I0038093

BORATE GLASSES, CRYSTALS & MELTS

2008

Proceedings of the Sixth International Conference on Borate Glasses, Crystals and Melts: Borate2008, held at Egret Himeji Convention Center, Himeji, Hyōgo Prefecture, Japan, on 18–22 August 2008.

Edited by
Norimasa Umesaki & Alex C. Hannon

Society of Glass Technology
Sheffield, 2010

This conference and proceedings are dedicated to
Professor Adrian C. Wright to honour his achievements in
Glass Science, and in particular Borate Glasses.

Borate2008
Proceedings of the Sixth International Conference on Borate Glasses, Crystals and Melts: new techniques and practical applications held at Himeji, Japan on 18–22 August 2008
A collected volume of papers from the Sixth International Conference, the papers were originally published in *European Journal of Glass Science and Technology:*
Physics and Chemistry of Glasses: European Journal of Glass Science and Technology Part B:
Glass Technology: European Journal of Glass Science and Technology Part A:

ISBN 978-0-900682-63-6

The objects of the Society of Glass Technology are to encourage and advance the study of the history, art, science, design, manufacture, after treatment, distribution and end use of glass of any and every kind. These aims are furthered by meetings, publications, the maintenance of a library and the promotion of association with other interested persons and organisations.

Officers of the Society
M. West (President)
J. Henderson (Honorary Secretary)
R. Duly (Honorary Treasurer)

Senior Editors
Dr R. J. Hand (Glass Technology)
Regional Editors
Professor J. M. Parker
Professor C. Rüssel
Professor L. Wondraczek
Professor A. Duran
Professor R. Vacher
Dr A. C. Hannon
Professor M. Liška
Professor S. Buddhudu
Professor Y. Yue

Managing Editor D. Moore
Assistant Editor S. Lindley

Society of Glass Technology
Unit 9, Twelve O'clock Court
21 Attercliffe Road
Sheffield S4 7WW, UK
Tel +44(0)114 263 4455
Fax +44(0)114 263 4411
Email info@sgt.org
Web http://www.sgt.org
The Society of Glass Technology is a registered charity no. 237438.

© Society of Glass Technology, 2010

Foreword

The *Sixth International Conference on Borate Glasses, Crystals and Melts: Borate2008* was held during the period 18–22 August, 2008 at the Egret Himeji Convention Center in Himeji, Hyōgo Prefecture, Japan. This followed previous conferences held at Alfred, USA (1977); Abingdon, United Kingdom (1996); Sofia, Bulgaria (1999); Cedar Rapids, USA (2002); and Trento, Italy (2005), and it was the first time that the Conference was held in Asia. The Sixth Conference was dedicated to Professor Adrian C. Wright to honour his achievements in glass science, and in particular borate glasses and neutron diffraction.

In total, 116 scientists from 16 countries gathered in Himeji to discuss recent advances in the study of borate glasses, crystals, and melts, and their applications. The participants came from: Armenia (2), Bulgaria (3), Canada (1), Czech Republic (1), France (2), Germany (2), Greece (2), Hungary (1), India (3), Italy (3), Japan (66), Poland (1), Romania (3), Russian Federation (4), United Kingdom (5), and United States of America (17). The first invited talk was by Francesco Rocca who summarised the contents of the previous Conference. The lecture in honour of Professor Adrian C. Wright, entitled "Adrian Wright: Glasses, Neutrons, Borates!", was presented by Alex Hannon, who described the key role played by Professor Wright in establishing the Borate Conference Series. This was followed by a plenary talk entitled "Borate Structures: Crystalline and Vitreous", given by Professor Wright. The other invited talks were by: M. Affatigato (USA), G. Dalba (Italy), S. A. Feller (USA), A. C. Hannon (UK), S. Kroeker (Canada), J. Kieffer (USA), G. Lucovsky (USA), A. Takada (Japan), M. Takata (Japan), T. Tanaka (Japan), N. M. Vedishcheva (Russia), S. Yamamoto (Japan), T. Yazawa (Japan), and S. Yoshida (Japan).

The conference structure involved the following sessions: Introduction (Chairmen K. Tadanaga and N. Umesaki), Structure and Glass Properties 1 (Chairman E. I. Kamitsos), Short and Intermediate Range Order (Chairman M. Tatsumisago), Thermodynamics and Glass Properties (Chairman A. C. Hannon), Spectroscopies and Local Structure (Chairman S. W. Martin), Structure and Glass Properties 2 (Chairman S. A. Feller), Structure and Modelling (Chairman T. Nanba), Novel Borate Glasses and Crystals (Chairman F. Rocca), Phase Separation and Inhomogeneities/Novel Borate Glasses and Crystals (Chairman S. Kroeker), Industrial and Technological Applications 1 (Chairman A. Takada), Industrial and Technological Applications 2 (Chairman N. Ohtori), Optical Properties and Materials (Chairman Y. B. Dimitriev), Physical Properties (Chairman N. M. Vedishcheva), and Structure and Glass Properties 3 (Chairman N. Umesaki). In total, there were 46 oral presentations. In addition, two poster sessions were held, with a total number of 47 posters.

The efforts and support of many people and organisations played a part in the success of the Conference. In particular I am most grateful to SPring8 (the world's largest third-generation synchrotron radiation facility) and to the Japan Synchrotron Radiation Research Institute (JASRI) for their support which greatly underpinned the conference.

Thanks are due to the members of the International Organising Committee: A. G. Clare (USA), G. Dalba (Italy), Y. B. Dimitriev (Bulgaria), D. Ehrt (Germany), S. A. Feller (USA), A. C. Hannon (UK), E. I. Kamitsos (Greece), N. M. Vedishcheva (Russia), A. C. Wright (UK) and J. W. Zwanziger (Canada). For logistic organisation of the Conference, thanks also to the members of the local committee: K. Hirao (Kyoto University), M. Tatsumisago (Osaka Prefecture University), T. Akai (AIST), Y. Daiko (University of Hyogo), K. Handa (Kyoto University), Y. Iwadate (Chiba University), A. Kajinami (Kobe University), M. Kodama (Sojo University), H. Ohno (JASRI/SPring-8), N. Ohtori (Niigata University), K. Tadanaga (Osaka Prefecture University), A. Takada (Asahi Glass Co. Ltd.), M. Takata (RIKEN/JASRI/SPring-8), S. Yamamoto (Nippon Electric Glass Co., Ltd.), and T. Yazawa (University of Hyogo). Special mention should be made of the careful work of Ms Inobe (SPring8) who diligently performed much of the detailed organisation and administration of many aspects of the Conference. Dr Katsumi Handa (Ritsumeikan University) is thanked for his help in the planning and organisation of the Conference and its website. Finally the role of Alex Hannon in the editorial process for the Conference Proceedings should be acknowledged.

The support of the sponsors of the Conference was most important, and I am grateful to: Asahi Glass Co. Ltd., Commemorative Organization for the Japan World Exposition '70, The Glass Division of The Ceramic Society of Japan, Himeji Convention & Visitors Bureau, Industrial Users Society of SPring-8, The Kansai Branch of the Iron and Steel Institute of Japan, The Kao Foundation for Arts and Science, The Murata Science Foundation, Nippon Electric Glass Co. Ltd., Nippon Sheet Glass Co. Ltd., Nippon Sheet Glass Foundation for Materials Science and Engineering, Rigaku Corporation, and Tsutomu Nakauchi Foundation. The Conference was also recognized as a satellite meeting of the XXI Congress of the International Union of Crystallography (held in Osaka, Japan), and the IUCr are gratefully thanked for the provision of support for students. Finally the support of the Society of Glass Technology in publishing the Conference Proceedings should be acknowledged.

The City of Himeji deserves special mention for providing a unique and very pleasant location for the conference. The conference venue had a spectacular view of Himeji Castle (sometimes known as White Egret Castle), which is the best preserved original Japanese castle, and is often described as the most beautiful castle in Japan. Himeji Castle was registered as a UNESCO world heritage site in 1993, and it was a pleasure to observe it in the breaks between scientific sessions, as well as in detail on the conference excursion. The Castle has previously achieved international fame in 1967 when Sean Connery visited to film a scene for the James Bond film *007 You Only Live Twice*. Then in 2003 Tom Cruise visited Himeji Castle to film part of his movie, *The Last Samurai*.

Two proposals for the location of the next Conference were considered by the International Organising Committee, and it was decided to hold the next conference in Dalhousie University, Halifax, Nova Scotia, Canada during 21–25 August 2011. Professor Josef W. Zwanziger will chair the *Seventh International Conference on Borate Glasses, Crystals and Melts*, which will be held in honour of Professor Tsutomu Minami to honour his achievements in Glass Science, and in particular Borate Glasses, and all of the participants present in Himeji are cordially invited to attend.

Norimasa Umesaki (JASRI/SPring-8), Conference Chairman.

The Borate2008 Conference Chairman,
Norimasa Umesaki, demonstrating how to use
a fan on a hot day at Himeji Castle

CONTENTS

Proceedings of the Sixth International Conference on Borate Glasses, Crystals and Melts: Borate2008

The journal references for the papers are given using these journal abbreviations:
PC=*Physics and Chemistry of Glasses: European Journal of Glass Science and Technology Part B;*
GT=*Glass Technology: European Journal of Glass Science and Technology Part A.*

SIXTH INTERNATIONAL CONFERENCE ON BORATE GLASSES, CRYSTALS AND MELTS: BORATE2008

Egret Himeji Convention Center – Himeji, Hyōgo Prefecture, Japan
18–22 August 2008

PROGRAMME

Monday, August 18th

15:00	*Registration*
18:00-20:00	*Welcome reception*

Tuesday, August 19th

Introductory section
Chairman: K. Tadanaga & N. Umesaki

9:20-9:25	WELCOME	
9:25-9:45	*Invited Talk*	
	F. Rocca	The 5th International Conference on Borate Glasses, Crystals and Melts: New Techniques and Practical Applications, Trento 10-14 July 2005
9:45-10:05	*Invited Talk*	
	A. C. Hannon	Adrian Wright: Neutron Diffraction Pioneer
10:05-10:55	*Plenary Talk*	
	A. C. Wright	Borate Structures: Crystalline and Vitreous
10:55-11:15	*Coffee break*	

Session 1: Structure and glass properties (1)
Chairman: E. I. Kamitsos

11:15-11:45	*Invited Talk*	
	S. A. Feller	A Review of Physical Properties of Borate Glasses Related to Atomic Structure
11:45-12:05	**Y. Matsuda**	Elastic Properties and Fragility of Lithium Borate Glasses

Session 2: Short and intermediate range order
Chairman: M. Tatsumisago

12:05-12:25	**S. Sen**	Atomic Structure of Beryllium Boroaluminate Glasses: A Multi-nuclear NMR Spectroscopic Study
12:25-12:45	**L. Cormier**	Temperature Effects on the Structure of Mixed-Alkali Borate Glasses and Melts by Neutron Diffraction and EPSR Modeling
12:45-14:00	*Lunch*	

Session 3: Thermodynamics and glass properties
Chairman: A. C. Hannon

14:00-14:30	*Invited Talk*	
	G. Lucovsky	Intermediate Phases (IPs) in Alkali Borosilicate Alloys: $(Na_2O)_X(B_2O_3)_{1-X}$ and Pyrex: $(SiO_2)_{0.81}(Na_2O)_{0.04}(Al_2O_3)_{0.02}(B_2O_3)_{0.13}$
14:30-15:00	*Invited Talk*	
	N. M. Vedishcheva	The Chemical Structure of Borate Glasses
15:00-15:20	**L. Koudelka**	WO_3-Doped Zinc Borophosphate Glasses
15:20-17:00	*Poster Session (A) & Coffee Break*	

Session 4: Spectroscopies and local structure
Chairman: S. W. Martin

17:00-17:30	*Invited Talk*	
	S. Kroeker	Phase Separation and Volatility in Model Nuclear Waste Glasses: Results from NMR
17:30-17:50	**N. S. Barrow**	Using Spin Diffusion ^{11}B NMR to Probe Medium Range Order in Lithium Diborate Crystals and Glasses
17:50-18:10	**T. Nanba**	Molecular Orbital Calculation of ^{29}Si NMR Chemical Shift in Borosilicates-the Effect of Boron Coordination to SiO_4 Unit
18:10-18:30	**E. R. Barney**	The Structure of Tellurium Borate Glasses, Neutron Diffraction, ^{11}B NMR and Density Studies

| 18:30-18:50 | **A. A. Osipov** | Structural Investigation of R_2O-B_2O_3 (R = Li, Na, K) Glasses and Melts by High-Temperature Raman Spectroscopy |

19:30-21:00 *Beer Party*

Wednesday, August 20th

Session 5: Structure and glass properties (2)
Chairman: S. A. Feller

9:00-9:30	*Invited Talk*	
	G. Dalba	What X-ray Absorption Spectroscopy Can Actually Say about the Short Range Order in Glasses
9:30-9:50	**D. Ehrt**	Electrical Conductivity and Viscosity of Borosilicate Glasses and Melts
9:50-10:10	**A. Khanna**	Investigation of Influence of Melt History on the Structure and Properties of Lead and Bismuth Borate Glasses
10:10-10:30	**T. S. Markova**	CSG Concept based on Vibrational Spectroscopy Analysis as Applied to BaO(SrO)–Al_2O_3–B_2O_3 Glassforming Systems

10:30-11:00 *Coffee Break*

Session 6: Structure and modeling
Chairman: T. Nanba

11:00-11:30	*Invited Talk*	
	A. Takada	Computer Modeling of B_2O_3-SiO_2 Glass Structures
11:30-12:00	*Invited Talk*	
	J. Kieffer	Prediction of New Crystalline Polymorphs in B_2O_3 and How This Explains the Behavior of Glass
12:00-12:20	**G. Ferlat**	First-principles Simulations of Liquid and Vitreous B_2O_3: Closing the Boroxol Controversy and Assessing Their Structural Role
12:20-12:40	**D. C. Möncke**	Thermal Poling of Sodium Borosilicate Glasses

12:40-14:00 *Lunch*

Session 7: Novel borate glasses and crystals
Chairman: F. Rocca

14:00-14:20	**D. M. Schubert**	Recent Studies of Polyborate Anions
14:20-14:40	**E. I. Kamitsos**	Structure of Lithium-Borosulphate Oxynitride Thin Film Amorphous Electrolytes
14:40-15:00	**Y. B. Dimitriev**	Boromolybadate Glasses Containing Rare-Earth Oxides
15:00-15:20	**S. W. Martin**	Mixed Glass Former Effects in Alkali Oxy- and Thio-Borophosphate Glasses

15:20-17:20 *Poster Session (B) & Coffee Break*

Session 8: Phase separation and in homogeneities/Novel borate glasses and crystals
Chairman: S. Kroeker

17:20-17:50	*Invited Talk*	
	T. Yazawa	Phase Separation of Borosilicate Glass-Mechanism and its Application
17:50-18:10	**R. M. Hovhannisyan**	Glass Formation & Crystallisation Behaviour of Yttrium Aluminum Tetraborate Glass Ceramic
18:10-18:30	**R. E. Youngman**	Composition and Structure/Property Studies of Novel Halide-Rich Borate Glasses
18:30-18:50	**H. Masai**	Crystallization Behaviour of MO_x-Doped CaO-Bi_2O_3-B_2O_3-Al_2O_3-TiO_2 Glass

Thursday, August 21st

Session 9: Industrial and technological applications (1)
Chairman: A. Takada

9:00-9:30	*Invited Talk*	
	S. Yamamoto	Borate and Borosilicate Glasses for Flat Panel Displays
9:30-10:00	*Invited Talk*	
	T. Tanaka	Fabrication of Low Temperature Foaming Glass by Hydrothermal Glass Chemistry
10:00-10:20	**S. Tomeno**	Effects of Rare Earth Oxides (La_2O_3, Gd_2O_3) on Optical and Thermal Properties in B_2O_3-La_2O_3-Based Glasses

10:20-10:40 *Coffee Break*

Session 10: Industrial and technological applications (2)
Chairman: N. Ohtori

10:40-11:00	**K. Sakaguchi**	Electrochemical Analysis of Bismuth-Doped Glasses
11:00-11:20	**O. A. Karas'**	Physical and Chemical Conditions of Boron Minerals Formation in the Dalnegorsky Borosilicate Deposit
11:20-11:40	**H. Takebe**	Viscosity and Sintering Behavior of $BaO–P_2O_5–B_2O_3$ Glasses and Their Powders
11:40-13:00	*Lunch*	
13:00-21:00	*Excursion & Banquet*	

Friday, August 22nd

Session 11: Optical properties and materials
Chairman: Y. B. Dimitriev

9:00-9:30	*Invited Talk*	
	M. Affatigato	Laser Desorption Time of Flight Mass Spectroscopy Applied to Heavy Metal Borates
9:30-9:50	**E. Fujinaka**	Luminescence Properties of $YBO_3:Eu^{3+}$ Crystallized from Borosilicate Glass
9:50-10:10	**Y. Yonesaki**	Space-selective Precipitation of Nonlinear Optical Crystals by Near-Infrared Femtosecond Laser Irradiation
10:10-10:30	**T. Honma**	Patterning of Ferro-Electric Crystals in Glass by Laser Irradiation
10:30-10:50	*Coffee Break*	

Session 12: Physical properties
Chairman: N. M. Vedishcheva

10:50-11:20	*Invited Talk*	
	S. Yoshida	Indentation-Induced Densification of Sodium Borate Glasses
11:20-11:40	**Y. Kato**	Effect of B_2O_3 on Crack Initiation in Na_2O-B_2O_3-SiO_2 Glass
11:40-12:00	**H. Yasunaga**	Pulse Laser Induced Nano-Structure and Thermal Properties of Bismuth Borate Glass
12:00-12:20	*Coffee Break*	

Session 13: Structure and glass properties (3)
Chairman: N. Umesaki

12:20-12:50	*Invited Talk*	
	A. C. Hannon	Tin Borate Glass Structure by Neutron Diffraction and NMR
12:50-13:20	*Invited Talk*	
	M. Takata	Challenge to Time Resolved Structural Visualization of the Phase Change of DVD-RAM Materials
13:20-13:30	*Closing Remarks*	
13:30-17:30	*SPring-8 Site Tour*	

Poster Session (A)

Tuesday, 19th August, 15:20-17:00

Neutron Diffraction and Reverse Monte Carlo Modelling Of B_2O_3 and Sodium Diborate Glasses
M. Fábián, E. Sváb, T. Proffen, E. Veress

Multilateral Analyses of Local Structure in B_2O_3-Li_2O Glasses
Y. Iwadate, *T. Harada*, H. Kuriyama, S. Nishiyama, M. Hatakeyama, T. Okubo, K. Itoh, T. Fukunaga, M. Misawa, N.Ohtori, A. Kajinami, S. Deki, H. Matsuura, N. Umesaki

Brillouin Scattering Study of Elastic Properties in Sodium Borate Binary Glasses
Y. Fukawa, Y. Matsuda, M. Kawashima, S. Kojima, M. Kodama

Al Substitution in Borophosphosilicate Glasses
B. G. Aitken, *R. E. Youngman*

FTIR and Raman Study on Borate Glass in Ternary ZnO-TiO_2-B_2O_3 System
B. A. Sava, A. Diaconu, D. Ursu, M. Elisa

Physical Properties and Structural Determination of Alkaline Earth Borosilicates
T. Mullenbach, M. Franke, A. Ramm, M. Affatigato, S. Feller

Viscosity of Bi_2O_3-B_2O_3-SiO_2 Melts
S. Inaba, H. Tokunaga, C. Hwang, S. Fujino

First-principles Simulations of Liquid and Vitreous B_2S_3: Assessing the Superstructural Units
M. Micoulaut, **G. Ferlat**

Phase Separation and Crystallization Mechanism in Borate Glasses from BaO-TiO_2-B_2O_3 System
L. Boroica, *R. Medianu, D. Hülsenberg, I. S. Boroica*

Material Design to Create Porous Glass using Phase Separation in Multi-Component Borosilicate Glass
M. Suzuki, *T. Tanaka*

Simulation of Quadrupole Interactions in ^{10}B NMR Powder Patterns
J. Berkowitz, *M. Affatigato, S. Feller, D. Holland, T. Kemp, M. Smith*

The Mixed Glass Former Effect on the T_g, V_m and Density of Na_2S-B_2S_3-P_2S_5
M. J. Haynes, *A. Shaw, T. Kaufmann, S. W. Martin*

Correlation between Basicity and Coordination Structure in Borosilicate Glasses
Y. Tanaka, *S. Sakida, Y. Benino, T. Nanba, Y.Miura*

Elastic and Vibrational Properties of Potassium Borate Binary Glasses
M. Kawashima, *Y. Matsuda, Y. Fukawa, S. Kojima, M. Kodama*

Coloration and Local Structure Analysis of Borosilicate Glasses Containing Sulfur
T. Asahi, *S. Nakayama, T. Nanba, H. Kiyono, H. Yamashita, T. Maekawa*

In-situ Xafs Measurement of Transition Metal Ion in Borate Glass at High Temperature
A. Kajinami, *H. Matsuura, S. Deki, S. Fujiwara*

Rare Earth Oxides in a Borate Host Glass
J. A. E. Desa, *W. Vaz, B. De Souza, M. Singh*

Structural Effects of Europium on Lead Borate Glasses
A. Marquardt, *M. Roberts, S. Feller, M. Affatigato, S. Singleton*

Structure Property Relationship of Sodium Oxy-borophosphate Glasses
R. Christensen, *J. Byer, S. W. Martin*

Composite Materials based on Immiscible Borate Glass and Iron Spinel Nanoparticles
E. P. Kashchieva, *Y. Dimitriev, V. Ivanova, T. Merodiiska*

Bonding Differences in Modified Non-Crystalline Covalent Random Networks and Phase-Separated Nano-Composites
G. Lucovsky, *J. C. Phillips*

An XPS Study of O1s Binding Energy of Sodium Borate Glasses
H. Segawa, *T. Yano, S. Shibata*

XAFS Study on the Local Order of Pb in Borate Glasses
G. Dalba, P. Fornasini, **R. Grisenti**, *F. Rocca, M. Affatigato*

Effect of Three-body Interaction on Local Structure of Alkaline-earth Borate Glasses
M. Matsubara, *N. Ohtori, A. Kajinami, Y. Iwadate, N. Umesaki*

Poster Session (B)

Wednesday, 20th August, 15:20-17:20

Transformation of Boron Containing Hybrid Structures into Silicon Oxycarbide Glasses
Y. Ivanova, *Y. Vueva*

Boron Isotope Effect on the High Temperature Viscosity of Sodium Borosilicate Glasses
J. Matsuoka, *Y. Nishida, K. Kimura, S. Yoshida, T. Sugawara*

Thermal and Optical Properties of Bi_2O_3-GeO_2-B_2O_3 Glasses
N. Yamashita, *T. Suetsugu, T. Einishi, K. Fukumi, N. Kitamura, J. Nishii*

The Ecological Evaluation of Reusing Borosilicate Glass Cullet for Manufacturing Glass
A. Diaconu, *B. A. Sava, L.-D. Ursu, M.Elisa*

A New Generation of Twin-Roller Quenchers
A. Havel, *M. Affatigato, S. Feller, M. Karns, M. Karns*

High-Temperature Absorption Spectra and Magnetic Susceptibilities for Co(II) and Ni(II) in Alkali Borate Glasses and Melts
K. Kojima, *T. Sanada, H. Yano, H. Marusawa, K. Nakai, T. Hayashi, M. Makino, H. Mizuki, J. Matsuda*

Raman Scattering Study of Binary Potassium Germanate Glasses[1]
S. Mamiya, *Y. Matsuda, Y. Fukawa, M. Kawashima, S. Kojima*

---

1. Mamiya, S., Matsuda, Y., Fukawa, Y., Kawashima, M. & Kojima, S. *Phys. Chem. Glasses Eur. J. Glass Sci. Technol. B*, 2009, **50**, 321.

Proc. VI Int. Conf. Borate Glasses, Himeji, Japan, 18–22 August 2008   *Phys. Chem. Glasses: Eur. J. Glass Sci. Technol. B*, February 2010, **51** (1), 1–39

Plenary Lecture

Borate structures: crystalline and vitreous

Adrian C. Wright

J.J. Thomson Physical Laboratory, University of Reading, Whiteknights, Reading, RG6 6AF, UK

Manuscript received 2 March 2009
Revised version received 28 May 2009
Accepted 24 December 2009

A comprehensive survey is presented of the structures of the crystalline phases formed in M_2O–B_2O_3 and MO–B_2O_3 glass forming systems, with particular reference to those aspects of relevance to the structures of the corresponding glasses. Tables are given of the known crystalline phases in each system, together with the current status of knowledge concerning their structure. The crystalline phases for which the full structure has been determined are employed to investigate the compositional dependence of various structural parameters, viz. the content of the five different basic borate structural units, $B^{(n)}$, and of the various species of superstructural unit.

The information accumulated for the crystalline phases is employed as a guide to the structures of the equivalent glasses, and to the variation of the above structural parameters with composition in the vitreous state, especially the fraction, x_4, of the boron atoms in a given glass that are 4-fold co-ordinated. Topological models have been presented in the literature for the variation of x_4 with composition, and their predictions are compared with experimental data for both crystalline and vitreous samples. The simplest topological model assumes that $B\varnothing_4^-$ units cannot be immediate neighbours, due to electrostatic repulsion, but such an arrangement is found in several superstructural units, presumably as a result of the unfavourable electrostatic configuration being more than balanced by the superstructural unit stabilisation energy. Experimental evidence for the presence of superstructural units in borate glasses is therefore examined, together with their possible effect on the structural unit fractions, x_n. The paper concludes by discussing a composite structural model for borate glasses, and by posing 10 questions of importance to continuing research on borate systems.

1. Introduction

Borate glasses are an enigma, in view of the challenges they present to conventional ideas concerning the structure of network glasses, as embodied in the traditional random network theory. To fully understand borate structures, both vitreous and crystalline, it is necessary to explain why the initial addition of a network modifier to B_2O_3 leads to an increase in the coordination number of some of the boron atoms from 3 to 4, rather than to the formation of nonbridging oxygen atoms (NBO). Similarly, there is the question as to the important role played by superstructural units in the formation of borate networks although, as will be shown later, these two anomalous aspects of borate structures are in fact closely related.

As for other oxide systems, the formation of an alkali or alkaline earth borate liquid from a basic alkali or alkaline earth compound (usually the carbonate) and the acidic oxide B_2O_3 involves a chemical reaction. The chemistry is exactly the same, whether the liquid is subsequently slowly cooled to yield a crystalline phase or more rapidly quenched to form a glass. Thus it is to be expected that the bonding, and hence the local atomic configurations, in the crystal and glass are likely to be similar. This was the basis on which Zachariasen[1] proposed the random network theory, using structural principles elucidated from crystallography, but taking due regard of the additional degrees of topological freedom required by a disordered network appropriate to the vitreous state. The objective of the present paper is to employ crystallographic data for anhydrous alkali, alkaline earth and related glass forming borates to achieve a better understanding of the unique features of borate structures, both crystalline and vitreous. In this quest, the author is following in the pioneering footsteps of Krogh-Moe, but with the advantage of a greatly extended, although still inadequate, crystallographic database. Much useful guidance has also been obtained from the excellent review, entitled *Borate Glass Structure*, presented by Griscom[2] at the very first Borate Conference in 1977, to which the reader is referred for a more detailed exposition of the early history of borate glass structure research, together with an account of the associated experimental techniques and important results.

In the following sections, the crystalline compositions $mM_2O.nB_2O_3$ and $mMO.nB_2O_3$ are abbreviated $mM.nB$, where M is a network modifying cation. To avoid an excessively long list of references, those solely for the crystalline structures in the H_2O–B_2O_3 system and in the various glass forming systems considered in this paper are presented in tabular form

Email a.c.wright@reading.ac.uk

in Appendix 1 (Tables A1, A2 and A3). Polymorphs formed at different temperatures are conventionally indicated by Greek letters (α-mM.nB, β-mM.nB, etc., normally in order of increasing temperature, but there are some important exceptions, as may be seen from Tables A1 and A2), whilst those prepared under pressure are usually denoted by Roman numerals (mM.nB-I, mM.nB-II, etc, with increasing pressure). Note, however, that the polymorphs normally referred to as β-Li.B and γ-Li.B are in fact high pressure phases and should therefore have been designated Li.B-II and Li.B-III.

The convention employed here for the formulæ of (super)structural units differs from those employed previously and by other authors, in that the symbol $\varnothing$ is reserved solely for those bridging oxygen atoms that are shared between adjacent (super)structural units, whilst bridging oxygen atoms that are situated completely within a superstructural unit are denoted O, as are negatively charged nonbridging oxygen atoms that form part of a (super)structural unit. The advantages of this representation are that the number of $\varnothing$ indicates the (super)structural unit connectivity (i.e. the number of connections to the external network) and that it allows equations describing the various reactions/equilibria involving (super)structural units to be balanced, since $\varnothing$ is then equivalent to half of an (O) oxygen atom.

2. Brief history

The first anhydrous alkali or alkaline earth borate crystal structure, Ca.B, was determined by Zachariasen & Ziegler in 1932, the same year as Zachariasen's classic paper[1] on the random network theory, and comprises Ca^{2+} cations and infinite chains of $BO\varnothing_2^-$ ($B^{(2)}$) borate structural units (Figure 1C). $B\varnothing_3$ units (Figure 1B) were initially identified in vitreous B_2O_3 by Warren, Krutter & Morningstar[3] in 1936, whilst in 1938 Biscoe & Warren[4] showed that, for Na_2O–B_2O_3 glasses, the addition of Na_2O leads to the conversion of the $B\varnothing_3$ triangular into $B\varnothing_4^-$ tetrahedral structural units (Figure 1A). The Na^+ network modifying cations were envisaged as occupying the interstices in the borate network adjacent to the $B\varnothing_4^-$ tetrahedra, as represented schematically in two dimensions in Figure 2. In the anhydrous crystalline state, the conversion of $B\varnothing_3$ triangles into $B\varnothing_4^-$ tetrahedra was not reported until much later (1959),[5] although the latter had been observed in 1934 in crystalline BPO_4 and $BAsO_4$, which have the β-cristobalite structure with alternating $B\varnothing_4^-$ and either $P\varnothing_4^+$ or $As\varnothing_4^+$ tetrahedra.[6]

The structures of potassium and sodium metaborate, K.B and Na.B were elucidated in 1937 and 1938, respectively, and are both based on the cyclic metaborate anion, $B_3O_6^{3-}$, the first superstructural unit (cf. Section 4.2, Figure 7B) to be identified. Other early crystalline structures to be determined were those

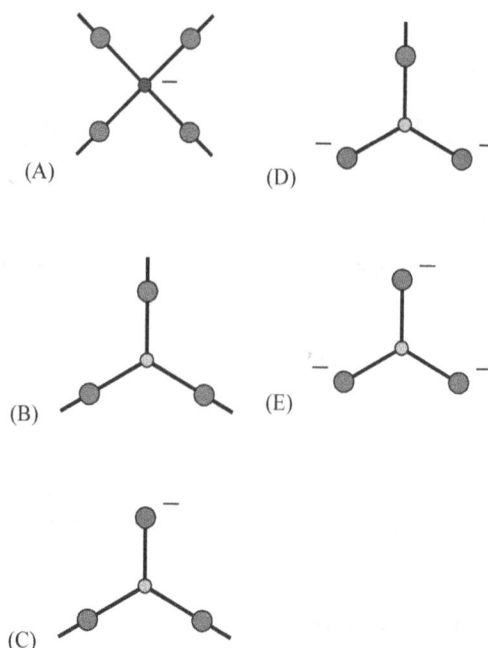

Figure 1. The five major basic $B^{(n)}$ structural units ($0 \le n \le 4$) found in vitreous and crystalline alkali, alkaline earth and related borates. A, $B\varnothing_4^-$; B, $B\varnothing_3$; C, $BO\varnothing_2^-$; D, $BO_2\varnothing^{2-}$ and E, BO_3^{3-}. (Key: blue, tetrahedral boron; cyan, trigonal boron; red, bridging oxygen atom and magenta, nonbridging oxygen atom)

of orthoboric acid, α-3H.B (H_3BO_3 molecules), and metaboric acid, H.B ($H_3B_3O_6$ molecules). The latter is particularly interesting in that the $H_3B_3O_6$ molecule incorporates the $B_3O_3\varnothing_3$ boroxol group (see later, Figure 7A), with the external bridging oxygen atoms, $\varnothing$, terminated by hydrogen atoms to form –OH groups. It was not until 1949 that the first anhydrous orthoborate structures were investigated (3Mg.B and 3Co.B[7]) and shown to contain BO_3^{3-} ($B^{(0)}$) orthoborate anions (Figure 1E) and, the following year, cobalt pyroborate,[8] 2Co.B, yielded the pyroborate anion, $B_2O_5^{4-}$ (Figure 3), which comprises two interconnected $BO_2\varnothing^{2-}$ ($B^{(1)}$) structural units (Figure 1D). Thus, by 1950, all five of the major $B^{(n)}$ borate structural units,

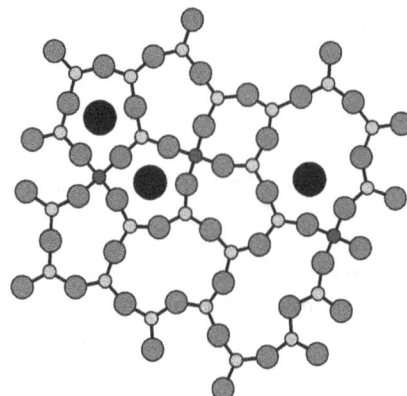

Figure 2. Two-dimensional schematic diagram representing the structure of vitreous Na_2O–B_2O_3 (after Biscoe & Warren[4]). Blue, tetrahedral boron; cyan, trigonal boron; red, bridging oxygen atom and black Na^+ cation

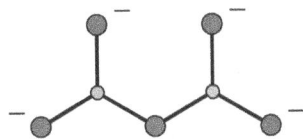

Figure 3. $B_2O_5^{4-}$ pyroborate anion (Key as Figure 1)

where n is the number of bridging oxygen atoms, Ø, ($0 \leq n \leq 4$), had been observed in either crystalline or vitreous anhydrous borate structures.

As indicated above, the conversion of $BØ_3$ triangles into $BØ_4^-$ tetrahedra was not observed in an anhydrous crystalline alkali or alkaline earth borate until 1959. This determination of the structure of crystalline K.5B and Rb.5B by Krogh-Moe[5] is also important in that it revealed the $BØ_4^-$ tetrahedra to be incorporated into a superstructural unit, viz. the $B_5O_6Ø_4^-$ pentaborate group (see later, Figure 8A), although the related bicyclic metaborate anion, $B_5Ø_6O_4^{5-}$ (see later, Figure 8E), had been found much earlier (1937) by Zachariasen[9] in the hydrate $K_2O.5B_2O_3.8H_2O$.

The advent of ^{11}B NMR spectroscopy allowed a much more accurate measurement of the fraction, x_4, of the boron atoms in borate glasses that are 4-fold coordinated (i.e. those that are present as $BØ_4^-$ tetrahedra) than was then possible from the average boron atom coordination number, $n_{B(O)}$, obtained by x-ray diffraction. In 1958, Silver & Bray[10] were the first to confirm the change in boron coordination number using NMR spectroscopy, and this was followed in 1963 by the famous paper of Bray & O'Keefe,[11] who determined the composition dependence of x_4 for all five alkali borate glass systems, as shown in Figure 4. As the mole fraction, x_M, of the network modifying oxide increases, x_4 first increases, then passes through a maximum and finally decreases, the variation with x_M being very similar for all five systems. Subsequent studies by the Bray group yielded x_4 data for a wide

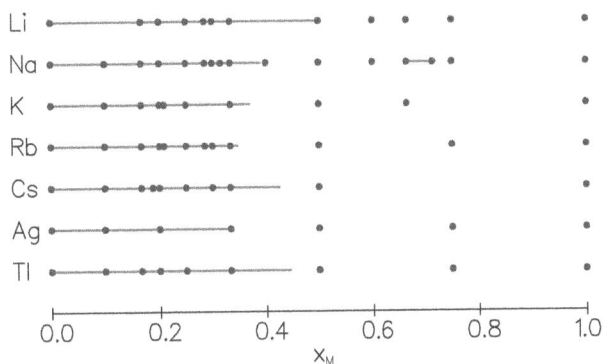

Figure 5. Conventional single phase glass forming regions (———) for binary M_2O–B_2O_3 systems, together with the corresponding crystalline phases (●)

variety of B_2O_3-containing glasses, such as those from the Tl_2O–B_2O_3[12] and PbO–B_2O_3[13] systems plotted later in Figures 29 and 32, respectively. The investigation of the compositional dependence of x_4 thus became a major objective of early NMR experiments on borate glasses. It is, however, perhaps unfortunate that structural studies of borate glasses have placed so much emphasis on the composition dependence of x_4, whilst ignoring that of the other borate structural unit fractions, which become increasingly important at higher x_M. In addition, the borate network is only one aspect of the structure. The network modifying cations also play a critical role (cf. Section 10.2), although this is frequently overlooked.

3. Glass formation and devitrification

Diboron trioxide is a glass former par excellence and forms binary glasses with a wide range of other oxides. The single phase glass forming regions obtained for binary M_2O–B_2O_3 and MO–B_2O_3 systems using conventional quenching techniques[14–16] are shown as a function of the mole fraction of the network modifying oxide, M_2O or MO, in Figures 5 and 6, respectively, together with the crystalline phases occurring in each system. For all of the M_2O–B_2O_3 systems in Figure 5, the single phase glass forming region commences at zero x_M, whereas, for all of the MO–B_2O_3 systems (Figure 6), single phase glasses are not formed at low modifier contents. Any glass formation in this region results in phase-separated glasses. For example, Shelby[17] records two glass transition temperatures for PbO–B_2O_3 glasses in the region $0.005 < x_{PbO} < 0.195$, corresponding to those for the two limiting compositions (B_2O_3, T_g=270°C and $PbO.4B_2O_3$, T_g=445°C). It is also interesting to note that there is no glass formation in the BeO–B_2O_3 system, whilst that in the MgO–B_2O_3 system is exceedingly narrow, suggesting that small divalent, and hence highly polarising, network modifying cations are not compatible with borate glass formation.

An early application of borate glass formation, and of the ability of B_2O_3 to act as a flux in dissolv-

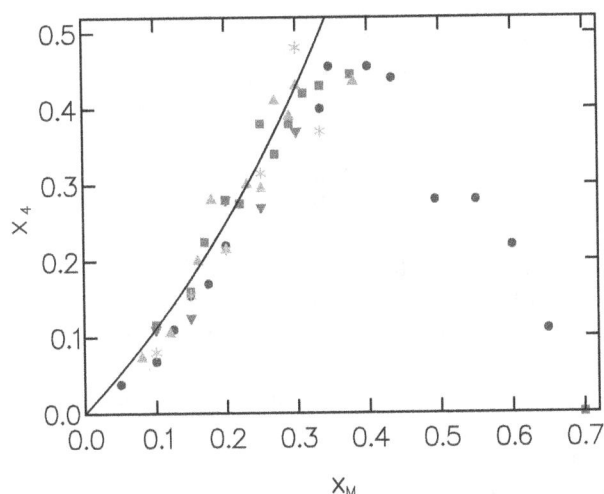

Figure 4. Bray & O'Keefe's x_4 data[11] for the five alkali borate glass systems. Blue filled circles, Li_2O–B_2O_3; red filled squares, Na_2O–B_2O_3; green filled triangles, K_2O–B_2O_3; magenta inverted filled triangles, Rb_2O–B_2O_3 and cyan asterisks, Cs_2O–B_2O_3

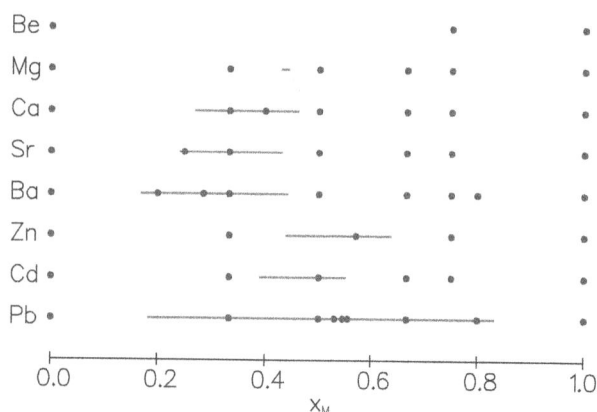

Figure 6. Conventional single phase glass forming regions (———) for binary MO–B_2O_3 systems, together with the corresponding crystalline phases (●)

ing other materials, can be found in the well known borax ($Na_2B_4O_5(OH)_4.8H_2O$) bead test, where borax is melted together with an unknown substance and quenched in a platinum wire loop to yield a glass bead of a characteristic colour. Borax is a hydrous sodium diborate with a structure based on the $[B_4O_5(OH)_4]^{2-}$ anion,[18,19] which consists of a diborate group with the four external bridging oxygen atoms (Ø) forming –OH groups (see Section 4.2, Figure 9B).

3.1. Larger groupings

The concept that glass formation is enhanced by the presence in the melt of stable groupings/aggregates that are larger than the basic structural units dates back to a paper by Tool & Hill.[20] The same idea was expressed by Hägg[21] following the publication of Zachariasen's paper expounding the random network theory.[1] Hägg's hypothesis was strongly criticised by Zachariasen,[22] but was taken up by Krogh-Moe,[23–26] who postulated the important role of superstructural units in vitreous borate networks, and it is perhaps amusing to note that several of Krogh-Moe's early papers (e.g. Ref. 27) were communicated to Arkiv för Kemi by Hägg!

The most well known example of larger groupings (superstructural units) inhibiting crystallisation and promoting glass formation is afforded by pure anhydrous B_2O_3, where crystallisation does not occur at ambient pressure due to the presence of boroxol groups in the melt. The ambient pressure crystalline modification of B_2O_3 does not contain boroxol groups (*cf.* Section 6), and so crystallisation from the melt requires that the boroxol groups are broken up, which is inhibited by their stabilisation energy. A crystal of B_2O_3-I seeded into the anhydrous supercooled melt does not grow, even over a period of several months,[28] indicating an extremely high activation energy for crystallisation.

A similar argument can be advanced for the binary glass forming systems in Figures 5 and 6. In the melt,

there will be a distribution of superstructural unit species formed as a result of the various equilibrium reactions that occur in the liquid state. This is true even for melts having the exact stoichiometry of a crystalline phase, which will therefore include superstructural units not found in this phase. Thus, in order for a borate melt to crystallise, it is necessary to break up these superstructural units and form those appropriate to the crystalline phase in the correct proportions, and then to rearrange them topologically to yield the correct crystalline structure. The activation energy for this process is higher than that for the crystallisation of systems without superstructural units, which explains the excellent glass forming ability of borate systems.

It is now generally accepted that the structures of binary and multicomponent glasses are nanoheterogeneous, in that there are local fluctuations in both density and composition that are present in the liquid and frozen in on quenching through the glass transition. Initially, the origin of such fluctuations was discussed in terms of compound formation and dissociation,[29] or of molecular groupings,[29] in glass forming melts, but more recently the thermodynamic modelling studies of Shakhmatkin & co-workers[30,31] have led to the concept of chemical groupings that have the stoichiometry of the crystalline phases that exist in the system in question. The chemical groupings are also assumed to exhibit a structural similarity to these crystalline phases, at least in terms of the distribution of (super)structural units present and that of the environment of the network forming cations.

3.2. Devitrification

The process of glass crystallisation is important in that devitrification products are often taken as a guide to the structure of the initial glass. However, great care must be taken to ensure that such products do indeed have structures that are relevant to that of the original glass. This requires an understanding of the process of devitrification, which involves an interplay between kinetics (atomic diffusion) and thermodynamics (equilibrium structure). First, it is important that the crystalline phase is homogeneously nucleated throughout the bulk of the glass, and not nucleated heterogeneously, or just at the glass surface; e.g. due to the presence of impurities, atmospheric/moisture attack, and/or a composition that differs from that of the bulk. Given homogeneous bulk nucleation, the devitrification temperature is of paramount importance. If the glass is devitrified at the lowest possible temperature, the crystallisation process is diffusion limited and so the (frequently thermodynamically metastable) crystalline phase formed will be that which requires the least diffusion; i.e. the one with a structure closest to that of the original glass. In terms of the nanoheterogeneous structure of the latter, the nucleation of the crystalline phase is most likely to occur at points where the frozen-in composition/density fluctuations lead to an atomic arrangement closest to that of the crystal. As the devitrification temperature is raised, diffusion becomes easier, and the structure of the resulting crystalline phase can deviate more readily from that of the glass. Finally, at the highest devitrification temperature, thermodynamics dominates and the crystalline phase formed is that which is stable (i.e. the equilibrium thermodynamic state) at the devitrification temperature. Note that the increased degrees of freedom at the glass surface means that structural rearrangement is easier than in the bulk, and hence that the structure of the crystalline phase formed by surface devitrification can more readily deviate from that of the glass.

4. Basic principles of borate structures

4.1. Structural units

For conventional silicate glasses, the various constituents are normally classified as either **network formers** or **network modifiers**, although some materials are intermediate in character and may change their role depending on the composition and/or the other constituents present. The classification of components into network formers and network modifiers arises from the fact that the nature of the bonding is different around the two types of cation. The network modifying cations are in general more ionic in character. The network formers contribute to the basic three-dimensional network, whereas the network modifiers are **network breakers**, leading to the formation of nonbridging oxygen atoms. The addition of a network modifier to B_2O_3, on the other hand, initially results in the conversion of $B\varnothing_3$ triangles into $B\varnothing_4^-$ tetrahedra

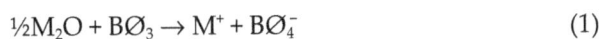

$$\tfrac{1}{2}M_2O + B\varnothing_3 \rightarrow M^+ + B\varnothing_4^- \tag{1}$$

or

$$MO + 2B\varnothing_3 \rightarrow M^{2+} + 2B\varnothing_4^- \tag{2}$$

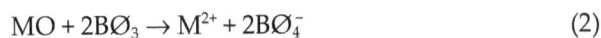

i.e. it acts as a **network compacter** via an increase in $n_{B(O)}$ from 3 to 4. In the absence of nonbridging oxygen atoms, the fraction of 4-fold coordinated boron atoms, x_4, is the same as that for hydrated borates[32] and is given by[25]

$$x_4 = x_M/(1-x_M) = m/n \tag{3}$$

where m and n refer to the crystalline compounds $mM_2O.nB_2O_3$ and $mMO.nB_2O_3$. The negative charge on the $B\varnothing_4^-$ tetrahedron is delocalised over the entire structural unit, which means that it is energetically unfavourable for $B\varnothing_4^-$ tetrahedra to be immediate neighbours.

Further additions of a network modifier eventually lead to the formation of negatively charged nonbridging oxygen atoms. The delocalised negative charge on the $B\varnothing_4^-$ tetrahedra means that, in systems containing network modifying cations with a high degree of ionic character, nonbridging oxygen atoms will tend to be confined to the $B\varnothing_nO_{3-n}^{n-3}$ triangles. (An exception occurs at high x_M for alkali borate glasses involving the heavier alkali cations, and in the crystal structure of 6Pb.5B (x_M=0·545), where the PbO is thought to be acting as a network former, such that the Pb–O bond has significant covalent character – see Section 12.4.) Thus in the case of an alkali oxide modifier, nonbridging oxygen atoms are generated by the following sequence of reactions

$$\tfrac{1}{2}M_2O + B\varnothing_3 \rightarrow M^+ + BO\varnothing_2^- \tag{4}$$

$$\tfrac{1}{2}M_2O + B\varnothing_4^- \rightarrow M^+ + BO_2\varnothing^{2-} \tag{5}$$

$$\tfrac{1}{2}M_2O + BO\varnothing_2^- \rightarrow M^+ + BO_2\varnothing^{2-} \tag{6}$$

and

$$\tfrac{1}{2}M_2O + BO_2\varnothing^{2-} \rightarrow M^+ + BO_3^{3-} \tag{7}$$

Since two negatively charged nonbridging oxygen atoms on the same boron atom ($B^{(1)}$ unit) are energetically less favourable than their being on separate boron atoms ($B^{(2)}$ units), $B^{(1)}$ units are unlikely to be formed in any number until this is unavoidable. Similarly, a $B^{(0)}$ unit, with three nonbridging oxygen atoms on the same boron atom, is even less favourable. Hence the reactions defined by Equations (6) and (7) will tend not to progress to any great extent until the number of nonbridging oxygen atoms exceeds the number of available trigonal boron atoms.

4.2. Superstructural units

Superstructural units comprise well defined arrangements of the basic borate structural units, with no internal degrees of freedom in the form of variable bond or torsion angles and all of those summarised in Table 1 have been found in at least one anhydrous crystalline borate structure. The naming convention for superstructural units used here follows that of Krogh-Moe and refers to the composition that would result for a borate containing only the unit in question. Thus the di-triborate group corresponds to the composition 2M.3B, etc.

In systems containing no nonbridging oxygen atoms, four different 3-membered rings are possible, as indicated in Figures 7A, 8B, 9D and 10B, and all four are found in crystalline borates. However, it is interesting to note that, despite the existence of the $B_3O_3\varnothing_3$ boroxol group (Figure 7A) and the $B_3O_6^{3-}$ cyclic metaborate anion (Figure 7B), neither of the intermediate 3-membered rings, incorporating one or two $BO\varnothing_2^-$ units (Figure 11), has ever been observed. The reason for this is unclear, especially given the presence of the $B_3O_4\varnothing_3^{2-}$ di-triborate (NBO)

Table 1. (Super)structural units and anions found in crystalline alkali, alkaline earth and related borates

Unit/Anion	Formula	x_M*	$B\emptyset_4^-$	$B\emptyset_3$	$BO\emptyset_2^-$	$BO_2\emptyset^{2-}$	BO_3^{3-}
B(3)	$B\emptyset_3$	0·000	-	1	-	-	-
Boroxol	$B_3O_3\emptyset_3$	0·000	-	3	-	-	-
Pentaborate	$B_5O_6\emptyset_4^-$	0·167	1	4	-	-	-
Triborate	$B_3O_3\emptyset_4^-$	0·250	1	2	-	-	-
Di-pentaborate	$B_5O_6\emptyset_5^{2-}$	0·286	2	3	-	-	-
Di-pentaborate (NBO)	$B_5O_7\emptyset_3^{2-}$	0·286	1	3	1	-	-
Diborate	$B_4O_5\emptyset_4^{2-}$	0·333	2	2	-	-	-
Tri-pentaborate	$B_5O_6\emptyset_6^{3-}$	0·375	3	2	-	-	-
Tri-pentaborate (NBO)	$B_5O_7\emptyset_4^{3-}$	0·375	2	2	1	-	-
Di-triborate	$B_3O_3\emptyset_5^{2-}$	0·400	2	1	-	-	-
Di-triborate (NBO)	$B_3O_4\emptyset_3^{2-}$	0·400	1	1	1	-	-
B(4)	$B\emptyset_4^-$	0·500	1	-	-	-	-
B(2)	$BO\emptyset_2^-$	0·500	-	-	1	-	-
Metaborate	$B_3O_3\emptyset_6^{3-}$	0·500	3	-	-	-	-
Metaborate (NBO)	$B_3O_4\emptyset_4^{3-}$	0·500	2	-	1	-	-
Cyclic metaborate	$B_3O_6^{3-}$	0·500	-	-	3	-	-
Bicyclic metaborate	$B_5\emptyset_6O_4^{5-}$	0·500	1	-	4	-	-
B(1)	$BO_2\emptyset^{2-}$	0·667	-	-	-	1	-
Pyroborate	$B_2O_5^{4-}$	0·667	-	-	-	2	-
B(0)	BO_3^{3-}	0·750	-	-	-	-	1

* Mole fraction of network modifying oxide for a structure containing only the unit/anion

group with a nonbridging oxygen atom (Figure 8D) in crystalline α-Na.2B.

Larger superstructural units are formed by combining two 3-membered rings, by sharing either one (pentaborate series) or two (diborate) $B\emptyset_4^-$ tetrahedra. As indicated in Section 4.1, the delocalised negative charge means that it is energetically unfavourable for $B\emptyset_4^-$ tetrahedra to be immediate neighbours. However, such an arrangement is found within superstructural units, presumably because this is more than compensated for by the superstructural unit stabilisation energy. In this respect, it is useful to group the superstructural units according to whether they incorporate no $B\emptyset_4^-$ tetrahedra (Figure 7), only one $B\emptyset_4^-$ tetrahedron (Figure 8), a pair of adjacent $B\emptyset_4^-$ tetrahedra (Figure 9), or three adjacent $B\emptyset_4^-$ tetrahedra (Figure 10).

Note that the so-called tetraborate group (Figure 12) does not qualify as a superstructural unit, since it comprises a triborate and a pentaborate group joined by a single bridging oxygen atom with variable bond and torsion angles. In addition, the combined unit is sometimes portrayed with a $B^{[3]}$–$\emptyset$–$B^{[3]}$ bridge between the constituent pentaborate and triborate groups, as in Figure 12, and at other times with a $B^{[3]}$–$\emptyset$–$B^{[4]}$ bridge, where $B^{[3]}$ represents a trigonal and $B^{[4]}$ a tetrahedral boron atom. The pyroborate anion (Figure 3), which consists of two interconnected $BO_2\emptyset^{2-}$ units with an unconstrained B–$\emptyset$–B linkage, is similarly not a superstructural unit. Superstructural

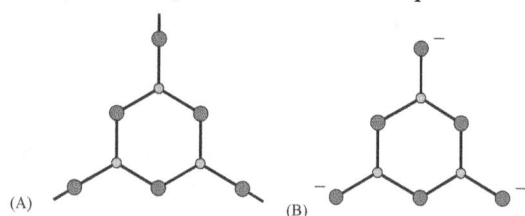

units also exist with more than two rings. The 3-ring $B_7O_9\emptyset_6^{3-}$ group (Figure 13A) is found in crystalline β-Tl.2B, whilst the structure of Ag.B is based on infinite $(B_2O_3\emptyset^-)_\infty$ chains of metaborate (NBO) groups sharing a $B^{(4)}$ unit with each of their neighbours (Figure 13B).

As discussed by Krogh-Moe,[33] the bond lengths and angles within a given superstructural unit vary in a systematic way from the average values for

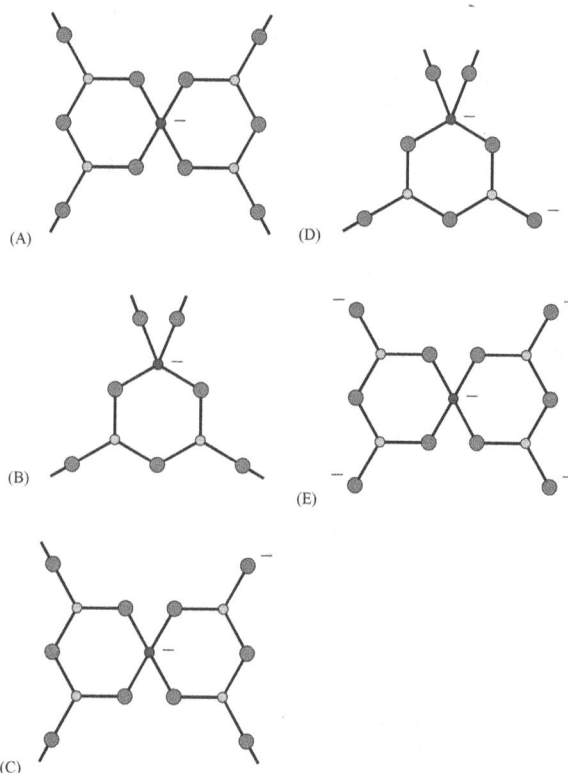

(A) (D) (B) (E) (C)

Figure 8. Superstructural units with one $B\emptyset_4^-$ tetrahedron: A, $B_5O_6\emptyset_4^-$ pentaborate group; B, $B_3O_3\emptyset_4^-$ triborate group; C, $B_5O_7\emptyset_3^{2-}$ di-pentaborate (NBO) group; D, $B_3O_4\emptyset_3^{2-}$ di-triborate (NBO) group and E, $B_5O_{10}^{5-}$ bicyclic metaborate anion (Key as Figure 1)

Figure 7. A, $B_3O_3\emptyset_3$ boroxol group and B, $B_3O_6^{3-}$ cyclic metaborate anion (Key as Figure 1)

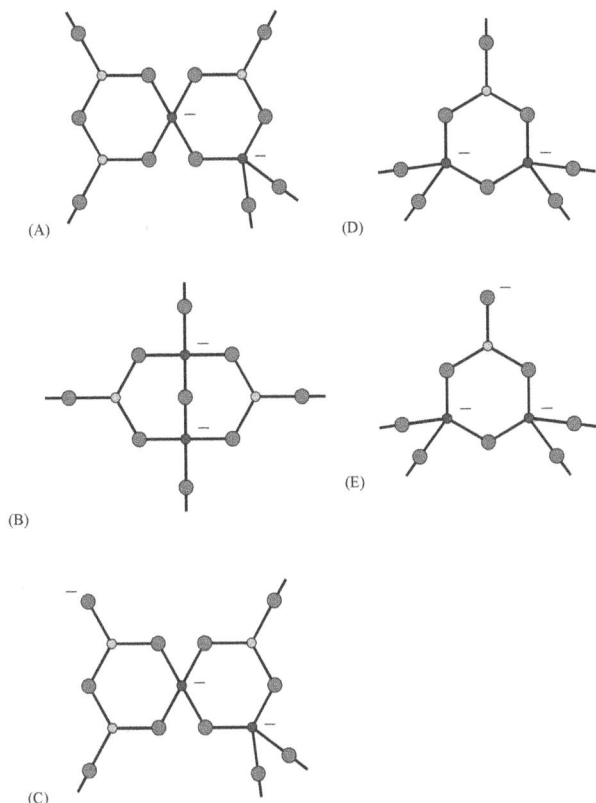

(A)

(B)

(C)

(D)

(E)

Figure 9. Superstructural units with two $B\emptyset_4^-$ tetrahedra: A, $B_5O_6\emptyset_5^{2-}$ di-pentaborate group; B, $B_4O_5\emptyset_4^{2-}$ diborate group; C, $B_5O_7\emptyset_4^{3-}$ tri-pentaborate (NBO) group; D, $B_3O_3\emptyset_5^{2-}$ di-triborate group and E, $B_3O_4\emptyset_4^{3-}$ metaborate (NBO) group (Key as Figure 1)

3-fold and 4-fold co-ordinated boron atoms, so that the internal atomic coordinates for the idealised form of each superstructural unit are specified by a minimum set of bond lengths and angles, as specified for the diborate group in Figure 14. The numbers of unique bond lengths and angles required for various superstructural units are given in Table 5 of Ref. 34, and a value for each parameter may be obtained from accurate structural data for crystals containing the superstructural unit in question, as discussed in Ref. 34.

The fact that borate crystallography up to the diborate composition is totally dominated by superstructural units indicates that their formation is not just random, due to the proximity of the equilibrium

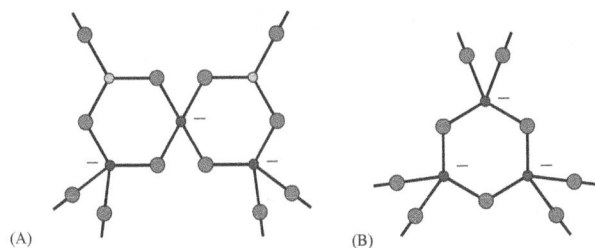

(A)

(B)

Figure 11. Non-existent superstructural units (3-membered rings) based on $B\emptyset_3$ and $BO\emptyset_2^-$ structural units: A, $B_3O_4\emptyset_2^-$ triborate (NBO) group and B, $B_3O_5\emptyset^{2-}$ di-triborate (2×NBO) group (Key as Figure 1)

B–Ô–B bond angle to that required for planar 3-membered rings, but that they must have a favourable formation/stabilisation energy. This is supported by the fact that the structures of many hydrated borates are also based on the same superstructural units, with their external bridging oxygen atoms, –Ø–, replaced by –OH groups.[32,35] In addition, it is interesting to note that, in the absence of a superstructural unit stabilisation energy, molecular dynamics simulations of borate systems do not generate large numbers of superstructural units, even when performed with interatomic potentials that yield the appropriate equilibrium B–Ô–B bond angles as, for example, in the case of pure vitreous B_2O_3.[36]

The stabilisation energy of the boroxol group has been determined from Raman studies of liquid B_2O_3 to be -6.4 ± 0.4 kcal/mol,[37] in good agreement with quantum chemistry calculations.[38] However, this stabilisation energy, and the planarity of the 3-membered ring of $B\emptyset_3$ triangles that form the boroxol group, do not arise from delocalised π-bonding, as demonstrated by the crystal structure of cæsium enneaborate (β-Cs.9B), which reveals that the B–Ø bonds within the 6-atom ring are longer than those outside the ring. The departures from planarity for the nine atoms that form each of the two symmetrically distinct boroxol groups in β-Cs.9B are small, being characterised by root mean square (rms) values of 0.077 and 0.088 Å, respectively.

For a planar 6-atom ring, the sum of the internal bond angles is $720°$. If the sum of the unconstrained equilibrium bond angles for the six atoms is less than $720°$, the ring will tend to be puckered but,

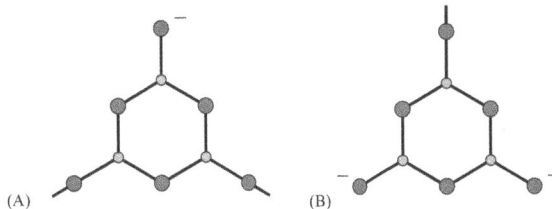

(A)

(B)

Figure 10. Superstructural units with three $B\emptyset_4^-$ tetrahedra: A, $B_5O_6\emptyset_6^{3-}$ tri-pentaborate group and B, $B_3O_3\emptyset_6^{3-}$ metaborate group (Key as Figure 1)

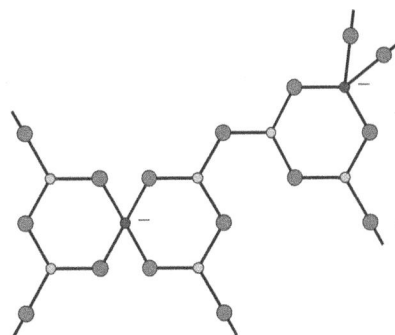

Figure 12. $B_8O_{10}\emptyset_6^{2-}$ tetraborate group, with internal $B^{[3]}-\emptyset-B^{[3]}$ bridge (Key as Figure 1)

(A)

(B)

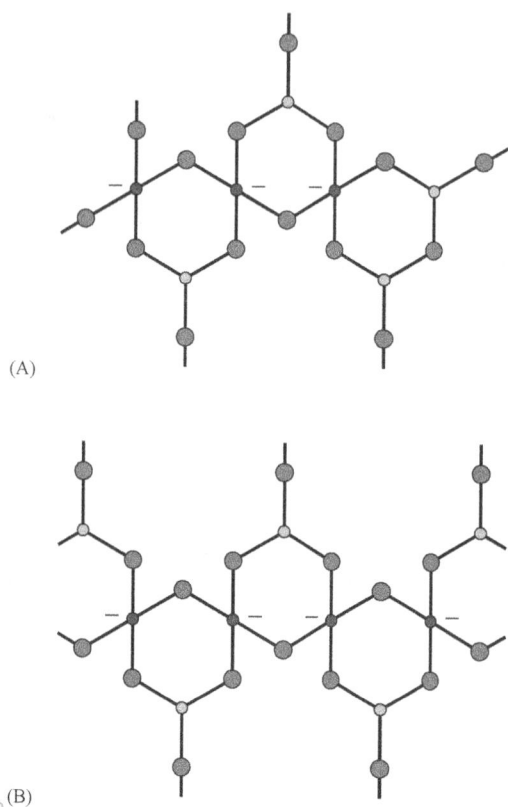

Figure 13. Species with more than two rings: A, 3-ring $B_7O_9\varnothing_6^{3-}$ group and B, infinite $(B_2O_3\varnothing^-)_\infty$ chain (Key as Figure 1)

if the sum is greater, this provides a driving force for the ring to be planar. The equilibrium B–Ô–B angle, not constrained by being incorporated into a superstructural unit, is ~130° (average for crystalline B_2O_3 is 130·7°), and the mean Ø–B̂–Ø angles for trigonal and tetrahedral boron atoms are 120° and ~110°, respectively. Thus, for a 3-membered ring of $B\varnothing_4^-$ tetrahedra (i.e. the metaborate unit in Figure 10B), the sum of the internal angles is ~720° and there is very little driving force for planarity, whereas the driving force increases with the sequence di-triborate (~730°)→triborate (~740°)→boroxol (~750°) group. A further driving force for planarity may also arise from the superstructural unit stabilisation mechanism, although it is interesting to note that the two nonexistent superstructural units in Figure 11 would both be naturally planar and so stabilisation cannot arise solely from planarity. Conversely, the frequent occurrence in crystalline borates of the diborate

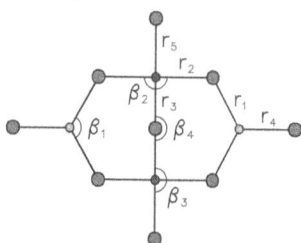

Figure 14. Definition of the unique bond lengths and angles for the diborate group[34] (Colour key as Figure 1)

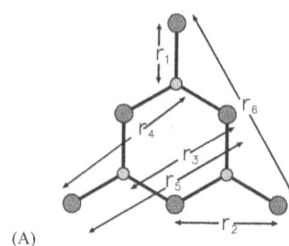

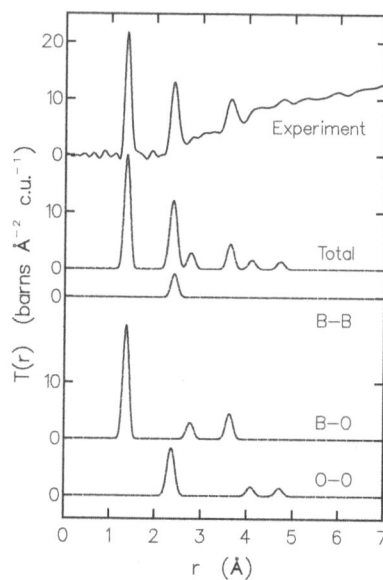

(A)

(B)

Figure 15. (A) Interatomic distances within the boroxol group (colour key as Figure 1) and (B) neutron total correlation function, $T(r)$, for vitreous B_2O_3 (solid line),[51] together with the contribution expected from boroxol groups (dashed lines)

group, which is based on two non-planar 6-atom rings, demonstrates that the stabilisation mechanism does not require such rings to be planar.

The situation for the metaborate group can be compared to that for vitreous GeO_2, where the equilibrium Ge–Ô–Ge angle is also ~130°. After constructing a ball-and-stick model of the latter, it was discovered that there were many 3-membered rings, but that almost none of them were planar.[39] Similarly, a comparison of the real-space total correlation function for vitreous GeO_2 with the contribution expected for *planar* 3-membered rings suggests that the concentration of such rings is not high.[39]

Irrefutable evidence for the existence of superstructural units in borate glasses is provided by optical spectroscopy; for example, the boroxol and triborate groups yield sharp lines in the Raman spectrum at ~808 and ~770 cm^{-1}, respectively.[40–45] Additional quantitative information is provided by NMR studies,[46,47] and potentially from inelastic neutron scattering measurements.[48–50] In diffraction studies, superstructural units lead to sharp features in the real space correlation function at higher interatomic distances, r, than found within the basic structural units and hence, to identify these features, it is neces-

Table 2. Crystal structures with monovalent network modifying cations*

Composition	x_M	Phase	Li	Na	K	Rb	Cs	Ag	Tl
M.9B	0·100	(α)		*	*	*	●	*	*
		β		*				◆	
		γ			*			◆	
M.5B	0·167	(α)	*	*	◆	◆	◆		◆
		β		*	◆	◆	◆		
		γ		*	◊	*	◆		
3M.13B	0·188						◆		
M.4B	0·200	(α)	◊	◆	*	*	*	◆	*
		β	◊	◆				◆	
5M.19B	0·208					◆	◆		
M.3B	0·250	(α)	◆	◆	◆	◆	◆		◆
		β		◆			◆		
		γ		*					
2M.5B	0·286		?	*			*		
3M.7B	0·300		◆	◆		◆	◆		
6M.13B	0·316			◆					
M.2B	0·333	(α)	◆	◆	◆	◆	◆	*	◊
		β		*					◆
		γ		*					*
2M.3B	0·400			*					
M.B	0·500	(α)	◆	◆	◆	◆	◆	◆	◆
		β	*	*			*		
		γ	◆						
3M.2B	0·600	(α)	◊	*					
		β	*						
2M.B	0·667	(α)	◊	◆	*				
		β	◊						
5M.2B	0·714			*					
3M.B	0·750	(α, I)	◆	◆			*	◆	◆
		β	◊						
		γ	◊						
		II							◆

* Key: ◆: Structure determined
●: Isostructural with, or related via a displacive transition to, a compound of known structure (Lattice parameters determined)
◊: Lattice parameters only
*: Polymorph exists (Gmelin Handbook[52] or reference in Appendix 1)
?: Polymorph possibly exists, or doubt has been expressed as to its existence

Table 3. Crystal structures with divalent network modifying cations*

Composition	x_M	Phase	Be	Mg	Ca	Sr	Ba	Zn	Cd	Pb
M.4B	0·200						◆			
M.3B						*				
2M.5B							*			
M.2B	0·333			◆	◆	◆	◆	◆	◆	●
2M.3B	0·400			◆						
M.B	0·500	(α, I)	*	◆	●	◆			◊	*
		β				◆				*
		γ								*
		II			◆					
		III			◆	●				
		IV			◆	◆				
9M.8B	0·529									*
6M.5B	0·545									◆
5M.4B	0·555									*
4M.3B	0·571						◆			
2M.B	0·667	(α)		◆	◆	◆	◊	◆		*
		β			*					*
3M.B	0·750		*	◆	◆	◆	*	◆	◆	
4M.B	0·800	(α)					*			*
		β								*

* Key as Table 2

binary glass forming systems of Figures 5 and 6 is summarised in Tables 2 and 3, and a complete list of references is given in the Appendix (Tables A1 and A2). Where no structural data have been reported, the existence of a given compound has been taken either from the Gmelin Handbook,[52] or from the reference given in Table A1 or A2. Table A3 lists the references for crystalline phases having two modifying cation species.

The lack of crystallographic information for binary borate systems can be seen from the $Na_2O–B_2O_3$ system, where a full crystal structure determination has only been performed for 10 out of the total of 24 compounds given in Table 2. The superstructural units or borate anions occurring in these ten polymorphs are summarised in Table 4, which shows the complexity of the crystalline borates, in that the first seven phases all have structures that include more than one superstructural unit, even when a single superstructural unit would yield a borate network of the correct stoichiometry. Similarly, the various diborate structures also exhibit a wide range of superstructural units, as may be seen from Table 5. Note that, unlike many diborates, α-Na.2B does not contain the diborate group, although this is found in α-Na.3B, which conversely does not have any triborate groups. The fact that the crystalline compounds frequently

sary to have good real space resolution by making measurements in reciprocal space to high scattering vector magnitudes, Q. This is illustrated in Figure 15B by the neutron total correlation function, $T(r)$, for vitreous B_2O_3,[51] which is compared to the contribution expected from boroxol groups. The sharp peak at 3·6 Å is due to the frequently occurring B–(3)O distance within the boroxol group (r_4 in Figure 15A).

5. Crystalline phases

The present state of knowledge concerning the structures of the crystalline phases that occur in the

Table 4. Distribution of (super)structural units and anions in $Na_2O–B_2O_3$ crystalline phases

Crystal	$B\emptyset_3$	$B_5O_6\emptyset_4^-$	$B_3O_3\emptyset_4^-$	$B_5O_6\emptyset_5^{2-}$	$B_4O_5\emptyset_4^{2-}$	$B_3O_3\emptyset_5^{2-}$	$B_3O_4\emptyset_3^{2-}$	$B\emptyset_4^-$	$B_3O_6^{3-}$	$B_2O_5^{4-}$	BO_3^{3-}
α-Na.4B	-	1/2	1/2	-	-	-	-	-	-	-	-
β-Na.4B	-	1/2	1/2	-	-	-	-	-	-	-	-
α-Na.3B	-	1/2	-	-	1/2	-	-	-	-	-	-
β-Na.3B	-	1/3	1/3	-	-	-	-	1/3	-	-	-
3Na.7B	1/3	-	1/3	-	-	1/3	-	-	-	-	-
6Na.13B	-	-	-	1/3	2/3	-	-	-	-	-	-
α-Na.2B	-	-	-	1/2	-	-	1/2	-	-	-	-
α-Na.B	-	-	-	-	-	-	-	-	1	-	-
2Na.B	-	-	-	-	-	-	-	-	-	1	-
3Na.B	-	-	-	-	-	-	-	-	-	-	1

Table 5. Distribution of (super)structural units in crystalline diborates

Diborate	$BØ_3$	$B_3O_3Ø_4^-$	$B_5O_6Ø_5^{2-}$	$B_5O_7Ø_3^{2-}$	$B_4O_5Ø_4^{2-}$	$B_3O_3Ø_5^{2-}$	$B_3O_4Ø_3^{2-}$	$BØ_4^-$	$n_{4(4)}$
Li, Mg, Zn, Cd	-	-	-	-	1	-	-	-	1
Na	-	-	$^1/_2$	-	-	-	$^1/_2$	-	$^2/_3$
K, Rb	$^1/_3$	-	-	-	$^1/_3$	$^1/_3$	-	-	1
Ca	-	$^1/_3$	-	-	$^1/_3$	-	-	$^1/_3$	$1^1/_2$
Sr, Pb	-	-	-	-	-	-	-	1*	-
Ba	-	-	$^1/_2$	-	-	$^1/_2$	-	-	$1^1/_2$
Li/K, Li/Rb	-	-	-	$^1/_2$	-	$^1/_2$	-	-	$^1/_2$

* Structure includes 3-fold co-ordinated oxygen atoms

involve a number of different superstructural units, even at the stoichiometric compositions for a single superstructural unit, indicates that this is likely to be even more true for the corresponding glasses, as indicated by NMR data.[46,47] For the crystalline diborates included in Figure 5, structures have not been reported for β- and γ-Na.2B, Cs.2B, Ag.2B and α- and γ-Tl.2B, whilst, in several non-glass forming systems, M^{2+} diborates exist as networks containing only diborate groups, including Mn.2B,[53,54] Fe.2B,[54] Co.2B,[54] Ni.2B[54] and Hg.2B.[54] The structures of vitreous Sr.2B and Pb.2B are unlikely to be similar to that of the corresponding (isostructural) crystalline polymorphs, since the latter contain 3-fold co-ordinated oxygen atoms, which violate Zachariasen's[1] criteria for glass formation[55] (see later, Section 9).

In the case of vibrational and NMR spectroscopy, the characteristic signature of a particular (super) structural unit is most easily determined from a crystalline phase having that unit as the only borate component. Table 6 lists those structures that are based on a single (super)structural unit. For those superstructural units that do not appear individually, crystal structures that contain them are listed, together with the other borate units present.

The different crystalline polymorphs formed with a given composition may be linked either by displacive transitions, as in the case of cæsium enneaborate (Cs.9B), or by reconstructive transitions, as exhibited by sodium triborate (Table 4). The former involve rapid atomic displacements at constant borate network topology, usually accompanied by a change in space group, whereas the latter are more sluggish and lead to a change in the borate network topology and in the superstructural units present. In the following sections, when calculating the number of different structures, compounds linked by displacive transitions are only counted once, whilst polymorphs resulting from reconstructive transitions are counted individually.

6. Pure B_2O_3

The fact that B_2O_3 does not crystallise from the pure anhydrous melt at ambient pressure makes it extremely difficult to grow single crystals of a size suitable for a structural study. Minute crystals of B_2O_3-I were first prepared by crystallisation from hydrous solution by Kracek et al,[28] and hence the initial structural study by Berger[56,57] employed x-ray

Table 6. Crystalline compounds containing superstructural units

Unit	Formula	Crystalline compound(s) Single unit	Multiple units
Structural units			
BO₃ triangle	$BØ_3$	B-I	-
BO₄ tetrahedron	$BØ_4^-$	γ-Li.B	-
BO₃ triangle (NBO)	$BOØ_2^-$	Li.B, Ca.B	-
Superstructural units			
Boroxol	$B_3O_3Ø_3$	-	Cs.9B (+ $B_3O_3Ø_4^-$)
Pentaborate	$B_5O_6Ø_4^-$	K.5B, Rb.5B, Cs.5B	-
Triborate	$B_3O_3Ø_4^-$	Li.3B, Cs.3B	-
Di-pentaborate	$B_5O_6Ø_5^{2-}$	Na.Cs.5B, Na.Tl.5B, K.Cs.5B	-
Di-pentaborate (NBO)	$B_5O_7Ø_3^{2-}$	-	Li.K.4B (+ $B_3O_3Ø_5^{2-}$), Li.Rb.4B (+ $B_3O_3Ø_5^{2-}$)
Diborate	$B_4O_5Ø_4^{2-}$	Li.2B, Mg.2B, Zn.2B	-
Tri-pentaborate	$B_5O_6Ø_6^{3-}$	-	2Ca.3B (+ $BØ_4^-$)
Tri-pentaborate (NBO)	$B_5O_7Ø_4^{3-}$	Na.2Ca.5B, K.2Sr.5B	-
Di-triborate	$B_3O_3Ø_5^{2-}$	-	3Li.7B (+ $B_3O_3Ø_4^-$ + $BØ_3$), K.2B & Rb.2B (+ $B_4O_5Ø_4^{2-}$ + $BØ_3$), Ba.2B (+ $B_5O_6Ø_5^{2-}$)
Di-triborate (NBO)	$B_3O_4Ø_3^{2-}$	-	α-Na.2B (+ $B_5O_6Ø_5^{2-}$)
Metaborate	$B_3O_3Ø_6^{3-}$	Ca.B-IV, Sr.B-IV	-
Metaborate (NBO)	$B_3O_4Ø_4^{3-}$	Li.4Ba.5B	-
Anions			
Cyclic metaborate	$B_3O_6^{3-}$	Na.B, K.B, α-Rb.B, Cs.B, Ba.B	-
Bicyclic metaborate	$B_5Ø_6O_4^{4-}$	3Na.2Ca.5B	-
Pyroborate (2×$BO_2Ø^{2-}$)	$B_2O_5^{4-}$	2Na.B, 2Mg.B, 2Ca.B, 2Sr.B	-
Orthoborate	BO_3^{3-}	α-3Li.B, 3Na.B, 3Ag.B-I, 3Tl.B, 3Mg.B, 3Ca.B, 3Sr.B, 3Zn.B	-

powder diffraction. Berger[56,57] concluded incorrectly that the structure is based on distorted BO_4 tetrahedra. This was shown to be incorrect by the NMR data of Kline et al,[58] and later electron[59] and x-ray[60] diffraction studies revealed that it is formed from planar ribbons of $BØ_3$ triangles. The structure of the high pressure modification, B_2O_3-II, is however based on highly distorted BO_4 tetrahedra, but also contains 3-fold coordinated oxygen atoms. The single crystals of B_2O_3-I studied by Gurr et al[60] were crystallised from the melt under pressures of 10 and 15 kbar, whereas Strong & Kaplow[59] indicate that the microscopic crystallite employed in their single grain electron diffraction study was prepared by solid state transformation from the glass, although no details are given as to how this was achieved.

There are various indications that the structure of vitreous B_2O_3 is very different from that of B_2O_3-I (cf. Section 3.1). For example, the density of B_2O_3-I is very much (41%[61]) higher than that of the glass. Vitreous B_2O_3 was the first glass for which the structural role of a superstructural unit (the $B_3O_3Ø_3$ boroxol group – Figure 7A) was discussed in detail. The presence of boroxol groups was first proposed by Goubeau & Keller,[62] and evidence for their existence is summarised in a review by Krogh-Moe.[63] Mozzi & Warren[64] compared their x-ray correlation function with the contribution expected for randomly linked boroxol groups and concluded that the major part of the glass is made up of boroxol groups. More recently, neutron diffraction[51,61] and NMR[65–68] and NQR[69] spectroscopy have yielded values for the fraction of the boron atoms in vitreous B_2O_3 that are in boroxol groups, x_B, of between 0.6 ± 0.2[61] and 0.85 ± 0.02.[69] (A value of 0.75 would correspond to equal numbers of boroxol groups and independent $BØ_3$ triangles.) The structure of vitreous B_2O_3 may thus be envisaged as a mixed random network of two corner sharing triangular units, one twice the size of the other, as illustrated schematically in two dimensions in Figure 1D of Ref. 61, the presence of boroxol groups being inextricably linked to the density and other anomalous properties of vitreous B_2O_3. Further evidence for the boroxol ring model, which explains both the anomalously low density of vitreous B_2O_3 and the extremely high activation energy for crystallisation from the melt, has been summarised elsewhere,[51,61,63] and will not be repeated here.

6.1. Crystalline boroxol B_2O_3 polymorph?

The question arises as to why the ambient pressure crystalline phase of B_2O_3 does not contain boroxol groups, or indeed why its structure is not solely based on boroxol groups. Clearly the presence of the latter in liquid B_2O_3, in the region of the melting point, and the fact that their concentration falls off with increasing temperature,[37] implies that their presence leads to a lower energy structure, at least in the liquid and vitreous states. The fact that pressure is required to crystallise B_2O_3-I from the anhydrous melt raises the question as to whether this is indeed the equilibrium crystalline phase at ambient pressure, since the application of pressure favours a more densely packed structure, whereas the larger size of the boroxol group compared to the basic $BØ_3$ triangle will lead to a more open structure as found in the glass, even in the presence of locally independent interpenetrating networks (cf. Sections 6.3 and 10.1). It is thus entirely possible that another crystalline polymorph is the stable phase at ambient pressure, and that this has a structure based on boroxol groups. The problem is how to access this polymorph, given that the high activation energy for crystallisation precludes its formation by crystallisation from the melt.

One possible method of preparing a boroxol-based ambient pressure crystalline polymorph of B_2O_3 might be by chemical synthesis, for example using sol-gel techniques and starting from an alkyl (R) substituted metaboric acid, $R_3B_3O_6$. Boroxol–boroxol linkages could be formed[70] via hydrolysis

$$R_2B_3O_5\text{–}OR + H_2O \rightleftharpoons R_2B_3O_5\text{–}OH + ROH \qquad (8)$$

followed by either alcohol condensation

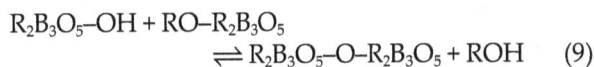

$$R_2B_3O_5\text{–}OH + RO\text{–}R_2B_3O_5$$
$$\rightleftharpoons R_2B_3O_5\text{–}O\text{–}R_2B_3O_5 + ROH \qquad (9)$$

or water condensation

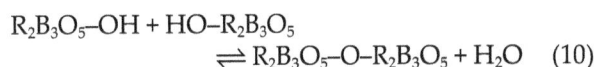

$$R_2B_3O_5\text{–}OH + HO\text{–}R_2B_3O_5$$
$$\rightleftharpoons R_2B_3O_5\text{–}O\text{–}R_2B_3O_5 + H_2O \qquad (10)$$

The resulting (amorphous?) B_2O_3 containing 100% boroxol groups could then be carefully crystallised to yield the required polymorph.

6.2. Mechanical coordination number change

As pointed out by Krogh-Moe,[71] the structures of crystalline Sr.2B and Pb.2B, which comprise a 3-dimensional network of corner sharing $BØ_4^-$ tetrahedra with one of the oxygen atoms in the asymmetric unit being 3-fold co-ordinated by boron (Section 5), demonstrate the precarious balance of boron between coordination numbers of 3 and 4, and the low energy required for activating states with 4-fold coordinated boron and 3-fold coordinated oxygen atoms. Hence such states can be mechanically induced, and this explains[23,72] the extremely low viscosity of liquid B_2O_3 compared to that of liquid SiO_2, even though the strengths of the B–Ø and Si–Ø bonds are closely similar. The same mechanism can be invoked[23,72] to account for the high compressibility of vitreous B_2O_3, which is nearly twice as large as that of vitreous SiO_2, despite the fact that the oxygen atom number density, $\rho_O{}^\circ$, for the former is larger (0.04704 cf. 0.04413 atoms Å^{-3}). The mechanism for the pressure-induced crystallisation of B_2O_3-I from the melt is also presumably based on the mechanically induced change in boron coordination number and transient 3-coordinated oxygen species.

An additional factor in the case of liquid and vitreous B_2O_3 is the planarity and relatively large size (cf. the basic $B\varnothing_3$ triangles) of the boroxol group. This means that there is a tendency for adjacent but not interconnected boroxol groups to align roughly parallel to each other, leading to regions that are locally layered, or to flattened network cages. Such "layers" can easily slide past each other, as was graphically illustrated by a ball and stick model of vitreous B_2O_3 constructed by the author's group[73] that, over a period of time, managed to "flow" over the edge of the (too-narrow) shelf on which it resided without breaking more than one or two bonds. Planarity also plays an important role in the structure of B_2O_3-I, which is based on sets of parallel planar ribbons of $B\varnothing_3$ triangles with three separate orientations.

6.3. Pressure compacted vitreous B_2O_3

In situ Raman spectroscopic studies[74] of vitreous B_2O_3 under pressure indicate that, with increasing pressure, the boroxol groups are broken up and essentially disappear at ~14 kbar, which is within the range of pressure used by Gurr et al[60] to crystallise B_2O_3-I from the anhydrous melt. On releasing the pressure, some boroxol groups are recreated almost immediately, but the glass density continues to relax over a period of several days at ambient pressure/temperature,[75] during which time the original boroxol ring fraction is restored. It is also interesting that, following the application and release of pressures up to 25 kbar, the resulting densified glass does not contain any $B\varnothing_4^-$ tetrahedra[76]. (Note that crystalline B_2O_3-II is prepared at high temperature (1100°C) and at very much higher pressure (65 kbar).)

In the liquid state, the above effects will occur at significantly lower pressures, and so the pressurised liquid from which Gurr et al[60] crystallised B_2O_3-I would have contained no boroxol groups, which explains why the same is true for crystalline product. In general, therefore, it can be concluded that borate systems are much more sensitive to pressure than their silicate counterparts. Another example of this pressure sensitivity concerns the possible role of pressure in determining the structure of rapidly quenched samples of vitreous Li.B, as discussed in Section 12.3.

7. Structural unit based models of borate glasses

The structural models that have been proposed for alkali and alkaline earth borate glasses are mainly of two types; those that predict the variation of x_4 with composition, and those that envisage the vitreous network as a random network of borate structural and/or superstructural units. These models are considered in greater detail in the following sections, in the light of the structural information provided by borate crystallography. In addition, those models that concentrate on x_4 have been extended to include the compositional dependence of the fractional content of all five major borate structural units.

Models that predict x_4 are based on the fact that the delocalised negative charge on the $B\varnothing_4^-$ tetrahedron means that adjacent $B\varnothing_4^-$ tetrahedra are energetically unfavourable (cf. Section 4.1); i.e. that $B^{[4]}-\varnothing-B^{[4]}$ bridges are unlikely to occur and that the structure will involve only $B^{[3]}-\varnothing-B^{[3]}$ and $B^{[3]}-\varnothing-B^{[4]}$ bridges. This is the basis of the Abe[77] and Beekenkamp[78] models discussed in Sections 7.1 and 7.2, respectively, whereas Gupta[79] has proposed a Random Pair Model (Section 7.3) involving pairs of adjacent $B\varnothing_4^-$ tetrahedra that are not incorporated into a superstructural unit. It is important to realise that all of these models do not include the effects of temperature. Hence they are only strictly valid at $T=0$, and so provide an upper limit for x_4.

When considering the various models that predict x_n ($0 \leq n \leq 4$), it is useful to divide the composition range into four different regions, according to the mix of basic structural units, $B^{(n)}$, present. In defining these regions, it is assumed that the repulsion between negatively charged nonbridging oxygen atoms on the same boron atom means that $B^{(1)}$ and $B^{(0)}$ units are not formed until it is unavoidable, consistent with the chosen model criteria. For each of the models, there are different scenarios, with increasing x_M, depending on the relative magnitudes of the energy required for the processes defined by Equations (1), (2) and (4)–(7), and on how these compare with the formation energy of any superstructural units present. For the standard Abe,[77] Beekenkamp[78] and Gupta[79] models, the overriding requirement is that x_4 has the maximum possible value, consistent with which it is assumed that the above processes occur with the following priority:

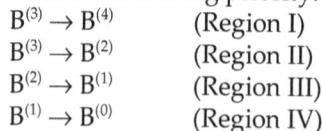

$B^{(3)} \rightarrow B^{(4)}$ (Region I)
$B^{(3)} \rightarrow B^{(2)}$ (Region II)
$B^{(2)} \rightarrow B^{(1)}$ (Region III)
$B^{(1)} \rightarrow B^{(0)}$ (Region IV)

Note that, in Regions II to IV, the minimum number of $B^{(4)}$ units are also lost in order to avoid violating the relevant model's criteria concerning adjacent $B^{(4)}$ units. Thus the nonbridging oxygen atoms that first appear in Region II arise from the conversion not only of the $B^{(3)}$ but also of some of the $B^{(4)}$ units present into $B^{(2)}$ units. The upper limit of Region II is at the metaborate composition ($x_M=\frac{1}{2}$), at which point the borate network comprises a combination of $B^{(4)}$ and $B^{(2)}$ units. In Region III, $B^{(2)}$ and the minimum number of $B^{(4)}$ units are converted into $B^{(1)}$ units, and Region III ends at the composition of a polyanion comprising a combination of $B^{(4)}$ and $B^{(1)}$ units and a structure consisting of these polyanions plus network modifying cations. Finally, in Region IV, the polyanions present at the end of Region III are converted into $B^{(0)}$ units, such that Region IV involves invert glasses[80] with the (poly)anions formed from $B^{(4)}$, $B^{(1)}$ and $B^{(0)}$ units. Region IV ends at the orthoborate composition (3M.B; $x_M = \frac{3}{4}$), with a structure consisting of BO_3^{3-} orthoborate anions and network modifying cations. The only alkali borate glasses formed in Region IV, by conventional quenching techniques, are from the Na_2O–B_2O_3 system, which has a second glass forming region at high x_M, although glasses can be prepared in other systems by rapid quenching.

7.1. Abe model

The Abe[77] model was the first to predict x_4 and incorporates the following assumptions:
(i) $B^{(4)}$ units cannot be immediate neighbours.
(ii) $B^{(3)}$ units cannot be bound to more than one $B^{(4)}$ unit.
(iii) Nonbridging oxygen atoms can only occur on trigonal boron atoms.

These assumptions are based on the concept of atomic groups comprising a $B\varnothing_4^-$ tetrahedron linked to four $B\varnothing_3$ triangles, and lead to the structural unit fractions

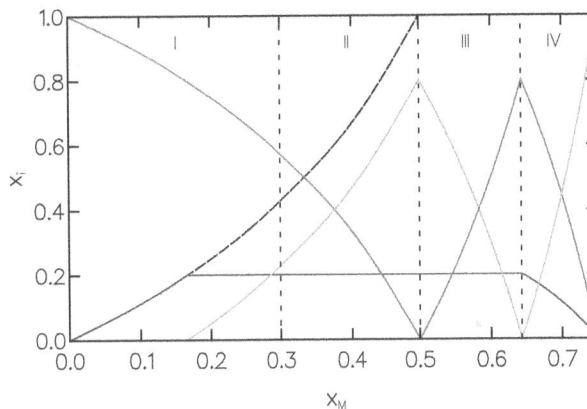

Figure 16. Variation of structural unit fractions, x_n, with composition for the Abe[77] model. Blue, x_4; red, x_3; green, x_2; magenta, x_1 and cyan, x_0. The black dashed line indicates the extension of Equation (3) and the vertical black short-dashed lines indicate the boundaries between Regions I, II, III and IV

shown in Figure 16 and Table 7. The upper limit of Region I is at $x_M = \frac{1}{6}$ (0·167) and x_4 remains constant at $x_4 = 0.2$ from this value of x_M until the boundary between Regions III and IV. The latter occurs at $x_M = \frac{9}{14}$ (0·643), at which point the structure consists of a combination of $[B(OBO_2)_4]^{9-}$ $(B_5O_{12}^{9-})$ polyanions (Figure

Table 7. Structural unit fractions for the Abe,[77] Beekenkamp,[78] Gupta[79] and Krogh-Moe[26]–Griscom[2] models

Fraction	Region I	Region II	Region III	Region IV
Abe[77] model				
x_4	$x_M/(1-x_M)$	$\frac{1}{5}$	$\frac{1}{5}$	$(3-4x_M)/6(1-x_M)$
x_3	$(1-2x_M)/(1-x_M)$	$(1-2x_M)/(1-x_M)$	-	-
x_2	-	$(6x_M-1)/5(1-x_M)$	$(9-14x_M)/5(1-x_M)$	-
x_1	-	-	$(2x_M-1)/(1-x_M)$	$2(3-4x_M)/3(1-x_M)$
x_0	-	-	-	$(14x_M-9)/6(1-x_M)$
Beekenkamp[78] model				
x_4	$x_M/(1-x_M)$	$(3-4x_M)/6(1-x_M)$	$(3-4x_M)/6(1-x_M)$	$(3-4x_M)/6(1-x_M)$
x_3	$(1-2x_M)/(1-x_M)$	$(1-2x_M)/(1-x_M)$	-	-
x_2	-	$(10x_M-3)/6(1-x_M)$	$(9-14x_M)/6(1-x_M)$	-
x_1	-	-	$(2x_M-1)/(1-x_M)$	$2(3-4x_M)/3(1-x_M)$
x_0	-	-	-	$(14x_M-9)/6(1-x_M)$
Gupta[79] model				
x_4	$x_M/(1-x_M)$	$(3-4x_M)/5(1-x_M)$	$(3-4x_M)/5(1-x_M)$	$(3-4x_M)/5(1-x_M)$
x_3	$(1-2x_M)/(1-x_M)$	$(1-2x_M)/(1-x_M)$	-	-
x_2	-	$3(3x_M-1)/5(1-x_M)$	$(7-11x_M)/5(1-x_M)$	-
x_1	-	-	$(2x_M-1)/(1-x_M)$	$3(3-4x_M)/5(1-x_M)$
x_0	-	-	-	$(11x_M-7)/5(1-x_M)$
Krogh-Moe[26]–Griscom[2] model (½)*				
x_4	$x_M/(1-x_M)$	$(1-2x_M)/(1-x_M)$	-	-
x_3	$(1-2x_M)/(1-x_M)$	$(1-2x_M)/(1-x_M)$	-	-
x_2	-	$(3x_M-1)/(1-x_M)$	$(2-3x_M)/(1-x_M)$	$(3-4x_M)/(1-x_M)$
x_1	-	-	$(2x_M-1)/(1-x_M)$	$(3-4x_M)/(1-x_M)$
x_0	-	-	-	$(3x_M-2)/(1-x_M)$
Krogh-Moe[26]–Griscom[2] model (⅔)*				
x_4	$x_M/(1-x_M)$	$(2-3x_M)/3(1-x_M)$	$(2-3x_M)/3(1-x_M)$	-
x_3	$(1-2x_M)/(1-x_M)$	$(2-3x_M)/3(1-x_M)$	$(2-3x_M)/3(1-x_M)$	-
x_2	-	-	-	-
x_1	-	$(3x_M-1)/3(1-x_M)$	$(3x_M-1)/3(1-x_M)$	$(3-4x_M)/(1-x_M)$
x_0	-	-	-	$(3x_M-2)/(1-x_M)$
Krogh-Moe[26]–Griscom[2] model (¾)*				
x_4	$x_M/(1-x_M)$	$(3-4x_M)/5(1-x_M)$	$(3-4x_M)/5(1-x_M)$	$(3-4x_M)/5(1-x_M)$
x_3	$(1-2x_M)/(1-x_M)$	$(3-4x_M)/5(1-x_M)$	$(3-4x_M)/5(1-x_M)$	$(3-4x_M)/5(1-x_M)$
x_2	-	-	-	-
x_1	-	-	-	-
x_0	-	$(3x_M-1)/5(1-x_M)$	$(3x_M-1)/5(1-x_M)$	$(3x_M-1)/5(1-x_M)$

* Value in brackets is upper x_M limit for diborate groups

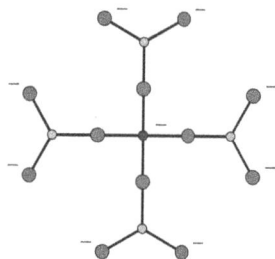

Figure 17. $[B(OBO_2)_4]^{9-}$ ($B_5O_{12}^{9-}$) polyanion (Key as Figure 1)

17) and network modifying cations. Note that rule (ii) implies that there must be a minimum of two $B\varnothing_3$ triangles between any two $B\varnothing_4^-$ tetrahedra.

7.2. Beekenkamp model

As may be seen from a comparison of Figures 4 and 16, the Abe[77] model predicts a maximum value of x_4 that is too low and is reached at too low a value of x_M. This was realised by Beekenkamp,[78] who eliminated rule (ii), to allow two $B\varnothing_4^-$ tetrahedra to be separated by only one $B\varnothing_3$ triangle. The relaxing of this rule leads to the structural unit fractions in Figure 18 and Table 7. Region I ends at $x_M=^3/_{10}$, which corresponds to a network of alternating $B^{(3)}$ and $B^{(4)}$ units. As for the Abe[77] model, Region III terminates at $x_M=^9/_{14}$ (0·643) with a structure involving $B_5O_{12}^{9-}$ polyanions and network modifying cations. Throughout Regions II to IV, x_4 is given by

$$x_4=(3-4x_M)/6(1-x_M) \tag{11}$$

7.3. Gupta model

The position of the maximum in x_4 for the Beekenkamp[78] model is still at too low a value of x_M, which implies that rule (i) (Section 7.1) must also be violated. As a result, Gupta[79] has proposed a Random Pair Model, based on the following three rules:
(i) $B^{(4)}$ units occur in pairs joined by a bridging

oxygen atom with variable bond and torsion angles; i.e. as $B_2O\varnothing_6^{2-}$ units.
(ii) $B_2O\varnothing_6^{2-}$ units cannot be bound to each other.
(iii) Nonbridging oxygen atoms can only occur on trigonal boron atoms.

The structural unit fractions for the Gupta[79] model (Table 7) are plotted in Figure 19. The presence of the $B^{[4]}$–$\varnothing$–$B^{[4]}$ bridges linking the pairs of $B\varnothing_4^-$ tetrahedra means that the value of x_M at which the maximum in x_4, and the boundary between Regions I and II, occurs is increased to $^1/_3$ (0·333; the diborate composition), in better agreement with NMR data. The upper limit of Region III is at $x_M=^7/_{11}$ (0·636) and corresponds to a $B_8O_{19}^{14-}$ polyanion, similar to that in Figure 17 but with a pair of $B\varnothing_4^-$ tetrahedra at the centre. The value of x_4 throughout Regions II to IV is higher than for the Beekenkamp[78] model, being

$$x_4=(3-4x_M)/5(1-x_M) \tag{12}$$

7.4. Alternative scenario for Region IV

For all of the above models, in Region IV, the need to maximise x_4 takes preference over the requirement that $B^{(0)}$ units are not formed until absolutely necessary. If the situation is reversed, the borate structure initially comprises a mixture of either $B_5O_{12}^{9-}$ (Beekenkamp[78]) or $B_8O_{19}^{14-}$ (Gupta[79]) polyanions, together with pyroborate ($B_2O_5^{4-}$) anions, and x_4 goes to zero at $x_M=^2/_3$ (0·667; the pyroborate composition), at which point the structure involves only pyroborate anions and network modifying cations. For both models

$$x_4=(2-3x_M)/(1-x_M) \tag{13}$$

and

$$x_1=(2x_M-1)/(1-x_M) \tag{14}$$

whilst, for $^2/_3 \leq x_M \leq ^3/_4$

$$x_1=(3-4x_M)/(1-x_M) \tag{15}$$

and

$$x_0=(3x_M-2)/(1-x_M) \tag{16}$$

Note that, for the latter range of x_M, the $B^{(1)}$ units still exist in pairs as pyroborate anions. This alternative behaviour for Region IV is denoted by the dashed lines in Figures 18 and 19.

8. Crystal structures

The structural role of $B\varnothing_4^-$ tetrahedra in ambient pressure borate crystals is summarised in Table 8, which details the location and interconnection of $B\varnothing_4^-$ tetrahedra, listing the number of crystalline structures having the given arrangement. The hypothesis that $B^{[4]}$–$\varnothing$–$B^{[4]}$ bridges not within a superstructural unit are energetically unfavourable is supported by

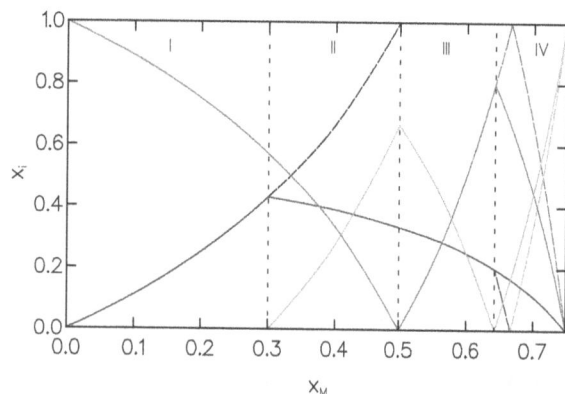

Figure 18. Variation of structural unit fractions, x_n, with composition for the Beekenkamp[78] model. The alternative pyroborate model for Region IV (Section 7.4) is denoted by the coloured dashed lines (Colour key as Figure 16)

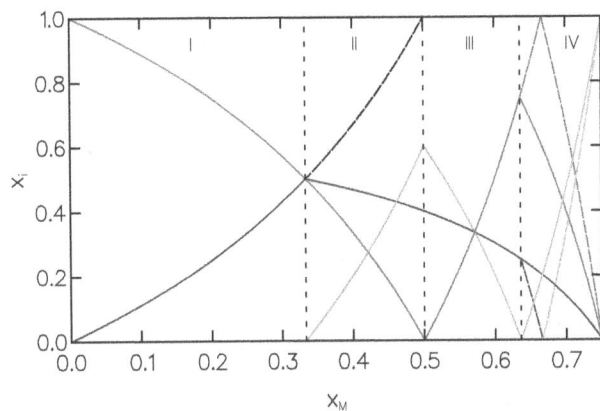

Figure 19. Variation of structural unit fractions, x_i, with composition for the Gupta[79] model and alternative pyroborate model for Region IV (Key as Figure 18)

crystallographic studies, in that only $B^{[3]}$–$\varnothing$–$B^{[3]}$ and $B^{[3]}$–$\varnothing$–$B^{[4]}$ bridges are found outside superstructural units in the crystalline state until $B^{[4]}$–$\varnothing$–$B^{[4]}$ bridges are unavoidable, which occurs at x_M=0·3 if there are also none within superstructural units.

8.1. Region I

At the lowest values of x_M, the addition of a network modifier leads to the formation of $B^{(4)}$ units, according to Equation (1) or (2). Hence this region is characterised by a combination of $B^{(3)}$ and $B^{(4)}$ units; x_4 has the value given by Equation (3) and

$$x_3=1-x_4=(1-2x_M)/(1-x_M) \qquad (17)$$

This applies to all of the crystalline phases with $x_M<^1/_3$ (diborate composition), and to all of the single

modifier diborates except α-Na.2B, Sr.2b and Pb.2B. The diborate composition is also that at which superstructural units incorporating three $B^{(4)}$ units (Figure 10) are first observed.

Table 8 reveals that, for every crystalline phase with $x_M\leq^1/_5$ (tetraborate composition), all of the $B\varnothing_4^-$ tetrahedra are incorporated into superstructural units, and hence that their creation is facilitated by the superstructural unit formation energy. Independent $B^{(4)}$ units (i.e. those that are not part of a superstructural unit) first appear at $x_M=^5/_{24}$ (0·208; 5M.19B) and pairs of $B^{(4)}$ units within a superstructural unit (Figure 9) begin at $x_M=^1/_4$, with the diborate group in α-Na.3B. External $B^{[4]}$–$\varnothing$–$B^{[4]}$ bridges (i.e. those not within a superstructural unit) are not observed until $x_M=^3/_{10}$.

It is interesting to note that the structure of crystalline Cs.9B is based on boroxol and triborate groups in the ratio 2:1. The fact that the triborate group is by far the most commonly occurring superstructural unit in crystalline structures (*cf.* Table 9) suggests that it may be the one with the largest stabilisation energy. This is supported by the fact that it is possible to conceive of an alternative structure for cæsium enneaborate consisting of alternating boroxol and pentaborate groups in the ratio 4 boroxol to 3 pentaborate (topologically equivalent to a Si_3N_4 network). This does not form, presumably due to the fact that it is energetically less favourable although, in this respect, it would be extremely interesting to know whether the structures of other enneaborates (Na, K, Rb, Ag and Tl) are similar to that of Cs.9B. On the other hand, all of the pentaborates studied to date have network structures based solely on pentaborate groups and so the difference in stabilisation energy

Table 8. Distribution of $B^{(4)}$ species in ambient pressure borate phases*

Formula	x_M	$B^{(4)}$ (ssu)	$2\times B^{(4)}$ (ssu)	$3\times B^{(4)}$ (ssu)	$B^{(4)}$ (indep)	$B^{[4]}$–$\varnothing$–$B^{[4]}$	NBO	Total
M⁺ modifying cation								
M.9B	0·100	1	-	-	-	-	-	1
M.5B	0·167	4	-	-	-	-	-	4
3M.13B	0·188	1	-	-	-	-	-	1
M.4B	0·200	4	-	-	-	-	-	4
5M.19B	0·208	2	-	-	2	-	-	2
M.3B	0·250	7	1	-	1	-	-	7
3M.7B	0·300	3	3	-	1	1	-	3
6M.13B	0·316	-	1	-	-	1	-	1
M.2B	0·333	1	4	1	-	-	1	5
M.B	0·500	-	-	-	2	2	6	7
M²⁺ modifying cation								
M.4B	0·200	1	-	-	-	-	-	1
M.2B	0·333	1	5	-	1	2	-	5
2M.3B	0·400	-	-	1	1	1	-	1
M.B	0·500	-	-	-	-	-	2	2
Two modifying cations								
M.5B	0·167	1	-	-	-	-	-	1
M.4B	0·200	1	-	-	-	-	-	1
M.3B	0·250	5	-	-	-	-	-	5
2M.5B	0·286	-	3	-	-	-	-	3
3M.7B	0·300	1	-	-	1	1	-	1
M.2B	0·333	2	2	-	-	-	2	2
3M.5B	0·375	-	2	-	-	-	2	2
M.B	0·500	1	1	-	-	-	2	2

* Superstructural unit and nonbridging oxygen atom are denoted ssu and NBO, respectively, and the $B^{[4]}$-O-$B^{[4]}$ bridges are between, or external to, superstructural units

Table 9. Frequency of occurrence of superstructural units in ambient pressure borate phases

Unit/Anion	Formula	No. of occurrences
Boroxol	$B_3O_3\varnothing_3$	1
Pentaborate	$B_5O_6\varnothing_4^-$	16
Triborate	$B_3O_3\varnothing_4^-$	27
Di-pentaborate	$B_5O_6\varnothing_5^{2-}$	8
Di-pentaborate (NBO)	$B_5O_7\varnothing_3^{2-}$	2
Diborate	$B_4O_5\varnothing_4^{2-}$	9
Tri-pentaborate	$B_5O_6\varnothing_6^{3-}$	2
Tri-pentaborate (NBO)	$B_5O_7\varnothing_4^{3-}$	2
Di-triborate	$B_3O_3\varnothing_5^{2-}$	7
Di-triborate (NBO)	$B_3O_4\varnothing_3^{2-}$	1
Metaborate	$B_3O_3\varnothing_6^{3-}$	0
Metaborate (NBO)	$B_3O_4\varnothing_3^{3-}$	1
Cyclic metaborate	$B_3O_6^{3-}$	5
Bicyclic metaborate	$B_5\varnothing_6O_4^{5-}$	1

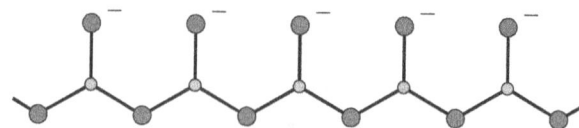

Figure 20. $(BO\varnothing_2^-)_\infty$ metaborate chain (Key as Figure 1)

between triborate and pentaborate groups is likely to be small.

All of the tetraborates so far investigated have borate networks based on the so-called tetraborate group. As explained in Section 4.2, this does not qualify as a superstructural unit but consists of a combination of a pentaborate and a triborate group. The way in which the latter are interconnected differs between the various tetraborates, but all of the structures comprise two independent interpenetrating networks and have equal numbers of pentaborate and triborate groups. In α-Na.4B and Ba.4B, the pentaborate and triborate groups alternate, whereas the other tetraborates include both pentaborate–pentaborate and triborate–triborate linkages.

8.2. Region II

In the crystalline state, nonbridging oxygen atoms ($B^{(2)}$ units) do not occur until $x_M = 1/3$, as part of the di-triborate (NBO) group (Figure 8D) in the structure of α-Na.2B (Table 4), and in di-pentaborate (NBO) groups (Figure 8C) in Li.K.4B and Li.Rb.4B. Thus, in the ambient pressure crystalline state, nonbridging oxygen atoms are relatively rare for $x_M < 1/2$ (metaborate composition) and *only* occur in superstructural units, *never* as independent $B^{(2)}$ units. Hence, as with the $B^{(4)}$ units in Region I, their initial introduction is facilitated by the superstructural unit stabilisation energy. Similarly, the fact that nonbridging oxygen atoms are not formed until $x_M = 1/3$ is presumably linked to the non-existence of the superstructural units in Figure 11, so that there would be no contribution to their creation (Equation (4)) from superstructural unit formation.

The Ca.2B and Ba.2B structures have a $B^{[3]}$–$\varnothing$–$B^{[3]}$ bridge within a superstructural unit (triborate and di-pentaborate, respectively), and hence require external $B^{[4]}$–$\varnothing$–$B^{[4]}$ bridges, suggesting that the superstructural unit stabilisation energies more than compensate for the increased energy due to these external $B^{[4]}$–$\varnothing$–$B^{[4]}$ bridges. Note that the di-pentaborate composition ($x_M = 2/7 = 0.286$) does not require the presence of any $B^{[4]}$–$\varnothing$–$B^{[4]}$ bridges, and that crystalline

polymorphs (e.g. Na.Tl.5B) do exist with networks comprising only di-pentaborate groups (*cf.* Table 6 and Figure 9A). On the other hand, the presence of the diborate group in crystal structures with $x_M < 0.3$ (e.g. α-Na.3B), and its common occurrence as the only superstructural unit in many crystalline diborates, suggests that it has a particularly favourable formation energy.

The only crystalline compositions found between the diborate and metaborate compositions are 3M.5B and 2M.3B. The former has only been found in mixed modifier systems (Na.2Ca.5B and K.2Ca.5B), and the latter only in the sodium and calcium borate systems. The structures of Na.2Ca.5B and K.2Ca.5B are based on the tri-pentaborate (NBO) superstructural unit shown in Figure 9C. The structure of 2Na.3B is unknown, whilst the network in 2Ca.3B is formed from equal numbers of tri-pentaborate groups (Figure 10A) and $B^{(4)}$ units. This structure also has external $B^{[4]}$–$\varnothing$–$B^{[4]}$ bridges between the tri-pentaborate groups and $B^{(4)}$ units, but not between pairs of independent $B\varnothing_4^-$ tetrahedra or between tri-pentaborate groups.

All but one (Tl.B) single modifier crystalline metaborate formed at ambient pressure have structures based either on infinite chains of $B^{(2)}$ units (Figure 20; Li, Ca, Sr) or on the cyclic metaborate anion (Figure 7B; Na, K, Rb, Cs, Ba). These structures yield values of x_4 and x_2 of zero and unity, respectively. The atoms forming the metaborate chains in crystalline Li.B, Ca.B-I and Sr.B-I are constrained by the periodic symmetry to be almost coplanar whereas, in a glass, similar chains would be expected to exhibit a range of bond and torsion angles. Tl.B also has a chain structure but, in this case, the chains are formed from equal numbers of $B^{(4)}$ and $B^{(2)}$ units ($x_4 = x_2 = 0.5$), whilst Li.4Ba.5B incorporates chains formed from metaborate (NBO) groups (Figure 9E) and independent $B^{(2)}$ units ($x_4 = 0.4$; $x_2 = 0.6$). $B^{(4)}$ units are also found in 3Na.2Ca.5B, as part of the bicyclic metaborate anion (Figure 8E), and in 3Li.Na.4B[81], although the structure of the latter remains to be determined. The application of pressure leads to the formation of independent $B^{(4)}$ units, as in the case of the high pressure polymorphs of Li.B and Ca.B. The structures of Ca.B-III and Ca.B-IV both include 3-membered rings of three $B\varnothing_4^-$ tetrahedra (metaborate group; Figure 10B).

8.3. Region III

It is interesting to note that for none of the M_2O–B_2O_3 or alkaline earth borate systems does the conventional glass forming region extend into Region III. Bray and O'Keefe's NMR data[11] for lithium borate glasses do, however, include two compositions in Region III that were prepared by rapid quenching. The only systems that exhibit conventional glass formation in this region are ZnO–B_2O_3, CdO–B_2O_3 and PbO–B_2O_3, and crystalline phases occur in both the ZnO–B_2O_3 (4Zn.3B) and PbO–B_2O_3 (9Pb.8B, 6Pb.5B and 5Pb.4B) systems. The structure of 4Zn.3B is unusual in that it has an isolated O^{2-} ion that is not involved in the borate network, and hence its formula is better written $Zn_4O(BO_2)_6$. More interesting, however, is the fact that, in the Li_2O–B_2O_3 and Na_2O–B_2O_3 systems, crystalline phases are found at the 3:2 stoichiometry (α- and β-3Li.2B and 3Na.2B), whereas this is not the case for any of the other systems in Figures 5 and 6. Unfortunately, none of the 3M.2B crystal structures are known.

8.4. Region IV

Crystalline orthoborates occur in the Li_2O–B_2O_3, Na_2O–B_2O_3, Rb_2O–B_2O_3, Ag_2O–B_2O_3 and Tl_2O–B_2O_3 systems, and in all of the MO–B_2O_3 systems except PbO–B_2O_3. All of the crystalline orthoborates so far investigated, including those with two network modifier species, have structures based on the BO_3^{3-} orthoborate anion. Region IV also includes the pyroborate composition (2M.B), at which all but one (Be.Ca.B – see Section 13.6) of the known crystalline structures incorporate the $B_2O_5^{4-}$ pyroborate anion (Figure 3; Tables 2 and 3). The second glass forming region for the Na_2O–B_2O_3 system[14,15] commences very close to the pyroborate composition, at x_M=0·665. The compound 5Na.2B (x_M=0·714) is similarly found in the sodium borate system, at the upper limit of the second glass forming region (x_M=0·715), but no structural data have been reported.

9. Superstructural unit models

The idea that the structure of borate glasses involves a random network composed of superstructural units is due to Krogh-Moe,[26] and further support for this model has been provided by the work of Bray.[82,83] Gupta[79] has rejected the suggestion by Krogh-Moe[23] and Bray[84] that superstructural units play an important role in borate glass networks, and referred to their approach as a modified crystallite model. This is, however, incorrect in that Krogh-Moe and Bray merely stated that the vitreous network incorporates superstructural units. There is no concept of periodicity in their model, nor the assertion that the interconnection and relative orientation of these superstructural units are restricted to those found in the crystalline state, and hence their model cannot be considered a crystallite model. Indeed, Krogh-Moe[26] described the resulting structure as a three-

dimensional disordered framework, and Bray[82] referred to it as a modified Warren–Zachariasen network, in which the superstructural units (rather than individual borate structural units) are connected randomly to each other. This, Bray[83] argues, can be viewed as a natural refinement of the continuous random network model.

The major difference between Gupta's random pair model,[79] and that of Krogh-Moe[26] and Bray,[82,83] is that the $B^{[4]}$–$\varnothing$–$B^{[4]}$ bridge linking his pairs of $B\varnothing_4^-$ tetrahedra is unconstrained, and therefore incompatible with their being incorporated into a superstructural unit. However, the fact that $B^{(4)}$ units are only found as part of a superstructural unit at low x_M ($x_M \leq 1/5$), and that $B^{[4]}$–$\varnothing$–$B^{[4]}$ bridges do not occur outside superstructural units in the crystalline state for $x_M < 3/10$, strongly suggests that the great majority of any pairs of linked $B\varnothing_4^-$ tetrahedra present in Region I will indeed be part of a superstructural unit with a constrained internal $B^{[4]}$–$\varnothing$–$B^{[4]}$ bridge. In this connection, it is interesting to observe that one of the possible $B\varnothing_4^-$ clusters (No. IV) considered by Gupta[79] consists of a 3-membered ring of two $B\varnothing_4^-$ tetrahedra and one $B\varnothing_3$ triangle; i.e. a di-triborate group! It should also be noted that Gupta's formulæ[79] (Table 7) for the variation of x_4 with x_M are unaffected by his pairs of $B\varnothing_4^-$ tetrahedra being incorporated into superstructural units with no external $B^{[4]}$–$\varnothing$–$B^{[4]}$ bridges.

The question thus arises as to how borate superstructural units can be connected together topologically to achieve a disordered glass network of the appropriate number density with no broken bonds, except for nonbridging oxygen atoms, and what are the criteria for their interconnection. Zachariasen's criteria[1] for oxide glass formation state that:

"(1) an oxygen atom is linked to not more than two atoms A;

"(2) the number of oxygen atoms surrounding atoms A must be small;

"(3) the oxygen polyhedra share corners with each other, not edges or faces;"

and, for a three-dimensional network,

"(4) at least three corners in each oxygen polyhedron must be shared."

These criteria are based on a network formed by $A\varnothing_n$ structural units, where A is the network forming cation, but can easily be extended to apply to a network formed by superstructural units. The first criterion then refers to the oxygen atoms shared between adjacent superstructural units, whilst the second involves the connectivity of the superstructural unit in question, which should be small (i.e. 3 or 4 for a 3-dimensional network (criterion 4)). It is therefore interesting to note that this is true for the connectivities of many of the commonly occurring superstructural units in the crystalline state (Table 9); those of the pentaborate and triborate groups (Figures 8A and 8B), are both 4, and that of the boroxol group (Figure 7A) is 3. The connectivity of the diborate group (Figure 9B) is also 4, but those of the di-pentaborate and di-triborate groups (Figures 9A and 9D), which occur with similar frequency in the crystalline state, are 5, suggesting that the latter are less likely to play such an important structural role in glasses. The third criterion mitigates against edge or face sharing, and indicates that there should only be one shared bridging oxygen atom between any two adjacent superstructural units.

All of the interconnecting (external) B–Ø- bonds for the boroxol and pentaborate groups involve trigonal boron atoms, whereas the triborate and diborate groups have two $B^{[3]}$–Ø- and two $B^{[4]}$–Ø- interconnecting bonds. The fact that the latter have equal numbers of $B^{[3]}$–Ø- and $B^{[4]}$–Ø- interconnections means that they can individually or together form a network having only $B^{[3]}$–Ø–$B^{[4]}$ bridges, which occurs at the maximum value of x_4 without external $B^{[4]}$–Ø–$B^{[4]}$ bridges. The four interconnecting bonds for the pentaborate, triborate and diborate groups project towards the vertices of a (distorted) tetrahedron and hence, in their relationship to Zachariasen's criteria,[1] these groups may be considered as topologically equivalent to a $BØ_4^-$ or $SiØ_4$ tetrahedron.

The increased size of the superstructural units, compared to the basic $BØ_3$ and $BØ_4^-$ structural units, means that a network formed solely by the former will have a very open, low density structure. As discussed later in Section 10.1, this may lead to locally independent, interpenetrating networks, as demonstrated by many borate crystal structures. In addition, the requirement for efficient packing means that the network will almost certainly include independent $B^{(n)}$ structural units. The structure of borate glasses may thus be envisaged as being based on a mixed Zachariasen–Warren network that includes both superstructural units and independent basic structural units connected randomly to each other.

9.1. Krogh-Moe model

Based on an analysis of melting point depression in the Na_2O–B_2O_3 system, Krogh-Moe[26] has predicted the distribution of superstructural units in sodium borate glasses as a function of x_M, up to the diborate composition ($x_M = 1/3$):

$0.05 \leq x_M \leq 0.20$: Boroxol + tetraborate (+ diborate) groups.
$0.20 \leq x_M \leq 0.26$: (Boroxol +) tetraborate + diborate groups.
$0.30 \leq x_M \leq 0.33$: Tetraborate + diborate groups.

All of these superstructural units are consistent with Zachariasen's criteria,[1] as expounded above, in that they have a connectivity of 3 or 4, if the tetraborate group is considered in terms of its constituent pentaborate and triborate groups. For the first two regions, the group in brackets results from a 10% dissociation of the tetraborate groups, not into triborate plus pentaborate groups, but into boroxol

plus diborate groups, according to the equilibrium reaction[26]

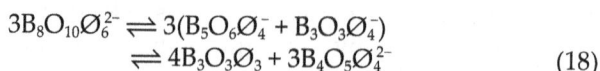

$$3B_8O_{10}Ø_6^{2-} \rightleftharpoons 3(B_5O_6Ø_4^- + B_3O_3Ø_4^-)$$
$$\rightleftharpoons 4B_3O_3Ø_3 + 3B_4O_5Ø_4^{2-} \qquad (18)$$

Further evidence in support of Krogh-Moe's model[26] can be found in the review by Griscom[2].

Krogh-Moe's arguments[26] concerning the presence of tetraborate and diborate groups are based on the fact that the corresponding crystalline phases are not dissociated to any great extent ($\leq 10\%$) in the melt. In addition, in a later paper,[85] he concluded that sodium tetraborate melts with little decomposition, and hence that the melt does not contain separate pentaborate and triborate groups, but rather tetraborate groups; i.e. that a triborate group must therefore (nearly) always appear in such melts associated with a pentaborate group, presumably due to some form of association energy. Thus the above distribution relies on the assumption that the network structure of crystalline sodium tetraborate is based solely on tetraborate groups, and that of crystalline sodium diborate on diborate groups. Unfortunately, however, unlike that of crystalline Li.2B, which was then known to Krogh-Moe, the structure of crystalline α-Na.2B was later (1974) found (by Krogh-Moe) *not* to involve diborate groups, but to comprise layers formed from di-pentaborate and di-triborate (NBO) groups.

On the other hand, there is evidence concerning the presence of diborate groups in sodium borate glasses close to the diborate composition. For example, in 1938, Hibben[86] noted the similarity in the Raman spectra of borax (see Section 3) and sodium diborate glass and concluded that their structures must be related. Similarly, Krogh-Moe[25,87] discussed the role of diborate groups in the structures of vitreous Li.2B and Na.2B, based on a comparison of their IR spectra with that for crystalline Li.2B. The increased connectivity (5) of the di-pentaborate and di-triborate groups, compared to that for the diborate group (4), would also suggest that the latter is more likely to occur in vitreous systems. A possible resolution of this apparent inconsistency involves the structures of the high temperature polymorphs β-Na.2B and γ-Na.2B. It is entirely conceivable that the structure of one (or both) of these polymorphs includes diborate groups and that the transition that links them to α-Na.2B is reconstructive, as is the case for the transition linking α- and β-Na.3B. Note that it is γ-Na.2B that is thermodyamically stable just below the melting point, and hence it is this phase that is responsible for the melting point depression.

The fact that, at low x_M, $B^{(4)}$ units are always involved in superstructural units suggests that Equations (1) and (2) should be rewritten in terms of superstructural units. Thus, for an alkali–oxide modifier, boroxol groups can be converted into triborate

groups according to the reaction

$$\tfrac{1}{2}M_2O + B_3O_3\varnothing_3 \rightarrow M^+ + B_3O_3\varnothing_4^- \tag{19}$$

In the vitreous state, the number of degrees of freedom can be increased, and independent $B\varnothing_3$ triangles introduced, by the equilibrium reaction

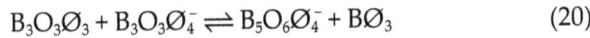

$$B_3O_3\varnothing_3 + B_3O_3\varnothing_4^- \rightleftharpoons B_5O_6\varnothing_4^- + B\varnothing_3 \tag{20}$$

This introduces pentaborate groups, which can combine with a triborate group to form tetraborate groups. Alternatively, pentaborate groups can be formed by triborate groups combining with two independent $B\varnothing_3$ triangles

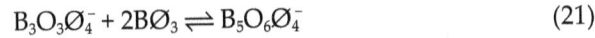

$$B_3O_3\varnothing_4^- + 2B\varnothing_3 \rightleftharpoons B_5O_6\varnothing_4^- \tag{21}$$

The tetraborate association energy would then provide a driving force to the right for Equations (20) and (21). In the presence of larger superstructural units, the independent $B\varnothing_3$ triangles aid network packing, and hence increase the specific network density (Section 12.5). With the addition of further alkali oxide, the triborate groups of Equation (19) can react to form di-triborate groups

$$\tfrac{1}{2}M_2O + B_3O_3\varnothing_4^- \rightarrow M^+ + B_3O_3\varnothing_5^{2-} \tag{22}$$

However, as indicated above, the increased connectivity of the di-triborate group suggests that, in the vitreous state, diborate groups are more likely to form; e.g. either directly via the reaction

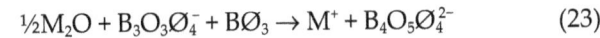

$$\tfrac{1}{2}M_2O + B_3O_3\varnothing_4^- + B\varnothing_3 \rightarrow M^+ + B_4O_5\varnothing_4^{2-} \tag{23}$$

or from initial di-triborate groups as a result of the equilibrium

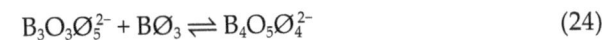

$$B_3O_3\varnothing_5^{2-} + B\varnothing_3 \rightleftharpoons B_4O_5\varnothing_4^{2-} \tag{24}$$

The $B\varnothing_3$ units required for these reactions are likely to be present in the melt as a result of various reactions/equilibria, such as that of Equation (20).

9.2. Krogh-Moe–Griscom model

Griscom[2] has employed Krogh-Moe's predictions[26] to formulate the superstructural unit model in Figure 21. An idealised model, which ignores the 10% dissociation, is denoted by the dashed lines. Above $x_M = 1/3$, the diborate contribution is linearly extrapolated to zero at $x_M = 1/2$. The solid line represents a more realistic model that includes the 10% dissociation and rounds off the sharp changes in gradient. In addition, the diborate contribution is linearly extrapolated to zero at $x_M = 2/3$, to better represent experimental NMR data.

There are two problems with the KroghMoe[26]–Griscom[2] model. The first is that the borate network of pure vitreous B_2O_3 is assumed to be composed of 100% boroxol groups ($2/3$ of a boroxol group per B_2O_3 unit), whereas both NMR and neutron diffraction data indicate approximately equal numbers of boroxol groups and independent $B\varnothing_3$ triangles. In

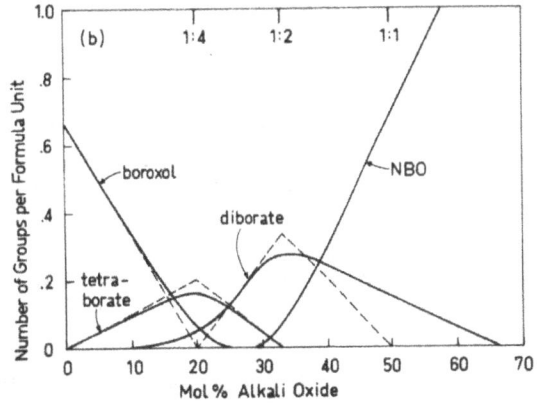

Figure 21. Variation of superstructural unit content with composition, as predicted for the Na_2O–B_2O_3 system by Krogh-Moe[26] and plotted by Griscom[2]

this connection, it is interesting to note that Osipov[88] has proposed an alternative form of Equation (18) that is consistent with these data

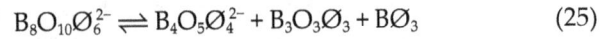

$$B_8O_{10}\varnothing_6^{2-} \rightleftharpoons B_4O_5\varnothing_4^{2-} + B_3O_3\varnothing_3 + B\varnothing_3 \tag{25}$$

A second problem with the more realistic version of the Krogh-Moe[26]–Griscom[2] model is that, as shown in Figure 21, it is internally inconsistent. If nonbridging oxygen atoms are substituted for the $B^{(4)}$ units in diborate groups, it is also necessary to include $B^{(3)}$ units to maintain the correct stoichiometry. The structural unit fractions for the idealised model are plotted as the dashed lines in Figure 22, whereas the solid lines represent a variation of the idealised model in which the number of diborate groups is extrapolated linearly to zero at $x_M = 2/3$, rather than at $x_M = 1/2$; i.e. to the same value of x_M as for the more realistic model, but without the 10% dissociation or the arbitrary rounding off of the sharp peaks. The structural unit fractions for both models are included in Table 7. It is interesting to note that, for $x_M > 1/3$, the first variant

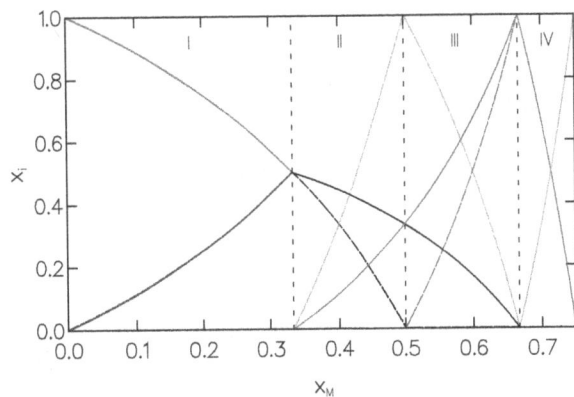

Figure 22. Structural unit fractions, x_n, for the idealised Krogh-Moe[26]–Griscom[2] model (dashed lines) and for the modified version with the diborate contribution extrapolated to zero at $x_M = 2/3$ (solid lines). Blue, x_4; red, x_3; green, x_2; magenta, x_1 and cyan, x_0. The black lines denote overlapping x_4 and x_3 contributions

leads to a structure comprising diborate groups and $B^{(2)}$ units, whilst the second implies a combination of diborate groups and $B^{(1)}$ units, although some of the latter could dissociate into equal numbers of $B^{(2)}$ and $B^{(0)}$ units, as discussed later in Section 12.2 (Equation (35)). Extrapolating the diborate content to zero at $x_M = ¾$ would require a mixture of diborate groups and isolated BO_3^{3-} anions ($B^{(0)}$ units). The value of x_4 in this case is identical to that for the Gupta[79] model (Table 7).

In the light of the discussion concerning diborate groups in Section 9.1, it is interesting to compare Figure 21 with the chemical structure of sodium borate glasses, as predicted by the model of associated solutions.[89] As may be seen from Figure 23, the major contributions to the chemical structure are indeed from B_2O_3, Na.4B and Na.2B, in agreement with Krogh-Moe's predictions.[26] This is, of course, to be expected, since Krogh-Moe's model is also based on thermodynamics (melting point depression). Krogh-Moe's conclusions should, therefore, be taken to refer to chemical groupings, rather than to specific superstructural units. In Figure 23, there is also a minor contribution from Na.5B at low x_M, indicating a slight excess of pentaborate groups, over those needed to form tetraborate groups in association with triborate groups, and a significant presence of Na.3B chemical groupings around the triborate composition ($x_M = 0·25$). These extra contributions replace Krogh-Moe's 10% dissociation. The diborate contribution starts at $x_M \sim 0·2$ and extends to $x_M \sim 0·6$. In addition to those from the diborate chemical grouping (di-triborate (NBO) groups), nonbridging oxygen atoms ($B^{(2)}$ units) are introduced via cyclic metaborate anions (Na.B chemical grouping), commencing at $x_M \sim 0·25$, and by pyroborate anions ($B^{(1)}$ units), which first appear at $x_M \sim 0·4$. Although not present in α-Na.2B, some diborate groups are contributed via the α-Na.3B chemical grouping.

10. Stereochemical constraints

10.1. Network density

In addition to the stabilisation energy, there are also stereochemical considerations that influence the detailed distribution of superstructural units in a given glass. As pointed out by Krogh-Moe,[85] the bond angles and superstructural units of a single borate network do not in general permit efficient atom packing, and hence many crystalline borates containing superstructural units exist as two independent, interpenetrating networks, suggesting that, in the presence of significant numbers of superstructural units, the corresponding glasses are locally similar, in order to achieve a sufficiently high borate network density. Alternatively, as in 5K.19B, a normal borate network density can be obtained by the incorporation of larger numbers of independent $BØ_3$ triangles and/

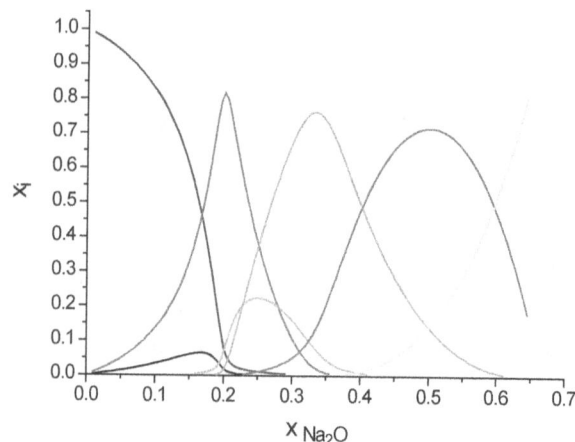

Figure 23. Chemical structure of Na_2O–B_2O_3 glasses at 800 K.[89] Blue, B_2O_3; black, Na.5B; red, Na.4B; cyan, Na.3B; green, Na.2B; magenta, Na.B and yellow, 2Na.B

or $BØ_4^-$ tetrahedra. Even with locally independent, interpenetrating networks, some independent $BØ_3$ triangles and/or $BØ_4^-$ tetrahedra are almost certainly necessary to ensure more efficient packing and the necessary degrees of freedom for glass formation. In the absence of such units, excessive network strain and/or numbers of broken bonds would be likely to be "frozen in" on passing through the glass transition and would be reflected in a relatively high heat of crystallisation. However, the heats of crystallisation of binary borate glasses are typically in the range $\sim$−2 to −4 kcal/mol,[90] indicating the absence of excessive network or (super)structural unit deformation relative to the crystalline state.

It should be noted that one of the main objections to the boroxol ring model for vitreous B_2O_3 has been that it is extremely difficult to construct a random network model with a sufficiently high number density, $\rho°$. As noted in Section 6.2, the oxygen atom number density, $\rho_O°$, for vitreous B_2O_3 is in fact greater than that for vitreous SiO_2, even though the boron atom coordination number is lower than that for silicon, which is the reverse of what might be expected. The problem with all of the random network models of vitreous B_2O_3 to date is that they have been constructed with the preconceived notion that the structure comprises a *single* network, rather than locally independent interpenetrating networks, as is almost certainly the case for the real material. Krogh-Moe[91] was the first to suggest that the structure of vitreous B_2O_3 consists of interpenetrating networks and that, in the glass phase, the disorder is probably enhanced by a network being joined with its twin at random positions within its structure. In effect, therefore, in the bulk, the structure as a whole comprises a single network, but locally within a small region, such as the volume of a typical ball-and-stick cluster model or the unit cell of a molecular dynamics simulation, there are likely to be two, or even more, separate interpenetrating network segments that are not interconnected. However, in the case of molecular dynamics simulations, it is unclear as to the effect periodic boundary conditions will have on the ease of formation of interpenetrating networks and the frequency of crosslinking. Krogh-Moe[91] similarly proposed a structure for vitreous lithium and sodium diborates mainly consisting of interpenetrating networks of diborate groups. The absence of two perfectly defined sub-networks (i.e. the presence of occasional crosslinking) is reminiscent of the problem of the absence of antiferromagnetism in glasses, such as Fe_2O_3–P_2O_5,[92] where the lack of two clearly independent magnetic sub-networks leads to frustration and speromagentism (short range antiferromagnetic ordering).

10.2. Network modifying cations

Another important stereochemical parameter is the effective radius and charge of the network modifying cations and the degree of ionicity in their bonding. Simple radius ratio considerations mean that, in the melt and during vitrification or crystallisation, the network modifying cations will attempt to restructure their local environments in order to achieve their optimum first coordination shell. This determines the minimum dimension of the cavities in the borate network in which these ions reside, and hence influences the borate network number density. However, it is characteristic of borate crystal structures that the coordination number, $n_{M(O)}$, of a given network modifying cation varies for different crystalline phases and frequently between different sites in the same structure, as may be seen for the Na^+ ion from Table 10. Similarly, the oxygen polyhedra surrounding the network modifying cations are often considerably distorted, with a wide range of M–O distances. Krogh-Moe[33] ascribed this to the effect of the rigid superstructural units, and hence similar distorted network modifying cation polyhedra will exist in the vitreous state. In addition, the lack of periodicity in the latter will lead to a considerable variation in the network modifying cation environment for a given glass, both in terms of the distortion and coordination number. Thus, regular coordination polyhedra do not usually form around the network modifying cations in borate structures where there are significant numbers of superstructural units. The distorted oxygen polyhedron surrounding network modifying cations also means that it is frequently difficult to define an exact coordination number, especially for the larger cations. A criterion that is sometimes employed is to count as neighbours all of the surrounding oxygen atoms with an M–O bond length less than the shortest M-B distance. The irregular network modifying cation polyhedra also explain why isostructural series of crystalline compounds can have different cation coordination numbers.

Table 10. Sodium ion coordination in crystalline Na_2O–B_2O_3 phases

Crystal	4	5	6	7	8	9
α-Na.4B	-	-	-	-	1	-
β-Na.4B	-	¼	¼	-	¼	¼
α-Na.3B	-	⅓	⅔	-	-	-
β-Na.3B	-	-	⅓	⅓	⅓	-
3Na.7B	-	⅓	⅔	-	-	-
α-Na.2B	-	-	¾	¼	-	-
α-Na.B	-	-	-	1	-	-
2Na.B	-	½	½	-	-	-
3Na.B	⅓	⅔	-	-	-	-

Whereas the information obtained from earlier broad line NMR experiments was limited to the fractions of the various structural unit species present in the vitreous network, more recent high resolution and double resonance studies have yielded data concerning the environment and spatial distribution of the network modifying cations in both borate and silicate systems. For the latter, clustering of the network modifying cations, together with their associated nonbridging oxygen atoms, leads to their being concentrated into channels, or into two-dimensional boundary regions between silica rich clusters, as represented by the modified random network model.[93] However, in the case of Na_2O–B_2O_3 glasses, NMR data indicate a near homogeneous spatial distribution of the Na^+ ions for $x_M \geq 0.15$, with "some degree" of clustering only being found at lower cation concentrations.[94] In addition, in systems containing two network modifying cations (Li^+/Na^+ and Na^+/K^+), the data indicate a statistical mixing of the cations with no evidence of unlike-cation pairing or like-cation segregation.[94,95]

The network modifying cation distribution in borate glasses described in the previous paragraph is at variance with earlier conclusions concerning a non-random distribution of (heavy) network modifying cations (Cs^{+}[27], Ba^{2+}[96], Cd^{2+}[97], Sr^{2+}[98] and Tl^{+}[99]) and, in particular, that these ions exist in pairs or as larger clusters, with a definite interionic spacing of ~4–5 Å that is independent of concentration. These conclusions were reached on the basis of the x-ray heavy element technique, according to which prominent maxima in the electronic differential radial distribution function were identified as M–M interactions. However, Prins[100] has pointed out that this is by no means always correct, as may be seen by comparing the areas in units of electrons^2 (el^2) under the M–M peak for a pair of such cations having z_M electrons ($A_{M(M)} \sim z_M^2$) with that for $n_{M(O)}$ oxygen atom neighbours around a single cation ($A_{M(O)} \sim 2n_{M(O)}z_Mz_O$). Thus, even for the highest atomic number cation in the above list, Tl^+, assuming a value of 8 for $n_{Tl(O)}$ yields $A_{Tl(O)}$=10240 el^2, which is significantly greater than $A_{Tl(Tl)}$=6400 el^2. That the strong peak at ~4–5 Å in the electronic differential radial distribution function may not be evidence of pairs of M^+ or M^{2+} cations is also supported by the fact that the strongest peak for

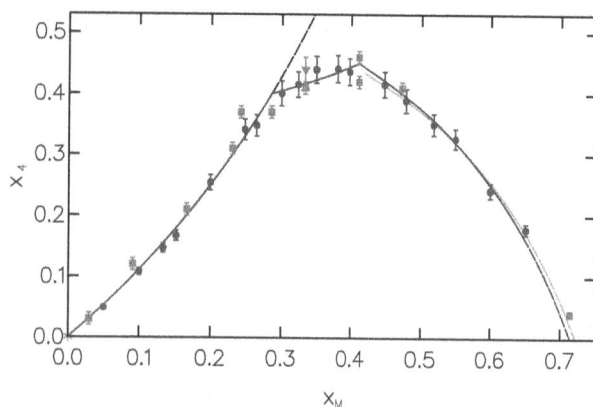

Figure 24. Fraction of 4-fold coordinated boron atoms in lithium borate glasses. Blue closed circles, Jellison et al[102] (NMR); red closed squares, Kroeker and co-workers[103,104] (NMR); red closed triangle, Ratai et al[95] (NMR) and red inverted closed triangle, Cormier et al[105] (neutron diffraction). The green and blue solid lines are high x_M[102] and three section[47] fits, respectively, to the data of Jellison et al[102] and the black dashed line is the continuation of Equation (3). The green and blue dashed lines indicate the extrapolation of the fits to zero x_4

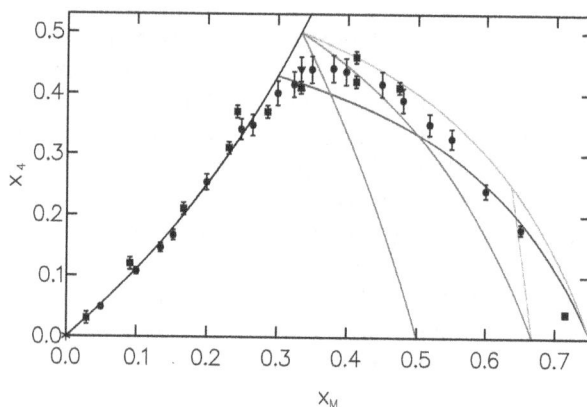

Figure 25. Comparison of the Beekenkamp[78] (blue solid line), Gupta[79] (green solid line) and Krogh-Moe[26]–Griscom[2] (red solid line, idealised model and magenta solid line modified model with diborate extrapolated to zero at $x_M = 2/3$) models with the experimental data of Figure 24 (black point symbols – key as Figure 24). The black line denotes Equation (3) and the cyan dashed line indicates the alternative Region IV scenario for the Beekenkamp[78] and Gupta[79] models discussed in Section 7.4

pure vitreous B_2O_3, and for a sequence of lithium, sodium and potassium borate glasses, is also situated between 4 and 5 Å.[101]

11. Comparison of models with experiment

11.1. Compositional dependence of x_4

A composite plot of NMR[95,102–104] and neutron diffraction[105] data for the variation of x_4 with x_M for the $Li_2O–B_2O_3$ glass system is shown in Figure 24. In contrast to the models of Figures 18, 19 and 22, the NMR data follow a smooth curve, without a discontinuity at $x_M = 3/10$ or $1/3$. The maximum value of x_4 (~0·44) occurs at $x_M \sim 0·37$. The green solid line in Figure 24 is the result of a linear fit to the data of Jellison et al[102] when plotted against

$$R = x_M/(1-x_M) \quad (26)$$

such that, for $x_M > 0·42$ ($R = 0·724$)

$$x_4 = -(0·23 \pm 0·03)R + (0·60 \pm 0·02) \quad (27)$$

This fit extrapolates to zero at $x_M = 0·722$ (green dashed line), close to the orthoborate composition ($x_M = 3/4$). In a later paper, Feller et al[47] specify x_4 using three linear regions (blue solid line), the first of which is equivalent to Equation (3)

$$x_4 = R \quad 0·000 \leq x_M < 0·286; \ (0·0 \leq R < 0·4) \quad (28)$$

$$x_4 = 1/3 + R/6 \quad 0·286 \leq x_M < 0·412; \ (0·4 \leq R < 0·7) \quad (29)$$

$$x_4 = 5/8 - R/4 \quad 0·412 \leq x_M < 0·650; \ (0·7 \leq R < 1·86) \quad (30)$$

Equation (30) extrapolates to zero at $x_M = 5/7$ (0·714; $R = 2·5$), as indicated by the blue dashed line.

The lithium borate system is that for which the experimental data are the most comprehensive and cover the widest range of x_M. In addition, the $Li_2O–B_2O_3$ system is not complicated by carbonate retention at high x_M,[106] and so the experimental data of Figure 24 are compared to the predictions of the various models in Figure 25. It should be noted, however, that the Krogh-Moe[26]–Griscom[2] model is strictly based on data for the $Na_2O–B_2O_3$ system. On the other hand, unlike crystalline α-Na.2B, crystalline Li.2B does have a structure based on the diborate group. Although they have not yet been determined, the structures of Li.5B and of the polymorphs of Li.4B are likely to be based on pentaborate and tetraborate groups, respectively, as is the case for all of those pentaborates and tetraborates whose structures have so far been investigated. Li.5B is, however, very unstable and rapidly decomposes to give the tetraborate.[107] Therefore, unlike the case of the Na.5B chemical grouping in Figure 23 (Section 9.2), the Li.5B chemical grouping is unlikely to make a significant contribution to the chemical structure of the corresponding glasses (see Section 12.3, Figure 30). Sodium is the only alkali metal for which both tetraborate and diborate structures are established, although those of all three sodium pentaborate polymorphs remain to be determined.

The absence of discontinuities in the NMR data may be explained as follows: First, it should be realised that a perfect network of alternating $B^{(4)}$ and $B^{(3)}$ units (Beekenkamp;[78] $x_M = 3/10$), which implies only even-membered rings of $B^{(n)}$ structural units (and hence no superstructural units), or of diborate groups with only external $B^{[3]}–\emptyset–B^{[4]}$ bridges ($x_M = 1/3$), represents an extremely small region of configuration

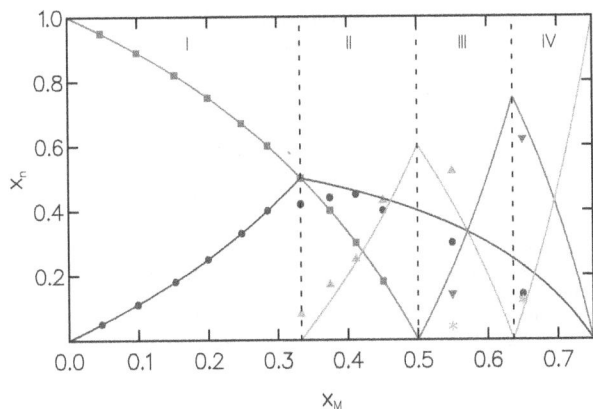

Figure 26. Comparison of the structural unit fractions for Li_2O–B_2O_3 glasses obtained by NMR spectroscopy[110] (point symbols) with those predicted by the Gupta[79] model (solid lines). Blue line and solid circles, x_4; red line and solid squares, x_3; green line and solid triangles, x_2; magenta line and solid inverted triangles, x_1 and cyan line and asterisks, x_0

space, which will not be accessible to a glass obtained by conventional quenching techniques. (A similar argument has been employed by Wright & Desa[108] to explain the absence of glass formation for the stoichiometric III–V and II–VI analogues of vitreous silica.) Thus a real network will contain some odd-membered rings, formed by incorporating a $B^{(2)}$ unit, and/or adjacent $B^{(4)}$ units. Any $B^{(2)}$ units or $B^{[4]}$–$\varnothing$–$B^{[4]}$ bridges are likely to be present *within* superstructural units (which indeed are ***based on*** odd-membered (3-membered) rings), particularly above and just below the diborate composition ($x_M = {}^1/_3$). Hence x_4 for the NMR data starts to deviate from the value given by Equation (3) close to the di-pentaborate composition ($x_M = 0.286$), due to the formation of nonbridging oxygen atoms, and gradually falls below the Equation (3) value as x_M approaches $^3/_{10}$.

It is interesting to note that the inclusion of superstructural units incorporating nonbridging oxygen atoms into the structures of crystalline α-Na.2B, Li.K.4B and Li.Rb.4B (*cf.* Section 8.2) reduces the value of x_4 from that of the Gupta[79] model and other crystalline diborates ($x_4 = \frac{1}{2}$) to $x_4 = ^3/_8$, which is slightly lower than those for the alkali borate glasses. This strongly supports the idea that the initial deviation of x_4 for the glasses from the value given by Equation (3) is also due to the appearance of superstructural units having nonbridging oxygen atoms.

Except for the highest values of x_M ($x_M \geq 0.6$), the experimental data mainly lie between the lines defining the Beekenkamp[78] and Gupta[79] models, suggesting that not all of the $B^{(4)}$ units exist as pairs. In Region IV, the extrapolated fits (green and blue dashed lines in Figure 24) and the highest x_M data point[104] suggest that x_4 falls to zero between the pyroborate and orthoborate compositions, but closer to the latter ($x_M \sim 0.72$), indicating that the x_4 cut-off predicted

by the pyroborate model of Section 7.4 is too sharp. The value of x_M at which x_4 becomes zero is also too low both for the idealised Krogh-Moe[26]–Griscom[2] model, and for the revised model with the diborate contribution extending to the pyroborate composition. For the other alkali borate glass systems, it should be noted that the NMR data of Zhong & Bray[109] indicate that, in the region $0.14 \leq x_M \leq 0.45$, the value of x_4 falls progressively further below that for the Li_2O–B_2O_3 system with increasing alkali atomic number, and hence the agreement of the models with experiment will vary as a function of both the alkali cation and x_M.

11.2. Structural unit fractions

The fractional content of all five major structural unit species in Li_2O–B_2O_3 glasses has been investigated as a function of composition by both NMR[47,104,110,111] and optical[112,113] spectroscopy. The NMR data from Ref. 110 are compared to the Gupta[79] model in Figure 26, and it can be seen that, whilst the agreement between the model and experiment is not perfect, the model does predict the order of the appearance and disappearance of the various structural unit species and the approximate magnitude of their fractions, x_n. A similar optical spectroscopy study of the compositional dependence of the structural unit fractions for binary and pseudo-binary Na_2O–B_2O_3 glasses has been published by Kamitsos & Karakassides.[114] (To cover a continuous range of x_M it was necessary to add 5–7 mol% Al_2O_3 in the gap between the two sodium borate glass forming regions (*cf.* Figure 5).) The various distributions are qualitatively similar to those for the lithium borate system, although x_4 peaks much earlier, and more $B^{(4)}$ units appear to be present at the highest values of x_M.

11.3. Superstructural units

Bray and co-workers have published a detailed study of lithium[47,110,111] and sodium[46,115] borate glasses in which they have performed a fit to their data incorporating different sites for both tetrahedral and trigonal boron atoms that are based on the (super)structural units specified by the Krogh-Moe[26] model, and are summarised in Table 11. The data for the Li_2O–B_2O_3

Table 11. Tetrahedral and trigonal boron atom sites in Li_2O–B_2O_3 glasses identified by NMR spectroscopy[47]

Site	Unit	$n_{B(O)}$	(Super)structural unit or anion
B^3	$B\varnothing_3$	3	Boroxol
T^3	$B\varnothing_3$	3	Tetraborate
T^4	$B\varnothing_4^-$	4	Tetraborate
D^3	$B\varnothing_3$	3	Diborate
D^4	$B\varnothing_4^-$	4	Diborate
L^4	$B\varnothing_4^-$	4	Independent
M^3	$BO\varnothing_2^-$	3	-
P^3	$BO_2\varnothing^{2-}$	3	Pyroborate
O^3	BO_3^{3-}	3	Orthoborate

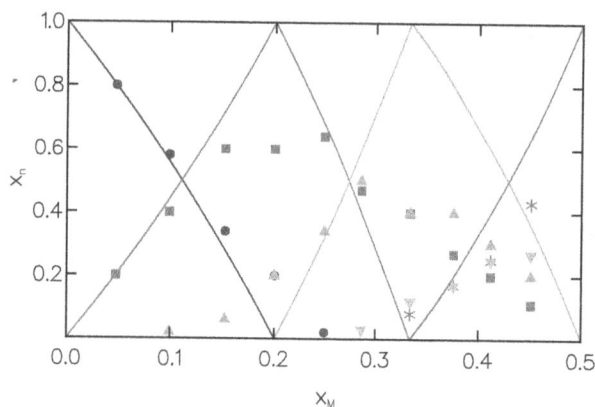

Table 12. (Super)structural units in binary and pseudo-binary sodium borate glasses[114]

(Super)structural unit	Composition range
Boroxol	$0.00 \leq x_M \leq 0.25$
Tetraborate (pentaborate + triborate)	$0.00 < x_M \leq 0.35$
Di-triborate and/or di-pentaborate	$0.30 < x_M < 0.75$
Diborate (interconnected)	$0.15 \leq x_M < 0.45$
Diborate (independent)	$0.45 < x_M < 0.67$
$B\varnothing_4^-$ (independent)	$0.45 < x_M < 0.67$
Network NBO	$0.05 \leq x_M < 0.71$
Cyclic metaborate	$0.35 \leq x_M < 0.71$
Pyroborate	$0.45 \leq x_M < 0.75$
Orthoborate	$0.61 \leq x_M \leq 0.75$

Figure 27. Superstructural unit fractions for Li_2O–B_2O_3 glasses obtained from NMR spectroscopy[110] (point symbols) and as predicted by the idealised Krogh-Moe[26]–Griscom[2] model (solid lines). Blue line and solid circles, boroxol (x_B); red line and solid squares, tetraborate (x_{Te}); green line and solid triangles, diborate (x_D); cyan inverted triangles, independent (loose) $B^{(4)}$ (x_I) and magenta line and asterisks, "metaborate" (x_2)

system are the more comprehensive and are plotted in Figure 27. The boron atom fractions for tetraborate (x_{Te}) and diborate (x_D) groups have been obtained as the sum of the relevant tetrahedral and trigonal boron atom fractions, whereas the "metaborate" fraction, x_2, does not distinguish between independent $B^{(2)}$ units and those incorporated into cyclic metaborate anions and other superstructural units, Similarly, the $B^{(1)}$ units are all assigned to pyroborate anions although, in practice, some of them will act as network terminators, as the network is broken into fragments with increasing x_M.

The data analysis that led to Figure 27 is based on the superstructural units predicted to be present in the Na_2O–B_2O_3 system by Krogh-Moe[26]. In practice, however, the structural unit environments specified in Table 11 may include contributions from other (super)structural units not considered by Krogh-Moe;[26] e.g. the boroxol fraction is now known to include independent $B^{(3)}$ units in addition to those in boroxol groups. Nevertheless, the data highlight several important aspects of the glass structure:

(i) The tetraborate (pentaborate + triborate) peak is lower and broader than for the idealised Krogh-Moe[26]–Griscom[2] model. This contribution presumably includes all of the pentaborate and triborate groups present, whether or not they are involved in tetraborate pairs.

(ii) The diborate (plus other superstructural units with two adjacent $B^{(4)}$ units (Figure 9)) fraction commences much earlier ($x_M \sim 0.1$) than for the model and, together with the boroxol fraction, is responsible for the reduction in the height of the tetraborate peak (cf. Krogh-Moe's 10% dissociation[26]).

(iii) $B^{(3)}$ units (from both tetraborate and diborate groups) exist well into Region II, and x_3 appears

to go to zero at the Region II/III boundary, as predicted by the Beekenkamp[78] and Gupta[79] models; i.e. there are no $B^{(3)}$ units in Region III, as would be required for the realistic version of the Krogh-Moe[26]–Griscom[2] model in Figure 21.

(iv) There is a significant contribution from independent $B^{(4)}$ units throughout Regions II and III.

The predictions of Krogh-Moe's superstructural unit model[26] should also be compared to the optical spectroscopy data of Kamitsos et al[116] for the Li_2O–B_2O_3 system. These authors conclude that, with increasing Li_2O content, there is systematic formation and then destruction of pentaborate, tetraborate, diborate, pyroborate and orthoborate groups, which were found to be the main constituents of the borate network. The cyclic metaborate anion was also detected, but in smaller amounts. It is interesting to note the presence of pentaborate groups, as predicted for the Na_2O–B_2O_3 system by the thermodynamic modelling studies of Figure 23. In a later paper, Kamitsos et al[112] show the relative boroxol ring content obtained from Raman spectroscopy decreasing linearly from its value for pure vitreous B_2O_3 to zero at $x_M \sim 0.26$, in excellent agreement with the NMR data for the Li_2O–B_2O_3 system in Figure 27. This value should be compared to $x_M = 0.20$ for the idealised Krogh-Moe[26]–Griscom[2] model (dashed line in Figure 21). However, for the more realistic version of the model (solid line), the boroxol rings disappear at $x_M = 0.25$.

Kamitsos & Karakassides[114] have performed a more detailed investigation of the Na_2O–B_2O_3 system (see Section 11.2), from which they derive the regions of x_M within which various superstructural units are present, as summarised in Table 12. Note the presence of di-triborate and/or di-pentaborate groups, which both include an adjacent pair of $B\varnothing_4^-$ tetrahedra, as is the case for the diborate group. The di-triborate group is found in crystalline 3Na.7B and the di-pentaborate group in crystalline 6Na.13B and α-Na.2B. The upper limit of x_M for boroxol groups (0.25) is very close to that for Li_2O–B_2O_3 glasses (~ 0.26[112]).

11.4. Structural unit coordination numbers

For a given $B^{(i)}$ unit, the average coordination number for neighbouring $B^{(j)}$ units, $n_{i(j)}$, defines the relative

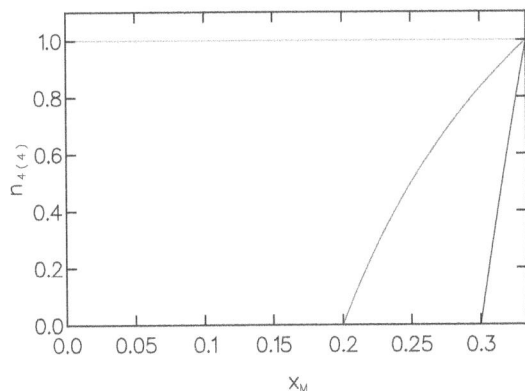

Figure 28. Structural unit coordination number, $n_{4(4)}$, in Region I for the extended Beekenkamp[78] (blue), Gupta[79] (green) and idealised Krogh-Moe[26]–Griscom[2] (red) models. The extension of the Beekenkamp[78] model beyond $x_M = {}^3/_{10}$ assumes the minimum number of $B^{[4]}-\emptyset-B^{[4]}$ linkages (i.e. pairs of adjacent $B\emptyset_4^-$ tetrahedra)

numbers of $B^{[i]}-\emptyset-B^{[j]}$ bridges present in a given glass. Hence, in Region I, the average coordination numbers for $B^{(4)}$ and $B^{(3)}$ units can be used to distinguish between the various structural models. This is particularly true for $n_{4(4)}$, which is zero for the Beekenkamp[78] model ($x_M \leq {}^3/_{10}$) and unity for the Gupta[79] model ($x_M \leq {}^1/_3$). If, on the other hand, it is assumed that pairs of $B^{(4)}$ units only occur (in superstructural units) when absolutely necessary, $n_{4(4)}$ is the same as for the Beekenkamp[78] model for $x_M \leq {}^3/_{10}$ and then, in the region ${}^3/_{10} \leq x_M \leq {}^1/_3$

$$n_{4(4)} = (10x_M - 3)/x_M \qquad (31)$$

For the idealised Krogh-Moe[26]–Griscom[2] model (dashed lines in Figure 21), $n_{4(4)}$ is zero for $x_M \leq {}^1/_5$, whereas, in the range ${}^1/_5 \leq x_M \leq {}^1/_3$, it is given by

$$n_{4(4)} = (5x_M - 1)/2x_M \qquad (32)$$

The compositional dependence of $n_{4(4)}$ for all of the above models in Region I is plotted in Figure 28.

12. Discussion

12.1. Glass formation region

The lack of single phase glass formation and crystalline compounds for the $MO-B_2O_3$ systems at low modifier contents (Figure 6) is extremely interesting and is almost certainly connected with the fact that two $B\emptyset_4^-$ tetrahedra are required in close proximity to balance the charge on an M^{2+} cation. This is supported by the lack (Be^{2+}) or very small range (Mg^{2+}) of glass formation for the smallest network modifying cations. Conversely, single phase glasses are formed at the lowest modifier contents for the largest and most easily polarised cations (Ba^{2+} and Pb^{2+}), the lowest of all being for Ba^{2+}, which is the only system for which a crystalline tetraborate has been reported. As indicated in Section 3, the tetraborate composition is also the lowest at which single phase glasses occur

in the $PbO-B_2O_3$ system.[17] The increased size of the coordination shell surrounding these large cations means that the two $B\emptyset_4^-$ tetrahedra required to balance their 2+ charge can be separated by at least one $B\emptyset_3$ triangle. In addition, the lack of glass formation strongly suggests that diborate groups are not present in significant numbers in borate networks at low x_M, in contrast to Krogh-Moe's suggested mechanism[26] for the 10% dissociation in this region.

12.2. Borate network

The structural unit fractions for several of the models discussed in Sections 8 and 9 exhibit sharp maxima at the boundaries between adjacent regions. The absence of the constraints imposed by the periodicity of a crystalline material means that, for a glass, the available region of configuration space can be extended by disproportionation reactions of the form

$$B\emptyset_4^- \rightleftharpoons BO\emptyset_2^- \qquad (33)$$

$$2B\emptyset_4^- \rightleftharpoons 2BO\emptyset_2^- \rightleftharpoons B\emptyset_3 + BO_2\emptyset^{2-} \qquad (34)$$

and

$$B_2O_5^{4-} \rightleftharpoons 2BO_2\emptyset^{2-} \rightleftharpoons BO\emptyset_2^- + BO_3^{3-} \qquad (35)$$

In the vitreous state, such reactions only need to occur to a limited extent, rather than in every unit cell, as would be the case for a crystalline compound. Similarly, for a glass, the number of $B^{[4]}-\emptyset-B^{[4]}$ bridges may exceed the minimum number required by the composition, in order to attain the degrees of topological freedom necessary for the formation of an experimentally accessible vitreous network.

Most of the glass forming systems in Figures 5 and 6 exhibit a maximum in x_4 in Region II. The peak value is usually around 0·5 or less, whereas crystalline phases exist with $x_4 = 1$. However, the latter either include 3-fold coordinated oxygen atoms (Sr.2B and Pb.2B), or are formed under pressure. A value of 0·5 for x_4 corresponds to a network of diborate groups, as found in several crystalline diborates (Table 5). The highest x_4 so far observed for an ambient pressure crystalline phase (2Ca.3B) without 3-fold coordinated oxygen atoms is ${}^2/_3$.

$Tl_2O-B_2O_3$ glasses are an exception to the general rule[12] in that, at low x_M, x_4 exceeds the value given by Equation (3), as may be seen from Figure 29. Having eliminated the possibility that Tl^+ is oxidised to Tl^{3+}, Baugher & Bray[12] also attribute this to the formation of 3-fold coordinated oxygen atoms, which is in violation of Zachariasen's first criterion[1] (Section 9). They suggest that, in the range $0·05 \leq x_M \leq 0·20$, each added Tl_2O unit gives rise to the transformation of three $B\emptyset_3$ triangles into $B\emptyset_4^-$ tetrahedra whereas, above $x_M = 0·20$, only two $B\emptyset_3$ triangles are converted. This model is denoted by the green line in Figure 29, whereas the blue line, which is a much better fit to the data,

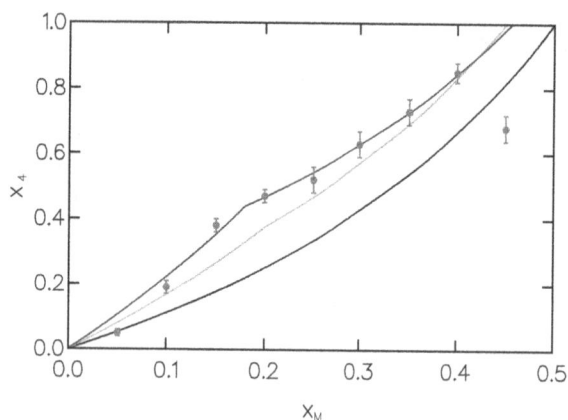

Figure 29. *Fraction of 4-fold coordinated boron atoms in $Tl_2O–B_2O_3$ glasses. Red closed circles, NMR data[12]; black line, Equation (3); green line, Baugher & Bray[12] model and blue line, revised model in which 4 $B\emptyset_3$ triangles are converted below $x_M=0.18$ and $1^1/_3$ are converted above this value*

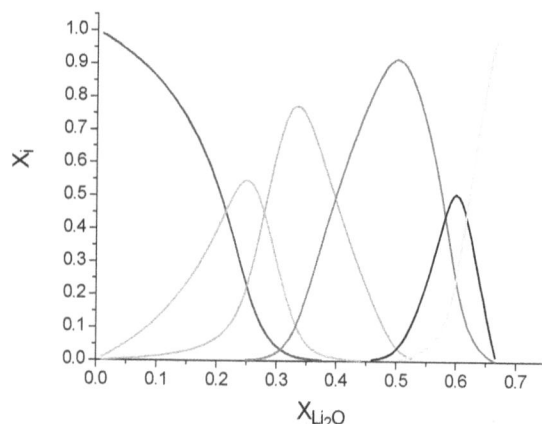

Figure 30. *Chemical structure of $Li_2O–B_2O_3$ glasses at 800 K (approx. average T_g).[117] Blue, B_2O_3; cyan, Li.3B; green, Li.2B; magenta, Li.B; black, 3Li.2B and yellow, 2Li.B*

corresponds to four $B\emptyset_3$ triangles being converted at low x_M ($x_M \leq 0.18$) and $1^1/_3$ at higher values. The maximum in x_4 (0.85 ± 0.03) occurs at $x_M \sim 0.4$, and is still in excess of the value ($^2/_3$) predicted by Equation (3) (black line). Given the behaviour of x_4 for $Tl_2O–B_2O_3$ glasses, it is interesting to note that crystalline Tl.5B (pentaborate groups), Tl.3B (triborate groups) and β-Tl.2B (tri-pentaborate + 3-ring $B_7O_9\emptyset_6^{3-}$ groups) all conform to Equation (3), having structures composed only of $B^{(4)}$ and $B^{(3)}$ structural units. Crystalline Tl.B, on the other hand, has a chain structure based on equal numbers of $B^{(4)}$ and $B^{(2)}$ units ($x_4=0.5$). This value of x_4 is consistent with an extrapolation of x_4 for the corresponding glasses to $x_M=0.5$, suggesting that vitreous Tl.B would also have a similar chain/network structure.

12.3. Lithium metaborate anomaly

The finite value of x_4 (~0.38) for (rapidly quenched) vitreous lithium metaborate is in stark contrast to that for the ambient pressure crystalline polymorph α-Li.B ($x_4=0$; $x_2=1$). Since an increase in temperature favours the formation of $B^{(2)}$ over $B^{(4)}$ units, it is to be expected that the structure of the ambient pressure (supercooled) liquid will also be mainly based on $B^{(2)}$ units, with minor contributions from other structural unit species formed via the disproportionation reactions of Equation (33) to (35), as indeed is predicted by thermodynamic modelling[117] (Figure 30). However, the value of x_4 for the rapidly quenched glass is clearly incompatible with the above, if it is assumed that the structure of the supercooled melt is simply frozen in on quenching through the glass transition. Conversely, if the resulting glass is instantaneously raised to its fictive temperature, T_f, then, within the period of one or two atomic vibrations, the resulting (supercooled) liquid should have its equilibrium

structure with the same value of x_4 as the original glass, but such a liquid appears not to exist at ambient pressure.

In the crystalline state, the application of pressure leads to the conversion of $B^{(2)}$ into $B^{(4)}$ units, as may be seen from the high pressure polymorphs of Li.B, Ag.B, Ca.B and Sr.B. Hence it appears that the process of rapid quenching does not just freeze in the supercooled liquid structure, but is equivalent to the application of pressure for the crystalline state. A possible explanation is that the rapid solidification and contraction of the outer surfaces that accompanies rapid quenching, and/or the pressure applied during roller quenching, results in pressure being applied to the still liquid interior. If this scenario is indeed correct, the structure of the fictive temperature liquid should be that which is (meta)stable at the same pressure, and hence it is important to investigate x_4 for (supercooled) liquid Li.B as a function of both temperature and pressure.

The above scenario is supported by the IR reflection data of Machowski et al,[118] who investigated the variation of x_4, as a function of depth from the surface, for a sample of vitreous 3Na.7B, by removing successive surface layers. The data reveal that x_4 is a minimum at the sample surface and increases with depth up to ~200 mm; i.e. that there is a progressive transformation of $B^{(2)}$ units into $B^{(4)}$ units. Further evidence for the role of pressure is the fact that the composition unit (c.u.) number density of Li.B glass ($\rho°=0.0270\pm0.003$ c.u. $Å^{-3}$)[119] is very slightly larger than that of α-Li.B (0.02689 c.u. $Å^{-3}$) whereas, when a glass and crystal have similar structures, the glass number density is usually a few percent less than that of the crystal due to the former's inherent disorder.

In addition to α-Li.B, lithium metaborate has at least two high pressure phases. The structure of γ-Li.B ($\rho°=0.03489$ c.u. $Å^{-3}$), grown under a pressure of 15 kbar at 950°C, consists of a network of independent $B\emptyset_4^-$ tetrahedra plus Li^+ cations ($x_4=1$; $x_2=0$). An intermediate high pressure phase, β-Li.B, synthesised directly at 15 kbar and 230°C, has been reported by Chang & Margrave,[120] who also prepared γ-Li.B directly at 18 kbar and 370°C. Although the structure of β-Li.B has not been determined, IR spectroscopy[120] reveals the presence of both $B^{(4)}$ and $B^{(2)}$ units.

Chryssikos et al[121] have investigated the crystallisation of vitreous Li.B as a function of devitrification temperature, T_x (cf. Section 3.2). At the highest temperatures, $450 \leq T_x \leq 850$°C, the thermodynamically stable, ambient pressure phase, α-Li.B, is formed whilst, at $T_x < 410$°C, the devitrification product is γ-Li.B, the highest pressure polymorph. Crystallisation at intermediate temperatures ($410 \leq T_x \leq 450$°C) yielded a mixture of α-Li.B and another phase that Chryssikos et al[121] denote as β'-Li.B, because its x-ray powder diffraction pattern does not match that for β-Li.B published by Chang & Margrave.[120] Chryssikos et al[121] suggest that the structure of β'-Li.B comprises a network of alternating $BO\varnothing_2^-$ triangles and $B\varnothing_4^-$ tetrahedra, which is the crystalline (i.e. ordered) equivalent of the structure predicted by the Beekenkamp[78] model for the structure of vitreous Li.B ($x_4 = 1/3$; $x_2 = 2/3$). The fact that devitrification at the lowest temperatures yields a high pressure phase is again indicative of the role of pressure during rapid quenching. However, in the light of the discussion of Section 3.2, it is interesting to note that the polymorph formed at the lowest temperatures is γ-Li.B ($x_4 = 1$), rather than β- or β'-Li.B, which have values of x_4, and almost certainly also average number densities, closer to that of the original rapidly quenched glass. Presumably, the transformation from the network of the glass to that of γ-Li.B is easier (involves less diffusion) than transformation to the network of β'-Li.B, since the mechanism for the former only involves the conversion of $B^{(2)}$ into $B^{(4)}$ units, whilst the mechanism for the latter involves a rearrangement of $B^{(4)}$ and $B^{(2)}$ units, to yield the correct (alternating) topology, and/or the conversion both of $B^{(4)}$ into $B^{(2)}$ and $B^{(2)}$ into $B^{(4)}$ units.

12.4 High x_M region

The structural unit models of Section 7 all predict the glasses in Region IV to be invert; i.e. that they comprise a mixture of polyanions and network modifying cations, and have a network dimensionality of zero. This is also true of the extension of the Krogh-Moe[26]–Griscom[2] model to high x_M in Section 9.2. (Note that Krogh-Moe[26] only considered compositions up to $x_M = 1/3$.) NMR[47,104,110,111] and optical spectroscopy[112–114,116,122] both indicate that, for lithium and sodium borate glasses above $x_M = 1/2$ (Regions III and IV), the borate network is gradually broken up until, in Region IV, the glasses become invert, and x_4 reduces to zero close to the orthoborate composition ($x_M = 3/4$). Kamitsos & Chryssikos[113] associate this

complete disruption of the borate network with the high charge density of the Li^+ and Na^+ cations. In lithium borate crystals, the Li^+ cation is either 4-fold (distorted tetrahedron) coordinated by oxygen, or exhibits a 4+1 arrangement, with the fifth oxygen atom at a larger distance. The glass structure might therefore be envisaged as a network of alternate LiO_4 tetrahedra and BO_3 triangles, with the Li–O bonds having considerable ionic character.

The mechanism of glass formation for invert glasses differs from that for network systems. For the latter, the high activation energy associated with the breaking of network bonds leads to rapidly increasing viscosity and relaxation time with decreasing temperature such that, in the glass transition region, the metastable equilibrium supercooled liquid structure can no longer be maintained, and so the liquid structure corresponding to the fictive temperature, T_f, is kinetically frozen-in. Glass formation for invert glasses, on the other hand, relies on the "confusion principle"; i.e. the presence of a random mixture of a number of different ions of varying shapes/sizes. To crystallise such a mixture, it is necessary to segregate those ions appropriate to a particular crystalline structure and/or, in the case of borate polyanions, to interconvert them via the equilibrium reactions occurring in the melt, to yield a local nucleus of the required species, which must then be rearranged into the correct crystalline configuration. Glass formation will occur if the segregation/interconversion/rearrangement time is in excess of that needed to rapidly quench through the supercooled liquid and glass transition regions. Note that, in the case of binary alkali borate systems, glass formation relies on there being a number of polyanion species present, whereas the original silicate invert glasses of Trap & Stevels[80] contained three or more network modifying cations. As indicated by Chryssikos et al,[122] the excess configurational entropy of the vitreous state, over that of related crystalline phases, is essentially the entropy of mixing of the various ionic species.

The situation for the Rb_2O–B_2O_3 and Cs_2O–B_2O_3 systems is very different,[113] in that the large and polarisable Rb^+ and Cs^+ ions do not favour the reversal of $n_{B(O)}$ from 4 to 3. Hence, as $x_M \to 3/4$, x_4 approaches unity due to the formation of $BO_2\varnothing_2^{3-}$ tetrahedra (Figure 31A) that, in the ionic approximation, have two bridging and two nonbridging oxygen atoms; i.e. the equilibrium reaction

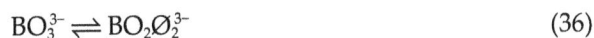

$$BO_3^{3-} \rightleftharpoons BO_2\varnothing_2^{3-} \qquad (36)$$

is displaced very much to the right. The $BO_2\varnothing_2^{3-}$ structural units have a connectivity of 2 and hence will tend be present as chains (Figure 31B), or possibly as $B_3O_9^{9-}$ cyclic orthoborate anions (Figure 31C), together with network modifying cations. If it exists, the cyclic orthoborate anion would be isostructural with the $Si_3O_9^{6-}$ cyclic metasilicate anion found in crystalline

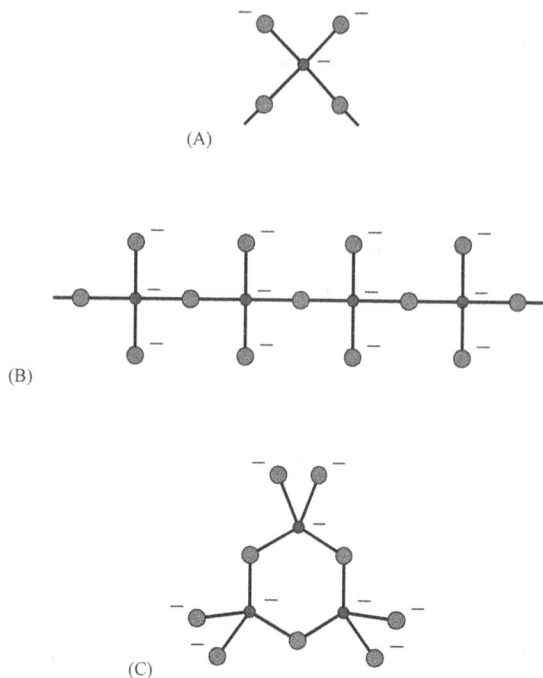

Figure 31. A, $BO_2\varnothing_2^{3-}$ tetrahedral structural unit; B, $(BO_2\varnothing_2^{3-})_\infty$ chain and C, $B_3O_9^{9-}$ cyclic orthoborate anion (Key as Figure 1)

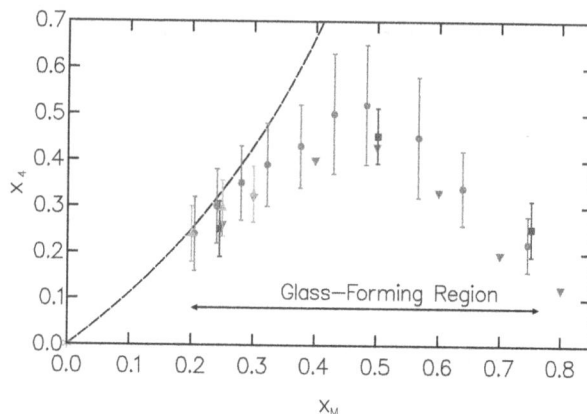

Figure 32. Fraction of 4-fold co-ordinated boron atoms in PbO–B_2O_3 glasses, as a function of composition.[124] Red circles, Bray et al[13] (NMR); magenta inverted triangles, Takaishi et al[125] (NMR; no uncertainty quoted); green triangles, Vedishcheva et al[55,126] (thermodynamic model) and blue squares, Wright & Sinclair[124] (neutron diffraction). The black dashed line denotes Equation (3) and the conventional single phase glass forming region[14] is indicated by the horizontal bar

potassium metasilicate, $K_2O.SiO_2$.[123] Vitreous 3K.B is intermediate in character,[122] but with a structure predominantly based on $BO_2\varnothing_2^{3-}$ tetrahedra.

Lead borate glasses are also formed in Region IV with very much higher values of x_4 than predicted by any of the x_4 models (Figure 32). However, there is considerable evidence from crystallography that, at high x_M, the Pb^{2+} ion changes its role from a network modifying to a network forming cation, as may be seen from the crystal structure of 6Pb.5B (x_M=0·545), which has a large isolated borate anion consisting of two diborate groups connected together by a chain of two $B^{(2)}$ units. This includes a $B^{[4]}$–O^- bond (i.e. a $BO\varnothing_3^{2-}$ tetrahedron with a nonbridging oxygen atom in the ionic modifier approximation – see later, Figure 35A) on each diborate group, but examination of the Pb–O bonding reveals the presence of considerable covalent character in that the Pb atoms are present as PbO_3 and PbO_4 pyramids with the oxygen atoms forming the base and the Pb atom the apex. It is therefore likely that similar groupings or superstructural units also occur in lead borate glasses at high PbO concentrations, as demonstrated by both NMR and neutron diffraction data.[55,124] In addition, the fact that x_4 is non-zero at x_M=¾ means that, at this composition, there must be borate tetrahedra present with nonbridging oxygen atoms, and that the simple ionic modifier model does not apply. The differential correlation functions, $D(r)$, for two lead borate glasses, $3PbO.B_2O_3$ and $PbO.B_2O_3$, are compared with the contribution expected from diborate groups in Figure 33, and it can be seen that, whilst the peak alignment is not perfect, the structure in $D(r)$ is entirely consistent

with the presence of diborate groups in both glasses. Small peak shifts are to be expected due to interaction with other features, and to the increased covalency of the Pb–O bond. (The parameters for the diborate group[34] (Figure 14) were estimated from crystalline Li.2B, α-Na.3B and K.2B.)

12.5. Density and glass transition

The so-called **borate** (or **boron oxide**) **anomaly** refers to the fact that the physical properties, such as the density, of borate glasses do not vary monotonically

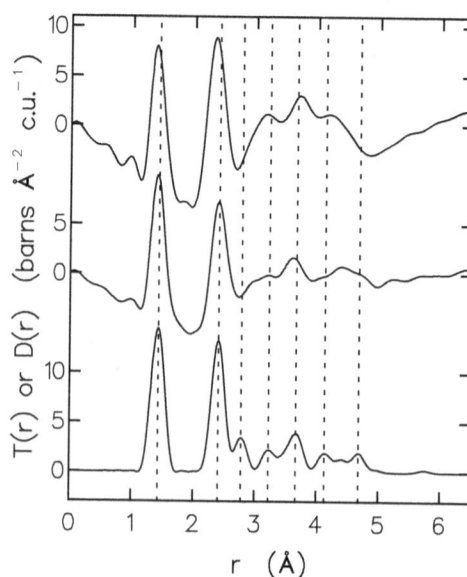

Figure 33. Comparison of the differential correlation functions, $D(r)$, for vitreous 3Pb.B (top) and Pb.B (centre) with the contribution ($T(r)$) expected from diborate groups[124] (bottom)

with x_M, but exhibit a maximum or minimum at a specific composition. Glass properties are clearly a consequence of the underlying atomic structure, and hence early attempts to explain the borate anomaly concentrated on the most unusual aspect of borate glass structure; *viz.* the change in boron atom coordination number. Thus various attempts[127] have been made to correlate the compositional dependence of physical properties with that of x_4, which is available from NMR studies. However, whilst the change in average coordination number, $n_{B(O)}$, is clearly relevant, it is extremely important to consider the glass structure as a whole, including the role played by superstructural units (the other unusual aspect of borate structures) and that by the network modifying cations. The interplay between these different factors is illustrated by the compositional dependence of the density and glass transition temperature, T_g, for the alkali borate systems. Note, however, that in the case of the former much confusion has arisen as a result of considering the mass density, rather than the average number density, ρ°.

Kodama's data[128] for all five alkali borate systems are plotted in Figure 34A in terms of the specific network number density, $\rho_N^\circ(x_M)$; i.e. the number of borate structural units per unit volume, which is equal to the average boron atom number density, ρ_B°. If the borate network is considered in isolation, starting from pure vitreous B_2O_3, the conversion of $B\varnothing_3$ triangles into $B\varnothing_4^-$ tetrahedra will lead to a compaction of the network and a consequent increase in ρ_N° that might be expected to reach a maximum at the composition corresponding to the maximum in x_4. However, this is to ignore the role of the network modifying cations, which must be incorporated into the cages formed by the borate network. Within a disordered network, there will be a distribution of cage sizes, in an exactly analogous way to the distribution of canonical or Bernal holes in a random close packing,[129] and the average cage size will be reduced with increasing network compaction (increasing x_4). For a given network modifying cation, at the lowest values of x_M, there may well be cages available of the necessary size into which the cation can be inserted. With increasing x_M, however, all of these cages will become occupied and the cations must then be inserted into smaller cages with the result that the network will be stretched (in practice, of course, a less compact network will be formed), leading to a reduction in ρ_N°. Thus ρ_N° will first increase, with increasing x_M, pass through a maximum and then decrease. In addition, the larger the network modifying cation, the lower the value of x_M at which the maximum should occur, as may be seen from Figure 34A. Indeed, the data for the caesium borate system do not exhibit a maximum at all, but decrease immediately from the value at $x_M=0$, suggesting that there are no cages in the initial borate (i.e. vitreous B_2O_3) network that are

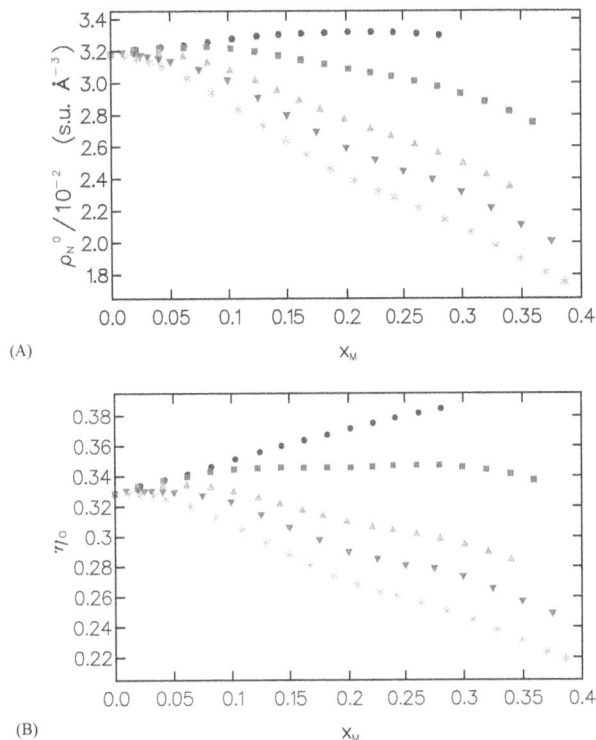

(A)

(B)

Figure 34. A, Specific network number density, ϱ_N°, and B, apparent oxygen atom packing density, η_O', for the five alkali borate glass systems, calculated from the mass density data of Kodama.[128] Blue filled circles, $Li_2O–B_2O_3$; red filled squares, $Na_2O–B_2O_3$; green filled triangles, $K_2O–B_2O_3$; magenta inverted filled triangles, $Rb_2O–B_2O_3$ and cyan asterisks, $Cs_2O–B_2O_3$

naturally large enough to contain a Cs^+ ion.

An alternative way of investigating the network density is to consider the apparent packing density[130] of the oxygen atoms, since the addition of a network modifier leads to extra oxygen atoms. The oxygen atom radius, r_O, can be estimated in the approximation that the oxygen atoms of a $B\varnothing_3$ triangle are "in contact." Assuming a bond length of 1·365 Å, yields $\rho_O=1·18$ Å, and the oxygen atom packing density is given by

$$\eta_O'=4\pi r_O^3\rho_O^\circ/3 \tag{37}$$

where ρ_O° is the oxygen atom number density. Note that η_O' is directly proportional to ρ_O° and so the fractional variation with x_M is the same for both functions, and is not affected by the value of r_O. For a $B\varnothing_4^-$ tetrahedron, the delocalisation of the negative charge leads to increased repulsion between the oxygen atoms and hence a slightly larger B–Ø bond length and Ø–Ø distance. The data of Figure 34A are replotted in Figure 34B in terms of η_O', and it can be seen that the positions of the maxima are moved to higher x_M. Indeed, the data for the $Li_2O–B_2O_3$ system do not reach a maximum within the x_M range of Kodama's data.[128]

The subtle variations in the gradient of both ρ_N° and η_O' in the region $0·15{\leq}x_M{\leq}0·25$ are qualitatively similar for all of the systems except $Li_2O–B_2O_3$, which suggests that they are characteristic of the borate network, rather than a consequence of the size of the network modifying cation. The thermal expansion coefficients for alkali (Li, Na, K and Rb)[131] and silver[132] borate glasses all exhibit a minimum in the centre of this region (x_M~0·2), and Shelby[131] has pointed out that it is exactly at x_M=0·2 where a lot is taking place in the idealised Krogh-Moe[26]–Griscom[2] model (disappearance of boroxol groups, first appearance of diborate groups and a maximum in the concentration of tetraborate groups). The glass transition temperatures, T_g, and the dilatometric softening temperatures for the same glasses[131,132] all have maxima at x_M~0·27. This is very close to the point at which x_4 begins to deviate from the value given by Equation (3), due to the formation of nonbridging oxygen atoms. These break up the network, leading to a lowering of both T_g and the softening temperature.

13. Composite structural model

Since the formulation of the models considered in Sections 7 and 9, a considerable body of experimental data concerning the structure of both borate crystals and glasses has been published in the literature. Hence it is necessary to revisit these models and consider how they can be revised in the light of modern knowledge, to arrive at a composite model to predict the (super)structural unit fractions that acknowledges the important role played by both superstructural units and the network modifying cations. The data clearly indicate that the (super)structural unit species present, and the way in which their fractions vary with composition, differ for the various M_2O–B_2O_3 and MO–B_2O_3 systems, and hence that it is not possible to arrive at a universal quantitative model that applies to every system.

In the vitreous state, the absence of a periodic structure means that a given structural feature (nonbridging oxygen atom, superstructural unit, etc.) can initially occur at a very small concentration, as opposed to the crystalline state where such a feature must be present in every unit cell, and hence in significant numbers or not at all. This means that the boundaries between Regions I–IV are not sharply defined, as might be inferred from the models of Sections 7 and 9. Similarly, glass formation requires a reasonable region of configuration space to allow the degrees of freedom necessary for the formation of a disordered, but relatively strain free, network (cf. Section 11.1), which again will blur the inter-region boundaries. However, the concept of different regions is still useful in describing the evolution of borate glass structure with increasing x_M. Based on the foregoing discussion, and the models of Sections 7 and 9, the Region I/II and II/III boundaries should be set at the diborate (x_M=$^1/_3$) and metaborate (x_M=½) compositions, respectively. The location of the Region

III/IV boundary is more difficult to establish, but perhaps the best position would be at the pyroborate composition (x_M=$^2/_3$). In this case, the second Na_2O–B_2O_3 glass forming region commences at the Region III/IV boundary and lies completely within Region IV.

As may be seen from Figures 23 and 30, thermodynamic modelling demonstrates that the contributions from a given chemical grouping, and hence its associated structural features, start to appear at values of x_M that are significantly lower than its own stoichiometry, and persist to correspondingly higher x_M. Thus, while it is reasonable to assume that crystal structure data can be employed to predict the order of the appearance (and disappearance) of structural features with increasing x_M, it is necessary to resort to either modelling or experiment to investigate the range of x_M over which they make a significant contribution to the glass structure; i.e. are present in concentrations higher than those associated with defects.

13.1. Region I

Commencing at x_M=0, the structure of pure vitreous B_2O_3 is now known to contain approximately equal numbers of boroxol groups and independent $B\text{\O}_3$ triangles (Section 6). At low x_M, single phase glass formation only occurs in the M_2O–B_2O_3 systems (Figure 5), and not in any of the MO–B_2O_3 systems of Figure 6. On adding a network modifier to vitreous B_2O_3, crystal structure data suggest that boroxol groups are initially converted into triborate groups (Equation (19)). Pentaborate groups can then be formed by the reactions of Equations (20) and (21). The way in which the balance between the fractions of triborate and pentaborate groups varies with x_M for each M_2O–B_2O_3 system is an important, but unanswered question, since the NMR data for the Li_2O–B_2O_3[47,110,111] and Na_2O–B_2O_3[46,115] systems identify a single tetrahedral boron site for the tetraborate group, and do not distinguish between the sites in the constituent triborate and pentaborate groups. This balance is clearly complicated, as may be deduced from the known crystal structures. Thus the structure of the only enneaborate investigated to date includes both boroxol and triborate groups, all of the pentaborates are based solely on pentaborate groups, and all of the tetraborates involve equal numbers of triborate and pentaborate groups, paired together to form tetraborate groups. Note that Krogh-Moe's model[26] is based on the Na_2O–B_2O_3 system for which none of the enneaborate or pentaborate structures have been determined.

Region I can conveniently be divided into two parts at the tetraborate composition (x_M=0·2). It is at this value of x_M that several important features of the Krogh-Moe[26]–Griscom[2] model occur, and that the subtle variations in the gradient of $\eta_O{'}$ are centred (Section 12.5). For x_M<0·2, the major structural components are boroxol, pentaborate and triborate groups, together with independent $BØ_3$ triangles. Crystallographic data suggest that increasing x_M above 0·2 leads to the introduction of independent $BØ_4^-$ tetrahedra ($B^{(4)}$ units) and of superstructural units incorporating two adjacent $BØ_4^-$ tetrahedra (Figure 9), such as the diborate group (cf. Krogh-Moe[26] model). As indicated above, the absence of periodicity and a unit cell means that initially only small numbers of such units may occur, even below x_M=0·2 (cf. Figure 27), and that this, together with residual boroxol groups and/or independent $BØ_3$ triangles, may be responsible for rounding off the maximum in the tetraborate component at x_M=0·2, as in the realistic version of the Krogh-Moe[26]–Griscom[2] model. An alternative explanation of the reduced tetraborate component in the region of the maximum may be an excess of either pentaborate or triborate groups, as predicted by the thermodynamic modelling of the chemical structure of Na_2O–B_2O_3 glasses in Figure 23. Note that, in the case of the Li_2O–B_2O_3 system (Figure 30), there are no significant contributions to the chemical structure from either Li.5B or Li.4B, but only from Li.3B.

13.2. Region I/II (diborate) boundary

Region I for the Gupta[79] model ends at the diborate composition (x_M=$1/3$). At this value of x_M, two borate networks are possible involving a single superstructural unit from Figure 9 and no external $B^{[4]}$–$Ø$–$B^{[4]}$ bridges. The first, and most obvious, is a network of diborate groups linked only by $B^{[3]}$–$Ø$–$B^{[4]}$ bridges, as found in the many common crystalline diborate structures (cf. Table 5) and in the idealised Krogh-Moe[26]–Griscom[2] model. Alternatively, such a network can be formed from equal numbers of di-triborate groups (Figure 9D) and independent $BØ_3$ triangles, again with only external $B^{[3]}$–$Ø$–$B^{[4]}$ bridges, although the increased connectivity of the di-triborate group would mitigate against this arrangement (cf. Section 9). As indicated in Section 11.1, a vitreous network having only diborate groups with $B^{[3]}$–$Ø$–$B^{[4]}$ bridges represents an extremely small region of configuration space, and hence other (super)structural units will also be present, even at the diborate composition.

NMR data (e.g. Figures 4, 24 and 25) indicate that x_4 first starts to deviate from the value given by Equation (3) at x_M~0·28, close to the di-pentaborate composition, due to the formation of nonbridging oxygen atoms. The fact that, in the crystalline state, nonbridging oxygen atoms first appear incorporated into superstructural units, suggests that the same may be true for borate glasses, especially in the Na_2O–B_2O_3 system, where the nonbridging oxygen atoms in crystalline $α$-Na.2B are included in di-triborate (NBO) groups (Figure 8D). The available region of configuration space approaching the diborate composition

may thus be extended by converting diborate groups into di-triborate (NBO) groups plus $B^{(3)}$ units via the equilibrium reaction

$$B_4O_5Ø_4^{2-} \rightleftharpoons B_3O_4Ø_3^{2-} + BØ_3 \qquad (38)$$

or into di-triborate groups plus $B^{(3)}$ units by driving the equilibrium of reaction (24) to the left.

13.3. Region II

The upper limits of the conventional glass forming regions of all of the M_2O–B_2O_3 and alkaline earth borate systems (except BeO–B_2O_3, which does not form glasses via conventional quenching) are in Region II, although glass formation can be extended to higher x_M using rapid quenching techniques. As the value of x_M is increased above $1/3$, the formation of $B^{(2)}$ units will lead to the diborate (or di-triborate) groups being broken up and/or converted into superstructural units incorporating $B^{(2)}$ units, and also to an increase in the number of independent $B^{(4)}$ units (Figure 27), if pairs of adjacent $B^{(4)}$ units are postulated to be energetically unfavourable outside superstructural units. One possible reaction involves the conversion of a diborate group into a di-triborate (NBO) group plus an independent $BØ_4^-$ tetrahedron

$$½M_2O + B_4O_5Ø_4^{2-} \rightarrow M^+ + B_3O_4Ø_3^{2-} + BØ_4^- \qquad (39)$$

For the Li_2O–B_2O_3 system, x_4 reaches a maximum at x_M~0·37 (Section 11.1; Figures 24 and 25), after which it decreases due to the conversion of $B^{(4)}$ into $B^{(2)}$ units. The dearth of crystalline phases between the boundary compositions of Region II makes it extremely difficult to predict the borate constituents of the structure of conventionally quenched glasses, although it should be stressed that, in the few ambient pressure crystalline structures that are known, nonbridging oxygen atoms only occur within superstructural units.

Once glass formation requires rapid quenching, the apparent equivalence with the application of pressure (cf. Section 12.3) would suggest that, with increasing x_M, superstructural units are gradually broken up to yield a network of independent $B^{(4)}$ and $B^{(2)}$ units at the metaborate composition. The presence of independent $B^{(4)}$ units in Li_2O–B_2O_3 glasses in Region II is apparent from Figure 27. Note, however, that the realistic version of the Krogh-Moe[26]–Griscom[2] model predicts the presence of diborate groups, and hence $B^{(3)}$ units, up to the pyroborate composition. This implies the presence of $B^{(1)}$ units to compensate for the excess of $B^{(3)}$ units over the value given by Equation (17) (cf. Section 9.2).

13.4. Region II/III (metaborate) boundary

At the metaborate composition (Region II/III boundary), both the Beeekenkamp[78] and Gupta[79] models predict a mixed network of $B^{(4)}$ and $B^{(2)}$ units. Since alkali and alkaline earth borate glasses of this composition are prepared by rapid quenching, it is likely that the majority of the structural units will be independent, although some of them may be incorporated into superstructural units such as metaborate (NBO) or metaborate groups, or independent cyclic or bicyclic metaborate anions. For example, Raman studies reveal the presence of cyclic metaborate anions in vitreous Li_2O–B_2O_3[116] and $Cs_2O.B_2O_3$.[43] As noted above, if any superstructural units containing $B^{(3)}$ units remain at the metaborate composition, the $B^{(3)}$ units must be compensated for by an equal number of $B^{(1)}$ units, which will act as network terminators. An argument in favour of the presence of superstructural units in the form of metaborate (NBO) or metaborate groups is that they incorporate adjacent $B^{(4)}$ units that would otherwise be energetically unfavourable, and therefore less likely to occur in significant numbers.

13.5. Region III

Assuming a mixed network of $B^{(4)}$ and $B^{(2)}$ units at the Region II/III boundary, the Beekenkamp[78] and Gupta[79] models predict that a further increase in x_M will lead to the conversion of the $B^{(2)}$ and a minimum number of the $B^{(4)}$ units into $B^{(1)}$ units. The latter will either act as network terminators or form pyroborate anions. Thus the network will gradually be broken up such that, at some point, there will no longer be a single network but an increasing number of network fragments. Finally, approaching the Region III/IV boundary, the structure will become invert, comprising a mixture of polyanions and network modifying cations (Section 12.4). An interesting semantic question is how small does a network fragment have to be before it is considered a polyanion?

13.6. Region IV

The conventional picture for Region IV is that the glasses are invert and that, with increasing x_M, the larger polyanions are gradually converted, first into pyroborate, and then into orthoborate anions. This scenario, and that of Section 13.5, are appropriate for the Li_2O–B_2O_3 and Na_2O–B_2O_3 systems but, as discussed in Section 12.4, at high x_M the larger alkali cations favour the formation of $BO_2\emptyset_2^{3-}$ tetrahedra with two bridging and two nonbridging oxygen atoms (Figure 31A). The detailed way in which the structures of K_2O–B_2O_3, Rb_2O–B_2O_3, and Cs_2O–B_2O_3 glasses evolve from the metaborate to the orthoborate composition requires further study, especially since, of the latter three systems, only Rb_2O–B_2O_3 exhibits a crystalline compound of the orthoborate composition, and only potassium, which is intermediate in character, forms a pyroborate. (Neither structure has been determined.) A possible key to this evolution

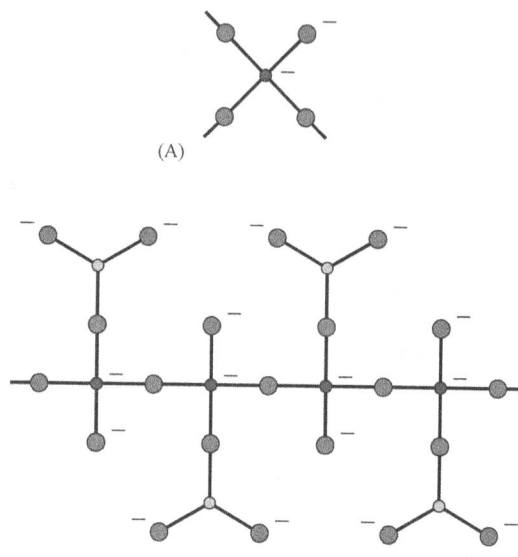

Figure 35. A, $BO\emptyset_3^{2-}$ tetrahedral structural unit and B, $(B_2O_4\emptyset_2^{4-})_\infty$ chain (Key as Figure 1)

can be found in the structure of crystalline Be.Ca.B, the only pyroborate so far investigated that does not have a structure based on pyroborate anions (Figure 3). Instead, in the ionic modifier approximation, the borate component of the structure involves equal numbers of $BO_2\emptyset^{2-}$ triangles ($B^{(1)}$ units – half of a pyroborate anion) and $BO\emptyset_3^{2-}$ tetrahedra with one nonbridging oxygen atom (Figure 35A) that are combined to form $(B_2O_4\emptyset_2^{4-})_\infty$ chains (Figure 35B). A comparison of Figures 31B and 35B shows that the former can be converted into the latter by replacing one nonbridging oxygen atom on each $BO_2\emptyset_2^{3-}$ tetrahedron with a $B^{(1)}$ unit. Hence the structure can evolve from the pyroborate to the orthoborate composition by gradually replacing $B^{(1)}$ units with nonbridging oxygen atoms. In Region III, the transformation from metaborate to pyroborate would then involve the conversion of $B^{(2)}$ into $B^{(1)}$ units, and the organisation of the tetrahedral structural units into chains. In respect of the crystal structure of Be.Ca.B, it is also interesting to note that the Be atoms are present as distorted BeO_4 tetrahedra, which are incorporated into 3-membered rings together with a $BO\emptyset_3^{2-}$ tetrahedron and a $BO_2\emptyset^{2-}$ triangle.

14. Conclusions

From the preceding sections, it is apparent that a study of the appropriate crystalline structures can yield considerable insight into the structures of borate glasses, provided suitable allowance is made for the extra degrees of network freedom necessary for glass formation. However, the information that can be inferred is limited by the fact that many key crystalline borates have unknown structures. Similarly, in the case of the glasses, far fewer structural data are available for MO–B_2O_3 systems than for M_2O–B_2O_3

systems. A number of criteria have been discussed relevant to the formation of vitreous borates containing superstructural units and used to further refine the various structural models proposed in the literature. As in the case of Zachariasen's criteria for glass formation, these criteria should not be taken as hard and fast rules, but rather as general guidelines. The fact that a given feature does not occur in the crystalline state should not be taken to infer that it will never be found in the glass, but rather that it is energetically unfavourable and hence will only be likely to occur with a low probability. As with crystalline defects, a small concentration of an unfavourable configuration may be desirable for entropic reasons.

Since the seminal work of Krogh-Moe, considerable progress has been made both in borate crystallography and in structural studies of borate glasses. However, there remain some important questions, the answers to which would contribute greatly to current understanding concerning borate structures, both crystalline and vitreous. The author's "top ten" such questions are:

14.1. General questions

1. **What is the origin of the superstructural unit stabilisation energy, and what are the relative stabilisation energies for the various superstructural units?** The answer to this question should explain the frequency of occurrence of the various superstructural units in the crystalline state (Table 9), and hence suggest their possible distribution in glasses. The relative stabilisation energies will also determine the direction of the various equilibrium reactions between superstructural units. An associated question is whether the stabilisation mechanism promotes planarity of the 6-atom rings that form the basis of superstructural units. Clearly, the 6-atom rings do not have to be planar, as evidenced by the frequent occurrence of the diborate group (Table 9). As indicated in Section 4.2, the stabilisation energy and planarity of the boroxol group do not arise from delocalised π-bonding.

2. **Why are the triborate (NBO) and di-triborate (2×NBO) superstructural units in Figure 11 not formed?** The reason for the non-formation of these superstructural units should provide a fundamental explanation as to why $B\varnothing_4^-$ tetrahedra and not nonbridging oxygen atoms are formed on the initial introduction of a network modifier.

14.2. Crystallography

3. **How many of the crystalline phases up to and including the tetraborate composition ($x_M \leq 0.2$), indicated by * or ? in Table 2 ($M_2O–B_2O_3$ phases), actually exist and, if they do exist, what are their** structures? The structures of those phases that do exist should provide guidance as to which superstructural units are likely to be present in the corresponding glasses and, in the case of the various tetraborates, the strength of the association energy between pentaborate and triborate groups. (Note that only two tetraborates in Table 2 have known structures and that the network modifying cations involved (Na^+ and Ag^+) are of not too dissimilar size ($r_{Na}=1.02$ Å and $r_{Ag}=1.15$ Å [133]).

4. **What are the structures of certain key crystalline phases in Regions II (2Na.3B, β- and β'-Li.B and 3Li.Na.4B), III (α- and β-3Li.2B and 3Na.2B) and IV (2K.B, 5Na.2B and 3Rb.B)?** A knowledge of these crystal structures is of paramount importance in understanding glass formation, or the lack of it, in Regions II to IV, together with the structures of the resulting glasses.

5. **What are the structures of the many $M_2O–B_2O_3$ high temperature polymorphs that remain to be determined, and are the transitions that link them to the corresponding low temperature polymorph displacive or reconstructive?** Devitrification of a given glass at the lowest possible temperature tends to yield the polymorph that has a structure closest to that of the initial glass, which may well be a high temperature polymorph. Of particular interest, in connection with the Krogh-Moe[26] model, are the structures of β-Na.2B and γ-Na.2B (Section 9.1). Thermodynamic and other modelling studies also require structural data for high temperature polymorphs.

14.3. Vitreous and liquid borates

6. **What is the balance between the fractions of the boron atoms in pentaborate, x_P, and triborate, x_{Tr}, groups for the different M^+ cations in Region I, and particularly for $x_M \leq 0.2$?** Thermodynamic modelling, for example, suggests that this should be very different for the $Li_2O–B_2O_3$ and $Na_2O–B_2O_3$ systems whilst, for the latter system, the Krogh-Moe[26] model predicts $x_P=x_{Tr}$.

7. **At what values of x_M do significant numbers of the various structural and superstructural units first appear, and then finally disappear, for the different $M_2O–B_2O_3$ and $MO–B_2O_3$ glass systems?** It is important to establish to what extent there is a qualitative difference in the evolution of structure with increasing x_M for the different systems (e.g. for light and heavy alkali cations close to the orthoborate composition) and to what extent the differences are merely quantitative, influencing the relative fractions of the various (super)structural unit species. The absence of a periodic structure and the increased degrees of freedom in the vitreous state mean that small

numbers of energetically less favourable configurations can occur earlier than in the crystalline state, where they have to appear in every unit cell.

8. **Are the initial $B^{(2)}$ units that appear in the conventional glass forming region, and especially those present around the diborate composition, independent or incorporated into superstructural units?** The mechanism governing the initial introduction of $B^{(2)}$ units will provide further enlightenment concerning the role of superstructural units, and their associated stabilisation energies, in determining the evolution of the glass structure with increasing modifier content.

9. **How does $n_{4(4)}$ vary with x_M, especially in Region I?** As discussed in Section 11.2, the variation of the average value of $n_{4(4)}$ with composition clearly differentiates the various models that predict x_4. In Regions II to IV, it also provides information as to the types of superstructural unit that may be present (i.e. those incorporating one, two or three $B^{(4)}$ units).

10. **What are the structural unit fractions in (supercooled) liquid Li.B as a function of both temperature and pressure?** Understanding the origin of the $B^{(4)}$ units in rapidly quenched vitreous Li.B and other metaborate glasses would greatly increase current knowledge concerning the rapid quenching process, and of the relationship between the structures of the resulting glasses and of those formed by conventional quenching.

Acknowledgement

It is first necessary to acknowledge the seminal work of Jan Krogh-Moe that laid the foundation of borate structural science, and upon which this paper so greatly depends. The author would also like to thank Robert L. Snyder and the late Roger. N. Sinclair for their help in searching various crystallographic databases, Rimma S. Bubnova and Marcel Touboul for complete listings and reprints of their borate crystal structure papers, Masao Kodama for providing his numerical alkali borate glass density data and Natalia M. Vedishcheva for supplying Figures 23 and 30, and for helpful discussions.

References

1. Zachariasen, W. H. *J. Am. Chem. Soc.*, 1932, **54**, 3841.
2. Griscom, D. L. In: *Borate Glasses: Structure, Properties, Applications*, Eds L. D. Pye, V. D. Fréchette & N. J. Kreidl, Plenum, New York, 1978, p. 11.
3. Warren, B. E., Krutter, H. & Morningstar, O. *J. Am. Ceram. Soc.*, 1936, **19**, 202.
4. Biscoe, J. & Warren, B. E. *J. Am. Ceram. Soc.*, 1938, **21**, 287.
5. Krogh-Moe, J. *Arkiv Kemi*, 1959, **14**, 439.
6. Schulze, G. E. R. *Z. Phys. Chem. (Leipzig)*, 1934, **B24**, 215.
7. Berger, S. V. *Acta Chem. Scand.*, 1949, **3**, 660.
8. Berger, S. V. *Acta Chem. Scand.*, 1950, **4**, 1054.
9. Zachariasen, W. H. *Z. Krist.*, 1937, **98**, 266.
10. Silver, A. H. & Bray, P. J. *J. Chem. Phys.*, 1958, **29**, 984.
11. Bray, P. J. & O'Keefe, J. G. *Phys. Chem. Glasses*, 1963, **4**, 37.
12. Baugher, J. F. & Bray, P. J. *Phys. Chem. Glasses*, 1969, **10**, 77.
13. Bray, P. J., Leventhal, M. & Hooper, H. O. *Phys. Chem. Glasses*, 1963, **4**, 47.
14. Rawson, H. *Inorganic glass-forming systems*, Academic Press, London, 1967.
15. Mazurin, O. V., Streltsina, M. V. & Shvaiko-Shvaikovskaya, T. P. *Svoistva Steklo i Stekloobrazuyushchikh Rasplavov (The properties of Glasses and Glass-Forming Systems)*, Vol. 2. Nauka, Leningrad, 1975 and Vol. 4, Part 1, 1980.
16. Wright, A. C., Vedishcheva, N. M. & Shakhmatkin, B. A. In: *Borate Glasses, Crystals and Melts*, Eds A. C. Wright, S. A. Feller & A. C. Hannon, Society of Glass Technology, Sheffield, 1997, p. 80.
17. Shelby, J. E. *J. Non-Cryst. Solids*, 1982, **49**, 287.
18. Morimoto, N. *Miner. J. (Jpn)*, 1956, **2**, 1.
19. Levy, H. A. & Lisensky, G. C. *Acta Cryst.*, 1978, **B34**, 3502.
20. Tool, A. Q. & Hill, E. E. *J. Soc. Glass Technol.*, 1925, **9**, T185.
21. Hägg, G. *J. Chem. Phys.*, 1935, **3**, 42.
22. Zachariasen, W. H. *J. Chem. Phys.*, 1935, **3**, 162.
23. Krogh-Moe, J. *Phys. Chem. Glasses*, 1960, **1**, 26.
24. Krogh-Moe, J. *Acta Cryst.*, 1962, **15**, 190.
25. Krogh-Moe, J. *Phys. Chem. Glasses*, 1962, **3**, 1.
26. Krogh-Moe, J. *Phys. Chem. Glasses*, 1962, **3**, 101.
27. Krogh-Moe, J. *Arkiv Kemi*, 1959, **14**, 451.
28. Kracek, F. C., Morey, G. W. & Merwin, H. E. *Am. J. Sci.*, 1938, **35A**, 143.
29. Morey, G. W. & Bowen, N. L. *J. Soc. Glass Technol.*, 1925, **9**, T226.
30. Vedishcheva, N. M., Shakhmatkin, B. A., Shultz, M. M. & Wright, A. C. In: *Borate Glasses, Crystals and Melts*, Ed. A. C. Wright, S. A. Feller & A. C. Hannon, Society of Glass Technology, Sheffield, 1997, p. 215.
31. Wright, A. C., Shakhmatkin, B. A. & Vedishcheva, N. M. *Glass Phys. Chem.*, 2001, **27**, 97.
32. Edwards, J. O. & Ross, V. J. *Inorg. Nucl. Chem.*, 1960, **15**, 329.
33. Krogh-Moe, J. *Acta Cryst.*, 1972, **B28**, 168.
34. Wright, A. C., Shaw, J. L., Sinclair, R. N., Vedishcheva, N. M., Shakhmatkin, B. A. & Scales, C. R. *J. Non-Cryst. Solids*, 2004, **345&346**, 24.
35. Schubert, D. M. & Knobler, C. B. *Phys. Chem. Glasses Eur. J. Glass Sci. Technol. B*, 2009, **50**, 71.
36. Park, B. & Cormack, A. N. In: *Borate Glasses, Crystals and Melts*, Eds A. C. Wright, S. A. Feller & A. C. Hannon, Society of Glass Technology, Sheffield, 1997, p. 443.
37. Walrafen, G. E., Samanta, S. R. & Krishnan, P. N. *J. Chem. Phys.*, 1980, **72**, 113.
38. Snyder, L. C. *Bull. Am. Ceram. Soc.*, 1978, **57** (), 825.
39. Wright, A. C. unpublished work.
40. Kristiansen, L. A. & Krogh-Moe, J. *Phys. Chem. Glasses*, 1968, **9**, 96.
41. Bronswijk, J. P. & Strijks, E. *J. Non-Cryst. Solids*, 1977, **24**, 145.
42. Galeener, F. L. & Geissberger, A. E. *J. Phys. Colloque*, 1982, **C9**, 343.
43. Kamitsos, E. I., Karakassides, M. A. & Chryssikos, G. D. *Phys. Chem. Glasses*, 1989, **30**, 229.
44. Chryssikos, G. D., Kamitsos, E. I. & Karakassides, M. A. *Phys. Chem. Glasses*, 1990, **31**, 109.
45. Meera, B. N. & Ramakrishna, J. *J. Non-Cryst. Solids*, 1993, **159**, 1.
46. Jellison Jr., G. E. & Bray, P. J. *J. Non-Cryst. Solids*, 1978, **29**, 187.
47. Feller, S. A., Dell, W. J. & Bray, P. J. *J. Non-Cryst. Solids*, 1982, **51**, 21.
48. Sinclair, R. N., Wright, A. C., Wanless, A. J., Hannon, A. C., Feller, S. A., Mayhew, M. T., Meyer, B. M., Royle, M. L., Wilkerson, D. L., Williams, R. B. & Johanson, B. C. In: *Borate Glasses, Crystals and Melts*, Eds A. C. Wright, S. A. Feller & A. C. Hannon, Society of Glass Technology, Sheffield, 1997, p. 140.
49. Sinclair, R. N., Stone, C. E., Wright, A. C., Polyakova, I. G., Vedishcheva, N. M., Shakhmatkin, B. A., Feller, S. A., Johanson, B. C., Venhuizen, P., Williams, R. B. & Hannon, A. C. *Phys. Chem. Glasses*, 2000, **41**, 286.
50. Sinclair, R. N., Haworth, R., Wright, A. C., Parkinson, B. G., Holland, D., Taylor, J. W., Vedishcheva, N. M., Polyakova, I. G., Shakhmatkin, B. A., Feller, S. A., Rijal, B. & Edwards, T. *Eur. J. Glass Sci. Technol B: Phys. Chem. Glasses*, 2006, **47**, 405.
51. Hannon, A. C., Grimley, D. I., Hulme, R. A., Wright, A. C. & Sinclair, R. N. *J. Non-Cryst. Solids*, 1994, **177**, 299.
52. Heller, G. *Gmelin Handbuch der Anorganischen Chemie*, Vol. 28, Bor-verbindungen Part 7, Springer-Verlag, Berlin, 1975.
53. Abrahams, S. C., Bernstein, J. L., Gibart, P., Robbins, M. & Sherwood, R. C. *J. Chem. Phys.*, 1974, **60**, 1899.
54. Krogh-Moe, J. *Acta Cryst.*, 1972, **B28**, 3089.
55. Vedishcheva, N. M., Shakhmatkin, B. A., Wright, A. C., Grimley, D. I.,

Etherington, G. & Sinclair, R. N. In: *Fundamentals of Glass Science and Technology*, Stazione Sperimentale del Vetro, Venice, 1993, p. 459.

56. Berger, S. V. *Acta Cryst.*, 1952, **5**, 389.

57. Berger, S. V. *Acta Chem. Scand.*, 1953, **7**, 611.

58. Kline, D., Bray, P. J. & Kriz, H. M. *J. Chem. Phys.*, 1968, **48**, 5277.

59. Strong, S. L. & Kaplow, R. *Acta Cryst.*, 1968, **B24**, 1032.

60. Gurr, G. E., Montgomery, P. W., Knutson, C. D. & Gorres, B. T. *Acta Cryst.*, 1970, **B26**, 906.

61. Johnson, P. A. V., Wright, A. C. & Sinclair, R. N. *J. Non-Cryst. Solids*, 1982, **50**, 281.

62. Goubeau, J. & Keller, H. *Z. Anorg. Allg. Chem.*, 1953, **272**, 303.

63. Krogh-Moe, J. *J. Non-Cryst. Solids*, 1969, **1**, 269.

64. Mozzi, R. L. & Warren, B. E. *J. Appl. Crystallogr.*, 1970, **3**, 251.

65. Jellison Jr., G. E., Panek, L. W., Bray, P. J. & Rouse Jr., G. B. *J. Chem. Phys.*, 1977, **66**, 802.

66. Youngman, R. E. & Zwanziger, J. W. *J. Non-Cryst. Solids*, 1994, **168**, 293.

67. Youngman, R. E. & Zwanziger, J. W. *J. Am. Chem. Soc.*, 1995, **117**, 1397.

68. Joo, C. G., Werner-Zwanziger, U. & Zwanziger, J. W. *Phys. Chem. Glasses*, 2000, **41**, 317.

69. Gravina, S. J. & Bray, P. J. *J. Magn. Reson.*, 1990, **89**, 515.

70. Brinker, C. J. *J. Non-Cryst. Solids*, 1988, **100**, 31.

71. Krogh-Moe, J. *Acta Chem. Scand.*, 1964, **18**, 2055.

72. Krogh-Moe, J. *Arkiv Kemi*, 1959, **14**, 553.

73. Wright, A. C., Sumner, D. J. & Clare, A. G. In: *The Structure of Non-Crystalline Materials 1982*, Eds P. H. Gaskell, J. M. Parker & E. A. Davis, Taylor & Francis, London, 1983, p. 395.

74. Grimsditch, M., Polian, A. & Wright, A. C. *Phys. Rev.*, 1996, **B54**, 152.

75. Wright, A. C., Stone, C. E., Sinclair, R. N., Umesaki, N., Kitamura, N., Ura, K., Ohtori, N. & Hannon, A. C. *Phys. Chem. Glasses*, 2000, **41**, 296.

76. Bray, P. J., Kline, D. & Poch, W. *Glastechn. Ber.*, 1966, **39**, 175.

77. Abe, T. *J. Am. Ceram. Soc.*, 1952, **35**, 284.

78. Beekenkamp, P. In: *Physics of Non-Crystalline Solids*, Ed. J. A. Prins, North-Holland, Amsterdam, 1965, p. 512.

79. Gupta, P. K. *Collected Papers XIV Int. Congr. Glass*, 1986, p. 1.

80. Trap, H. J. L. & Stevels, J. M. *Glastechn. Ber.*, 1959, **32K**, VI/31.

81. Chryssikos, G. D., Kapoutsis, J. A., Kamitsos, E. I., Patsis, A. P. & Pappin, A. J. *J. Non-Cryst. Solids*, 1994, **167**, 92.

82. Bray, P. J. *J. Non-Cryst. Solids*, 1985, **75**, 29.

83. Bray, P. J. *Mater. Res. Soc. Symp. Proc.*, 1986, **61**, 121.

84. Bray, P. J. *J. Non-Cryst. Solids*, 1985, **73**, 19.

85. Krogh-Moe, J. *Acta Cryst.*, 1965, **18**, 77.

86. Hibben, J. H. *Am. J. Sci.*, 1938, **35A**, 113.

87. Krogh-Moe, J. *Arkiv Kemi*, 1958, **12**, 475.

88. Osipov, A. A. Private communication, 2008.

89. Vedishcheva, N. M., Shakhmatkin, B. A. & Wright, A. C. In: *Glass – The Challenge for the 21st Century* (Adv. Mater. Res. **39-40**), Ed. M. Liška, D. Galusek, R. Klement and V. Petrušková, Trans Tech, Stafa-Zurich, 2008, p. 103.

90. Shultz, M. M., Vedishcheva, N. M. & Shakhmatkin, B. A. In: *Fizika i Khimiya Silicatov (Physics and Chemistry of Silicates)*, Eds M. M. Shultz & R. G. Grebenshchikov, Nauka, Leningrad, 1987, p. 5.

91. Krogh-Moe, J. *Phys. Chem. Glasses*, 1965, **6**, 46.

92. Shaw, J. L., Wright, A. C., Sinclair, R. N., Marasinghe, G. K., Holland, D., Lees, M. R. & Scales, C. R. *J. Non-Cryst. Solids*, 2004, **345&346**, 245.

93. Greaves, G. N. *J. Non-Cryst. Solids*, 1985, **71**, 203.

94. Ratai, E. -M., Janssen, M., Epping, J. D., Chan, J. C. C. & Eckert, H. *Phys. Chem. Glasses*, 2003, **44**, 45.

95. Ratai, E., Chan, J. C. C. & Eckert, H. *Phys. Chem. Chem. Phys.*, 2002, **4**, 3198.

96. Krogh-Moe, J. *Phys. Chem. Glasses*, 1962, **3**, 208.

97. Block, S., Molin, J. & Piermarini, G. *J. Am. Ceram. Soc.*, 1963, **46**, 557.

98. Block, S. & Piermarini, G. *J. Phys. Chem. Glasses*, 1964, **5**, 138.

99. Krogh-Moe, J. & Jürine, H. *Phys. Chem. Glasses*, 1965, **6**, 30.

100. Prins, J. A. In: *Physics of Non-Crystalline Solids*, Ed. J. A. Prins, North-Holland, Amsterdam, 1965, p. 39.

101. Grjotheim, K. & Krogh-Moe, J. *Kgl. Norske Vidensk. Selsk. Forh.*, 1956, **29**, 24.

102. Jellison Jr., G. E., Feller, S. A. & Bray, P. J. *Phys. Chem. Glasses*, 1978, **19**, 52.

103. Kroeker, S., Aguiar, P. M., Cerqueira, A., Okoro, J., Clarida, W., Doerr, J., Olesiuk, M., Ongie, G., Affatigato, M., & Feller, S. A. *Eur. J. Glass Sci. Technol. B: Phys. Chem. Glasses*, 2006, **47**, 393.

104. Aguiar, P. M. & Kroeker, S. *J. Non-Cryst. Solids*, 2007, **353**, 1834.

105. Cormier, L., Calas, G. & Beuneu, B. *J. Non-Cryst. Solids*, 2007, **353**, 1779.

106. Kamitsos, E. I., Karakassides, M. A. & Patsis, A. P. *J. Non-Cryst. Solids*, 1989, **111**, 252.

107. Bétourné, E. & Touboul, M. *J. Alloys Compd.*, 1997, **255**, 91.

108. Wright, A. C. & Desa, J. A. E. *Phys. Chem. Glasses*, 1978, **19**, 140.

109. Zhong, J. & Bray, P. J. *J. Non-Cryst. Solids*, 1989, **111**, 67.

110. Feller, S. A. *PhD Thesis*, Brown Univ., 1980.

111. Yun, Y. H. & Bray, P. J. *J. Non-Cryst. Solids*, 1981, **44**, 227.

112. Kamitsos, E. I., Patsis, A. P., Karakassides, M. A. & Chryssikos, G. D. *J. Non-Cryst. Solids*, 1990, **126**, 52.

113. Kamitsos, E. I. & Chryssikos, G. D. *J. Mol. Struct.*, 1991, **247**, 1.

114. Kamitsos, E. I. & Karakassides, M. A. *Phys. Chem. Glasses*, 1989, **30**, 19.

115. Jellison Jr., G. E. *PhD Thesis*, Brown Univ., 1977.

116. Kamitsos, E. I., Karakassides, M. A. & Chryssikos, G. D. *Phys. Chem. Glasses*, 1987, **28**, 203.

117. Vedishcheva, N. M. & Wright, A. C. *Eur. J. Glass Sci. Technol. B: Phys. Chem. Glasses*, in preparation.

118. Machowski, P. M., Varsamis, C. P. E. & Kamitsos, E. I. *J. Non-Cryst. Solids*, 2004, **345&346**, 213.

119. Feller, S. A., Kottke, J., Welter, J., Nijhawan, S., Boekenhauer, R., Zhang, H., Feil, D., Parameswar, C., Budhwani, K., Affatigato, M., Bhatnagar, A., Bhasin, G., Bhowmik, S., Mackenzie, J., Royle, M., Kambeyanda, S., Pandikuthira, P. & Sharma, M. In: *Borate Glasses, Crystals & Melts*, Eds A. C. Wright, S. A. Feller & A. C. Hannon, Society of Glass Technology, Sheffield, 1997, p. 246.

120. Chang, C. H. & Margrave, J. L. *Mater. Res. Bull.*, 1967, **2**, 929.

121. Chryssikos, G. D., Kamitsos, E. I., Patsis, A. P., Bitsis, M. S. & Karakassides, M. A. *J. Non-Cryst. Solids*, 1990, **126**, 42.

122. Chryssikos, G. D., Kamitsos, E. I., Patsis, A. P. & Karakassides, M. A. *Mater. Sci. Eng.*, 1990, **B7**, 1.

123. Werthmann, R. & Hoppe, R. *Rev. Chim. Minér.*, 1981, **18**, 593.

124. Wright, A. C. & Sinclair, R. N. Submitted to *Proc. 4th Balkan Conf. on Glass Science and Technology and 16th Conf. on Glass and Ceramics*, Varna, Sept. 27–Oct. 1 2008.

125. Takaishi, T., Jin, J., Uchino, T. & Yoko, T. *J. Am. Ceram. Soc.*, 2000, **83**, 2543.

126. Vedishcheva, N. M., Shakhmatkin, B. A., Wright, A. C., Sinclair, R. N. & Grimley, D. I. *Bol. Soc. Esp. Ceram. Vid.*, 1992, **31-C**, 3, 41.

127. Weinberg, M. C., Uhlmann, D. R. & Poisl, W. H. In: *Borate Glasses, Crystals & Melts*, Eds A. C. Wright, S. A. Feller & A. C. Hannon, Society of Glass Technology, Sheffield, 1997, p. 63.

128. Feller, S. A., Affatigato, M., Giri, S. K., Basu, A. & Kodama, M. *Phys. Chem. Glasses*, 2003, **44**, 117.

129. Bernal, J. D. In: *Liquids: Structure, Properties and Solid Interactions*, Ed. T. J. Hugel, Elsevier, Amsterdam, 1965, p. 25.

130. Wright, A. C., Etherington, G., Desa, J. A. E., Sinclair, R. N., Connell, G. A. N. & Mikkelsen Jr., J. C. *J. Non-Cryst. Solids*, 1982, **49**, 63.

131. Shelby, J. E. *J. Am. Ceram. Soc.*, 1983, **66**, 225.

132. Piguet, J. L. & Shelby, J. E. *J. Am. Ceram. Soc.*, 1985, **68**, 450.

133. Shannon, R. D. & Prewitt, C. T. *Acta Cryst.*, 1969, **B25**, 925.

Appendix 1

The following tables list the references for the structures and/or lattice parameters of the crystalline phases considered in this paper. A reference is also given for the existence of those phases not listed in the Gmelin Handbook.[52] In the case of compounds having polymorphs stable at different temperatures, the convention is to label these polymorphs α-, β-, etc., in order of increasing temperature. Where this is not the case, or there is confusion as to the labelling, the low, intermediate and high temperature polymorphs are indicated by (LT), (IT) and (HT), respectively. Similarly, where Greek letters have been employed for high pressure polymorphs, these are denoted (HP).

Table A1. mM_2O.nB_2O_3 *crystal structure references*

Compound	Reference(s)
Boron oxide	
B-I	Strong, S. L. & Kaplow, R. *Acta Cryst.*, 1968, **B24**, 1032.
	Gurr, G. E., Montgomery, P. W., Knutson, C. D. & Gorres, B. T. *Acta Cryst.*, 1970, **B26**, 906.
	Strong, S. L., Wells, A. F. & Kaplow, R. *Acta Cryst.*, 1971, **B27**, 1662.
	Borisova, N. V., Vedishcheva, N. M. & Pivovarov, M. M. *Rus. J. Inorg. Chem.*, 1978, **23**, 388.
B-II	Prewitt, C. T. & Shannon, R. D. *Acta Cryst.*, 1968, **B24**, 869.
Hydrogen/Deuterium	
α-H.B	Tazaki, H. *J. Sci. Hiroshima Univ. Ser. A*, 1940, **10**, 37.
	Tazaki, H. *J. Sci. Hiroshima Univ. Ser. A*, 1940, **10**, 55.
	Peters, C. R. & Milberg, M. E. *Acta Cryst.*, 1964, **17**, 229.
β-H.B	Tazaki, H. *J. Sci. Hiroshima Univ. Ser. A*, 1940, **10**, 37.
	Zachariasen, W. H. *Acta Cryst.*, 1963, **16**, 385.
γ-H.B	Zachariasen, W. H. *Acta Cryst.*, 1963, **16**, 380.
3H.B	Zachariasen, W. H. *Z. Krist.*, 1934, **88**, 150.
	Zachariasen, W. H. *Acta Cryst.*, 1954, **7**, 305.
3D.B	Craven, B. M. & Sabine, T. M. *Acta Cryst.*, 1966, **20**, 214.
Lithium	
Li.5B	Bétourné, E. & Touboul, M. *J. Alloys Compd.*, 1997, **255**, 91.
Li.4B	Bétourné, E. & Touboul, M. *J. Alloys Compd.*, 1997, **255**, 91.
α-Li.4B	Mathews, M. D., Tyagi, A. K. & Moorthy, P. N. *Thermochim. Acta*, 1998, **320**, 89.
β-Li.4B	Mathews, M. D., Tyagi, A. K. & Moorthy, P. N. *Thermochim. Acta*, 1998, **319**, 113.
Li.3B	König, H. & Hoppe, R. *Z. Anorg. Allgem. Chem.*, 1978, **439**, 71.
	Ihara, M., Yuge, M. & Krogh-Moe, J. *Yogyo-Kyokai-Shi*, 1980, **88**, 179.
	Zhao, S., Huang, C. & Zhang, H. *Cryst. Growth*, 1990, **99**, 805.
	Wei, L., Guiqing, D., Qingzhen, H., An, Z. & Jingkui, L. *J. Phys. D: Appl. Phys.*, 1990, **23**, 1073.
	Radaev, S. F., Genkina, E. A., Lomonov, V. A., Maksimov, B. A., Pisarevskii, Yu. V., Chelokov, M. N. & Simonov, V. I. *Sov. Phys. Cryst.*, 1991, **36**, 803.
	Radaev, S. F., Maximov, B. A., Simonov, V. I., Andreev, B. V. & D'yakov, V. A. *Acta Cryst.*, 1992, **B48**, 154.
	Mathews, M. D., Tyagi, A. K. & Moorthy, P. N. *Thermochim. Acta*, 1998, **320**, 89.
	Shepelev, Yu. F., Bubnova, R. S., Filatov, S. K., Sennova, N. A. & Pilneva, N. A. *J. Solid State Chem.*, 2005, **178**, 2987.
3Li.7B	Aidong, J., Shirong, L., Qingzhen, H., Tianbin, C. & Deming, K. *Acta Cryst.*, 1990, **C46**, 1999.
	Mathews, M. D., Tyagi, A. K. & Moorthy, P. N. *Thermochim. Acta*, 1998, **320**, 89.
Li.2B	Krogh-Moe, J. *Acta Cryst.*, 1962, **15**, 190.
	Krogh-Moe, J. *Acta Cryst.*, 1968, **B24**, 179.
	Natarajan, M., Faggiani, R. & Brown, I. D. *Cryst. Struct. Commun.*, 1979, **8**, 367.
	Radaev, S. F., Muradyan, L. A., Malakhova, L. F., Burak, Ya. V. & Simonov, V. I. *Sov. Phys. Cryst.*, 1989, **34**, 842.
	Mathews, M. D., Tyagi, A. K. & Moorthy, P. N. *Thermochim. Acta*, 1998, **320**, 89.
	Sennova, N., Bubnova, R., Shepelev, J., Filatov, S. & Yakovleva, O. *J. Alloys Compd.*, 2007, **428**, 290.
α-Li.B	Zachariasen, W. H. *Acta Cryst.*, 1964, **17**, 749.
	Will, G., Kirfel, A. & Josten, B. *J. Less-Common Met.*, 1981, **82**, 255.
	Kirfel, A., Will, G. & Stewart, R. F. *Acta Cryst.*, 1983, **B39**, 175.
	Bétourné, E. & Touboul, M. *Powder Diffr.*, 1997, **12**, 155.
	Mathews, M. D., Tyagi, A. K. & Moorthy, P. N. *Thermochim. Acta*, 1998, **320**, 89.
β-Li.B (HP)	Chang, C. H. & Margrave, J. L. *Mater. Res. Bull.*, 1967, **2**, 929.
	Chryssikos, G. D., Kamitsos, E. I., Patsis, A. P., Bitsis, M. S. & Karakassides, M. A. *J. Non-Cryst. Solids*, 1990, **126**, 42.
γ-Li.B (HP)	Marezio, M. & Remeika, J. P. *J. Phys. Chem. Solids*, 1965, **26**, 2083.
	Marezio, M. & Remeika, J. P. *J. Chem. Phys.*, 1966, **44**, 3348.
α-3Li.2B	Mathews, M. D., Tyagi, A. K. & Moorthy, P. N. *Thermochim. Acta*, 1998, **320**, 89.
α-2Li.B	Mathews, M. D., Tyagi, A. K. & Moorthy, P. N. *Thermochim. Acta*, 1998, **320**, 89.
β-2Li.B	Mathews, M. D., Tyagi, A. K. & Moorthy, P. N. *Thermochim. Acta*, 1998, **320**, 89.
α-3Li.B	Stewner, F. *Acta Cryst.*, 1971, **B27**, 904.
	Mathews, M. D., Tyagi, A. K. & Moorthy, P. N. *Thermochim. Acta*, 1998, **320**, 89.
β-3Li.B	Stewner, F. *Acta Cryst.*, 1971, **B27**, 904.
γ-3Li.B	Stewner, F. *Acta Cryst.*, 1971, **B27**, 904.
Sodium	
α-Na.4B (HT)	Krogh-Moe, J. *Acta Cryst.*, 1957, **10**, 435.
	Hyman, A., Perloff, A., Mauer, F. & Block, S. *Acta Cryst.*, 1967, **22**, 815.
	Bubnova, R. S., Shepelev, Ju. F., Sennova, N. A. & Filatov, S. K. *Z. Krist.*, 2002, **217**, 444.
β-Na.4B (LT)	Penin, N., Touboul, M. & Nowogrocki, G. *J. Solid State Chem.*, 2002, **168**, 316.
α-Na.3B	Krogh-Moe, J. *Acta Cryst.*, 1974, **B30**, 747.
β-Na.3B	Krogh-Moe, J. *Acta Cryst.*, 1972, **B28**, 1571.
3Na.7B	Penin, N., Touboul, M. & Nowogrocki, G. *J. Alloys Compd.*, 2004, **363**, 104.
6Na.13B	Penin, N., Touboul, M. & Nowogrocki, G. *J. Solid State Chem.*, 2005, **178**, 671.
α-Na.2B	Krogh-Moe, J. *Acta Cryst.*, 1974, **B30**, 578.
	Borisova, N. V., Vedishcheva, N. M. & Pivovarov, M. M. *Rus. J. Inorg. Chem.*, 1978, **23**, 388.
Na.B	Cole, S. S., Scholes, S. R. & Amberg, C. R. *J. Am. Ceram. Soc.*, 1935, **18**, 58.
	Fang, S. -M. *Z. Krist.*, 1938, **99**, 1.
	Marezio, M., Plettinger, H. A. & Zachariasen, W. H. *Acta Cryst.*, 1963, **16**, 594.
	Borisova, N. V., Vedishcheva, N. M. & Pivovarov, M. M. *Rus. J. Inorg. Chem.*, 1978, **23**, 388.
2Na.B	König, H., Hoppe, R. & Jansen, M. *Z. Anorg. Allgem. Chem.*, 1979, **449**, 91.
3Na.B	König, H. & Hoppe, R. *Z. Anorg. Allgem. Chem.*, 1977, **434**, 225.
Potassium	
α-K.5B (HT)	Krogh-Moe, J. *Arkiv Kemi*, 1959, **14**, 567.
	Krogh-Moe, J. *Acta Cryst.*, 1972, **B28**, 168.
	Bubnova, R. S., Polyakova, I. G., Anderson, Yu. E. & Filatov, S. K. *Glass Phys. Chem.*, 1999, **25**, 183.

(Table A1 continued)

β-K.5B (LT) Krogh-Moe, J. *Arkiv Kemi*, 1959, **14**, 439.
Krogh-Moe, J. *Acta Cryst.*, 1965, **18**, 1088.
Bubnova, R. S., Polyakova, I. G., Anderson, Yu. E. & Filatov, S. K. *Glass Phys. Chem.*, 1999, **25**, 183.

γ-K.5B (IT) Krogh-Moe, J. *Acta Cryst.*, 1961, **14**, 68.
Bubnova, R. S., Polyakova, I. G., Anderson, Yu. E. & Filatov, S. K. *Glass Phys. Chem.*, 1999, **25**, 183.

5K.19B Krogh-Moe, J. *Acta Cryst.*, 1974, **B30**, 1827.

K.3B Krogh-Moe, J. *Acta Cryst.*, 1961, **14**, 68.
Bubnova, R. S., Fundamenskii, V. S., Filatov, S. K. & Polyakova, I. G. *Dokl. Physical Chem.*, 2004, **398**, 249.

K.2B Krogh-Moe, J. *Acta Cryst.*, 1972, **B28**, 3089.

K.B Zachariasen, W. H. *J. Chem. Phys.*, 1937, **5**, 919.
Schneider, W. & Carpenter, G. B. *Acta Cryst.*, 1970, **B26**, 1189.

Rubidium
α-Rb.5B Bubnova, R. S., Polyakova, I. G., Anderson, Yu. E. & Filatov, S. K. *Glass Phys. Chem.*, 1999, **25**, 183.

β-Rb.5B Krogh-Moe, J. *Arkiv Kemi*, 1959, **14**, 439.
Bubnova, R. S., Polyakova, I. G., Anderson, Yu. E. & Filatov, S. K. *Glass Phys. Chem.*, 1999, **25**, 183.
Krzhizhanovskaya, M. G., Bubnova, R. S., Filatov, S. K., Belger, A. & Paufler, P. *Z. Krist.*, 2000, **215**, 740.
Penin, N., Seguin, L., Touboul, M. & Nowogrocki, G. *J. Solid State Chem.*, 2001, **161**, 205.

5Rb.19B Krzhizhanovskaya, M. G., Bubnova, R. S., Bannova, I. I. & Filatov, S. K. *Crystallogr. Rep.*, 1999, **44**, 187.

α-Rb.3B Bubnova, R. S., Krzhizhanovskaya, M. G., Trofimov, V. B., Polyakova, I. G. & Filatov, S. K. Abstracts VII Conf. on *Crystal Chemistry of Inorganic and Co-ordination Compounds*, St. Petersburg, Russia, June 27-30 1995, p. 97.
Bubnova, R. S., Krzhizhanovskaya, M. G., Polyakova, I. G., Trofimov, V. B. & Filatov, S. K. *Inorg. Mater.*, 1998, **34**, 1119.
Krzhizhanovskaya, M. G., Kabalov, Yu. K., Bubnova, R. S., Sokolova, E. V. & Filatov, S. K. *Crystallogr. Rep.*, 2000, **45**, 572.

β-Rb.3B Bubnova, R. S., Krzhizhanovskaya, M. G., Trofimov, V. B., Polyakova, I. G. & Filatov, S. K. Abstracts VII Conf. on *Crystal Chemistry of Inorganic and Co-ordination Compounds*, St. Petersburg, Russia, June 27-30 1995, p. 97.
Krzhizhanovskaya, M. G., Bubnova, R. S., Fundamenskii, V. S., Bannova, I. I., Polyakova, I. G. & Filatov, S. K. *Crystallogr. Rep.*, 1998, **43**, 21.
Bubnova, R. S., Krzhizhanovskaya, M. G., Polyakova, I. G., Trofimov, V. B. & Filatov, S. K. *Inorg. Mater.*, 1998, **34**, 1119.

2Rb.5B Toledano, P. *Bull. Soc. Chim. Franc.*, 1966, **7**, 2302.

3Rb.7B Bubnova, R. S., Polyakova, I. G., Krzhizhanovskaya, M. G. & Filatov, S. K. Abstracts Int. Conf. on *Powder Diffraction and Crystal Chemistry*, St. Petersburg, Russia, June 20-23 1994, p. 80.
Bubnova, R. S., Krivovichev, S. V., Shakhverdova, I. P., Filatov, S. K., Burns, P. C., Krzhizhanovskaya, M. G. & Polyakova, I. G. *Solid State Sci.*, 2002, **4**, 985.

Rb.2B Krzhizhanovskaya, M. G., Bubnova, R. S., Bannova, I. I. & Filatov, S. K. Abstracts VII Conf. on *Crystal Chemistry of Inorganic and Co-ordination Compounds*, St. Petersburg, Russia, June 27-30 1995, p. 28.
Krzhizhanovskaya, M. G., Bubnova, R. S., Bannova, I. I. & Filatov, S. K. *Crystallogr. Rep.*, 1997, **42**, 226.

α-Rb.B Schneider, W. & Carpenter, G. B. *Acta Cryst.*, 1970, **B26**, 1189.
Schmid, S. & Schnick, W. *Acta Cryst.*, 2004, **C60**, i69.

3Rb.B Bubnova, R. S., Krzhizhanovskaya, M. G., Trofimov, V. B., Polyakova, I. G. & Filatov, S. K. Abstracts VII Conf. on *Crystal Chemistry of Inorganic and Co-ordination Compounds*, St. Petersburg, Russia, June 27-30 1995, p. 97.

Cæsium
α-Cs.9B Haworth, R., Wright, A. C., Sinclair, R. N., Knight, K. S., Vedishcheva, N. M., Polyakova, I. G. & Shakhmatkin, B. A. *Eur. J. Glass Sci. Technol. B: Phys. Chem. Glasses*, 2006, **47**, 352.

β-Cs.9B Krogh-Moe, J. & Ihara, M. *Acta Cryst.*, 1967, **23**, 427.
Wright, A. C., Sinclair, R. N., Stone, C. E., Knight, K. S., Polyakova, I. G., Vedishcheva, N. M. & Shakhmatkin, B. A. *Phys. Chem. Glasses*, 2003, **44**, 197.
Penin, N., Touboul, M. & Nowogrocki, G. *J. Solid State Chem.*, 2003, **175**, 348.
Haworth, R., Wright, A. C., Sinclair, R. N., Knight, K. S., Vedishcheva, N. M., Polyakova, I. G. & Shakhmatkin, B. A. *Eur. J. Glass Sci. Technol. B: Phys. Chem. Glasses*, 2006, **47**, 352.

γ-Cs.9B Haworth, R., Wright, A. C., Sinclair, R. N., Knight, K. S., Vedishcheva, N. M., Polyakova, I. G. & Shakhmatkin, B. A. *Eur. J. Glass Sci. Technol. B: Phys. Chem. Glasses*, 2006, **47**, 352.

α-Cs.5B (HT) Krogh-Moe, J. *Arkiv Kemi*, 1959, **14**, 451.
Penin, N., Seguin, L., Touboul, M. & Nowogrocki, G. *J. Solid State Chem.*, 2001, **161**, 205.
Bubnova, R. S., Fundamensky, V. S., Anderson, J. E. & Filatov, S. K. *Solid State Sci.*, 2002, **4**, 87.
Filatov, S., Bubnova, R., Shepelev, Yu., Anderson, J. & Smolin, Yu. *Cryst. Res. Technol.*, 2005, **40**, 65.

β-Cs.5B (LT) Bubnova, R., Dinnebier, R. E., Filatov, S. & Anderson, J. *Cryst. Res. Technol.*, 2007, **42**, 143.

γ-Cs.5B Penin, N., Seguin, L., Touboul, M. & Nowogrocki, G. *J. Solid State Chem.*, 2001, **161**, 205.

3Cs.13B Penin, N., Seguin, L., Touboul, M. & Nowogrocki, G. *Solid State Sci.*, 2002, **4**, 67.

Cs.3B Krogh-Moe, J. *Acta Cryst.*, 1960, **13**, 889.
Krogh-Moe, J. *Acta Cryst.*, 1974, **B30**, 1178.
Touboul, M., Pénin, N. & Seguin, L. *Powder Diffr.*, 1999, **14**, 234.

3Cs.7B Nowogrocki, G., Penin, N. & Touboul, M. *Solid State Sci.*, 2003, **5**, 795.

Cs.B Schläger, M. & Hoppe, R. *Z. Anorg. Allgem. Chem.*, 1994, **620**, 1867.
Schneider, W. & Carpenter, G. B. *Acta Cryst.*, 1970, **B26**, 1189.

Silver
α-Ag.4B (HT) Krogh-Moe, J. *Acta Cryst.*, 1965, **18**, 77.

β-Ag.4B (LT) Penin, N., Touboul, M. & Nowogrocki, G. *Solid State Sci.*, 2003, **5**, 559.

Ag.B Jansen, M. & Brachtel, G. *Naturwiss.*, 1980, **67**, 606.
Brachtel, G. & Jansen, M. *Z. Anorg. Allgem. Chem.*, 1981, **478**, 13.

3Ag.B Jansen, M. & Scheld, W. *Z. Anorg. Allgem. Chem.*, 1981, **477**, 85.

3Ag.B-II Jansen, M. & Brachtel, G. *Z. Anorg. Allgem. Chem.*, 1982, **489**, 42.

Thallium
Tl.5B Touboul, M. *C. R. Acad. Sci. Paris Sér. C*, 1973, **277**, 1025.
Touboul, M. & Nowogrocki, G. *J. Solid State Chem.*, 1998, **136**, 216.
Touboul, M., Penin, N. & Seguin, L. *Powder Diffr.*, 1999, **14**, 237.

Tl.3B Touboul, M. *C. R. Acad. Sci. Paris Sér. C*, 1973, **277**, 1025.

(Table A1 continued)

	Touboul, M. Bétourné E, & Nowogrocki, G. *J. Solid State Chem.*, 1997, **131**, 370.
	Touboul, M., Penin, N. & Seguin, L. *Powder Diffr.*, 1999, **14**, 234.
α-Tl.2B	Touboul, M. & Amoussou, D. *C. R. Acad. Sci. Paris Sér. C*, 1977, **285**, 145.
β-Tl.2B	Touboul, M. & Amoussou, D. *C. R. Acad. Sci. Paris Sér. C*, 1977, **285**, 145.
	Penin, N., Seguin, L., Gérand, B., Touboul, M. & Nowogrocki, G. *J. Solid State Chem.*, 2001, **160**, 139.
Tl.B	Touboul, M. & Amoussou, D. *J. Less-Common Metals*, 1977, **56**, 39.
	Touboul, M. & Amoussou, D. *Rev. Chim. Minér.*, 1978, **15**, 223.
3Tl.B	Marchand, R., Piffard, Y. & Tournoux, M. *C. R. Acad. Sci. Paris Sér. C*, 1973, **276**, 177.
	Piffard, Y., Marchand, R. & Tournoux, M. *Rev. Chim. Minér.*, 1975, **12**, 210.

Table A2. mMO.nB_2O_3 *crystal structure references*

Compound	Reference(s)
Magnesium	
Mg.2B	Kuzel, H. -J. *Neues Jahrb. Mineral. Monatsh.*, 1964, 357.
	Bartl, H. & Schuckmann, W. *Neues Jahrb. Mineral. Monatsh.*, 1966, 142.
2Mg.B	Takéuchi, Y. *Acta Cryst.*, 1952, **5**, 574.
	Block, S., Burley, G., Perloff, A. & Mason Jr., R. D. *J. Res. NBS*, 1959, **62**, 95.
3Mg.B	Berger, S. V. *Acta Chem. Scand.*, 1949, **3**, 660.
	Effenberger, H. & Pertlik, F. *Z. Krist.*, 1984, **166**, 129.
Calcium	
Ca.2B	Kindermann, B. *Z. Krist.* **146**, 1977, 61.
	Zayakina, N. V. & Brovkin, A. A. *Sov. Phys. Cryst.*, 1977, **22**, 156.
2Ca.3B	Zayakina, N. V. & Brovkin, A. A. *Sov. Phys. Cryst.* **21**, 1976, 277.
Ca.B-I	Zachariasen, W. H. & Ziegler, G. E. *Z. Krist. Kristallgeom. Kristallphys.*, 1932, **83**, 354.
	Marezio, M., Plettinger, H. A. & Zachariasen, W. H. *Acta Cryst.*, 1963, **16**, 390.
	Kirfel, A. *Acta Cryst.*, 1987, **B43**, 333.
Ca.B-II	Marezio, M., Remeika, J. P. & Dernier, P. D. *Acta Cryst.*, 1969, **B25**, 965.
	Dernier, P. D. *Acta Cryst.*, 1969, **B25**, 1001.
	Shashkin, D. P., Simonov, M. A. & Belov, N. V. *Sov. Phys. Dokl.*, 1971, **15**, 1003.
Ca.B-III	Marezio, M., Remeika, J. P. & Dernier, P. D. *Acta Cryst.* **B25**, 1969, 955.
	Dernier, P. D. *Acta Cryst.* **B25**, 1969, 1001.
Ca.B-IV	Marezio, M., Remeika, J. P. & Dernier, P. D. *Acta Cryst.*, 1969, **B25**, 965.
	Dernier, P. D. *Acta Cryst.* **B25**, 1969, 1001.
2Ca.B	Schäfer, U. L. *Neues Jahrb. Mineral. Monatsh.*, 1968, 75.
	Ji, Y., Liang, J., Xie, S., Zhu, N. & Li, Y. *Acta Cryst.*, 1993, **C49**, 78.
3Ca.B	Schäfer, U. L. *Neues Jahrb. Mineral. Monatsh.*, 1968, 75.
	Schuckmann, W. *Neues Jahrb. Mineral. Monatsh.*, 1969, 142.
	Vegas, A., Cano, F. H. & García-Blanco, S. *Acta Cryst.*, 1975, **B31**, 1416.
	Vegas, A. *Acta Cryst.*, 1985, **C41**, 1689.
Strontium	
Sr.3B	Chenot, C. F. *J. Am. Ceram. Soc.*, 1967, **50**, 117.
Sr.2B	Krogh-Moe, J. *Acta Chem. Scand.*, 1964, **18**, 2055.
	Krogh-Moe, J. *Nature*, 1965, **206**, 613.
	Block, S., Perloff, A. & Weir, C. E. *Acta Cryst.*, 1964, **17**, 314.
	Perloff, A & Block, S. *Acta Cryst.*, 1966, **20**, 274.
Sr.B-I	Block, S., Perloff, A. & Weir, C. E. *Acta Cryst.*, 1964, **17**, 314.
	Dernier, P. D. *Acta Cryst.*, 1969, **B25**, 1001.
Sr.B-III	Dernier, P. D. *Acta Cryst.*, 1969, **B25**, 1001.
Sr.B-IV	Dernier, P. D. *Acta Cryst.*, 1969, **B25**, 1001.
	Ross, N. L. & Angel, R. J. *J. Solid State Chem.*, 1991, **90**, 27.
2Sr.B	Bartl, H. & Schuckmann, W. *Neues Jahrb. Mineral. Monatsh.*, 1966, 253.
3Sr.B	Hata, H., Adachi, G. & Shiokawa, J. *Mater. Res. Bull.*, 1977, **12**, 811.
	Richter, L. & Müller, F. *Z. Anorg. Allgem. Chem.*, 1980, **467**, 123.
Barium	
Ba.4B	Krogh-Moe, J. *Acta Chem. Scand.*, 1960, **14**, 1229.
	Krogh-Moe J. & Ihara, M. *Acta Cryst.*, 1969, **B25**, 2153.
2Ba.5B	Hübner, K. -H. *Neues Jahrb. Mineral. Monatsh.*, 1969, 335.
Ba.2B	Block, S., Perloff, A. & Weir, C. E. *Acta Cryst.*, 1964, **17**, 314.
	Block, S. & Perloff, A. *Acta Cryst.*, 1965, **19**, 297.
	Filatov, S. K., Nikolaeva, N. V., Bubnova, R. S. & Polyakova, I. G. *Glass Phys. Chem.*, 2006, **32**, 471.
α-Ba.B (HT)	Block, S., Perloff, A. & Weir, C. E. *Acta Cryst.*, 1964, **17**, 314.
	Mighell, A. D., Perloff, A. & Block, S. *Acta Cryst.*, 1966, **20**, 819.
β-Ba.B (LT)	Hübner, K. -H. *Neues Jahrb. Mineral. Monatsh.*, 1969, 335.
	Liebertz, J. & Stähr, S. *Z. Krist.*, 1983, **165**, 91.
	Fröhlich, R. *Z. Krist.*, 1984, **168**,109.
	Ito, K., Marumo, F., Ohgaki, M. & Tanaka, K. *Rep. Res. Lab. Eng. Mater. Tokyo Inst. Technol.*, 1990, (15), 1.
	Filatov, S. K., Nikolaeva, N. V., Bubnova, R. S. & Polyakova, I. G. *Glass Phys. Chem.*, 2006, **32**, 471.
2Ba.B	Hübner, K. -H. *Neues Jahrb. Mineral. Monatsh.*, 1969, 335.
4Ba.B	Hübner, K. -H. *Neues Jahrb. Mineral. Monatsh.*, 1969, 335.
Zinc	
Zn.2B	Martínez-Ripoll, M. & García-Blanco, S. *Anal. Fis.*, 1970, **66**, 209.
	Martínez-Ripoll, M., Martinez-Carrera, S. & García-Blanco, S. *Acta Cryst.*, 1971, **B27**, 672.
4Zn.3B	Smith, P., García-Blanco, S. & Rivoir, L. *Anal. Real Soc. Esp. Fis. Quim. Ser. A* **57A**, 1961, 263.
	Smith, P., García-Blanco, S. & Rivoir, L. *Z. Krist.*, 1961, **115**, 460.
	Smith, P., García-Blanco, S. & Rivoir, L. *Z. Krist.*, 1964, **119**, 375.

(Table A2 continued)

	Bondareva, O. S., Egorov-Tismenko, Yu. K., Simonov, M. S. & Belov, N. V. *Sov. Phys. Dokl.*, 1978, **23**, 529.
	Smith-Verdier, P. & García-Blanco, S. *Z. Krist.*, 1980, **151**, 175.
3Zn.B	García-Blanco, S. & Fayos, J. *Z. Krist.*, 1968, **127**, 145.
	Baur, W. H. & Tillmanns, E. *Z. Krist.*, 1970, **131**, 213.
Cadmium	
Cd.2B	Hand W. D. & Krogh-Moe, J. *J. Am. Ceram. Soc.*, 1962, **45**, 197.
	Ihara, M. & Krogh-Moe, J. *Acta Cryst.*, 1966, **20**, 132.
Cd.B	Sokolova, E. V., Mel'nikov, O. K., Ivashchenko, A. N., Simonov, M. A. & Belov, N. V. *Inorg. Mater.*, 1983, **19**, 1349.
	Laureiro, Y., Pico, C. & Jerez, A. *Synth. React. Inorg. Met. -Org. Chem.*, 1988, **18**, 119.
2Cd.B	Hand, W. D. & Krogh-Moe, J. *J. Am. Ceram. Soc.*, 1962, **45**, 197.
	Sokolova, E. V. Simonov M. A. & Belov, N. V. *Sov. Phys. Dokl.*, 1979, **24**, 524.
	Sokolova, E. V., Mel'nikov, O. K., Ivashchenko, A. N., Simonov, M. A. & Belov, N. V. *Inorg. Mater.*, 1983, **19**, 1349.
3Cd.B	Weir, C. E. & Schroeder, R. A. *J. Res. NBS*, 1964, **68A**, 465.
	Laureiro, Y., Veiga, M. L., López, M. L., García-Martín, S., Jerez, A. & Pico, C. *Powder Diffr.*, 1991, **6**, 28.
Lead	
Pb.2B	Block, S., Perloff, A. & Weir, C. E. *Acta Cryst.*, 1964, **17**, 314.
	Perloff, A. & Block, S. *Acta Cryst.*, 1966, **20**, 274.
6Pb.5B	Krogh-Moe, J. & Wold-Hansen, P. S. *Acta Cryst.*, 1973, **B29**, 2242.

Table A3. References for crystal structures with two modifying cation species

Compound	Reference(s)
Mixed pentaborate	
Cs.2Ag.15B	Wiesch, A. & Bluhm, K. *Z. Naturf.*, 1998, **53b**, 157.
Mixed tetraborate	
0·4Na.0·6Ag.4B	Krogh-Moe, J. *Acta Cryst.*, 1965, **18**, 77.
Mixed triborate	
Li.Cs.6B	Sasaki, T., Mori, Y., Kuroda, I., Nakajima, S., Yamaguchi, K. & Watanabe, S. *Acta Cryst.*, 1995, **C51**, 2222.
	Tu, J.-M. & Keszler, D. A. *Mater. Res. Bull.*, 1995, **30**, 209.
Li.2Sr.9B	Penin, N., Seguin, L., Touboul, M. & Nowogrocki, G. *Int. J. Inorg. Mater.*, 2001, **3**, 1015.
Li.2Ba.9B	Penin, N., Seguin, L., Touboul, M. & Nowogrocki, G. *Int. J. Inorg. Mater.*, 2001, **3**, 1015.
	Pushcharovsky, D. Yu., Gobetchia, E. R., Pasero, M., Merlino, S. & Dimitrova, O.V. *J. Alloys Compd.*, 2002, **339**, 70.
Na.2(Na/K).9B	Bubnova, R., Albert, B., Georgievskaya, M., Krzhizhanovskaya, M., Hofmann, K. & Filatov,S. *J. Solid State Chem.*, 2006, **179**, 2954.
Na.2K.9B	Bubnova, R., Albert, B., Georgievskaya, M., Krzhizhanovskaya, M., Hofmann, K. & Filatov,S. *J. Solid State Chem.*, 2006, **179**, 2954.
(Na/K).2K.9B	Bubnova, R., Albert, B., Georgievskaya, M., Krzhizhanovskaya, M., Hofmann, K. & Filatov,S. *J. Solid State Chem.*, 2006, **179**, 2954.
Na.2Ba.9B	Penin, N., Seguin, L., Touboul, M. & Nowogrocki, G. *Int. J. Inorg. Mater.*, 2001, **3**, 1015.
Mixed di-pentaborate	
Na.Cs.5B	Tu, J.-M. & Keszler, D. A. *Inorg. Chem.*, 1996, **35**, 463.
Na.Tl.5B	Penin, N., Touboul, M. & Nowogrocki, G. *J. Alloys Compd.*, 2004, **363**, 104.
K.Cs.5B	Tu, J.-M. & Keszler, D. A. *Inorg. Chem.*, 1996, **35**, 463.
Mixed tri-heptaborate	
Ag.2Sr.7B	Wiesch, A. & Bluhm, K. *Z. Naturf.*, 1997, **52b**, 227.
Mixed diborate	
Li.K.4B	Ono, Y., Nakaya, M., Kajitani, T., Sugawara, T., Watanabe, N., Shiraishi, H. & Komatsu, R. *Acta Cryst.*, 2000, **C56**, 1413.
	Ono, Y., Nakaya, M., Sugawara, T., Watanabe, N., Siraishi, H., Komatsu, R. & Kajitani, T. *J. Cryst. Growth*, 2001, **229**, 472.
Li.Rb.4B	Ono, Y., Nakaya, M., Kajitani, T., Sugawara, T., Watanabe, N., Shiraishi, H. & Komatsu, R. *Acta Cryst.*, 2000, **C56**, 1413.
	Ono, Y., Nakaya, M., Sugawara, T., Watanabe, N., Siraishi, H., Komatsu, R. & Kajitani, T. *J. Cryst. Growth*, 2001, **229**, 472.
Mixed tri-pentaborate	
Na.2Ca.5B	Fayos, J., Howie, R. A. & Glasser, F. P. *Acta Cryst.*, 1985, **C41**, 1394.
K.2Sr.5B	Tu, J.-M. & Keszler, D. A. *Acta Cryst.*, 1995, **C51**, 341.
Mixed metaborate	
Li.4Ba.5B	Smith, R. W. & Keszler, D. A. *Mater. Res. Bull.*, 1989, **24**, 725.
3Na.2Ca.5B	Fayos, J., Howie, R. A. & Glasser, F. P. *Acta Cryst.*, 1985, **C41**, 1396.
Mixed pyroborate	
Be.Ca.B	Schaffers, K. I. & Keszler, D. A. *Acta Cryst.*, 1993, **C49**, 647.
Mg.Ca.B	Yakubovich, O. V., Yamnova, N. A., Shchedrin, B. M., Simonov, M. A. & Belov, N. V. *Sov. Phys. Dokl.*, 1976, **21**, 294.
Mixed orthoborate	
Li.2Zn.B	Bondareva, O. S., Simonov, M. A., Egorov-Tismenko, Yu. K. & Belov, N. V. *Sov. Phys. Cryst.*, 1978, **23**, 269.
Li.2Cd.B-I	Kazanskaya, E. V., Sandomirskii, P. A., Simonov, M. A. & Belov, N. V. *Sov. Phys. Dokl.*, 1978, **23**, 108.
	Sokolova, E. V., Mel'nikov, O. K., Ivashchenko, A. N., Simonov, M. A. & Belov, N. V. *Inorg. Mater.*, 1983, **19**, 1349.
Li.2Cd.B-II	Sokolova, E. V., Boronikhin, V. A., Simonov, M. A. & Belov, N. V. *Sov. Phys. Dokl.*, 1979, **24**, 417.
	Sokolova, E. V., Mel'nikov, O. K., Ivashchenko, A. N., Simonov, M. A. & Belov, N. V. *Inorg. Mater.*, 1983, **19**, 1349.
2Na.Rb.B	Schläger, M. & Hoppe, R. *Aust. J. Chem.*, 1992, **45**, 1427.
2Na.Cs.B	Schläger, M. & Hoppe, R. *Aust. J. Chem.*, 1992, **45**, 1427.
2Be.Sr.B	Schaffers, K. I. & Keszler, D. A. *J. Solid State Chem.*, 1990, **85**, 270.
Mg.2Sr.B	Diaz, A. & Keszler, D. A. *Chem. Mater.*, 1997, **9**, 2071.
	Chen, G.-J., Wu, Y.-C. & Fu, P.-Z. *Acta Cryst.*, 2007, **E63**, i175.
Mg.2Ba.B	Akella, A. & Keszler, D. A. *Mater. Res. Bull.*, 1995, **30**, 105.
Ca.2Ba.B	Akella, A. & Keszler, D. A. *Main Group Met. Chem.*, 1995, **18**, 35.
Ba.2Zn.B	Smith, R. W. & Keszler, D. A. *J. Solid State Chem.*, 1992, **100**, 325.

Proc. VI Int. Conf. Borate Glasses, Himeji, Japan, 18–22 August 2008   *Phys. Chem. Glasses: Eur. J. Glass Sci. Technol. B*, February 2010, **51** (1), 40–51

Adrian C. Wright: Glasses, Neutrons, Borates!

Alex C. Hannon[1]

ISIS Facility, Rutherford Appleton Laboratory, Chilton, Didcot, Oxon OX11 0QX, U.K.

Manuscript received 18 March 2009
Revised version received 28 June 2009
Accepted 6 July 2009

Professor Adrian C Wright of the Physics Department at Reading University has a long and illustrious scientific career, based on the use of neutron scattering to study the atomic structure of glasses. He and Roger Sinclair, his longstanding collaborator, have been key figures in the development of pulsed neutron diffraction as an experimental technique for structural studies of glass. One of his outstanding areas of work has been the investigation of the structure of vitreous B_2O_3, showing clear support for a model which involves a large proportion of boroxol groups. A brief outline of the life and scientific career of Adrian Wright is given. Emphasis is given to his early work on glasses, neutron scattering and vitreous B_2O_3, showing how this led to the development of the International Conference on Borate Glasses, Crystals and Melts into a successful conference series.

1. Introduction

The Sixth International Conference on Borate Glasses, Crystals and Melts (Borate2008) took place on 18–22 August 2008 in Himeji, Japan, under the chairmanship of Dr Norimasa Umesaki, of the SPring-8 synchrotron radiation facility. The Borate Conference has always been held in honour of a scientist who has made an outstanding contribution to the study of borate materials (see Table 1), ever since the first Borate Conference,[1] held on 3–8 June 1977 at Alfred University, USA, in honour of Professor Jan Krogh-Moe[2] from the Norwegian Institute of Technology in Trondheim, for the great insight that he brought to the investigation of borates. This tradition has continued until the Sixth Borate Conference, which was held in honour of Professor Adrian C. Wright (see Figure 1) of Reading University, UK, for his achievements in glass science, and in particular borate glasses.

Adrian Wright is a most appropriate person to honour at the Borate Conference because, in addition to his great contributions to the study of the structure of borate materials, he has played the key role in making the Borate Conference into a successful conference

Email a.c.hannon@rl.ac.uk

Figure 1. *Professor Adrian Wright, before giving the Scholes lecture at the New York State College of Ceramics, Alfred University, USA, on 27th April, 2006*

series. In this article I will describe the events around Adrian's earlier contributions to the study of borates, and the circumstances which led to the development of the Borate Conference. It should be noted, however, that Adrian has made a major contribution to the

Table 1. *Historical details of the Borate Conference*

Conference	Dates	Conference title	Location	Conference chairmen	Honouree
Borate-I[1]	3–8 June 1977	Boron in Glass and Glass Ceramics	Alfred, USA	David Pye & Norbert Kreidl	Jan Krogh-Moe
Borate-II[4]	22–25 July 1996	Borate Glasses, Crystals and Melts	Abingdon, UK	Adrian Wright & Steve Feller	Philip J. Bray
Borate-III[5]	4–9 July 1999	Borate Glasses, Crystals and Melts: Structure and Applications	Sofia, Bulgaria	Yanko Dimitriev	Evgenii A. Porai-Koshits
Borate-IV[6]	14–18 July 2002	Borate Glasses, Crystals and Melts	Cedar Rapids, USA	Mario Affatigato & Steve Feller	David L. Griscom
Borate-V[7]	10–14 July 2005	Borate Glasses, Crystals and Melts: New Techniques and Practical Applications	Trento, Italy	Giuseppe Dalba & Francesco Rocca	Werner Vogel
Borate-VI	18–22 August 2008	Borate Glasses, Crystals and Melts: Borate2008	Himeji, Japan	Norimasa Umesaki	Adrian C. Wright

Figure 2. Adrian Wright in November 1996, receiving the George W Morey award of the American Ceramic Society at their Glass & Optical Materials Division Fall Meeting in San Antonio, USA, from Cate Simmons

study of the structure of non-crystalline materials with a wide range of compositions, extending far wider than just those materials which contain boron and oxygen. For example, see the book chapter on *Neutron and X-ray Amorphography*,[3] one of the two publications which were cited when Adrian was awarded the George W. Morey award of the American Ceramic Society (see Figure 2).

Adrian Wright's enthusiasm for his subject and tenaciousness as an organiser have been vital for the development of the Borate Conference. He has also been a superb editor of the proceedings of the conference.[4–7] But anyone who gets to know Adrian will soon find that, as well as his interest in glass and neutron scattering, he also takes a great interest in the people in the field, so that he has many friends, collaborators and colleagues amongst his fellow researchers. In my view, it is Adrian's large network of contacts that has been of the greatest importance to his successful development of the Borate Conference and so, as well as discussing Adrian himself, I

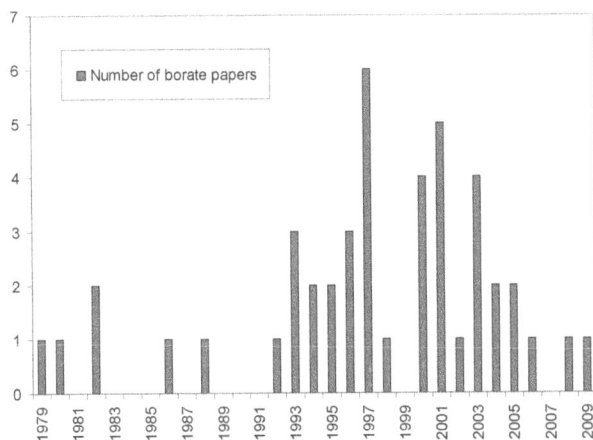

Figure 3. The annual number of borate-related publications by Adrian Wright (see Table 2)

Figure 4. Adrian Wright as a young boy (exact age unknown, but thought to be between 5 and 11)

will discuss some of the other people who have been important figures in Adrian's studies of borates.

2. The beginning

Adrian has published many borate-related scientific papers, as listed in Table 2 and shown by the histogram in Figure 3. His borate paper with the biggest impact is his 1982 neutron diffraction study of the structure of vitreous boron trioxide, v-B_2O_3,[8] which has so far been cited 163 times. But it is interesting to discover how he came to produce this paper, and to go on to develop the Borate Conference into a regular, long running series.

Adrian was born on 1st February, 1944, near Sittingbourne in Kent, UK (Figure 4). When he was two years old his parents divorced, and not long after that his father went to live in Canada at first, and later in the USA. Consequently, Adrian was brought up by his mother and maternal grandparents. At that time his grandfather was the head gardener at Stubbings House, a large country house near Maidenhead, Berkshire, UK, where his grandmother also worked, and they lived in a cottage in the grounds of Stubbings House that was provided in connection with their employment. Stubbings House was the home of Queen Wilhelmina of the Netherlands during the Second World War, but in 1947 it had been bought by Sir Thomas Merton, a spectroscopist in the Clarendon Laboratory (the physics department) at Oxford University. Sir Thomas was a man of considerable wealth, and at Stubbings House he maintained what was probably the last private physics laboratory in Britain. He was noted for his work on spectroscopy and diffraction gratings, and also as an art collector and for the inventions he contributed to the British war effort during the Second World War. When Adrian was 11 years old his grandfather retired, and he was able to buy a house in Maidenhead using his savings which a previous employer had helped him to invest wisely. At about that time Adrian passed an important academic examination, the 11-plus, which won him a place at Maidenhead Grammar School – a kind of school with the purpose of educating the most

Table 2. Adrian Wright's scientific publications on borates

Publication	Reference
Neutron-diffraction study of vitreous B_2O_3 (Wright, A C, Johnson, P A V and Sinclair, R N)	(72)
Neutron-diffraction studies of amorphous solids (Sinclair, R N, Desa, J A E, Etherington, G, Johnson, P A V and Wright, A C)	(73)
A neutron-diffraction investigation of the structure of vitreous boron trioxide (Johnson, P A V, Wright, A C and Sinclair, R N)	(8)
Random network models of vitreous B_2O_3 containing boroxol rings (Wright, A C, Sumner, D J and Clare, A G)	(74)
Neutron scattering studies of the structure and dynamics of vitreous B_2O_3 (Wright, A C, Hannon, A C, Grimley, D I, Sinclair, R N, Galeener, F L, Elliott, S R and Gladden, L F)	(75)
Phonon spectra of vitreous B_2O_3 (Hannon, A C, Sinclair, R N, Blackman, J A, Wright, A C and Galeener, F L)	(59)
Inelastic neutron scattering studies of the vibrational modes of vitreous B_2O_3 (Hannon, A C, Sinclair, R N and Wright, A C)	(60)
A combined thermodynamic and neutron scattering investigation of glasses in the system $PbO-B_2O_3$ (Vedishcheva, N M, Shakhmatkin, B A, Wright, A C, Sinclair, R N and Grimley, D I)	(54)
Structural order in lead borate glasses: a test of recent thermodynamic predictions (Vedishcheva, N M, Shakhmatkin, B A, Wright, A C, Grimley, D I, Etherington, G and Sinclair, R N)	(55)
The vibrational modes of vitreous B_2O_3 (Hannon, A C, Sinclair, R N and Wright, A C)	(61)
Inelastic neutron scattering and modelling studies of vitreous B_2O_3 (Hannon, A C, Wright, A C and Sinclair, R N)	(62)
Boroxol groups in vitreous boron oxide : New evidence from neutron diffraction and inelastic neutron scattering studies (Hannon, A C, Grimley, D I, Hulme, R A, Wright, A C and Sinclair, R N)	(32)
The thermodynamic properties of oxide glasses and glass-forming liquids and their chemical structure (Shakhmatkin, B A, Vedishcheva, N M, Shultz, M M and Wright, A C)	(76)
Vitreous borate networks containing superstructural units: a challenge to the random network theory? (Wright, A C, Vedishcheva, N M and Shakhmatkin, B A)	(69)
The vibrational modes of vitreous B_2O_3: inelastic neutron scattering and modelling studies (Hannon, A C, Wright, A C, Blackman, J A and Sinclair, R N)	(63)
Irreversible structural changes in vitreous B_2O_3 under pressure (Grimsditch, M, Polian, A and Wright, A C)	(77)
Borate glasses, superstructural units and the random network theory (Wright, A C, Sinclair, R N, Grimley, D I, Hulme, R A, Vedishcheva, N M, Shakhmatkin, B A, Hannon, A C, Feller, S A, Meyer, B M, Royle, M L and Wilkerson, D L)	(78)
The thermodynamic modelling of glass properties: A practical proposition? (Vedishcheva, N M, Shakhmatkin, B A, Shultz, M M and Wright, A C)	(79)
Borate glasses, crystals and melts (Eds Wright, A C, Feller, S A and Hannon, A C)	(80)
Some paradoxes of borate glasses and melts (Porai-Koshits, E A and Wright, A C)	(81)
The interrelationship between the structures of borate glasses and crystals (Wright, A C, Vedishcheva, N M and Shakhmatkin, B A)	(47)
Inelastic neutron scattering techniques for studying superstructural units in borate glasses (Sinclair, R N, Wright, A C, Wanless, A J, Hannon, A C, Feller, S A, Mayhew, M T, Meyer, B M, Royle, M L, Wilkerson, D L, Williams, R B and Johanson, B C)	(48)
A proposed relationship between the position of the first neutron diffraction peak for alkali borate glasses and the separation between the modifying cations as found from density measurements (Wanless, A J, Wright, A C, Sinclair, R N, Feller, S A, Williams, R B, Johanson, B C and Mayhew, M T)	(56)
A crystallographic guide to the structure of borate glasses (Wright, A C, Vedishcheva, N M and Shakhmatkin, B A)	(70)
A thermodynamic approach to the structural modeling of oxide melts and glasses: Borate and silicate systems (Vedishcheva, N M, Shakhmatkin, B A and Wright, A C)	(82)
Inelastic neutron scattering studies of superstructural units in borate glasses and crystalline phases (Sinclair, R N, Stone, C E, Wright, A C, Polyakova, I G, Vedishcheva, N M, Shakhmatkin, B A, Feller, S A, Johanson, B C, Venhuizen, P, Williams, R B and Hannon, A C)	(64)
Structure of bismuth borate glasses (Stone, C E, Wright, A C, Sinclair, R N, Feller, S A, Affatigato, M, Hogan, D L, Nelson, N D, Vira, C, Dimitriev, Y B, Gattef, E M and Ehrt, D)	(57)
The structure of pressure-compacted vitreous boron oxide (Wright, A C, Stone, C E, Sinclair, R N, Umesaki, N, Kitamura, N, Ura, K, Ohtori, N and Hannon, A C)	(45)
Influence of fictive temperature on the structure of borate glasses: a thermodynamic prediction (Vedishcheva, N M, Shakhmatkin, B A and Wright, A C)	(83)
Thermodynamic modelling of the structure of glasses and melts: single-component, binary and ternary systems (Vedishcheva, N M, Shakhmatkin, B A and Wright, A C)	(67)
The chemical structure of oxide glasses: A concept consistent with neutron scattering studies? (Wright, A C, Shakhmatkin, B A and Vedishcheva, N M)	(84)
The structure of vitreous boron sulphide (Sinclair, R N, Stone, C E, Wright, A C, Martin, S W, Royle, M L and Hannon, A C)	(85)
Borate Glasses, Crystals & Melts: Structure & Applications (Dimitriev, Y B and Wright, A C)	(5)
Evgenii Alexandrovich Porai-Koshits: Glass scientist extraordinaire (Vedishcheva, N M and Wright, A C)	(86)
Neutron diffraction studies of binary network glasses containing PbO and Bi_2O_3 (Wright, A C, Etherington, G, Feller, S A, Gnappi, G, Grimley, D I, Montenero, A, Sinclair, R N and Stone, C E)	(87)
Borate Glasses, Crystals & Melts 2002: New Techniques and Applications (Affatigato, M, Feller, S A, Vedishcheva, N M and Wright, A C)	(6)
Structure of crystalline caesium enneaborate (Wright, A C, Sinclair, R N, Stone, C E, Knight, K S, Polyakova, I G, Vedishcheva, N M and Shakhmatkin, B A)	(88)
A neutron diffraction investigation of the structure of caesium borate glasses (Shaw, J L, Wright, A C, Sinclair, R N, Frueh, J R, Williams, R B, Nelson, N D, Affatigato, M, Feller, S A and Scales, C R)	(58)
Thermodynamic modelling of the structure of sodium borosilicate glasses (Vedishcheva, N M, Shakhmatkin, B A and Wright, A C)	(89)
The use of crystallographic data in interpreting the correlation function for complex glasses (Wright, A C, Shaw, J L, Sinclair, R N, Vedishcheva, N M, Shakhmatkin, B A and Scales, C R)	(90)
The structure of sodium borosilicate glasses: thermodynamic modelling vs. experiment (Vedishcheva, N M, Shakhmatkin, B A and Wright, A C)	(91)
Thermodynamic modelling of the structure and properties of glasses in the sytems $Na_2O-B_2O_3-SiO_2$ and $Na_2O-CaO-SiO_2$ (Vedishcheva, N M, Shakhmatkin, B A and Wright, A C)	(92)
Superstructural units in vitreous and crystalline caesium borates (Haworth, R, Shaw, J L, Wright, A C, Sinclair, R N, Knight, K S, Taylor, J W, Vedisheheva, N M, Polyakova, I G, Shakhmatkin, B A, Feller, S A and Winslow, D W)	(65)
The polymorphs of crystalline caesium enneaborate (Haworth, R, Wright, A C, Sinclair, R N, Knight, K S, Vedishcheva, N M, Polyakova, I G and Shakhmatkin, B A)	(68)
Neutron spectroscopic studies of caesium borate crystals and glasses (Sinclair, R N, Haworth, R, Wright, A C, Parkinson, B G, Holland, D, Taylor, J W, Vedishcheva, N M, Polyakova, I G, Shakhmatkin, B A, Feller, S A, Rijal, B and Edwards, T)	(66)
Thermodynamic modelling of the real-space correlation function for four sodium borosilicate glasses (Wright, A C)	(93)
Borate glasses, crystals and melts: New techniques and practical applications (Dalba, G, Rocca, F, Vedishcheva, N M and Wright, A C)	(7)
Borate structures: Crystalline and vitreous (Wright, A C)	(71)

Figure 5. Professor Alan Leadbetter, Adrian Wright's PhD supervisor, relaxing at home in 1968. (Thanks to Alan Leadbetter for providing this picture)

academically capable children, which is sadly now abolished in most parts of the UK.

In 1962, after completing school, Adrian went to study Chemistry at Bristol University. After three years he passed his examinations and was awarded a First Class BSc Special Honours degree in Chemistry – the highest grade that can be achieved. It was always his intention to participate in scientific research towards a PhD, but first he spent a year training as a teacher and gaining a Postgraduate

Figure 6. Adrian Wright in academic dress, on the occasion of his PhD graduation at Bristol University in 1970

Certificate in Education. When Adrian started his Ph.D. studies, Sir Thomas Merton gave him a generous gift of £200 (a large sum at that time) to help with his education. Adrian's PhD studies were performed under the supervision of Alan Leadbetter (see Figure 5), who was then a young university lecturer who had established himself studying the heat capacity and vibrational spectra of glasses. In relation to the Borate Conference, it is interesting to note here that, although Alan Leadbetter worked on many different glasses during his career, he never published any work on v-B_2O_3 or other borates. Adrian was set to work initially on the effect of radiation damage on the density of glasses. This topic was not sufficient for a PhD, and Adrian soon began to apply x-ray diffraction in the laboratory to study the structure of glasses, using a modified Phillips diffractometer with a molybdenum tube. In 1969 Adrian completed his PhD, with a thesis entitled The Structure of Five Simple Glasses,[9] describing x-ray diffraction studies of v-SiO_2, irradiated v-SiO_2, GeO_2, BeF_2 and vitreous carbon (see Figure 6). Early neutron diffraction and inelastic neutron scattering measurements on BeF_2 glass were also reported in Adrian's thesis.

I myself obtained my own PhD[10] under Adrian's supervision at the University of Reading, but did not meet Alan Leadbetter until I applied for a position at the ISIS Facility pulsed neutron source in 1986, at which time he was the head of ISIS. In my job interview, Alan Leadbetter put me at my ease by introducing himself as my "scientific grandfather". In this view, Adrian Wright is my "scientific father", but as I shall show below, he is also the father of the Borate Conference.

After his PhD, Adrian moved to the Physics Department of Reading University, where he worked from 1969 to 1971 as an ICI Research Fellow, and then as a Senior Research Physicist, until 1973 when he was taken on as a permanent member of the academic staff. It was a time of growth for the study of amorphous materials, and the head of department, Bill Mitchell (later Professor Sir E. W. J. Mitchell FRS), had the foresight to see that a chemist was just what was needed for a physics department to become active in this field. Also, Adrian's teaching contribution was of great value to the Reading Physics Department for the strong chemical physics degree course which was then taught at Reading University.

3. Roger Sinclair – An enduring partnership

Towards the end of his PhD, Adrian had started to become involved in the use of neutron scattering to study glasses, and in fact his first scientific paper was a report of inelastic neutron scattering studies of BeF_2 in both its vitreous and crystalline forms.[11] The work on BeF_2 was performed using a spectrometer at a nuclear reactor called HERALD (Highly Enriched

Figure 7. (above) The participants in the Harwell Laboratory neutron scattering summer school in 1966. Adrian Wright is to the left of the picture, as indicated by a rectangle, and shown to the left. (Thanks to Professor Terry Willis of Oxford University for providing this picture.) Both Adrian and Alan Leadbetter can be seen in the group photo of the 1970 summer school, recently published by Lander[13]

Reactor at ALDermaston),[12] which was located at the Aldermaston nuclear weapons laboratory. However, there was a larger neutron scattering programme being developed at the Harwell Laboratory, where there were two reactors being used for neutron scattering, and Adrian attended two of the excellent Harwell neutron scattering summer schools, in 1966 and 1970[13] (see Figure 7). These summer schools provided the starting point for the scientific careers of many of the leading figures in neutron scattering.[13] Adrian first met Ed Lorch (whose name is well known in the field of glass diffraction studies, due to the modification function which he devised[14]) at the Harwell summer school in 1966, and Ed can be

seen at fourth from the left in the front row in Figure 7.

After moving to Reading, Adrian began to apply neutron diffraction more widely to the study of glass structure, using the Aldermaston and Harwell reactors, and whilst at Harwell in the spring of 1970 Adrian heard talk of a researcher called Roger Sinclair (see Figures 8 and 9), who was starting to use the first Harwell electron linac neutron source as a means of measuring neutron diffraction data to very high values of momentum transfer. High momentum transfer, $\hbar Q$, ($Q=4\pi\sin\theta/\lambda$) is a kind of holy grail for the study of the structure of disordered materials, because it leads to high resolution in real-space,[15] and so Adrian quickly met and began to work with Roger Sinclair.

Figure 8. Adrian Wright (left) and Roger Sinclair (right) during the 1970s whilst performing an experiment on the D4 neutron diffractometer at the Institut Laue Langevin nuclear reactor, in Grenoble, France (photo courtesy of Prof Erwin Desa of Goa University, India)

Figure 9. Adrian Wright (left) and Roger Sinclair (right) in 2005 during the Fifth International Conference on Borate Glasses, Crystals and Melts: New Techniques and Practical Applications at Trento, Italy. (Thanks to Professor Ladislav Koudelka of University of Pardubice, Czech Republic, for providing this picture)

This meeting was the start of a great friendship, and a scientific partnership of exceptional strength and duration (37 years), during which Adrian and Roger worked together continually, producing a great number of scientific papers on neutron scattering and the structure of glass, and only ending when Roger passed away in 2007.[16] Roger was a great friend and scientist who is missed by all who knew him.

Originally neutron scattering had been performed solely using neutrons which were obtained from nuclear reactors, but at the end of the 1960s accelerator-based neutron sources began to be used as well. Roger was at the forefront of developing techniques for the study of materials using neutrons obtained from an electron linac, the first type of accelerator-based neutron source to be developed. (Subsequently the electron linac has been superseded by the spallation source[17] as the preferred type of accelerator-based neutron source.) Almost all accelerator-based sources produce neutrons in pulses, and the time-of-flight (t-o-f) measurement technique is essential for performing diffraction measurements on such pulsed neutron sources. During the 1960s there were reports in the literature of the use of the t-o-f technique with a reactor source to make neutron diffraction measurements on single crystal and polycrystalline samples. Likewise the first feasibility studies for the use of the t-o-f technique to perform neutron diffraction on glass samples were performed at the HERALD reactor for SiO_2 and GeO_2 by Lorch during his PhD,[18] but were not published in the literature. However, for non-crystalline materials the virtues of using the t-o-f technique are greater if an accelerator-based neutron source is employed, due to the availability of high energy neutrons (i.e. energies greater than ~300 meV) which are required for high Q-values to be achieved. The first report in the literature of a diffraction measurement on a non-crystalline sample using the t-o-f technique was made in 1973 by the Japanese group who used the Tohoku electron linac to measure the structure factor of liquid bromine.[19] However, this publication was quickly followed in 1974 by a paper from Adrian, Roger and their collaborators[15] who used Roger's original version of the total scattering spectrometer (TSS) to study the structure of GeO_2 glass (and also liquid nitrogen). Roger had built this diffractometer for the small cost of £2000 (in comparison, a modern total scattering neutron diffractometer costs several million pounds), and in his work at the Harwell electron linac he built the prototypes of most of the different types of neutron scattering instrument that are in use today at much more modern and powerful pulsed neutron sources.[17,20,21] Roger was thus a pioneer of the pulsed neutron technique.

The paper published in 1974[15] was the first report in the literature of a t-o-f diffraction measurement for a glass, and was the first to show a Fourier transform

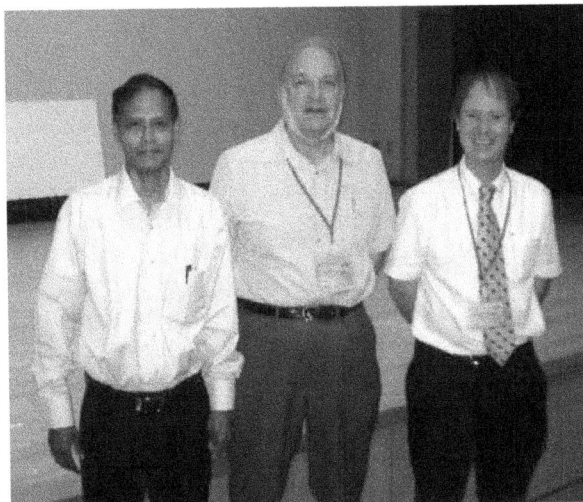

Figure 10. Adrian Wright (centre) during the Sixth Borate Conference in Himeji, Japan, with Professor Erwin Desa (left) of Goa University, India, and the author (right). (Thanks to Professor Desa for this photograph)

of such data – a Fourier transform of the data should be performed if precise, detailed information about the structure is to be extracted (the Japanese group had studied a liquid, not a glass, and had not shown a real-space Fourier transform of their data). This paper is thus a major milestone for the study of glass structure by neutron diffraction methods. Before Adrian and Roger started their work together on glass, there had been a few previous reports of neutron diffraction measurements on glasses. For example, the first reports of true neutron diffraction measurements on glasses (as opposed to mere conference abstracts, or transmission measurements) had been reported in 1964 at the first Physics of Non-Crystalline Solids conference.[22,23] These first measurements had been performed on the archetypal glass v-SiO_2 and, as was the case for most reports during the 1960s, served more as demonstration 'proof of principle' experiments than as detailed structural studies. Thus Adrian and Roger were at the forefront of the development of neutron diffraction as a significant experimental probe which can reveal detailed structural information about a wide range of glassy materials, as is shown by Adrian's many scientific publications since that time.[3] Their first t-o-f neutron diffraction report[15] was followed by a workshop about the use of pulsed neutron sources to study condensed matter[24] which they organised at Reading University. It was thus fitting that when Roger retired from Harwell Laboratory he was appointed as a visiting professor in the Physics Department of Reading University.

Adrian and Roger worked together closely, with Roger concentrating more on the development of instrumentation and techniques for t-o-f neutron scattering, whilst Adrian's efforts were more on the development of neutron scattering methods for the study of glass structure; thus Adrian and Roger were the first to report the application of several different neutron diffraction techniques for glass studies. In 1977 a paper was published[25] which describes neutron diffraction isotopic substitution measurements on $K_2O.TiO_2.2SiO_2$ glass. By measuring the diffraction pattern for samples with different titanium isotopic compositions, which are otherwise identical, element-specific information about the Ti environment was obtained. This was the first ever use of neutron diffraction isotopic substitution to study a glass, and this technique has now come to be extensively used for glass structure investigations. Adrian was also the first to propose and use a special application of this technique, known as neutron diffraction double-null isotopic substitution.[26] Another related technique which Adrian was the first to develop for glass structure studies is the neutron diffraction anomalous dispersion technique,[27] which in some cases provides an alternative means of obtaining element-specific information. This work was performed with Professor Erwin Desa (see Figure 10), now of Goa University, during his PhD studies in Reading under Adrian's supervision, when he used the technique to investigate the environment of samarium in a Sm_2O_3–Al_2O_3–GeO_2 glass. A third technique which Adrian was also first to apply to the study of the structure of non-metallic glasses is magnetic neutron diffraction.[28] In this case neutron diffraction data were recorded for an iron phosphate glass at two different temperatures above and below the magnetic short range ordering transition region. The only major change in the diffraction pattern arises from the change in magnetic Fe–Fe correlations, and so the difference between the two datasets provides both structural information on the Fe–Fe distances in the glass, and magnetic information on the interactions between the magnetic ions (i.e. it shows whether they align ferromagnetically or antiferromagnetically).

Neutron diffraction is a very powerful tool for probing glass structure, but it must be acknowledged that it suffers from several difficulties, notably the way in which the results contain distance information from all possible pairs of elements in the glass. In view of this difficulty, Windsor has likened the determination of glass structure to trying to measure the period of a handclap by listening to applause.[29] A common theme of the techniques which Adrian first applied to glass structure is that they seek to overcome this difficulty by providing element-specific diffraction information. Unfortunately, none of these techniques are of practical application to pure B_2O_3, although they are now starting to be applied by other workers to the study of alkali borate glasses.[30]

4. Neutron scattering from v-B_2O_3

The basic model for the structure of vitreous B_2O_3, proposed in 1932 by Zachariasen,[31] is probably

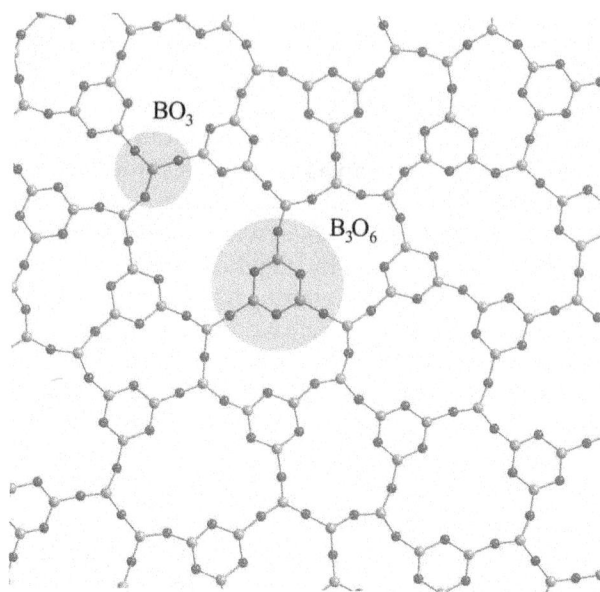

Figure 11. A two-dimensional representation of the boroxol ring model for the structure of v-B_2O_3; a randomly organised network of boroxol groups and independent BO_3 triangles. Shaded areas indicate a typical boroxol group (B_3O_6) and independent triangle (BO_3), respectively. The figure is an updated version of the cover art for the proceedings for the Second International Conference on Borate Glasses, Crystals and Melts[4]

universally accepted. In this model the glass is considered to be a three-dimensional continuous random network, constructed by corner sharing between BO_3 structural units, each of which is formed by an equilateral triangle with an oxygen at each vertex and a boron at the centre. However, there has long been considerable controversy[32] whether there is a high frequency of occurrence of a larger well-defined structural unit known as a boroxol group. The B_3O_6 boroxol group is composed of three corner sharing BO_3 triangles which form a very highly planar ring, and Figure 11 illustrates a network in which a high proportion of the atoms are in boroxol groups.

In 1976 Adrian participated in the First International Conference on the Structure of Non-Crystalline Materials, organised by Phil Gaskell in Cambridge, England.[33] A paper was presented by Al Cooper of Case Western Reserve University in Cleveland, Ohio, in which it was concluded, on the basis of x-ray diffraction results, that boroxol groups do not play an important role in the structure.[34] Al Cooper suggested to Adrian that neutron diffraction might be able to help to resolve the controversy about boroxol groups. He also invited Adrian to participate in a conference on the subject of "Boron in Glass and Glass Ceramics" which was then being planned to be held at the New York State College of Ceramics, Alfred, USA, in 1977. Adrian was most enthusiastic to participate in this conference, which is the first in the Borate Conference series, but needed an official letter

Figure 12. Frank Galeener and his wife, Janet, in front of his beloved citrus trees at their Palo Alto home in 1985

Figure 13. Adrian Wright, digging the ground in Frank Galeener's garden in 1985

of invitation. Unfortunately the letter from Al Cooper went astray, so that Adrian was unable to attend the conference, and thus Steve Feller of Coe College, Iowa, USA, is the only person to have attended all of the Borate Conferences from the first to the sixth.

By this time Roger Sinclair's neutron diffractometer, TSS, had moved to the new HELIOS[35] electron linac neutron source at Harwell Laboratory, and Peter Johnson, a PhD student, worked with Adrian and Roger to use the TSS to study the structure of v-B_2O_3. The experiment was reported in a lengthy paper,[8] which includes a comprehensive review of the boroxol controversy until 1982, and is Adrian's third most highly cited paper. The essential conclusion of this paper is that 60% of the boron atoms in v-B_2O_3 are in boroxol groups. This conclusion confirmed the results of Warren's x-ray diffraction work[36] and of Bray's pioneering NMR work[37] and, with refinement,[32] the conclusion remains until the present day. It is telling to note that Adrian embarked on his study expecting, on the basis of Al Cooper's work,[34] to find that boroxol groups are not an important structural feature in v-B_2O_3; the fact that the conclusion was the opposite of the initial expectation tends to make it even more convincing. Personally, I have found it surprising that the boroxol controversy has remained

so heated,[38–40] subsequent to the publication of such clear experimental evidence.

In 1979 and 1980 Adrian spent a sabbatical year at the Xerox Palo Alto Research Lab where he came into contact with Frank Galeener (see Figures 12 and 13). Frank had made a major impact studying the vibrational spectra of glasses, and he emphasised a point which is not always appreciated by those studying the structure by diffraction and modelling, that the rings are highly planar,[41] i.e. a ring of three BO_3 units is not a boroxol group unless there is a high degree of coplanarity. Inspired by Frank's Raman scattering work,[42] Adrian and his group went on to study v-B_2O_3, and later other borate glasses, by inelastic neutron scattering, which has helped to shed further light on the presence of boroxol groups in v-B_2O_3, and of other superstructural units in borate glasses. Unfortunately, Frank was taken from us prematurely in 1993,[43] or otherwise there can be little doubt that he would have been one of the key figures to be involved in the Borate Conference over the years. Frank's loss at a relatively young age was felt deeply in the glass structure community, and was marked by a special symposium on Amorphous Insulators.[43]

5. Scientific visitors to Reading – re-starting the Borate Conference

Over the years, a number of scientists from other countries have been attracted to Reading for a long term scientific visit, to work with Adrian in studying the structure of glasses. It is amusing to note that subsequently almost all of these visitors have gone on to chair the Borate Conference at their own home institution.

The first of Adrian's scientific visitors to Reading

Figure 14. Norimasa Umesaki and his wife, Atsuko, at the ISIS Facility in 1992

Figure 16. In his role as President of the Society of Glass Technology, Adrian Wright welcomes Steve Feller (right) as a Fellow of the Society in 2003

was Dr Norimasa Umesaki, then of the Welding Research Institute, Osaka University, Japan, who visited Reading University for a sabbatical year in 1987 and 1988 (see Figure 14). At that time Norimasa worked on germanate glasses,[44] but he later worked with Adrian on densified B_2O_3 glass,[45] and most recently has chaired the Sixth International Conference on Borate Glasses, Crystals and Melts, in Himeji, Japan, at which Adrian was honoured for his achievements in glass science, and in particular borate glasses.

In April and May 1991, Professor Yanko Dim-

itriev (see Figure 15) of the University of Chemical Technology and Metallurgy, Sofia, Bulgaria, made a scientific visit to Adrian in Reading. Subsequently he collaborated on scientific studies with Adrian, and in 1999 he chaired the Third Borate Conference in Sofia, Bulgaria, at which Evgenii Porai-Koshits was honoured in recognition of his outstanding contribution to the science of borate glasses.

The most significant visitor to Reading, as far as the Borate Conference is concerned, is Professor Steve Feller (see Figure 16) of Coe College, Cedar Rapids, Iowa, USA. Steve had been developing techniques for making high purity borate glasses over the widest range in composition, and for careful studies of their density from a structural point of view. This work complemented Adrian's own interest in studying the structure of borate glasses by means of neutron scattering, and a most fruitful collaboration between Steve and Adrian has developed over the years, starting in the early 1990s. Steve first made a brief visit to Reading in October 1993, and during this visit, whilst walking back from lunch in the Senior Common Room of Reading University, he suggested holding another Borate Conference. Adrian seized upon this suggestion and brought his tenacious organisational and editorial abilities to bear on the subject. In June 1994, Steve returned to Reading for a week to begin making detailed plans for re-starting the Borate

Figure 15. Yanko Dimitriev (left), together with Doris Ehrt (centre) of the Otto-Schott-Institut, Friedrich-Schiller-Universität, Jena, Germany, and Adrian Wright during the conference excursion at the Third Borate Conference

Figure 17. Steve Feller at the conference dinner of the Second Borate Conference, with Phil Bray (centre) and Evgenii Porai-Koshits (right)

Figure 18. The Russian participants at the Second Borate Conference with the organisers, in front of the Cosener's House (back: Stanislav Filatov, Natalia Vedishcheva, Adrian Wright, Evgenii Porai-Koshits, Steve Feller, Nikolai Bokov, Galina Sycheva; front: Valerii Golubkov, Boris Shakhmatkin)

Conference. It was planned that in 1996 Steve would make a sabbatical visit to Reading from January to August, culminating in the Borate Conference.

With the help of the Society of Glass Technology, especially Jill Costello, a detailed plan was put together to hold the second Borate Conference from 22nd to 25th July 1996. The Cosener's House, which belongs to Rutherford Appleton Laboratory, was selected as a venue; this is a big, old house constructed on the site of the kitchens (the word "Cosener" is derived from the French word for kitchen) of the former abbey in Abingdon. It is a delightful site, next to the river Thames (or the river Isis as it is otherwise known), which always enhances the experience of attending a meeting, and helps to make it a success.

The field of borates is a small, specialised field, and it was uncertain whether there would be sufficient participants. In the event 96 experimental and theoretical scientists from 18 countries came – there were too many for the accommodation available at the Cosener's House, the staff became concerned that the audience in the lecture theatre would exceed the number permitted by safety rules, and it was necessary to turn away some late applicants. A notable presence were the scientists from Russia (see Figure 18); a large number of Russian scientists participated, despite the difficulties that still existed in 1996, a short time after the end of the Cold War. Since then, Adrian has had a close scientific collaboration, especially on

borates, with Natalia Vedishcheva and others from the Institute of Silicate Chemistry in Saint Petersburg, Russia, and recently Adrian became Visiting Director of Glass Research at this institute. There was considerable controversy at the conference about the importance of boroxol groups in v-B_2O_3, with most of the experimentalists claiming strong evidence for a large proportion of boroxol groups in the glass,[46–50] whilst the computer simulators reported far fewer boroxol groups in their simulations,[51–53] and expressed strong doubts as to their importance in the structure. It is amusing to relate how, at the Third Borate Conference, there was a serious blowout of a tyre on the bus taking the experimentalists to see the Roman amphitheatre in the Bulgarian city of Plovdiv – without the driver's skill in keeping the bus on the road, the boroxol controversy might now have taken a very different course, with the anti-boroxol camp in the ascendancy.

Since 1996, the Borate Conference has become a regular feature in the scientific calendar. It was 19 years between the first and second conferences in the series, but since then it has been held exactly every 3 years. Approximately 100 scientists have participated in every conference, and the published proceedings show that the quality of the presentations has been surprisingly high for such a small, specialised field. In short, the Borate Conference has become a success, and this has been due to the efforts and influence of Adrian Wright more than any other.

The fourth conference in the series was organised in 2002 by Steve Feller and his colleague, Mario Affatigato, when they held it at Coe College, Cedar Rapids, Iowa, USA, in honour of David Griscom in recognition of his outstanding contributions to the science of borate glasses and his distinguished career in scientific research.

6. Recent and current work on borates

As shown by Figure 3, there has been a steady growth in the number of borate-related papers published by Adrian since the Second Borate Conference in 1996. These studies can be considered to involve four different topics of interest, although it must be acknowledged that there is considerable inter-relation between these topics: boroxol groups; superstructural units; four-coordinated boron and the borate anomaly; crystallography.

Subsequent to the study of pure v-B_2O_3 by neutron diffraction (see Section 4), Adrian has extended this work to the study of the structure of a number of different borate glass systems, such as lead borates,[54,55] alkali borates,[56] bismuth borates,[57] and caesium borates.[58] As well as neutron diffraction, Adrian and Roger Sinclair have made extensive use of inelastic neutron scattering to obtain further evidence on the presence of boroxol groups in v-B_2O_3,[32,59–64] and other superstructural units in borate glasses.[48,64–66] In addition to his neutron diffraction and inelastic neutron scattering studies, Adrian has been involved with his Russian collaborators in using thermodynamics to successfully predict the temperature-dependence of the boroxol ring fraction in molten B_2O_3 (see Figure 6 in the paper by Vedishcheva et al[67]).

Adrian has used crystallographic techniques to show the surprising result that in the crystal structure of β-caesium enneaborate, the B–O bonds in the boroxol rings are longer than outside the rings.[68] Adrian has also used results from crystallographic studies of borates as a source of concepts for the understanding of the structure of borate glasses. For example, this has led to a proposal that the structure of borate glasses involves independent, interpenetrating networks,[69] and in this way models of a realistic density can be constructed. Adrian has also used information from crystal structures to help to understand the evolution of four-coordinated boron in borate glasses. My own personal favourite in this area is the paper, *A crystallographic guide to the structure of borate glasses*,[70] which contains a great deal of insight into both the evolution of four-coordinated boron in borate glasses and the presence of superstructural units in borate glasses. The reader is also recommended to read Adrian's plenary paper in these proceedings,[71] which is a more recent study in the same general area. It will be interesting to find out (perhaps at the next Borate Conference in Nova Scotia) what will be Adrian's next contribution to the study of borates!

Acknowledgements

This paper is a tribute to Adrian Wright, to whom I am greatly indebted for a superlative education in glass structure and neutron scattering. I am also grateful to Roger Sinclair who was a great mentor in the field of neutron scattering. Adrian's help in the preparation of this paper was invaluable, and so too was Steve Feller's help. Finally, many thanks to Norimasa Umesaki for providing the rare opportunity to praise a great glass scientist.

References

1. Pye, L. D., Fréchette, V. D. & Kreidl, N. J., Eds. *Borate Glasses: Structure, Properties, Applications.* Materials Science Research. Vol. 12. Plenum Press, New York, 1978.
2. Mackenzie, J. D. & Urnes, S. *J. Non-Cryst. Solids*, 1975, **17**, 1.
3. Wright, A. C. In: *Experimental Techniques of Glass Science*, Eds. C. J. Simmons & O. H. El-Bayoumi, American Ceramic Society, Westerville, 1993, p. 205.
4. Wright, A. C., Feller, S. A. & Hannon, A. C., Eds. *Borate glasses, crystals and melts*. The Society of Glass Technology, Sheffield, 1997.
5. Dimitriev, Y. B. & Wright, A. C., Eds. *Borate Glasses, Crystals & Melts: Structure & Applications*. The Society of Glass Technology, Sheffield, 2001.
6. Affatigato, M., Feller, S. A., Vedishcheva, N. M. & Wright, A. C., Eds. *Borate Glasses, Crystals & Melts 2002: New Techniques and Applications*. The Society of Glass Technology, Sheffield, 2003.
7. Dalba, G., Rocca, F., Vedishcheva, N. M. & Wright, A. C., Eds. *Borate glasses, crystals and melts: New techniques and practical applications*. The Society of Glass Technology, Sheffield, 2008.
8. Johnson, P. A. V., Wright, A. C. & Sinclair, R. N. *J. Non-Cryst. Solids*, 1982, **50**, 281.
9. Wright, A. C., PhD Thesis, 1969, *The structure of five simple glasses*, University of Bristol, UK.
10. Hannon, A. C., PhD Thesis, 1989, *Neutron scattering and modelling studies of the atomic and magnetic structure of the metallic glass Dy_7Ni_3 and of the atomic structure and dynamics of vitreous B_2O_3*, University of Reading, UK.
11. Leadbetter, A. J. & Wright, A. C. *J. Non-Cryst. Solids*, 1970, **3**, 239.
12. McEnhill, J. J., Rodgers, A. L. & Todd, M. C. *J. J. Br. Nucl. Energ. Soc.*, 1965, **4**, 344.
13. Lander, G. H. *Neutron News*, 2004, **15** (3), 32.
14. Lorch, E. *J. Phys. C*, 1969, **2**, 229.
15. Sinclair, R. N., Johnson, D. A. G., Dore, J. C., Clarke, J. H. & Wright, A. C. *Nucl. Instr. Meth.*, 1974, **117**, 445.
16. Wright, A. C. *Neutron News*, 2007, **18** (4), 34.
17. Fender, B. E. F., Hobbis, L. C. W. & Manning, G. *Phil. Trans. R. Soc. B*, 1980, **290**, 657.
18. Lorch, E., PhD Thesis, 1967, University of Birmingham, UK.
19. Misawa, M., Fukushima, Y., Suzuki, K. & Takeuchi, S. *Phys. Lett. A*, 1973, **45**, 273.
20. Mason, T. E., Abernathy, D., Anderson, I., Ankner, J., Egami, T., Ehlers, G., Ekkebus, A., Granroth, G., Hagen, M., Herwig, K., Hodges, J., Hoffmann, C., Horak, C., Horton, L., Klose, F., Larese, J., Mesecar, A., Myles, D., Neuefeind, J., Ohl, M., Tulk, C., Wang, X. L. & Zhao, J. *Physica B*, 2006, **385–86**, 955.
21. Nagae, T. *Nucl. Phys. A*, 2008, **805**, 486C.
22. Egelstaff, P. A. In: *The Physics of Non-Crystalline Solids*, Ed. J. A. Prins, North-Holland, Amsterdam, 1965, p. 127.
23. Carraro, G., Domenici, M. & Zucca, T. In: *The Physics of Non-Crystalline Solids*, Ed. J. A. Prins, North-Holland, Amsterdam, 1965, p. 152.
24. Wright, A. C. & Sinclair, R. N. *Condensed matter research using pulsed neutron sources*, Rutherford Laboratory Report RL-74-051, 1974.
25. Wright, A. C., Yarker, C. A., Johnson, P. A. V. & Wedgwood, F. A. In: *The Physics of Non-Crystalline Solids*, Ed. G. H. Frischat, Trans. Tech., Aedermannsdorf, 1977, p. 118.
26. Wright, A. C., Hannon, A. C., Sinclair, R. N., Johnson, W. L. & Atzmon, M. *J. Phys. F*, 1984, **14**, L201.
27. Wright, A. C., Etherington, G., Desa, J. A. E. & Sinclair, R. N. *J. de Phys. Coll.*, 1982, **C9**, 31.
28. Wedgwood, F. A. & Wright, A. C. *J. Non-Cryst. Solids*, 1976, **21**, 95.
29. Windsor, C. G. *Pulsed neutron scattering*. Taylor and Francis, London, 1981.
30. Majerus, O., Cormier, L., Calas, G. & Beuneu, B. *J. Phys. Chem. B*, 2003, **107**, 13044.
31. Zachariasen, W. H. *J. Am. Chem. Soc.*, 1932, **54**, 3841.
32. Hannon, A. C., Grimley, D. I., Hulme, R. A., Wright, A. C. & Sinclair, R. N. *J. Non-Cryst. Solids*, 1994, **177**, 299.
33. Gaskell, P. H., Ed. *The structure of non-crystalline materials*. A Physics and Chemistry of Glasses conference volume. Taylor & Francis, London, 1977.
34. Dunlevey, F. M. & Cooper, A. R. In: *The Structure of Non-Crystalline Materials*, Ed. P. H. Gaskell, Taylor and Francis Ltd, London, 1977, p. 211.
35. Lynn, J. E. *Contemp. Phys.*, 1980, **21**, 483.
36. Mozzi, R. L. & Warren, B. E. *J. Appl. Cryst.*, 1970, **3**, 251.
37. Jellison, G. E., Panek, L. W., Bray, P. J. & Rouse, G. B. *J. Chem. Phys.*, 1977, **66**, 802.
38. Umari, P. & Pasquarello, A. *Phys. Rev. Lett.*, 2005, **95**, 137401.
39. Swenson, J. & Borjesson, L. *Phys. Rev. Lett.*, 2006, **96**, 199701.
40. Umari, P. & Pasquarello, A. *Phys. Rev. Lett.*, 2006, **96**, 199702.
41. Galeener, F. L., Lucovsky, G. & Mikkelsen, J. C. *Phys. Rev. B*, 1980, **22**, 3983.
42. Galeener, F. L. & Thorpe, M. F. *Phys. Rev. B*, 1983, **28**, 5802.
43. Thorpe, M. F. & Wright, A. C. *J. Non-Cryst. Solids*, 1995, **182**, vii.
44. Umesaki, N., Brunier, T. M., Wright, A. C., Hannon, A. C. & Sinclair, R. N. *Physica B*, 1995, **213-214**, 490.
45. Wright, A. C., Stone, C. E., Sinclair, R. N., Umesaki, N., Kitamura, N., Ura, K., Ohtori, N. & Hannon, A. C. *Phys. Chem. Glasses*, 2000, **41**, 296.
46. Zwanziger, J. W., Youngman, R. E. & Braun, M. In: *Borate glasses, crystals and melts*, Eds. A. C. Wright, S. A. Feller, and A. C. Hannon, The Society of Glass Technology, Sheffield, 1997, p. 21.
47. Wright, A. C., Vedishcheva, N. M. & Shakhmatkin, B. A. In: *Borate glasses, crystals and melts*, Eds. A. C. Wright, S. A. Feller, and A. C. Hannon, The Society of Glass Technology, Sheffield, 1997, p. 80.
48. Sinclair, R. N., Wright, A. C., Wanless, A. J., Hannon, A. C., Feller,

S. A., Mayhew, M. T., Meyer, B. M., Royle, M. L., Wilkerson, D. L., Williams, R. B. & Johanson, B. C. In: *Borate Glasses, Crystals and Melts*, Eds. A. C. Wright, S. A. Feller, and A. C. Hannon, The Society of Glass Technology, Sheffield, 1997, p. 140.

49. Hübert, T., Harder, U., Mosel, G. & Witke, K. In: *Borate glasses, crystals and melts*, Eds. A. C. Wright, S. A. Feller, and A. C. Hannon, The Society of Glass Technology, Sheffield, 1997, p. 156.

50. Ota, R., Fukunaga, J. & Wakasugi, T. In: *Borate glasses, crystals and melts*, Eds. A. C. Wright, S. A. Feller, and A. C. Hannon, The Society of Glass Technology, Sheffield, 1997, p. 199.

51. Teter, M. In: *Borate Glasses, Crystals and Melts*, Eds. A. C. Wright, S. A. Feller, and A. C. Hannon, The Society of Glass Technology, Sheffield, 1997, p. 407.

52. Swenson, J. & Börjesson, L. In: *Borate glasses, crystals and melts*, Eds. A. C. Wright, S. A. Feller, and A. C. Hannon, The Society of Glass Technology, Sheffield, 1997, p. 425.

53. Park, B. & Cormack, A. N. In: *Borate glasses, crystals and melts*, Eds. A. C. Wright, S. A. Feller, and A. C. Hannon, The Society of Glass Technology, Sheffield, 1997, p. 443.

54. Vedishcheva, N. M., Shakhmatkin, B. A., Wright, A. C., Sinclair, R. N. & Grimley, D. I. *Bol. Soc. Esp. Ceram. V.*, 1992, **31-C,3**, 41.

55. Vedishcheva, N. M., Shakhmatkin, B. A., Wright, A. C., Grimley, D. I., Etherington, G. & Sinclair, R. N. In: *Fundamentals of Glass Science and Technology (ESG, Venice)*, 1993, p. 459.

56. Wanless, A. J., Wright, A. C., Sinclair, R. N., Feller, S. A., Williams, R. B., Johanson, B. C. & Mayhew, M. T. In: *Borate glasses, crystals and melts*, Eds. A. C. Wright, S. A. Feller, and A. C. Hannon, Society of Glass technology, Sheffield, 1997, p. 506.

57. Stone, C. E., Wright, A. C., Sinclair, R. N., Feller, S. A., Affatigato, M., Hogan, D. L., Nelson, N. D., Vira, C., Dimitriev, Y. B., Gattef, E. M. & Ehrt, D. *Phys. Chem. Glasses*, 2000, **41**, 409.

58. Shaw, J. L., Wright, A. C., Sinclair, R. N., Frueh, J. R., Williams, R. B., Nelson, N. D., Affatigato, M., Feller, S. A. & Scales, C. R. *Phys. Chem. Glasses*, 2003, **44**, 256.

59. Hannon, A. C., Sinclair, R. N., Blackman, J. A., Wright, A. C. & Galeener, F. L. *J. Non-Cryst. Solids*, 1988, **106**, 116.

60. Hannon, A. C., Sinclair, R. N. & Wright, A. C. In: *The Physics of Non-Crystalline Solids*, Eds. L. David Pye, W. C. LaCourse, and H. J. Stevens, Taylor and Francis, London, 1992, p. 67.

61. Hannon, A. C., Sinclair, R. N. & Wright, A. C. *Physica A*, 1993, **201**, 375.

62. Hannon, A. C., Wright, A. C. & Sinclair, R. N. *suppl. Riv. della Staz. Sper. del Vetro*, 1993, **XXIII**, 479.

63. Hannon, A. C., Wright, A. C., Blackman, J. A. & Sinclair, R. N. *J. Non-Cryst. Solids*, 1995, **182**, 78.

64. Sinclair, R. N., Stone, C. E., Wright, A. C., Polyakova, I. G., Vedishcheva, N. M., Shakhmatkin, B. A., Feller, S. A., Johanson, B. C., Venhuizen, P., Williams, R. B & Hannon, A. C. *Phys. Chem. Glasses*, 2000, **41**, 286.

65. Haworth, R., Shaw, J. L., Wright, A. C., Sinclair, R. N., Knight, K. S., Taylor, J. W., Vedisheheva, N. M., Polyakova, I. G., Shakhmatkin, B. A., Feller, S. A. & Winslow, D. W. *Phys. Chem. Glasses*, 2005, **46**, 477.

66. Sinclair, R. N., Haworth, R., Wright, A. C., Parkinson, B. G., Holland, D., Taylor, J. W., Vedishcheva, N. M., Polyakova, I. G., Shakhmatkin, B. A., Feller, S. A., Rijal, B. & Edwards, T. *Phys. Chem. Glasses: Eur. J. Glass Sci. Technol. B*, 2006, **47**, 405

67. Vedishcheva, N. M., Shakhmatkin, B. A. & Wright, A. C. *J. Non-Cryst. Solids*, 2001, **293**, 312.

68. Haworth, R., Wright, A. C., Sinclair, R. N., Knight, K. S., Vedishcheva, N. M., Polyakova, I. G. & Shakhmatkin, B. A. *Phys. Chem. Glasses: Eur. J. Glass Sci. Technol. B*, 2006, **47**, 352.

69. Wright, A. C., Vedishcheva, N. M. & Shakhmatkin, B. A. *J. Non-Cryst. Solids*, 1995, **192–193**, 92.

70. Wright, A. C., Vedishcheva, N. M. & Shakhmatkin, B. A. *Mater. Res. Soc. Symp. Proc.*, 1997, **455**, 381.

71. Wright, A. C. *Phys. Chem. Glasses: Eur. J. Glass Sci. Technol. B*, 2010, **51**, 1.

72. Wright, A. C., Johnson, P. A. V. & Sinclair, R. N. *Am. Ceram. Soc. Bull.*, 1979, **58**, 384.

73. Sinclair, R. N., Desa, J. A. E., Etherington, G., Johnson, P. A. V. & Wright, A. C. *J. Non-Cryst. Solids*, 1980, **42**, 107.

74. Wright, A. C., Sumner, D. J. & Clare, A. G. In: *The structure of non-crystalline materials 1982*, Eds. P. H. Gaskell, J. M. Parker, and E. A. Davis, Taylor and Francis, London, 1982, p. 395.

75. Wright, A. C., Hannon, A. C., Grimley, D. I., Sinclair, R. N., Galeener, F. L., Elliott, S. R. & Gladden, L. F. *Ceram. Bull.*, 1986, **65**, 1372.

76. Shakhmatkin, B. A., Vedishcheva, N. M., Shultz, M. M. & Wright, A. C. *J. Non-Cryst. Solids*, 1994, **177**, 249.

77. Grimsditch, M., Polian, A. & Wright, A. C. *Phys. Rev. B*, 1996, **54**, 152.

78. Wright, A. C., Sinclair, R. N., Grimley, D. I., Hulme, R. A., Vedishcheva, N. M., Shakhmatkin, B. A., Hannon, A. C., Feller, S. A., Meyer, B. M., Royle, M. L. & Wilkerson, D. L. *Fiz. Khim. Stekla*, 1996, **22**, 268.

79. Vedishcheva, N. M., Shakhmatkin, B. A., Shultz, M. M. & Wright, A. C. *J. Non-Cryst. Solids*, 1996, **196**, 239.

80. Wright, A. C., Feller, S. A. & Hannon (Eds.), A. C., Eds. *Borate glasses, crystals and melts*. Society of Glass Technology, Sheffield. The Society of Glass Technology, Sheffield, 1997.

81. Porai-Koshits, E. A. & Wright, A. C. In: *Borate, glasses crystals and melts*, Eds. A. C. Wright, S. A. Feller, and A. C. Hannon, Society of Glass Technology, Sheffield, 1997, p. 51.

82. Vedishcheva, N. M., Shakhmatkin, B. A. & Wright, A. C. *Glass Phys. Chem.*, 1998, **24**, 308.

83. Vedishcheva, N. M., Shakhmatkin, B. A. & Wright, A. C. *Phys. Chem. Glasses*, 2000, **41**, 260.

84. Wright, A. C., Shakhmatkin, B. A. & Vedishcheva, N. M. *Glass Phys. Chem.*, 2001, **27**, 97.

85. Sinclair, R. N., Stone, C. E., Wright, A. C., Martin, S. W., Royle, M. L. & Hannon, A. C. *J. Non-Cryst. Solids*, 2001, **293-295**, 383.

86. Vedishcheva, N. M. & Wright, A. C. In: *Borate Glasses, Crystals & Melts: Structure & Applications*, Eds. Y. B. Dimitriev and A. C. Wright, The Society of Glass Technology, Sheffield, 2001, p. xvii.

87. Wright, A. C., Etherington, G., Feller, S. A., Gnappi, G., Grimley, D. I., Montenero, A., Sinclair, R. N. & Stone, C. E. *Glass Sci. Technol.*, 2002, **75**, 139.

88. Wright, A. C., Sinclair, R. N., Stone, C. E., Knight, K. S., Polyakova, I. G., Vedishcheva, N. M. & Shakhmatkin, B. A. *Phys. Chem. Glasses*, 2003, **44**, 197.

89. Vedishcheva, N. M., Shakhmatkin, B. A. & Wright, A. C. *Phys. Chem. Glasses*, 2003, **44**, 191.

90. Wright, A. C., Shaw, J. L., Sinclair, R. N., Vedishcheva, N. M., Shakhmatkin, B. A. & Scales, C. R. *J. Non-Cryst. Solids*, 2004, **345&346**, 24.

91. Vedishcheva, N. M., Shakhmatkin, B. A. & Wright, A. C. *J. Non-Cryst. Solids*, 2004, **345&346**, 39.

92. Vedishcheva, N. M., Shakhmatkin, B. A. & Wright, A. C. *Phys. Chem. Glasses*, 2005, **46**, 99.

93. Wright, A. C. *Phys. Chem. Glasses: Eur. J. Glass Sci. Technol. B*, 2006, **47**, 497.

Proc. VI Int. Conf. Borate Glasses, Himeji, Japan, 18–22 August 2008   *Phys. Chem. Glasses: Eur. J. Glass Sci. Technol. B*, June 2009, **50** (3), 224–228

Packing in alkali and alkaline earth borosilicate glass systems

Saurav Bista, Anthony O'Donovan-Zavada, Tyler Mullenbach, Maranda Franke, Mario Affatigato & Steve Feller*

Physics Department, Coe College, 1220 First Avenue NE, Cedar Rapids, IA 52402, USA.

Manuscript received 18 August 2008
Revised version received 21 December 2008
Accepted 12 January 2009

Packing fraction is defined as the ratio of ionic volume to molar volume, and is a useful parameter for analysing structural changes with composition. Alkali borosilicate glasses, $RM_2O.B_2O_3.KSiO_2$ (M=Li, Na, K, Rb or Cs), were studied using fixed K while varying R. This was done for several suitable K values for each alkali used. Packing fractions were determined using experimental density data and Shannon radii, and were then compared with the packing fractions for binary borate and silicate alkali glasses. Further, two alkaline earth borosilicate glass families, $RMO.B_2O_3.KSiO_2$ (M=Ca or Ba), were studied and the packing fractions were also compared with the packing fractions for binary borate and silicate alkali glasses. Alkali ions having volumes larger than oxygen (K, Rb and Cs) dominate the packing at all K values; we define this as ionic packing. For alkalis smaller than oxygen (Li and Na), the packing is controlled by the covalent oxygen network (borate and silicate units); this is defined as covalent packing. The alkaline earth borosilicates ($RCaO.B_2O_3.KSiO_2$ and $RBaO.B_2O_3.KSiO_2$) have packing that is intermediate between the ionic and covalent packing trends of the alkali borosilicate systems.

1. Introduction

In our previous work we used the packing fraction to describe the structure of binary alkali and alkaline earth oxide glasses[1,2] and observed three different packing trends. Ionic packing trends were found for ions much greater in size than that of oxygen (K, Rb, and Cs). These trends were marked by monotonic rises in packing fraction followed by an asymptotic approach to the packing fraction for a dense random packing of equally sized hard spheres, 0·64. Another trend, covalent packing, was observed for ions smaller than oxygen (Li and Na), where structural changes in the oxide network (i.e. borate and silicate short range units) dominated the packing. The packing trends of alkaline-earth glasses were intermediate between the ionic and covalent trends.

In this paper we analyse the packing fractions from alkali and alkaline earth borosilicate glasses and compare them to previous binary systems. As before, we have chosen Shannon radii and diffraction data,[3–7] adjusted for the coordination number of the ions in crystals, for calculating the ionic volume. Table

*Corresponding author. Email SFELLER@coe.edu

1 lists the relevant radii, coordination, ionic volume, and estimated uncertainties for each ion used in the packing calculation.

2. Calculation of packing fractions

The packing fraction (PF) is given by

$$PF = \frac{\frac{4}{3}\pi \sum r_i^3 N_i}{V_f} \quad (1)$$

where the *i*-summation is taken over the elements in the formula unit, r_i and N_i are the radii and number of each ion per formula unit, and V_f is the volume of a formula unit of glass. V_f is found by the ratio of the mass of the formula unit to the density of the glass. Density data used in this study were taken from the literature.[8–15]

3. The Dell, Bray, and Xiao structural model

The Shannon radii that we used to calculate the ionic volume are different for three or four-coordinated boron. Modified borosilicate glasses share added oxygen to form two glass subnetworks: borate and

Table 1. Ion coordination, radii, and volumes[1–7]

	Ca	Ba	Li	Na	K	Rb	Cs	O	Si	^{III}B	^{IV}B
Coordination	7-8	9	4	6	8	9	10	2	4	3	4
Radius (Å)	1·23	1·61	0·73	1·16	1·65	1·77	1·95	1·21	0·4	0·15	0·25
Radial uncertainty (Å)	0·05	0·05	0·05	0·03	0·02	0·02	0·02	0·01	0·01	0·01	0·01
Volume (Å³)	7·80	17·48	1·63	6·54	14·71	19·16	31·06	7·42	0·25	0·01	0·07
Volume uncertainty(Å³)	0·95	1·63	0·34	0·51	0·6	0·8	1·0	0·37	0·02	0·003	0·008
Fractional volume uncertainty	0·12	0·09	0·21	0·08	0·04	0·04	0·03	0·05	0·08	0·3	0·11

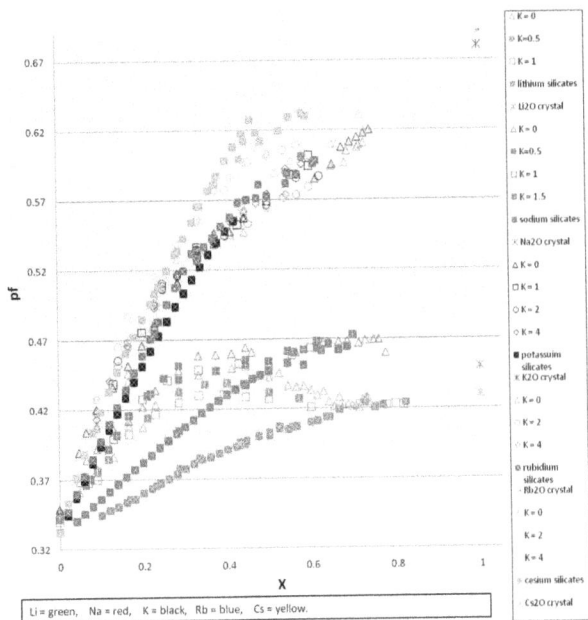

Figure 1. Packing fraction of alkali borate, silicate and borosilicate glasses with several K (molar ratio of silicate to borate) values as a function of the molar fraction of alkali oxide, X

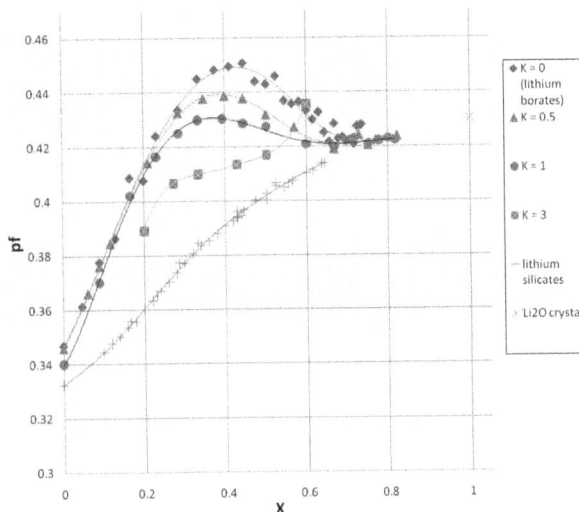

Figure 2. Packing fraction in lithium borate, silicate and borosilicate glasses as a function of molar fraction of Li₂O, X. K is the molar ratio of borate to silicate. The lines through the data are guides to the eye

4. Packing in borosilicate glass systems

(A) Alkali borosilicate glasses

As with binary alkali borate glasses,[1] we observe that packing in alkali borosilicate glass systems follows two general trends, see Figure 1. Glasses with modifying ions larger than oxygen (K, Rb and Cs) follow ionic packing as defined earlier, and glasses with ions smaller than oxygen (Li and Na) exhibit covalent packing controlled by the borate and the silicate structural changes. The packing fraction was plotted against the molar fraction of alkali oxide, which is denoted by X

$$X = \frac{R}{R+K+1} \tag{6}$$

Figure 1 and Table 2 show the PFs of the respective metal oxides. In general, as X approaches 1, the PF of the glass approaches that of the corresponding metal oxide.

In the lithium borosilicate glasses, as K decreases the PF displays more pronounced maxima, indicative of changes in the borate structure and the formation of tetrahedral borons.[16] We note that the packing curve of the lithium silicate glasses demonstrates a monotonic rise with the addition of alkali oxide (Figure 2). This happens as there is no coordination change in silicon; it remains tetrahedral with the addition of alkali oxide. What we observe in Figure 2 is that the PF roughly follows the prediction of the Dell model, where added alkali initially associates principally with the borate part of the glass structure

silicate. The sharing is model dependent. In this paper, we use "silicate" and "borate" to indicate these subnetworks. The Dell model[16] was used to find the fractions of three (N_3) and four (N_4) coordinated boron in each glass since the boron radius is dependent on coordination.

The glass composition is $RM_2O.B_2O_3.KSiO_2$ for alkali borosilicate glasses and $RMO.B_2O_3.KSiO_2$ for alkaline earth borosilicate glasses. The Dell model gives the N_4 fraction as a function of R[11,16] according to in Equations 2(a,b) through 4(a,b)

For $0 \leq R \leq R_{max}$ (where $R_{max}=1/2+K/16$)

$$N_4 = R \tag{2}$$

For $R_{max} \leq R \leq R_{d1}$ (where $R_{d1}=1/2+K/4$)

$$N_4 = R_{max} \tag{3}$$

For $R_{d1} \leq R \leq R_{d3}$ (where $R_{d3}=2+K$)

$$N_4 = \frac{8+K}{12} - \frac{R}{24+12K} \tag{4}$$

For all values of R

$$N_3 = 1 - N_4 \tag{5}$$

The PF for borosilicate glasses was calculated by using Equation (1), together with Equations (2) to (5) for the Dell model, to take into account the effect of coordination on the boron radius.

Table 2. Density and packing of alkali and alkaline earth oxides[17]

	Li₂O	Na₂O	K₂O	Rb₂O	Cs₂O	CaO	BaO
Density (g/cm³)	2·013	2·27	2·35	4·0	4·65	3·34	5·72
PF	0·43	0·45	0·68	0·69	0·69	0·54	0·56
Structure	cubic	amorphous	cubic	cubic	hexagonal	cubic	cubic

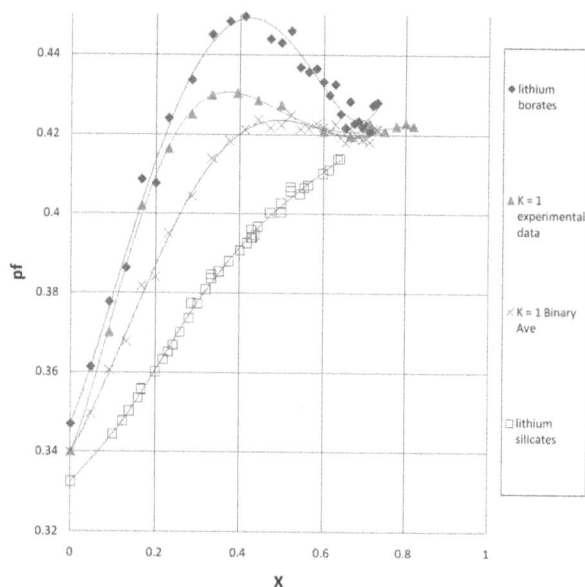

Figure 3. Packing fraction in lithium borate, silicate and (K=1) borosilicate glasses compared with averages from the respective binary systems as a function of the molar fraction of Li$_2$O, X. K is the molar ratio of borate to silicate. The lines through the data are guides to the eye

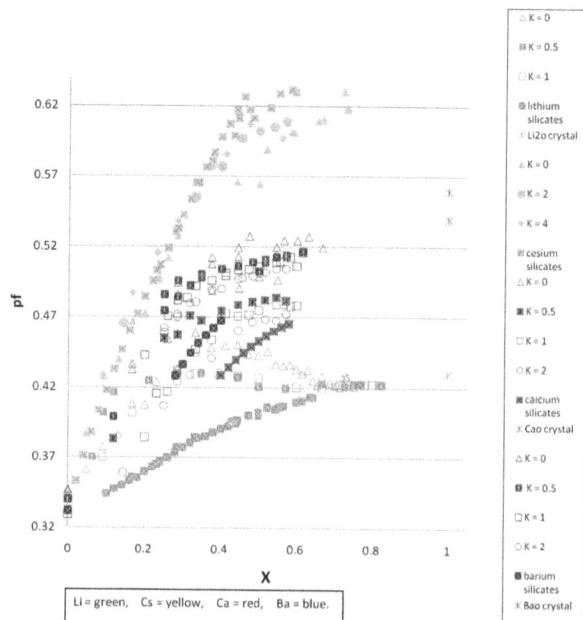

Figure 5. Packing fraction in alkali and alkaline earth borate, silicate and borosilicate glasses with several K (molar ratio of borate to silicate) values. The Cs and Li systems are shown for comparison to the Ca and Ba system

and the PF is close to the binary borate value. The $K=1$ and $0·5$ systems follow the $K=0$ binary borate glass curve until about $X=0·2$ and $0·25$, respectively, where they diverge from the binary curve. The $K=3$ system only forms glass for $X≥0·2$ and already has deviated from the binary borate glass curve. At higher lithia contents, all systems converge near the packing fraction, $0·43$, of lithium oxide ($X=1$).

Figure 3 shows a detailed plot of the PF data for

the $K=1$ and binary systems for lithium borosilicate glasses. Again it is found that the $K=1$ F data are closer to those for borate glasses than those for silicate glasses, as is made apparent by the simple binary average shown in the figure. We believe this is because the alkali is preferentially associated with the borate part of the network, as discussed above, and also because as K increases, the Dell model[16] tells us that there is a larger fraction of four-coordinated borons, N_4, in the borosilicates compared with the binary lithium borate glasses. The increase in N_4 causes the PF to be closer to that for borates than a simple average would dictate.

Figure 4 shows the packing trends of K, Rb and Cs borosilicate glasses. We observe that they do not exhibit peaks as for Li borosilicates. The packing fraction of K, Rb and Cs borosilicate glasses rise towards $0·64$, which is the packing fraction for a dense random packing of equally sized hard spheres. At $0·64$, these packing curves appear to level off. The K, Rb and Cs ions are much larger than oxygen, and because of this the boron coordination change from three-fold to four-fold does not appear to have an effect on the packing fraction trend.

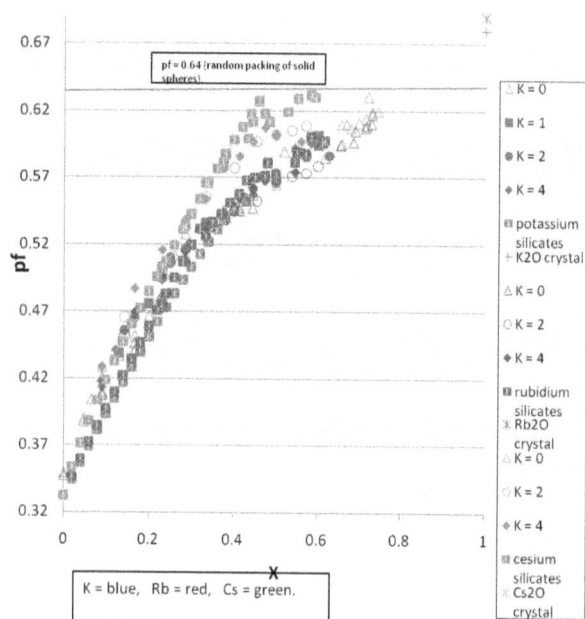

Figure 4. Packing fraction in potassium, rubidium and cesium borate, silicate and borosilicate glasses with several K (molar ratio of borate to silicate) values as a function of molar fraction of alkali oxide, X

(B) Alkaline earth borosilicate glasses

The calcium and barium borosilicate glass packing fraction curves lie in between the ionic (e.g. Cs) and covalent (e.g. Li) packing curves of the alkali borosilicate glasses as shown in Figure 5. Hence, we do not see a distinct covalent or ionic trend in these systems. Since the Ca and Ba ions are as large or

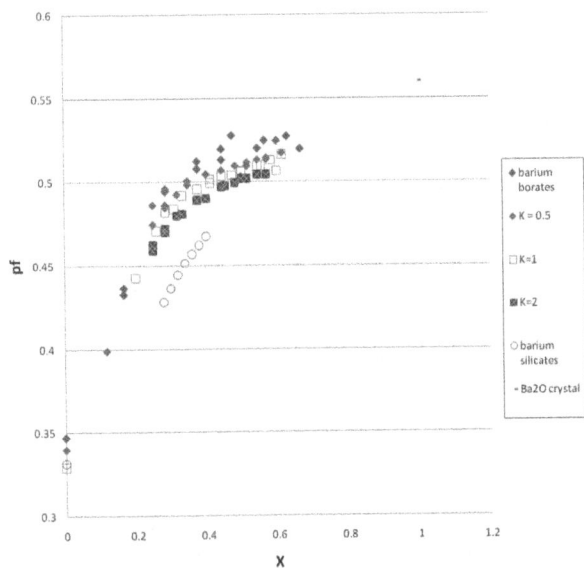

Figure 6. Packing fraction in barium borate, silicate and borosilicate glasses with several K (molar ratio of borate to silicate) values as a function of molar fraction of BaO, X

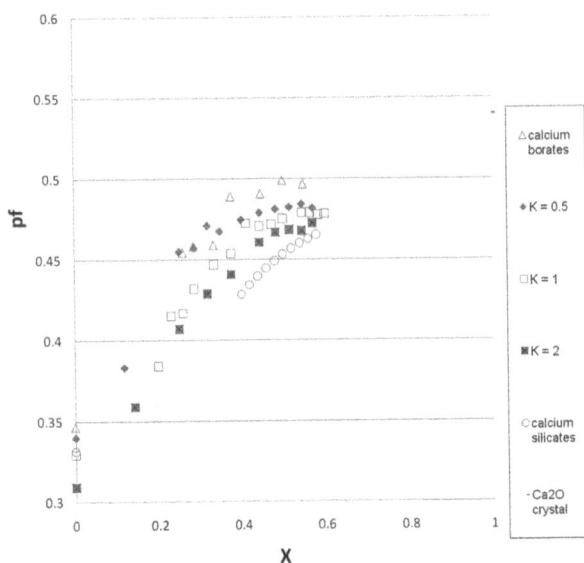

Figure 8. Packing fraction in barium borate, silicate and (K=1) borosilicate glasses compared with averages from the respective binary systems as a function of the molar fraction of BaO, X. K is the molar ratio of borate to silicate. The lines through the data are guides to the eye

larger than the oxygen, we might expect to see the ionic trend in the PFs in these glasses, but due to their bivalency, oxygen is comparatively more abundant than in the alkali borosilicate glass systems. This oxygen abundance suppresses the dominant effect of the larger metal ions on the PFs (Figures 6 and 7). As with the alkali system, the PFs in barium and calcium borosilicate glasses rise towards the PF of the respective oxides as X approaches 1.

Figure 8 and Figure 9 show a detailed plot of the PF data for the K=1 and binary systems for barium and calcium, respectively. The data for barium follow a trend similar to those for lithium and sodium, while calcium follows a different trend. We note

that up to about X=0·3 the K=1 barium borosilicate PF data are much closer to the binary borate data than they are to the binary silicate data. For larger values of X, the K=1 barium borosilicate PF data become increasingly close to the average of the borate and silicate data (Figure 8). Contrastingly the K=1 calcium borosilicate PF data are close to the average of the borate and silicate data for all values of X (Figure 9). There is no observed range in X for which the K=1 calcium borosilicate PF data lie close to the binary borate data.

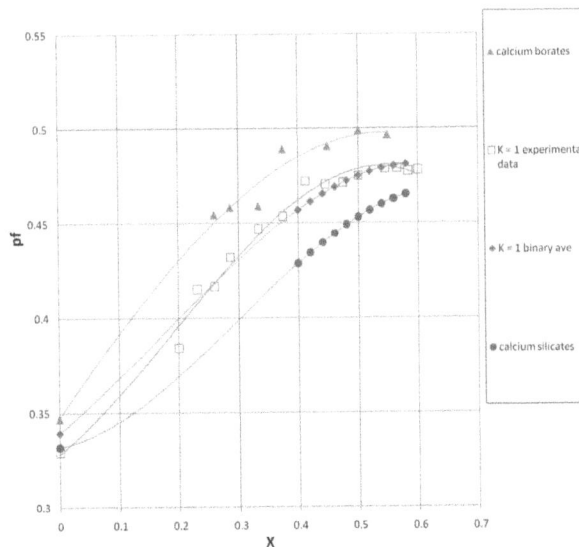

Figure 7. Packing fraction in calcium borate, silicate and borosilicate glasses with several K (molar ratio of borate to silicate) values as a function of molar fraction of CaO, X

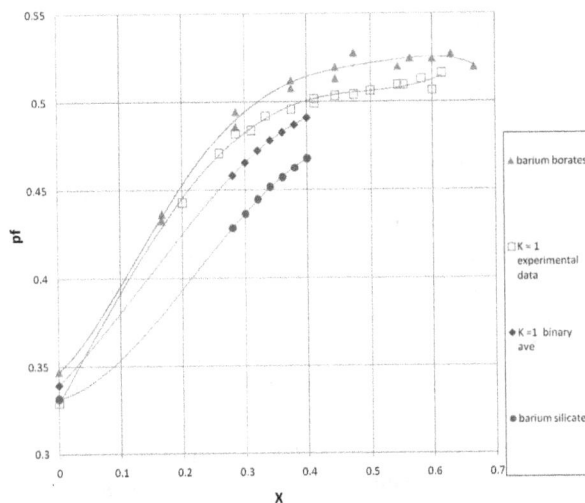

Figure 9. Packing fraction in calcium borate, silicate and (K=1) borosilicate glasses compared with averages from the respective binary systems as a function of the molar fraction of CaO, X. K is the molar ratio of borate to silicate. The lines through the data are guides to the eye

Conclusions

The packing fractions (PFs) of alkali and alkaline earth borosilicate glasses were studied and three distinct trends were observed. Packing in alkali borosilicate glasses having modifying ions smaller than oxygen (Li and Na) is dominated by the structural changes in the borate parts of the network. The PFs for lithium and sodium borosilicate ternary glasses are much closer to the PF for the respective borate binary glass than they are to the PF for the silicate binary glass (see Figure 3). For alkali borosilicate glasses having ions larger than oxygen (K, Rb and Cs), the packing is dominated by the modifying ions and the borate structural change does not have an apparent effect on the PF. Packing in alkaline earth borosilicate glasses is intermediate between the ionic and covalent packing trends observed for the alkali systems. This happens even though the modifying ions are as large as, or larger than oxygen, because there are half as many modifying ions added per oxygen in alkaline earth borosilicate glasses compared to alkali borosilicate glasses (MO versus M_2O).

Acknowledgements

This work was performed with the assistance of the National Science Foundation under grant DMR 0502051. Coe College is thanked for providing housing and stipends to the students.

References

1. Giri, S., Gaebler, C., Helmus, J., Affatigato, M., Feller, S. & Kodama, M. *J. Non-Cryst. Solids*, 2004, **347**, 87.
2. Burgess, M., McClarnon, D., Affatigato, M. & Feller, S. *J. Non-Cryst. Solids*, 2008, **354**, 3491.
3. Shannon, R. D. *Acta Cryst. A*, 1976, **32**, 751.
4. Mozzi, R. L. & Warren, B. E. *J. Appl. Cryst.*, 1970, **3**, 251.
5. Wright, A. C. *private communication*,
6. Zachariasen, W. H. *Acta Cryst.*, 1964, **17**, 749.
7. Shaw, J. L. PhD Thesis, *A neutron scattering study of glasses suitable for nuclear waste vitrification*, The University of Reading, UK, 2003.
8. Doweidar, H., Feller, S., Affatigato, M., Tischendorf, B., Ma, C. & Hammarsten, E. *Phys. Chem. Glasses*, 1999, **40**, 339.
9. Peters, A. M., Alamgir, F. M., Messer, S. W., Feller, S. A. & Loh, K. L. *Phys. Chem. Glasses*, 1994, **35**, 212.
10. Boekenhauer, R. *The density of lithium borosilicate glasses related to the atomic arrangement*, Honors Paper, Coe College, USA, 1991.
11. Feil, D. *The density of sodium borosilicate glasses related to atomic arrangements*, Honors Paper, Coe College, USA, 1989.
12. Budhwani, K. *Master recipe book*, Coe College, USA, 1990.
13. Kottke, J. A. *An examination of the physical properties of the rubidium and cesium borosilicate glasses and the development of the quantitative model for the alkali borosilicate glass system*, Honors Paper, Coe College, USA, 1995.
14. Bansal, N. & Doremus, R. H. *Handbook of glass properties*, Academic Press, Orlando, 1986, p.54–56.
15. Mullenbach, T., Franke, M., Ramm, A., Betzen, A. R., Kapoor, S., Lower, N., Munhollen, T., Berman, M., Affatigato, M. & Feller, S. A. *Phys. Chem. Glasses Eur. J. Glass Sci. Technol. B*, 2009, **50** (2), 89–94.
16. Dell, W. J., Bray, P. J. & Xiao, S. Z. *J. Non-Cryst. Solids*, 1983, **58**, 1.
17. Weast, R. *Handbook of Chemistry and Physics*, 66th edition. CRC, Boca Raton, 2001–2002, p.4–44.

Proc. VI Int. Conf. Borate Glasses, Himeji, Japan, 18–22 August 2008   *Phys. Chem. Glasses: Eur. J. Glass Sci. Technol. B, December 2009, **50**(6), 367–371*

Elastic properties and fragility of lithium borate glasses

Yu Matsuda,[*1] *Yasuteru Fukawa,*[1] *Mitsuru Kawashima,*[1] *Seiichi Mamiya,*[1] *Masao Kodama*[2] *&*
Seiji Kojima[1]

[1] *Graduate School of Pure and Applied Sciences, University of Tsukuba, Tsukuba, Ibaraki 305-8573, Japan*

[2] *Department of Applied Chemistry, Sojo University, Kumamoto 860-0082, Japan*

Manuscript received 18 August 2008
Revised version received 8 March 2009
Accepted 14 July 2009

The elastic properties of lithium borate glasses, $xLi_2O.(100-x)B_2O_3$, over a wide composition range up to x=64 mol% Li_2O have been investigated by Brillouin scattering spectroscopy (BSS). The longitudinal (V_L) and transverse (V_T) sound velocity and elastic moduli, i.e. longitudinal modulus (L), shear modulus (G), bulk modulus (K) and Young's modulus (E), have been determined as a function of the composition. They monotonically increase in the composition range $0 \leq x \leq 28$, and then take nearly constant values for x>40 mol% Li_2O. The correlation between the elastic properties of the glassy state and the fragility of the liquid state has been investigated. It is found that, except for the lower and higher composition ranges, there is a linear correlation of V_L, V_T, L and G to the fragility.

1. Introduction

Binary lithium borate glasses (LiBO), $xLi_2O.(100-x)$ B_2O_3 where x denotes the Li_2O composition in mol%, have attracted much attention for several decades, because the physical properties vary drastically with alkali metal oxide content, and often exhibit maxima or minima in their composition dependences, a behaviour known as the borate anomaly.[1] For instance, the composition dependences of thermal expansion coefficient,[1] density,[2,3] sound velocity,[2,4,5] boson peak,[6] and non-Debye relaxational parameter[7–9] of LiBO do not change monotonically, but exhibit maxima or minima against composition.

The elastic properties, such as sound velocity and elastic moduli, are of great interest to investigate these compositional variations, since they are sensitive to, and indicative of, the modification of the glass structures induced by the addition of Li_2O. Kodama *et al*[2] studied the composition dependences of sound velocity and elastic moduli by means of the ultrasonic pulse-echo overlap method, using a 10 MHz frequency transducer, in the composition range $0 \leq x \leq 28$ mol% Li_2O, which is the bulk glass forming range suitable for ultrasonic measurement. The composition dependence was discussed on the basis of the coordination change of the boron atoms. In the present paper, we have used Brillouin scattering spectroscopy (BSS) to investigate the elastic properties of LiBO over a wide composition range, extended by means of the plate quenching technique. The application and power of BSS for the measurement of hypersonic velocity of glasses has already

been reported.[4,5] This paper is the sequel of a series of papers on the elastic properties of LiBO,[2,4,5] but this is the first investigation to report the elastic properties for a composition range extending as high as 64 mol% Li_2O.

Another challenging topic in glass science is to understand a mechanism for the dramatic evolution of transport coefficients, such as viscosity or α-relaxation time, τ, in the supercooled liquid state when the temperature approaches the glass transition temperature, T_g. The temperature dependence of τ is characterised by the concept of fragility, introduced by Angell.[10] In this concept, a strong liquid exhibits an Arrhenius temperature dependence, while a fragile liquid shows a notable deviation from Arrhenius behaviour, involving a rapid temperature evolution of τ near T_g. The origin of the fragility is not yet fully understood. The fragility is quantified using the fragility index, m, or steepness index defined by

$$m = \lim_{T \to T_g} \left| \frac{d \log \tau}{d(T_g/T)} \right| \tag{1}$$

which is equivalent to the slope of $\log \tau$ versus T_g/T at $T=T_g$ in the Arrhenius plot.

Chryssikos *et al*[11] showed that the fragility of the LiBO system increases monotonically as Li_2O is added in the composition range $0 \leq x \leq 25$ mol% Li_2O. For compositions greater than 25 mol% Li_2O, we have reported the composition dependence of the fragility up to 64 mol% Li_2O by observing the frequency dependent dynamic heat capacity near T_g, using temperature modulated DSC.[12,13] The results show that the fragility has a single, broad maximum at around x=40–50 mol% Li_2O, corresponding closely to the variation in the boron coordination number.[13]

* Corresponding author. Email s-matsuda@ims.tsukuba.ac.jp, hiiro_s721@yahoo.co.jp

Very recently, the correlation of the fragility, which is a property of the liquid, with the elastic properties of its glassy state, has been hotly debated.[14,15] Knowledge of the fragility, which characterises the temperature evolution of transport properties in the supercooled liquid state near T_g, is of great importance for glass manufacturing processes. However, the correlation itself is still unclear.

In this paper, we have investigated the correlation between the fragility and elastic properties, and showed that the longitudinal and transverse sound velocities (or moduli) are linearly correlated with the fragility, associated with the formation of 4-coordinated boron atoms.

2. Experimental

2.1 Sample preparation

The detailed description of the sample preparation has been given previously.[2,8] In order to investigate the inherent nature of the binary system, highly homogeneous samples were prepared by the solution method. The elastic properties in a lower composition range, $0 \le x \le 28$, were previously determined accurately by the ultrasonic pulse echo overlap method.[2] In this study, samples in the higher composition range $30 \le x \le 64$ were newly prepared.

Analytical reagent grade $LiOH.H_2O$ and H_3BO_3 were used as the starting materials, and these were initially made to react in an aqueous solution in order to achieve high homogeneity. After the complete evaporation of water in a dry box kept at 140°C, the chemically synthesised powders, $xLi_2O.(100-x)B_2O_3$, were obtained. The powders were fused for 1·5 h in a Pt crucible at about 950 to 1300°C, depending on the composition. The glasses were then obtained by plate quenching. The quenched samples were used without any annealing because they are quite easy to crystallise.

2.2 Brillouin scattering spectroscopy

Brillouin scattering (BS) is inelastic light scattering induced by thermally excited acoustic phonons. Hypersonic velocity and elastic moduli can be determined by measuring the frequency shift of scattered light from a transparent sample such as a glass.

The experimental setup of BSS apparatus has been reported elsewhere.[4,5] A Sandercock-type 3+3 pass tandem Fabry-Perot interferometer was used. Scattering was excited by a single frequency green YAG laser with a wavelength of 532 nm. A conventional photon counting system and a multichannel analyser were used to observe the signals. The free spectral range (FSR) and scanning range were 30 GHz and ±25 GHz, respectively.

The BS spectra were measured in right angle (90A) geometry[5,16] at room temperature. The feature of

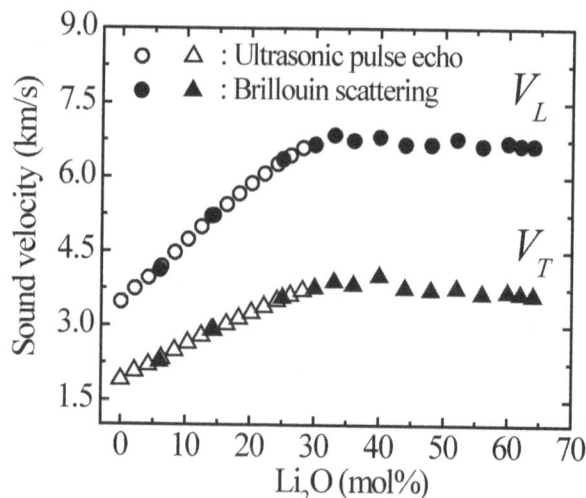

Figure 1. Composition dependences of longitudinal sound velocity (V_L) and transverse sound velocity (V_T) for lithium borate glasses. Open symbols denote the experimental results obtained using the ultrasonic pulse echo method by Kodama et al.[2] Closed symbols denote the results obtained by Brillouin scattering spectroscopy

this geometry is that the sound velocity, V, can be determined by the Brillouin shifts, Δv, without any information on the refractive index of the sample using the equation

$$V_k = \frac{\Delta v_{90A}^k \lambda}{\sqrt{2}} \qquad (2)$$

where λ is the wavelength of the incident light, k=Longitudinal (L) or Transverse (T).

3. Results

Examples of the acquired BS spectra of LiBO containing both longitudinal acoustic (LA) and transverse acoustic (TA) modes have been presented previously.[5] Figure 1 shows the composition dependences of longitudinal sound velocity (V_L) and transverse sound velocity (V_T) determined by Equation (2). Open symbols denote the experimental results obtained using the ultrasonic pulse echo method by Kodama et al,[2] while closed symbols denote the results obtained by BSS.

The elastic moduli, i.e. longitudinal modulus (L), shear modulus (G), bulk modulus (K) and Young's modulus (E) can be calculated by

$$L = \rho V_L^2 \qquad (3)$$

$$G = \rho V_T^2 \qquad (4)$$

$$K = L - \frac{4}{3}G \qquad (5)$$

$$E = \frac{G(3L-4G)}{L-G} \qquad (6)$$

where ρ is the density of the samples. To calculate the

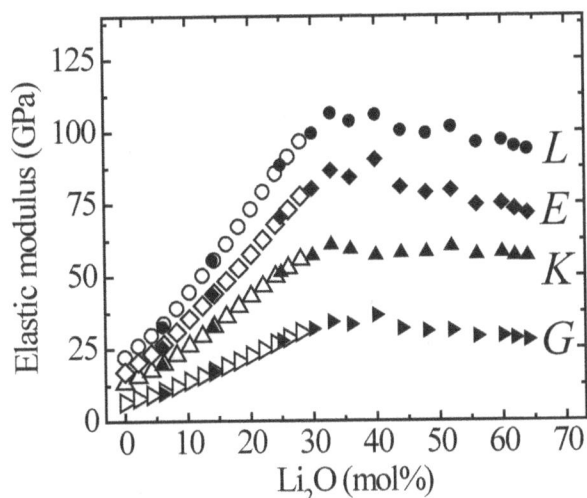

Figure 2. Composition dependences of the elastic moduli, longitudinal modulus (L), shear modulus (G), bulk modulus (K) and Young's modulus (E) for lithium borate glasses. The notation for the open and closed symbols is the same as for Figure 1

elastic moduli, knowledge of the density of LiBO is required; the composition dependence of the density has been reported by Kodama et al[2] and Feller et al.[3]

Figure 2 shows the composition dependences of the elastic moduli calculated by Equations (3) to (6). The meaning of the open and closed symbols is the same as for Figure 1. The composition range studied by the previous ultrasonic method was limited to $0 \leq x \leq 28$ mol% Li_2O, which is equivalent to the bulk glass forming range, since the method requires bulk homogeneous samples with typical size 15 mm in diameter and 15 mm in length.[2] In contrast, BS is light scattering using a visible laser beam, and thus the sound velocities and elastic moduli of a thin sample whose thickness is less than 1 mm can be measured (in 90A geometry). This is a unique advantage of BSS which is a powerful tool to probe the elastic properties of both bulk and rapidly quenched glasses. In the present study the compositional variations have been revealed over a wide composition range up to 64 mol% Li_2O.

4. Discussion

4.1 Elastic properties

The composition dependence of the sound velocity and elastic moduli show similar trends; the values increase linearly with increasing composition up to about 30 mol% Li_2O, and then reach approximately constant values. More precisely, because of the experimental uncertainty, it is not clear whether for V_T and V_L the composition dependences have a broad maximum at about 40 mol% Li_2O. Meanwhile each elastic modulus surely shows a broad maximum at about 40 mol% Li_2O, as can be seen from Figure 2. This is due to the effect of the compositional varia-

tions of density.

The structures of LiBO have been studied by NMR[17,18] and vibrational spectroscopy.[19,20] The structure of pure B_2O_3 glass is constructed from 6-membered boroxol rings consisting of $BØ_3$ triangular units, where the boron atoms are 3-coordinated with bridging oxygen atoms ("Ø" denotes a bridging oxygen). Further additions of Li_2O up to 40 mol% Li_2O induce a gradual change of the boron coordination number from 3 to 4, and structural units including the $BØ_4$ tetrahedron are formed. As has been discussed previously,[2] in the case of LiBO, the structural conversion from $BØ_3$ to $BØ_4$ results in an increase of the elastic properties because of the efficient tetrahedral arrangement in the glass network in the composition range $0 \leq x \leq 28$ mol% Li_2O. This is also supported by the density measurements of Affatigato et al.[3]

Above 40 mol% Li_2O, the fraction of 4-coordinated boron atoms decreases again. Structural units composed of 3-coordinated boron atoms and nonbridging oxygen atoms, are formed, such as $BØ_2O$, $BØO_2$ and BO_3, where "O" denotes a nonbridging atom. From a consideration of the boron coordination number and the microscopic structure of the glasses, one might expect that the sound velocity and elastic moduli decrease drastically for high Li_2O content, because the glassy network connectivity is reduced by the formation of nonbridging oxygen atoms. Actually, the glass transition temperature[9] and density[3] decrease markedly with increasing Li_2O content above $x=40$ mol% Li_2O. In the case of the elastic properties, surprisingly, the present results show that the values of the elastic moduli above 40 mol% Li_2O show an approximately constant value, or a very slight decrease, indicating that the composition dependence of the elastic properties cannot be explained in the same way as that of the glass transition temperature and density. Although the exact structure for high Li_2O content is not clear at the present stage, it is suggested that the nonbridging oxygen atoms and lithium ions play an important role in the elastic properties of LiBO with high x by stabilising the glass network, and these results provide experimental evidence which may help to provide an understanding of the role of nonbridging oxygens and lithium ions.

4.2 Correlation of the elastic property with the fragility

As mentioned in the Introduction, recently the correlation of the elastic properties in the glassy state with the fragility (which is a property of the melt) has been hotly argued. In Ref. 14, it was found that the fragility index, m, is linearly correlated with the ratio of the bulk and shear moduli, K/G, of the glassy state, by the equation

$$m = 29(K/G - 0.41) \qquad (7)$$

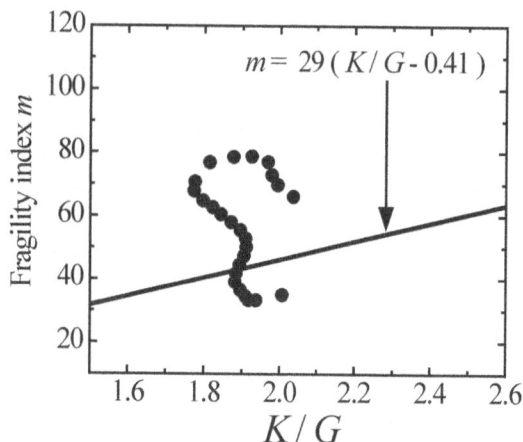

Figure 3. Fragility index, m, as a function of the ratio of bulk (K) and shear (G) moduli. The data for the fragility are taken from Refs 11 and 13

Yannopoulos et al[15] presented a detailed analysis of the correlation given by Equation (5) for various kinds of glass former, such as inorganic, organic and metallic. They pointed out that the correlation of Equation (7) does not hold for some polymeric and metallic glasses. Very recently, Sokolov et al[21] have discussed the possible main problem of Equation (7) for metallic glasses, in which the free electron contribution to the bulk modulus K is important, and for polymeric glasses, in which a specific intermolecular effect, of long polymer chains is important.

Here, in order to investigate the correlation for the binary borate system, we present the relationship between the fragility of LiBO and the elastic properties. The compositional dependence of m up to x=64 mol% Li$_2$O has already been presented elsewhere,[12,13] and also the fragility up to x=25 mol% Li$_2$O has previously been considered by Chryssikos et al,[11] on the basis of heat capacity and viscosity data. Very recently, the values of m have been determined by observing the frequency dependence dynamic heat capacity near T_g by temperature modulated DSC.[13]

Figure 3 shows the fragility as a function of K/G in the same way as in Refs 14 and 15. As can be seen, there is no simple correlation for LiBO. The reason for the failure may relate to the compositional variations of K/G., which are connected to the composition dependence of Poisson's ratio, σ, of the glasses. As has been discussed previously,[5] σ shows a quite complicated behaviour; σ as a function of the composition exhibits a very broad minimum at around x=10 mol% Li$_2$O, a broad maximum at around x=20 mol% Li$_2$O, and another broad minimum at around x=40 mol% Li$_2$O. The trend for σ may be explained by the coexistence of different superstructural units (such as the boroxol ring and tetraborate, pentaborate and diborate units etc.) which are composed of both 3-coordinated and 4-coodinated boron atoms, which may have different σ values. Such complicated topology of the glass network leads to the violation of the

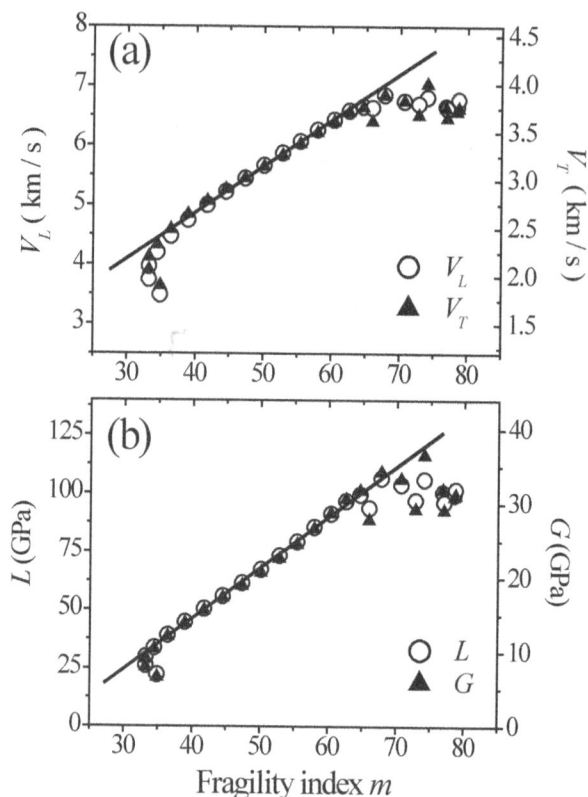

Figure 4. (a) Longitudinal (V$_L$) and transverse (V$_T$) sound velocities as a function of the fragility. (b) Longitudinal (L) and shear (G) moduli as a function of the fragility. The straight lines are a guide for the eye. In both (a) and (b), the linear correlation of V$_L$, V$_T$, L and G to the fragility clearly hold except for the lower and higher composition ranges

correlation in Equation (7). An understanding of the relationship between the microscopic glass structure of LiBO and σ will give more clear insight into the correlation of Equation (7).

In contrast to Poisson's ratio, the sound velocities and elastic moduli vary in a rather simple way (Figures 1 and 2). Figure 4(a) shows V_L and V_T as a function of the fragility index, m; Figure 4(b) shows the elastic moduli L and G. Except for the lower and higher composition ranges, the linear correlation of V_L, V_T, L and G to the fragility clearly holds. As was discussed earlier, V_L and V_T monotonically increase in the range 0≤x≤28 mol% Li$_2$O, due to the direct conversion of structural units from BØ$_3$ to BØ$_4$. On the other hand, the fragility of LiBO, the behaviour of which can be explained as arising from the fluctuation of the boron coordination number,[12,13] also monotonically increases with increasing Li$_2$O content in the range 0≤x≤28 mol% Li$_2$O. Thus, the coordination change of the boron atoms plays an important role in the linear correlation of the sound velocity and the fragility. At least for LiBO, one might easily estimate the fragility, which gives information on the temperature evolution of the transport properties for the supercooled liquid near T_g, by just measuring one of V_L, V_T, L and G, if the fragility is in the range of about 35≤m≤65.

5. Conclusions

The elastic properties of lithium borate glasses, xLi$_2$O.$(100-x)$B$_2$O$_3$, over a wide composition range up to $x=64$ mol% Li$_2$O have been investigated by Brillouin scattering spectroscopy (BSS). BSS is a powerful tool for exploring the elastic properties of both bulk and rapidly quenched thin glassy samples. The elastic properties can be determined using the 90A geometry of BSS without any information on refractive index of the sample.

The longitudinal (V_L) and transverse (V_T) sound velocity and elastic moduli, longitudinal modulus (L), shear modulus (G), bulk modulus (K) and Young's modulus (E), monotonically increase in the composition range $0 \leq x \leq 28$ mol% Li$_2$O, and then reach nearly constant values for $x > 40$ mol% Li$_2$O. The behaviour for the high compositions cannot be explained only by considering the coordination change of boron atoms, and thus the nonbridging oxygens and lithium ions may play an important role in the elastic properties for high Li$_2$O content. The present results, which cover a wide composition range, can give an experimental basis for a development of the understanding of the relationship between the intermediate structure and elastic properties of borate glasses.

The correlation of the elastic properties of the glassy state and the fragility of the liquid state has been investigated. A clear correlation of V_L, V_T, L and G with the fragility has been found, except for the lower and higher composition ranges, whereas the reported linear correlation of the ratio of K and G (relating to Poisson's ratio, σ) with the fragility was not found.

Acknowledgements

Y. M. is thankful for a JSPS Research Fellowship for Young Scientists. This research was partially supported by MEXT, the Grant-in-Aids for 19·574.

References

1. Shelby, J. E. *J. Am. Ceram. Soc.*, 1983, **66**, 501.
2. Kodama, M., Matsushita T. & Kojima, S. *Jpn. J. Appl. Phys.*, 1994, **34**, 2570.
3. Affatigato, M., Feller, S., Khaw, E. J., Feil, D., Teoh, B. & Mathews, O. *Phys. Chem. Glasses*, 1990, **31**, 19.
4. Ike, Y., Matsuda, Y., Kojima, S. & Kodama, M. *Jpn. J. Appl. Phys.*, 2006, **45**, 4474.
5. Fukawa, Y., Matsuda, Y., Ike, Y., Kodama, M. & Kojima, S. *Jpn. J. Appl. Phys.*, 2008, **47**, 3833.
6. Kojima, S., Novikov, V. N. & Kodama, M. *J. Chem. Phys.*, 2000, **113**, 6344.
7. Matsuda, Y., Matsui, C., Ike, Y., Kodama, M. & Kojima, S. *J. Therm. Anal. Cal.*, 2006, **85**, 725.
8. Matsuda, Y., Fukawa, Y., Matsui, C., Ike, Y., Kodama, M. & Kojima, S. *Fluid. Phase Eq.*, 2007, **256**, 127.
9. Matsuda, Y., Fukawa, Y., Ike, Y., Kodama, M. & Kojima, S. *J. Phys. Soc. Jpn.*, 2008, **77**, 084602.
10. Angell, C. A. *J. Non-Cryst. Solids*, 1991, **131–133**, 13.
11. Chryssikos, G. D., Duffy, J. A., Hutchinson, J. M., Ingram, M. D., Kamitsos, E. I. & Pappin, A. J. *J. Non-Cryst. Solids*, 1994, **172–174**, 378.
12. Matsuda, Y., Fukawa, Y., Kawashima, M. & Kojima, S. *AIP Conf. Proc.*, 2008, **982**, 207.
13. Matsuda, Y., Fukawa, Y., Kawashima, M., Mamiya, S. & Kojima, S. *Solid State Ionics*, 2008, **179**, 2424.
14. Novikov, V. N. & Sokolov, A. P. *Nature*, 2004, **431**, 961.
15. Yannopoulus, S. N. & Johari, G. P. *Nature*, 2006, **442**, E7.
16. Kruger, J. K., Marx, A., Peetz, L., Roberts, R. & Unruh, H.-G. *Colloid Polymer Sci.*, 1986, **264**, 403.
17. Bray, P. J. & O'Keefe, J. G. *Phys. Chem. Glasses*, 1963, **4**, 37.
18. Jellison, G. E., Feller, S. A. & Bray, P. J. *Phys. Chem. Glasses*, 1978, **19**, 52.
19. Kamitsos, E. I., Karakassides, M. A. & Chryssikos, G. D. *Phys. Chem. Glasses*, 1987, **28**, 203.
20. Kamitsos, E. I., Patsis, A. P., Karakassides, M. A. & Chryssikos, G. D. *J. Non-Cryst. Solids*, 1990, **126**, 52.
21. Sokolov, A. P., Novikov, V. N. & Kisliuk, A. *Phil. Mag.*, 2007, **87**, 613.

Proc. VI Int. Conf. Borate Glasses, Himeji, Japan, 18–22 August 2008   *Phys. Chem. Glasses: Eur. J. Glass Sci. Technol. B, August 2009, **50** (4), 262–266*

Atomic structure of BeO–B$_2$O$_3$–Al$_2$O$_3$ glasses: ^{11}B and ^{27}Al MAS NMR spectroscopy at 21·8 Tesla

S. Sen,[*1] E. L. Gjersing,[1] H. Maekawa,[2] Y. Noda,[2] M. Ando,[2] M. Tansho,[3] T. Shimizu,[3] V. P. Klyuev[4] & B. Z. Pevzner[5]*

[1] *Department of Chemical Engineering and Materials Science, University of California, Davis, California 95616, USA*
[2] *Graduate School of Engineering, Tohoku University, Aoba-ku, Sendai, 980-8579, Japan*
[3] *National Institute for Materials Science, 3-13, Sakura, Tsukuba, Ibaraki 305-0003, Japan*
[4] *Grebenshchikov Institute of Silicate Chemistry of the Russian Academy of Science, Nab. Makarova 2, St. Petersburg 199034, Russia*
[5] *Laboratory of Glass Properties, ul. Varshavskaya, 5A, St. Petersburg 196084, Russia*

Manuscript received 21 December 2008
Revised version received 2 April 2009
Accepted 18 May 2009

The structures of BeO–Al$_2$O$_3$–B$_2$O$_3$ glasses with a range of BeO/Al$_2$O$_3$ ratios at fixed BeO/B$_2$O$_3$ ratios of 1:1·5 and 1:1 have been investigated using ^{11}B and ^{27}Al MAS NMR spectroscopy at ultrahigh magnetic field (21·8 T). These ^{11}B and ^{27}Al MAS NMR spectra are characterised by remarkably superior resolution and thus provide more accurate results for B and Al speciation compared to the low field (11·7 T) NMR results reported in previous studies. All glasses are characterised by a BO$_3$:BO$_4$ ratio of ~86:14, and approximately 70% of all Al atoms are found to be five- and six-fold coordinated. Such large concentrations of BO$_3$, AlO$_5$ and AlO$_6$ species are consistent with the highest field strength of Be among all alkali and alkaline earth metals. The compositional variation of cation speciation, when combined with bond valence calculations, indicates that most of the oxygens in these glasses would have to be 3-coordinated and shared between B^{III}–Al^{VI}–Be^{IV}, B^{III}–Be^{IV}–Be^{IV}, B^{IV}–Al^{IV}–Be^{IV}, Al^{IV}–Al^{IV}–Be^{IV} or Al^{IV}–Al^{V}–Be^{IV} coordination polyhedra. This bonding scenario would be expected to give rise to a dense structure with increased rigidity, consistent with the remarkably high T$_g$ values reported in the literature for these glasses.

1. Introduction

The structure of a variety of alkaline earth boroaluminate glasses has been studied in the past using ^{11}B, ^{27}Al and ^{17}O nuclear magnetic resonance (NMR) spectroscopic techniques.[1–11] These glasses are characterised by an unusually wide variety of Al and B coordination environments with Al being 4-, 5- and 6-coordinated (Al^{IV}, Al^V and Al^{VI}, respectively) to oxygen, and B being 3- and 4-coordinated (B^{III} and B^{IV}, respectively) to oxygen. The relative fractions of these coordination environments have been found to be strongly dependent on composition and thermal history of the glass.[2,3,11] One of the most interesting aspects of the compositional dependence is the remarkable effect of the field strength (charge:radius ratio) of the modifier alkaline earth cation (R) on the relative ratios of various B and Al species in the structure. For example, for identical RO:Al$_2$O$_3$:B$_2$O$_3$ molar ratios, the high field strength (HFS) cations, such as Mg^{2+}, prefer higher concentrations of B^{III}, Al^V and Al^{VI} species, compared to those in glasses where R=Ca^{2+}, Sr^{2+} or Ba^{2+}. Moreover, HFS cations are also known to reduce the compositional dependence of the relative fractions of various Al and B species, as well as that of various thermophysical properties, such as thermal expansion coefficient, α_s, and glass

transition temperature, T_g.[2,7,8,11,12]

The Be^{2+} cation is characterised by the highest field strength and electronegativity among all alkali and alkaline earth metals. Although several detailed structural studies of Mg, Ca, Sr and Ba boroaluminates over a wide range of composition have been reported over the last two decades, the structure of beryllium boroaluminate glasses has been investigated for the first time only recently by us, using ^{11}B, ^{27}Al and ^{9}Be NMR spectroscopic techniques.[13] The compositional range of glass formation in the BeO–B$_2$O$_3$–Al$_2$O$_3$ system is rather limited compared to that for other alkaline earth cations. However, the results of this study clearly confirmed the abovementioned effects of HFS cations on glass structure.[13] Accordingly, the concentrations of B^{III}, Al^V and Al^{VI} species in BeO–B$_2$O$_3$–Al$_2$O$_3$ glasses were found to be significantly higher compared to those in other alkali and alkaline earth boroaluminate glasses reported previously in the literature.[13] However, the unusually distorted geometry of the B–O and Al–O coordination polyhedra in these glasses resulted in significant second order quadrupolar broadening of the ^{11}B and ^{27}Al magic angle spinning (MAS) NMR line shapes at the magnetic field of 11·7 T that was used in this study.[13] Such broadening made NMR line shape simulation, and hence quantitative determination of the relative fractions of various B and

*Corresponding author. Email sbsen@ucdavis.edu

Al species in the glass structure, difficult and less accurate. The high resolution technique triple quantum MAS (3QMAS) NMR was shown to be successful in resolving the different B and Al sites in these glasses.[13] Nevertheless, such techniques are not fully quantitative, especially for relative quantification of sites with significantly different quadrupolar coupling constants, C_Q. The reduction of the second order quadrupolar broadening with increasing magnetic field can be used to obtain more quantitative NMR spectra with higher resolution and sensitivity under simple magic angle spinning.[14] We report here the results of a high resolution, ultrahigh field (21·8 T) ^{11}B and ^{27}Al MAS NMR spectroscopic characterisation of the atomic structure of beryllium boroaluminate glasses with a range of BeO/Al$_2$O$_3$ ratios at fixed BeO/B$_2$O$_3$ ratios of 1:1·5 and 1:1 (see Table 1). The Al and B speciation results and quadrupolar parameters are compared with those reported in our previous study at a lower magnetic field of 11·7 T.[13] Finally, a qualitative structural model of these glasses in terms of the connectivity between various cation–oxygen coordination polyhedra is proposed.

2. Experimental

2.1. Sample preparation

Samples were synthesised from reagent grade BeCO$_3$, H$_3$BO$_3$ and Al(OH)$_3$. The batch materials were mixed in stoichiometric proportions, decarbonated, dehydrated and melted in 20 to 30 g batches in platinum crucibles at 1400 to 1500°C for 1 h. Melts were poured into steel moulds, and the resulting glasses were annealed at temperatures near the respective T_g. The nominal compositions of the glasses investigated in this study are listed in Table 1. The chemical compositions of all glasses were analysed with inductively coupled plasma mass spectrometry (ICP-MS) and were found to be in good agreement (within ±0·5 mol%) with the nominal compositions. Moreover, Be contents were independently checked with quantitative ^{9}Be MAS NMR using phenakite (Be$_2$SiO$_4$) crystal as a standard, and were also found to be in good agreement with the nominal values reported in Table 1.

2.2. NMR spectroscopy

NMR spectra were acquired using a JNM-ECA930 (JEOL) spectrometer equipped with a JEOL magnet

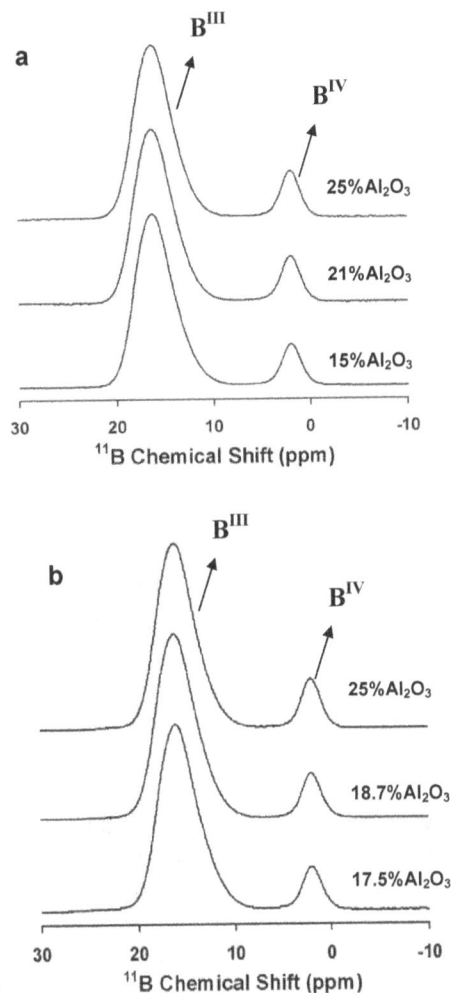

Figure 1. ^{11}B MAS NMR spectra of beryllium boroaluminate glasses with BeO:B$_2$O$_3$=1:1 (a) and 1:1·5 (b) collected at 21·8 T. The corresponding Al$_2$O$_3$ contents are shown alongside each spectrum

(21·8 T) operating at Larmor frequencies of 298·2 and 242·2 MHz for ^{11}B and ^{27}Al, respectively. All MAS NMR spectra were acquired using a solid 10° pulse (1·15 and 0·84 μs for ^{11}B and ^{27}Al, respectively). Samples were placed in ZrO$_2$ rotors, and were spun at 14 kHz in a 4 mm JEOL MQ/MAS probe for ^{11}B and at 22 kHz in a 3·2 mm JEOL MQ/MAS probe for ^{27}Al. The latter probe contains a sharp Gaussian background signal at ~15 ppm for ^{27}Al. A total of 4 (64) scans with a recycle delay of 60 (2) s was used to obtain each ^{11}B (^{27}Al) MAS NMR spectrum. The chemical shifts for ^{11}B and ^{27}Al were externally referenced to BF$_3$.Et$_2$O and 1 M AlCl$_3$ aqueous solution, respectively.

Table 1. Compositions (mol%) of BeO–Al$_2$O$_3$–B$_2$O$_3$ glasses and corresponding B and Al speciation

BeO	Al$_2$O$_3$	B$_2$O$_3$	BeO/B$_2$O$_3$	BeO/Al$_2$O$_3$	%B^{IV*}	%Al^{IV**}	%Al^{V**}	%Al^{VI**}
35·2	18·7	46·1	0·67	1·88	13·7	29	48	23
33·0	17·5	49·5	0·67	1·88	12·7	28	47	25
30·0	25·0	45·0	0·67	1·20	14·6	25	48	27
42·5	15·0	42·5	1·00	2·83	13·6	30	47	23
39·5	21·0	39·5	1·00	1·88	14·4	34	46	20
37·5	25·0	37·5	1·00	1·50	14·8	30	47	23

* Typical error is ~0·5%. ** Typical error is ~4%.

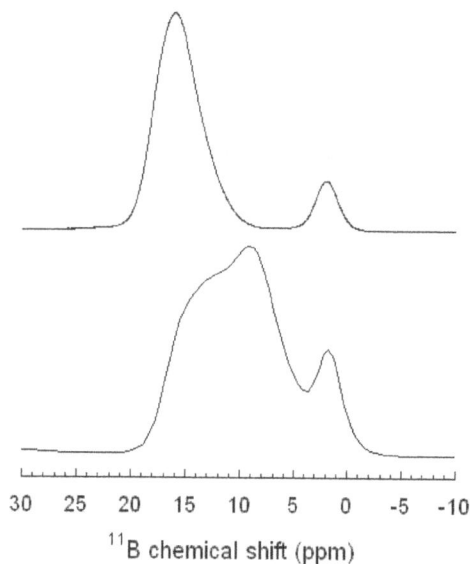

Figure 2. Comparison of the ^{11}B MAS NMR spectra of $33BeO.17\cdot5Al_2O_3.49\cdot5B_2O_3$ glass collected at 21·8 T (top) and at 11·7 T (bottom)

3. Results

3.1. ^{11}B NMR

The ^{11}B MAS NMR spectra of beryllium boroaluminate glasses collected at 21·8 T are shown in Figure 1. These spectra are characterised by a broad, nearly symmetric peak, centred at 16·2 ppm, corresponding to the B^{III} sites, and a small, relatively narrow, symmetric peak, centred at 2 ppm, corresponding to the B^{IV} sites. The peak positions and line shapes do not show any significant change with glass composition. A comparison of the ^{11}B MAS NMR spectra collected at 11·7 and at 21·8 T for the same glass is given in Figure 2, clearly displaying the remarkably superior resolution of the peaks corresponding to the two B sites at the higher magnetic field. The relative fractions of B^{III} and B^{IV} sites can be immediately obtained from the areas under the corresponding peaks in the ^{11}B MAS NMR spectra in Figure 1, and are reported in Table 1. These B speciation results are of significantly higher accuracy than those reported in our previous study based on the poorly resolved ^{11}B MAS NMR spectra collected at 11·7 T.[13] The results show that the relative fraction of B^{IV} sites in all glasses varies within a small range of ~12·7 to 14·7%. A comparison of the centre of gravity of the B^{III} peak in the ^{11}B MAS NMR spectra collected previously at 18·8 and at 11·7 T for one glass sample with composition $33BeO.17\cdot5Al_2O_3.49\cdot5B_2O_3$[13] with that collected in this study at 21·8 T allows us to estimate the quadrupolar coupling constant, C_Q, and isotropic chemical shift, δ_{iso}, for this site with good accuracy.[15] Such calculations yield a C_Q value of 2·8±0·2 MHz, and a δ_{iso} value of 18±0·5 ppm when the asymmetry parameter, η, is taken to be 0·5. Similar calculations for the B^{IV} site yield a C_Q value of 1·0±0·1

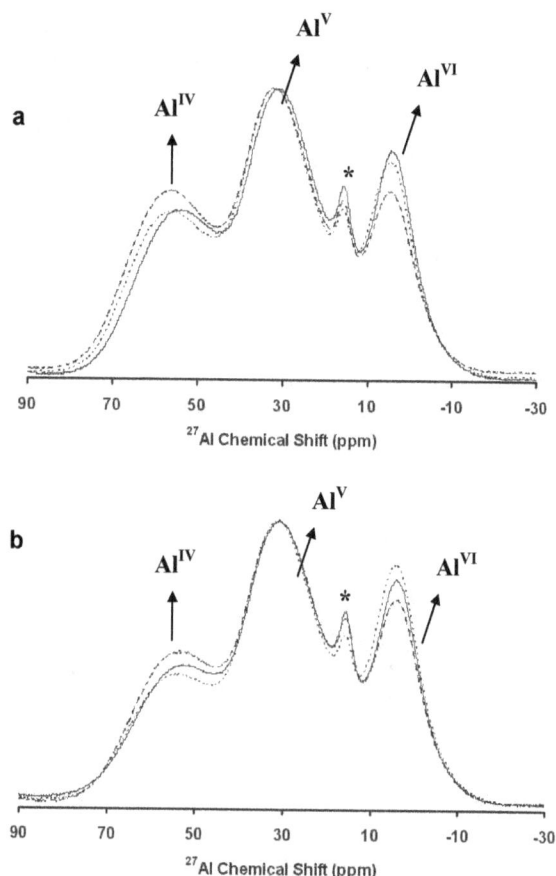

Figure 3. ^{27}Al MAS NMR spectra of beryllium boroaluminate glasses in the two compositional series with $BeO:B_2O_3$=1:1 (a) and 1:1·5 (b) collected at 21·8 T. The solid, dashed and dotted lines correspond to progressively increasing Al_2O_3 content in each series. The sharp peak near ~15 ppm denoted by an asterisk is from the probe background

MHz, and a δ_{iso} of 2±0·1 ppm. These values of C_Q for the B^{III} and B^{IV} sites are relatively high compared to the C_Q values of ~2·5 to 2·6 (0·3 to 0·5) MHz typical of B^{III} (B^{IV}) sites in other boron containing glasses, and possibly implies significant asymmetry or distortion of such sites in beryllium boroaluminates. The δ_{iso} of 18±0·5 ppm for the B^{III} sites is characteristic of either BO_3 species with one nonbridging oxygen or triborate and/or boroxol ring type environments with all bridging oxygens.[16] Although the relatively large C_Q value of ~2·8 MHz for these B^{III} sites possibly discards the existence of boroxol rings in these glasses, such values are consistent with those characteristic of B^{III} sites in triborate rings.[16]

3.2. ^{27}Al NMR

The ^{27}Al MAS NMR spectra of beryllium boroaluminate glasses collected at 21·8 T are shown in Figure 3. These spectra are characterised by three clearly resolved symmetric peaks centred at ~55, 31 and 4 ppm that can be readily assigned to Al^{IV}, Al^V and Al^{VI} sites, respectively.[2–7,11,15] A comparison of the ^{27}Al

MAS NMR spectra collected at 11·7 and at 21·8 T for the same glass is given in Figure 4, demonstrating the excellent resolution of the peaks corresponding to the three Al sites at the higher magnetic field. The peak positions and line shapes of these high field ^{27}Al MAS NMR spectra do not show any significant change with glass composition, and can be simulated well with three symmetric peaks with 50% Gaussian and 50% Lorentzian components (Figure 3). The relative fractions of Al^{IV}, Al^V and Al^{VI} sites thus obtained from these simulations are listed in Table 1. Once again, these high field Al speciation results are of significantly higher accuracy than those reported in our previous study, based on the poorly resolved ^{27}Al MAS NMR spectra collected at 11·7 T.[13] The results show that the relative fraction of Al^V sites in all glasses is nearly constant at ~47%, while the Al^{VI}:Al^{IV} ratio shows small but systematic changes with composition (Table 1). A combination of the centres of gravity of the Al^{IV}, Al^V and Al^{VI} line shapes obtained previously at 11·7 T (~45, 24 and 0 ppm, respectively) with those obtained in this study at 21·8 T allows estimation of the values of C_Q and δ_{iso} for these sites with good accuracy.[15] Such calculations, with the approximation that $\eta=0·5$, yield C_Q values of 9·5, 7·9 and 5·9 (±10%) MHz, and δ_{iso} values of 59 (±2), 34 (±1) and 5·6(±1) ppm for the Al^{IV}, Al^V and Al^{VI} sites, respectively. These C_Q values are significantly higher than the values typical of such sites in other alkali and alkaline earth aluminoborate glasses,[2,11] and are indicative of significant distortion of the Al–O coordination polyhedra in beryllium boroaluminates.

Previous ^{27}Al NMR studies of alkali and alkaline earth boroaluminate glasses have shown that the concentrations of high coordinated Al^V and Al^{VI} species increase in the presence of HFS modifier cations.[2,11] The beryllium boroaluminate compositions studied here are characterised by some of the highest concentrations of the Al^V (~47%) and Al^{VI} (~20–27%) species reported so far for any alkali or alkaline earth boroaluminate glass of comparable composition (Table 1). This result is consistent with the highest field strength of the Be^{2+} ions among all alkali and alkaline earth cations.

4. Discussion

When taken together, the ^{11}B and ^{27}Al MAS NMR spectroscopic results indicate relatively small compositional variation of B and Al speciation in these beryllium boroaluminate glasses, with the structure being dominated by B^{III} and high coordinated Al^V and Al^{VI} species (Table 1). However, these high resolution NMR spectra also reveal that in these glasses the B^{IV}:B^{III} and Al^{IV}:Al^{VI} ratios increase simultaneously and systematically with decreasing B$_2$O$_3$ content at a constant BeO:Al$_2$O$_3$ ratio of 1·88:1 (Table 1). On the other hand, no systematic compositional effect of

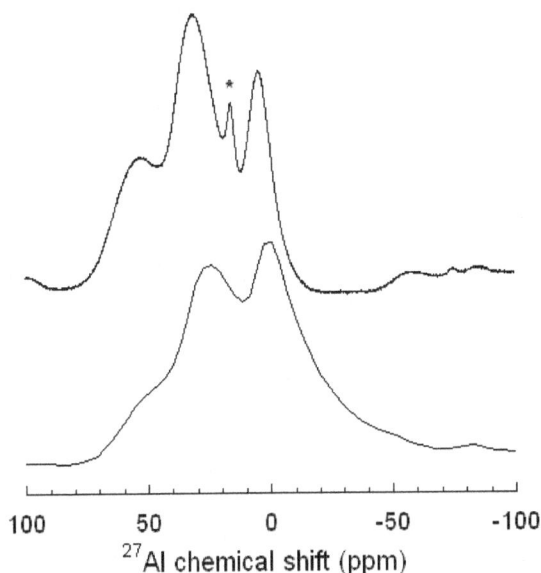

Figure 4. Comparison of the ^{27}Al MAS NMR spectra of 33BeO.17·5Al$_2$O$_3$.49·5B$_2$O$_3$ glass collected at 21·8 T (top) and at 11·7 T (bottom). The sharp peak near ~15 ppm denoted by an asterisk in the top spectrum is from the probe background

Al$_2$O$_3$ content on B and Al speciation is apparent in glasses with constant BeO:B$_2$O$_3$ ratios.

Previous structural studies of alkaline earth boroaluminate glasses have suggested on the basis of simple bond valence calculations that the bonding requirements of the oxygens are best satisfied when the sum of the bond strengths to oxygen from the surrounding cations is +2, i.e. the oxygen is charge neutral.[4] Be is known to be 4-coordinated to oxygen and to form regular BeO$_4$ tetrahedra in these glasses.[13] Therefore, the Brown–Shannon bond strength for Be–O bonds in a BeO$_4$ tetrahedron (Be^{IV}) is 0·5 bond valence units, which is identical to that for Al–O bonds in an AlO$_6$ octahedron (Al^{VI}).[17] Bond valence calculations would then indicate that oxygens in BeO$_4$ tetrahedra would have to be 3-coordinated and shared by a few possible combinations of three cation sites in order to be neutral or near neutral.[17] These combinations are: B^{III}–Al^{VI}–Be^{IV}, B^{III}–Be^{IV}–Be^{IV}, B^{IV}–Al^{IV}–Be^{IV}, Al^{IV}–Al^{IV}–Be^{IV} or Al^{IV}–Al^V–Be^{IV} sites. Such coordination environments imply the preferential association of B^{III} and Al^{VI}, or of B^{IV} and Al^{IV} in the structure. This scenario is consistent with the abovementioned correlated increase in B^{IV} and Al^{IV} fractions in beryllium boroaluminate glasses with decreasing B$_2$O$_3$ content at constant BeO:Al$_2$O$_3$ ratio of 1·88:1 (Table 1). These triply coordinated oxygen environments may also result in significant distortion of the B^{III}, Al^{IV} and Al^V coordination polyhedra, a prediction that is consistent with the characteristically large C_Q values associated with these sites in beryllium boroaluminate glasses, as already discussed. Moreover, it is worth noting here that for nearly constant B_2O_3 content of ~45 mol%, the glass with BeO:Al_2O_3 ratio of 1·9:1 is characterised by a higher Al^{IV}:Al^{VI} ratio (1·2:1) compared to that (0·9:1) in the glass with BeO:Al_2O_3 ratio of 1·2:1 (Table 1). This result may indicate a preference for Be to be associated with Al^{IV} sites in Al^{IV}–Al^{IV}–Be^{IV} or Al^{IV}–Al^{V}–Be^{IV} type of linkages. Similar to oxygen triclusters (oxygen shared by three AlO_4 tetrahedra), the 3-coordinated oxygen in the Al^{IV}–Al^{IV}–Be^{IV} environment would be expected to give rise to a dense structure with increased rigidity.[18] This hypothesis is corroborated by the fact that the glass transition temperatures of beryllium boroaluminate glasses are significantly higher (712–733°C) than those characteristic of other alkali or alkaline earth boroaluminate glasses of comparable composition, which typically range between 400 and 670°C.[12,19]

4. Summary

Highly accurate B and Al speciation results have been obtained for beryllium boroaluminate glasses using high resolution ^{11}B and ^{27}Al MAS NMR spectroscopy at 21·8 T. These results indicate that all the beryllium boroaluminate glasses studied here are characterised by nearly constant BO_3:BO_4 ratio of ~86:14 and (Al^{V}+Al^{VI}):Al^{IV} ratio of ~70:30. The relatively large values of these ratios are consistent with the high field strength of the Be^{2+} cation. Simple bond valence calculations indicate that the oxygens in the BeO_4 tetrahedra in these glasses would have to be 3-coordinated and shared between B^{III}–Al^{VI}–Be^{IV}, B^{III}–Be^{IV}–Be^{IV}, B^{IV}–Al^{IV}–Be^{IV}, Al^{IV}–Al^{IV}–Be^{IV} and Al^{IV}–Al^{V}–Be^{IV} sites. These oxygen bonding environments are consistent with the unusually large distortion and C_Q (quadrupolar coupling constant) values of the B–O and Al–O coordination polyhedra, as well as with the high T_g values which are characteristic of these glasses. Future studies based on ^{17}O NMR and ^{11}B–^{27}Al two-dimensional NMR correlation spectroscopy would be crucial to establish the nature of such connectivities between various cation–oxygen coordination polyhedra in the structure of these glasses.

Acknowledgment

This work was partially supported by an NSF grant DMR-0603933 to S. Sen.

References

1. Hahnert, M. & Hallas, E. Z. Chem., 1986, **26**, 144.
2. Bunker, B. C., Kirkpatrick, R. J., Brow, R. K., Turner, G. L. & Nelson, C. J. Am. Ceram. Soc., 1991, **74**, 1430.
3. Sen, S., Xu, Z. & Stebbins, J. F. J. Non-Cryst. Solids, 1998, **226**, 29.
4. Baltisberger, J. H. Xu, Z. Stebbins, J. F. Wang, S. Pines, A. J. Am. Chem. Soc., 1996, **118**, 7209.
5. Bertmer, M., Züchner, L., Chan, J. C. C. & Eckert, H. J. Phys. Chem. B, 2000, **104**, 6541.
6. Chan, J. C. C., Bertmer, M. & Eckert, H. J. Am. Chem. Soc., 1999, 121, 5238.
7. van Wüllen, L., Gee, B., Züchner, L., Bertmer, M. & Eckert, H. Ber. Bunsen-Gesell. Phys. Chem., 1996, **100**, 1539.
8. van Wüllen, L., Züchner, L., Müller-Warmouth, W. & Eckert, H. Solid State Nucl. Magn. Reson., 1996, **6**, 203.
9. Brow, R. K., Tallant, D. R. & Turner, G. L. J. Am. Ceram. Soc., 1997, **80**, 1239.
10. Hémono, N., Rocherullé, J., Le Floch, M., Bureau, B. & Benard-Rocherullé, P. J. Mater. Sci., 2006, **41**, 445.
11. Du, L.-S. & Stebbins, J. F. Solid State Nucl. Magn. Reson., 2005, **27**, 37.
12. Klyuev, V. P. & Pevzner, B. Z. J. Non-Cryst. Solids, 2007, **353**, 2008.
13. Sen, S., Yu, P., Klyuev, V. P. & Pevzner, B. Z. J. Non-Cryst. Solids, 2008, **354**, 4005.
14. Kroeker, S., Neuhoff, P. S. & Stebbins, J. F. J. Non-Cryst. Solids, 2001, **293–295**, 440.
15. Samoson, A. Chem. Phys. Lett., 1985, **119**, 29.
16. Sen, S. Molec. Simulat., 2008, **34**, 1115.
17. Bunker, B. C., Kirkpatrick, R. J. & Brow, R. K. J. Am. Ceram. Soc., 1991, **74**, 1425.
18. MacDowell, J. F. & Beall, G. H. J. Am. Ceram. Soc., 1969, **52**, 17.
19. Klyuev, V. P. & Pevzner, B. Z. Glass Phys. Chem., 2002, **28**, 207.

Proc. VI Int. Conf. Borate Glasses, Himeji, Japan, 18–22 August 2008   *Phys. Chem. Glasses: Eur. J. Glass Sci. Technol. B, June 2009, **50** (3), 195–200*

Quantification of boron coordination changes between lithium borate glasses and melts by neutron diffraction

Laurent Cormier,[1]* *Georges Calas*[1] *& Brigitte Beuneu*[2]

[1] *Institut de Minéralogie et de Physique des Milieux Condensés, CNRS UMR 7590, Université Pierre et Marie Curie - Paris 6, Université Denis Diderot - Paris 7, IPGP, 140 rue de Lourmel, 75015 Paris, France*

[2] *Laboratoire Léon Brillouin, C.E. Saclay, 91191 Gif-sur-Yvette, France*

Manuscript received 12 September 2008
Revised version received 14 November 2008
Accepted 6 March 2009

The structure of lithium borate glasses and liquids (xLi$_2$O.$(1-x)$B$_2$O$_3$, with x=0·1, 0·20, 0·25 and 0·33) has been investigated by neutron diffraction coupled with Li isotopic substitution at temperatures up to 1173–1273 K. Both the short range order around B and Li is changed, with the conversion of BO$_4$ to BO$_3$ and a subsequent increase of NBOs (nonbridging oxygens) around Li as temperature increases. The quantitative boron coordination change has been assessed by a Gaussian fitting of the B–O pair contribution. The increasing presence of NBOs results in the opening of the borate network and the disappearance of superstructural borate rings. The reorganisation of the structure above the glass transition temperature provides a structural mechanism for the modifications of the thermodynamic and configurational properties of borate liquids.

The temperature driven structural changes of oxide glasses and melts affect transport properties, such as diffusion, conductivity, and viscosity. These changes are often subtle because they affect the medium range order (MRO).[1,2] Alkali borate glasses possess a peculiar behaviour as their short range order (SRO) may also be modified. Indeed a boron coordination change, corresponding to the partial conversion of BO$_4$ tetrahedra to BO$_3$ triangles as temperature increases, has been reported by ^{11}B NMR,[3,4] Raman[5–7] and x-ray[8] and neutron diffraction[9,10] studies.

In Angell's classification,[11] liquids are separated between "strong" liquids that exhibit an Arrhenius dependence of the viscosity, η, over a broad range of temperature, and "fragile" liquids that display pronounced deviations from this behaviour. Fragility, m, reflects the ease with which the short and medium range structure can be modified by going from the solid to the liquid state across the glass transition temperature, T_g. It varies with temperature T according to the relation

$$m = \left[\frac{d \log \eta}{d \left(T_g / T \right)} \right]_{T=T_g} \tag{1}$$

Lithium borate glasses are considered to be strong glass formers, and they exhibit an increasing fragility as the alkali oxide content increases.[12,13] This fragile behaviour indicates a rapid increase in the configurational entropy above T_g, contrary to what is observed in the strong B$_2$O$_3$ liquid. However, the way in which

the structural mechanisms involved in the configurational entropy increase is not well understood, partly due to the lack of data on the structural reorganisation when going from glass to liquid.

The purpose of this paper is to better assess the structural reorganisation of lithium borate glasses with increasing temperature, as a function of their Li$_2$O content. We present new neutron diffraction data from room temperature up to the liquid state, coupled with Li isotopic substitution, in order to gain information on the lithium environment.

Experimental

Sample preparation

Glasses of composition xLi$_2$O.$(1-x)$B$_2$O$_3$, with x=0·1, 0·2, 0·25 and 0·33, were prepared from dried reagent grade powders of boron oxide and lithium carbonate, Li$_2$CO$_3$. Boron was isotopically enriched in [11]B (99·62%) to avoid the high neutron absorption cross section of the [10]B isotope. For the Li$_2$O.2B$_2$O$_3$ and Li$_2$O.3B$_2$O$_3$ composition, two glasses have been synthesised using [6]Li$_2$CO$_3$ (95·7% [6]Li) for the first glass and [7]Li$_2$CO$_3$ (99·94% [7]Li) for the second glass. The starting powders were mixed and melted for 1 h at 900°C, quenched by immersion of the bottom of the Pt crucible in water, then crushed and remelted for 15 min at 1050°C and quenched again, to ensure glass homogeneity. The composition and nomenclature for the glasses studied are listed in Table 1, along with the glass transition temperatures and the densities of the glasses expected from previous studies.[14–17]

* Corresponding author. Email cormier@impmc.jussieu.fr

Table 1. *Molar compositions, glass transition temperatures,* T_g, *and densities for the glasses investigated*

Glass	B_2O_3 (mol%)	Li_2O (mol%)	R (alkali/boron)	T_g (K)	d_{glass} (g cm^{-3})	d_{liquid} (g cm^{-3})
LB9	90	10	0·11	616[a]	1·972[b]	1·747[c]
LB4	80	20	0·25	750[d]	2·1[a]	1·916[c]
LB3	75	25	0·33	762[a]	2·19[a]	1·921[c]
LB2	66.67	33.33	0·50	773[a]	2·253[a]	1·952[c]

[a] Ref. 14, [b] Ref. 15, [c] Ref. 16, [d] Ref. 17

Neutron diffraction

Neutron diffraction experiments have been carried out on the 7C2 diffractometer at the Orphée reactor of the Laboratoire Léon Brillouin (Saclay, France). This instrument uses hot neutrons of wavelength 0·697 Å, giving access to a Q-range of 0·5–15·6 Å^{-1}. Glasses were powdered just before the measurement in order to avoid hydration, and set in a cylindrical vanadium cell, at the centre of a cylindrical vanadium furnace. Measurements were performed on the glass and melt, at room temperature and 1273 K for LB2 and LB3, and room temperature and 1173 K for LB4 and LB9. Complementary measurements of the scattering from the empty cans in the furnace at 300 K and at 1273 K, from the empty furnace at 300 K, and from the background were performed for use in the data analysis procedure. The data were corrected as previously explained.[9,18] Data reduction yields the total structure factor, $F(Q)$, which is the distinct scattering term in the total neutron differential cross section. The differential correlation function, $D(r)$, is obtained by Fourier transformation of $F(Q)$, and corresponds to the neutron weighted sum of all the partial pair distribution functions, $g_{\alpha\beta}(r)$ (the notation is as used in Ref. 19)

$$D(r) = 4\pi r \rho_0 \sum_{\alpha,\beta=1}^{n} c_\alpha c_\beta \bar{b}_\alpha \bar{b}_\beta \left[g_{\alpha\beta}(r) - 1 \right] \quad (2)$$

where c_α and $\bar{b}_\alpha$ are the atomic concentration and neutron scattering length for each element.

The isotopic substitution technique[20] involves subtracting the total structure factors measured for the samples enriched in ^{6}Li and ^{7}Li. This allows the cancellation of the identical terms (B–B, B–O and O–O) and gives only a weighted sum of the Li-centred partial structure factors (Li–O, Li–B and Li–Li). By Fourier transforming this differential structure factor, $F_{Li}(Q)$, one obtains a correlation function, $D_{Li}(r)$, which is the sum of all Li centred partial correlation functions.[21]

Results

Total structure factors and total correlation functions

The total structure factors of the investigated glasses at room temperature are shown in Figures 1 and 2. The high Q range ($Q>5$ Å^{-1}) is very similar for all glasses and describes the SRO. The low Q region

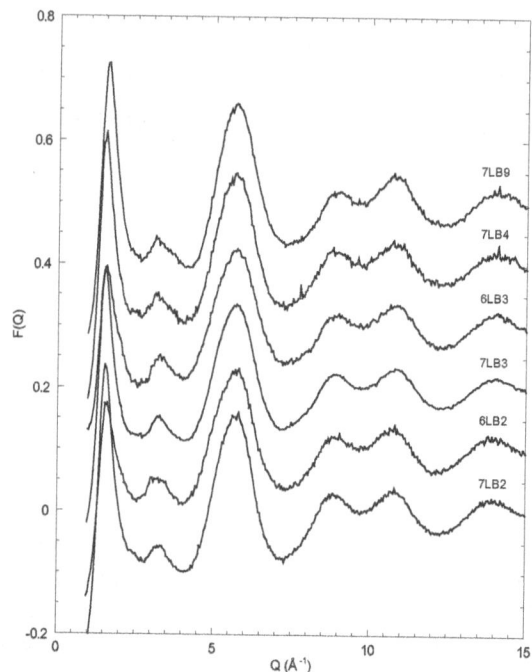

Figure 1. *Neutron total structure factors for the different lithium borate glasses. Some curves have been displaced vertically for clarity*

is mainly characterised by a sharp peak, Q_p, near 1·55 Å^{-1}, related to medium range correlations (4–20 Å), involving preferential arrangements between borate units and/or alkali ions. The intensity of peak Q_p decreases for the samples enriched in ^{6}Li, for which ^{6}Li has a positive neutron scattering length, as compared to the ^{7}Li enriched samples. This peak does

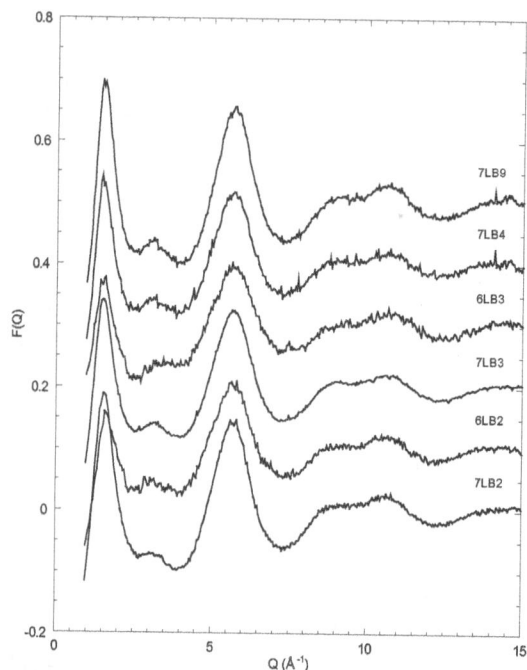

Figure 2. *Neutron total structure factors for the different lithium borate liquids at 1273 K (LB2 and LB3) and 1173 K (LB4 and LB9). Some curves have been displaced vertically for clarity*

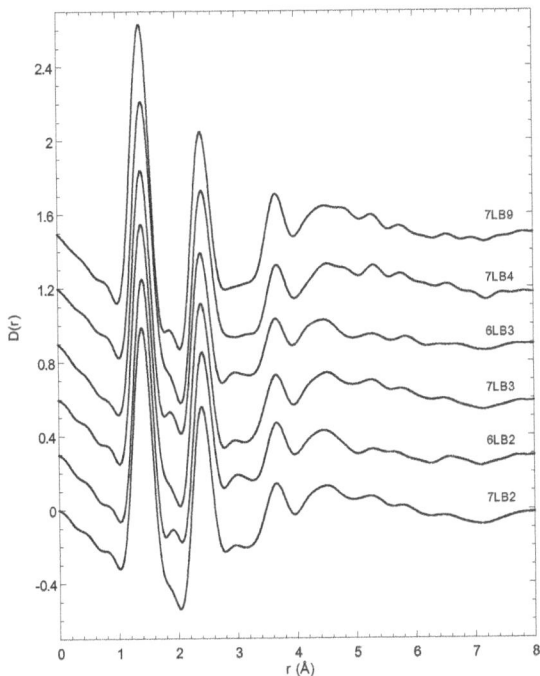

Figure 3. Neutron total correlation functions, D(r), for the different lithium borate glasses. Some curves have been displaced vertically for clarity

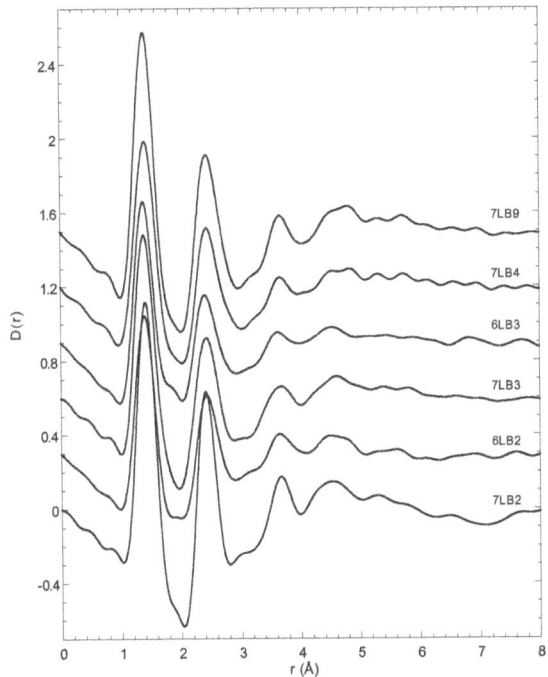

Figure 4. Neutron total correlation functions, D(r), for the different lithium borate liquids at 1273 K (LB2 and LB3) and 1173 K (LB4 and LB9). Some curves have been displaced vertically for clarity

not change significantly for the ^{7}LB2, ^{7}LB3 and ^{7}LB4 samples but it is shifted to higher Q values for ^{7}LB9 (1.53 ± 0.02 Å^{-1} for ^{7}LB2 and 1.58 ± 0.02 Å^{-1} for ^{7}LB9). This implies a structural organisation with a shorter correlation distance ($D=2\pi/Q_p$) when decreasing the alkali content. The dampening of the oscillations at high temperature (Figure 2) is due to both static and thermal disorder. The peak Q_p decreases in intensity relative to the peak at 5.7 Å^{-1}. The shift, observed for the ^{7}LB9 sample at room temperature, is no longer discernible at high temperature but its width remains smaller than for the other samples.

The differential correlation functions are plotted in Figures 3 and 4 for the glasses and melts. $D(r)$ functions are characterised by two strong contributions at 1.4 Å and 2.42 Å. The first peak is due to the

B–O pairs and the second to mainly O–O pairs and small contributions from B–B pairs. The Li–O pair is discernible at ~2 Å as a difference in intensity due to the different neutron scattering lengths for ^{6}Li and ^{7}Li isotopes. At higher r-values, the contributions can be mainly ascribed to the borate network. With increasing temperatures, the B–O contribution is shifted to lower r-values by at least ~0.01 Å, which corresponds to a change in the mean boron coordination. We have previously shown that changes in the first B–O peak appear only above the glass transition temperature,[9] in agreement with Raman data.[6] The peaks at medium range distance are reduced in intensity and the O–O peak at 2.42 Å is slightly shifted by 0.01 Å to higher r.

The quantification of trigonal and tetrahedral boron contributions has been obtained by Gaussian fitting of the first B–O peak in the total correlation functions (Table 2). Two Gaussian functions were used to fit the glass data, except for the LB9 sample where two functions cannot be justified, resulting from the low amount of BO$_4$ tetrahedra. A single Gaussian function was used at high temperature due to the lower resolution. The mean coordination and mean distance allow the determination of the fraction of boron occurring in BO$_4$ tetrahedra, N_4.[9] The accuracy of the derived parameters (mean B–O distance, R_i, Debye–Waller factor, σ, and coordination number, CN) has been carefully evaluated previously.[9] The average B–O distance determined in the liquid state decreases by about 0.01 Å compared with the mean distance in the glassy state, except for

Table 2. Gaussian parameters used to fit the B–O peak in the total correlation functions. The subscripts 1 and 2 correspond to BO$_3$ and BO$_4$ in the glassy state, respectively. R_i is the B–O distance, σ is the Debye–Waller factor, CN is the coordination number. N_4 is given by CN$_2$/(CN$_1$+CN$_2$)

Glass	R_1 ±0.01 Å	CN_1 ±0.07	σ_1 ±0.01 Å	R_2 ±0.01 Å	CN_2 ±0.07	σ_2 ±0.01 Å	N_4 ±0.02	$N_4{}^a$
LB9	1.403	3.11	0.009				0.20	0.12
LB4	1.380	2.00	0.07	1.485	1.33	0.075	0.33	0.26
LB3	1.380	1.89	0.08	1.470	1.52	0.070	0.37	0.31
LB2	1.380	1.62	0.060	1.4790	1.84	0.096	0.46	0.41
Liquid	R ±0.01 Å	CN ±0.12	σ ±0.01 Å				N_4 ±0.05	
LB9	1.405	3.05	0.105				0.10	
LB4	1.418	3.05	0.118				0.15	
LB3	1.395	3.28	0.100				0.28	
LB2	1.415	3.30	0.115				0.30	

^{a} NMR values from Ref. 23

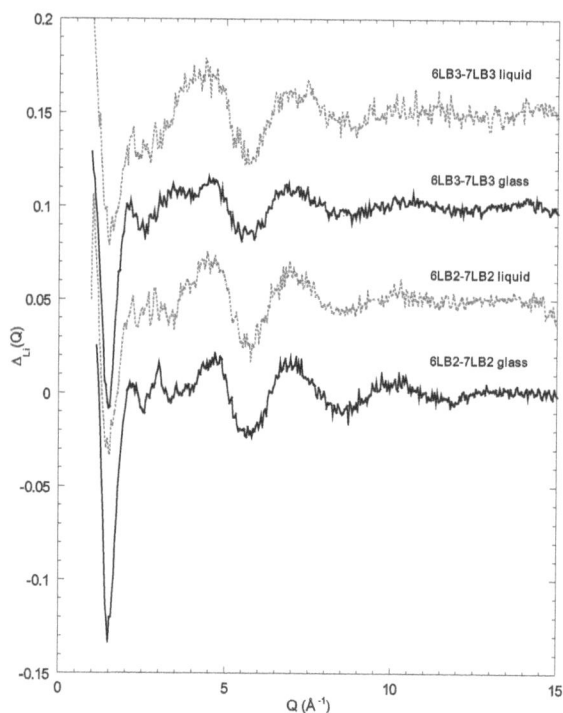

Figure 5. First difference structure factors for the LB2 and LB3 glasses and melts, giving a sum of all the Li centred partial structure factors. Some curves have been displaced vertically for clarity

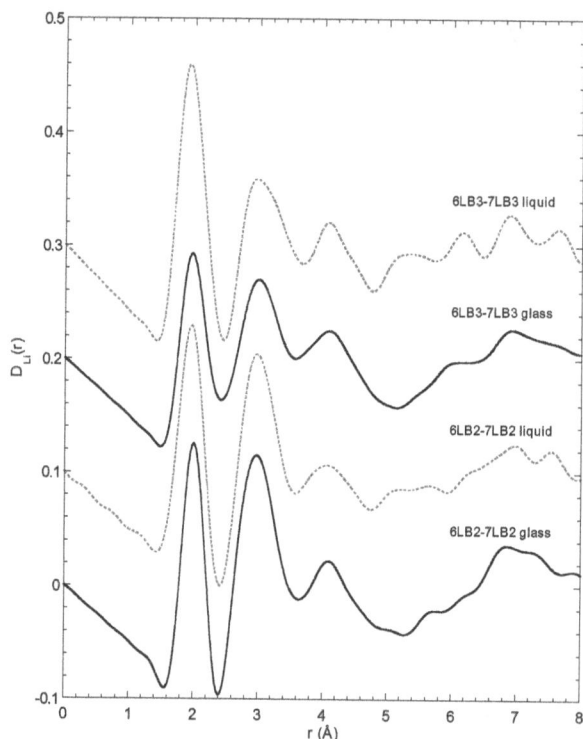

Figure 6. First difference correlation functions for the LB2 and LB3 glasses and melts, giving a sum of all the Li centred partial structure factors. Some curves have been displaced vertically for clarity

the [7]LB9 sample. The shortening of the B–O distance results from the competition between a decrease of the N_4 ratio that reduces the mean B–O distance and an increase due to the thermal expansion of the B–O bond. However, this latter effect would yield a B–O bond length increase by 0·004 Å and 0·005 Å at 1273 K, considering the mean linear expansion coefficients for the [3]B–O and the [4]B–O bonds (4×10^{-6} K^{-1} and $5·3 \times 10^{-6}$ K^{-1}, respectively[22]). Such variation is undetectable in our neutron data and explains why the [7]LB9 sample does not show variation in the mean B–O bond length, since most of boron atoms are in three-fold coordination in the glass. For the other samples, the decrease of boron coordination explains the variation of the B–O distance.

First difference neutron functions: environment around Li

The first difference functions, $F_{Li}(Q)$ and $D_{Li}(r)$, are plotted in Figures 5 and 6 for the LB2 and LB3 glasses and liquids. The real space functions, $D(r)$ and $D_{Li}(r)$, are compared in Figure 7 for the LB2 and LB3 samples, showing their respective amplitudes. These functions are the sum of all Li-centred pairs, but the $D_{Li}(Q)$ and $D_{Li}(r)$ functions are mainly dominated by Li–O and Li–B pairs, as the Li–Li weighting factor is almost negligible. Both functions are very similar for the two glasses, showing that the Li environment is similar for these compositions. Such difference functions are difficult to extract for the LB4 and LB9

samples, due to the lower Li_2O content.

The differential structure factor has oscillations at high Q values ($Q>8$ Å^{-1}) due to the short Li–O distances and exhibits a first negative sharp peak at $1·45 \pm 0·02$ Å^{-1}. This corresponds to the peak Q_p in the total structure factors (Figure 1), which indicates that the MRO of both the B–O network and the Li surrounding are related. This negative peak is responsible for the change in intensity of the peak Q_p as the Li neutron scattering length varies. The [6]Li isotope has a positive neutron scattering length ($\bar{b}=2$ fm), which reduces the intensity of the peak Q_p in the total structure factors. On the contrary, [7]Li has a negative scattering length ($\bar{b}=-2·22$ fm) and the $F_{Li}(Q)$ function then contributes to a decrease in the intensity of the peak Q_p. By comparison with the diborate ($Li_2B_4O_{10}$) crystal structure, this peak is interpreted as a result of a chemical packing of the alkalis, similar to that existing in $Li_2B_4O_{10}$ crystal.[24] The $F_{Li}(Q)$ function also shows a weak contribution between 2 and 4 Å^{-1}, which indicates strong anti-phase interferences between Li–O and Li–B contributions. This supports alternate ordering of Li and B/O atoms. At high temperature, the peak at 1·44 Å^{-1} is less intense and broader than at room temperature, indicating a smaller extension of the medium range structure. A peak around 4·3 Å^{-1} is more pronounced in the liquid state.

The first peak in the $D_{Li}(r)$ functions corresponds to the Li–O pair and the second peak at 2·95 Å is as-

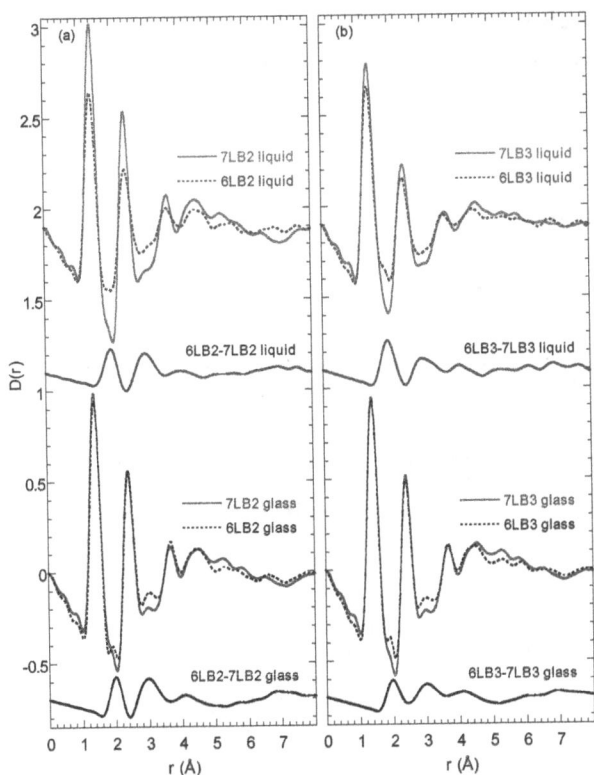

Figure 7. Comparison between the D(r) and $D_{Li}(r)$ functions for (a) the LB2 glasses and liquids and (b) the LB3 glasses and liquids

signed mainly to the first Li–B correlation. Beyond, two broad peaks are observed around 4·1 and 7 Å but they are difficult to interpret due to overlapping Li-centred pairs. The same contributions are visible in the liquid, indicating that both SRO and MRO are maintained around Li ions above T_g. The main variation upon increasing the temperature is a shift in position of the first Li–O peak by ~0·04 Å to lower r, for both the LB2 and LB3 samples.

Discussion

With the neutron diffraction data, we observe the progressive formation of BO_4 tetrahedra as the Li_2O content increases in the investigated glasses, in agreement with NMR results.[23,25] In the liquid state, there is a clear conversion of BO_4 tetrahedra into BO_3 triangles. Previous studies using Raman scattering indicate that this changes occurs in the supercooled liquid between the glass transition and the liquidus temperatures.[6] A direct quantification of the boron coordination change can be assessed by a Gaussian analysis of the first B–O peak (Table 2). The results indicate a decrease of N_4 for all compositions, with 40 to 50% less BO_4 tetrahedra in the liquids compared to the glasses. In particular, the decrease in N_4 at high temperature which we observe is not as severe as has been reported from a high temperature Raman spectroscopy study of sodium borates,[6] which has been used to determine N_4 values of less than 0·1

for melts with 25 and 30 mol% Na_2O at comparable temperatures. The temperature dependence found in our study mimics more closely the temperature and composition dependence predicted by a model of associated solutions for sodium borate melts.[26] The reason might be that the Raman peak in the region 1500–1300 cm^{-1}, used for N_4 quantification, is not directly related to the proportion of each species. However, the changes in the relative abundance of the high frequency bands, and the disappearance of bands in the mid-frequency region associated with borate rings containing BO_4 tetrahedra, are consistent with a change in boron coordination in the liquid state. *In situ* ^{11}B NMR measurements also support a coordination change, but a quantitative determination of N_4 is difficult due to averaging of the two isotropic shifts in the liquid.[3,4]

Modifications beyond the first peak are observed in the total correlation functions (Figure 4). The peak at 2·44 Å is shifted by +0·01 Å and decreases in intensity at high temperature. This peak is dominated by O–O with an additional small contribution from B–B correlations. The decrease in intensity results from the thermal disorder, and the decrease in the O–O coordination results from the $^{[4]}B$ to $^{[3]}B$ conversion. The shift in position can be due to longer B–B distances in the liquids as recently observed in an EPSR study.[27] EPSR modelling indicates that the B–B distance distribution presents an important tail at high r values in the liquids. This change can be interpreted as an opening of the borate rings, with larger mean B–Ô–B angles. This also agrees with Raman spectra which show a decreasing intensity of the mid-frequency bands in liquid borates, associated with various borate rings containing BO_4 units.[10]

A consequence of the BO_4 to BO_3 conversion is the formation of NBOs (nonbridging oxygens) above T_g, which thus depolymerise the borate network and affect melt properties. Isothermal viscosities at low temperatures are maximum near 25 mol% of alkali oxide in borate liquids,[28] which is close to the maximum of formation of BO_4 tetrahedra and thus to the highest connectivity of the network. At high temperature, the disruption of the network associated with the formation of NBOs results in the disappearance of the maximum in the viscosity isotherms, and their progressive decrease with increasing alkali content.

The formation of NBO also affects the alkali environment, which is clearly seen in the first difference functions determined for Li (Figure 6). Indeed, the $D_{Li}(r)$ function shows a shift of the Li–O distance by −0·04 Å from the glass to the liquid. Given the mean linear expansion coefficient of 16×10^{-6} K^{-1} for the Li–O bond in tetrahedra,[22] the Li–O distances should increase by ~0·02 Å at 1300 K. The observed decrease of the Li–O distances has thus been interpreted as an increasing proportion of Li–NBO bonds in the liquid, which are shorter than Li–BO (where BO denotes bridging oxygen) ones.[18] The increasing number of NBOs, resulting from the boron coordination change, indicates that the Li environment changes from a charge compensating role near BO_4 tetrahedra in the glass, to a modifying role associated with NBOs in the liquid.

The structural changes observed between glassy and liquid borates can be related to transport and thermodynamic properties, as they are involved in the configurational entropy appearing above T_g. The enthalpy of boroxol ring rupture has been estimated as 20·9 kJ mol⁻¹ in B_2O_3,[29] while the enthalpy of boron coordination change is in the range 12–37 kJ mol⁻¹, as estimated from the present neutron diffraction data. These two structural mechanisms will thus contribute equally to the heat capacity, an important thermodynamic value that is directly related to configurational changes. The heat capacity, C_p, associated with the boron coordination conversion for LB2 has been reassessed as 16 J mol⁻¹K⁻¹, considering that 0·16 moles of B are converted and normalising by the number of moles of B per mole of glass. This estimate is lower than the previous erroneous estimation of 75 J mol⁻¹K⁻¹,[9] and lower than the experimental calorimetric difference in C_p (71 J mol⁻¹K⁻¹ for LB2),[30] showing that the configuration entropy changes are due to both SRO and MRO. Modifications in the Li environment can also participate in the heat capacity change, though an estimate of this contribution is difficult. The stabilisation of boron in tetrahedral sites is energetically costly but this species could lower the energy of the glass by favouring the formation of a well defined MRO, characterised by the borate superstructual units, similar to the stabilisation of B_2O_3 glass by the presence of boroxol rings.[31]

Conclusions

This study reports the structural modifications induced by temperature between lithium borate glasses and liquids, using neutron diffraction coupled with Li isotopic substitution. We have quantitatively determined the fraction, N_4, of BO_4 tetrahedra in glasses and liquids, showing a decrease of N_4 when either the alkali content decreases or the temperature increases. Associated with the BO_4 to BO_3 conversion with temperature, the neutron data indicate a modification in the Li environment for diborate and triborate compositions, due to more NBOs around Li in the liquid state. These structural changes are important to better assess the configurational properties of melts.

Acknowledgments

We are grateful to Jonathan Stebbins for fruitful discussion regarding the calculation of the thermodynamic values. This is IPGP Contribution No. 2504.

References

1. Majérus, O., Cormier, L., Calas, G. & Beuneu, B. *Chem. Geol.*, 2004, **213**, 89.
2. Fischer, H. E., Barnes, A. C. & Salmon, P. S. *Rep. Prog. Phys.*, 2006, **69**, 233.
3. Stebbins, J. F. & Ellsworth, S. E. *J. Am. Ceram. Soc.*, 1996, **79**, 2247.
4. Sen, S. *J. Non-Cryst. Solids*, 1999, **253**, 84.
5. Akagi, R., Ohtori, N. & Umesaki, N. *J. Non-Cryst. Solids*, 2001, **293–295**, 471.
6. Yano, T., Kunimine, N., Shibata, S. & Yamane, M. *J. Non-Cryst. Solids*, 2003, **321**, 147.
7. Cormier, L., De Sousa Meneses, D., Neuville, D. R. & Echegut, P. *Phys. Chem. Glasses: Eur. J. Glass Sci. Technol. B*, 2006, **47**, 430.
8. Herms, G. & Sakowski, J. *Phys. Chem. Glasses*, 2000, **41**, 309.
9. Majérus, O., Cormier, L., Calas, G. & Beuneu, B. *Phys. Rev. B*, 2003, **67**, 024210.
10. Cormier, L., Majérus, O., Neuville, D. R. & Calas, G. *J. Am. Ceram. Soc.*, 2006, **89**, 13.
11. Angell, C. A. *J. Non-Cryst. Solids*, 1991, **131–133**, 13.
12. Chryssikos, G. D., Kamitsos, E. J. & Yainnopoulos, Y. D. *J. Non-Cryst. Solids*, 1996, **196**, 244.
13. Kodama, M. & Kojima, S. *J. Therm. Anal. Calorim.*, 2002, **69**, 961.
14. Feller, S. A., Kottke, J., Welter, J., Nijhawan, S., Boekenhauer, R., Zhang, H., Feil, D., Parameswar, C., Budhwani, K., Affatigato, M., Bhatnagar, A., Bhasin, G., Bhowmik, S., Mackenzie, J., Royle, M., Kambeyanda, S., Pandikuthira, P. & Sharma, M. In: *Borate Glasses, Crystals and Melts*, Eds. A. C. Wright, S. A. Feller and A. C. Hannon, Society of Glass Technology, Sheffield, 1997, p. 246.
15. Carini, G., Carini, G., D'Angelo, G. & Tripodo, G. *Phys. Rev. B*, 2005, **72**, 014201.
16. Shi, X. M., Wang, Q., Li, C. X., Niu, X. J., Wang, F. P. & Lu, K. Q. *J. Cryst. Growth*, 2006, **290**, 637.
17. Shelby, J. E. *J. Am. Ceram. Soc.*, 1983, **66**, 225.
18. Majérus, O., Cormier, L., Calas & G., Beuneu, B. *J. Phys. Chem. B*, 2003, **107**, 13044.
19. Keen, D. A. *J. Appl. Cryst.*, 2001, **34**, 172.
20. Cormier, L., Calas, G. & Gaskell, P. H. *Chem. Geol.*, 2001, **174**, 349.
21. Majérus, O., Cormier, L., Calas, G. & Soper, A. K. *Physica B*, 2004, **350**, 258.
22. Brown, G. E. Jr., Farges, F. & Calas, G. *Rev. Mineral.*, 1995, **32**, 317.
23. Kroeker, S., Aguiar, P.L., Cerquiera, A., Okoro, J., Clarissa, W., Doerr, J., Olesiuk, M., Ongie, G., Affatigato, M. & Feller, S. A. *Phys. Chem. Glasses: Eur. J. Glass Sci. Technol. B*, 2006, **47**, 393.
24. Gaskell, P. H. In: *Borate Glasses, Crystals and Melts*, Eds. A. C. Wright, S. A. Feller and A. C. Hannon, Society of Glass Technology, Sheffield, 1997, p. 71.
25. Zhong, J. & Bray, P.J. *J. Non-Cryst. Solids*, 1989, **111**, 67.
26. Vedishcheva, N. M., Shakhmatkin, B. A., Schultz, M. M. & Wright, A.C. In: *Borate Glasses, Crystals and Melts*, Eds. A. C. Wright, S. A. Feller and A. C. Hannon, The Society of Glass Technology, Sheffield, 1997, p. 215.
27. Cormier, L., Calas, G. & Beuneu, B. submitted.
28. Visser, T. J. M. & Stevels, J. M. *J. Non-Cryst. Solids*, 1972, **7**, 376.
29. Walrafen, G. E., Hokambadi, M. S., Krishnan, P. N., Guha, S. & Munro, R. G. *J. Chem. Phys.*, 1983, **79**, 3609.
30. Uhlmann, D. R., Kolbeck, A. G. & de Witte, D. L. *J. Non-Cryst. Solids*, 1971, **5**, 426.
31. Ferlat, G., Charpentier, T., Seitsonen, A. P., Takada, A., Lazzeri, M., Cormier, L., Calas, G., Mauri, F. *Phys. Rev. Lett.*, 2008, **101**, 65504.

Proc. VI Int. Conf. Borate Glasses, Himeji, Japan, 18–22 August 2008   *Phys. Chem. Glasses: Eur. J. Glass Sci. Technol. B, August 2009,* **50** (4), 243–248

WO₃-doped zinc borophosphate glasses

Jiří Šubčík, Ladislav Koudelka,* Petr Mošner

*Department of General and Inorganic Chemistry, Faculty of Chemical Technology, University of Pardubice, 532 10
Pardubice, Czech Republic*

Ivan Gregora

Institute of Physics, Academy of Sciences of the Czech Republic, Na Slovance 2, 18221 Prague, Czech Republic

Lionel Montagne & Bertrand Revel

*Laboratoire de Cristallochimie et Physicochemie du Solide, Ecole Nationale Superieure de Chemie de Lille, BP107
59652, Villeneuve d'Ascq cedex, France*

Manuscript received 18 August 2008
Revised version received 17 April 2009
Accepted 12 June 2009

*The effect of WO₃ on the properties and structure of zinc borophosphate glasses was investigated and the role of WO₃ in
these glasses was considered. Homogeneous glasses of the composition (100–x)[0·5ZnO.0·1B₂O₃.0·4P₂O₅].xWO₃ were
obtained within the concentration range of x=0–40 mol% WO₃, whereas glasses with x=45 and 50 contain microinclusions of β-WO₃. Their glass transition temperature increases with increasing WO₃ content, whereas the crystallisation
temperature, and the thermal stability of the glasses reveal a maximum at ~10–15 mol% WO₃. The structure of the
glasses was studied by Raman and infrared spectroscopy, combined with ³¹P and ¹¹B MAS NMR spectroscopy. The
Raman spectra are characterised by a strongly polarised band at 941–930 cm⁻¹ and a depolarised band at 832–809 cm⁻¹,
ascribed to vibrations of WO₆ and W–O–W bonds, respectively. Compositional changes in the ¹¹B MAS NMR spectra
reveal a partial conversion of BO₄ to BO₃ units with increasing WO₃ content, and the replacement of B–O–P bridges
by B–O–W bridges.*

1. Introduction

Alkali phosphate glasses are able to incorporate large amounts of tungsten oxide.[1–7] The addition of WO₃ improves the chemical resistance of alkali phosphate-based glasses against atmospheric moisture.[2,3] Tungsten oxide based materials are known for their electrochromic and photochromic properties,[8] resulting in a wide range of applications, such as smart windows, display devices and sensors. Also Poirier *et al*[9,10] have reported photochromic behaviour in tungstate–phosphate glasses, where the absorption coefficient and refractive index can be modified under laser exposure.

Tungsten oxide based glasses are mostly dark blue in colour, which is ascribed to the presence of some tungsten atoms in a lower valence as W(V).[3] Several EPR (electron paramagnetic resonance) studies have been devoted to the effect of glass composition[5] and preparation conditions[6] on the proportion of W⁵⁺ ions in the glasses. In some WO₃-containing glasses, a broad EPR signal from W⁵⁺ ions at about 3800 G (g=1·70) is accompanied by a narrow signal at 3400 G (g=1·95) due to Mo⁵⁺ ions present in WO₃ as an impurity.[3,5,7]

Raman and IR spectroscopy,[2,6,7,11] NMR spectroscopy,[2] and x-ray absorption spectroscopy[11,12] have been applied for structural studies of alkali phosphate glasses. XANES (x-ray absorption near edge structure) studies of local order in NaPO₃–BaF₂–WO₃ glasses showed that tungsten atoms are surrounded by a distorted octahedron of oxygen atoms which, for high WO₃ concentrations, form clusters of WO₆ octahedra with W–O–W bridges, as identified by Raman spectra.[11,12] Nevertheless, some authors suggested the formation of WO₄ tetrahedra,[13,14] or mixed coordination of WO₄ and WO₆ structural units.[15,16]

This paper is devoted to the study of the formulation of zinc borophosphate glasses with tungsten oxide WO₃, and of the effect of WO₃ on their structure and properties. The interaction of WO₃ with zinc borophosphate was investigated in the compositional series of (100–*x*)[0·5ZnO.0·1B₂O₃.0·4P₂O₅].*x*WO₃. We have studied glass formation, basic physical properties and thermal behaviour of these glasses. For structural studies we have applied Raman, infrared and ³¹P and ¹¹B MAS NMR spectroscopy.

2. Experimental

The investigated glasses were prepared from reagent grade ZnO, WO₃, H₃BO₃ and H₃PO₄ (85%) in batches of 20 g. First the reaction mixture was heated slowly in a platinum crucible up to 600°C, with final calcination at the maximum temperature for 2 h to remove

* Corresponding author. Email ladislav.koudelka@upce.cz

water. After calcination, the reaction mixture was slowly heated up to 1260°C. After 20 min heating at the maximum temperature, the obtained melt was cooled by pouring into a graphite mould of 30×30 mm^2 dimensions to form a suitable glass block. The obtained glasses were separately annealed for 30 min at a temperature close to their T_g, and then slowly cooled to room temperature.

The glass transition temperatures, T_g, and crystallisation temperatures, T_c, of the glasses were obtained from DTA curves (Netzsch DTA 404 PC) at a heating rate of 10 K min^{-1}. The thermal expansion coefficient, a, was obtained from the dilatation curves between 150–250°C. The dilatation measurements were carried out on glass cubes of 5×5×5 mm^3, at a heating rate of 5 K min^{-1}.

The glass density, ρ, was determined at 25°C by Archimedes' method, using toluene as the immersion liquid. The molar volume, V_M, was calculated as $V_M = \overline{M}/\rho$, where $\overline{M}$ is the average molar weight of the glass composition.

The chemical durability of the glasses was evaluated from their dissolution rate in distilled water at 25°C. Glass cubes of 5×5×5 mm^3 were immersed in 100 cm^3 of water for 10 h. Before and after the dissolution test, the samples were dried at 105°C. The dissolution rate (DR) was calculated from the expression $DR = \Delta\omega/St$, where $\Delta\omega$ is the weight loss (g), S is the sample area (cm^2) before the dissolution test, and t is the dissolution time (min).

The Raman spectra were measured on bulk samples at room temperature, using a Renishaw RM 1000 Raman microscope. The spectra were recorded in back scattering geometry under excitation with Ar-laser radiation (514·5 nm) at a power of 5 mW. The spectral slit width was 1·5 cm^{-1} and the total integration time was 100 s.

IR spectra were recorded at spectral resolution of 2 cm^{-1}, using a Nicolet Protege 460 FT-IR spectrometer in the range of 400–4000 cm^{-1} taking 32 scans. For the measurement, the powdered samples were mixed and homogenised with spectroscopically pure KBr, and pressed into pellets.

^{31}P MAS NMR spectra were measured using a Bruker Avance 400 spectrometer with a 4 mm probe. The spinning speed was 12·5 kHz. The pulse length was 1·2 µs ($\pi/4$), and the recycle delay was 60 s, which was sufficient to enable relaxation at this field strength. ^{11}B MAS NMR spectra were measured on a Bruker Avance 400 spectrometer with a 4 mm probe. The spinning speed was 12·5 kHz. A rotor synchronised echo was used with selective pulse lengths of 20 µs and 40 µs for the first and second pulses respectively, with 10 s recycling delay. The Larmor frequencies were 162·3 MHz and 128·4 MHz for ^{31}P and ^{11}B, respectively. The chemical shifts of ^{11}B nuclei are given relative to BPO$_4$ at −3·6 ppm, those of ^{31}P are relative to 85% H$_3$PO$_4$ at 0 ppm.

Figure 1. Microinclusions of crystals of orthorhombic β-WO$_3$, observed by electron microscopy in the 50[0·5ZnO.0·1B$_2$O$_3$.0·4P$_2$O$_5$].50WO$_3$ glass

Results

By slow cooling of the melt in air, we have obtained glassy samples of the (100−x)[0·5ZnO.0·1B$_2$O$_3$.0·4P$_2$O$_5$]. xWO$_3$ series within the concentration region of x=0–50 mol% WO$_3$, which were of blue colour and had x-ray diffraction diagrams characteristic of glasses. Nevertheless, under microscopic examination we observed microinclusions in the samples with 45 and 50 mol% WO$_3$ (see Figure 1). Their size was smaller than 20 µm and gave no diffraction lines under x-ray diffraction analysis.

The density and molar volume values for the studied glasses are given in Table 1. The density of the glasses steadily increases with increasing WO$_3$ content from 3·13 g/cm^3 up to 5·02 g/cm^3 within the studied concentration region. On the other hand, the molar volume remains nearly constant whatever the composition. All the prepared glasses reveal relatively good chemical durability, according to the dissolution rate, DR, in distilled water. DR does not change drastically within the studied concentration region, and its values vary between 3·3–14·3 g/cm^2 min (see Table 1).

Table 1. Composition, density (ρ), molar volume (V_M), dissolution rate (DR), crystallisation temperature (T_c) and thermal stability (T_c–T_g) of (100−x) [0·5ZnO.0·1B$_2$O$_3$.0·4P$_2$O$_5$].xWO$_3$ glasses

x(WO$_3$) [mol%]	$\rho\pm0.05$ [g/cm^3]	$V_M\pm0.5$ [cm^3/mol]	DR×10$^7\pm0.5$ [g/cm^2 min]	$T_g\pm2$ [°C]	$T_c\pm5$ [°C]	T_c–T_g [°C]
0	3·13	33·4	7·3	481	627	146
2	3·20	33·4	5·3	488	640	152
5	3·33	33·3	4·3	500	666	166
10	3·54	33·1	3·3	508	690	182
15	3·75	33·0	8·2	521	702	181
20	3·93	33·0	11·5	534	702	168
25	4·10	33·3	11·0	544	679	135
30	4·31	33·1	10·1	548	687	139
35	4·48	33·3	12·2	559	680	121
40	4·68	33·2	13·3	563	673	110
45	4·87	33·3	14·3	569	636	67
50	5·02	33·5	15·1	571	629	58

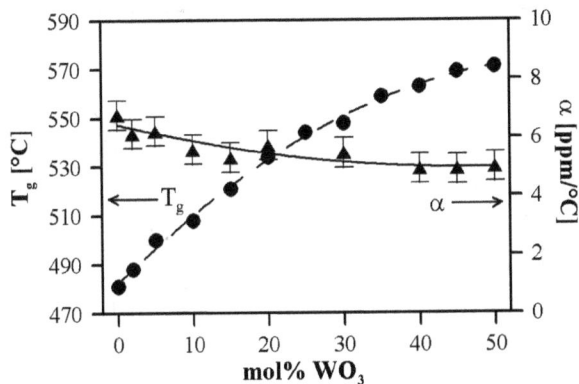

Figure 2. Compositional dependences of the glass transition temperature (circles), T_g, and the thermal expansion coefficient (triangles), α, for the glass series $(100-x)$ $[0.5ZnO.0.1B_2O_3.0.4P_2O_5].xWO_3$. The error in the values of T_g is within the point symbol. The lines are only a guide for the eye

Thermoanalytical studies yielded glass transition temperatures, T_g (see Figure 2), which increase by ~90°C as the WO_3 content is increased. On the other hand, the thermal expansion coefficient, a, slightly decreases with increasing WO_3 content (see Figure 2). On heating, the glasses reveal mostly two crystallisation peaks in their thermoanalytical curves. The values of the crystallisation temperature, T_c, taken as the onset of the crystallisation peak, are given in Table 1. The thermal stability of the glasses, evaluated as T_c-T_g (see Table 1), reveals a maximum at the glass composition containing 10–15 mol% WO_3. This maximum arises mainly from the high crystallisation temperature of the glasses in this compositional range.

We have annealed glassy samples at temperatures above their second crystallisation peak, and used x-ray diffraction analysis on powder samples to identify the crystallisation products. The results are summarised in Table 2.

As can be seen in Table 2, glasses with a low WO_3 content give rise mostly to zinc diphosphate, $Zn_2P_2O_7$,

Table 2. Compounds identified on x-ray diffraction diagrams of products of crystallisation of the studied $(100-x)[0.5ZnO.0.1B_2O_3.0.4P_2O_5].xWO_3$ glasses, annealed at the given temperature

$x(WO_3)$ [mol%]	T_{anneal} [°C]	Crystallisation products
0	670	$Zn_2P_2O_7$, BPO_4, $Zn_3(PO_4)_2$
	750	$Zn_2P_2O_7$, BPO_4
5	750	$Zn_2P_2O_7$, BPO_4
10	750	$Zn_2P_2O_7$, BPO_4
15	750	$Zn_2P_2O_7$, $W_{18}O_{49}$
20	750	$Zn_2P_2O_7$, d-WO_3, $W_{12}PO_{38.5}$, BPO_4
25	750	d-WO_3, $Zn_2P_2O_7$, $W_{12}PO_{38.5}$
30	750	$W_{12}PO_{38.5}$, PW_8O_{26}, $Zn_2P_2O_7$
35	750	PW_8O_{26}, $W_{12}PO_{38.5}$, $Zn_2P_2O_7$, $Zn_{0.06}WO_3$
40	690	$Zn_{0.3}WO_3$
	750	PW_8O_{26}, $W_{12}PO_{38.5}$, $Zn_2P_2O_7$
45	750	d-WO_3, $W_{12}PO_{38.5}$, $Zn_2P_2O_7$
50	690	PW_8O_{26}, $Zn_2P_2O_7$

and boron phosphate, BPO_4. Diffractograms of some samples also contained weak diffraction lines of zinc orthophosphate, $Zn_3(PO_4)_2$.

Diffraction patterns of samples with $x>10$ mol% WO_3 contain also diffraction lines of tungsten compounds. Besides regular tungsten oxide δ-WO_3, a reduced form of tungsten oxide $W_{18}O_{49}$ was also identified among the crystallisation products; this form usually appears in the process of production of metallic tungsten from WO_3,[17] and sometimes is also given by the formula $WO_{2.72}$. It is possible that the formation of this deficient oxide is supported by the presence of W^{5+} ions in the glasses. The presence of W^{5+} ions is indicated by the blue colour of the glasses and electron spin resonance measurements showing the characteristic signal of W^{5+} ions.[5,6]

Glasses with $x>20$ mol% WO_3 gave crystalline products containing also the compound $W_{12}PO_{38.5}$, which corresponds to the stoichiometric ratio of $1P_2O_5/24WO_3$. This compound can be prepared by thermal decomposition of $H_3[PW_{12}O_{40}].6H_2O$.[18] The structure of this phosphotungstate acid consists of twelve WO_6 octahedra in four tritungstate groups in which WO_6 octahedra share their edges, and belongs to Keggin-type compounds.[19] Another polytungstate compound PW_8O_{26} was identified in samples with $x>30$ mol% WO_3. This compound can be also prepared by thermal decomposition of phosphotungstate acid.[18]

In some samples we tried to study the difference in crystallisation products after the first and the second crystallisation peak. In the sample with 30 mol% WO_3, after the first crystallisation at 690°C, the only product was cubic $Zn_{0.06}WO_3$ (prepared recently[20] by an electrochemical insertion of Zn into monoclinic WO_3), but after the second crystallisation at ~750°C, thermodynamically more stable phases $W_{12}PO_{38.5}$ and PW_8O_{26} were formed.

Raman spectra of the glass series $(100-x)$ $[0.5ZnO.0.1B_2O_3.0.4P_2O_5].xWO_3$ are shown in Figure 3. The spectrum of the parent $0.5ZnO.0.1B_2O_3.0.4P_2O_5$ borophosphate glass shows one broad band in the high frequency region, with a maximum at 1165 cm^{-1}, which is ascribed to the symmetric stretching vibration of nonbridging oxygen atoms in diphosphate units,[21] and two medium bands in the middle frequency region at 659 and 749 cm^{-1}, which are ascribed to the vibrations of oxygen atoms in P–O–P bridges between metaphosphate (Q^2) and diphosphate (Q^1) units, respectively.[21] The band at 659 cm^{-1} vanishes within the concentration region of $x=0$–5 mol% WO_3. The band at 749 cm^{-1}, characteristic of the vibrations of bridging oxygen between diphosphate (Q^1) units gradually decreases within the concentration region of $x=0$–20 mol% WO_3. The dominant band of stretching vibrations of PO_4 units shifts steadily from 1165 cm^{-1} to lower wavenumbers, and its relative intensity decreases in comparison with strong bands appearing within the wavenumber range 800–1000 cm^{-1}. A new

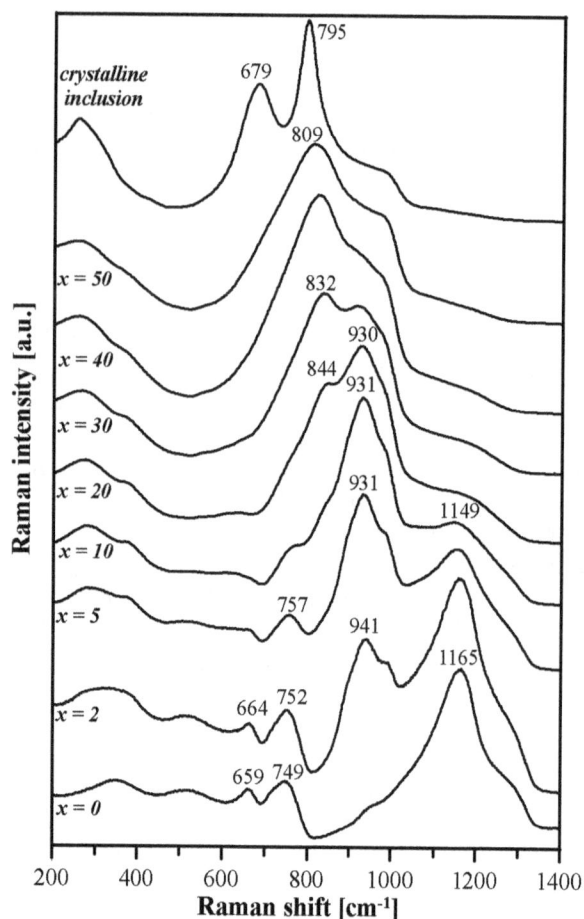

Figure 3. Raman spectra of (100–x)[0·5ZnO. 0·1B₂O₃.0·4P₂O₅].xWO₃ glasses. The Raman spectrum of microinclusions (orthorhombic β-WO₃) taken from the glass with 50 mol% WO₃ is given at the top

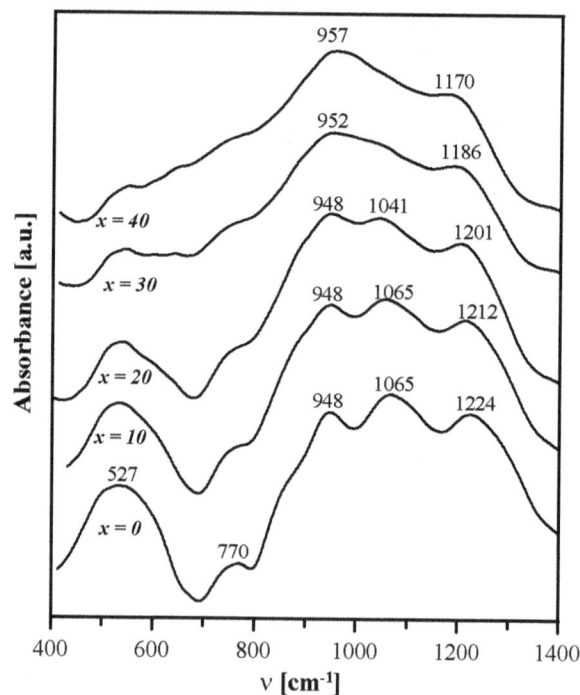

Figure 4. Infrared spectra of (100–x)[0·5ZnO. 0·1B₂O₃.0·4P₂O₅].xWO₃ glasses

strong broad band, peaking at 941–930 cm⁻¹, appears in the Raman spectrum of the glass with 2 mol% WO₃, the maximum of which shifts to lower wavenumbers with increasing WO₃ content. In the glasses with $x>20$ mol% WO₃, another strong broad band appears at 844 cm⁻¹, the intensity of which increases with increasing WO₃ content, and it dominates the Raman spectra of the glasses with $x\geq30$ mol% WO₃. Its maximum shifts to lower wavenumbers with increasing WO₃ content.

Infrared spectra of (100–x)[0·5ZnO.0·1B₂O₃.0·4P₂O₅]. xWO₃ glasses are shown in Figure 4. Infrared spectrum of the parent glass, 0·5ZnO.0·1B₂O₃.0·4P₂O₅, contains three vibrational bands in the high frequency region at 1224, 1065 and 948 cm⁻¹. The band at 948 cm⁻¹ can be ascribed to asymmetric stretching vibration of bridging oxygen atoms in P–O–P bridges, ν_{as}(P–O–P), the band at 1065 cm⁻¹ to the vibrations of diphosphate units, ν_{as}(PO₃), and the band at 1224 cm⁻¹ to the asymmetric stretching vibrations of nonbridging oxygen atoms in metaphosphate units, ν_{as}(PO₂)[22]. The medium band at 770 cm⁻¹ can be ascribed to the symmetric stretching vibration of bridging oxygen atoms in P–O–P bridges, ν_s(P–O–P).[23] The broad band at 527 cm⁻¹ lies within the region characteristic

of deformation vibrations of PO₄ units.[23] With increasing WO₃ content, the high frequency part of the spectrum gradually evolves in such a way that the ν_{as}(PO₂) band shifts to lower wavenumbers, which indicates the phosphate network depolymerisation. With increasing WO₃ content, vibrational bands broaden and develop gradually into two dominant bands at 957 and 1170 cm⁻¹ in the infrared spectrum

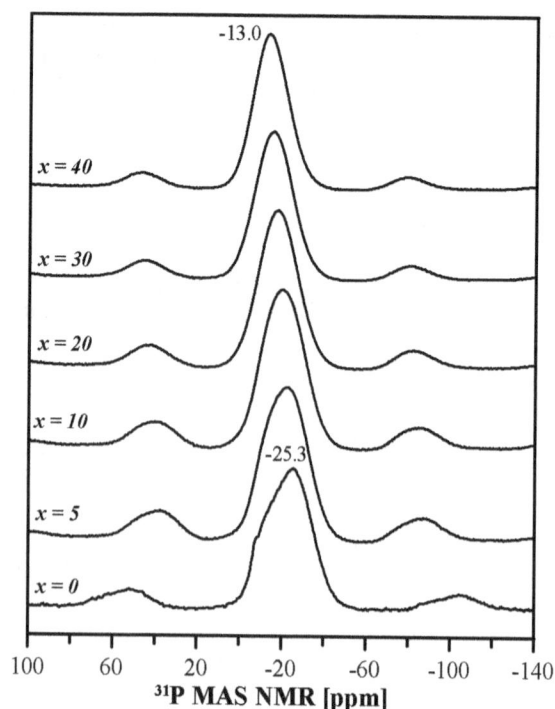

Figure 5. ³¹P MAS NMR spectra of (100–x)[0·5ZnO. 0·1B₂O₃.0·4P₂O₅].xWO₃ glasses

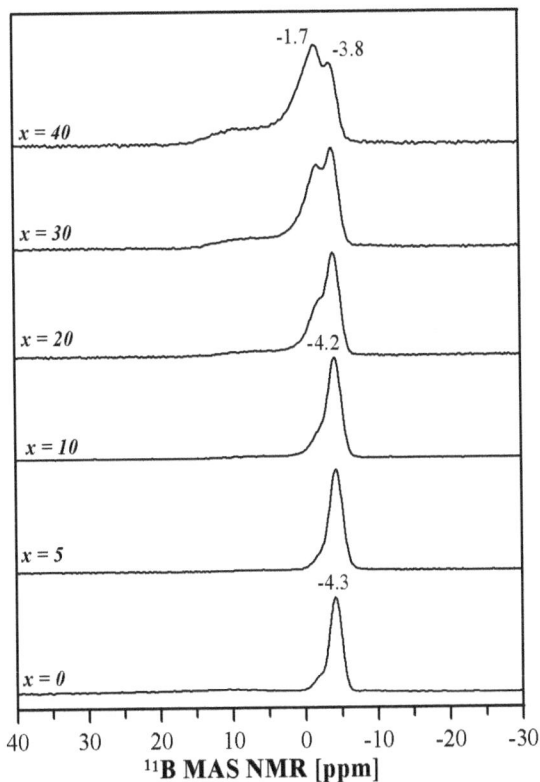

Figure 6. ^{11}B MAS NMR spectra of $(100-x)[0\cdot5ZnO.$ $0\cdot1B_2O_3.0\cdot4P_2O_5].xWO_3$ glasses

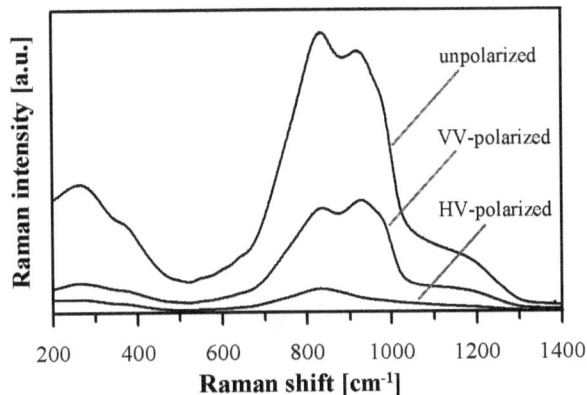

Figure 7. Polarised Raman spectra of the $70[0\cdot5ZnO.$ $0\cdot1B_2O_3.0\cdot4P_2O_5].30WO_3$ glass

of the glass containing 40 mol% WO_3. They are assigned to the stretching vibration of W–O bonds in WO_6 octahedra.[6,7]

The ^{31}P MAS NMR spectra of the $(100-x)$ $[0\cdot5ZnO.0\cdot1B_2O_3.0\cdot4P_2O_5].xWO_3$ glasses (Figure 5) show two overlapping signals at −25·3 and −16 (shoulder) ppm for the base glass, and their shape evolves with increasing WO_3 content into one band peaking at δ=−13 ppm. These changes in the NMR spectra are associated with the depolymerisation of the borophosphate network, and the formation of diphosphate units (Q^1 sites) bonded to WO_6 polyhedra.

The ^{11}B MAS NMR spectra of the $(100-x)$ $[0\cdot5ZnO.0\cdot1B_2O_3.0\cdot4P_2O_5].xWO_3$ glasses are shown in Figure 6. The spectrum for the base glass reveals one symmetrical signal at −4·3 ppm, characteristic of BO_4 units coordinated by four phosphorus atoms.[24] With increasing WO_3 content, the structure of the NMR signal changes very significantly, with a further BO_4 resonance appearing in the region of −(1·7–2·0) ppm.

Discussion

Tungsten oxide has a relatively high solubility in zinc borophosphate glasses from the series $(100-x)$ $[0\cdot5ZnO.0\cdot1B_2O_3.0\cdot4P_2O_5].xWO_3$, and homogeneous bulk glasses containing up to 40 mol% WO_3 can be obtained. The incorporation of tungsten oxide into the glass structure is especially strongly manifested in the Raman spectra and ^{11}B MAS NMR spectra. In the Raman spectra there are two strong bands, peaking

at 941–930 cm^{-1} and 832–809 cm^{-1}, which increase in strength with increasing WO_3 content, and can be ascribed to the vibrations of tungstate structural units. The first band is similar to that observed for lithium tungsten phosphate glasses by Boudlich et al,[7] who ascribed it to the W–O stretching vibration in WO_6 units, and for tungstate fluorophosphate glasses by Poirier et al,[11] who ascribed it also to the vibrations of W–O$^-$ or W=O terminal bonds in WO_6 octahedra. In the Raman spectra of glasses with a low WO_3 content, this band has a shoulder at 982 cm^{-1}, the position of which is close to the dominant band at 968 cm^{-1} in crystalline $Li_2W_2O_7$,[25] which contains edge sharing WO_6 octahedra. The second band appearing in the Raman spectra of glasses with $x>20$ mol% WO_3 is characteristic of stretching vibrations of W–O–W bridges between WO_6 octahedra,[2,11] and points to a larger number of W–O–W bridges in the glasses with a higher WO_3 content. Tungstate polyhedra have a high Raman scattering cross section, and thus in the WO_3-rich glasses Raman bands characteristic of vibrations of W–O bonds dominate the Raman spectra and suppress the bands characteristic of vibrations of phosphate units.

Polarised Raman spectra measurements on the glass with 30 mol% WO_3 showed that the character of the two dominant vibrations is different (see Figure 7). The band at 930 cm^{-1} is polarised, i.e. it should be ascribed to a totally symmetric stretching vibration, whereas the band at 832 cm^{-1} is depolarised. Therefore we conclude that the band at 930 cm^{-1} should be ascribed to the symmetric stretching vibration of WO_6 octahedra or tetrahedra, rather than to the vibration of W–O or W=O bonds only.

Raman spectroscopy helped also with the identification of microcrystalline inclusions observed under the microscope in the samples with x=45 and 50 mol% WO_3 (see Figure 1), despite the fact that x-ray diffraction analysis provided a diffraction pattern characteristic of an amorphous material. Microscopic EDX (energy dispersive x-ray spectroscopy) analysis indicated the presence of tungsten and oxygen in these microinclusions, and the Raman spectrum of these microinclusions (see Figure 3) was compared with the Raman spectra of WO_3 modifications.[26] The obtained Raman spectrum is similar to that of the high temperature modification of orthorhombic β-WO_3. As we used γ-WO_3 for the preparation of the glass, it means that these inclusions are formed during cooling of the melt, at temperatures high above T_g of these glasses.

The depolymerisation of phosphate chains is manifested not only in the Raman spectra, by the shift of the dominant band of stretching vibrations of PO_4 units steadily from 1165 cm^{-1} to lower wavenumbers, but also in the infrared spectra, by the decreasing intensity of the symmetrical stretching vibration of bridging oxygen atoms in P–O–P bridges ν_s(P–O–P),[23] due to a decrease of the number of P–O–P bridges, and their replacement by P–O–W bridges, as the WO_3 content increases.

In the ^{11}B MAS NMR spectra the formation of W–O–B bonds is reflected by the second resonance in the region of $-(1\cdot7$–$2\cdot0)$ ppm, characteristic of BO_4 units.[24] As its strength increases with increasing WO_3 content, we ascribe it to the formation of mixed $B(OP)_{4-x}(OW)_x$ units. In the NMR spectra of WO_3-rich glasses, a broad resonance characteristic of BO_3 units appears in the region of positive chemical shift,[27] which reveals a change in the coordination of some boron atoms from tetrahedral in BO_4 units, to three-coordinated in BO_3 units. Thus information on boron structural units can only be obtained from the NMR spectra, because Raman scattering of borate units is much weaker than phosphate units, and their vibrational band can only be observed in the IR spectra for borophosphate glasses with a high B_2O_3 content.[28]

Conclusions

A broad glass forming region was observed in the system $(100-x)[0\cdot5ZnO.0\cdot1B_2O_3.0\cdot4P_2O_5].xWO_3$ for x=0–50 mol% WO_3, but glasses with 45–50 mol% WO_3 contain microinclusions of β-WO_3 crystallites which are not detected by x-ray diffraction. The chemical durability and thermal stability of the glasses is very good. The highest thermal stability, evaluated as the difference between crystallisation temperature and glass transition temperature, is for glasses containing 10–15 mol% WO_3. According to the studies of Raman spectra, tungsten oxide incorporates into the structural network most probably in the form of WO_6 octahedra. These octahedra form clusters via W–O–W bonds, the number of which increases with increasing WO_3 content. The incorporation of WO_6 octahedra into the structural network is associated with the depolymerisation of phosphate chains, and

the formation of diphosphate and orthophosphate groups, and with the formation of P–O–W and also B–O–W bonds, as revealed from NMR spectra.

Acknowledgements

The Czech authors are grateful for financial support from research project No. 0021627501 of the Ministry of Education of the Czech Republic, grant No. 104/07/0315 of the Grant Agency of the Czech Republic, and research project KAN301370701 of the Academy of Sciences of the Czech Republic. LM thanks the Feder, the USTL, the Région Nord Pas de Calais and the CNRS for funding of NMR spectrometers.

References

1. Bynton, P. L., Rawson, H. & Stanworth, J. E. *Nature*, 1956, **178**, 910.
2. de Araujo, C. C., Strojek, W., Zhang, L., Eckert, H., Poirier, G., Riberio, S. J. L. & Messaddeq, Y. *J. Mater. Chem.*, 2006, **16**, 3277.
3. Studer, F., Rih, H. & Raveau, B. *J. Non-Cryst. Solids*, 1988, **107**, 101.
4. Chowdari, B. V. R., Tan, K. L. & Chia, W. T. *Mater. Res. Soc. Symp. Proc.*, 1993, **293**, 325.
5. Bih, L., Abbas, L., Azrour, M., Amraoui, Y. L. & Nadiri, A. *J. Therm. Anal. Cal.*, 2005, **81**, 57.
6. Poirier, G., Poulain, M., Messaddeq, Y. & Ribeiro, S. J. L. *J. Non-Cryst. Solids*, 2005, **351**, 293.
7. Boudlich, D., Bih, L., Archidi, M. El H., Haddad, M., Yacoubi, A., Nadiri, A. & Elouadi, B. *J. Am. Ceram. Soc.*, 2002, **85**, 623.
8. Granqvist, C.G. *Sol. Energy Mater. Sol. Cells*, 2000, **60**, 201.
9. Poirier, G., Nalin, M., Cescato, L., Messaddeq, Y. & Ribeiro, S. J. L. *J. Chem. Phys.*, 2006, **125**, 161101.
10. Poirier, G., Nalin, M., Messaddeq, Y. & Ribeiro, S. J. L. *Solid State Ionics*, 2007, **178**, 871.
11. Poirier, G., Messaddeq, Y., Ribeiro, S. J. L. & Poulain, M. *J. Solid State Chem.*, 2007, **178**, 1553.
12. Poirer, G., Cassanjes, F.C., Messaddeq, Y., Ribeiro, S. J. L., Michalowicz, A. & Poulain, M. *J. Non-Cryst. Solids*, 2005, **351**, 3644.
13. Shaltout, I., Tang, Y., Braunstein, R. & Shaisha, E. E. *J. Phys. Chem. Solids*, 1996, **57**, 1223.
14. Shaltout, I., Tang, Y., Braunstein, R. & Abu-Elazm, A. M. *J. Phys. Chem. Solids*, 1995, **56**, 141.
15. Tatsumisago, M., Kowada, Y., Minami, T. & Adachi, H. *Phys. Chem. Glasses*, 1994, **35**, 89.
16. Chowdari, B. V. R. & Kumari, P. *Mater. Res. Bull.*, 1999, **34**, 327.
17. Haubner, R., Schubert, W. D., Lassner, E., Schreiner, M. & Lux, B. *J. Refract. Hard Metals*, 1983, **2**, 108.
18. Varfolomeev, M. B., Burljaev, V. V., Toporenskaja, T. A., Lunk, H. J., Wilde, W. & Hilmer, W. *Z. Anorg. Allg. Chem.*, 1981, **472**, 185.
19. Keggin, J. F. *Proc. R. Soc.*, 1934, **A144**, 75.
20. Martinez-de la Cruz, A., Torres-Martinez, L., Garcia-Alvarado, F., Moran, E. & Alario-Franco, M. *J. Mater. Chem.*, 1998, **8**, 1805.
21 Fawcett, V., Long, D. A. & Taylor, L. H. *Proc. Fifth Int. Conf. Raman Spectrosc.*, Freiburg 1976, Hans Ferdinand Schulz Verlag, Freiburg/Br., 1976, p. 112.
22. Bobovich, Ya. S. *Opt. Spektrosk.*, 1962, **13**, 459.
23. Nyquist R. A., Putzig C. L. & Leugers, M. A. *The Handbook of Infrared and Raman Spectra of Inorganic Compounds and Organic Salts*, Vol 1, Academic Press, San Diego 1997, p.25.
24. Brow, R. K. & Tallant, D. R. *J. Non-Cryst. Solids*, 1997, **222**, 396.
25. Sekiya, T., Mochida, N. & Ogawa, S. *J. Non-Cryst. Solids*, 1994, **176**, 105.
27. Ducel, J. F., Videau, J. J., Suh, K. S. & Senegas, J. *Phys. Chem. Glasses* 1994, **35**, 10.
28. Koudelka, L., Mošner, P., Zeyer, M. & Jäger, C. *Phys. Chem. Glasses* 2002, **43C**, 102.

Proc. VI Int. Conf. Borate Glasses, Himeji, Japan, 18–22 August 2008   *Phys. Chem. Glasses: Eur. J. Glass Sci. Technol. B*, August 2009, **50** (4), 249–252

Caesium volatilisation in borosilicate glasses: a multinuclear magnetic resonance study

Vladimir K. Michaelis & Scott Kroeker*

Department of Chemistry, University of Manitoba, Winnipeg, Manitoba, R3T 2N2, Canada

Manuscript received 7 November 2008
Revised version received 17 June 2009
Accepted 29 June 2009

The evaporation of Cs from caesium borosilicate glasses with variable melting times has been quantified using inductively coupled plasma optical emission spectroscopy and [133]Cs NMR spin echo intensities. Caesium is shown to be lost as an oxide or in elemental form, not as a borate or silicate compound. The composition change is associated with a decrease in the fraction of four-coordinated boron, as measured by ^{11}B MAS NMR, and regular changes in the [133]Cs NMR peak position. Distinct behaviour of the two tetrahedral boron NMR peaks suggests that some degree of phase separation may also occur after long heating times. These results on a simplified model for a nuclear waste glass show that NMR can play a role in determining structural changes due to elemental volatility, and provide valuable information on which solutions to this problem may be based.

Introduction

Borosilicate glass constitutes the base network glass of materials used to immobilise high level liquid radioactive waste in many countries. Due to the critical environmental and health considerations demanded of such materials over very long periods of time, such glasses have attracted intense and sustained research interest. In addition to extensive engineering tests, basic science has played an important role in the development of nuclear waste glasses. For such laboratory based studies to be effective, it is often necessary to simplify the complex glasses used in industry.

Sodium borosilicate glasses have been extensively studied by many techniques due to their applications as laboratory and kitchen glassware, for example. [11]B nuclear magnetic resonance (NMR) spectroscopy has contributed to the well defined and detailed models of structural evolution with composition currently in use, with the most comprehensive articulation found in the "Dell & Bray" model of Ref. 1. While some of the details of this model have recently come under reconsideration,[2–4] the main features involve exclusive formation of four-coordinated boron with sodium loading up to a critical composition, determined by the ratio of SiO_2/B_2O_3, above which depolymerisation of the silicate units occurs, followed by depolymerisation of trigonal boron.

More relevant to nuclear waste glasses is a borosilicate glass containing caesium, due to its incorporation as a long lived fission product ubiquitous in high level liquid waste streams from fuel reprocessing. Although it was once thought that all alkali metals behaved similarly as modifiers incorporated into borate[5] and borosilicate glasses, it has been shown that significant differences exist between light and heavy alkali cations at high alkali loadings.[6,7] Hence,

our recent work has sought to define the behaviour of different alkalis in borate based network glasses.[7,8]

From the standpoint of processing properties, caesium volatility at the high processing temperatures required to achieve sufficient viscosity is a known problem that creates auxiliary decommissioning problems. Recent work has shown that many elements evaporate at typical vitrification temperatures,[9] and it has been proposed that elemental loss occurs in the form of borate phases.[10,11] The associated structural changes have been probed by NMR and Raman spectroscopy in a mixed alkali borosilicate nuclear waste model glass,[3,9,12] and include a reduction in both four-coordinated boron and silicon Q^3 units, which are expected to have an impact on the durability of such materials. Finding ways to reduce caesium volatility and maintain the integrity of the final waste product will be aided by detailed studies of such microstructural changes and reliable methods of study.

In this paper, we quantify compositional and structural changes that occur within a caesium borosilicate glass as a function of heating time. In addition to the common ^{11}B MAS NMR method, we find [133]Cs MAS NMR to be a useful probe of these changes.

Experimental procedures

Sample preparation

A caesium borosilicate glass of nominal composition (in mol%) $14\cdot3Cs_2O$–$28\cdot6B_2O_3$–$57\cdot1SiO_2$ ($K=SiO_2/B_2O_3=2$ and $R=Cs_2O/B_2O_3=0\cdot5$) was prepared from B_2O_3 (dehydrated from commercially available H_3BO_3), SiO_2 and Cs_2CO_3. Appropriate amounts were measured and ground in an agate mortar and pestle for five minutes to ensure homogenous mixing. The sample was decarbonated in a box furnace at 600°C for three hours; complete CO_2 loss was verified by weight loss measurements. This product was re-ground, sepa-

*Corresponding author. Email Scott_Kroeker@umanitoba.ca

rated into smaller portions in zirconia grain stabilised Pt crucibles, and heated to 1350°C in a vertical tube furnace under an inert argon atmosphere. Individual samples were quenched after the specified heating times, and examined under a polarising microscope for evidence of crystallinity.

X-ray diffraction

Glass samples were powdered using an agate mortar and pestle and assessed using a PANalytical X'Pert Pro Bragg-Brentano powder x-ray diffractometer employing Bragg-Brentano geometry, a Cu K_α radiation source, an X'Celerator detector and a Ni-filter diffracted beam. All data were acquired at room temperature with a 2θ range of 5°–80° at 0·0167° increments using a time per step of 75 s. The samples were mounted with grease on a single crystal quartz (SiO_2) zero background sample holder.

Inductively coupled plasma optical emission spectroscopy (ICP-OES)

Glass samples (100 mg) heated for 3, 15 and 60 min were digested using concentrated HF and analysed for Cs and B by inductively coupled plasma optical emission spectroscopy (ICP-OES) on a Varian Liberty 200 instrument. Because of the volatility of SiF_4 formed under these work up conditions, Si concentrations obtained by this method are not reliable.

Electron microprobe analysis (EMPA)

Samples were powdered and mounted on one-inch diameter Perspex disks using epoxy, polished, carbon coated and probed with a Cameca SX-100 EMP. The collections used wavelength dispersion mode with an excitation voltage of 15 kV, specimen current of 20 nA, a 10 μm beam size, a peak count time of 20 s and a background count time of 10 s. Five points on each sample were analysed.

Nuclear magnetic resonance spectroscopy

NMR data were obtained on a Varian UNITY^{INOVA} 600 NMR spectrometer (magnetic field B_o=14·1 T) using a 3·2 mm double resonance Chemagnetics magic angle spinning (MAS) probe. Sample amounts ranging from 30 to 38 mg were placed in 22 μl ZrO_2 rotors. ^{133}Cs MAS NMR spectra were observed using a Hahn-echo pulse sequence employing the extended phase cycling scheme of Kunwar & Oldfield,[13] with a radiofrequency (rf) field of 50 kHz. Sixteen transients were co-added, with recycle delays of 500 to 1000 s between transients. CsCl (aq, 1·0 M) was used as a primary reference, set to 0 ppm. ^{11}B MAS NMR spectra were observed using a Bloch-decay sequence with short pulses corresponding to a 22° tip angle, with an

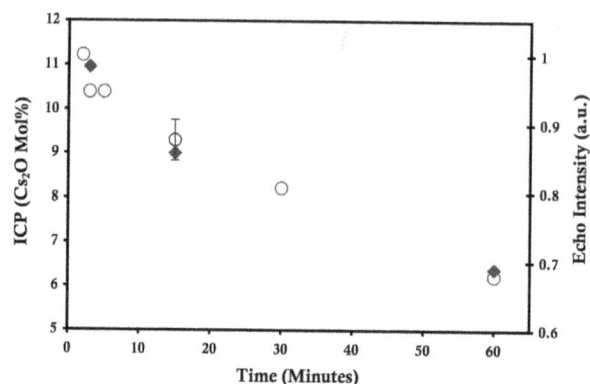

Figure 1. Cs concentration measured by ICP-OES (◆) and monitored by ^{133}Cs NMR spin-echo (○) measurements as a function of glass heating time. ICP error bars are within the size of the symbol

rf field of 69 kHz. Sixteen transients were co-added with recycle delays of 10 to 20 s. The primary reference for boron is BF_3-Et_2O. However, boric acid (aq, 0·1 M) was used as a secondary reference, set to 19·6 ppm.

Results and discussion

The absolute molar quantities of Cs and B were measured by ICP-OES in samples heated for 3, 15 and 60 min (Figure 1). Even after three minutes of heating, the amount of Cs_2O dropped from 14 to 11 mol%, whereas the B content remained stable. This shows that the volatile species is not an alkali borate compound, as previously proposed.[10] Expressed as the ratio of alkali modifier to boron, (R=Cs/B), the R value decreases by over 50% after 60 min heating, from 0·5 to 0·24. This change would be expected to have a marked impact on the resulting glass structure. While accurate quantification of Si by ICP-OES is challenging, due to the extreme volatility of the tetrafluorosilicon compound formed during HF digestion, electron microprobe analysis reveals that the silicon concentration in these samples does not change with heating time. This is not unexpected, given the high boiling temperature of SiO_2 (2950°C),[14] and relatively high melting temperatures of alkali silicate phases compared to caesium oxide (T_m=490°C).[14] Hence, the elemental loss is likely in the form of Cs_2O, or possibly elemental Cs. Mass loss measurements support this observation, as the mass difference between the decarbonated sample and vitrified products can be approximately accounted for by the loss of Cs_2O.

This compositional change can also be monitored by comparing the ^{133}Cs spin echo NMR intensities of these closely related samples collected under identical experimental conditions. Figure 1 depicts the fractional decline in echo intensity with heating time. Reasonable agreement between the data points highlights the reliability of the NMR approach for quantifying spins (i.e. relative atomic concentrations). However, absolute Cs contents can only be measured by comparison with a

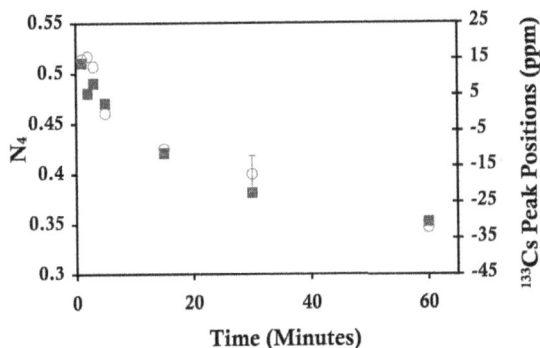

Figure 2. Fraction of four-coordinated boron, N_4 (■), and ^{133}Cs NMR peak positions (○) as a function of heating time. Error bars for N_4 data are within the symbol

standard with an accurately known Cs content.

^{11}B MAS NMR is a convenient way of accurately measuring the fractions of three- and four-coordinated boron, N_4, and reflects the Cs loss as a decrease in N_4 (Figure 2). The traditional model for the behaviour of N_4 for alkali borosilicates for this composition ($K=2$) predicts that it will scale linearly as $N_4=R$ up to $R=0.62$.[1] However, the present data have $N_4=0.35$ at $R=0.24$. Although the details of the traditional model have recently come under criticism,[2,4,9] the origin of the observed discrepancy may instead lie in some degree of phase separation at long heating times (see below).

Figure 3 shows ^{11}B MAS NMR spectra of the same nominal composition after various heating times. Subtle changes in the trigonal boron lineshapes are suggestive of a shift in the balance of ring and non-ring boron toward greater ring concentrations after longer heating times, consistent with the measured reduction of alkali cations. Two distinct tetrahedral boron peaks

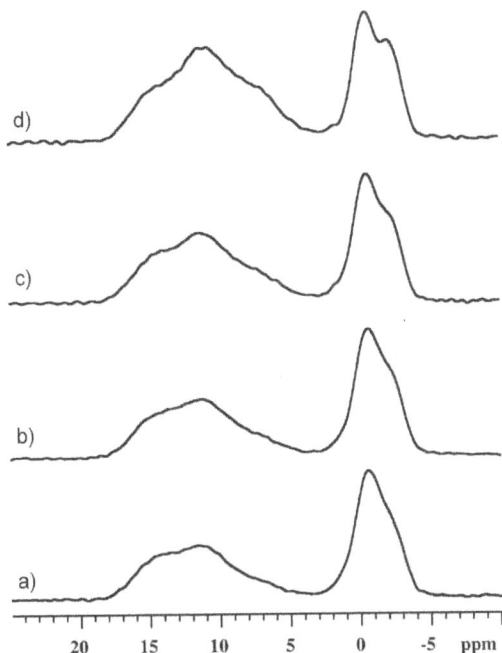

Figure 3. ^{11}B MAS NMR spectra of glasses heated for (a) 3, (b) 5, (c) 15 and (d) 60 min

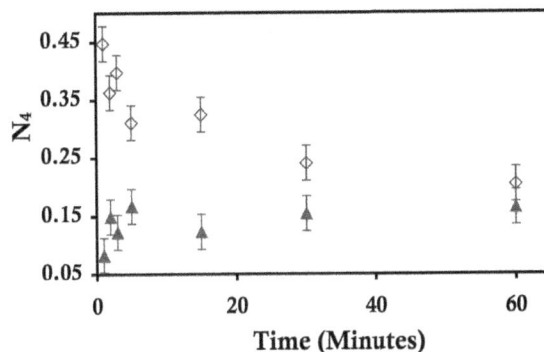

Figure 4. N_4 contributions from individual tetrahedral boron peaks, $\delta \approx -2$ ppm (▲) and $\delta \approx 0$ ppm (◇)

are observed, likely representing $BO_{4/2}$ bonded to four Si (chemical shift $\delta \approx -2$ ppm) and $BO_{4/2}$ bonded to three Si and one $BO_{3/2}$ ($\delta \approx 0$ ppm).[15,16] Careful integration and peak fitting of the entire ^{11}B NMR spectrum reveals that the higher frequency peak decreases with heating time, while the absolute intensity of the lower frequency signal remains essentially constant (Figure 4). (Note that the first data point must be discarded due to incomplete reaction; see below for discussion.) The different behaviour of these two $BO_{4/2}$ units indicates that Cs loss is site specific, or that interspecies equilibria are rapidly re-established in the melt. It appears that Cs is preferentially lost from sites near boron tetrahedra which are bonded to the boron network, and is retained where the neighbouring boron tetrahedron is surrounded by silicon tetrahedra. Both of these observations (i.e. tetrahedral boron speciation and the ring/non-ring populations) differ from what is found for related sodium borosilicate compositions,[15] which could be due to a different alkali cation, or to specific effects of volatilisation.

If the $BO_{4/2}$ (1B, 3Si) contribution is plotted against the R-value determined by ICP-OES, a slope of 1.02 is obtained, with an intercept near zero (0.008) and a high correlation coefficient ($R^2=0.997$) (Figure 5). Although only three compositions were analysed by ICP, this is precisely the relationship predicted by the Dell & Bray model, and suggests that one putative phase of

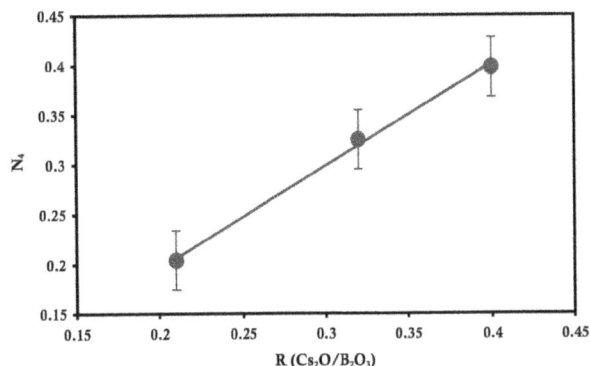

Figure 5. Relationship between intensity of high frequency tetrahedral boron peak and analysed composition. Error bars for the R-value are within the size of the symbol

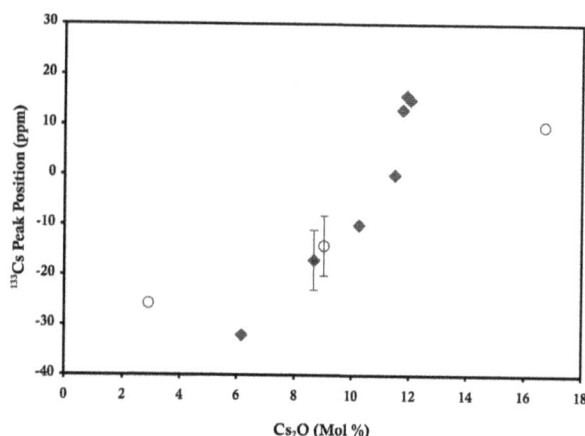

Figure 6. ^{133}Cs *NMR centre-of-gravity peak positions as a function of analysed Cs_2O content* ($\blacklozenge$), *overlaid with previous data on binary caesium borate glasses* ($\circ$)[7]

the glass is being modified in the normal way with caesium evaporation, while a second phase appears to retain a constant composition. While there may be other explanations for this interesting observation, the liquid–liquid immiscibility region and documented occurrence of phase separation for similar alkali borosilicate compositions renders this plausible.[15]

^{133}Cs MAS NMR can also be used to infer structural changes in the glasses. While no lineshape or linewidth changes accompany the loss of caesium from the glasses, the peak position shifts to more negative values with heating time, mirroring the N_4 data (Figure 2). This relationship was observed in binary caesium borate glasses and interpreted as a decrease in coordination number with increasing alkali loading.[7] While the addition of a second network former (and possibly phase separation) complicates this simple explanation, the results show the same general trend as the binary system (Figure 6). In the ternary however, a smaller change in the total Cs content is required to effect a large change in the peak position, as seen previously,[3] implying that silicon influences the ^{133}Cs shift.

Figures 1, 2, 5 and 6 show some scatter in the data for short heating times (≤ 3 min). Both x-ray diffraction and ^{133}Cs MAS NMR indicate the presence of a minor unidentified crystalline phase (<2%) in the sample quenched after one minute. While this amount is too low to have a measurable impact on the glass results, it suggests the presence of incompletely reacted components that have not had enough time to melt and fuse. Despite this scatter, it appears that most of the volatilisation occurs during the first few minutes of heating, perhaps precisely because of incomplete mixing.

Conclusions

Compositional analysis of a caesium borosilicate glass subjected to variable melting times shows that Cs is lost as an oxide or in elemental form, while

the other cations remain intact. NMR spectroscopic studies provide alternate ways to track these changes non-destructively, and offer additional structural information not available from inductively coupled plasma optical emission spectroscopy or electron microprobe analysis. In particular, the observation of multiple tetrahedral boron peaks opens the door to a much more detailed analysis only hinted at in the present work. Confounding a simple explanation is the likelihood of phase separation, which will require further study to confirm.

The significance of volatility during high level nuclear waste vitrification can hardly be overstated. Finding ways to assess its occurrence and understand its impact on glass structure is a first step toward seeking solutions to mitigate this problem. While the high degree of caesium loss in this work can be attributed to smaller sample sizes and higher melting temperatures than are used in industrial vitrification plants, the results can be considered analogous because of the longer heating times employed industrially. The composition chosen in this study is a reasonable model for such materials, and indicates that NMR can play a role in laboratory studies of this challenging issue.

Acknowledgements

The authors would like to thank Mr M. Hancock and Dr M. Bieringer (Chemistry, University of Manitoba) for access to x-ray diffraction facilities, Mr G. Mordan (Geological Sciences, University of Manitoba) for the ICP analysis, and Mr A. Lussier (Geological Sciences, University of Manitoba) for EMPA. This research is supported by NSERC and CFI grants to SK, and an NSERC scholarship to VKM.

References

1. Dell, W. J., Bray, P. J. & Xiao, S. Z. *J. Non-Cryst. Solids*, 1983, **58**, 1.
2. Martens, R. & Müller-Warmuth, W. *J. Non-Cryst. Solids*, 2000, **265**, 167.
3. Parkinson, B. G., Holland, D., Smith, M. E., Howes, A. P. & Scales, C. R. *J. Non-Cryst. Solids*, 2005, **351**, 2425.
4. Wang, S. & Stebbins, J. F. *J. Non-Cryst. Solids*, 1998, **231**, 286.
5. Bray, P. J. & O'Keefe, J. G. *Phys. Chem. Glasses*, 1963, **4**, 37.
6. Kroeker, S., Aguiar, P. M., Cerquiera, A., Clarida, W. J., Doerr, J., Olesiuk, M., Ongie, G., Affatigato, M. & Feller, S. A. *Phys. Chem. Glasses: Eur. J. Glass Sci. Technol. B*, 2006, **47**, 393.
7. Michaelis, V. K., Aguiar, P. M. & Kroeker, S. *J. Non-Cryst. Solids*, 2007, **353**, 2582.
8. Aguiar, P. M. & Kroeker, S. *J. Non-Cryst. Solids*, 2007, **353**, 1834.
9. Parkinson, B. G., Holland, D., Smith, M. E., Howes, A. P. & Scales, C. R. *J. Non-Cryst. Solids*, 2007, **353**, 4076.
10. Delorme, L. *Mécanismes de volatilité des verres et des fontes borosilicatés d'intérêt nucleaire, Influence de la structure*, Thesis, Université d'Orléans, France, 1998.
11. Archakov, I. Y., Stolyarova, V. L. & Shultz, M. M. *Rapid Commun. Mass Spectr.*, 1998, **12**, 1330.
12. Parkinson, B. G., Holland, D., Smith, M. E., Howes, A. P. & Scales, C. R. *J. Phys.: Condens. Matter*, 2007, **19**, 415144.
13. Kunwar, A. C., Turner, G. L. & Oldfield, E. *J. Magn. Reson.*, 1986, **69**, 124.
14. Lide, D. R. *CRC Handbook of Chemistry and Physics*, 82nd edition, CRC Press, Baton Rouge, 2001.
15. Du, L. S. & Stebbins, J. F. *J. Phys. Chem. B*, 2003, **107**, 10063.
16. Du, L. S. & Stebbins, J. F. *J. Non-Cryst. Solids*, 2003, **315**, 239.

Proc. VI Int. Conf. Borate Glasses, Himeji, Japan, 18–22 August 2008   *Phys. Chem. Glasses: Eur. J. Glass Sci. Technol. B,* June 2009, **50** (3), 201–204

Developing ^{11}B solid state MAS NMR methods to characterise medium range structure in borates

Nathan S. Barrow,[a],* *Sharon E. Ashbrook,*[b] *Steven P. Brown*[a] *& Diane Holland*[a]

[a] *Physics Department, University of Warwick, Coventry, CV4 7AL, UK*
[b] *School of Chemistry and EaStCHEM, University of St. Andrews, North Haugh, St. Andrews, KY16 9ST, Scotland*

Manuscript received 29 October 2008
Revised version received 10 February 2009
Accepted 11 February 2009

^{11}B *magic angle spinning solid state NMR spectra of polycrystalline lithium diborate are presented. Second order quadrupolar broadened resonances corresponding to three- (B3) and four-coordinated (B4) boron nuclei overlap for a magnetic field of 7·05 T, but are resolved for 14·1 T. Cross peaks linking the B3 and B4 resonances are observed in two-dimensional homonuclear spin diffusion spectra. There is a need to develop analytical methods for the characterisation of medium range structure in borate glasses. In this respect, such solid state NMR methods have potential for identifying the presence of specific superstructural units.*

1. Introduction

Determining the microscopic structure of a material is essential if the macroscopic properties of the material are to be rationalised. This is a particular challenge for disordered materials, which lack the long range ordering and symmetry that crystals possess. For such disordered materials, the lack of long range order means that it is difficult to extract distance information from diffraction data beyond the first or possibly second coordination sphere, even in simple compositions.

It has been suggested that borate glasses contain medium range order superstructural groups,[1] see Figure 1. These have well defined boron–boron distances in crystals and the presence of sharp peaks in the Raman spectra of the corresponding glasses indicates that these distances should still be distinct even in the disordered material.[2] Even using Raman, neutron and x-ray spectroscopies, the quantities of different superstructural units present in a borate glass are currently difficult to directly measure.[3]

This paper presents ^{11}B solid state NMR results for a model crystalline compound lithium diborate ($Li_2O.2B_2O_3$). ^{11}B is 80% naturally abundant and is a spin-3/2 nucleus. As a half integer quadrupolar nucleus, there is the complication that resonances in a ^{11}B solid state NMR spectrum are broadened, even under magic angle spinning (MAS),[4] due to the second order quadrupolar interaction of the nuclear electric quadrupole moment. This moment interacts with the electric field gradient at the site of the nucleus, which is a consequence of the surrounding electronic environment, i.e. the valence of the boron atom. Early experiments were able to quantify the ratio of three- to four-coordinated boron in static powders of alkali

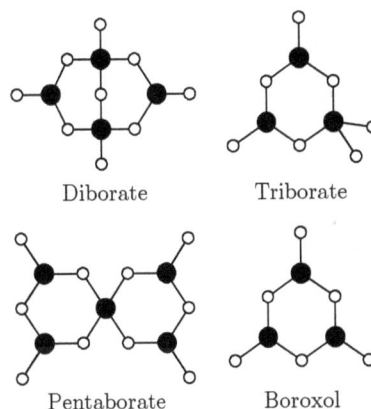

Figure 1. *Superstructural units found in borate crystals and glasses, showing bridging oxygen (white circles) and boron (black circles)*

borates.[5,6] Under MAS and at higher static magnetic fields, the resolution was found to be sufficient for ^{11}B solid state NMR to provide valuable structural information for various borate crystals and glasses.[7]

The strength of the magnetic dipole coupling between two nuclei, *j* and *k*, is given (in rad s^{-1}) by the dipolar coupling constant

$$b_{jk} = -\frac{\mu_0}{4\pi}\frac{\hbar\gamma_j\gamma_k}{r^3} \tag{1}$$

where γ_j is the gyromagnetic ratio of the *j*th nucleus, and *r* is the internuclear distance. Solid state NMR methods that probe dipolar couplings hence provide valuable atomic level structural information in the form of homonuclear distances.

This article considers solid state NMR experiments that probe homonuclear dipolar interactions between ^{11}B nuclei using a longitudinal magnetisation exchange experiment, whose use (in solution state NMR) to probe dipolar relaxation in a two spin

*Corresponding author. Email n.s.barrow@warwick.ac.uk

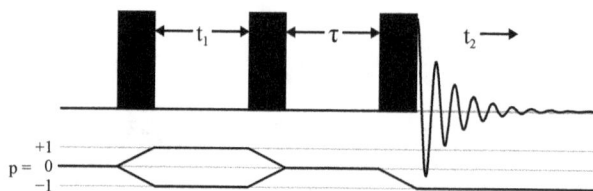

Figure 2. The pulse sequence and coherence transfer pathway diagram[18] for the NOESY-like experiment employed to observe homonuclear spin diffusion (see text for details)

system was first reported by Ernst and coworkers in 1979.[8] In solid state NMR, such experiments have been widely applied for the case of spin-1/2 nuclei, for example to provide structural constraints so as to determine the three-dimensional structure of a protein.[9] By comparison, for half-integer quadrupolar nuclei, there have been fewer publications applying this approach.[10,11,12,13] In addition, double quantum homonuclear dipolar recoupling methods have also been reported.[14,15]

2. Experimental details

2.1 Sample preparation

Polycrystalline lithium diborate, $Li_2O.2B_2O_3$ was prepared by mixing Li_2CO_3 and B_2O_3 in the correct stoichiometric ratio, followed by melting at 1000°C in a platinum–rhodium crucible for 20 min, and subsequent pouring onto a room temperature steel plate. It is noted that if the melt is left to cool in the crucible, crystallisation does not occur.[16] Confirmation of crystallinity and phase purity was provided by powder x-ray diffraction.

2.2 ^{11}B MAS NMR

^{11}B solid state NMR experiments were carried out with magnetic fields of 7·05 T or 14·1 T using a Varian/Chemagnetics Infinity+ or Bruker Avance II+ spectrometer, respectively. In both cases, MAS experiments were performed using a Bruker wide bore 4 mm MAS probe. All spectra were recorded on the same sample of lithium diborate, at a MAS frequency of 10 kHz. ^{11}B chemical shifts were calibrated using BPO_4 (−3·3 ppm) or $NaBH_4$ (−42·06 ppm) as an external reference.

Two-dimensional spin diffusion MAS spectra were recorded using the NOESY-like (Nuclear Overhauser Effect Spectroscopy) pulse sequence shown in Figure 2. Experiments were performed in a rotor-synchronised fashion, i.e. the spectral width was set equal to the MAS frequency in both dimensions. Sign discrimination was achieved in the indirect, F_1, dimension using the States method.[17] A 2-step (with change in coherence order $\Delta p = \pm 1$ on the first $\pi/2$ pulse) or 8-step (with $\Delta p = \pm 1$ on the first (2 steps) and $\Delta p = -1$ on the last (4 steps) $\pi/2$ pulses, respectively) phase cycle was used.[18]

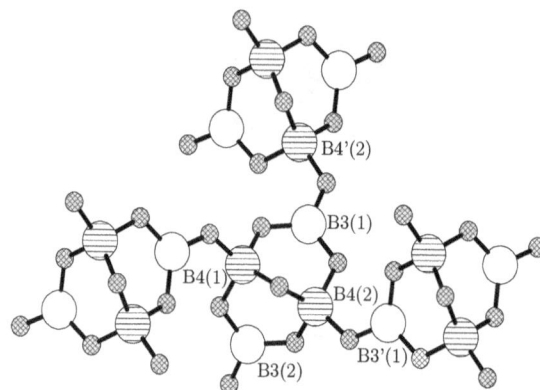

Figure 3. A representation of the crystal structure of lithium diborate, $Li_2O.2B_2O_3$.[19] The figure shows bridging oxygen atoms (grey cross-hatched circles), three-coordinated boron atoms (plain white circles), four-coordinated boron atoms (horizontal lined white circles). Lithium atoms are not shown. The superstructural diborate groups contain two three-coordinated boron atoms and two four-coordinated boron atoms and are clearly discernible

3. Results

3.1 Medium range order in polycrystalline lithium diborate

As shown in Figure 3, crystalline lithium diborate is comprised solely of superstructural diborate units plus charge balancing Li^+ ions.[19] Each diborate unit consists of two three-coordinated boron atoms and two four-coordinated boron atoms. All oxygen in the structure is bridging. $Li_2O.2B_2O_3$ is not unique in containing diborate superstructural units; $ZnO.2B_2O_3$ and $MgO.2B_2O_3$ contain only diborate units whilst $K_2O.2B_2O_3$ and $Rb_2O.2B_2O_3$ contain a mixture of diborate units and BO_3 triangles.[1,20]

Boron–boron separations and the corresponding homonuclear dipolar coupling constants (as calculated using Equation (1)) are presented in Table 1. The closest pair of boron nuclei in the structure is B4(1)–B4(2). By contrast, B3 nuclei are much further away from each other and hence have a low dipolar coupling constant. B3–B4 dipolar couplings are strong and prevalent in the structure.

Table 1. Boron–boron distances and corresponding dipolar coupling constant between different boron atoms in the diborate group. Bracketed numbers are used to differentiate between different boron atoms in the same superstructural group. Each B3 has a connectivity to a B4 in a different superstructural group and vice versa. See Figure 3

Nuclei	Atomic separation (Å)	Dipolar coupling constant $(b_{jk}/2\pi)$ (Hz)
B4(1)–B4(2)	2·36	−937
B3(1)–B4(1)	2·44	−848
B4(2)–B3(2)	2·44	−848
B3(1)–B4(2)	2·49	−801
B3(2)–B4(1)	2·49	−801
B3(1)–B4'(2)	2·50	−791
B3'(2)–B4(1)	2·50	−791
B3(1)–B3(2)	3·58	−269

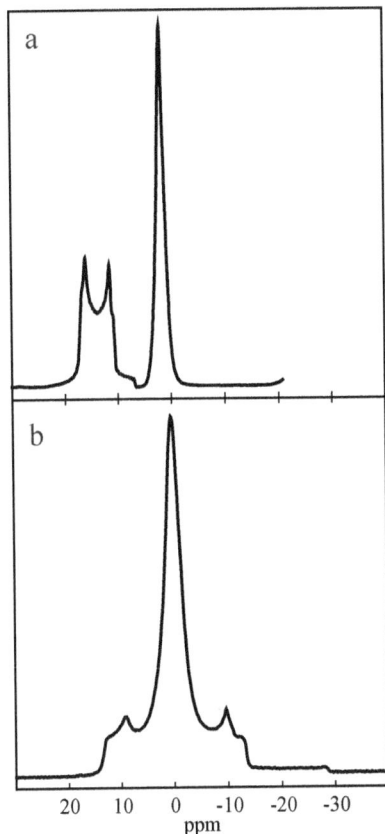

Figure 4. 1D ^{11}B MAS (10 kHz) NMR spectra of polycrystalline lithium diborate recorded at (a) 14·1 T and (b) 7·05 T

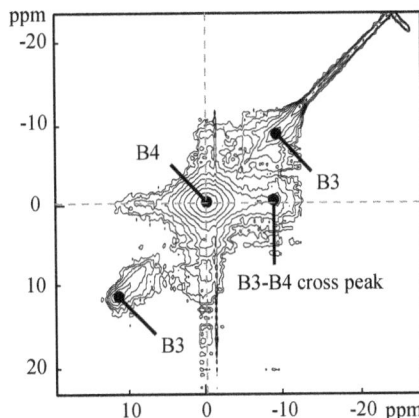

Figure 5. 2D ^{11}B MAS (10 kHz) NMR spectrum of polycrystalline lithium diborate recorded at 7·05 T using the pulse sequence shown in Figure 2 with a mixing time of 114 ms. Two transients were co-added for each of 128 t_1 slices. The recycle delay was 18 s

3.2 One-dimensional ^{11}B MAS NMR spectra

One pulse ^{11}B spectra of lithium diborate, Li$_2$O.2B$_2$O$_3$, recorded at two different magnetic fields are shown in Figure 4. The resonances exhibit second order quadrupolar broadening that is proportional to the square of the quadrupolar frequency, C_Q, and inversely proportional to the Larmor frequency. At the higher static magnetic field, the B3 and B4 lineshapes are resolved. At the lower static magnetic field, the increased broadening of the resonances, which is more pronounced for the B3 site, results in an overlap of the lineshapes. A fit of the high field spectrum gives the quadrupolar coupling parameter, C_Q, as 2600±100 kHz for the trigonal planar B3 site, 570±100 kHz for the tetrahedral B4 site, and the quadrupolar asymmetry parameter, η_Q as 0·14±0·5 for the B3 site. The chemical shifts are 18·5±0·5 ppm and 2·1±0·5 ppm for the B3 and B4 site, respectively. These values are in agreement with those previously published in the literature.[21]

3.3 Two-dimensional ^{11}B MAS NMR spin diffusion spectra

Spin diffusion between ^{11}B nuclei is probed using the pulse sequence shown in Figure 2. Such a pulse sequence corresponds to the commonly employed solution state NMR NOESY experiment[8] that has been widely used in solid state NMR, e.g. ^{1}H–^{1}H

experiments that probe proton–proton proximities.[22]

The pulse sequence is easily modified for use on quadrupolar nuclei by using selective pulses that only affect the central transition. The first pulse in the sequence, and subsequent t_1 time, allows the single quantum (SQ) magnetisation to precess at a characteristic frequency (i.e. that of the B3 or B4 chemical site) and is similar to a one pulse experiment. The second pulse converts the SQ coherence into a population state. This state is maintained for a duration called the mixing time, τ, as it is during this period that the individual spins can exchange magnetisation. For zero mixing time it is expected that there will be no magnetisation exchange and for long mixing times it is expected that the magnetisation will reach an equilibrium between the different chemical sites. After the mixing time a third pulse is applied and the spins once again precess at their characteristic frequencies. After a Fourier transformation of both t_1 and t_2 time dimensions is performed, a two-dimensional spectrum is obtained. Magnetisation that has not been exchanged will have precessed at the same characteristic frequency both times and will appear along the diagonal of the spectrum. Magnetisation that has been exchanged will have precessed at one frequency during t_1 and another during t_2. This will appear off-diagonal in the spectrum and at the F_2 chemical shift of the spin where the magnetisation has finally resided.

^{11}B MAS NMR spin diffusion spectra recorded at 7·05 T (300 MHz proton frequency) and 14·1 T (600 MHz proton frequency) are shown in Figures 5 and 6, respectively. At 7·05 T, the narrower B4 resonance overlaps the second order quadrupolar broadened B3 resonance. As indicated by dashed lines in Figure 5, B3–B4 cross peaks are observed to the left, right, above and below the B4 diagonal peak. As has been discussed previously for a spin diffusion experiment involving the spin I=5/2 nucleus, ^{27}Al, this case of

a) 0 ms

b) 114 ms

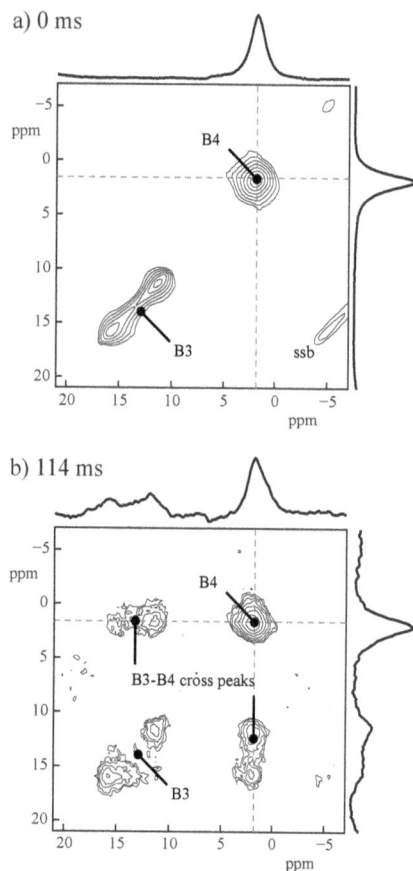

Figure 6. 2D ^{11}B MAS (10 kHz) NMR spectrum of poly-crystalline lithium diborate recorded at 14·1 T using the pulse sequence shown in Figure 2 with a mixing time, τ, of (a) 0 ms and (b) 114 ms. Eight transients were co-added for each of 256 t_1 slices. The recycle delay was 15 s. All pulses were of 20 μs duration, with this having been calibrated to give maximum B3 signal in a one-pulse experiment. The one-dimensional spectra are slices taken at the B4 resonance (1·8 ppm), as indicated by the dashed lines, and clearly show the cross peak intensity in Figure 6(b)

overlapping resonances corresponds to the n=0 rotational resonance condition. (Rotational resonance refers to the case where the difference in isotropic chemical shifts of two resonances is a multiple of the MAS frequency, i.e. $n=\Delta\omega_{iso}/\omega_r$.[11] The n=0 rotational resonance condition corresponds to the special case of the overlap of resonance lines due to distinct nuclei, i.e. since $\Delta\omega_{iso}$ is zero, n is zero, regardless of the MAS frequency, ω_r.) At 14·1 T, cross peak intensity between different parts of the second order quadrupolar broadened B3 lineshape is also evident – it has been shown previously that the analysis of such lineshapes enables the determination of the relative orientation of the quadrupole tensors.[13] At high field the B3 and B4 lineshapes are resolved, and there is no overlap corresponding to an n=0 rotational resonance condition. Nevertheless, clear B3–B4 cross peaks are also observed at long mixing times, indicating that a spin diffusion mechanism is still active.

4. Discussion

Spin diffusion between different ^{11}B chemical sites in crystalline $Li_2O.2B_2O_3$ has been observed under magic angle spinning (MAS) NMR at 7·05 T and 14·1 T. For the lower magnetic field, B3–B4 cross peaks were observed as a consequence of n=0 rotational resonance, due to the overlap of the second order quadrupolar broadened lineshapes. For the higher magnetic field, where the lineshapes do not overlap, B3–B4 cross peaks were still observed. These observations suggest that ^{11}B MAS spin diffusion experiments can be used to probe distinct boron–boron proximities. Since different superstructural borate units are characterised by different boron–boron proximities, ^{11}B MAS spin diffusion experiments have potential as probes of medium range order in borate glasses.

5. Acknowledgements

This work was supported by EPSRC grant EP/D080355/1. The Borate 2008 conference organisers are acknowledged for the provision of a generous travel grant, as well as allowing this work to be presented at the conference in Himeji, Japan. Helpful discussions with Dr Ivan Hung are acknowledged.

References

1. Wright, A. C., Vedishcheva, N. M. & Shakhmatkin, B. A. *J. Non-Cryst. Solids*, 1995, **192–193**, 92.
2. Meera, B. N. & Ramakrishna, J. *J. Non-Cryst. Solids*, 1993, **159**, 1.
3. Wright, A. C., Shaw, J. L., Sinclair, R. N., Vedishcheva, N. M., Shakhmatkin, B. A. & Scales, C. R. *J. Non-Cryst. Solids*, 2004, **345–346**, 24.
4. Lowe, I. J. *Phys. Rev. Lett.*, 1959, **2**, 285.
5. Jellison, G. E., Feller, S. A. & Bray, P. J. *Phys. Chem. Glasses*, 1978, **19**, 52.
6. Bray, P. J. *J. Non-Cryst. Solids*, 1987, **95–96**, 45.
7. Eckert, H. *Prog. Nucl. Magn. Reson. Spectrosc.*, 1992, **24**, 159.
8. Jeener, J., Meier, B. H., Bachmann, P. & Ernst, R. R. *J. Chem. Phys.*, 1979, **71**, 4546.
9. Castellani, F., van Rossum, B., Diehl, A., Schubert, M., Rehbein, K. & Oschkinat, H. *Nature*, 2002, **420**, 98.
10. Hartmann, P., Jager, C. & Zwanziger, J. W. *Solid State Nucl. Magn. Reson.*, 1999, **13**, 245.
11. Nijman, M., Ernst, M., Kentgens, A. P. M. & Meier, B. H. *Mol. Phys.*, 2000, **98**, 161.
12. Eden, M. & Frydman, L. *J. Chem. Phys.*, 2001, **114**, 4116.
13. Dowell, N. G., Ashbrook, S. E. & Wimperis, S. *J. Phys. Chem. A*, 2002, **106**, 9470.
14. Mali, G., Fink, G. & Taulelle, F. *J. Chem. Phys.*, 2004, **120**, 2835.
15. Mali, G., Kaucic, V. & Taulelle, F. *J. Chem. Phys.*, 2008, **128**, 204503.
16. Goktas, A. A., Neilson, G. F. & Weinberg, M. C. *J. Mater. Sci.*, 1992, **27**, 24.
17. States, D. J., Haberkorn, R. A. & Ruben, D. J. *J. Magn. Reson.*, 1982, **48**, 286.
18. Bodenhausen, G., Kogler, H. & Ernst, R. R. *J. Magn. Reson.*, 1984, **58**, 370.
19. Radaev, S. F., Muradyan, L. A., Malakhova, L. F., Burak, Y. V. & Simonov, V. I. *Kristallografiya*, 1989, **34**, 1400.
20. Sinclair, R. N., Haworth, R., Wright, A. C., Parkinson, B. G., Holland, D., Taylor, J. W., Vedishcheva, N. M., Polyakova, I. G., Shakhmatkin, B. A., Feller, S. A., Rijal, B. & Edwards, T. *Phys. Chem. Glasses: Eur. J. Glass Sci. Technol. B*, 2006, **47**, 405.
21. Hansen, M. R., Vosegaard, T., Jakobsen, H. J. & Skibsted, J. *J. Phys. Chem. A*, 2004, **108**, 586.
22. Brown, S. P. *Prog. Nucl. Magn. Reson. Spectrosc.*, 2007, **50**, 199.

Proc. VI Int. Conf. Borate Glasses, Himeji, Japan, 18–22 August 2008   *Phys. Chem. Glasses: Eur. J. Glass Sci. Technol. B*, October 2009, **50** (5), 301–304

Molecular orbital calculation of the ^{29}Si NMR chemical shift in borosilicates: the effect of boron coordination to SiO$_4$ units

T. Nanba,[1] *Y. Asano, Y. Benino*

Graduate School of Environmental Science, Okayama University, 3-1-1, Tsushima-Naka, Okayama 700-8530, Japan

S. Sakida

Environmental Management Center, Okayama University, 3-1-1, Tsushima-Naka, Okayama 700-8530, Japan

Y. Miura

University of Shiga Prefecture, 2500, Hassaka-Cho, Hikone-City, Shiga 522-8533, Japan

Manuscript received 1 September 2008
Revised version received 28 February 2009
Accepted 15 September 2009

Borosilicate cluster models were constructed, and the ^{29}Si NMR chemical shift was estimated by molecular orbital calculations. For Q^4 species (an SiO$_4$ unit consisting of four bridging oxygen atoms), a higher frequency shift was confirmed due to the replacement of the surrounding SiO$_4$ with BO$_4$ units, and for Q^4 species associated with more than one BO$_4$ unit, the chemical shifts were nearly identical to those for Q^3 species (an SiO$_4$ unit including one nonbridging oxygen) in alkali silicates. The chemical shifts of Q^4 species in borosilicates were interpreted in terms of the change in bond angle of Si–O–(Si,B4) bridges. A change in chemical shift anisotropy of Q^4 species was also found. The Q^4 species associated with two BO$_4$ units indicated an anisotropy maximum, which was smaller than the anisotropy of Q^3 species.

Borosilicate glasses have high thermal shock resistance and high chemical durability, and hence they have been used as laboratory and heat resistant glasses. Recently, they have also been used in glass solidification of nuclear wastes. Borosilicate glasses are also known for phase separation, and they have been used as separation membranes, catalyst supports and an alternative to SiO$_2$ (Vycor®) glass. Structural studies have been extensively performed by various experimental techniques, such as ^{11}B and ^{29}Si NMR, XPS, IR and Raman spectroscopies, and structural models have also been proposed.[1–4] Among the models, a model proposed by Dell *et al*[1] has been commonly accepted.

Nanba *et al*[5,6] found a disagreement between the amounts of nonbridging oxygen (NBO) estimated from ^{11}B and ^{29}Si NMR analyses for borosilicate glasses with low alkali content; ^{11}B NMR suggested that all alkali ions were associated with BO$_4$ units as a charge compensator with no NBOs in the low alkali glasses, but ^{29}Si NMR suggested the presence of Q^3 species (SiO$_4$ units including one NBO) in a measureable amount. In the ^{29}Si MAS (magic angle spinning) NMR spectrum of 0·2Na$_2$O.B$_2$O$_3$.0·5SiO$_2$ glass,[6] an NMR peak was successfully extracted at −90 ppm. According to a conventional peak assignment for alkali binary silicate glasses,[7] the −90 ppm peak is assigned to Q^3 species. In aluminosilicates, the ^{29}Si NMR signal shifts to the higher frequency side (lower magnetic field side) as the number of AlO$_4$ units surrounding an SiO$_4$ unit increases, and Q^4 species associated with three AlO$_4$ units appear as a peak at around −90 ppm in ^{29}Si NMR.[8] It was hence considered that the conventional assignments were not applicable to alkali borosilicate glasses, and Nanba *et al*[6] proposed an interpretation of the −90 ppm component in the ^{29}Si NMR spectra of borosilicates;[6] according to their molecular orbital (MO) calculations, Si2p and O1s orbital energies of Si and O atoms in B4–O–Si bridges (where B4 indicates four-fold coordinated boron) with small bond angle (~125°) were close to those of Si and O atoms in Q^3 species and NBOs in alkali silicates. It was finally proposed that in borosilicate glasses, Si atoms in such B4–O–Si bridges appeared in the −90 ppm component of the ^{29}Si NMR spectra.

In the present study, various structural models were constructed changing the number of BO$_m$ units surrounding an SiO$_4$ unit, and the ^{29}Si NMR chemical shift of the Si atom in the central SiO$_4$ unit was calculated by using MO calculations to investigate the effect of boron coordination to SiO$_4$ units.

Computational

Cluster models were constructed by two methods. In the first method, the objective structural units were extracted from various crystals. In the second method, the objective structures were initially hand built, and geometrical optimisations were finally performed by using the Gaussian03 program[9] to

[1] Corresponding author. Email tokuro_n@cc.okayama-u.ac.jp

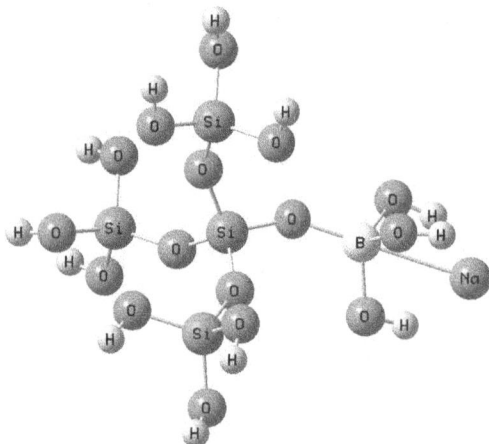

Figure 1. A cluster model of a Q^4 species surrounded by three SiO_4 units and one BO_4 unit

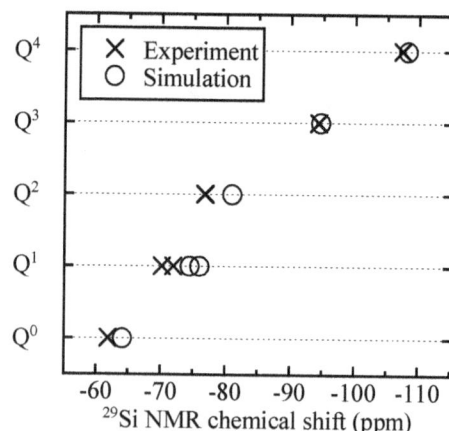

Figure 2. The ^{29}Si NMR chemical shift of Si atoms in Q^n species (Q^n are SiO_4 units containing n bridging oxygens) in silicate crystals (Q^4 in SiO_2, Q^3 in $Na_2Si_2O_5$, Q^2 in Na_2SiO_3, Q^1 in $Ca_3Si_2O_7$, Q^0 in Mg_2SiO_4)

stabilise the structures. In the cluster models, SiO_4 units were centrally positioned and four SiO_4, BO_4 or BO_3 units were additionally placed around the central SiO_4 unit. Q^{0-3} species including NBOs were also constructed. Sodium ions were introduced to compensate the negative charges of BO_4 units and NBOs. Hydrogen atoms were also added to the terminal oxygen atoms to reduce bond termination effects.

A cluster model of a Q^4 species surrounded by three SiO_4 units and one BO_4 unit is shown in Figure 1. Geometry optimisation was performed by using the Gaussian03 program at the HF6-31G(d) level. In the geometry optimisation, hydrogen atoms were firstly optimised, and secondly the central SiO_4 unit was fixed and the surrounding atoms, including hydrogen and alkali atoms, were optimised. Lastly the limitations were removed, and all atoms in a cluster were optimised, obtaining a final structure.

NMR parameters were obtained by making calculations with the Gaussian03 program at the HF/6-311+G(2df,p) level. Magnetic shielding parameters were calculated by using the Gauge-Independent Atomic Orbital (GIAO) method.[10] Chemical shifts, δ, for ^{29}Si NMR (in ppm) were estimated from the following equation

$$\delta(cluster) = \sigma(TMS) - \sigma(cluster) \qquad (1)$$

where σ is the ^{29}Si magnetic shielding (in ppm) obtained from MO calculations. Tetra methyl silane (TMS) was used as a reference. The TMS cluster geometry was optimised at the HF/6-31G(d) level, and the shielding parameter was calculated at the HF/6-311+G(2df,p) level.

Results

Firstly, Q^n structures were extracted from various silicate crystals, and the ^{29}Si NMR chemical shifts for Si atoms in each of the Q^n species were estimated for the purpose of evaluating the validity of the cluster models and the calculation conditions. As shown in

Figure 2, the ^{29}Si NMR chemical shift along with the change in Q^n structure is successfully reproduced, indicating that the calculation methods used are applicable to borosilicate structures.

Figure 3 shows the ^{29}Si NMR chemical shift calculated from the borosilicate clusters $Q^4(kBm)$ (Q^4 species surrounded by k BO_m units). High frequency shift with increasing number of BO_4 units surrounding the central Q^4 species is commonly confirmed, both in the cluster models extracted from the borosilicate crystals, and the hand built clusters after geometry optimisation. The chemical shifts of the Q^4 species associated with two and three BO_4 units ($k=2$, 3) are slightly different, but the other Q^4 species ($k=0$, 1, 4) are in good agreement between the cluster models constructed in different ways. It is noted that the chemical shifts of the Q^4 species associated with more than one BO_4 unit ($k=2$, 3, 4) are nearly identical to

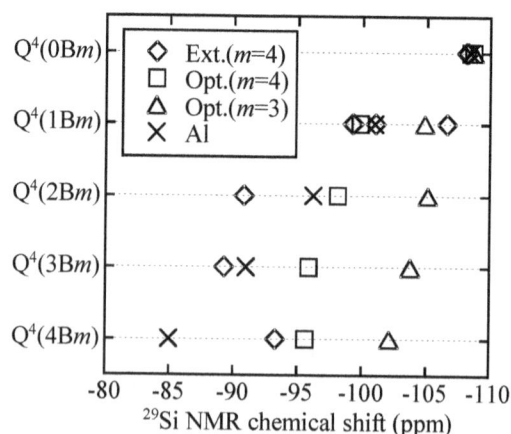

Figure 3. The ^{29}Si NMR chemical shift of Si atoms in various Q^4 species surrounded by BO_m units. $Q^4(kBm)$ indicates Q^4 species surrounded by k BO_m units. Ext.: cluster models extracted from borosilicate crystals (k=0: SiO_2, 1: $KNa_2B_3Si_{12}O_{30}$, $KBSi_3O_8$, $NaBSi_3O_8$, 2: $KBSi_2O_6$, 3: $CaB_2Si_2O_8$, 4: $NaBSiO_4$). Opt.: hand built cluster models after geometry optimisation. Al: experimental chemical shift of Q^4 species surrounded by AlO_4 units in aluminosilicate crystals[8]

those of the Q^3 species in alkali silicates. Figure 3 also shows the chemical shifts of Q^4 species associated with BO_3 units (m=3). Borosilicate crystals consisting of such structures could not be found, and hence the cluster models consisting of Si–O–B3 bonds were constructed by the hand built method. As shown in Figure 3, the ^{29}Si NMR chemical shifts due to B3 coordination to an SiO_4 unit are smaller than those due to B4 coordination. It is consequently concluded that the −90 ppm component in the ^{29}Si NMR signal of borosilicate glasses can be attributed to Q^4 species associated with more than one BO_4 unit.

Discussion

Lippmaa et al[8] reported the ^{29}Si NMR chemical shift for various aluminosilicates. For comparison, the chemical shift of Q^4 species in aluminosilicates is also illustrated in Figure 3. A linear change in the chemical shift occurs as the number of AlO_4 units associated with the SiO_4 unit increases. It is noted that the calculated chemical shifts of the $Q^4(kB4)$ species are quite similar to the experimental shifts of the $Q^4(kAl)$ species, except for k=4. If B4 and Al coordination to SiO_4 unit actually results in a similar ^{29}Si NMR chemical shift, this seems a curious phenomenon because B and Al atoms have different size and electronegativity.

According to the simplest interpretation, the NMR chemical shift depends on the electron population of objective atoms; when the electron population at outer shells increases, the core orbitals such as Si 1s, 2s and 2p expand and increase in orbital energy. As a result, Si nuclei are less shielded, and the ^{29}Si NMR signal shifts to the higher frequency (lower magnetic field) side. Based on this interpretation, the electron population of Si atoms in Q^4 species should increase with increasing number of surrounding BO_4 or AlO_4 units. Pauling's electronegativity for B, Al and Si atoms is 2·0, 1·6 and 1·8, respectively. Therefore, it is reasonable that more electrons are populated on Si in Si–O–Al bonds, whilst less electrons are populated on Si in Si–O–B bonds. However, this is inconsistent with the ^{29}Si NMR chemical shift predicted by the MO calculations.

The electronic states of Si, such as electron population and core orbital energy, were then examined. According to Mulliken population analysis,[11] the electron population on Si is given by $Q_{Si}=Q_{SiSi}+\frac{1}{2}\Sigma Q_{SiO}$, where Q_{SiSi} and Q_{SiO} are the populations of nonbonding electrons localised on Si and electrons shared between Si and neighbouring oxygens, respectively. In this paper, the total electron population around a Si nucleus was estimated as $Q_{SiSi}+\Sigma Q_{SiO}$, without dividing the shared electrons between Si and O. As shown in Figure 4(a), a linear correlation between total electron population on Si and ^{29}Si NMR chemical shift is confirmed in the Q^n species extracted from silicate crystals. In the case of borosilicate cluster models, however, it is difficult to recognise the changes in continuous

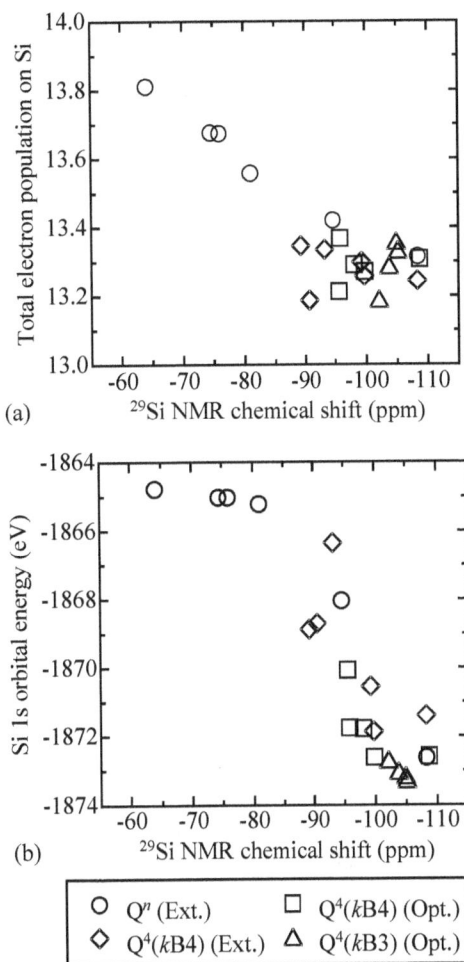

(a)

(b)

| $\bigcirc$ | Q^n (Ext.) | $\square$ | $Q^4(kB4)$ (Opt.) |
| $\diamond$ | $Q^4(kB4)$ (Ext.) | $\triangle$ | $Q^4(kB3)$ (Opt.) |

Figure 4. Correlation of the calculated ^{29}Si NMR chemical shift with (a) total electron population on the Si atom, and (b) Si1s orbital energy. Q^n (Ext.) and $Q^4(kB4)$ (Ext.) represent the cluster models extracted from silicate crystals (see Figure 2) and borosilicate crystals (see Figure 3), respectively. $Q^4(kB4)$ (Opt.) and $Q^4(kB3)$ (Opt.) indicate the hand built cluster models after geometry optimisation

sequence in electron population. Especially, in the case of B3 coordination, the change in electron population is completely opposite. As for the Si1s orbital energy, shown in Figure 4(b), a correlation which is not linear, but has a downward sloping trend, is commonly observed even in the borosilicate cluster models, and similar correlations are also confirmed against Si2s and Si2p orbital energies. It is suggested from these results that the electron population is not the decisive factor in determining the ^{29}Si NMR chemical shift.

It is known that ^{29}Si NMR chemical shift is also dependent on bond angle, and Xue & Kanzaki[12,13] reported the angular dependence for Si–O–T bonds (T=Si, Al) by using MO calculations. Hence the angular dependence for Si–O–T bonds (T=Si, B3, B4) in borosilicate structures was also examined. Figure 5 shows the angular dependence of the ^{29}Si NMR chemical shift, where the bond angles of central Si–O bonds to the surrounding Si, B3 and B4 atoms are averaged. A good correlation is recognised between the ^{29}Si

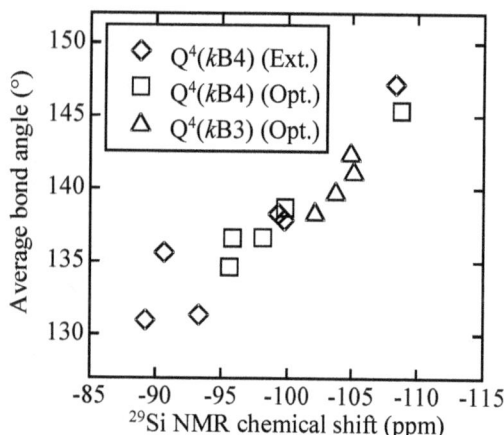

Figure 5. The correlation between the calculated ^{29}Si NMR chemical shift and the average bond angle for Si–O–(Si,B) bonds. Q^4(kB4) (Ext.) represents the cluster models extracted from borosilicate crystals (see Figure 3). Q^4(kB4) (Opt.) and Q^4(kB3) (Opt.) indicate the hand built cluster models after geometry optimisation

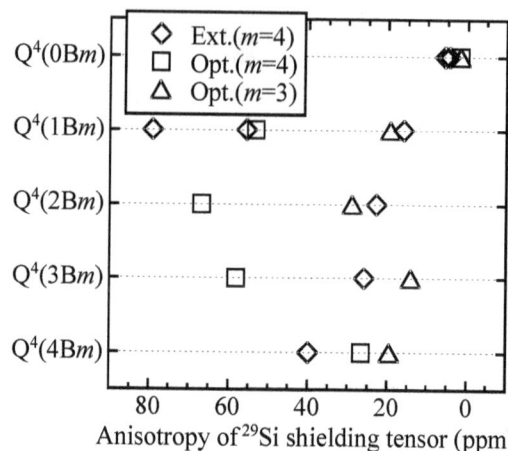

Figure 6. The anisotropy of the ^{29}Si shielding tensor of Si atoms in various Q^4 species surrounded by BO_m units, Q^4(kBm). Ext.: cluster models extracted from borosilicate crystals (see Figure 3). Opt.: hand built cluster models after geometry optimisation

NMR chemical shift and the average bond angle for the Si–O–(Si,B) bonds. It is consequently suggested that the boron coordination to SiO_4 unit results in a decrease in the Si–O–B bond angle, and hence leads to a higher frequency shift of the ^{29}Si NMR signal.

As discussed, boron coordination to SiO_4 units resulted in a change of bond angle and ^{29}Si NMR chemical shift to the higher frequency side. The change in bond angle may also induce chemical shift anisotropy, and hence the change in anisotropy of the ^{29}Si shielding tensor was investigated. As shown in Figure 6, the change in anisotropy of Si atoms in cluster models extracted from borosilicate crystals is not systematic. In the case of Si atoms in cluster models after geometry optimisation, however, maxima in anisotropy are confirmed for Si atoms surrounded by two BO_4 or BO_3 units, and larger anisotropy is caused by the coordination of BO_4 units. The anisotropy of the ^{29}Si shielding tensor of a Si atom in a cluster model of Q^3 structure after geometry optimisation is 94·6 ppm, which is much larger than that of Q^4(2B4) of 66·8 ppm. It is concluded from these results that discrimination between Si atoms in Q^3 and Q^4(kB4) species is almost impossible by ^{29}Si MAS NMR, but for ^{29}Si static NMR, however, it may be possible based on the asymmetric parameters of the NMR signal.

Conclusions

^{29}Si NMR chemical shifts were examined by using molecular orbital calculations. Various Q^4 structural models were constructed, changing the number of BO_m units surrounding the central SiO_4 unit, and the effect of boron coordination to SiO_4 units on the ^{29}Si NMR chemical shift was investigated. As the number of BO_4 units surrounding the central SiO_4 unit increased, chemical shifts to the higher frequency side were observed, both for cluster models extracted from

borosilicate crystals, and for hand built clusters after geometry optimisation. The chemical shifts of the Q^4 species associated with more than one BO_4 unit were nearly identical to those of the Q^3 species in alkali silicates, leading to the conclusion that the –90 ppm component observed in the ^{29}Si NMR signal in borosilicate glasses may also be attributed to Q^4 species associated with more than one BO_4 unit. In the case of Q^n species in silicates, the change in the electron population on Si atoms which is caused by the formation of NBOs in SiO_4 units is the predominant factor for the ^{29}Si NMR chemical shift. In the case of Q^4 species in borosilicates, however, the change in the bond angle for Si–O–(Si,B4) bridges is the most significant factor determining the chemical shift. Discrimination between Q^3 and BO_4-associated Q^4 species seems to be impossible by using ^{29}Si MAS NMR, but these species have different anisotropy of the ^{29}Si shielding tensor, suggesting the possibility that they can be distinguished by using ^{29}Si static NMR measurements.

References

1. Dell, W. J., Bray P. J. & Xiao S. Z. *J. Non-Cryst. Solids*, 1983, **58**, 1.
2. Zhong, J., Wu, X., Liu, M. L. & Bray, P. J. *J. Non-Cryst. Solids*, 1988, **107**, 81.
3. MacKenzie, J. W., Bhatnagar, A., Bain, D., Bhowmik, S., Parameswar, C., Budhwani, K., Feller, S. A., Royle, M. L. & Martin, S. W. *J. Non-Cryst. Solids*, 1994, **177**, 269.
4. Miura, Y., Kusano, H., Nanba, T. & Matsumoto, S. *J. Non-Cryst. Solids*, 2001, **290**, 1.
5. Nanba, T. & Miura, Y. *Phys. Chem. Glasses*, 2003, **44**, 244.
6. Nanba, T., Nishimura, M. & Miura, Y. *Geochim. Cosmochim. Ac.*, 2004, **68**, 5103.
7. Maekawa, H., Maekawa, T., Kawamura, K. & Yokokawa, T. *J. Non-Cryst. Solids*, 1991, **127**, 53.
8. Lippmaa, E., Maegi, M., Samoson, A., Tarmak, M. & Engelhardt, G. *J. Am. Chem. Soc.*, 1981, **103**, 4992.
9. GAUSSIAN 03, Gaussian, Inc., Pittsburgh, PA, 2003.
10. Cheeseman, J. R., Trucks, G. W., Keith, T. A. & Frisch, M. J. *J. Chem. Phys.*, 1996, **104**, 5497.
11. Mulliken R. S. *J. Chem. Phys.*, 1955, **23**, 1833.
12. Xue, X. & Kanzaki, M. *Phys. Chem. Miner.*, 1998, **26**, 14.
13. Xue, X. & Kanzaki, M. *J. Phys. Chem. B*, 1999, **103**, 10816.

Proc. VI Int. Conf. Borate Glasses, Himeji, Japan, 18–22 August 2008   *Phys. Chem. Glasses: Eur. J. Glass Sci. Technol. B, June 2009, **50** (3), 156–164*

A multi-technique structural study of the tellurium borate glass system

Emma R. Barney,[1*] *Alex C. Hannon*[1] *& Diane Holland*[2]

[1] *ISIS Facility, Rutherford Appleton Laboratory, Chilton, Didcot OX11 0QX, UK*
[2] *Physics Department, University of Warwick, Coventry CV4 7AL, UK*

Manuscript received 28 January 2009
Revised version received 1 April 2009
Accepted 1 April 2009

A series of tellurium borate, $xTeO_2.(100-x)B_2O_3$, glasses has been studied using neutron diffraction, ^{11}B nuclear magnetic resonance (NMR), Raman spectroscopy and density measurements to investigate how the environments of tellurium and boron change with composition. It was necessary to determine the sample composition by quantitative NMR, after which ^{11}B NMR and neutron diffraction gave good agreement for the B–O coordination number. The effects of phase separation were studied by making samples with compositions above and below the reported low TeO_2 limit of single phase glass formation, 74 mol% TeO_2. It was found that samples prepared with 70 mol% TeO_2 or less were phase separated, with a clear glass component of composition ~80 mol% TeO_2. A linear relationship was established between the density and the fraction of four-coordinated boron, N_4, and previous literature data also support this relationship. However, the previous data have glass compositions which are inconsistent with the present study, and the current work shows that the previous compositions are incorrect, highlighting the problems of phase separation and the importance of careful compositional analysis. With the addition of B_2O_3, no shift in the primary Te–O bond length to shorter distances, such as is seen in alkali tellurites, is observed. The measured Te–O coordination number decreased with the addition of B_2O_3, due to a reduction in the number of longer Te–O bonds, indicating that the total number of $[TeO_4]$ units decreases, to be replaced by $[TeO_3]$ units. The range over which the Te–O distribution was measured was limited by the presence of the O_B–O_B peak. Therefore, though the observed tellurium coordination numbers are smaller than predicted by a literature model for the structure of these glasses, further work is needed to determine the full extent of the Te–O distribution which overlaps with the O–O peaks.

1. Introduction

This study of tellurium borate glasses is part of an ongoing investigation into the microstructural changes in the environment of boron with composition in a wide range of binary and ternary glasses. It is common for binary borate glasses to exhibit an anomaly in the thermophysical properties, characterised by maxima or minima at a specific composition. This *borate anomaly* is associated with a maximum in the fraction, N_4, of boron atoms which are four-coordinated (though the maximum in N_4 may not necessarily occur at exactly the same composition as for a particular property). For example, for alkali borate glasses N_4 does not exceed a value of about 0·45 at a composition of about 40 mol% A_2O (where A is an alkali element such as Li).[1] However, several previous magic angle spinning (MAS) ^{11}B NMR studies[2,3] have shown that the boron environment in the $xTeO_2.(100-x)B_2O_3$ system exhibits unusual behaviour for a binary borate glass, in that there is no maximum for N_4. Sekiya *et al*[4] demonstrated, using both NMR and Raman spectroscopy, that both the tellurium coordination number and N_4 increase linearly over the glass forming range,[2] with N_4 appearing to approach one for small B_2O_3 contents. However,

although N_4 increases with increasing TeO_2 content, $N_4^*=[BO_4]^-/(B_{tot}+Te_{tot})$ (the number of 4-coordinated borons as a fraction of the total number of cations, B or Te) decreases for more than 70 mol% TeO_2. It is thought that the absence of the borate anomaly is due to the complex relationship between the boron and tellurium structural units.[4]

In alkali borate glasses, $xM_2O.(100-x)B_2O_3$, the addition of the modifier, M_2O, provides extra oxygen atoms to the glass network, converting $[BO_3]$ units into $[BO_4]^-$ units, up to a composition where it becomes more favourable to form nonbridging oxygens (NBOs); above this composition N_4 then decreases again. TeO_2 is a conditional glass former, containing tellurium in oxidation state 4+, and if it is incorporated into the borate glass network as $[TeO_4]$ units, it does not provide additional oxygen atoms to facilitate the change in boron coordination. However, many studies of alkali tellurite glasses have shown that the tellurium structural units are $[TeO_3]^{2-}$ (with three NBOs) at high alkali content, progressively changing to $[TeO_4]$ as the concentration of TeO_2 in the glass is increased.[5–9] Thus, in the case of $xTeO_2.(100-x)B_2O_3$ glasses, it has been proposed[2,4] that the oxygen required to change the $[BO_3]$ units into $[BO_4]^-$ units is supplied by a concurrent change in the tellurite network from $[TeO_4]$ to $[TeO_3]^+$ units (which have

* Corresponding author. Email emma.barney@stfc.ac.uk

three bridging oxygens). Therefore, by measuring N_4 in a tellurium borate glass, the expected average tellurium coordination number can be inferred. This study aims to test this model by measuring both the B–O and Te–O coordination numbers using neutron diffraction.

As well as the unusual behaviour of the boron and tellurium environments, there are discrepancies in the literature concerning the range in composition for glass formation for the tellurium borate system. The glass forming range was first reported by Stanworth[10,11] to have an upper limit of 95 mol% TeO_2, beyond which the melt crystallises very easily on cooling. Subsequent work indicated that the lower boundary for single phase glass formation is imposed by a liquid–liquid phase separation, which has been reported to occur below 74·5[12] or 73·6[13] mol% TeO_2. Below this composition, two phases form; one region is a clear TeO_2 rich glass, whilst the other is an opaque, less dense, milky-white amorphous phase.[12] Compositions of tellurium borate glasses published in later studies using NMR,[2] density measurements,[4,14] Raman spectroscopy[4] and Fourier transform infrared (FTIR) spectroscopy[15] extend far below this limit. An aim of this study is to investigate the final compositions of glasses which have starting compositions on both sides of the reported phase separation boundary, and to examine the reliability of compositions reported previously.

2. Neutron diffraction theory

A neutron diffraction experiment measures the differential cross section, $d\sigma/d\Omega$, which is equivalent to the total scattering from the sample, $I(Q)$ where $\hbar Q$ is the magnitude of the momentum transfer.[16] The total scattering is the sum of the self scattering, $I^s(Q)$ (the interference between scattered waves from the same nucleus), and the distinct scattering, $i(Q)$ (the interference between scattered waves from different nuclei)

$$\frac{d\sigma}{d\Omega} = I(Q) = I^s(Q) + i(Q) \tag{1}$$

The self scattering can be calculated within an approximation for a glass of known composition, and is subtracted from the data to give the distinct scattering in reciprocal space, which can then be Fourier transformed to give a real space correlation function

$$T(r) = T^0(r) + \frac{2}{\pi} \int_0^\infty Q i(Q) M(Q) \sin(rQ) dQ \tag{2}$$

where $T^0(r)$ is the average density contribution to the correlation function, and $M(Q)$ is a modification function which is used to reduce termination ripples in the Fourier transform.

The result from a diffraction experiment is not element specific, and $T(r)$ is a weighted sum of all possible partial correlation functions, $t_{ll'}(r)$, broadened by convolution with the experimental resolution in real space $P(r)$

$$T(r) = \sum_{ll'} c_l \bar{b}_l \bar{b}_{l'} t_{ll'}(r) \tag{3}$$

where $\bar{b}_l$ and $\bar{b}_{l'}$ are the coherent scattering lengths for elements l and l' respectively, and c_l is the atomic fraction of element l. The summation is over all the pairwise combinations of elements in the sample. For a peak in $T(r)$ due to a particular correlation between atoms of element j and k, the coordination number, n_{jk}, can be calculated from the area, A_{jk}, and position, r_{jk}, of the peak and the coefficient for $t_{ll'}(r)$ according to

$$n_{jk} = \frac{r_{jk} A_{jk}}{(2 - \delta_{jk}) c_j \bar{b}_j \bar{b}_k} \tag{4}$$

3. Experimental detail

3.1 Sample preparation

Nine samples were prepared with nominal compositions of 90, 80, 70, 60 (×2), 50 (×3), and 40 mol% TeO_2. The chemicals used to produce the samples were TeO_2 and $^{11}B_2O_3$. Boron enrichment is necessary for neutron scattering studies to remove the problems which arise from the high neutron absorption of the ^{10}B isotope. The samples were heated in Pt/Rh crucibles to 800°C and held at temperature for 15 min before being splat quenched between copper sheets.

Analysis of neutron diffraction data requires accurate knowledge of the chemical and isotopic composition of the sample. The isotopic composition of the enriched B_2O_3 was checked using secondary ion mass spectroscopy (SIMS) and the ^{11}B content was not significantly different from that given by the supplier (99·62%). In addition, weight loss measurements were carried out on a 50 mol% TeO_2 glass. This sample was heated to 800°C and was held at that temperature. At intervals, over a period of 5 h, the sample was removed from the furnace and allowed to cool before being weighed and then returned to the furnace. Only a small weight loss (~0·37%) was measured, which occurred on the first heating of the sample and was attributed to H_2O loss. However, it should be emphasised that only clear portions of the resultant glassy products were selected for further analysis and, as explained below, this resulted in compositional variations.

3.2 Density measurements

Density measurements were carried out using a Quantachrome micropycnometer with helium as a working fluid. The densities of selected samples were re-measured using a Micromeritics Accupyc 1330 pycnometer to estimate the error. No significant discrepancies between the measurements were observed.

3.3 ^{11}B MAS NMR

Quantitative ^{11}B MAS NMR spectra were recorded at an applied field of 14·1 T on a Bruker Avance II+ spectrometer operating at a frequency of 192·04 MHz. The samples were loaded into a 3·2 mm MAS rotor and spun at 12–15 kHz. Acquisitions were taken with a pulse delay of 10 s and a 0·7 μs pulse width. Samples were referenced against crystalline BPO_4, which has a chemical shift of −3·3 ppm with respect to the primary reference $Et_2O:BF_3$ at 0 ppm.[17] The B_2O_3 content of the sample was determined by comparison of the spectral intensity from a known mass of sample with that from a known mass of $^{11}B_2O_3$.

3.4 Raman spectroscopy

Raman spectra were obtained on all glasses using a Renishaw Invia Raman spectrometer equipped with a 20 mW argon ion laser operating at 514·5 nm. The measurements were performed with 10 mW incident laser power and a ×50 objective, across the range 100–3200 cm^{-1}. The spectrometer resolution was approximately 2 cm^{-1}. Background corrections were carried out by fitting an exponential to the high wavenumber (>1200 cm^{-1}) part of the spectra. Repeated measurements were taken across the surface of the glass to ensure that the data were representative of the bulk sample.

3.5 Neutron diffraction

The neutron diffraction data were taken using the GEM diffractometer[18] at the ISIS pulsed neutron source, Rutherford Appleton Laboratory. 8·3 mm diameter vanadium cans with walls of 25 μm thickness were used to contain the samples, in the form of glass fragments. The experimental data, in both reciprocal and real space, are available from the ISIS Disordered Materials Database.[19]

4. Results

4.1 Density measurements

Figure 1 shows the densities of the clear glass samples, plotted against the nominal composition, compared with literature data. The densities for samples containing less than 80 mol% TeO$_2$ deviate from the linear fit through the published density data, giving clear evidence of phase separation, which is consistent with the glass forming ranges reported in previous studies.[12,13] The selected clear glass material has higher TeO$_2$ content than expected from the nominal compositions, and a comparison of the density values in Table 1 and Figure 1 with those reported in the literature indicates that three distinctly different compositions have been made successfully. The 90 and 80 mol% TeO$_2$ samples have not phase separated, whilst the majority of the samples made

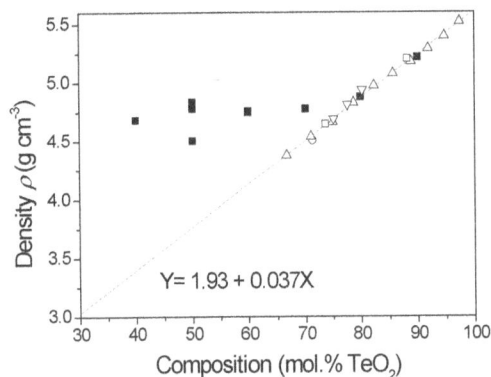

Figure 1. The measured densities of the glasses using nominal compositions (closed squares) plotted with densities reported in the literature[4,11,13,14] (open symbols: up-triangle[4]; circle[11]; down-triangle[13]; square[14]). The dotted line is a linear fit to the literature data

with nominal compositions of 70 mol% TeO$_2$ or less have densities that are very close to that measured for the 80 mol% TeO$_2$ sample. One of the samples, with a nominal composition of 50 mol% TeO$_2$, has a lower density, and it is thus probable that this glass has a lower TeO$_2$ content.

4.2 ^{11}B NMR

Quantitative NMR was carried out to determine the B_2O_3 content of each of the glasses used in this study. Figure 2 shows the ^{11}B NMR spectra for the nine samples. The data confirm one of the conclusions drawn from the density data; three distinctly different compositions have been made in this study. These compositions are given in Table 1, and they confirm that the 90 and 80 mol% TeO$_2$ samples have not changed composition (within the error of ±2 mol% TeO$_2$ determined by repeated measurements) and that the remaining samples have phase separated, with the majority of selected clear glass regions having compositions of between 80 and 83 mol% TeO$_2$. The nominal 50 mol% TeO$_2$ sample, which has the

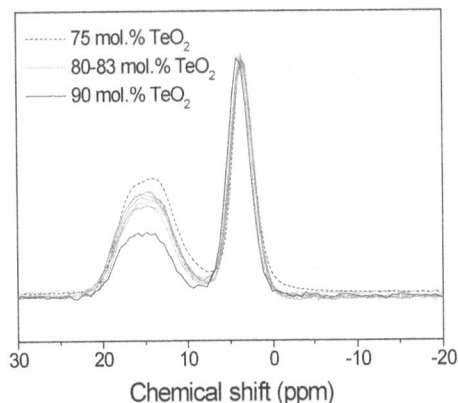

Figure 2. ^{11}B NMR spectra, plotted to show the three-coordinated peak, between 10 and 20 ppm, and the four-coordinated peak at ~4 ppm. All the data are normalised to the height of the [BO$_4$] peak. The compositions are those determined by ^{11}B NMR

Table 1. Density, N_4 and N_4^* values for the tellurium borate samples, and compositions measured from quantitative ^{11}B NMR. For duplicate measured compositions, only one sample was used in the neutron diffraction analysis, indicated by †

Nominal composition (mol% TeO₂)	Density (g cm⁻³) (±0·005)	N_4 (±0·02)	N_4^* (±0·01)	Compositions from quantitative ^{11}B NMR (mol% TeO₂) (±2)
90	5·243	0·59	0·10	90
80	4·877	0·51	0·15	82
70	4·777	0·48	0·17	81
60 (#1)	4·76	0·48	0·16	80 †
60 (#2)	4·75	0·45	0·15	80
50 (#1)	4·78	0·48	0·16	81 †
50 (#2)	4·835	0·51	0·15	83
40	4·7093	0·45	0·15	80
50 (#3)	4·505	0·41	0·17	75

lowest N_4 value and density, was determined to have a composition of 75 mol% TeO₂. For the remainder of this paper all quoted compositions are the analysed values from quantitative ^{11}B NMR.

4.3 Raman spectroscopy

The Raman scattering data for the nine samples are shown in Figure 3(a), normalised to the height of the 660 cm⁻¹ peak. The spectra were fitted with seven peaks, using wavenumber values obtained by Sekiya et al[4] as starting parameters. When fitting the data, the widths of the four peaks between 600 and 800 cm⁻¹ were constrained to be the same, generating a final fit which had a similar appearance to published fits to Raman data.[4,20] Figure 3(b) shows the ratio between the areas of the 660 cm⁻¹ peak (reported to be due to vibrations of [TeO₄]⁻ and [TeO₃₊₁]⁻ structural units[9,21,22]) and the 710 cm⁻¹ peak (attributed variously to [TeO₃]⁻ units[20] and [TeO₄]⁻ units[4,9]) as measured in both this work and in previous studies.[4,20]

4.4 Neutron diffraction

The distinct scattering curves, $i(Q)$, after full correction using the ATLAS suite of programs[23] are plotted in Figure 4(a), in which the compositions determined by quantitative ^{11}B NMR are given. There is very

good agreement between the $i(Q)$ data for the three 80 mol% TeO₂ samples and similarly good agreement between the two 81 mol% TeO₂ samples. Further processing will only be described for one glass of each composition (for duplicate measured compositions, the samples used are labelled in Table 1 using †). Oscillations were clearly visible up to $Q=40$ Å⁻¹. A quadratic of the form $A+BQ^2$ was fitted to the data at low Q to extrapolate the data back to $Q=0$. A Fourier transform was carried out on the distinct scattering data using a maximum momentum transfer, Q_{max}, of 35 Å⁻¹ and the Lorch modification function[24] to reduce termination ripples. Figure 4(b) shows the resultant total correlation functions, $T(r)$.

The neutron data $T(r)$ exhibit four clear physical peaks below 3·0 Å. The first peak, at ~1·4 Å, and the third peak, at ~2·4 Å, decrease in height with the addition of TeO₂ and so can be attributed to the borate part of the network, arising from B–O and O_B–O_B correlations. The second peak, at ~1·9 Å, and the fourth peak, at ~2·8 Å, increase in height with increasing TeO₂ content and so are attributed to Te–O and O_{Te}–O_{Te} correlations.

In order to extract the B–O and Te–O coordination numbers, the correlation functions were fitted with Gaussian peaks convoluted with the resolution function of the measurement[25] (see Figure 5). This function is appropriate for a single well defined interatomic distance. With the exception of the 90 mol% TeO₂ sample, the B–O correlation could not be accurately fitted with a single peak, due to the presence of both [BO₄]⁻ and [BO₃] units which have different B–O bond lengths, leading to an asymmetric peak shape. A simultaneous two peak fit was used and the positions of the peaks were constrained using the bond lengths calculated by the bond valence method,[26] assuming all of the B–O bond lengths within a unit are equal (leading to $r_{BO}=1·370$ Å for [BO₃] units and $r_{BO}=1·477$ Å for [BO₄]⁻ units). Table 2 gives the B–O coordination numbers for each of the samples, as measured by both NMR and neutron diffraction, and these two measurements agree within error. Accurate coordination numbers can only be extracted from neutron diffraction data if the

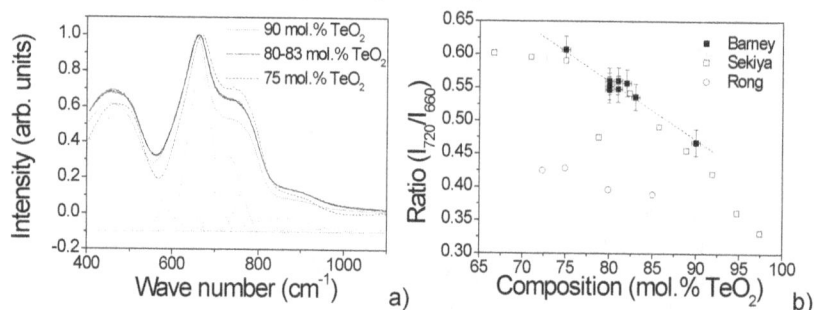

Figure 3. (a) Raman spectra measured for the nine samples, normalised to the height of the 660 cm⁻¹ peak. The five peak fit to the Te–O vibrational peaks, and the two peak fit to the B–O vibrational peaks for the 90 mol% TeO₂ glass are shown as a light dotted line, and displaced by −0·1. (b) The ratio of the areas of the two fitted Gaussians at 660 and 720 cm⁻¹, as measured in this study (closed squares), and as reported by Sekiya et al[4] (open squares) and Rong et al[20] (open circles). The dashed line is a linear fit to our data

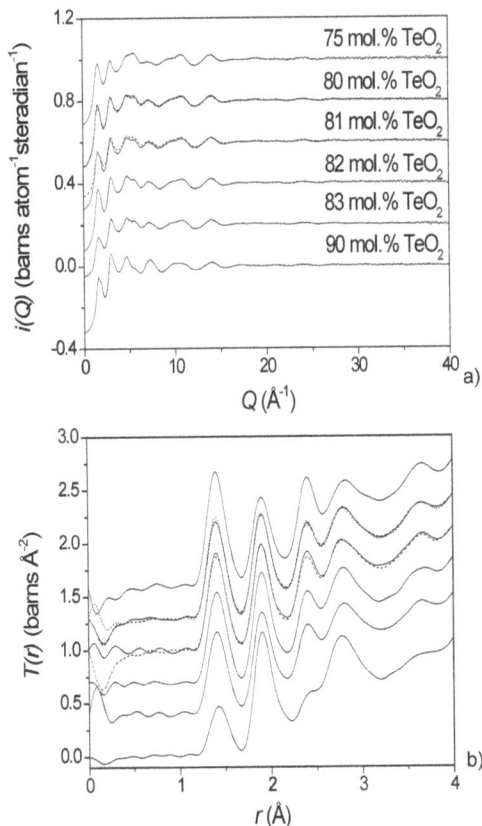

Figure 4. (a) The distinct scattering, i(Q), and (b) the total correlation function, T(r), for the nine tellurium borate glasses. The compositions given are those determined by ^{11}B NMR. The order of the samples in part (a) is repeated in part (b). Dashed and dotted lines show data from samples with repeat compositions

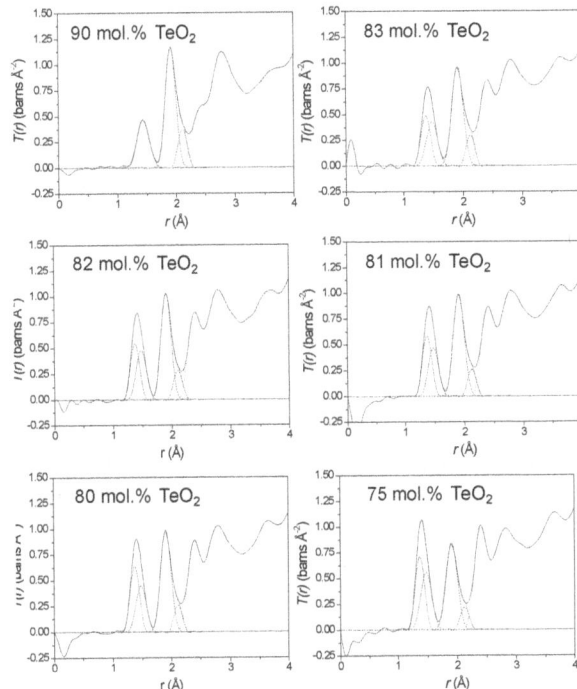

Figure 5. Selected correlation function fits, spanning the range of compositions measured. (For compositions where more than one sample was studied, the sample identified in Table 1 was used)

correct sample composition is used (due to both the coefficient $c_j \bar{b}_j \bar{b}_k$ in Equation (4), and the average density term, $T^0(r)$), and hence this agreement is an indication that the compositions used to analyse the neutron diffraction data are correct. The differences between the coordination numbers from ^{11}B NMR and from neutron diffraction are of order 1%, and coordination number errors of this magnitude are the best that can currently be achieved by neutron diffraction.[27]

The Te–O correlation in $T(r)$ is also asymmetric and requires a two peak fit. The main Te–O peak at ~1·9 Å was fitted using only the leading edge of the peak, by using a fitting range with a maximum distance of 1·95 Å, just past the peak maximum. The residual

clearly showed a second Te–O peak at ~ 2·11 Å, prior to the O_B–O_B correlation, and both these peaks in the residual were simultaneously fitted in order to model the longer Te–O contributions. Table 2 gives the individual coordination numbers for the two fitted Te–O peaks and the total coordination number (i.e. the sum of the two contributions). Figure 6 shows that, as the concentration of TeO_2 in the glass is reduced, the total tellurium coordination number decreases because the coordination number for the second component at ~ 2·11 Å declines, whilst the coordination number for the first component remains essentially constant. Table 2 also gives the expected tellurium coordination number, n'_{TeO}, calculated from the NMR B–O coordination number according to Sekiya et al's model.[4] This model is based on implicit assumptions, firstly that a pure TeO_2 glass would consist entirely of $[TeO_4]$ units connected by bridging oxygens, and secondly that the incorporation of B_2O_3 into the network involves only bridging oxygens,

Table 2. Coordination numbers measured for the B–O and Te–O peaks in T(r) along with the Te–O coordination calculated using the model proposed by Sekiya et al.[4] The B–O bond lengths, used to fit T(r), were constrained, using bond–valence calculations, to be 1·37 and 1·47 Å

Measured composition mol% TeO_2 (±2 mol% TeO_2)	Neutron diffraction n_{BO}	NMR n_{BO}	$r_{Te-O(1)}$ (Å) (±0·05)	$r_{Te-O(2)}$ (Å) (±0·05)	$n_{TeO(1)}$	$n_{TeO(2)}$	Total n_{TeO}	Calculated n'_{TeO}	Δn_{TeO} (calculated– measured)
90	3·57(3)	3·59(2)	1·91	2·11	2·62(2)	0·80(1)	3·42(2)	3·87(3)	0·45(4)
83	3·53(4)	3·51(2)	1·91	2·12	2·58(1)	0·65(2)	3·23(4)	3·79(4)	0·56(5)
82	3·54(3)	3·51(2)	1·91	2·11	2·59(2)	0·67(1)	3·25(2)	3·78(4)	0·53(4)
81	3·48(4)	3·48(2)	1·91	2·12	2·65(1)	0·58(1)	3·23(1)	3·77(4)	0·54(4)
80	3·45(2)	3·48(2)	1·90	2·12	2·64(3)	0·58(1)	3·22(3)	3·76(4)	0·54(5)
75	3·41(3)	3·41(2)	1·90	2·12	2·62(2)	0·56(2)	3·18(3)	3·73(4)	0·54(5)

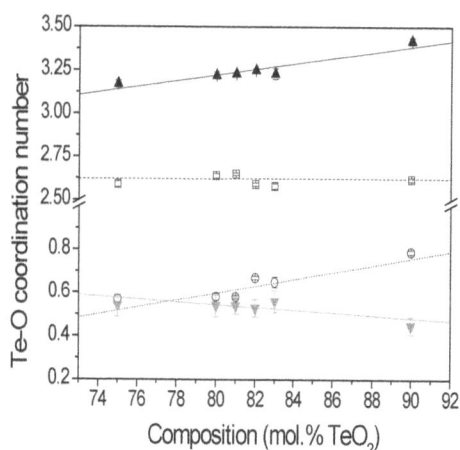

Figure 6. Change in total coordination number of the Te atom (black triangles and solid black line), comprising the main Te–O peak (open squares and dashed line) and the asymmetric shoulder (open circles and dotted line). The missing Te–O coordination for each sample, Δn_{TeO}, calculated in Table 2, is also shown (grey triangles and grey line)

with the result that n'_{TeO} is given by

$$n'_{TeO}=(600-2x-2(100-x)n_{BO})/x \qquad (5)$$

The coordination number measured by the fitting approach used in this study is significantly less than this expected value, suggesting that the number of $[TeO_3]^+$ units grows much faster than the number of $[BO_4]^-$ units.

5. Discussion

5.1 Sample compositions, N_4 and density

The results from ^{11}B NMR, Raman spectroscopy and density measurements presented in this work are consistent with the reported lower single phase glass forming limit for tellurium borate glasses, given in the literature as 74 mol% TeO_2.[12,13] This contradicts several other reports of structural studies,[2,4,14,15] which describe results for samples with significantly less than 74 mol% TeO_2, but these reports do not give evidence that compositional analysis has been carried out. For a structural study of glasses, it is important that the composition is known reliably, for which the use of several experimental techniques is advantageous, as also shown, for example, by our studies of tin borate glasses.[28]

Sekiya *et al* have published Raman scattering and density data for ten different tellurium borate glass compositions,[4] and N_4 values from ^{11}B NMR measurements for seven of these samples.[2] The samples with more than 95 mol% TeO_2 have not been considered here as they are outside the range of glass formation[10,11] and the NMR peaks are very narrow, which may indicate crystallisation. Combining the NMR[2] and density data,[4] we have performed a linear fit to obtain the following relationship

$$N_4=0.2529\rho-0.718 \qquad (6)$$

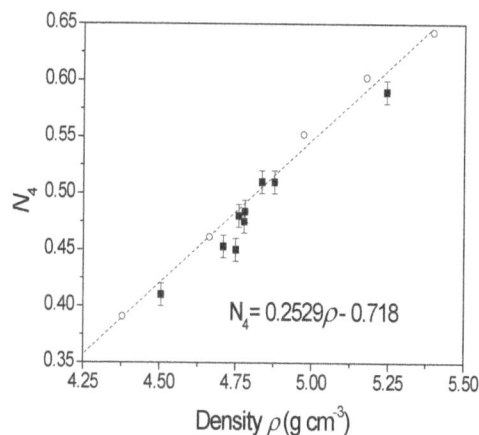

Figure 7. The relationship between density and N_4 calculated using the data published by Sekiya et al[2,4] (open circles and dotted line) compared with the data measured in the current study (closed squares)

Figure 7 shows both the density–N_4 data reported by Sekiya *et al*[4] and the linear relationship, along with the values measured in this study. The two data sets agree well, confirming the linear relationship between density and N_4. However, the compositions reported by Sekiya *et al*[2,4] do not agree with the compositions measured in this study for samples with similar density and N_4 values. Figure 8 shows the composition dependence of the N_4 values measured in the present work, together with the values measured by Sekiya *et al*[2,4] and Goring *et al*;[3] the difference in relationship between N_4 and composition in the two major studies (Sekiya *et al*[2,4] and the current work) is illustrated with guides to the eye. At 90 mol% TeO_2, the two lines agree within error, but the deviation between the two lines becomes greater as the concentration of TeO_2 is reduced. The two samples reported by Sekiya *et al*[2,4] which produce such strong disagreement with this study, have nominal compositions of 67 mol% TeO_2 (inside the phase separation region) and 75 mol% TeO_2 (on the phase separation boundary). The lack of compositional analysis in the report by Sekiya *et al*,[2,4] in conjunction with the range of compositions which are given (extending far into the phase separation region[12,13]), suggest that the correct compositions for their samples with lowest TeO_2 content differ significantly from the nominal compositions. Comparisons with samples in the present study which have the same N_4 values, within error, indicate that Sekiya *et al's*[2,4] samples with nominal compositions of 67 and 75 mol% TeO_2 actually had correct compositions of 75 and 80 mol% TeO_2, respectively. In contrast, Goring *et al*[3] have published NMR data for three samples with analysed compositions and there is close agreement between the N_4 fractions which they report for samples with 75 and 77 mol% TeO_2 and the data presented in the present work.

Usually a linear density versus composition curve, such as that shown in Figure 1 and parameterised by Equation (6), is an indication that the coordination

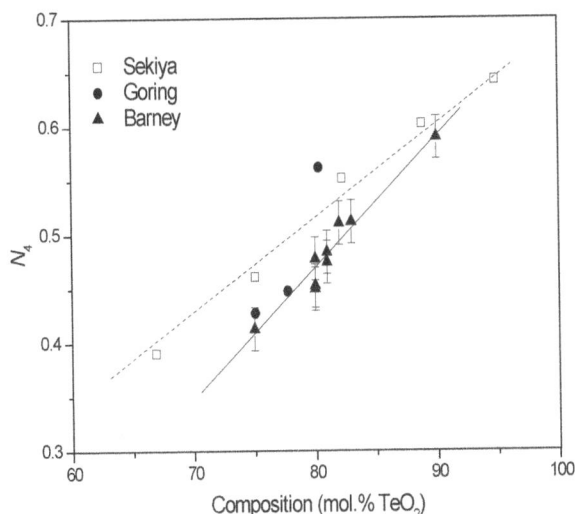

Figure 8. The N_4 values from data published by Sekiya et al[2,4] (open squares and dashed line guide to the eye) and Goring et al[3] (closed circles), compared to the values measured in this study (closed triangles and continuous line guide to the eye)

number for the network former does not change with composition.[29] For example, the change in B–O coordination number that occurs for most borate glasses is accompanied by a nonlinear density curve. [1] However, tellurium borates are an exceptional case, for which the density curve is linear due to the presence of two network formers, for both of which there is a change in coordination number.

5.2 Raman spectra

Unambiguous determination of the origin of vibrational peaks in Raman spectra from tellurites is difficult. Studies of pure TeO_2 crystal phases[21,22] have shown that the features in the 400–500 cm^{-1} region of the spectrum are due to symmetric Te–O–Te vibrations, indicating a polymerised network. Higher frequency features (~600–800 cm^{-1}) are the result of vibrations of both terminal Te–O bonds and asymmetric Te–O–Te bridges (for example, the Te–$_{eq}$O$_{ax}$–Te bridges present in crystalline α-TeO_2). The exact frequency of peaks in the 600–800 cm^{-1} region is, as shown in alkali tellurite crystal studies,[9] dependent on the Te–O bond lengths, and so attributing the vibrations in tellurite crystals to specific structural units is not possible. Studies of tellurite glasses also disagree in assigning the vibrational origins of the peaks. All tellurite glasses have a broad peak at ~450 cm^{-1}, which indicates that there is a continuous Te–O network with symmetric Te–O–Te vibrations. In addition, there is a peak manifold at 600–800 cm^{-1}, with the most intense peak at ~660 cm^{-1}. Both previous Raman studies of tellurium borate glasses[4,20] have applied five peak fits to model the spectra, yielding similar peak positions and widths. The two studies disagree in the precise assignment of the peak

origins, but agree that a decrease in the intensity of the 660 cm^{-1} peak relative to the 720 cm^{-1} peak indicates a reduction in the number of 4-coordinated Te atoms (whether considered as [TeO_4] or [TeO_{3+1}] units).[4,20] The assignment of the 660 cm^{-1} peak in terms of 4-coordinated Te is made because it corresponds to the Te–$_{eq}$O$_{ax}$–Te asymmetric vibration associated with [TeO_4] units in crystalline α-TeO_2.

The ratio of the areas of the peaks at 720 cm^{-1} and 660 cm^{-1} was determined from the fits to the Raman spectra, and is plotted in Figure 3(b) together with the ratios reported by Sekiya et al[4] and Rong et al. [20] Rong et al observed only a slight change in peak area ratio with composition and concluded that there is very little change in the tellurium environment, whereas Sekiya et al[4] observed a decrease in the intensity of the 660 cm^{-1} peak as the amount of TeO_2 in the glass is reduced. The data in the present study show a linear change in the peak area ratio with composition, in good agreement with Sekiya et al[4] for samples of similar composition (note that the nominal compositions have been used to plot the data of Sekiya et al[4] in our Figure 3(b), and our work above shows that these points should be shifted to the right to account for problems with the composition, particularly for lower TeO_2 contents).

The Raman spectroscopy work of Sekiya et al[4] clearly demonstrates the problems of phase separation in the tellurium borate system. Their three samples with nominal compositions at or below the reported limit of glass formation ($\leq$75 mol% TeO_2) have very similar peak area ratios (see Figure 3(b)), indicating that the samples have phase separated, producing clear homogeneous glass material of the same composition (~75 mol% TeO_2). Furthermore, considerations of composition based on Raman spectroscopy, which only samples a small sample volume, and bulk measurements (density and NMR) lead to inconsistent conclusions. For example, the Raman peak area ratio reported by Sekiya et al[4] for a sample with nominal 79 mol% TeO_2 indicates a true composition of about 87 mol% TeO_2 (according to the dashed line in Figure 3(b)), whereas the composition correction which we have established in the previous section on the basis of density and NMR data indicates a true composition of 83 mol% TeO_2 for this sample. This finding suggests the possibility that for some samples microscopic phase separation may occur for compositions above the reported phase separation boundary. A study of phase separation on this scale is beyond the scope of this work, but has been considered in other studies using electron microscopy. Dimetriev et al[12] found evidence for microheterogeneity in clear glasses formed on cooling from melts containing between 74·5 and 94–95 mol% TeO_2. However, a more recent study by Bürger et al found no evidence to support this, and concluded that TeO_2-rich glasses are microhomogeneous.[13]

5.3 The tellurium environment

Using density measurements, Raman spectroscopy and ^{11}B MAS NMR, the importance of compositional analysis in the tellurium borate system has been demonstrated, especially for samples with initial compositions near the phase separation boundary. Equations (3) and (4) show that knowledge of the atomic fraction of each element in the sample is critical in the analysis of neutron diffraction data, particularly if accurate coordination numbers are required. The quality of the compositional analysis performed in this study is verified by a comparison of the boron coordination number measured by neutron diffraction with the value derived from the N_4 fraction measured by ^{11}B NMR. The coordination numbers measured by the two methods are listed in Table 2 and they agree within the experimental errors. The results extracted for the peak position and coordination number for the Te–O correlation are therefore reliable over the distance range considered.

A comparison of the tellurium environment in the tellurium borate system with that reported for the potassium tellurite system shows that the addition of B_2O_3 to TeO_2 has a very different effect on the tellurium bonding than K_2O.[30] The addition of an alkali to tellurite glass causes a shortening of the primary bond length in the Te–O distribution,[7,8,30] but the Te–O(1) peak positions given in Table 2 show that there is almost no change in the primary peak position with the addition of B_2O_3 to the glass. This reflects the difference between the [TeO_3] units which form on adding these two different oxides to the tellurite structure. B_2O_3 forms part of the glass network, introducing no excess oxygen to the structure so that the 3-coordinated Te atoms are in [TeO_3]$^+$ units, regardless of the concentration of B_2O_3, and the Te–O bond length does not change. However, the addition of K_2O, which acts as a modifier, contributes excess oxygen atoms to the network, allowing the formation of NBOs around the network cations, and thus shortening the Te–O bond lengths. The higher the concentration of K_2O, the more NBOs are formed, with a limiting situation in which all the Te cations are in [TeO_3]$^{2-}$ units.

Figure 6 shows that the average coordination number for tellurium increases as TeO_2 is added to the glass. This is not due to the change in coordination under the primary peak in the correlation function, but due to the increase in the number of longer Te–O bonds (which were modelled by fitting a second peak at ~2·11 Å to the residual from fitting the first peak). The increase in the average Te–O coordination number with TeO_2 content, and the increase in the number of longer Te–O bonds, is consistent with the observation from Raman scattering that there is an increase in the proportion of 4-coordinated tellurium atoms with increasing TeO_2 content. However, the average coordination number measured using this two peak model for the asymmetric Te–O bond distribution is significantly less than is expected from model proposed by Sekiya et al[4] (the difference is calculated in Table 2 and shown in Figure 6).

The Te–O contribution to the correlation function, $T(r)$, is asymmetric and broad. This is clear evidence that there is a distribution of Te–O bond lengths, rather than a single well defined Te–O bond length. In order to determine a value for the total Te–O coordination number, we have obtained the area under the Te–O peak in the correlation function by a procedure that involves fitting two peaks which are appropriate for single well defined distances.[25] It must be acknowledged that, whilst this is a reasonably satisfactory way of parameterising the peak, it is not proof that there are two well defined Te–O bond lengths, short and long, in these glasses. Instead, it appears that there is a narrower and more symmetric distribution of shorter bonds, and a broader and more asymmetric distribution of longer bonds. In particular, our fit treats the distribution of longer bonds as symmetric, with the consequence that we may have underestimated the longer bond coordination number to some extent. In addition, whilst the Te–O coordination number we have determined is reliable over the distance range for which the fit is defined, the measurement is limited by overlap with the large O_B–O_B peak at ~2·4 Å. Coordination numbers are only of meaning for a given range in interatomic distance.[31,32] For the B–O peak in the correlation function the definition of a distance range for the extraction of a coordination number is clear, because the correlation function falls close to zero after the B–O peak. This is due to the narrow distribution of B–O bond lengths for the two borate structural units. However, such a clear definition of distance range is never possible for Te–O coordination numbers in tellurite glasses because the asymmetric longer bond distribution extends to longer distance so that it overlaps with the shortest O_{Te}–O_{Te} contribution from the tellurite units themselves.[7]

For the range in Te–O distances which we have been able to consider, our results show that, as the TeO_2 content is reduced, there is a decline in the value of $n_{TeO(2)}$ (see Table 2), so that the Te–O distribution becomes narrower and more symmetric. This indicates that, as B_2O_3 is added, the number of longer Te–O bonds becomes smaller, and that there is an increase in the number of [TeO_3] units. Thus our results imply that when B_2O_3 is added to the glass the number of [TeO_3] units increases more than is predicted by the model of Sekiya et al,[4] suggesting that the assumptions of this model may not be correct. To further investigate the Te–O coordination number it is necessary to reduce the effect of the overlap of the Te–O and O–O contributions, and an x-ray diffraction study will be performed on the samples used in this study, from which a second set of correlation functions can be extracted. This will allow the Te–O peak to be modelled without a major interference from the O–O peaks and, by fitting the two correlation functions simultaneously, it will be possible to model the Te–O distribution more accurately. If there is significant additional Te–O coordination at longer distance then it may be that Sekiya *et al*'s model[4] will be proved correct. Determining the validity of this model will show whether NBOs are present in tellurite units in pure TeO_2 and this is fundamental to the understanding of the structure of tellurite glasses.

6. Conclusions

A multi-technique (^{11}B NMR, Raman spectroscopy, density and neutron diffraction) structural study has been performed on glass samples from the tellurium borate system. The sample composition was analysed by quantitative NMR, and the results are consistent with reports of phase separation for melts with a nominal composition of less than 74 mol% TeO_2. A linear correlation was found between the fraction of 4-coordinated borons, N_4, and the density which was of great use in interpreting the results of other studies in the literature in which the sample composition was not analysed. This approach shows strong evidence that claims to have studied single phase glass samples containing less than 74 mol% TeO_2 are not reliable.

A correct knowledge of the sample composition is a prerequisite for a reliable analysis of neutron diffraction data. For the work reported here, a close agreement was obtained between the B–O coordination numbers measured by neutron diffraction and by ^{11}B NMR, showing that the composition and neutron diffraction data are reliable. The Te–O coordination number decreases from 3·42, for a sample with 90 mol% TeO_2, to 3·18 for a sample with 75 mol% TeO_2, over the distance range for which the coordination number could be accurately measured. This result is consistent with the Raman spectra, which show a decline in the proportion of 4-coordinated Te as the TeO_2 content is reduced. The neutron diffraction results show evidence for two components in the distribution of Te–O bond lengths: There is a narrow distribution of short bonds, of length about 1·91 Å, and the average coordination number at these distances is essentially constant at about 2·6 for all compositions. There is also a broad distribution of longer bonds, of length about 2·11 Å, and the change in total coordination number with TeO_2 content arises from changes in the coordination number associated with this second distribution. The observed Te–O coordination numbers are not consistent with a previous literature model in which there is a simple balance between the numbers of 3-coordinated telluriums and 4-coordinated borons. However, further work is needed to determine the full extent of the Te–O distribution and if there is a significant contribution beyond ~2·3 Å the model may be correct.

Acknowledgements

The sample preparation and characterisation was carried out at the University of Warwick Physics Department. The authors would like to acknowledge the assistance of Dr Richard Morris for carrying out the SIMS for this study, and Dr Andrew Howes and Dr Tom Kemp for their assistance in collecting the NMR data. Dr Solveig Felton, Dr Robin Cruddace and the Warwick Diamond group are also thanked for their assistance in using the Raman spectrometer. This work was funded by EPSRC and the Centre for Materials Physics and Chemistry, grant number CMPC04108.

References

1. Hannon, A. C. & Holland, D. *Phys. Chem. Glasses : Eur. J. Glass Sci. Technol. B*, 2006, **47**, 449.
2. Sekiya, T., Mochida, N., Ohtsuka, A., Soejima, A., Yasumori, A. & Yamane, M. *Glastech. Ber. Glass Sci. Technol.*, 1993, **66**, 15.
3. Goring, R., Burger, H., Nass, H. & Schnabel, B. *Phys. Status Solidi A*, 1981, **68**, K29.
4. Sekiya, T., Mochida, N., Ohtsuka, A. & Soejima, A. *J. Non-Cryst. Solids*, 1992, **151**, 222.
5. Neov, S., Gerasimova, I., Kozhukharov, V., Mikula, P. & Lukas, P. *J. Non-Cryst. Solids*, 1995, **192-3**, 53.
6. McLaughlin, J. C., Tagg, S. L., Zwanziger, J. W., Haeffner, D. R. & Shastri, S. D. *J. Non-Cryst. Solids*, 2000, **274**, 1.
7. McLaughlin, J. C., Tagg, S. L. & Zwanziger, J. W. *J. Phys. Chem.*, 2001, **105**, 67.
8. Iwadate, Y., Mori, T., Hattori, T., Nishiyama, S., Fukushima, K., Umesaki, N., Akagi, R., Handa, K., Ohtori, N., Nakazawa, T. & Iwamoto, A. *J. Alloys Compd.*, 2000, **311**, 153.
9. Sekiya, T., Mochida, N., Ohtsuka, A. & Tonokawa, M. *J. Non-Cryst. Solids*, 1992, **144**, 128.
10. Stanworth, J. E. *J. Soc. Glass Technol.*, 1952, **36**, 217.
11. Stanworth, J. E. *J. Soc. Glass Technol.*, 1954, **38**, 425.
12. Dimitriev, Y. & Kashchieva, E. *J. Mater. Sci.*, 1975, **10**, 1419.
13. Bürger, H., Vogel, W., Kozhukharov, V. & Marinov, M. *J. Mater. Sci.*, 1984, **19**, 403.
14. Vogel, W., Burger, H., Winterstein, G., Ludwig, C. & Jaekel, W. *Silikattechnik*, 1974, **25**, 209.
15. Rada, S., Culea, M. & Culea, E. *J. Non-Cryst. Solids*, 2008, **354**, 5491.
16. Wright, A. C. In: *Experimental Techniques of Glass Science*, Eds. C. J. Simmons and O. H. El-Bayoumi, American Ceramic Society, Westerville, 1993, p. 205.
17. Turner, G. L., Smith, K. A., Kirkpatrick, R. J. & Oldfield, E. *J. Magn. Reson.*, 1986, **67**, 544.
18. Hannon, A. C. *Nucl. Instrum. Meth. A*, 2005, **551**, 88.
19. Hannon, A. C. ISIS Disordered Materials Database, http://www.isis.rl.ac.uk/disordered/Database
20. Rong, Q. J., Osaka, A., Nanba, T., Takada, J. & Miura, Y. *J. Mater. Sci.*, 1992, **27**, 3793.
21. Mirgorodsky, A. P., Merle-Mejean, T., Champarnaud, J. C., Thomas, P. & Frit, B. *J. Phys. Chem. Solids*, 2000, **61**, 501.
22. Noguera, O., Merle-Mejean, T., Mirgorodsky, A. P., Smirnov, M. B., Thomas, P. & Champarnaud, J. C. *J. Non-Cryst. Solids*, 2003, **330**, 50.
23. Hannon, A. C., Howells, W. S. & Soper, A. K. *Inst. Phys. Conf. Ser.*, 1990, **107**, 193.
24. Lorch, E. *J. Phys. C*, 1969, **2**, 229.
25. Hannon, A. C., Grimley, D. I., Hulme, R. A., Wright, A. C. & Sinclair, R. N. *J. Non-Cryst. Solids*, 1994, **177**, 299.
26. Hannon, A. C. & Parker, J. M. *Phys. Chem. Glasses*, 2002, **43C**, 6.
27. Hannon, A. C., Di Martino, D., Santos, L. F. & Almeida, R. M. *J. Phys. Chem. B*, 2007, **111**, 3342.
28. Hannon, A. C., Barney, E. R. & Holland, D. *Phys. Chem. Glasses: Eur. J. Glass Sci. Technol. B*, 2009, submitted.
29. Barney, E. R. PhD Thesis, 2007, *The structural role of lone pair ions in novel glasses*, University of Warwick, United Kingdom.
30. Hannon, A. C., Umesaki, N. & Tatsumisago, M. *unpublished work*,
31. Black, P. J. & Cundall, J. A. *Acta Cryst.*, 1965, **19**, 807.
32. Mikolaj, P. G. & Pings, C. J. *Phys. Chem. Liq.*, 1968, **1**, 93.

Proc. VI Int. Conf. Borate Glasses, Himeji, Japan, 18–22 August 2008   *Phys. Chem. Glasses: Eur. J. Glass Sci. Technol. B*, December 2009, **50** (6), 343–354

Structure of lithium borate glasses and melts: investigation by high temperature Raman spectroscopy

A. A. Osipov* & L. M. Osipova

Institute of Mineralogy UB RAS, Miass, Chelyabinsk region, Russian Federation

Manuscript received 30 October 2008
Revised version received 12 January 2009
Accepted 29 September 2009

Raman spectra of xLi$_2$O.(1–x)B$_2$O$_3$ glasses and melts (x=0·05, 0·1, 0·15, 0·25 and 0·4) have been measured and analysed in the temperature range from 20 to 1000°C. Modelling of mid frequency (650–900 cm^{-1}) and high frequency (1200–1650 cm^{-1}) spectra ranges has been performed to determine quantitative information about the structural changes in the samples, depending on composition and temperature. Concentrations of the fundamental structural units were determined as a result of modelling of the high frequency spectra contours. The fraction of four-fold coordination boron atoms, N$_4$, decreases with temperature increase. Dynamic equilibrium between structural units in the melts is given by (BØ$_{4/2}$)$^-$ ⇌ BØ$_{2/2}$O$^-$ (Ø – bridging oxygen; O$^-$ –nonbridging oxygen). A dominating mechanism of the structural change of the glasses with x≤0·15 as a function of heating and fusion is a process of destruction of the boroxol rings. Structural changes in the intermediate range order for x>0·15 are not limited by the dissociation of boroxol rings, but are also connected with the transformations between other superstructural groupings and their destruction.

Introduction

Borate glasses and melts are a unique subject for research because boron atoms can be present in the structure both in three- and four-fold coordination. Also, in the structure of borate glasses there are larger groups than the fundamental structural units – so called superstructural groups. Quite a number of works have allowed determination of the main regularities of the composition dependence of the formation of the glass network, both in terms of short range order (SRO) and intermediate range order (IRO).[1–5] The structure of borate melts has been investigated much less, in contrast to widely investigated borate glasses. This is most likely connected with the technical difficulties of experiments at high temperatures. Nevertheless, the knowledge of the structure of melts, and their structural change depending on composition and temperature, are very important, both from a scientific and a practical point of view.

Raman spectroscopy is a useful method for direct investigation of melt structure at high temperature. Information can be obtained by modelling of the Raman spectra as a superposition of Gaussian bands. The efficiency of such an approach has been shown in a number of works devoted to the study of silicate glasses and melts.[6–9] However, this method has not been used practically for the structural investigation of borate glasses and melts, until recently. Significant progress in this direction has been achieved in the

works of Japanese researchers,[10–12] where mechanisms for the structural changes of Na$_2$O–B$_2$O$_3$ glasses and melts have been considered, and the concentrations of the fundamental structural units are calculated on the basis of modelling of the Raman spectra obtained at different temperatures. The application of deconvolution procedures to Raman spectra of glasses and melts from alkali borate systems are thus of great interest. This is why we have carried out the high temperature investigation of Li$_2$O–B$_2$O$_3$ glasses and melts.

Experiment

Samples of the glasses were prepared from boron oxide (special purity grade) and lithium carbonate (reagent grade). Initial reagents were carefully dried before weighing. Then they were weighed in the appropriate ratios and mixed in a mortar with an alcohol. The batch was dried in a furnace at temperature t=120°C for 2 h, and melted in a platinum crucible at t=1100°C. The weight of the batch was 15 g in all experiments. The obtained homogeneous melt was poured into a small platinum crucible and cooled in air at room temperature.

A specially designed high temperature setup, based on a DFS-24 monochromator, was used for the Raman spectra measurements (Figure 1). The small platinum crucible containing the sample (7 on the figure) was placed into a compact electrical furnace, which was placed on the adjustment table (8) which

*Corresponding author. Email armik@mineralogy.ru

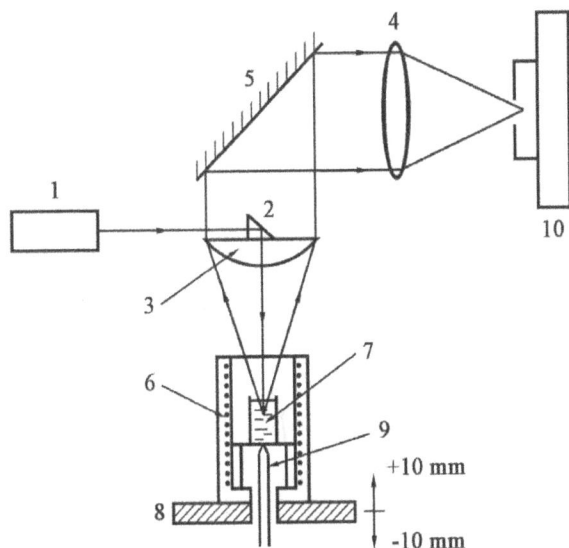

Figure 1. Schematic illustration of the optic arrangement for Raman scattering measurement. 1 - LTI – 701 pulse laser (λ=532 nm, <P>=1 W, τ=2 μs), 2 – prism, 3 - focusing/collecting lens, 4 - focusing lens, 5 - flat mirror, 6 - electrical furnace, 7 - Pt crucible with a sample, 8 - table for adjustment, 9 – thermocouple, 10 - DFS – 24 double monochromator

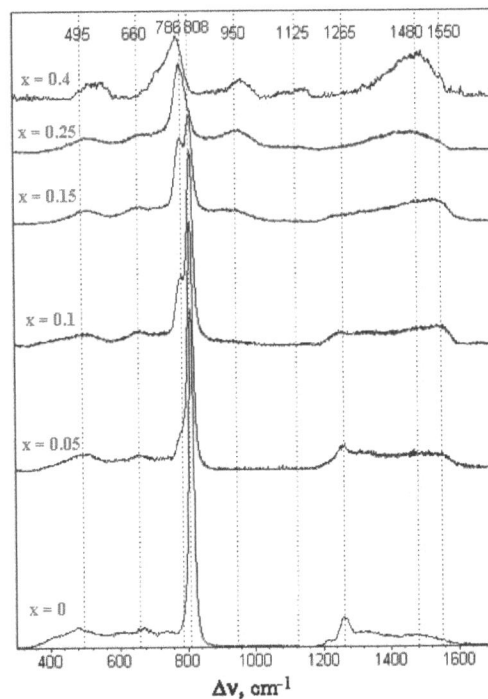

Figure 2. Raman spectra of the lithium borate glasses with various Li_2O contents. All spectra are measured at room temperature

Results

has a vertical degree of freedom. A lens (3) with a small prism (2) was used both for focusing of laser light on the sample and for collection of scattered light. The parallel beam of the scattered light reflected by a flat mirror (5) was collected on the entrance slit of the DFS-24 monochromator by a lens (4). The laser ray was focused into a sample volume and not on the surface by vertical movement of the furnace. Thus the spectra were recorded in 180° geometry. An uncooled FEU-79 photomultiplier was used for detection of the Raman signal. The temperature of the sample was controlled within ±1°C by a platinum thermocouple placed in immediate proximity to the crucible bottom. The Raman spectra were excited by LTI-701 Nd:YAG laser (λ=532 nm, <P>=1 W) operating at a modulation frequency of 8·7 kHz. The pulse duration of the acousto-optic switch was equal to 2 μs. The spectral width of the slit was equal to 6 cm^{-1} in all experiments.

Results

The Raman spectra of the lithium borate glasses are shown in Figure 2. All spectra were normalised so that the integrated intensities are the same. The most essential changes are observed in the 700–820 cm^{-1} region and are connected with the occurrence of a new line at about 770–780 cm^{-1} for small concentrations of lithium oxide. As the modifier oxide content increases, the intensity of this band increases. This is accompanied by decreases in the intensity of the strongest band in the spectrum of g-B_2O_3 at 810 cm^{-1}. For x=0·25, the bands in the 700–820 cm^{-1} region

merge and form a single line with a maximum near 780 cm^{-1}. The maximum of this line shifts toward low frequency, and its shape becomes asymmetric for the further increase of the Li_2O content. This asymmetry is apparently caused by the formation of a new unresolved line at 750–760 cm^{-1}.

Changes are observed in the 900–1200 cm^{-1} range which depend on the glass composition. A band appears near 950 cm^{-1} on addition of alkali oxide. The intensity of this band monotonically increases with alkali content over all of the investigated range of compositions, and its maximum gradually shifts towards high frequency. Besides this band, it is necessary to note the appearance of a weak wide line near 1125 cm^{-1}, but this band is distinctly observed only in the spectrum of 0·4Li_2O.0·6B_2O_3 glass. The form of the spectral envelope in the 1200–1650 cm^{-1} region also changes depending on glass composition. First and foremost, we note the decrease of intensity of the band with its maximum near 1250–1260 cm^{-1}, and some increase of intensity of the high frequency component of this envelope. As a result, the 1200–1650 cm^{-1} contour becomes practically flat at x=0·05, and for further increase of Li_2O concentration a maximum near 1530–1550 cm^{-1} is formed. The intensity of this band is not monotonically dependant on glass composition. Initially its intensity increases and reaches a maximum at x≈0·15. Then the intensity of this band decreases, and the high frequency envelope as a whole becomes dome-shaped. A maximum near 1480–1500 cm^{-1} appears for further increase of the alkali oxide content.

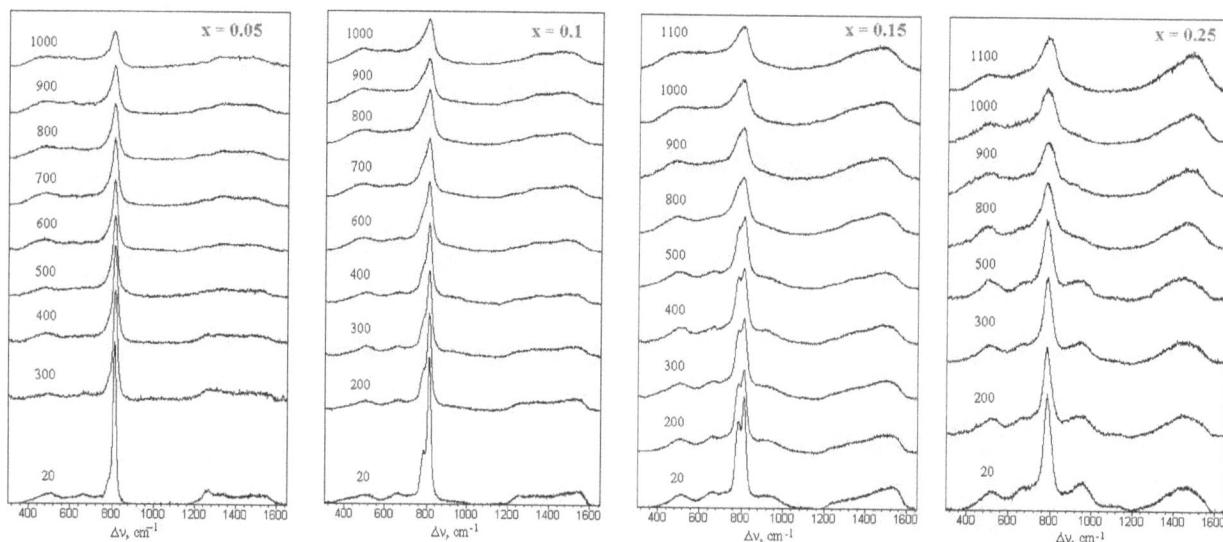

Figure 3. The high temperature Raman spectra obtained for the glasses and melts. Numbers near spectra are temperatures of samples in °C

The Raman spectra of Li_2O–B_2O_3 glasses and melts measured at different temperatures are shown in Figure 3. Raman spectra of $0.4Li_2O.0.6B_2O_3$ melts with a good signal/noise ratio were not obtained. All spectra were corrected for the thermal population of the vibrational levels

$$I_{red} = \frac{v}{(v_0 + v)^4}\left[1 - \exp\left(-\frac{hv}{kT}\right)\right] \times I_{obs} \qquad (1)$$

where v, I_{obs}, v_0, k and T represent the observed Raman shift, the observed Raman intensity, the wavenumber of the excitation source (cm^{-1}), Boltzmann constant and temperature, respectively. As seen from Figure 3, the increase of temperature of low alkali samples results in a significant decrease of the band at 810 cm^{-1}, while the high frequency envelope (1200–1650 cm^{-1}) varies insignificantly. The reduction of peak intensity of the line at 810 cm^{-1} is accompanied by an increase of its width and the width of the neighbouring band with its maximum near 770–780 cm^{-1}. As a result, these bands become unresolved, and the 700–820 cm^{-1} frequency range is represented by one asymmetric line in the spectra of the melts. The decrease of intensity with increasing temperature is also typical for the Raman bands located in the 900–1200 cm^{-1} frequency range. These bands are present in the spectra of the samples with $x > 0.15$. There is more and more obvious change of the form of the high frequency contour in the higher alkali composition range. The increase of temperature results in an increase of the Raman signal at 1480–1500 cm^{-1}.

Discussion

Raman spectra and structure of Li_2O–B_2O_3 glasses

Three frequency ranges of the Raman spectra are of great interest for interpretation: 700–820, 900–1150 and 1200–1650 cm^{-1}. Raman scattering in the 700–820

cm^{-1} region is caused by vibration of six-membered rings with various numbers of $BØ_4^-$ tetrahedra (Ø – bridging oxygen atom). Here we select the following bands:

(a) a band at 810 cm^{-1} due to the breathing vibration of boroxol rings[4,5,14-21]

(b) a band near 770–780 cm^{-1} due to the breathing vibration of six-membered rings with one $BØ_4^-$ tetrahedron (i.e. triborate, pentaborate and tetraborate)[5,18,20-22]

(c) a peak at about 750–760 cm^{-1} due to the vibration of six-membered rings with two $BØ_4^-$ tetrahedra (i.e. ditriborate or dipentaborate groups).[5,18,20-22]

The appearance of a band at 950 cm^{-1} in the spectra also indicates the formation of $BØ_4^-$ tetrahedra, both included and not included in superstructural groupings.[4,5,19,21,23-26] A weak band near 1125 cm^{-1} was assigned to the vibrational mode of diborate groups.[18,19,26] Lastly, Raman scattering in the high frequency region was connected with various vibrational modes of symmetric ($BØ_3$) and asymmetric ($BØ_2O^-$) triangles (O$^-$ – nonbridging oxygen). Vibrations of symmetric $BØ_3$ units give a set of bands in the 1200–1600 cm^{-1} region (1210, 1250–1260, 1325–1335 and 1470–1475 cm^{-1}), and a band with a maximum near 1480–1500 cm^{-1} corresponds to vibrations of nonbridging oxygen atoms in metaborate triangles.[3-5,16-21,23-26] There is no information in the literature concerning the band at 1530–1550 cm^{-1}. The intensity of this band depends non-monotonically on the glass composition, and reaches its maximum intensity at $0.15 < x < 0.25$. The origin of this band cannot be connected with the formation of nonbridging oxygens in the structure of glasses, as in this case the intensity of this band should increase monotonically up to the metaborate composition. It is known that the addition of a small amount of alkali oxide leads to the formation of $BØ_4^-$ tetrahedra in the structure of

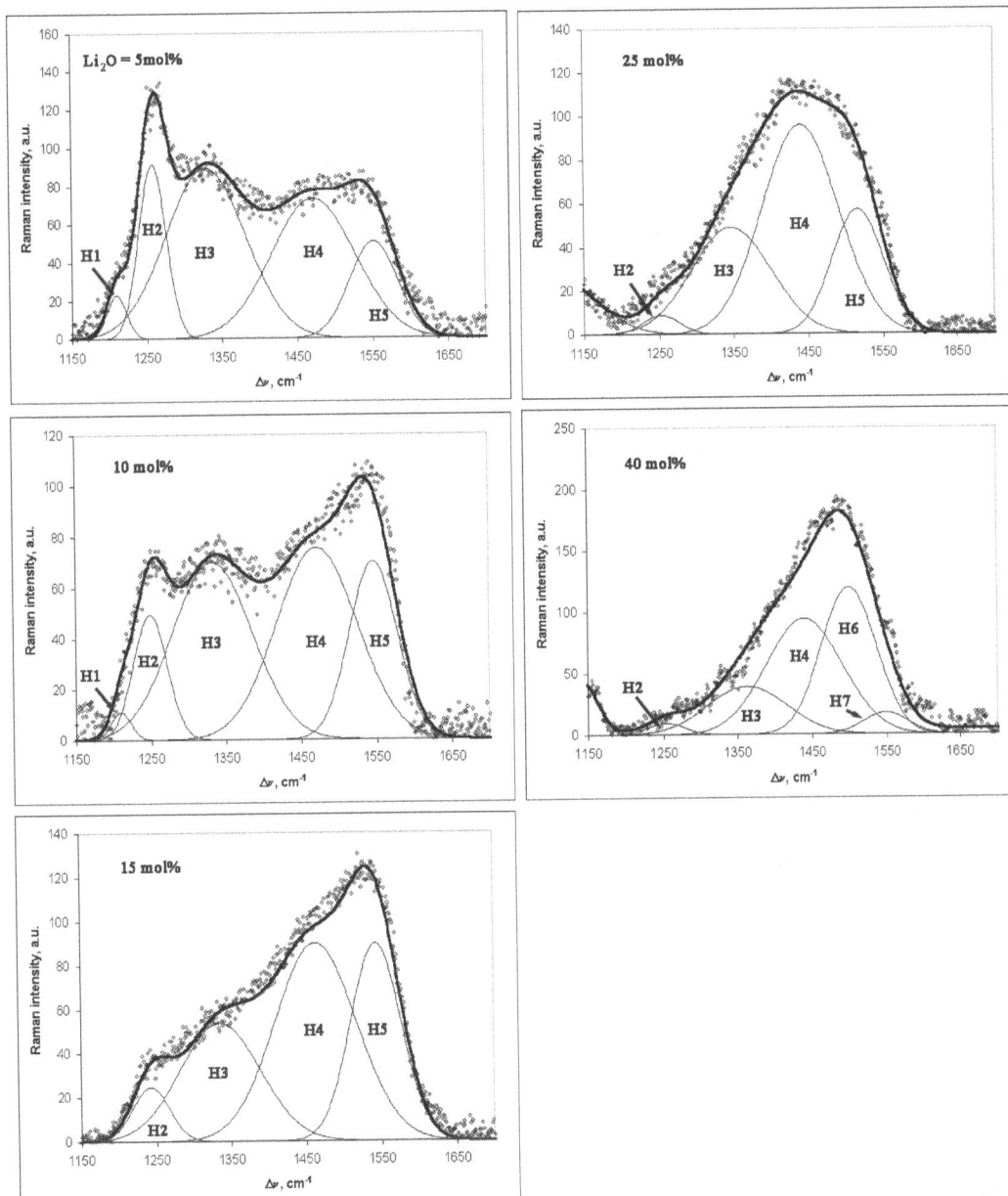

Figure 4. Examples of the band deconvolution of the high frequency region of the Raman spectra of lithium borate glasses. All spectra are measured at room temperature

borate glasses. Therefore we have assumed that the Raman scattering near 1530–1500 cm^{-1} is due to the vibrational mode of BØ$_3$ symmetric triangles having one bridging bond with a BØ$_4^-$ tetrahedron.

The analysis of the high frequency envelope of the Raman spectra of the glasses, within the framework of this interpretation, shows that gradual increase of modifier oxide results only in the formation of BØ$_4^-$ tetrahedra in the structure of lithium borate glasses, at least up to $x=0.25$. Metaborate BØ$_2$O$^-$ units are absent in this composition region. The band near 1480–1500 cm^{-1}, corresponding to vibrations of these units, is observed only in the spectrum of 0.40Li$_2$O.0.6B$_2$O$_3$ glass. Changes of the Raman scattering intensity in the 700–820 cm^{-1} region show that the structural changes in the intermediate range are connected with a gradual transformation of the boroxol rings

into triborate, and then ditriborate groups, in the investigated range of compositions.

Yano *et al* have reported high temperature Raman spectroscopy structural investigations of sodium borate glasses and melts,[10–12] and have interpreted the spectra quantitatively in terms of a superposition of Gaussian bands. This method has provided not only a qualitative description of the structure of glasses and melts, but also has given quantitative information about the structure, such as concentrations of fundamental structural units, and values which are proportional to the concentrations of various type of six-membered borate rings. Here we have performed deconvolution of the obtained spectra of Li$_2$O–B$_2$O$_3$ glasses based on this technique.

Examples of modelling of the high frequency envelope of Raman spectra of the lithium borate glasses

Table 1. The peak positions and FWHM of the partial bands (H1–H7) of high frequency contour of Raman spectra. Peak position and FWHM are given in cm^{-1}

t, °C	H1	H2	H3	H4	H5	H6	H7
	$5Li_2O.95B_2O_3$						
20	1210/32	1256/44	1329/125	1471/129	1551/74	-	-
300	1192/44	1251/52	1322/130	1464/130	1547/74	-	-
400	1190/49	1246/56	1321/131	1460/130	1550/78	-	-
500	1182/53	1234/65	1322/131	1458/132	1536/83	-	-
600	1179/58	1230/65	1322/132	1455/132	1532/87	-	-
700	1175/58	1226/65	1322/132	1455/132	1532/88	-	-
800	1175/58	1223/64	1321/128	1451/129	1532/88	-	-
900	1150/68	1214/64	1320/132	1449/132	1527/92	-	-
1000	1142/78	1214/66	1316/130	1445/132	1527/97	-	-
	$10Li_2O.90B_2O_3$						
20	1209/38	1251/52	1332/125	1470/127	1547/74	-	-
200	1203/36	1247/55	1332/125	1467/127	1547/78	-	-
300	1199/37	1243/60	1331/129	1464/131	1543/79	-	-
400	1192/42	1234/63	1331/128	1463/130	1540/83	-	-
600	1181/45	1226/68	1329/129	1457/130	1536/88	1491/72	-
700	1170/48	1216/70	1329/130	1454/132	1528/93	1491/72	-
800	1161/49	1214/71	1324/133	1452/132	1527/95	1486/72	-
900	-	1203/77	1323/134	1447/134	1528/97	1486/76	-
1000	-	1196/86	1323/134	1441/134	1527/102	1482/82	-
	$15Li_2O.85B_2O_3$						
20	-	1243/58	1336/125	1468/127	1543/78	-	-
200	-	1237/61	1334/128	1460/126	1539/82	-	-
300	-	1232/65	1329/129	1458/130	1537/84	-	-
400	-	1225/66	1329/129	1454/130	1532/88	-	-
500	-	1228/69	1332/129	1453/130	1525/89	-	-
800	-	1209/74	1325/130	1447/133	1524/93	1494/75	-
900	-	1208/79	1324/134	1441/134	1522/97	1489/79	-
1000	-	1171/82	1305/141	1426/142	1517/104	1487/85	1616/92
1100	-	1167/94	1309/141	1429/142	1517/102	1483/90	1590/96
	$25Li_2O.75B_2O_3$						
20	-	1248/62	1348/125	1443/122	1521/79	-	-
200	-	1244/60	1341/125	1443/123	1524/82	-	-
300	-	1244/60	1340/125	1440/123	1523/84	-	-
500	-	1236/62	1332/124	1433/127	1516/86	-	-
800	-	1214/66	1327/128	1436/128	1516/92	1495/79	1597/89
900	-	1200/67	1326/128	1434/131	1517/99	1495/83	1594/89
1000	-	1192/74	1316/132	1429/135	1516/103	1494/88	1593/95
1100	-	1173/77	1314/137	1428/140	1516/108	1496/98	1581/99
	$40Li_2O.60B_2O_3$						
20	-	1250/62	1362/125	1434/122	-	1498/75	1549/71

are shown in Figure 4. As seen from this figure, the original spectra could be well reproduced by using seven bands over the full range of investigated compositions. However, five lines are sufficient for the modelling of each individual spectrum. The peak positions and FWHM (full width at half maximum) of these lines are shown in Table 1, which also indicates the notation used to identify the lines.

The gradual increase of Li_2O content results in a monotonic decrease in intensity of the H1 and H2 bands, an increase in their width, and some shift of the peak positions towards low frequency. The H1 band disappears in the spectrum of $0.15Li_2O.0.85B_2O_3$ glass, whilst the H2 line is present even in the deconvolution of the spectrum of $0.4Li_2O.0.6B_2O_3$ glass. The intensity of the H3 band also decreases with increasing modifier oxide concentration, but, unlike the H1 and H2 bands, the peak position of this band is shifted towards high frequency, and the width of this band does not depend on composition within the limits of experimental error. The intensity of the H4 band varies little, and its peak position shifts

towards low frequency as x increases. The H1–H4 bands are inherited from the deconvolution of the pure g-B_2O_3 spectrum. These bands were assigned to the vibrational modes of $B\varnothing_3$ symmetric triangles, both included and not included in super-structural groups.[16] The H1 and H2 bands were assigned to B–O' stretching of planar $B\varnothing_3$ triangles incorporated in boroxol rings, where O' are not in the ring, following Walrafen et al[14] and Hassan et al.[16] The origin of the H3 band has been connected to the vibration of independent $B\varnothing_3$ units, whilst the H4 band is due to the vibration of $B\varnothing_3$ triangle units both in boroxol rings and independent units.[10,16] However, the presence of only H1–H4 lines is not enough for correct reproduction of the experimental envelope, even for small modifier oxide concentration. Therefore, there is a necessity for the introduction of a new H5 line. For reasons given above, this band has also been assigned to the vibrations of $B\varnothing_3$ units, but having a bridging bond with a $B\varnothing_4^-$ tetrahedron. The peak position of the H5 line also depends on composition. As the alkali oxide concentration increases, this band

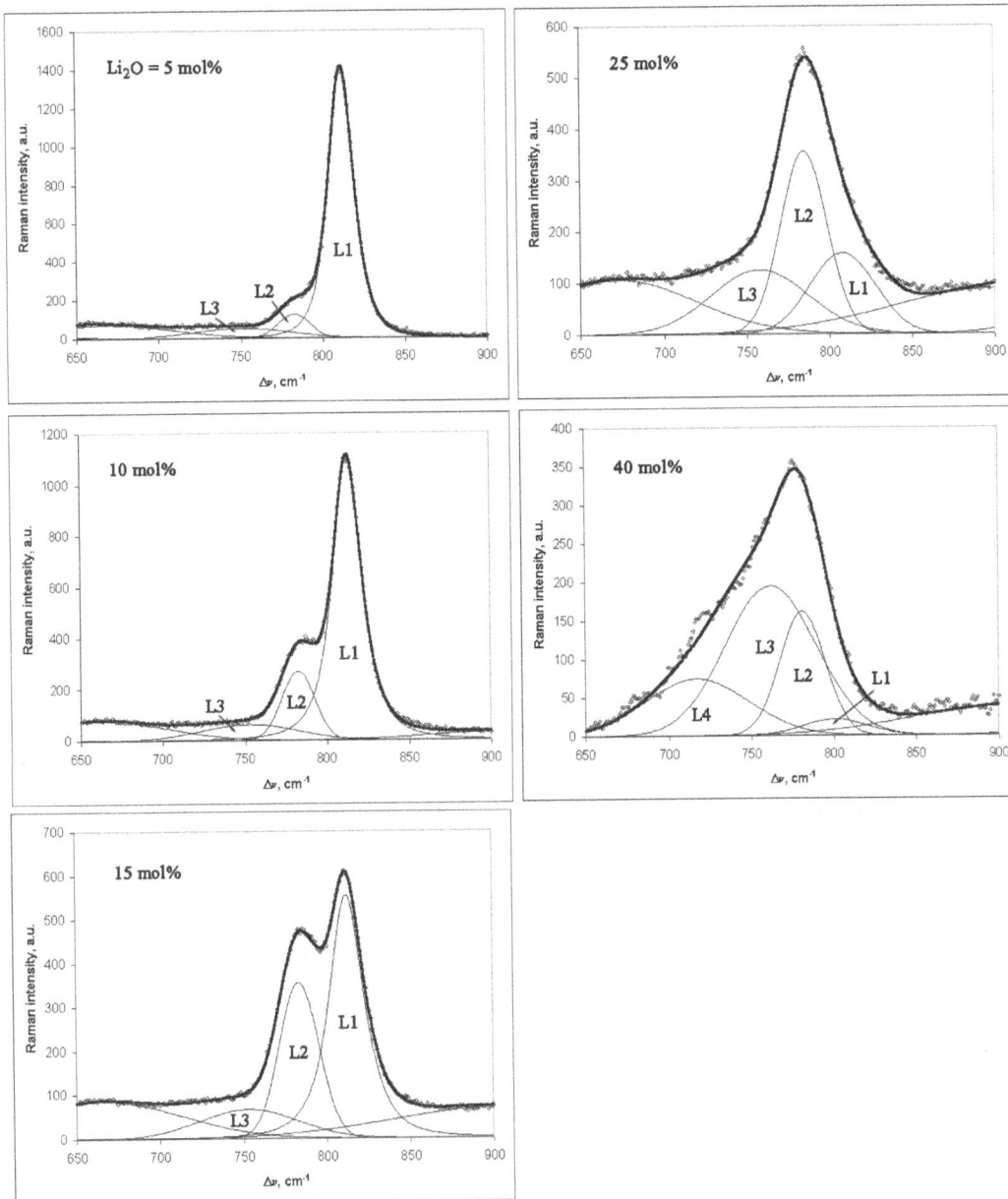

Figure 5. Examples of the band deconvolution of the middle frequency region of the Raman spectra. All spectra are measured at room temperature

shifts towards low frequency. The observed changes of positions of the H1–H5 lines are consistent with the data of Yano *et al.*[10]

Two new H6 and H7 bands appear in the deconvolution of the high frequency envelope of the $0.4Li_2O.0.6B_2O_3$ glass spectrum. The maximum of the H6 line is located near 1498 cm^{-1}, i.e. it is within the frequency interval corresponding to the vibrations of $B\emptyset_2O^-$ units. The H7 line has low intensity, but nevertheless it is necessary for modelling of the high frequency envelope. It is assumed that the H7 band is due to another vibrational mode of $B\emptyset_2O^-$ asymmetric triangles, distinct from the H6 line. Thus, this band is also assigned to the vibration modes of $B\emptyset_2O^-$ triangles, following Ref. 10.

It is assumed that the intensities of the H1–H7 lines are proportional to the concentration of the corresponding structural units in the glasses, i.e.

$$I_{H1} = C_{H1}N_{3s}, \quad I_{H2} = C_{H2}N_{3s}, \ldots, \quad I_{H5} = C_{H5}N_{3s} \quad (2)$$
$$I_{H6} = C_{H6}N_{3a}, \quad I_{H7} = C_{H7}N_{3a}$$

Thus the following expression can be obtained

$$\frac{I_{H6} + I_{H7}}{\sum\limits_{k=1}^{5} I_{Hk}} = \frac{(C_{H6} + C_{H7})N_{3a}}{N_{3s}\sum\limits_{k=1}^{5} C_{Hk}} = A\frac{N_{3a}}{N_{3s}}, \quad (3)$$

$$A = \frac{C_{H6} + C_{H7}}{\sum\limits_{k=1}^{5} C_{Hk}}$$

where N_{3a} and N_{3s} are the concentrations of $B\emptyset_2O^-$ metaborate units and $B\emptyset_3$ symmetrical triangles, respectively, and C_{Hk} are proportionality coefficients.

It is a matter of general observation that the addition of alkali oxide into B_2O_3 only converts $BØ_3$ triangles to $BØ_4^-$ tetrahedra or $BØ_2O^-$ units in the investigated range of compositions. Pyroborate ($BØO_2^{2-}$) and orthoborate (BO_3^{3-}) units are absent in structure of the investigated glasses. Therefore we can obtain two more equations connecting the concentrations of structural units

$$N_{3a} + N_4 = \frac{x}{1-x} \tag{4}$$

$$N_{3s} + N_{3a} + N_4 = 1 \tag{5}$$

Equations (3) to (5) allow us to calculate the concentrations of structural units in the glasses by using the H1–H7 bands and the coefficient A, which was determined using the data on local structure of the glasses from Refs 27 to 29. Following these studies, we have accepted that $N_{3a}/N_{3s}=0.67$ for $x=0.4$ and obtained $A=1.06$. Since the intensities I_{H6} and I_{H7} are equal to zero for $0 \leq x \leq 0.25$, there is no necessity to use the factor A for the determination of the concentrations of structural units. Here, the required concentrations are easily determined from Equations (4) and (5), provided that $N_{3a}=0$. As will be shown, $N_{3a}\neq0$ for melts with $x \leq 0.25$. In this case, the coefficient A allows us to calculate the concentrations of the structural units in the melts at high temperatures.

In addition, the intensity of the H1–H5 lines can be used to study the glass structure in the intermediate range. The boroxol ring, triborate group, and di-triborate group contain strictly specific numbers of $BØ_3$ triangles (3, 2 and 1, respectively). Therefore the ratio of the lines intensity corresponding to the vibrations of a particular type of six-membered rings to the total intensity of the H1–H5 lines should reflect the distribution of these structural units over borate rings or the fractions of these rings in the structure of glasses and melts. Modelling of the 650–900 cm^{-1} interval as a superposition of Gaussian lines has been carried out for determination of intensities of the lines corresponding to vibrations of borate rings of various types. Examples of the deconvolution of this interval of the Raman spectra are shown in Figure 5. According to the accepted interpretation of the Raman spectra, the L1–L3 bands are connected to vibrations of boroxol, triborate and di-triborate rings respectively, and the L4 band is connected to bending vibrations of B–O–B bridges in metaborate chains.[5,30]

The dependences of the normalised intensities of the L1–L3 lines on the glass composition, calculated according to Equation (6) are shown in Figure 6

$$I_{Li}^{normal} = \frac{I_{Li}}{\sum\limits_{k=1}^{5} I_{Hk}}, \quad i=1-3 \tag{6}$$

Here I_{Li} is the intensity of the Li line, and I_{Hk} are the intensities of the H1–H5 lines in the high frequency envelope. It is clear that the changes of the normal-

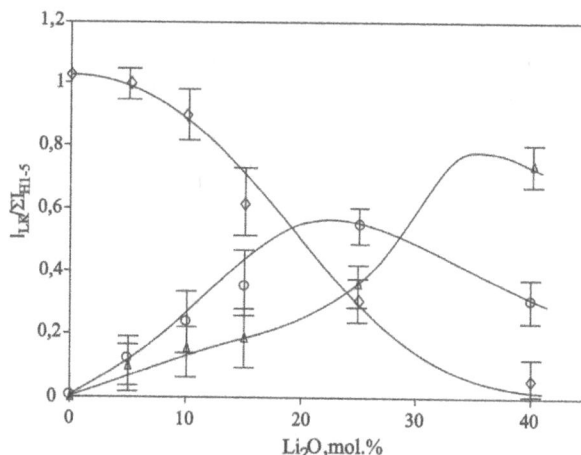

Figure 6. The normalised intensities of the L1–L3 bands, depending on glass composition. Solids lines are drawn as guides for the eye, taking into account Krogh-Moe's model. ◇ - L1, ○ – L2, △ – L3

ised intensities of the middle frequency lines are in good agreement with modern work on the structural modification of alkali borate glasses depending on composition. However, our results show a higher value of x at which the concentration of boroxol rings becomes equal to zero in comparison with the data of other authors.[26] The reason is that they determined the concentration of boroxol rings as a function x by the peak height of the corresponding Raman band on the condition that the bandwidths do not change with glass composition. Our data show that the bandwidths of the L1 and L2 bands grow with increasing x. The width increase of the L1–L3 lines is consistent with Yano et al's data.[10] As these lines are close to each other, even a weak change of their width will result in merging these bands, and it can seem that the L1 band is absent in the spectrum. We believe that the increase of the L1 band width is caused by deformation of boroxol rings due to the formation of B–O–B bridging bonds near these rings on addition of Li_2O.

Raman spectra and structure of Li_2O–B_2O_3 glasses and melts

The decrease of the band intensity in the 700–1200 cm^{-1} frequency range shows that there is significant change with temperature of the structure of all the investigated glasses in the intermediate range. It is of great interest to use Equation (6) to study the transformation of the glass structure in the intermediate range as the temperature changes. The weak dependence of the intensity of the high frequency bands on temperature, for the samples with $x<0.15$, indicates a significant similarity of local structure of glasses and melts in this composition range. The intensity of the Raman bands in the 1480–1500 cm^{-1} region increases with increasing temperature in the spectra of the samples with $x>0.15$. This shows that

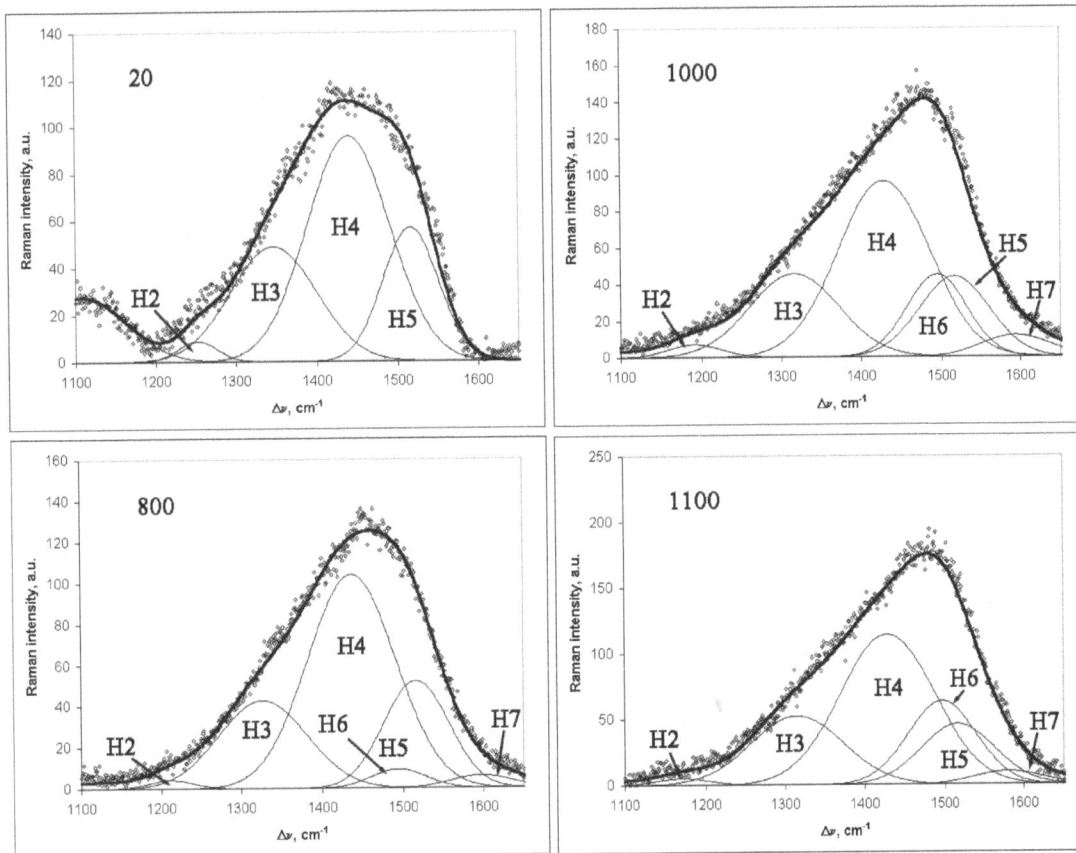

Figure 7. Examples of the band deconvolution of the high frequency region of the Raman spectra for x=0·25. Numbers near spectra are temperatures of samples in °C

the concentration of $B\emptyset_2O^-$ units in the melts is higher than in glasses with the same composition.

Modelling of the high frequency envelope of all measured spectra was carried out to determine how the concentrations of structural units depend on temperature. The same set of H1–H7 lines was used for the first approach. Examples of the deconvolution of the high frequency envelope for the sample with x=0·25 are shown in Figure 7. The peak positions and FWHM of the H1–H7 bands are given in Table 1. The values found for the ratios of line intensities are given in Table 2. As seen from these tables, there is no necessity for the use of the H6–H7 lines at any temperature for the glass with x=0·05. Here the local structure of glass and melt are similar within the

limits of experimental error, and consist of only $B\emptyset_3$ and $B\emptyset_4^-$ units. For all other samples, the H1–H5 lines are not sufficient for the correct reproduction of the spectra. Besides the H1–H5 lines, it is necessary to use the H6–H7 bands in the deconvolution procedure of the high temperature spectra. The occurrence of these last two bands, and the increase of their intensity, indicate the formation of $B\emptyset_2O^-$ units in the structure of melts with increase of temperature.

The concentrations of structural units have been calculated for all investigated samples according to

Table 2. The $\Sigma I_{H6,7}/\Sigma I_{H1-5}$ values obtained from deconvolution of the high frequency envelope of the Raman spectra

t, °C	$\Sigma I_{H6,7}/\Sigma I_{H1-5}$			
	10%Li$_2$O	15%Li$_2$O	25%Li$_2$O	40%Li$_2$O
20	0	0	0	0·691
200	0	0	0	-
300	0	0	0	-
400	0	0	-	-
500	-	0	0	-
600	0·004	-	-	-
700	0·014	-	-	-
800	0·020	0·011	0·053	-
900	0·030	0·019	0·118	-
1000	0·037	0·022	0·215	-
1100	-	0·036	0·262	-

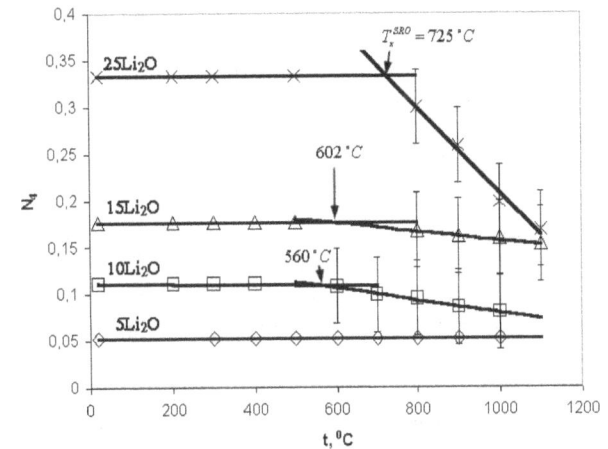

Figure 8. The fraction of $B\emptyset_4^-$ units depending on temperature

$$\ln\left(\frac{N_{3a}}{N_4}\right) = -\frac{4987.9}{T} + 2.9615$$

10Li$_2$O

$$\Delta H = 41 \pm 5 \frac{kJ}{mol}$$

$$\ln\left(\frac{N_{3a}}{N_4}\right) = -\frac{5113}{T} + 1.846$$

15Li$_2$O

$$\Delta H = 43 \pm 6 \frac{kJ}{mol}$$

$$\ln\left(\frac{N_{3a}}{N_4}\right) = -\frac{10959}{T} + 8.0755$$

25Li$_2$O

$$\Delta H = 91 \pm 13 \frac{kJ}{mol}$$

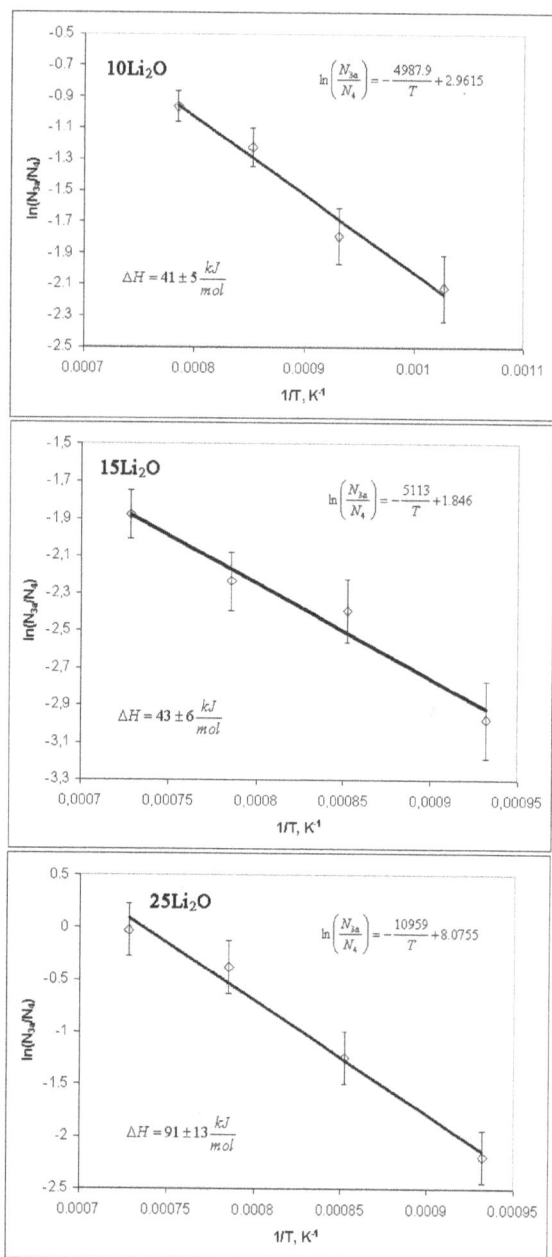

Figure 9. Arrhenius plots of N_{3a}/N_4 values for temperatures above T_x^{SRO}. The solid lines were obtained by least squares fitting

Equations (3) to (5). Here we assume that the coefficient A does not depend (or depends only slightly on temperature). Results of calculation of $B\varnothing_4^-$ tetrahedra concentrations are shown in Figure 8. It is obvious that the obtained data are well described by two direct lines with a break near to some temperature, T_x^{SRO} (SRO – short range order) at $x>0.05$. The values of T_x^{SRO} for each composition are shown in the same figure. The notation T_x^{SRO} is taken from Ref. 11, and corresponds to the temperature of "freezing" of the local structure of the melt as it is cooled. The local structure of glasses does not depend on temperature for temperatures lower than T_x^{SRO}. Here, I_{H6} and I_{H7} are equal to zero, and the concentration of $B\varnothing_4^-$ units is easily calculated from Equation (4) at $N_{3a}=0$. The

experimental data in Figure 8 are well described by straight lines for temperatures above T_x^{SRO}. The slope of these lines increases as the Li$_2$O concentration increases.

The obtained results confirm that a dynamic equilibrium between $B\varnothing_4^-$ tetrahedra and $B\varnothing_2O^-$ symmetric triangles exists in the melts

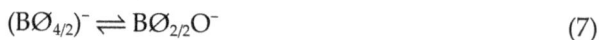

$$(B\varnothing_{4/2})^- \rightleftharpoons B\varnothing_{2/2}O^- \tag{7}$$

This equilibrium is shifted to the right as the temperature increases.

Ratios of $\ln(N_{3a}/N_4)$ as a function of $(1/T)$ at the temperatures above T_x^{SRO} are shown in Figure 9. The high temperature region could be least squares fitted

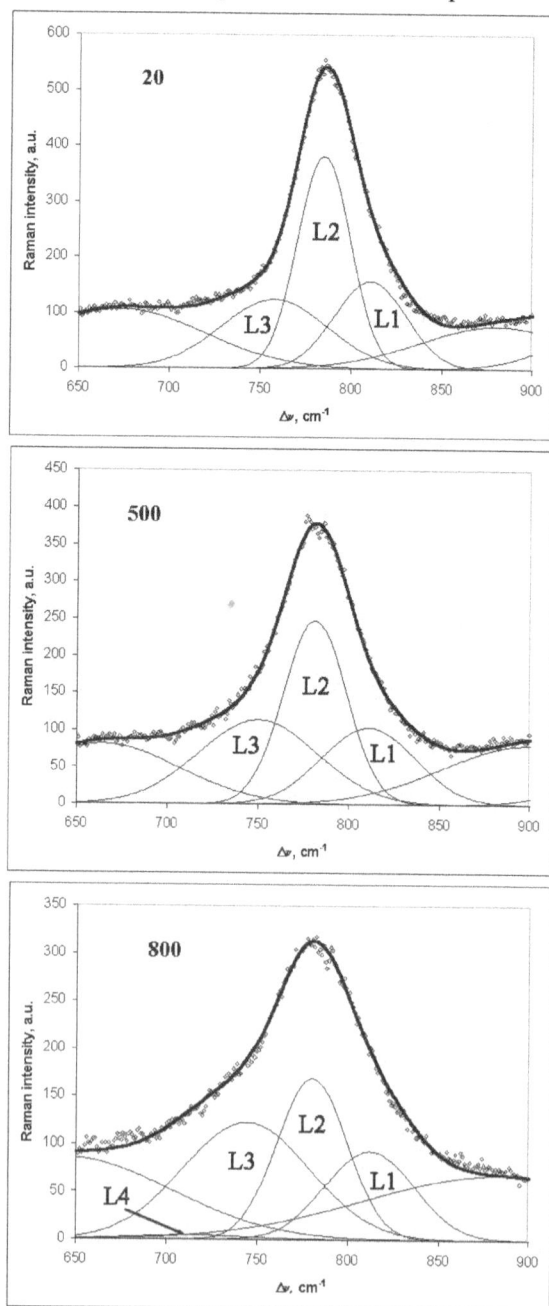

Figure 10. Examples of the deconvolution of the middle frequency region of the Raman spectra for $x=0.25$. Numbers near spectra are temperatures of samples in °C

well with a straight line. The slopes give the enthalpy, ΔH, for the equilibrium of Equation (7), according to the van't Hoff equation

$$\Delta H = -R \frac{d(\ln K)}{d(1/T)}, \quad K = \frac{N_{3a}}{N_4} \qquad (8)$$

Obtained ΔH values are shown in Figure 9. The ΔH values depend on the concentration of alkali oxide; the enthalpy increases with increasing Li_2O content.

Deconvolution of the middle frequency range of the measured spectra was carried out to study the temperature dependence of the structural changes in the intermediate range. We have used the same set of bands as used to model the spectra of the glasses. Examples of the deconvolution of the middle frequency envelope are shown in Figure 10 for $x=0.25$. New lines do not arise as the temperature increases, and only change of intensity of the lines takes place. The peak positions and FWHM of the L1–L4 lines are given in Table 3. The temperature dependences of the normalised intensities of the

Table 3. *The peak positions and FWHM of the partial bands (L1 – L4) of middle frequency envelope of the Raman spectra. Peak position and FWHM are given in cm^{-1}*

t, °C	L1	L2	L3	L4
	$5Li_2O.95B_2O_3$			
20	810/18	782/23	751/68	-
300	812/24	783/30	751/69	-
400	812/27	781/32	751/69	-
500	810/29	776/35	747/69	-
600	809/32	772/39	743/69	-
700	809/34	770/41	742/72	-
800	808/37	767/43	738/74	-
900	806/42	765/43	732/75	-
1000	804/44	761/47	728/80	-
	$10Li_2O.90B_2O_3$			
20	810/21	782/24	752/68	-
200	811/25	781/27	753/69	-
300	811/27	780/30	749/69	-
400	810/29	779/35	745/72	-
600	809/33	777/41	744/74	-
700	808/36	776/43	742/75	-
800	807/40	773/48	738/76	-
900	806/41	773/50	733/76	-
1000	805/44	770/52	730/81	-
	$15Li_2O.85B_2O_3$			
20	810/25	783/27	754/69	-
200	811/29	782/30	752/72	-
300	811/31	782/32	751/72	-
400	810/34	779/35	746/74	-
500	810/36	780/38	745/78	-
800	808/44	777/47	739/82	-
900	805/50	777/54	736/89	-
1000	805/55	771/58	729/94	-
1100	803/60	774/62	728/97	-
	$25Li_2O.75B_2O_3$			
20	808/34	784/33	758/69	-
200	810/37	782/36	755/69	-
300	809/41	783/38	751/73	-
500	810/48	781/41	750/76	-
800	810/54	780/45	744/82	714/82
900	806/63	779/55	744/87	712/86
1000	805/64	776/58	745/88	712/87
1100	802/64	778/63	746/88	704/84
	$40Li_2O.70B_2O_3$			
20	803/40	781/32	761/65	717/71

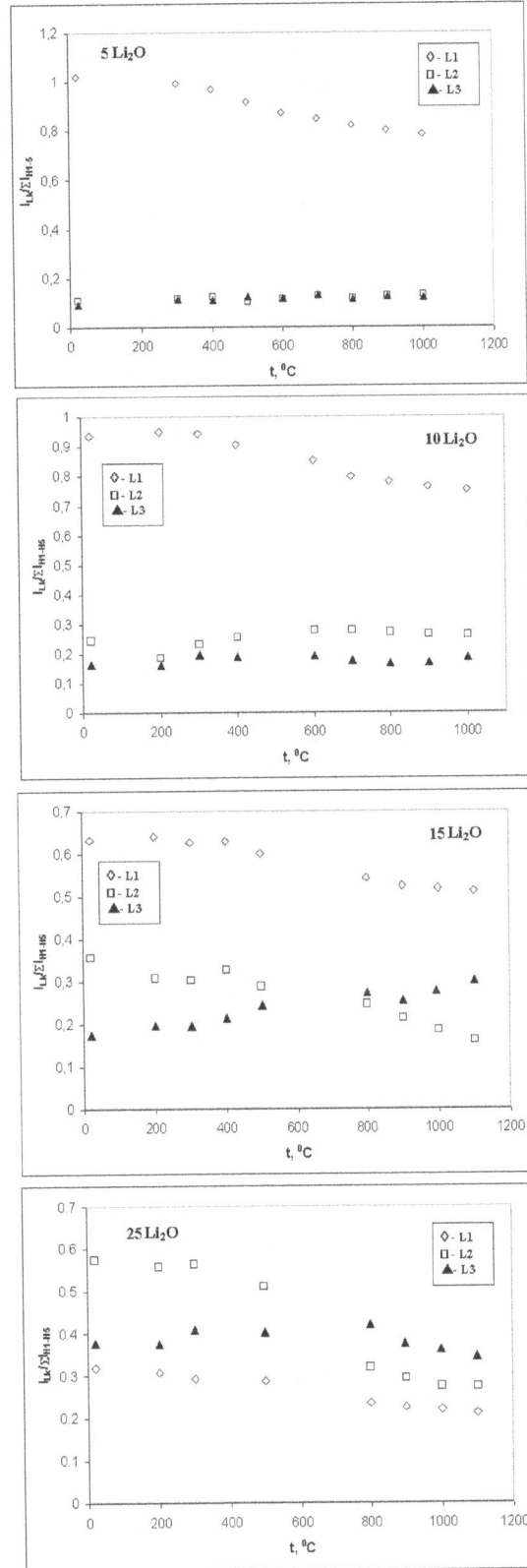

Figure 11. *The normalised intensities of the L1–L3 bands depending on the temperature*

L1–L3 lines according to Equation (6) are shown in Figure 11. The intensity of the L1 band decreases, but the intensities of the L2 and L3 bands depend little on temperature for $x \leq 0.15$. Walrafen *et al*[14] and Hassan *et al*[16] have shown that the decrease of the 810 cm^{-1} band corresponds to the destruction of boroxol rings. Our results show that the dominant process of structural changes of low alkali glasses on heating and fusion is still the dissociation of boroxol groups. Significant changes of the concentrations of triborate and ditriborate groups are not observed here. The destruction of boroxol rings becomes less pronounced as the Li$_2$O content increases (note the behaviour of L1(t) in Figure 11). Even so the general trend of dependence L1(t) remains unchanged. Yano et al[12] showed that the decrease of boroxol rings can be partly compensated by the destruction of larger superstructural groupings (diborate, pentaborate, etc.), forming boroxol groups. The intensity of the L2 line decreases and the L3 line increases for $x=0\cdot15$. It is possible to explain the behaviour of these bands by assuming that, as a result of the destruction of large groups (for example tetraborate or diborate groups), not only boroxol rings, but also ditriborate groups can be formed. Lastly a distinctive feature of the L2 and L3 bands for $x=0\cdot25$ is a decrease of the intensities of both lines for high temperatures above 800°C. It is probable that this decrease of intensity is due to structural transformations in the short range order. As a significant proportion of $BØ_4^-$ tetrahedra are in superstructural groups, a decrease in the number of these tetrahedra can result in the formation of nonbridging bonds inside superstructural groups and to their destruction as a consequence. Obviously, such a mechanism of destruction of superstructural groups can occur for any group which contains $BØ_4^-$ tetrahedra. Also, a decrease of the concentration of superstructural groups with $BØ_4^-$ tetrahedra could be connected with a more statistical distribution of the fundamental structural units in the melt structure in comparison with the glass.

The investigation of glasses and melts in the Na$_2$O–B$_2$O$_3$ system by Yano et al[11,12] established that T_x^{SRO}, T_x^{IRO} (the temperature of "freezing" of intermediate range structural reorganisation) and T_g are in the order

$$T_x^{IRO} \approx T_g < T_x^{SRO} \qquad (9)$$

The approximation of the L1(t) dependence by two lines was carried out to establish whether the Relation (9) is valid only for the Na$_2$O–B$_2$O$_3$ system, or if it is true for other alkali borate systems. The temperature T_x^{IRO} was obtained as the intersection point of these lines (Figure 12). The dependence L1(t) was chosen for the following reasons: first, this line is easily identified in the majority of the measured spectra, and

second, this band depends on temperature the most strongly in the investigated composition range. We approximated the experimental data by a constant for low temperatures, and by a line of the second degree at high temperature, as in Ref. 14. The obtained values of T_x^{IRO} are shown on Figure 12 and the T_g, T_x^{SRO} and T_x^{IRO} temperatures are given in Table 4. The Relation (9) is found to be valid the for lithium borate system within the limits of experimental error.

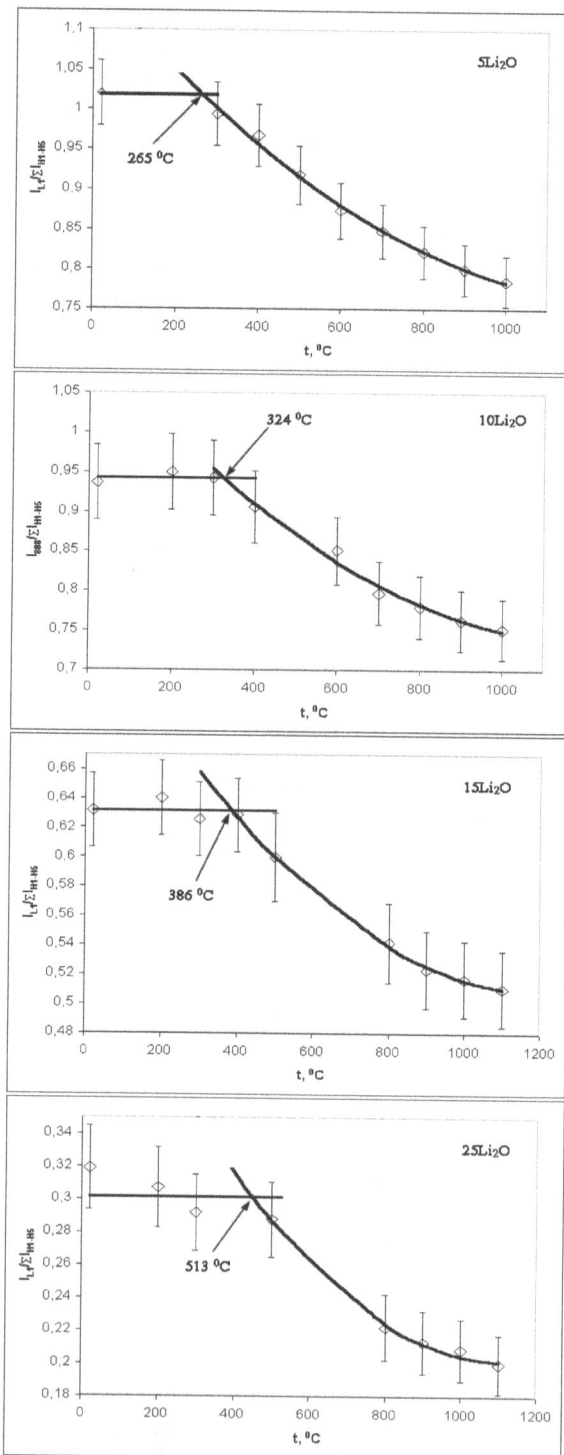

Figure 12. The normalised intensity of the L1 band depending on the temperature. The solid lines were obtained by least squares fitting

Table 4. The temperatures T_g, T_x^{SRO} and T_x^{IRO} for the investigated samples

x	T_g, °C	T_x^{IRO}, °C	T_x^{SRO}, °C
0·05	302	265	-
0·10	367	324	560
0·15	413	386	602
0·25	497	513	725

It is thus possible to distinguish the following probable mechanisms of structural modification of investigated glasses on heating and fusion: The main mechanism of structural transformation of g-B_2O_3 is a process of destruction of the boroxol rings. This mechanism dominates in samples with $x \leq 0.15$. Moreover, there is transformation between superstructural groups, and also destruction of superstructural groups due to the more statistical distribution of the fundamental structural units in the melt in comparison with glass. Also, the decrease of concentration of superstructural groups with $B\varnothing_4^-$ tetrahedra is due to the formation of nonbridging bonds inside superstructural groups for $T > T_x^{SRO}$. The composition region with $0.15 \leq x \leq 0.25$ is characterised by the greatest variety of superstructural groups coexisting in the glass structure. It is very likely that the variety of mechanisms of their transformation is also greatest here.

Conclusions

Raman spectra of glasses and melts in the Li_2O–B_2O_3 system were measured. The analysis of the obtained spectra has shown that the local structure of the investigated glasses consists only of $B\varnothing_3$ symmetric triangles and $B\varnothing_4^-$ tetrahedra, at least up to $x=0.25$. Asymmetrical $B\varnothing_2O^-$ units have been found only in the sample with $x=0.4$. The local structure of glasses and the relevant melts weakly depends on temperature for $x<0.15$. On the other hand, their structure in the intermediate range differs. The dominant process of structural transformation in this composition range is a thermal dissociation of boroxol rings. Here the concentration of six-membered borate rings with $B\varnothing_4^-$ tetrahedra depends weakly on temperature.

For $x>0.15$, increasing temperature results in a decrease of the coordination number of some boron atoms from 4 to 3. Dynamic equilibrium between the structural units in melts is described by $(B\varnothing_{4/2})^- \rightleftharpoons B\varnothing_{2/2}O^-$. This equilibrium is shifted to the left, and is frozen on cooling of the melts at a temperature T_x^{SRO}, which is greater than T_g. For $x>0.15$, changes of intermediate range structure are not limited to the dissociation of boroxol rings, but are also connected with transformation processes between other superstructural groupings, and their destruction. The destruction of superstructural groupings with

$(B\varnothing_{4/2})^-$ units can occur due to both the formation of nonbridging bonds inside them (with a modification of local structure), and to a more statistical distribution of the fundamental structural units in the melt structure compared to the glass (without a modification of local structure).

Acknowledgement

This work was supported by the RFBR (grant 07-05-96046-r_ural_a).

References

1. Krogh-Moe, J. *Phys. Chem. Glasses*, 1962, **3**, 1.
2. Bray, P. J. *J. Non-Cryst. Solids*, 1985, **75**, 29.
3. Kamitsos, E. I. & Karakassides, M. A. *Phys. Chem. Glasses*, 1989, **30**, 19.
4. Dwivedi, B. P. & Khanna, B. N. *J. Phys. Chem. Solids*, 1995, **56**, 39.
5. Konijnendijk, W. L. & Stevels J. M. *J. Non-Cryst. Solids*, 1975, **18**, 307.
6. Mysen, B. O., Finger, L. W., Virgo, D., Seifert, F. A. *Am. Miner.*, 1982, **67**, 686.
7. Mysen, B. O. & Frantz, J. D. *Chem. Geol.*, 1992, **96**, 321.
8. Bykov, V. N., Osipov, A. A. & Anfilogov V. N. *Rasplavy*, 1998, **6**, 86. (In Russian.)
9. Bykov, V. N., Osipov, A. A. & Anfilogov V. N. *Phys. Chem. Glasses*, 2000, **41**, 10.
10. Yano, T., Kunimine, N., Shibata, S. & Yamane, M. *J. Non-Cryst. Solids*, 2003, **321**, 137.
11. Yano, T., Kunimine, N., Shibata, S. & Yamane, M. *J. Non-Cryst. Solids*, 2003, **321**, 147.
12. Yano, T., Kunimine, N., Shibata, S. & Yamane, M. *J. Non-Cryst. Solids*, 2003, **321**, 157.
13. Bykov, V. N., Osipov, A. A. & Anfilogov V. N. *Rasplavy*, 1997, **4**, 28. (In Russian.)
14. Walrafen, G. E., Samanta, S. R. & Krishnan, P. N. *J. Chem. Phys.*, 1980, **72**, 113.
15. Galeener, F. L., Lucovsky, G. & Mikkelsen, J. C. *Phys. Rev. B*, 1980, **22**, 3983.
16. Hassan, A. K., Torell, L. M. & Börjesson, L. *Phys. Rev. B*, 1992, **45**, 12797.
17. Zhang, Z. & Soga, N. *Phys. Chem. Glasses*, 1990, **32**, 142.
18. Kamitsos, E. I., Karakassides, M. A. & Chryssikos, G. D. *J. Phys. Chem.*, 1987, **91**, 1073.
19. Kamitsos, E. I., Karakassides, M. A. & Chryssikos, G. D. *Phys. Chem. Glasses*, 1989, **30**, 229.
20. Maniu, D., Ardelean, I., Iliescu, T., Cinta, S. & Cozar, O. *J. Mol. Struct.*, 1997, **410&411**, 291.
21. Maniu, D., Iliescu, T., Ardelean, I., Ciceo-Lucacel, R., Bolboaca, M. & Kiefer, W. *Vibrat. Spec.*, 2002, **29**, 241.
22. Brill, T. W. *Philips Res. Rep. Suppl.*, 1976, **2**, 117.
23. Meera, B. N. & Ramankrishna, J. *J. Non-Cryst. Solids*, 1993, **159**, 1.
24. Yiannopoulos, Y. D., Chryssikos, G. D. & Kamitsos, E. I. *Phys. Chem. Glasses*, 2001, **42**, 164.
25. Akagi, R., Ohtori, N. & Umesaki, N. *J. Non-Cryst. Solids*, 2001, **293**, 471.
26. Kamitsos, E. I., Patsis, A. P., Karakassides, M. A. & Chryssikos, G. D. *J. Non-Cryst. Solids*, 1990, **126**, 52.
27. Varsamis, C.-P. E., Vegiri, A. & Kamitsos, E. I. *Condens. Matter Phys.*, 2001, **4**, 119.
28. Zhong, J. & Bray, P. J. *J. Non-Cryst. Solids*, 1989, **111**, 67.
29. Dingwell, D. B. *Rev. Miner.*, 1996, **33**, 331.
30. Chryssikos, G. D., Kamitsos, E. I., Patsis, A. P., Bitsis, M. S. & Karakassides M. A. *J. Non-Cryst. Solids*, 1990, **126**, 42.

Proc. VI Int. Conf. Borate Glasses, Himeji, Japan, 18–22 August 2008   *Phys. Chem. Glasses: Eur. J. Glass Sci. Technol. B, June 2009, **50** (3), 165–171*

Electrical conductivity and viscosity of borosilicate glasses and melts

Doris Ehrt* & Ralf Keding

Otto-Schott-Institut, Friedrich-Schiller-Universität Jena, Fraunhoferstr. 6, D-07743 Jena, Germany

Manuscript received 14 October 2008
Revised version received 12 February 2009
Accepted 1 April 2009

Simple sodium borosilicate and silicate glasses were melted on a very large scale (35 l Pt crucible) to prepare model glasses of optical quality in order to investigate various properties depending on their structure. The composition of the glass samples varied in a wide range: 3 to 33·3 mol% Na_2O, 0 to 62·5 mol% B_2O_3, and 25 to 85 mol% SiO_2. The glass samples were characterised by different methods. Refractive indices, density and thermal expansion were measured. Phase separation effects were investigated by electron microscopy. The electrical conductivity of glasses and melts were determined by impedance measurements in a wide temperature range (250 to 1450°C). The activation energies were calculated by Arrhenius plots in various temperature regions: below the glass transition temperature, T_g, above the melting point, T_l, and between T_g and T_l. Viscosity measurements were carried out with different methods from T_g to the melt. The measured data were fitted and the activation energies calculated. Simple exponential behaviour was found only in very narrow temperature ranges. The effect of B_2O_3 in sodium borosilicate glasses and melts is discussed in comparison with sodium silicate glasses and melts.

Introduction

Electrical conductivities and viscosities of solid glasses and melts are of great importance for a better understanding of the temperature dependence of the structure and also for industrial melting processes.

Borosilicate glasses based on the Na_2O–B_2O_3–SiO_2 system are of interest for various applications, ranging from chemically and thermal resistant technical glass to optical, sealing and nuclear waste glasses. The properties and structure are strongly dependent on the composition, mainly on the Na_2O/B_2O_3 ratio and the SiO_2 content. Many investigations exist and are published.[1-7] The glass former B_2O_3 affects the properties in a particular way, due to the possibility of forming $BØ_3$ triangles and $BØ_4^-$ tetrahedra with only bridging oxygen, and also borate groups with nonbridging oxygen. The study of the structure has been greatly influenced by the so-called borate anomaly: for example, the thermal expansion coefficient of binary sodium borate glasses decreases with increasing Na_2O content up to about 16 mol% Na_2O, after which it begins to increase again. The present work concentrates on investigations of electrical conductivity and viscosity of four sodium borosilicate model glasses and melts in comparison with two sodium silicate glasses and melts over a wide temperature range up to 1500°C. Phase separation effects were investigated by electron microscopy.

Experimental

The batch compositions and properties of the model glasses are shown in the tables and figures. High purity grade raw materials, Na_2CO_3, H_3BO_3 and SiO_2 were mixed and melted in an industrial scale, 35 l Pt crucible to prepare glasses to optical quality, with glass compositions which are specified as model glasses by Technical Committee TC03 of the International Commission on Glass. The melting temperature varied between 1300 and 1580°C. Glasses were obtained by pouring the melts into a mould. After annealing from T_g+50 K to room temperature at 3 K/min, the glasses were cut, ground and polished to produce samples for different measurements. The refractive indices in the visible range were measured with a refractometer from transparent homogenous samples which were obtained by remelting 200 g of glass in a Pt crucible, pouring in a mould and annealing at 5 K/min. The error was $\Delta n \pm 2 \times 10^{-5}$. The density was determined using Archimedes' principle with an error ±0·002. DTA (10 K/min) and dilatometry measurements (5 K/min) were carried out to obtain values for the thermal properties, T_g and thermal expansion coefficient (TEC). The viscosities as a function of temperature were determined with a rotating cylinder method in the η range, $10^{1.5}$–10^5 dPa s, and with the beam bending method in the range 10^9–10^{11} dPa s. The electrical conductivity of the glasses and melts was determined by impedance measurements over a wide temperature range. Solid samples were measured to a temperature where the viscosity was about 10^4 dPa s. Cylindrical glass samples with a diameter 10 mm and thickness ~5 mm were used. Low viscosity melts were measured with a high accuracy coaxial Pt crucible/Pt cylinder technique. The electrical resistance was measured depending on dipping depth of the cylinder in the melt and variation of the temperature. Impedance spectra were recorded with a Zahner IM5d electrochemical workstation, with an ac voltage of 20 or 50 mV, and frequencies in the range of

*Corresponding author. Email doris.ehrt@uni-jena.de

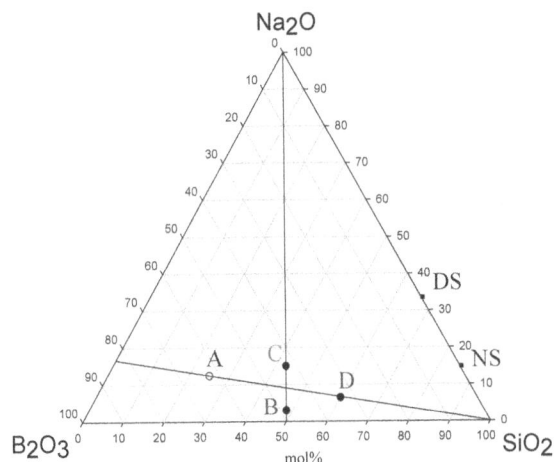

Figure 1. Binary sodium silicate NS, DS, and ternary sodium borosilicate NBS-A,-B,-C,-D model glasses in the ternary diagram: NBS-A and -D are on the borate anomaly line

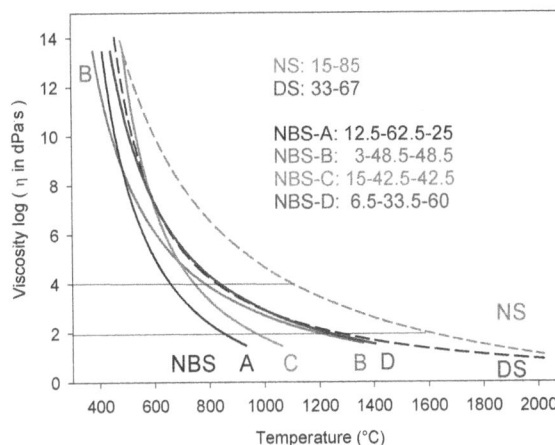

Figure 2. Viscosity curves of the four sodium borosilicate and two sodium silicate samples as global VFT-fits

0·1 Hz to 1 MHz, normally in steps of 20–50 K. The dc potentials were always 0 mV. The error of the resistance values was <5%. More details are given in Refs 8 to 11.

Results

This paper concentrates on investigations of the electrical conductivity and viscosity of four sodium borosilicate (NBS) model glasses and melts, in comparison with two sodium silicate glasses and melts, over a wide temperature range up to 1500°C. Phase separation effects were investigated by electron microscopy.

The composition of the NBS model glasses was varied in a wide range: 3 to 15 mol% Na_2O, 33·5 to 62·5 mol% B_2O_3, and 25 to 60 mol% SiO_2. Figure 1 shows the positions of the glasses in the ternary phase diagram: NBS-A and -D, with the molar ratio Na_2O/B_2O_3=0·2, and different SiO_2 content, 25 and 60 mol%, lie on the borate anomaly line; NBS-B and -C have a constant ratio B_2O_3/SiO_2=1 and different Na_2O content, 3 and 15 mol%. The two binary sodium silicate model glasses are NS with 15 mol% Na_2O and DS with 33 mol% Na_2O.

Table 1 gives an overview of the four NBS glasses in comparison with the two binary sodium silicate compositions, NS and DS, which were already used as model glasses in previous papers.[10,12,13] Figure 2 illustrates the different viscosity–temperature behaviours of the melts. Figures 3 to 6 summarise how the electrical conductivities depend on temperature from

solid glasses to liquid melts, as Arrhenius plots with calculated activation energies E_σ for different regions. Tables 2 to 6 give an overview of selected measured data and calculated values for electrical and viscosity properties. Figures 7(a)–(c) represent the comparison of measured (thick lines) and fitted (thin lines) viscosity data of NBS A–D samples in different temperature ranges. Figure 8 shows a graphical comparison of measured electrical conductivity and viscosity data of the sodium borosilicate melts NBS-A, -B, -C, -D, and the sodium silicate melts NS and DS, for constant temperatures in the range 1000–1500°C. Solid lines are guides for the eye. The errors of all experimental data in the figures are within the point symbols.

Discussion

Composition and characteristic properties

All four NBS glasses and also the NS glass can undergo sub-liquid phase separation by slow cooling of large melts or annealing below the binodal temperature.[1–3] It is well known and often investigated that NBS glasses form a nearly pure SiO_2 and a sodium borosilicate phase with various SiO_2 contents depending on composition, temperature and time. After slow cooling of these large melts the glass samples NBS-A and -C were translucent; NBS-B and -D, and NS white opaque. Electron microscopy investigations have shown the following results: NBS-A and -C are separated into small SiO_2 droplets, size ~100 nm, in a sodium borate rich matrix; NBS-B and -D consist of connected phases

Table 1. Composition and properties of glass sample

Glass (mol%)	NBS-A	NBS-B	NBS-C	NBS-D	NS	DS
Na_2O	12·5	3·0	15·0	6·5	15·0	33·3
B_2O_3	62·5	48·5	42·5	33·5	–	–
SiO_2	25·0	48·5	42·5	60·0	85·0	66·7
T_g (°C) ±5	415	380	490	445	485	460
TEC (ppm/K) ±0·3 100–300°C	8·5	6·4	8·7	5·3	7·5	16·0
Density (g/cm³) ±0·01	2·18	2·04	2·31	2·15	2·34	2·49
Refractive index* n_e±0·00002	1·49621	1·47151	1·50626	1·47595	1·48331	1·50832
Dispersion coeff.* v_e±0·3	62·5	62·4	64·2	64·2	61·0	55·0

* The values of refractive indices and dispersion coefficients were measured with homogeneous samples.

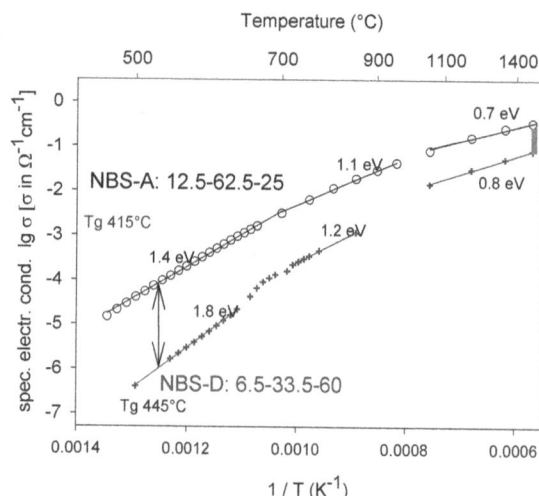

Figure 3. Temperature dependence of electrical conductivity of NBS-A and NBS-D from solid to liquid state: Solid lines represent Arrhenius linear fits with calculated activation energies

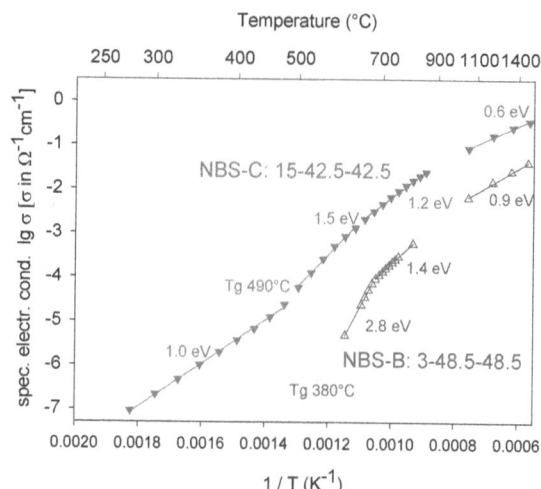

Figure 4. Temperature dependence of electrical conductivity of NBS-B and NBS-C from solid to liquid state: Solid lines represent Arrhenius linear fits with calculated activation energies

of μm size. The NS sample was separated into SiO_2 rich droplets and a sodium silicate matrix. It was possible to obtain homogenous glass samples by remelting of phase separated glass samples on a small scale with a faster cooling rate of the melts.

The temperatures for a viscosity $\log\eta = 4$ vary between 650°C for NBS A and 1100°C for NS; and for $\log\eta = 2$ between 850°C for NBS A and 1600°C for NS (Figure 2). These are very large differences, mainly affected by introducing B_2O_3. In contrast, the values for

Table 2. Measured data

Temperature T (°C)	Viscosity $\log\eta$ (η in dPa s)	Specific electrical conductivity			σ	Specific electrical resistance	
		$\ln\sigma$ (σ in $\Omega^{-1}m^{-1}$)	$\log\sigma$	$\log\sigma$ ($\Omega^{-1}cm^{-1}$)		ϱ (Ω cm)	$\log\varrho$ (ϱ in Ω cm)
NBS-A: $12.5Na_2O.62.5B_2O_3.25SiO_2$ T_g: 415°C T_{k100}: 290°C							
1500	−0·33	3·56	1·54	−0·45	0·35	2·84	0·45
1350	−0·03	3·27	1·42	−0·58	0·26	3·80	0·58
1200	0·36	2·79	1·21	−0·79	0·16	6·14	0·79
1050	0·88	2·09	0·91	−1·09	0·08	12·37	1·09
900	1·65	1·08	0·47	−1·53	0·03	33·88	1·53
800	2·39	0·15	0·06	−1·94	0·01	86·50	1·94
700	3·44	−1·14	−0·50	−2·50	3×10^{-3}	314·0	2·50
600	5·05	−2·97	−1·29	−3·29	5×10^{-4}	20×10^{2}	3·29
500	7· 85	−5·50	−2·39	−4·39	4×10^{-5}	$24·5\times10^{3}$	4·39
NBS-B: $3Na_2O.48.5B_2O_3.48.5SiO_2$ T_g: 380°C T_{k100}: 480°C							
1500	1·29	1·41	0·61	−1·39	0·04	24·42	1·39
1350	1·59	0·94	0·41	−1·59	0·02	39·07	1·59
1200	1·97	0·38	0·16	−1·84	0·01	68·39	1·84
1050	2·49	−0·43	−0·19	−2·19	6×10^{-3}	153·71	2·19
800	3·87	−2·85	−1·24	−3·24	5×10^{-4}	$17·2\times10^{2}$	3·34
700	4·78	−4·33	−7·88	−3·88	1×10^{-5}	$75·5\times10^{2}$	3·88
600	7·66	−7·66	−3·33	−5·33	5×10^{-6}	$21·2\times10^{4}$	5·33
NBS-C: $15Na_2O.42.5B_2O_3.42.5SiO_2$ T_g: 490°C T_{k100}: 220°C							
1500	0·17	3·58	1·55	−0·44	0·36	2·79	0·45
1350	0·49	3·27	1·42	−0·58	0·26	3·02	0·58
1200	0·92	2·82	1·22	−0·78	0·17	5·96	0·78
1050	1·52	2·15	0·93	−1·07	0·09	11·65	1·07
800	3·33	0·45	0·20	−1·80	0·02	63·55	1·80
700	4·70	−0·77	−0·34	−2·34	5×10^{-3}	217·6	2·34
600	7·02	−2·46	−1·07	−3·07	8×10^{-4}	$11·8\times10^{2}$	3·07
500	11·76	−5·16	−2·24	−4·24	6×10^{-5}	$17·4\times10^{3}$	4·24
400		−7·93	−3·44	−5·44	3×10^{-6}	$27·7\times10^{4}$	5·44
300		−10·74	−4·67	−6·67	2×10^{-7}	$46·4\times10^{5}$	6·67
NBS-D: $6.5Na_2O.33.5B_2O_3.60SiO_2$ T_g: 445°C T_{k100}: 410°C							
1500	1·33	2·09	0·91	−1·09	0·08	12·37	1·09
1350	1·66	1·67	0·72	−1·28	0·05	18·83	1·28
1200	2·09	1·12	0·49	−1·51	0·03	32·63	1·51
1050	2·68	0·36	0·16	−1·84	0·01	69·77	1·84
800	4·33	−2·71	−1·18	−3·18	$6·6\times10^{-4}$	$15·1\times10^{2}$	3·18
700	5·47	−4·41	−1·92	−3·92	$1·2\times10^{-4}$	$82·2\times10^{2}$	3·92
600	7·20	−6·99	−3·04	−5·04	$9·2\times10^{-6}$	$10·8\times10^{4}$	5·04
500	10·15	−10·11	−4·39	−6·39	$4·1\times10^{-7}$	$24·6\times10^{5}$	6·39

T_g are more similar. NBS-B with the lowest Na_2O content, 3 mol%, has the lowest T_g of 380°C due to the high content of $BØ_3$ and only a small content of $BØ_4^-$ units. NBS-C has the highest T_g of 490°C and should have a higher content of $BØ_4^-$ units. This T_g value is nearly the same as the T_g value 485°C for NS glass with the same Na_2O content, 15 mol%. The lowest TEC value, 5·3 ppm/K, is for NBS-D which lies on the anomaly line and has the highest SiO_2 content. NBS-B has the lowest density, 2·04 g/cm^3, and the lowest refractive index, n_e=1·47151 (at the e-line of Hg, λ_e=546·07 nm) caused by the lowest Na_2O and high B_2O_3 content.

Electrical properties

Electrical conductivity is the property of the greatest practical importance among the electrical properties. The strong change over large temperature ranges is of great interest.[8–11,14–17] As a global rule, the temperature dependence of a kinetic process may be described by a Boltzmann relation, and electrical conductivity σ and temperature are related by

$$\log\sigma = A - (E_\sigma/T) \tag{1}$$

The activation energies E_σ are nearly constant up to the glass transition temperature, T_g, in all cases measured (Figures 3–6); for NBS glasses E_σ~1·0 eV, for NS E_σ~0·8 eV, and for DS E_σ is only 0·6 eV.

The electrical conduction in these glasses is based on the transport of Na^+ ions through the glassy network. Thus, it should be dependent on the concentration of mobile Na^+ ions and their bond strength to the network. In silicate glasses, Na^+ ions are bonded to nonbridging oxygen. In borosilicate glasses, Na^+ ions prefer bonds to $BØ_4^-$ groups with one bridging oxygen which are stronger. This decreases the mobility

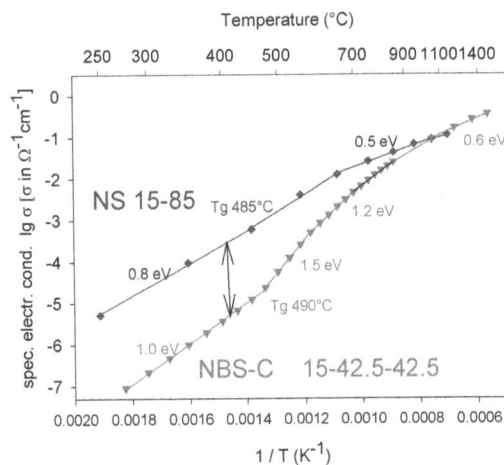

Figure 5. Temperature dependence of electrical conductivity of NBS-C in comparison with NS from solid to liquid state: Solid lines represent Arrhenius linear fits with calculated activation energies

of Na^+, thus the electrical conductivity, and increases the activation energy of the solid NBS glass samples. The electrical conductivity of the borosilicate glass NBS-C is about three orders lower than that of the silicate glass NS with the same Na_2O content, 15 mol% (Figure 5). With increasing temperature the conductivity of NBS-C approaches that of NS, and at ~1000°C the values are nearly the same.

There is a remarkably strong increase of the activation energy and electrical conductivity in the temperature range above T_g in the softening region of the glass network (Figures 3–6(a)). This increase is much stronger in NBS than in NS glasses (Figure 5). The highest activation energy, ~2·8 eV was measured for the NBS-B sample which has the lowest Na_2O content and a high B_2O_3 content (Figure 4). The E_σ values decrease with increasing Na_2O content and increasing

Table 3. Measured data

Temperature T (°C)	Viscosity $\log\eta$ (η in dPa s)	Specific electrical conductivity				Specific electrical resistance	
		$\ln\sigma$ (σ in $\Omega^{-1}m^{-1}$)	$\log\sigma$	$\log\sigma$ ($\Omega^{-1}cm^{-1}$)	σ	ϱ (Ω cm)	$\log\varrho$ (ϱ in Ω cm)
NS: $15Na_2O.85SiO_2$ T_g: 485°C T_{k100}: 120°C							
1500	2·11						
1350	2·78						
1200	3·61						
1050	4·65	2·16	0·94	−1·06	0·09	11·56	1·06
900	6·01	1·58	0·69	−1·31	0·05	20·51	1·31
800	7·16	1·10	0·48	−1·52	0·03	33·19	1·52
700	8·61	0·52	0·23	−1·77	0·02	59·16	1·77
600	10·47	−0·17	−0·08	−2·08	8×10^{-3}	1·2×10^2	2·07
500	12·95	−1·61	−0·70	−2·70	2×10^{-3}	5·0×10^2	2·70
400		−3·90	−1·69	−3·69	2×10^{-4}	4·9×10^3	3·69
300		−6·80	−2·95	−4·95	1×10^{-5}	9·0×10^4	4·95
200		−10·9	−4·73	−6·73	2×10^{-7}	5·4×10^6	6·73
DS: $33Na_2O.67SiO_2$ T_g: 460°C T_{k100}: 25°C							
1400	1·68	4·52	1·96	−0·04	0·92	1·09	0·04
1200	2·16	4·39	1·91	−0·09	0·81	1·24	0·09
1000	2·91	4·15	1·80	−0·20	0·63	1·58	0·20
900	3·47	3·64	1·58	−0·42	0·38	2·62	0·42
800	4·26	3·35	1·45	−0·54	0·28	3·51	0·54
700	5·43	cryst.					
600	7·36	cryst.					
500	11·19	cryst.					
400		−0·81	−0·35	−2·35	4×10^{-3}	2·25×10^2	2·35
300		−2·71	−1·18	−3·18	7×10^{-4}	1·50×10^3	3·18

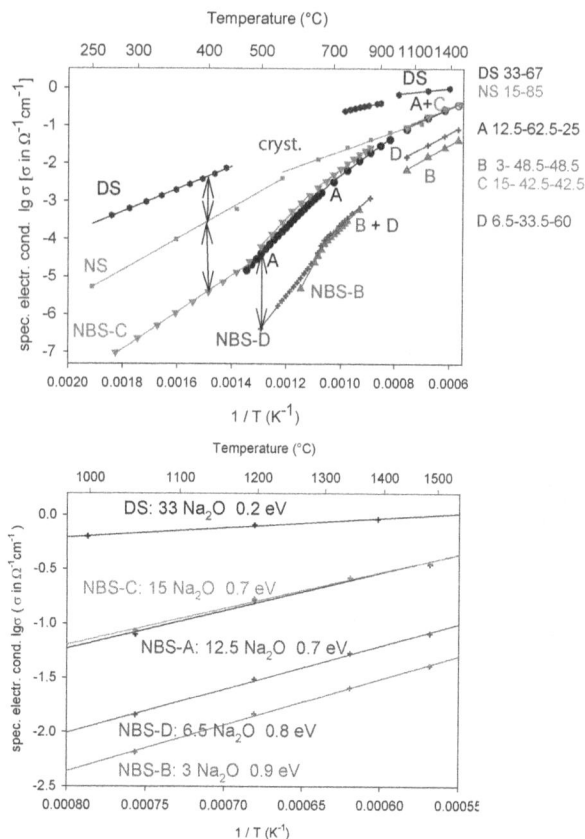

Figure 6. Overview of the electrical conductivities and activation energies (solid lines represent Arrhenius linear fits with calculated activation energies): (a) Glasses and melts, (b) Melts in more detail

temperature (Figure 6(a)). A superposition of different processes occurs: transport of mobile Na^+ ions and breaking the network which leads to viscous flow.

The electrical conductivities of the liquid melts in the temperature range 1000–1500°C correlate well with Arrhenius behaviour (Figure 6(b)). The conductivity increases and the activation energy decreases with increasing Na_2O content. Significant differences between silicate and borosilicate melts could not be recognised. Vogel–Fulcher–Tammann (VFT)-constants, A, B and T_0, for the conductivity and resistivity, according to

$$\log\sigma = A - B/(T - T_0) \qquad (2)$$

and

$$\log\rho = A - B/(T - T_0) \qquad (3)$$

were calculated for the NBS melts in the temperature

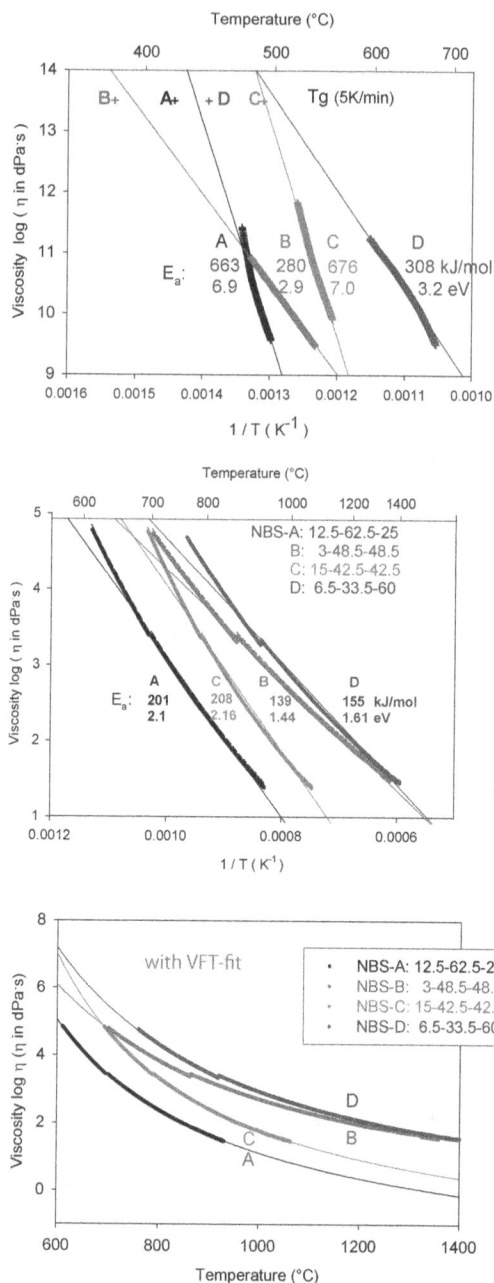

Figure 7. Comparison of measured (thick lines) and fitted (thin lines) viscosity data of NBS samples:(a) High viscosity region in comparison with T values at T_g determined by dilatometer; (b) Low viscosity region; (c) Measured data and VFT-fit in the temperature range 600 to 1400°C

range 1000–1500°C (Table 4). The electrical insulation of technical glasses is often defined by the temperature for electrical resistivity 10^8 Ω cm, denoted

Table 4. Calculated VFT-constants (VFT –fit only for the region 1000–1500°C)

	Viscosity VFT constants			Spec. electr. conductivity VFT constants			Spec. electr. resistance VFT constants		
	A	B	T_0	A	B	T_0	A	B	T_0
NBS-A	−2·5	2820·1	228·7	0·3	−793·6	493·7	−0·3	737·7	521·7
NBS-B	−1·1	3251·7	148·4	0·002	−1713·1	267·3	−0·08	1884·9	217·9
NBS-C	−2·0	2642·1	308·6	0·4	−874·9	447·0	−0·36	834·7	465·5
NBS-D	−1·2	3220·1	215·8	0·2	−1663·9	252·1	−0·25	1675·9	248·4
DS	−0·5	2391·3	295·0	0·16	−172·8	514·9	fit not possible		
NS	−4·4	10363·6	−98·4	1100°C					

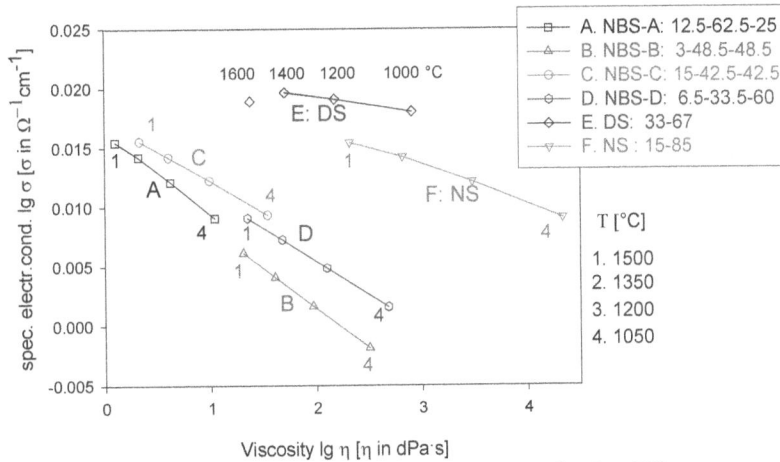

Figure 8. Electrical conductivity and viscosity of NBS, NS and DS samples for different temperatures: 1. 1500°C; 2. 1350°C; 3. 1200°C; 4. 1050°C (Solid lines are guides for the eye)

as T_{k100},[2] and in Tables 2 and 3, values of T_{k100} are also given. The lowest value, $T_{k100} \sim 25°C$, was found for DS glass, and the highest, $T_{k100} \sim 480°C$, for NBS-B. The NBS glasses have generally higher T_{k100} values than silicate glasses with the same Na_2O content.

Temperature dependence of viscosity

Structural changes of borate and borosilicate glasses have been intensively studies by different methods with regard to the borate anomaly.[1–4,6–7] It has been shown in previous papers[18–20] that melting and annealing conditions of two borosilicate model glasses and melts, NBS1 ($16Na_2O.10B_2O_3.74\,SiO_2$) and NBS2 ($4.3Na_2O.20.7B_2O_3.1Al_2O_3.74SiO_2$) affect their structure in a different way. The measured viscosity (η) temperature data were fitted by the VFT equation

$$\log\eta = A - B/(T - T_0) \qquad (4)$$

The VFT fit for the NBS1 sample was in good agreement with the experimental data and the effect of the melting and annealing conditions on the glass structure was very low. In contrast the NBS2 sample could not be fitted with the VFT equation. A strong deviation in the temperature range $T_g+(150–200)°C$ was found due to the effect of $B\emptyset_3$ groups. It can be shown by NMR, IR and Raman studies that in annealed glasses were more of $B\emptyset_3$ groups are linked

in boroxol rings, but in quenched glasses more $B\emptyset_3$ groups are connected with SiO_4 tetrahedra.[19–20] Differences in the thermal history of NBS2 samples result in glass structures with the same short range order (SRO) structural units while the medium range order (MRO) structure of the borosilicate network was found to be sensitive to changes in melt and cooling temperature.[20] Figure 7(a) shows the high viscosity region as Arrhenius plots with calculated activation energies, E_η, in comparison with a value, $\eta=10^{13.5}$ dPa s at T_g, determined by dilatometer measurements with a heating rate 5 K/min. Simple exponential behaviour was found in only very narrow ranges. NBS-C with the highest Na_2O content, 15 mol%, offers the best agreement with the value at T_g. The activation energy, $E_\eta=7.0$ eV, is very high; much higher than for the silicate sample NS, $E_\eta=3.0$ eV, with the same Na_2O content (Table 6). NBS-A, which has the highest B_2O_3 content, has a similar high value, $E_\eta=6.9$ eV. But NBS-A has a large deviation from the viscosity at T_g. Such large deviation near T_g was also detected for the samples NBS-B and -D. The activation energies for NBS-B and NBS-D, 2.9 eV and 3.2 eV, respectively, are much lower. The effect of phase separation on the viscosity data is relatively low.

Figure 7(b) shows the measured (thick lines) and calculated data for the low viscosity region.

Simple exponential behaviour was again found

Table 5. Activation energies of the electrical conductivity for various temperature ranges (calculated by Arrhenius plots)

Sample		$T<T_g$ E_σ 1 kJ/mol	eV	$T \sim T_g$ region E_σ 2 kJ/mol	eV	$T>T_g$ E_σ 3 kJ/mol	eV	$T>1000°C$ E_σ 4 kJ/mol	eV
NBS-A	E_A			135	1·4	100	1·05	65	0·7
	T (°C)			470–700		700–950		1050–1500	
NBS-B	E_A			265	2·75	130	1·35	80	0·85
	T (°C)			600–670		680–800		1050–1500	
NBS-C	E_A	95	1·0	145	1·5	110	1·15	60	0·65
	T (°C)	275–475		500–625		650–850		1050–1500	
NBS-D	E_A			175	1·8	120	1·25	80	0·8
	T (°C)			500–630		720–850		1050–1500	
NS 15-85	E_A	80	0·85			50	0·5		
	T (°C)	250–550				650–1150			
DS 33-67	E_A	60	0·6			30	0·3	15	0·15
	T (°C)	250–440				730–900		1000–1400	

Table 6. Activation energies of the viscosity for various temperature ranges (calculated by Arrhenius plots)

Sample	T_g (°C)	ΔT (°C)	$E_\eta 1$ (kJ/mol)	(eV)	ΔT (°C)	$E_\eta 2$ (kJ/mol)	(eV)
NBS-A	415	450–500	663	6·9	620–900	201	2·1
NBS-B	380	450–550	280	2·9	750–1400	139	1·4
			(279)	(2·9)			
NBS-C	490	490–580	676	7·0	700–1100	208	2·2
			(720)	(7·5)			
NBS-D	445	580–650	308	3·2	800–1400	155	1·6
			(377)	(3·9)			
NS 15-85	485	500–600	284	3·0	1200–1500	205	2·1
		(520–630)	(350)	(3·6)			
DS 33-67	460	480–520	551	5·7	850–1500	146	1·5

The values in the brackets are for phase separated samples

only in very narrow ranges. The calculated activation energies from Arrhenius plots are in the following order: NBS-C>-A>-D>-B.

Figure 7(c) shows the measured data and VFT-fits in the temperature range 600–1400°C. The experimental viscosity data are in good agreement with the VFT-fit in this temperature range. The calculated VFT constants are given in Table 4.

Electrical conductivity and viscosity of NBS, NS and DS melts for different temperatures

It is interesting to look for a correlation between data for the electrical conductivity and viscosity of the melts. It is clear that the mechanisms are different.

Figure 8 shows a graphical comparison of the measured electrical conductivity and viscosity data of the sodium borosilicate melts NBS-A, -B, -C, -D, and the sodium silicate melts NS and DS, for constant temperatures in the range 1000–1500°C. Solid lines are merely guides for the eye. It could be assumed that a nearly linear correlation exists between electrical conductivity, $\log\sigma$, and viscosity, $\log\eta$, dependent on composition and structure of the melts.

Conclusions

Four simple sodium borosilicate and two silicate model glasses were melted to investigate various properties of glasses and melts which depend on the structure. The composition of the samples varied over a wide range: 3 to 33·3 mol% Na_2O, 0 to 62·5 mol% B_2O_3, and 25 to 85 mol% SiO_2. The glass samples were characterized by several different methods.

The electrical conductivity of glasses and melts were determined by impedance measurements over a wide temperature range (250 to 1450°C). The activation energies were calculated by Arrhenius plots in various regions: below T_g, above the melting point, T_l, and between T_g and T_l.

Viscosity measurements were carried out with different methods from T_g to the melt. The measured data were fitted and the activation energies were calculated. Simple exponential behaviour was found only in very narrow temperature ranges.

The electrical conductivity is strongly dependent on the content of Na^+ ions and their mobility through the glassy network. Sodium borosilicate glasses ($T \leq T_g$) have much lower electrical conductivity than sodium silicate glasses due to the special borate units. The Na^+ ions are more strongly bonded to $B\varnothing_4^-$ tetrahedra with bridging oxygen than to SiO_3O^- tetrahedra with nonbridging oxygen.

Above T_g, in the softening region, the electrical conductivity of sodium borosilicate glasses increases strongly and approaches the values of sodium silicate samples near 1000°C. It is assumed that $B\varnothing_4^-$ units with bridging oxygen are transformed into $B\varnothing_2O^-$ units with nonbridging oxygen which increases the electrical conductivity drastically. Different mechanisms are superimposed.

The viscosity of sodium borosilicate melts is strongly dependent on the Na_2O/B_2O_3 ratio, the SiO_2 content and the temperature. A strong deviation from Arrhenius behaviour and a VFT-fit was found for $Na_2O/B_2O_3 \leq 0·2$ at $T \geq T_g$ depending on linking of $B\varnothing_3$ units in medium range order (MRO). The introduction of B_2O_3 in sodium silicate glasses strongly decreases the viscosity of the melts. It is found that a nearly linear correlation between the logs of electrical conductivity and viscosity exists for the melts.

Acknowledgement

The authors wish to thank R. Atzrodt, B. Keinert and R. Marschall for measurements and U. Kolberg from SCHOTT AG Mainz for collaboration and financial support. The samples investigated are also model glasses of Technical Committee TC03 of the International Commission on Glass.

References

1. Rawson, H. *Inorganic glass forming systems*, Academic Press, London, 1967.
2. Vogel, W. *Glass Chemistry*, Springer, Berlin, 1994.
3. Ehrt, D., Reiss, H. & Vogel, W. *Silikattechnik*, 1976, **27**, 304; 1977, **28**, 359.
4. Ehrt, D. *Glass Technol.*, 2000, **41**, 182.
5. Kloss, T., Lautenschläger, G. & Schneider, K. *Glass Technol.*, 2000, **41**, 177.
6. Ehrt, D. & Ebeling, P. *Glass Technol.*, 2003, **44**, 46.
7. Vedishcheva, N. M., Shakhmatkin, B. A. & Wright, A. C. *J. Non-Cryst. Solids*, 2004, **345&346**, 39.
8. Bach, H., Baucke, F. & Krause, D. *Electrochemistry of glasses and glass melts, including glass electrodes*, Springer, Berlin, 2001.
9. Schiefelbein, S. L. *Ceram. Trans*, Vol 92, *Proc. of the Electrochemistry of Glasses and Ceramics Symp.*, Ohio 1998, p.83.
10. Ravagnani, C., Keding, R. & Rüssel, C. *J. Non-Cryst. Solids*, 2003, **328**, 164.
11. Schirmer, H. & Rüssel, C. *J. Non-Cryst. Solids*, 2008, **354**, 889.
12. Ehrt, D., Leister, M. & Matthai, A. *Phys. Chem. Glasses*, 2001, **42**, 231.
13. Leister, M. & Ehrt, D. *Glastech. Ber. Sci. Technol.*, 1999, **72**, 153.
14. Pfeiffer, T. *Solid State Ionics*, 1998, **105**, 277.
15. Mazurin, O. W., Roskova, G. P. & Tschitsjakowa, E. B. *Silikattechnik*, 1973, **24**, 39.
16. Grandjean, A., Malki, M. & Simonnet, C. *J. Non-Cryst. Solids*, 2006, **352**, 2731.
17. Grandjean, A., Malki, M., Simonnet, C., Manara, D. & Penelon, B. *Phys. Rev. B*, 2007, **75**, 054112.
18. Möncke, D. & Ehrt, D. *Glass Sci. Technol.* 2002, **75**, 163.
19. Möncke, D., Ehrt, D., Eckert, H. & Mertens, V. *Phys. Chem. Glasses*, 2003, **44** (2), 113.
20. Möncke, D., Ehrt, D., Varsamis, C. E., Kamitsos, E. I. & Kalampounias, *Glass Technol.: Eur. J. Glass Sci. Technol. A*, 2006, **47** (5), 133.

Proc. VI Int. Conf. Borate Glasses, Himeji, Japan, 18–22 August 2008   *Phys. Chem. Glasses: Eur. J. Glass Sci. Technol. B, February 2010, **51** (1), 52–58*

Application of the constant stoichiometry grouping concept to the Raman spectra of BaO–Al$_2$O$_3$–B$_2$O$_3$ glasses

Irina Polyakova, Valentin Klyuev, Boris Pevzner

Institute of Silicate Chemistry, Russian Academy of Sciences, 29 Ul. Odoevskogo, St Petersburg, 199155 Russia

Vladimir Goncharuk

Institute of Chemistry of Far-Eastern Branch of Russian Academy of Sciences, 159 Pr. Stoletiya, Vladivostok, 690022 Russia

Oleg Yanush, Tatiana Markova,* Vladimir Maksimov & Vladimir Kabanov

St-Petersburg State Technological University of Plant Polymers, 4Ul. I. Chernykh, St Petersburg, 198095 Russia

Manuscript received 23 August 2008
Revised version received 18 October 2008
Accepted 4 December 2009

A new approach to glass structure, involving so-called constant stoichiometry groupings (CSGs) and based on Raman spectroscopy data, has been proposed to investigate the interactions of initial oxides in melts. The vibrational spectra of glasses are interpreted as a superposition of a relatively small number of unchangeable spectral forms belonging to the CSGs. The composition dependences of the concentrations of the CSGs afford an opportunity to explain, calculate and predict the compositional dependences of the properties of binary and ternary glasses, and the existence of new crystalline compounds. The present work shows the application of the CSG concept to the barium aluminoborate (BaAB) glass system, and demonstrates that systematic data on glass properties can give some information on the phase diagram and the localisation of compounds. The compositions obtained by use of the CSG concept for glasses were found to correspond to crystal stoichiometries on the BaAB phase diagram. Furthermore, it was shown that CSG localisation corresponds to extrema for the structural thermal expansion coefficient. We have found that the minima in T$_g$ versus composition for all series of glasses investigated coincide with the composition ranges where mixtures of CSGs occur.

1. Introduction

Borate based compositions are of interest for applications because of their piezoelectric, electro-optic and fast ionic properties, the second harmonic generation ability of poled glass samples, etc.[1] Recently, remarkable interest has been devoted to glass-ceramic materials containing polar phases characterised by nonlinear optical activity. These materials combine the advantages of glass technology and the special properties of borate crystals; nonlinear optical crystals of β-BaB$_2$O$_4$ exhibit second harmonic generation with intensity six times higher than for potassium dihydric phosphate crystals.[2] However, optimisation of the microphase properties of borate glasses is a very complicated technological problem.[3,4] The most interesting physical property in the system BaO–Al$_2$O$_3$–B$_2$O$_3$ is the low or negative thermal expansion coefficient (TEC). Alkaline earth aluminoborate glasses can be designed to have a thermal expansion match to battery pin material, to seal at relatively low temperatures, and to resist attack by both Li electrolyte and ambient conditions.

Vibrational spectroscopy is one of the most useful experimental techniques available for structural studies of glasses. This method is sensitive to small variation of the chemical bonds which determine the properties of glasses. The majority of investigators discuss assignments of vibrational spectra of borate glasses by focusing attention on the effects which conversion of triangular BO$_3$ structural units to BO$_4$ tetrahedra, and the formation of nonbridging oxygen atoms have on characteristic bands. It is conventional to explain the borate anomaly of some physical properties of borate glasses by this model. Alternatively, as early as 1962, Krogh-Moe[5] postulated that the structure of borate glasses can be described as a random network of large borate groups (such as diborate, triborate, pentaborate, etc). This 'molecular' approach is common practice for chemistry. It has thus been proposed[6,7] to interpret vibrational spectra (mainly Raman, but also IR) of glasses as a superposition of a relatively small number of unchangeable spectral forms (principal spectral components, PSCs[7]), which add to form the measured spectrum. These PSCs are interpreted as belonging to the stable products of the interaction of initial oxides, and are called constant

* Corresponding author. Email O.V.Yanush@inbox.ru

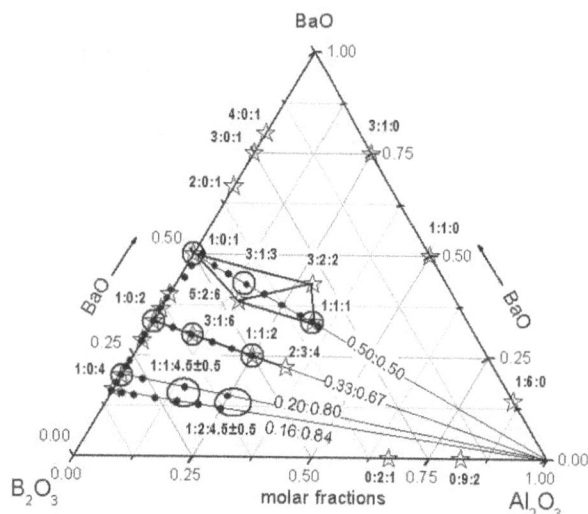

Figure 1. • – *Composition of BaO–Al₂O₃–B₂O₃ glasses used in this study.* ☆ – *known compounds (the [BaO]:[Al₂O₃]:[B₂O₃] stoichiometries are indicated on the figure).* ○ – *CSGs determined by Raman spectroscopy analysis in this paper*

stoichiometry groupings (CSGs). For the treatment of vibrational spectra, we hold to the hypothesis that CSGs are 'the smallest particles' which compose the glass, defining its properties.[7–9] Vibrational spectroscopy cannot deliver information concerning the size of CSGs. But in binary borate glasses with heavy modifier cations (Cs, Ba) cation clustering has been detected by x-ray diffraction.[10] An average Ba–Ba distance of 6·9 Å was obtained for barium tetraborate glass. In crystalline $BaO.4B_2O_3$, where barium atoms are arranged approximately in a body centred pseudo-tetragonal lattice, and each barium atom has two barium neighbours at 6·6 Å, and eight barium neighbours at 6·94 Å, the crystal distances coincide remarkably well with the average barium–barium distance of 6·9 Å for barium tetraborate glass. It should be emphasised that the Raman spectrum of crystalline $BaO.4B_2O_3$ resembles the corresponding PSC. This means that CSGs are of nanoscale range, and their real chemical formulae, for example, are not $BaO.4B_2O_3$ but $n(BaO.4B_2O_3)$, where $n≥1$. This point is important for the use of the Law of Mass Action.

The analysis of the intensities of the PSCs, which are extracted from the Raman spectra by the technique described in Ref. 7, makes it possible to calculate the composition dependences of the molar fraction of each CSG. This information makes it possible to explain, calculate and predict many properties of glasses in agreement with experimental data. It should be noted that the CSG concept is based on the experimentally measured vibrational spectra exclusively without using any supplementary data (for example, vibrational spectra of crystals, phase diagram, thermodynamic properties of crystals, constitutional relations of mineral phases, etc) in contrast to techniques developed, respectively, in Refs

11 to 13, and others. In the present work, we show the application of the CSG concept to the investigation of the barium aluminoborate glass system.

2. Experimental

The compositions of the barium aluminoborate glasses used in the present Raman spectroscopy study, and also the relevant crystal stoichiometries, are presented in Figure 1. The glasses were synthesised in a platinum crucible at temperatures in the range 1200–1250°C; the initial ingredients were reagent or spec-purity grade $BaCO_3$, $Al(OH)_3$ and H_3BO_3. Bulk glasses and powders were crystallised in electrical furnaces using appropriate temperature–time conditions. The phase composition of the crystallised samples was established by x-ray powder diffractometry.

Raman scattering was excited by an argon laser ($\lambda=0·488$ μm, 0·5 W). Polarised Raman spectra were measured with a 3 cm⁻¹ step at room temperature by a two beam spectrometer. The spectral width of the slit was no greater than 3 cm⁻¹. A sample made of fused silica was used as a reference, and the Raman scattering (VV) intensity at frequency 440 cm⁻¹ was taken as 1 (I^{SiO_2} (440 cm⁻¹)=1). Details of the measurement procedure are described elsewhere.[4,14]

3. Results

3.1. The $(1-y)(0·50BaO.0·50B_2O_3).yAl_2O_3$ glass series (a)

The Raman spectra (VV) of the $(1-y)(0·50BaO.0·50B_2O_3).yAl_2O_3$ series of glasses are shown in Figure 2(a). In the case of metaborate glasses, the band at 745 cm⁻¹ indicates the presence of six membered rings with two $BØ_4^-$ tetrahedra and one metaborate triangle $BØ_2O^-$.[15,16] The 630 cm⁻¹ band corresponds to metaborate triangles $BØ_2O^-$ arranged in metaborate rings ($B_3O_6^{3-}$). On adding Al_2O_3 up to 15–20 mol%, the bands at 745 and 630 cm⁻¹ vary slightly in intensity, and then at still higher Al_2O_3 concentrations they disappear (Figure 2(a)). In addition to these medium range order structures, other borate arrangements containing $BØ_4^-$ tetrahedra should be forming, as indicated by the bands developing at 500–530, 940 and 1100–1200 cm⁻¹.[16] As the Al_2O_3 content increases, the 500–530 cm⁻¹ band disappears, giving rise to a peak at 460 cm⁻¹ which indicates the formation of $AlØ_4^-$ species in the glass.[17] The band at 940 cm⁻¹ shifts monotonically to 1025 cm⁻¹, and the intensity of the band at 1100–1200 cm⁻¹ decreases monotonically (Figure 2(a)).

In the higher frequency range, the band at 1450 cm⁻¹ is assigned to the vibrations of terminal B–O⁻ bonds in large borate units containing both $BØ_4^-$ tetrahedra and metaborate triangles $BØ_2O^-$.[16] The 1450 cm⁻¹ band decreases in intensity and shifts to lower wavenumbers with increasing Al_2O_3 content (Figure 2(a)). In the low frequency region, the band at ~200

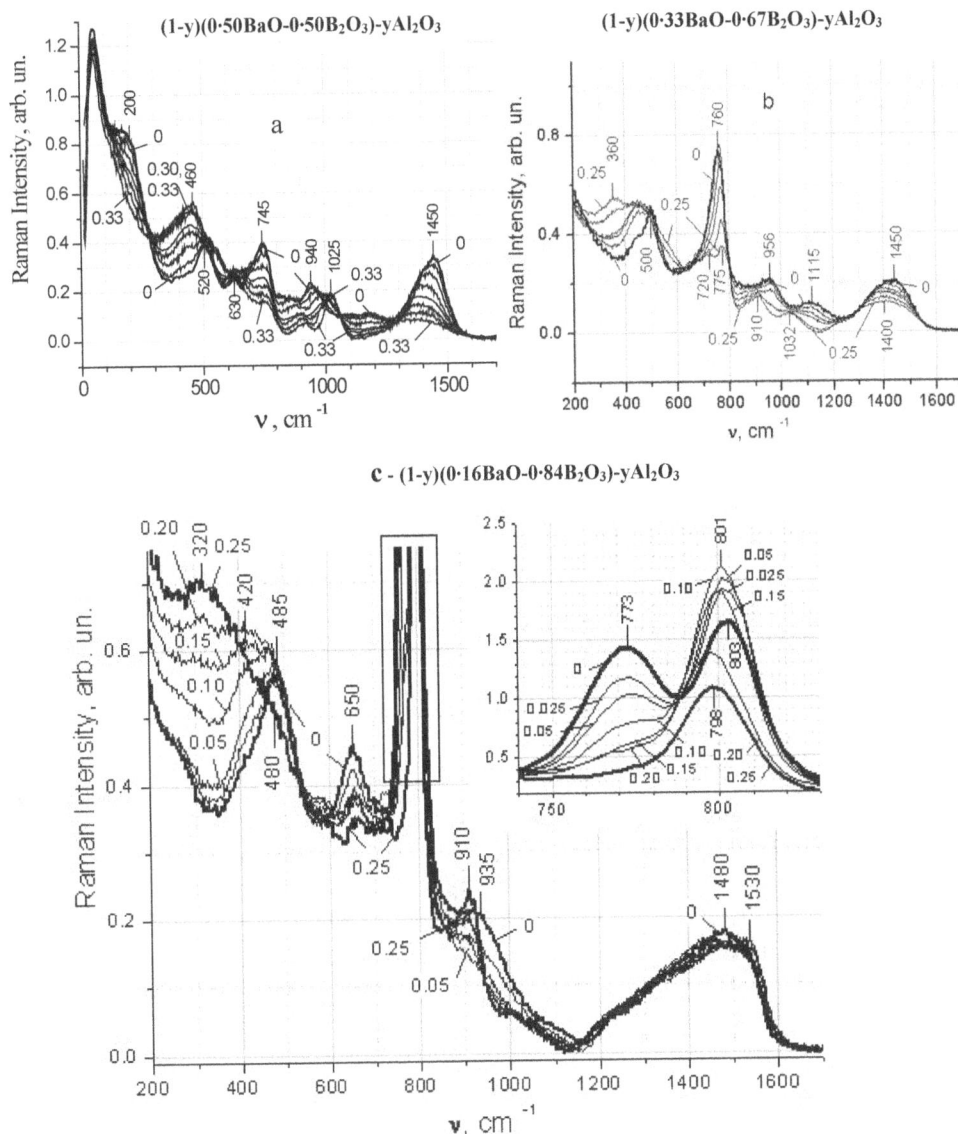

Figure 2. Raman spectra (VV) of all three glass series. (a) (1–y)(0·50BaO.0·50B₂O₃).yAl₂O₃; (b) (1–y)(0·33BaO.0·67B₂O₃). yAl₂O₃; (c) (1–y)(0·16BaO.0·84B₂O₃).yAl₂O₃). Horizontal numbers on the figures indicate Al₂O₃ mole fractions, whilst vertical numbers indicate peak positions

cm^{-1}, originating from the localised vibrations of metal ions (Ba^{2+}) in their equilibrium sites formed by oxygen atoms,[16,18] disappears with increasing Al$_2$O$_3$ content (Figure 2(a)), and this can be explained by the formation of Ba–O–Al linkages.

Analysis of the Raman spectra by the Wallace–Katz technique[19] shows that the experimental spectra of this series can be interpreted as the sum of only three PSCs. Examples of the deconvolution procedure and the constancy of the PSCs are illustrated in Figure 3(a) and Figure 4 (curve 3:1:3). All derived PSCs are demonstrated in Figure 5(a) (curves 1:0:1, 3:1:3 and 1:1:1). The composition dependences of the intensities of the PSCs (Figure 6(a-1)) allow us to determine the stoichiometry of every CSG: BaO.B$_2$O$_3$, 3BaO. Al$_2$O$_3$.3B$_2$O$_3$ and BaO.Al$_2$O$_3$.B$_2$O$_3$. In this figure the composition dependences of the Raman intensities of every PSC have been converted to give mole frac-

tions for each CSG of the series by using equations like the following

$$\frac{1}{2}\frac{I_{1:0:1}}{\sigma_{1:0:1}} + \frac{3}{7}\frac{I_{3:1:3}}{\sigma_{3:1:3}} + \frac{1}{3}\frac{I_{1:1:1}}{\sigma_{1:1:1}} = x_{B_2O_3} \qquad (1)$$

where I_{CSG} is the relative intensity of a certain PSC, and σ_{CSG} is its scattering cross section. The compositions of the CSGs are given in the form of [BaO]:[Al$_2$O$_3$]:[B$_2$O$_3$]. The coefficients in Equation (1) are consistent with the glass composition in molar fractions of initial oxides, whereby one mole of glass contains Avogadro's number of molecules of the initial oxides. If the glass, for example, is comprised of the CSG 3BaO.Al$_2$O$_3$.3B$_2$O$_3$ only, dissociation of which we ignore, 1 mole of glass will contain 3/7 mole of BaO, 1/7 mole of Al$_2$O$_3$, and 3/7 mole of B$_2$O$_3$, and correspondingly 1 mole of the CSG 3BaO. Al$_2$O$_3$.3B$_2$O$_3$.

Figure 4. Demonstration of the constancy of the PSCs belonging to the CSGs: $B_2O_3^{solv}$, $BaO.Al_2O_3.(4–5)B_2O_3$ $(1:1:(4–5))$, $3BaO.Al_2O_3.6B_2O_3$ $(3:1:6)$ and $3BaO.Al_2O_3.3B_2O_3$ $(3:1:3)$. Horizontal numbers to the right of each PSC indicate the Al_2O_3 compositional ranges used to determine it

decreases in intensity and shifts to lower frequencies with increasing Al_2O_3 content (Figure 2(a) and (b)).

Analysis by the Wallace–Katz technique shows that the experimental spectra of this series can also be interpreted as the sum of three PSCs belonging to the following CSGs: $BaO.2B_2O_3$, $3BaO.Al_2O_3.6B_2O_3$ and $BaO.Al_2O_3.2B_2O_3$ (curves 1:0:2, 3:1:6, and 1:1:2 in Figure 5(a)). Examples of the deconvolution procedure and the constancy of the PSCs are illustrated in Figure 3(b), and Figure 4 (curve 3:1:6). Using equations like Equation (1), the composition dependences of the mole fractions of the CSGs were obtained (Figure 6(b-1)).

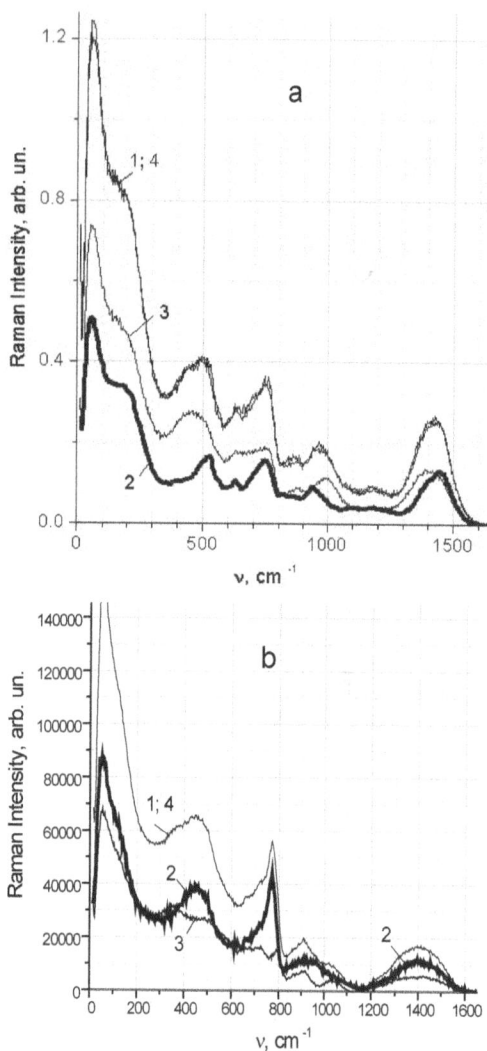

Figure 3. Examples of the deconvolution of the experimental Raman spectra. (1) (a) The spectrum for $0.90(0.50Ba$ $O.0.50B_2O_3).0.10Al_2O_3$ glass is deconvoluted into these PSCs: $BaO.B_2O_3$ (2) and $3BaO.Al_2O_3.3B_2O_3$ (3). (b) The spectrum for $0.80.(0.33BaO.0.67B_2O_3).0.20Al_2O_3$ glass is deconvoluted into these PSCs: $3BaO.Al_2O_3.6B_2O_3$ (2) and $BaO.Al_2O_3.2B_2O_3$ (3). In each case, curve (4) indicates the sum of the PSCs (2) and (3)

3.2. The $(1–y)(0.33BaO.0.67B_2O_3).yAl_2O_3$ glass series (b)

The Raman spectra (VV) of the $(1–y)(0.33BaO.0.67B_2O_3).$ yAl_2O_3 series of glasses are shown in Figure 2(b). As for the glasses in the previous series, the strong Raman band at 760 cm^{-1} decreases in intensity and shifts towards higher frequencies as the Al_2O_3 content increases (Figure 2(a) and (b)). Then, for $y>0.15$, the 760 cm^{-1} band splits into two (or three) bands at 775, 720, (and 660) cm^{-1} (Figure 2(b)). The intensities of the 500, 956 and 1115 cm^{-1} bands decrease monotonically in the composition range $0<y\leq0.15$, and then disappear, being replaced by some other type of structure, giving rise to the peaks at 455 and 360, 480, 910, and 1032 cm^{-1} for $y>0.15$ (Figure 2(b)). As for the glasses in the previous series, the peak at about 1450 cm^{-1}

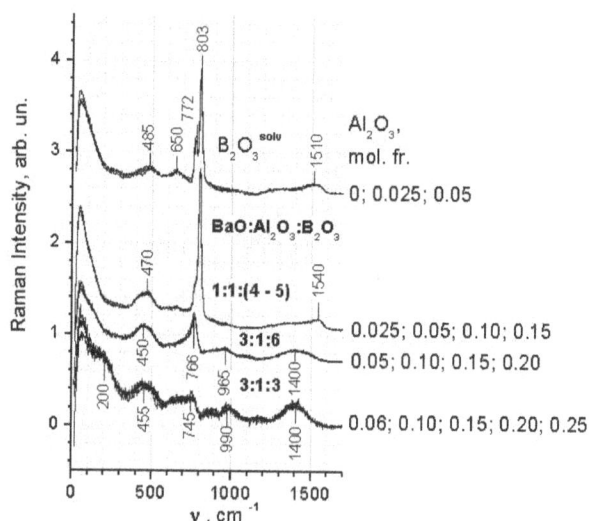

3.3. The $(1–y)(0.16BaO.0.84B_2O_3).yAl_2O_3$ glass series (c)

For the glasses in the $(1–y)(0.16BaO.0.84B_2O_3).yAl_2O_3$ series, a sharp, strong band at 803 cm^{-1} increases in intensity up to $y=0.10$, where it reaches a maximum and shifts to 801 cm^{-1}. At still higher Al_2O_3 concentrations, the 801 cm^{-1} band decreases in intensity and in frequency, and (at $y=0.25$) it shifts to 798 cm^{-1}. For this same composition, the band at about 770 cm^{-1} has completely disappeared (Figure 2(c)).

As the Al_2O_3 content increases, the 935 cm^{-1} band first decreases in intensity, and at approximately $0.05–0.10$ molar fraction Al_2O_3 it disappears, giving rise to the 910 cm^{-1} peak. The band at 650 cm^{-1} disappears on addition of Al_2O_3. With increasing Al_2O_3 content, the 485 cm^{-1} band is replaced by a peak at 420 cm^{-1}, and then by peaks at 320 and 490 cm^{-1} for $y=0.25$. The band at about 1450 cm^{-1} shifts to higher frequencies up to approximately 0.15 molar fraction Al_2O_3, and then occurs at lower frequency for $y=0.20$ and 0.25. The intensity of this band remains more

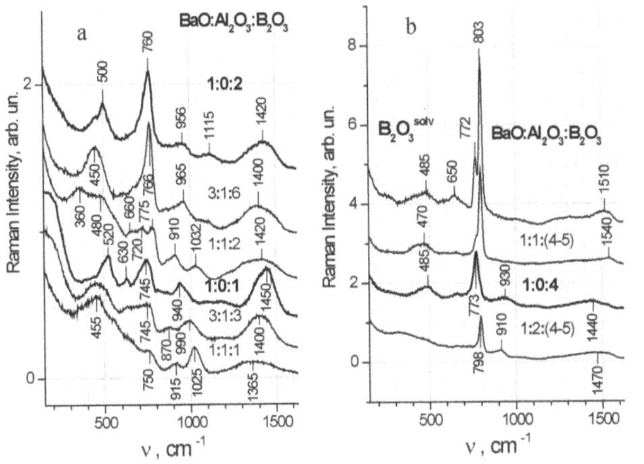

Figure 5. Illustration of all PSCs belonging to CSGs in glasses of all series investigated

or less constant across the whole composition range (Figure 2(c)).

Analysis by the Wallace–Katz technique shows that the experimental spectra of this series can be interpreted as the sum of four PSCs belonging to the following CSGs: $BaO.4B_2O_3$, $BaO.Al_2O_3.(4-5)B_2O_3$, $BaO.2Al_2O_3.(4-5)B_2O_3$ and $B_2O_3^{solv}$ (Figure 6(c-1)). Krogh-Moe[20] and Bril[15] attributed the 806 cm^{-1} peak

in the Raman spectra of vitreous boron oxide to the symmetrical breathing vibration of the boroxol ring. In Refs 8 and 22 it was assumed that these vibrations can shift to the frequency 803 cm^{-1} when boron oxide takes part in solvation of tetraborate groupings (modified B_2O_3). The CSG which is manifested by the band developing at 803 cm^{-1} is indicated in this work by $B_2O_3^{solv}$ (curve $B_2O_3^{solv}$ of Figure 5(b)). Demonstration of the constancy of the PSCs belonging to the CSGs $B_2O_3^{solv}$ and $BaO.Al_2O_3.(4-5)B_2O_3$ is shown in Figure 4 (curves $B_2O_3^{solv}$ and 1:1:(4–5)).

4. Discussion

Comparison of the stoichiometry of the CSGs obtained from Raman spectra analysis with the compositions of compounds[21] in the series $(1-y)(0·33BaO.0·67B_2O_3).yAl_2O_3$ shows their coincidence (1:0:2,[22] 3:1:6 and 1:1:2, (Figure 1)), and this confirms the validity of the CSG concept. The same coincidence was found between the CSGs for the series $(1-y)(0·50BaO.0·50B_2O_3).yAl_2O_3$ and the compounds of 1:0:1 and 1:1:1 composition. As for the CSG 3:1:3 obtained from Raman spectra analysis in this series, it does not correspond to any known

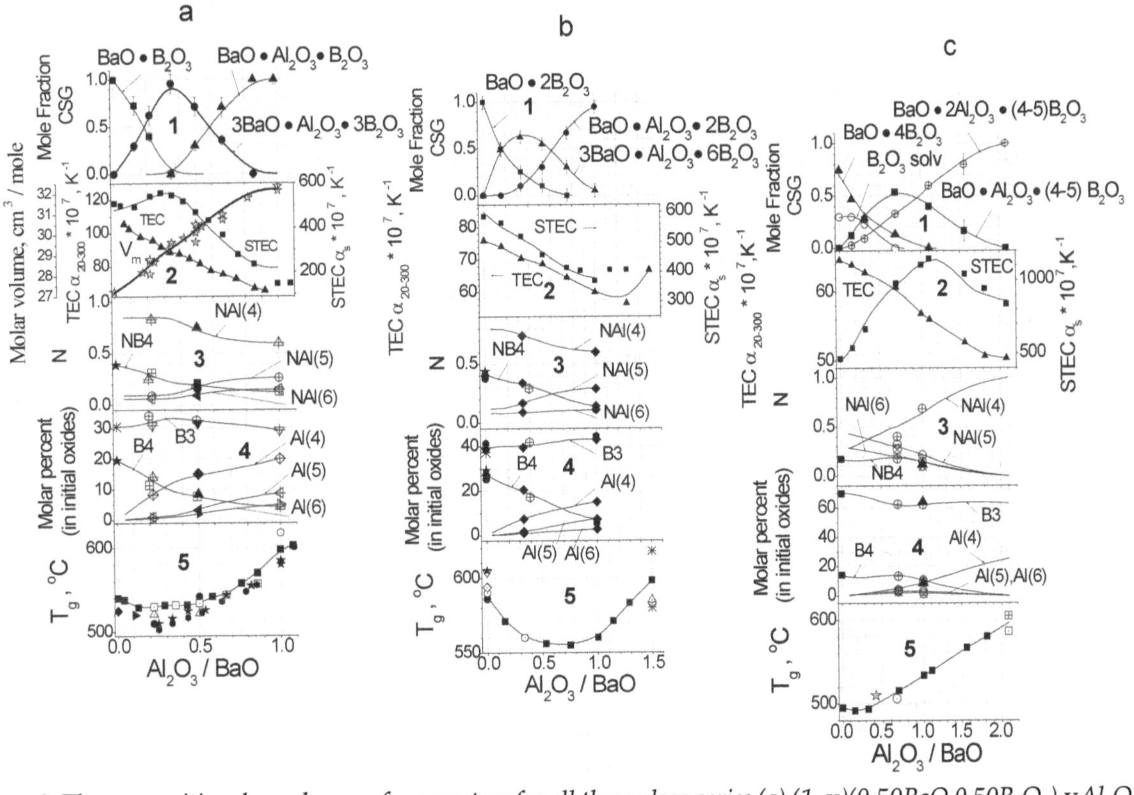

Figure 6. The composition dependences of parameters for all three glass series (a) $(1-y)(0·50BaO.0·50B_2O_3).yAl_2O_3$; (b) $(1-y)(0·33BaO.0·67B_2O_3).yAl_2O_3$; (c) $(1-y)(0·16BaO.0·84B_2O_3).yAl_2O_3$. (1) mole fraction of each CSG. (2) thermal expansion coefficient, α_{20-300}, structural thermal expansion coefficient, α_s, and molar volume, V_m. The fractions (3) and the molar concentrations, %, (4) of B and Al in different coordinations (filled symbols are experimental data obtained from NMR,[25–28] infrared spectroscopy,[16,18] neutron diffraction and molecular dynamics simulations,[29] and empty symbols are the result of their interpolation). (5) The glass transition temperature, T_g. Lines in 2–4 are calculations, and in 1 and 5 are guides to the eye. Points in 2 to 5 are experimental data from the literature[†(23,24)]

[†] Data references from SciGlass 7.0 software

compound. As can be seen from Figure 1, the series $(1-y)(0.50BaO.0.50B_2O_3).yAl_2O_3$ forms a boundary between two compounds, 5:2:6 and 3:2:2. It can be assumed that the CSG of 3:1:3 stoichiometry in reality is a mixture of CSGs of 5:2:6 and 3:2:2 compositions. This assumption should be checked by vibrational spectroscopy investigation of additional glasses belonging to the true pseudo-binary joint 5:2:6-3:2:2 (Figure 1). In the series $(1-y)(0.16BaO.0.84B_2O_3)$. yAl_2O_3 no compounds are yet known. For this reason, our results obtained from Raman spectra analysis concerning CSGs of 1:0:4, 1:1:(4–5), 1:2:(4–5) and $B_2O_3^{solv}$ have a tentative character.

An additional argument in favour of the results of the Raman spectra investigation is the coincidence of the extrema in the structural thermal expansion coefficient (STEC), α_{sr} [23,24] versus composition with the maxima of the CSG composition dependences in the series $(1-y)(0.50BaO.0.50B_2O_3)$. yAl_2O_3 (Figure 6(a-2) (curve STEC) and Figure 6(a-1) (curve $3BaO.Al_2O_3.3B_2O_3$)) and in the series $(1-y)(0.16BaO.0.84B_2O_3).yAl_2O_3$ (Figure 6(c-2) (curve STEC) and Figure 6(c-1) (curve $BaO.Al_2O_3.(4–5)B_2O_3$)). The CSG concept allows us to calculate the STEC curves using an additive approximation

$$\alpha_s = \sum_i v_i \alpha_i^s x_i^{CSG} \Big/ \sum_i v_i x_i^{CSG} \qquad (2)$$

where α_i^s is the partial STEC of the ith CSG, v_i is the partial molar volume of the ith CSG, and x_i^{CSG} is the mole fraction of the ith CSG. We assume here that the molar fractions of the CSGs do not change significantly within the narrow temperature interval of the STEC measurement. Among other glass properties that can be calculated in the additive approximation, only the molar volume V is extensively investigated[†]. Results of STEC and glass molar volume calculation according to

$$V = \sum_i v_i x_i^{CSG} \qquad (3)$$

are in good agreement with experimental data[†(23,24)] (part 2 of Figure 6(a) to (c)).

---

† Data references from SciGlass 7.0 software

Traditionally, the trend of the fraction of four-coordinated borons (NB4) is used to interpret the properties of borate glasses. The MAS (magic angle spinning) NMR technique provides valuable quantitative information about the fractions NB4, NAl(4), NAl(5) and NAl(6); however, there are very few data for barium aluminoborate glasses. These data, for all series investigated, are presented in parts 3 and 4 of Figure 6(a) to (c). The molar fractions of the CSGs resulting from the analysis of the Raman spectra make possible the calculation of the composition dependences of NB4, NAl(4), NAl(5) and NAl(6), assuming that the coordination of B and Al atoms belonging to a certain CSG do not depend on glass composition, and using equations like the following for the series $(1-y)(0.33BaO.0.67B_2O_3).yAl_2O_3$

$$NB4 = \frac{\frac{2}{3}N_{1:0:2}x^{1:0:2} + \frac{6}{10}N_{3:1:6}x^{3:1:6} + \frac{1}{2}N_{1:1:2}x^{1:1:2}}{\frac{2}{3}x^{1:0:2} + \frac{6}{10}x^{3:1:6} + \frac{1}{2}x^{1:1:2}} \qquad (4)$$

where x^{CSG} is the mole fraction of a CSG, N_{CSG} (for example $N_{3:1:6}$) is the NB4 value for the CSG itself (parts 3 and 4 (lines) of Figure 6(a) to (c) and partial values in Table 1). Though correlations with NB4 are observed for many properties of the binary barium borate system,[22] it is difficult to say that in the ternary system any of the properties (molar volume, TEC, STEC, and T_g) correlate to any of the B or Al coordinations (parts 2 to 5 of Figure 6(a) to (c)).

In Refs 17 and 30 the T_g for alkali boroaluminate glasses was shown to have a non-monotonic variation with a minimum. The results described above indicate that the minima in T_g versus composition for all series of glasses investigated coincide with the composition ranges where mixtures of CSGs take place (parts 1 and 5 of Figure 6(a) to (c)).

Traditionally it has been proposed that variations of the intensities of bands in the vibrational spectra depend on the concentrations of $B\varnothing_4^-$ and $B\varnothing_3$ units and 4-, 5- and 6-coordinated Al. In the context of this consider the most striking features of Raman spectra behaviour illustrated above.

The band at 1450 cm^{-1} rapidly decreases

Table 1. Partial properties of the CSGs in barium aluminoborate glasses

CSG	Partial properties of CSGs					
	The fraction of 4-coordinated borons (NB4)	The fraction of 4-coordinated Al (NAl(4))	The fraction of 5-coordinated Al (NAl(5))	The fraction of 6-coordinated Al(NAl (6))	Molar volume v_i, (cm³/mol)	Structural thermal expansion coefficient, STEC, $\alpha_s \times 10^7$ (K^{-1})
$B_2O_3^{solv}$	0	0	0	0	-	300
$BaO.4B_2O_3$	0.25	0	0	0	30.2*	500
$BaO.Al_2O_3.(4–5)B_2O_3$	0.28	0	0.6	0.4	-	1210
$BaO.2Al_2O_3.(4–5)B_2O_3$	0	1	0	0	-	860
$BaO.2B_2O_3$	0.41	0	0	0	27.6*	600
$3BaO.Al_2O_3.6B_2O_3$	0.316	0.8	0.12	0.08	-	500
$BaO.Al_2O_3.2B_2O_3$	0.13	0.6	0.3	0.11	-	370
$BaO.B_2O_3$	0.39	0	0	0	27.2	480
$3BaO.Al_2O_3.3B_2O_3$	0.21	0.85	0.095	0.06	29.4	570
$BaO.Al_2O_3.B_2O_3$	0.126	0.598	0.26	0.148	32.2	220

* data from Ref. 22

in intensity (by a factor of four) in the series $(1-y)(0\cdot50BaO.0\cdot50B_2O_3).yAl_2O_3$, decreases to a lesser extent in the series $(1-y)(0\cdot33BaO.0\cdot67B_2O_3).yAl_2O_3$, and remains more or less constant in intensity in the series $(1-y)(0\cdot16BaO.0\cdot84B_2O_3).yAl_2O_3$ (Figure 2(a) to (c)). In fact the composition dependences of molar concentration of B4 for the glasses in all series investigated were found to be similar to the behaviour of the intensity of the 1450 cm^{-1} Raman band (Figure 2(a) to (c) and parts 3 and 4 of Figure 6(a) to (c)). This correlation was also investigated in more detail in studies of the behaviour of the high frequency band in the IR spectra of alkaline earth borate glasses by Yiannopoulos et al.[16] Then the 1100–1200 cm^{-1} band was found to decrease in intensity for the series $(1-y)(0\cdot50BaO.0\cdot50B_2O_3).yAl_2O_3$ and $(1-y)(0\cdot33BaO.0\cdot67B_2O_3).yAl_2O_3$ in accordance with the B4 decrease, whilst this band is absent in the series $(1-y)(0\cdot16BaO.0\cdot84B_2O_3).yAl_2O_3$ where the B4 concentration is small (Figure 2(a) to (c) and parts 3 and 4 of Figure 6(a) to (c)).

The 900–1000 cm^{-1} envelope shifts towards higher frequencies and remains more or less constant in intensity as the Al_2O_3 content increases in the series $(1-y)(0\cdot50BaO.0\cdot50B_2O_3).yAl_2O_3$ (Figure 2(a)), and goes through a minimum in intensity and shifts to lower wavenumbers in the other series (Figure 2(b) and (c)). Thus in this case there is no vibrational evidence for any correlation with composition dependences of fractions and molar concentrations of B4.

In the context of the CSG concept, the 630 and 650 cm^{-1} Raman bands are due to the symmetric breathing vibrations of $BØ_2O^-$ triangles[16] that make up the CSGs $BaO.4B_2O_3$ and $BaO.B_2O_3$, but not the CSG $BaO.2B_2O_3$. It was shown that, for $xBaO.$ $(1-x)B_2O_3$ glasses at $x>0\cdot33$[16] and $x<0\cdot33$ (see Ref. 22 and references therein[28]) the bands at 630 and 650 cm^{-1} increase in intensity and follow the increase of the fraction of three coordinated boron atoms. With increasing Al_2O_3 content the bands at 630 and 650 cm^{-1} disappear for the glasses in the series investigated (Figure 2(a) and (c)) and thus it is evident that in the case of Al_2O_3 containing glasses there is no correlation with the increase of the concentration of B3, according to MAS NMR results,[25–28] (part 4 of Figure 6(a) and (c)). The 500–530 cm^{-1} band varies in intensity very slightly with increasing BaO content for $xBaO.$ $(1-x)B_2O_3$ glasses and this also signals the absence of correlation with the concentration of B4.[14,22] With increasing Al_2O_3 content this band disappears (despite the fact that the concentration of B4 decreases monotonically) in all series investigated (Figure 2(a) to (c), and parts 3 and 4 of Figure 6(a) to (c)).

Conclusions

Based on the vibrational spectra, the CSG (constant stoichiometry grouping) concept is demonstrated for the barium aluminoborate glass forming system. It was shown that CSG localisation corresponds to extrema for the structural thermal expansion coefficient (STEC). Calculated glass properties (molar volume, STEC, concentrations of B4, B3 and Al(4), Al(5), Al(6)) are in a good agreement with experimental data. Our results indicate that the minima in T_g versus composition of all series of glasses investigated coincide with the composition ranges where mixtures of CSGs take place. Some correlations between the behaviour of Raman bands and concentrations of B3, B4 and Al(4) are observed, but there are many exceptions.

Acknowledgement

This work was partially supported by a grant from RFBR (Project N 02-03-32499). This work was carried out with the support of the Committee on Science and Higher Education of the Government of Saint-Petersburg.

References

1. Affatigato, M., Feller, S. A., Vedishcheva, N. M. & Wright, A. C. Eds. Borate Glasses, Crystals & Melts 2002: New Techniques and Applications, Society of Glass Technology, Sheffield, 2003, p. 282.
2. Fedorov, P. P., Kokh, A. E. & Kononova, N. G. Uspekhi Khim., 2002, 71, 741. (In Russian.)
3. Gode, G. K. Dr Sc. Thesis, 1974. Riga. P 600. (In Russian.)
4. Lipovskii, A. A., Tagantsev, D. K., Vetrov, A. A. & Yanush, O. V. Opt. Mater., 2003, 21, 749.
5. Krogh-Moe, J. Phys. Chem. Glasses, 1962, 3, 101.
6. Kabanov, V. O. & Yanush, O. V. Fiz. Khim. Stekla, 1987, 13, 524. (In Russian.)
7. Yanush, O. V. In: Proceedings of the XVIII International Congress on Glass, San Francisco, 1998. p.75.
8. Yanush, O. V., Kabanov, V. O. & Mukhitdinova, I. A. Fiz. Khim. Stekla, 1988, 14, 330. (In Russian.)
9. Karapetyan, G. O., Maksimov, L. V. & Yanush, O. V. J. Non-Cryst. Solids, 1990, 126, 93.
10. Krogh-Moe, J. Phys. Chem. Glasses, 1962, 3, 208.
11. Vedishcheva, N. M., Shakhmatkin, B. A., Shultz, M. M. & Wright, A. C. In: Borate Glasses, Crystals and Melts, Eds. A. C. Wright, S. A. Feller & A. C. Hannon, 1997, Society of Glass Technology, Sheffield. p. 215.
12. Priven', A. I. Glass Phys. Chem., 2003, 29, 60.
13. Conradt, R. J. Non-Cryst. Solids, 2004, 345-346, 16.
14. Markova, T. S., Yanush, O. V., Polyakova, I. G., Pevzner, B. Z. & Klyuev, V. P. Glass Phys. Chem., 2005, 31, 721.
15. Bril, T. W. Philips Res. Repts. Suppl., 1976, 2.
16. Yiannopoulos, Y. D., Chryssikos, G. D. & Kamitsos, E. I. Phys. Chem. Glasses, 2001, 42, 164.
17. Chryssikos, G. D., Bitsis, M. S., Kapoutsis, J. A. & Kamitsos, E. I. J. Non-Cryst. Solids, 1997, 217, 278.
18. Kamitsos, E. I. Phys. Chem. Glasses, 2003, 44, 79.
19. Wallace, R. M. & Katz, S. M. J. Phys. Chem., 1964, 68, 3890.
20. Krogh-Moe, J. Phys. Chem. Glasses, 1965, 6, 46.
21. Hübner K.-H. Neues Jahrb. Mineral. Abhandl., 1969, 112, 150.
22. Markova, T. S., Yanush, O. V., Polyakova, I. G. & Pevzner, B. Z. Phys. Chem. Glasses: Eur. J. Glass Sci. Technol. B, 2006, 47, 476.
23. Klyuev, V. P. & Pevzner, B. Z. Glass Phys. Chem., 1998, 24, 372.
24. Klyuev, V. P. & Pevzner, B. Z. Glass Phys. Chem., 2001, 27, 54.
25. Bunker, B. C., Kirkpatrick R. J., Brow R. K., Turner, G. L. & Nelson C. J. Am. Ceram. Soc., 1991, 74, 1430.
26. Kim, K. S. & Bray, P. J. Phys. Chem. Glasses, 1974, 15, 47.
27. Pantano, C. G., Prabakar, S. & Mueller K. T. Phys. Chem. Glasses, 2003, 44, 125.
28. Brow, R. K. & Tallant, D. R. J. Non-Cryst. Solids, 1997, 222, 396.
29. Suzuki, Y., Ohtori, N., Takase, K., Handa, K., Itoh, K., Fukunaga, T. & Umesaki, N. Phys. Chem. Glasses, 2003, 44, 150.
30. Chryssikos, G. D., Kapoutsis, J. A., Bitsis, M. S., Kamitsos, E. I., Patsis, A. P. & Pappin, A.J. Bol. Soc. Esp. Ceram. Vid., 1992, 31-C, 27.

Proc. VI Int. Conf. Borate Glasses, Himeji, Japan, 18–22 August 2008   *Phys. Chem. Glasses: Eur. J. Glass Sci. Technol. B*, June 2009, **50** (3), 219–223

New geometrical modelling of B$_2$O$_3$ and SiO$_2$ glass structures

Akira Takada[1,2,3]

[1] *Research Center, Asahi Glass Co. Ltd., 1150 Hazawa-cho, Yokohama 221-8755, Japan*
[2] *University College London, Gower Street, London W1E 6BT, UK*
[3] *Institute of Industrial Sciences, University of Tokyo, 4-6-1, Komaba, Meguro-ku, Tokyo 153-8505, Japan*

Manuscript received 25 August 2008
Revised version received 16 January 2009
Accepted 29 January 2009

New geometrical modelling of glass structure employing 'local oxygen packing number (LOPN)' has been developed. This modelling enables the quantification of the variation in local structure in terms of oxygen packing. This new method was applied to analyse the structures of SiO$_2$, B$_2$O$_3$ and boric acid systems. The analysis of LOPN for the SiO$_2$ system shows that each structure can be classified into one of three different packing groups. Moreover, in terms of LOPN, the densification of silica glass is found to increase the fraction of coesite-like denser fragments taking the place of quartz-like and cristobalite-like fragments, although the packing group remains the same. B$_2$O$_3$ and boric acid systems can also be explained in terms of these three different packing groups. However, in contrast to the SiO$_2$ system, most of these borate structures have characteristics of two different groups. The differences in LOPN can be related to the difference in dimensionality of packing, as exhibited by: (1) coordination numbers of cations and anions, and (2) the existence of super-structural units such as the boroxol ring, or low dimensional structures such as chain packing. It is also revealed that the local oxygen packing can differ from place to place even in one structure, and the LOPN is a more powerful tool to investigate structures than the conventional concept of overall packing density.

1. Introduction

The structure of vitreous B$_2$O$_3$ is still the subject of controversy. The simpler random network model consists of planar BO$_3$ triangles, as proposed by Zachariasen.[1] However, several serious objections to this model have been raised, as reviewed in the volume of proceedings[2] and papers of Johnson *et al*[3] and Wright *et al*.[4] Goubeau & Keller[5] first suggested the existence of boroxol (3-membered ring) groups in B$_2$O$_3$ glass, and then Krogh-Moe[6] concluded that a random three-dimensional network of BO$_3$ triangles with a comparatively high fraction of boroxol rings gives the best explanation of the available data. From that time down to this day the boroxol ring model has been supported by other experiments.[7,8] The fraction of boroxol rings is usually defined by the fraction of boron atoms present in boroxol rings, estimated to be around 75%.[3,7–10]

In contrast, most ordinary MD (molecular dynamics) studies[11–22] have failed to reproduce the existence of a high fraction of boroxol rings. A coordination dependent model with three-body terms[23,24] or a polarisable model[25] has been able to confirm the existence of boroxol rings, but their fraction was still small (around 30%), although Takada *et al*[26] conducted a "computer (*in silico*) synthesis" and synthesised a structure that has a density close to that of vitreous B$_2$O$_3$ and a high fraction of boroxol rings (75%). More

improvements have been performed over the past few years. Takada[27] developed a hybrid MD/MC (Monte Carlo) method and produced a structure that has a 75% fraction of boroxol rings. Huang & Kieffer[28] also produced a structure with a high fraction of boroxol rings by melting from the caesium enneaborate structure. Furthermore, Ferlat *et al*[29] performed first principles MD calculations and showed that only a boroxol-rich model (75% boroxol fraction) can reproduce the full set of observables. It is also interesting to note that Takada[30] and Huang & Kieffer[28] investigated the effect of pressure on vitreous B$_2$O$_3$. The former discussed the densification mechanism, and the latter suggested thermomechanical anomalies that are analogous to those in the SiO$_2$ system and proposed new low density crystal structures.

On the other hand, many structural studies, using both experiments and computer simulations, have been performed for silica glass. The structures and properties of silica glass are reviewed in several papers[31–33] and monographs.[34,35] It should be emphasised that computer simulation has been proposing the microscopic mechanism of thermal and pressure effects on silica glass.

After performing MD and MC calculations, the structural changes in glasses have mainly been investigated in terms of coordination number, bond angles, torsional angles, and ring statistics. As an extension of this approach, the author[36–38] has recently proposed the 'structon model' which classifies local structures

Email akira-takada@agc.co.jp

into four structural units, and explains temperature induced structural changes in SiO_2, BeF_2, and GeO_2 glasses. On the other hand, there is another conventional approach called 'random close packing (RCP) of hard spheres' which was advanced by Bernal,[39] Scott,[40] and Finney.[41] RCP modelling has only been used for model glasses and simple metallic glasses. Elliot[42] reviews this sort of geometrical modelling in his book. In conclusion, there have been many studies on structural modelling, but no type of approach can cover the full variety of structural changes in crystals and glasses.

In this study, new geometrical modelling of glasses is reported. One new parameter, named "local oxygen packing number (LOPN)", is defined and applied to investigate crystal and glass structures of B_2O_3 and SiO_2. Finally, several similarities and dissimilarities in local structure among crystals and glasses are discussed.

2. Geometrical modelling

Useful information on glass structures has been obtained by the analysis of coordination number, bond angles, torsional angles, and ring statistics.[42,43] This method puts emphasis on chemical bonding. An alternative method is to emphasise the packing of atoms, such as in the random packing model.[39–41] This method works very well for monoatomic systems, but it is not easy to extend this model to a polyatomic system. On the other hand, by using a packing model for crystals, Dmitriev et al[44] showed that, when only the arrangement of oxygen atoms is considered, the crystal structures of SiO_2 are derived from a common parent disordered BCC (body centred cubic) structure having different fractional concentrations of SiO_2 molecules. This method, however, cannot be directly applied to amorphous systems. In general, crystal structures can be distinguished by symmetry. Each symmetry group holds its inherent number of surrounding atoms. When disordered structures are investigated, their symmetry group cannot be defined, but their number of surrounding atoms can be calculated by assuming a counting rule. In this study only structural information on oxygen atoms was investigated and the number of surrounding oxygen atoms around each oxygen was calculated. First, the relative arrangements of oxygen atoms in silica polymorphs were calculated and the data were projected onto BCC, FCC (face centred cubic), and HCP (hexagonal close packed) structures. The best fit showed that the number of surrounding oxygen atoms was 6 for cristobalite and 12 for stishovite. Next, a criterion of counting a surrounding number was investigated so that the surrounding numbers for cristobalite and stishovite can be reproduced as 6 and 12, and so that this criterion can also be applied to disordered systems. Finally, the following criterion was employed in this

study. A calculated surrounding number is termed 'local oxygen packing number' (LOPN)

$$f = \sum_i f_i$$

$$\begin{aligned} f_i &= 1 & r_i &< r_1 \\ &= (r_2 - r_i)/(r_2 - r_1) & r_1 &< r_i < r_2 \\ &= 0 & r_2 &< r_i \end{aligned}$$

where f is the LOPN, and r_1 and r_2 are 3·0 and 3·06 Å. The i summation is performed over all the other oxygens in the structure, each of which is a distance r_i from the oxygen under consideration. The same criterion is also applied to B_2O_3 and boric oxide system.

3. Results

3.1 Silica system

The distributions of local oxygen packing numbers for silica polymorphs are plotted in Figure 1. The LOPNs for cristobalite, quartz, and stishovite are 6, 6 and 12, respectively. Coesite has two LOPNs; 6 and 7. The LOPN for fibrous silica (silica W)[45,46] is 3. The enigmatic structure of fibrous silica is comprised of one-dimensional chains. The LOPN varies from 3 to 12 as silica structure changes from one-dimensional to three-dimensional and the packing density increases. The distributions of LOPNs for silica glass are plotted in Figure 2. For details of the silica glass simulation and its structure the author's previous paper[47] should be referred to. The LOPNs for an as-quenched glass and glasses decompressed from 25 and 40 GPa load are shown in Figure 2(a). The LOPNs for glasses under compressions of 25 and 40 GPa are also shown in Figure 2(b). The LOPNs for the as-quenched glass are 6 and 7 as for coesite, but the average LOPN for the glass is 6·1 and this value is closer to that of quartz and cristobalite (6·0) than that of coesite (6·5). Hence this as-quenched silica glass has a LOPN distribution which mixes features of quartz, cristobalite, and coesite. When this silica glass is compressed, its LOPN shifts to larger values (denser packing) and its average value approaches that of stishovite as the pressure increases. After decompression the densified glass has a larger average LOPN than that of the as-quenched glass. The

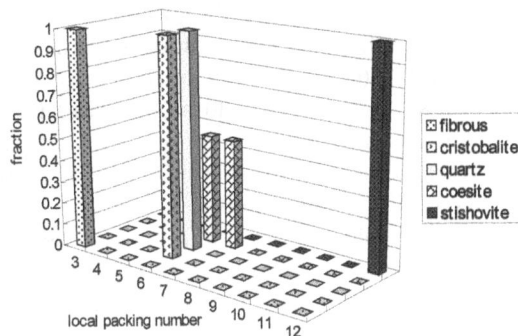

Figure.1 Distributions of local oxygen packing numbers for silica polymorphs

(a)

(b)

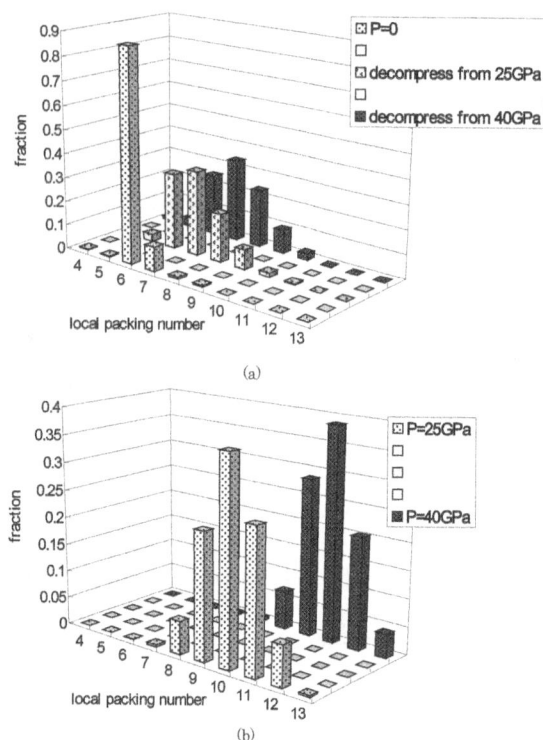

Figure 2. *Distributions of local oxygen packing numbers for silica glass. As-quenched, decompressed from 25GPa load, and decompressed from 40GPa load. Compressed under 25GPa load, and compressed under 40GPa load*

distributions of LOPN are broadened, ranging from 6 to 9. The average LOPNs for two densified glasses are 7·0 and 7·1 and are larger than that of coesite (6·5). To sum up, the effect of densification on silica glass is the transformation of an oxygen packing state from quartz-like and cristobalite-like to coesite-like through a denser packing like that of stishovite.

3.2 B_2O_3 system

The distributions of LOPNs for two B_2O_3 crystals, one borate crystal, and one B_2O_3 glass are plotted in Figure 3. For details of the B_2O_3 glass simulation and its structure the author's previous paper[27] should be referred to. Three crystals; B_2O_3-I (low pressure form), B_2O_3-II (high pressure form) and caesium enneaborate ($Cs_2O.9B_2O_3$) have different average LOPN values of 5·7, 9·3 and 4·3 respectively. B_2O_3-I is comprised of BO_3 units. One-third of oxygen atoms have a higher value of LOPN (6·6) than that of the other atoms (5·3), because the former are located at sites where two chains cross each other and the packing is denser than that for the latter two-thirds. B_2O_3-II is comprised of 4-fold coordinated BO_4 units. One-third of oxygen atoms are two-fold coordinated and the others are three fold-coordinated. The former and the latter have the LOPNs of 8 and 10. Caesium enneaborate crystal is comprised of two kinds of basic unit; the triborate group (containing a six-membered ring, but with one of the boron atoms coordinated tetrahedrally with

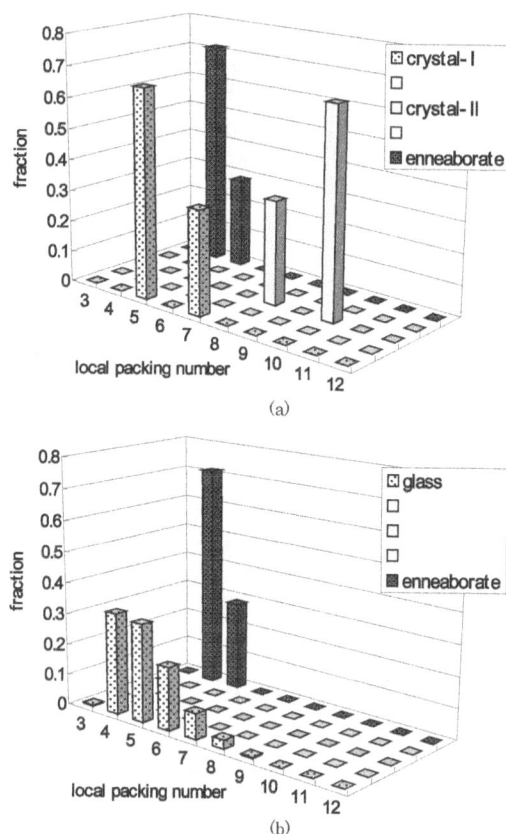

(a)

(b)

Figure 3. *Distributions of local oxygen packing numbers for B_2O_3-I, B_2O_3-II, $Cs_2O.9B_2O_3$ crystals, and B_2O_3 glass. (a) B_2O_3-I, B_2O_3-II, and $Cs_2O.9B_2O_3$. (b) B_2O_3 glass and $Cs_2O.9B_2O_3$*

oxygen atoms) and the boroxol group. The oxygen atoms surrounding 4-fold coordinated boron atoms have the higher LOPN value of 5 than that (4) of the oxygens coordinated to only 3-fold coordinated boron. B_2O_3 glass a LOPN distribution which mixes features of caesium enneaborate and B_2O_3-I. The oxygen atoms in boroxol rings are responsible for the LOPN value of 4, and oxygen atoms in independent BO_3 units are responsible for the other larger values.

To investigate the relation of LOPN with local structure in more detail, four crystal structures of boric acid, which are similar to B_2O_3, were analysed. The distributions of LOPNs for four boric acids are plotted in Figure 4. The structure of orthoboric acid, H_3BO_3, is built up by $B(OH)_3$ molecules which form endless layers. Its LOPN value of 4 is the same with that of caesium enneaborate. In caesium enneaborate the oxygen in boroxol rings is responsible for the LOPN value of 4, but H_3BO_3 has no boroxol rings. It is interesting to note that both structures have a common feature, that they are comprised of planar structural units, either boroxol rings or a layered structure. A key to explain the LOPN value of 4 seems to be the planarity of the structure. The structure of orthorhombic metaboric acid, HBO_2-III, is also layered. It is comprised of trimeric HBO_2 molecules which form a boroxol ring. Its LOPN values are 4

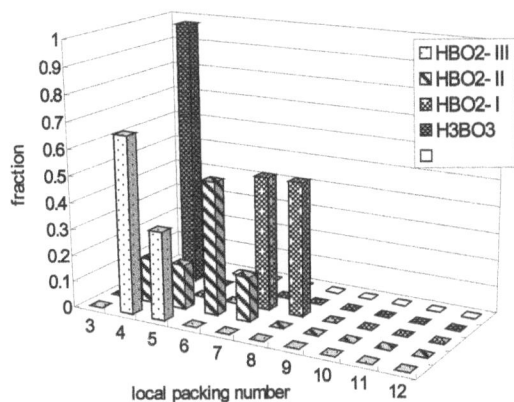

Figure 4. Distributions of local oxygen packing numbers for H_3BO_3, HBO_2-I, HBO_2-II, and HBO_2-III

Table 1. Classification of structures in terms of local oxygen packing number

Dimensionality of packing	Group-1 three-dimensional denser packing	Group-2 three-dimensional less dense packing	Group-3 planar-like or chain-like packing
Name of system	stishovite	cristobalite, quartz, coesite silica glass	fibrous silica
	B_2O_3-II	B_2O_3-I, B_2O_3-II, $Cs_2O.9B_2O_3$ B_2O_3 glass HBO_2-I, HBO_2-II, HBO_2-III	$Cs_2O.9B_2O_3$ B_2O_3 glass H_3BO_3, HBO_2-II, HBO_2-III
LOPN	10–12	5–9	3–4

(local oxygen packing number)

and 5. The oxygens inside and outside the ring are responsible for the former and the latter, respectively. The structure of cubic metaboric acid, HBO_2-I, forms a three-dimension network of BO_4 tetrahedra. Its LOPN values are 7 and 8. The oxygen atoms with the LOPN value of 7 have 2-fold coordination which is the same as one-third of the oxygen atoms in B_2O_3-II. The other two-thirds of oxygen atoms with the LOPN value of 8 have an additional hydrogen bonding. This hydrogen bonding seems to attract one additional oxygen atom and increase the LOPN value from 7 to 8. The structure of monoclinic metaboric acid, HBO_2-II, consists of endless zigzag chains of composition [B_3O_4(OH)(OH$_2$)]. Two-thirds of the boron atoms are in triangular and one-third in tetrahedral configuration. Its structure is comprised of chains like those found in B_2O_3-I crystal, and triborate groups like those found in caesium enneaborate. Its LOPNs are 4, 5, 6 and 7. The oxygen atoms which are located in an environment like that of enneaborate and HBO_2-I are responsible for the LOPN values of 4 and 5, and 7, respectively. The environment of the oxygen atoms with a LOPN value of 6 is similar to that of B_2O_3-I, but the degree of distortion of the chains is different. This difference in the distortion enables the LOPN value of HBO_2-II (6·0) to be different from those of B_2O_3-I (5·3 and 6·6).

It is interesting to note that most structures of both B_2O_3 and boric oxide crystals are comprised of two local fragments with different LOPNs, in contrast to those of all the SiO_2 crystals except coesite.

4. Discussion

Structures in silica, B_2O_3 and boric acid systems are compared in terms of LOPN in Table 1. The analysis of LOPN for oxygen in the silica system showed that each structure can be classified into one of three different packing groups. The first group contains stishovite. The LOPN of stishovite has a value of 12 which corresponds to denser packing structures such as FCC, HCP and icosahedron. The other crystals with lower density have smaller oxygen packing in terms of

LOPN. The second group contains quartz, cristobalite and coesite. Their occupation fraction is almost half compared with stishovite. Silica glass also belongs to the second group. Silica glass originally has the mixed features of quartz, cristobalite and coesite, but in its densified structure the fraction of coesite-like denser packing increases. The third group contains fibrous silica which is comprised of one-dimensional chains. In conclusion, one-dimensional or two-dimensional structures with less dense packing belong to the third group, although no two-dimensional planar structure of silica has been observed yet, in contrast to borate and boric acid systems.

B_2O_3 and boric acid systems can also be explained in terms of three packing groups. B_2O_3-II is in both the first and the second packing group. In contrast with stishovite, for which all the oxygen atoms are three-fold coordinated, B_2O_3-II has both two-fold and three-fold coordinated oxygen atoms. The fragments with different coordination number belong to different packing groups. B_2O_3-I belongs to the second group alone. Caesium enneaborate seems to be a mixture of the second and the third group, because there are two types of oxygen atoms; those bonded to 4-fold coordinated boron and those in boroxol rings. The latter belongs to the third group due to the planarity of the boroxol ring. B_2O_3 glass also has a mixed character of the second and the third group due to the mixture of independent BO_3 units and boroxol rings. The boric acid system has similar features to the B_2O_3 system, but the existence of hydrogen bonding produces a variety of combinations of LOPNs. Although the LOPN for densified B_2O_3 glass has not been analysed in this study, it is speculated that densification would transform the fragments in the second group into those in the third, because the decompressed glass still kept the 3-fold coordinated states, but the fraction of boroxol rings decreased significantly, as found in the previous study.[30]

The common feature in all systems investigated in this study is that all the systems can be classified into three groups depending on the manner of packing of oxygen atoms. Moreover, a common relation between the dimensionality of structure and the LOPN value is revealed, that is to say the lower dimensionality a structure has, the smaller value of LOPN its structure has. On the other hand, the structural difference between the SiO_2 and B_2O_3 systems is analysed as follows. All the silica crystals and glass can be classified into an individual group (1, 2 or 3). In contrast, B_2O_3-II crystal and B_2O_3 glass are in two different groups. The former is a mixture of two structural fragments with 2-fold and 3-fold coordinated oxygen atoms. The latter can be regarded to be an assembly of two-dimensional-like fragments comprised of boroxol rings, and three-dimensional-like fragments comprised of independent BO_3 units. This difference is assumed to be due to the more complex and diverse chemical bonding of B–O than that of Si–O.

Finally, the advantage of the concept of LOPN is that it can distinguish the difference in local structural environment very simply even for amorphous structures. Moreover, this new method, which is complementary to traditional methods such as bond length, bond angle, torsional angle, ring statistics, can provide new information on crystal and amorphous structures. In the previous papers,[36–38] the author developed 'structon' analysis, focusing on the local structural units in terms of chemical bonds. This structon can be named 'chemical structon.' In contrast, the structure fragments defined by the LOPN in this study can be named 'physical structon', because the focus here is on physical packing states. The applications of LOPN analysis to B_2O_3–SiO_2, silicate, and borate systems is in progress. They will be published elsewhere.

5. Conclusions

New geometrical modelling of glass structure employing 'local oxygen packing number (LOPN)' has been developed. This modelling enables the quantification of the variation in local structure in terms of oxygen packing. This new method was applied to analyse the structures of SiO_2, B_2O_3 and boric acid systems. The analysis of LOPN for the SiO_2 system showed that each structure could be classified into one of three different packing groups. Moreover, in terms of LOPN, the densification of silica glass was found to increase the fraction of coesite-like denser fragments, taking the place of quartz-like and cristobalite-like fragments. B_2O_3 and boric acid systems can also be explained in terms of these three different packing groups. However, in contrast to the SiO_2 system, most of these borate structures have characteristics of two different groups. The differences in LOPN can be related to the difference in dimensionality of packing, as exhibited by: (1) coordination numbers of cations and anions, and (2) the existence of superstructural units such as the boroxol ring, or low dimensional structures such as chain packing. It is also revealed that the local oxygen packing can differ from place to place even in one structure, and the LOPN is a more powerful tool to investigate structures than the conventional concept of overall packing density.

References

1. Zachariasen, W. H. *J. Am. Chem. Soc.*, 1932, **54**, 3841.
2. Pye, L. D., Frechette, V. D. & Kreidl, N. J. (Eds.), *Borate Glasses: Structure, properties, Applications*, 1978, Plenum Press, New York.
3. Johnson, P. A. V., Wright, A. C. & Sinclair, R. N. *J. Non-Cryst. Solids*, 1982, **50**, 281.
4. Wright, A. C., Vedishcheva, N. M. & Shakhmatkin, B. A. *J. Non-Cryst. Solids*, 1995, **192&193**, 92.
5. Goubeau, J. & Keller, H. *Z. Anorg. Chem.*, 1953, **272**, 303.
6. Krogh-Moe, J. *J. Non-Cryst. Solids*, 1969, **1**, 269.
7. Jellison, G. E. Jr, Panek, L. W., Bray, L. W. & Rouse, G. B. Jr *J. Chem. Phys.*, 1977, **66**, 802.
8. Hannon, A. C., Grimley, R. A., Hulme, A. C., Wright, A. C. & Sinclair, R. N. *J. Non-Cryst. Solids*, 1994, **177**, 299.
9. Youngman, R. E. & Zwanziger, J. W. *J. Non-Cryst. Solids*, 1994, **168**, 293.
10. Joo, C., Zwanziger, U. W. & Zwanziger, J. W. *Solid State NMR*, 2000, **16**, 77.
11. Soules, T. F. *J. Chem. Phys.*, 1980, **73**, 4032.
12. Soules, T. F. *J. Chem. Phys.*, 1979, **71**, 4570.
13. Soules, T. F. & Varshneya, A. K. *J. Am. Ceram. Soc.*, 1981, **64**, 145.
14. Soppe, W., van der Marel, C., van Gunsteren, W. F. & den Hartog, H. W. *J. Non-Cryst. Solids*, 1988, **103**, 201.
15. Soppe, W. & den Hartog, H. W. *J. Phys. C*, 1988, **21**, L689.
16. Soppe, W. & den Hartog, H. W. *J. Non-Cryst. Solids*, 1989, **108**, 260.
17. Amini, M., Mitra, S. K. & Hockney, R. W. *J. Phys. C.*, 1981, **14**, 3689.
18. Abramo, M. C. & Pizzimneti, G. *J. Non-Cryst. Solids*, 1986, **85**, 233.
19. Xu, Q., Kawamura, K. & Yokokawa, T. *J. Non-Cryst. Solids*, 1988, **104**, 206.
20. Hirao, K. & Soga, N. *J. Am. Ceram. Soc.*, 1985, **68**, 515.
21. Inoue, H., Aoki, N. & Yasui, I. *J. Am. Ceram. Soc.*, 1987, **70**, 622.
22. Vehoef, A. H. & den Hartog, H. W. *J. Non-Cryst. Solids*, 1992, **146**, 267.
23. Takada, A., Catlow, C. R. A. & Price, G. D. *J. Phys.: Condens. Matter*, 1995, **7**, 8659.
24. Takada, A., Catlow, C. R. A. & Price, G. D. *J. Phys.: Condens. Matter*, 1995, **7**, 8693.
25. Maranas, J. K., Chen, Y., Stillinger, D. K. & Stillinger, F. H. *J. Chem. Phys.*, 2001, **115**, 6578.
26. Takada, A., Catlow, C. R. A. & Price, G. D. *Phys. Chem. Glasses*, 2003, **44**, 147.
27. Takada, A. *Phys. Chem. Glasses Eur. J. Glass Sci. Technol. B*, 2006, **47**, 493.
28. Huang, L. & Kieffer, J. *Phys. Rev. B*, 2006, **45**, 4435.
29. Ferlat, G., Charpentier, T., Seitsonen, A. P., Takada, A., Lazzeri, M., Cormier L., Calas, G. & Mauri, F. *Phys. Rev. Lett.*, 2008, **101**, 065504.
30. Takada, A. *Phys. Chem. Glasses*, 2004, **45**, 156.
31. Brückner, R. *J. Non-Cryst. Solids*, 1970, **5**, 123.
32. Tomozawa, M. in *Silicon-Based Materials and Devices*, Ed. H. S. Nalwa, Vol. 1, 2001, Academic Press, New York, p.127.
33. Takada, A. & Cormack, A. N. *Phys. Chem. Glasses: Eur. J. Glass Sci. Technol. B*, 2008, **49**, 127.
34. Pacchini, G. (ed.) *Defects in SiO₂ and Related Dielectrics*, 2000, Academic Press, Dordrecht.
35. Devine, R. A. B., Durand, J.-P. & Dooryhee, E. (Eds.) *Structure and Imperfections in Amorphous and Crystalline Silicon Dioxide*, 2000, John Wiley & Sons, Chichester.
36. Takada, A., Richet, P., Catlow, C. R. A. & Price, G. D. *Phys. Chem. Glasses Eur. J. Glass Sci. Technol. B*, 2007, **48**, 182.
37. Takada, A., Richet, P., Catlow, C. R. A. & Price, G. D. *J. Non-Cryst. Solids*, 2007, **353**, 1892.
38. Takada, A., Richet, P., Catlow, C. R. A. & Price, G. D. *J. Non-Cryst. Solids*, 2008, **354**, 181.
39. Bernal, J. D. in *Liquids: Structure, Properties, Solid Interactions*, Ed. T.J. Hughel, Elsevier, Amsterdam, p.25.
40. Scott, G. D. & Kilgour, D. M. *J. Phy. D*, 1969, **2**, 263.
41. Finney, J. L. *Proc. Roy. Soc. A*, 1970, **319**, 479.
42. Elliot, S. R. *Physics of Amorphous Materials (second edition)*, 1990, Longman, Essex, Chapter 3.
43. Hobb, L. W., Jesurum, C. E. & Berger, C. E., in *Structure and Imperfections in Amorphous and Crystalline Silicon Dioxide* (eds.) R. A. B. Devine, J.-P. Durand & E. Dooryhee, 2000, John Wiley & Sons, Chichester, p.3.
44. Dmitriev, V., Torgashev, V., Toledano, P. & Salje, E. K. H. *Europhys. Lett.*, 1997, **37**, 553.
45. Weiss, A. & Weiss, A. *Naturwissenschaften*, 1954, **41**, 12.
46. Weiss, A. & Weiss, A. *Z. Anorg. Chem.*, 1954, **276**, 95.
47. Takada, A., Richet, P., Catlow, C. R. A. & Price, G. D. *J. Non-Cryst. Solids*, 2004, **345&346**, 224.

Proc. VI Int. Conf. Borate Glasses, Himeji, Japan, 18–22 August 2008   *Phys. Chem. Glasses: Eur. J. Glass Sci. Technol. B*, October 2009, **50** (5), 294–300

Understanding the anomalous behaviours of B$_2$O$_3$ glass

*John Kieffer**

Department of Materials Science and Engineering, University of Michigan, Ann Arbor, MI 48109, USA

Manuscript received 28 November 2008
Revised version received 30 July 2009
Accepted 6 September 2009

Anomalous thermomechanical properties in glasses include negative thermal expansion, the decrease of the elastic modulus upon compression, and the increase upon heating. The latter behaviour in particular has been experimentally observed for both four- and three-coordinated glass formers. Using MD simulations based on a charge transfer multiple-coordination potential, we are able to reproduce these anomalous behaviours, allowing us to identify the underlying mechanism. Accordingly, when compressing or expanding both tetrahedral and trigonal glasses, mechanically or thermally, these undergo reversible localised structural transitions by invoking buckling-like rotations of oxygen bridges, similar to those responsible for the transformations between different polymorphs of the crystalline counterparts of these materials. We predict that the anomalous behaviour observed in molten boron oxide resurfaces when the glass is subject to tensile deformation at room temperature. Furthermore, the increase of the elastic modulus upon expansion can be traced back to two previously unknown crystalline polymorphs for this compound that we discovered using simulation. Consequently, thermomechanical anomalies have the same origin for both three- and four-coordinated glasses, and this phenomenon appears to be universal for all major network formers.

1. Introduction

A major characteristic of glass forming oxides is the extended network that these systems develop, and which can be attributed to the divalent oxygen atoms establishing connectivity between the network cations, e.g. Si, Ge, B, Bi, Al, etc. These networks are the result of a pronounced need for these elements to satisfy their bonding requirement, a tendency that persists even under extreme conditions. Hence, these networks retain a large degree of integrity at high temperature,[1] under pressure, and even under tensile stress for as long as sufficient thermal activation is provided to facilitate bonding reconfiguration upon deformation.[2–4] A key feature of these structures is the lone pair of electrons located on the oxygen atom, which causes an equilibrium angle of significantly less than 180° to form between the two bonds of the X–O–X bridge, where X designates the network cation. In silica, for example, the archetypical tetrahedral network former, this kink in the network linkage has extraordinary effects on the dynamics[5] and the mechanical properties[6] of the network structure, i.e. it has a clear signature in the vibrational spectra, and is chiefly responsible for the anomalous thermo-mechanical properties of this substance. The present paper provides a synthesis of research findings with regard to the structure and properties of vitreous B$_2$O$_3$ that involved a number of co-workers over the past eight years. In particular, I would like to acknowledge the work by Drs Jason

Nicholas and Liping Huang. The combination and the comparison of data obtained by different experimental and computational methods of investigation provide new insights into the anomalous behaviours of networks based on trigonal units, i.e. a building block with fundamentally different symmetry and steric functionality, and put the observed behaviour into relationship with that of tetrahedral glasses. Discovering commonality between the responses of three- and four-coordinated structures provides additional support for the fact that the kink in the network bridges is ultimately responsible for the extraordinary properties of these materials.

2. Thermomechanical anomalies of network structures

Anomalous thermomechanical behaviours in network structures are manifest through properties such as negative thermal expansion, either under ambient conditions or while subject to compressive stress, density maxima in the liquid state, the increase of elastic moduli with increasing temperature, and the decrease of elastic moduli with increasing pressure. Such behaviours are best known for water and silica,[7,8] but other substances exhibit similar anomalies.[9–13] In fact, they seem to be common to all strong network glass formers, e.g. for SiO$_2$, GeO$_2$, and B$_2$O$_3$ the elastic modulus increases with increasing temperature, as can be seen in Figure 1.[1] Accordingly, thermomechanical anomalies also seem to be independent of the physical state of the substance;

---

* Corresponding author. Email kieffer@umich.edu

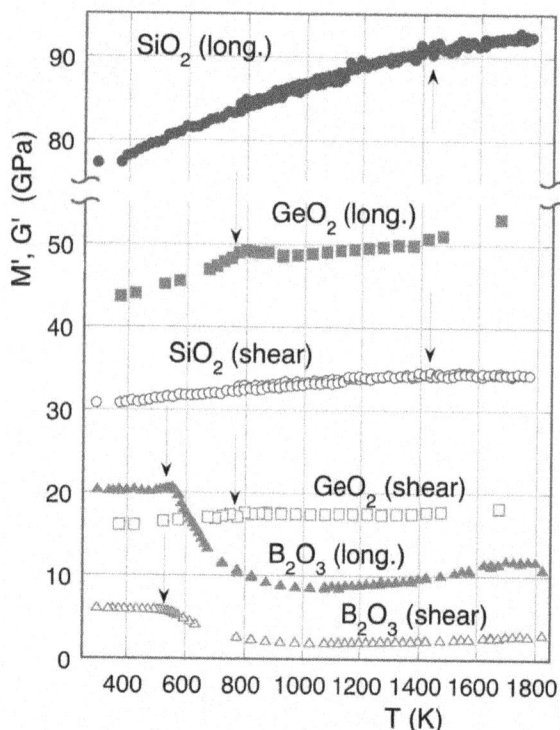

Figure 1. Longitudinal (solid symbols) and shear (open symbols) elastic moduli of SiO₂, GeO₂, and B₂O₃ glasses and melts, measured by Brillouin light scattering as a function of temperature. Glass transition temperatures are indicated by vertical arrows

they can occur in the glass, in liquid, or in both.[1] Models proposed to explain these phenomena are in one way or another based on the assumption that two or more energetically distinct amorphous states coexist in proportions that vary with pressure and temperature,[14,15] but until recently, little detail has been provided as to the structural entities that constitute these states and the atomic scale mechanisms that underlie the structural transitions between the states. Recent progress in this regard is largely due to the synergistic combination of experiments with atomic scale simulations, especially for research into the subject of amorphous–amorphous or polyamorphic transformations.[6,8,16–27]

For example, based on extensive molecular dynamics (MD) simulations, we showed that the anomalous temperature and pressure dependences of the mechanical properties of silica glass, a-SiO₂, are due to sudden localised structural rearrangements, similar to those that occur in crystalline silica upon transformation between high and low density modifications (α and β respectively).[8] These transitions involve the rotation of Si–O–Si bonds about the Si–Si axis by 90° or more. Contrary to previous beliefs, the Si–O–Si bond angle does not decrease much when the structure is compressed, or *vice versa*, increase upon expansion. Instead, when reaching a threshold strain the bond planes flip to release elastic energy, a process reminiscent of a buckling phenomenon. While in

cristobalite all bond planes rotate simultaneously by 90° during the stress induced α-to-β transformation, in a-SiO₂ the rotations occur sporadically at different pressure levels for different locations throughout the structure. They are reversible and gradually involve about 1/3 of all Si–O–Si bonds.

The elastic modulus of amorphous B₂O₃ at first decreases with increasing temperature, but above 750°C it increases, an anomalous behaviour that continues to more than 1000°C above the melting point of this compound (see Figure 1). This is extraordinary, because not only does the longitudinal elastic modulus increase with temperature, but also the structure supports shear deformational modes, a characteristic that should vanish in the liquid state. B₂O₃ is also one of the first materials in which polyamorphism has been observed.[28] Indeed, in a recent study we further developed the understanding of the underlying structural transitions using concurrent Brillouin and Raman light scattering measurements[22,29,30] and complementary MD simulations to interpret the experimental observations.[19] Our study revealed a rather unusual transformation in vitreous B₂O₃, which is continuous upon compression and discontinuous upon decompression. Upon compression the elastic modulus increases gradually. At relatively low pressures, the peak at 808 cm⁻¹ in the Raman spectrum, which has been assigned to the symmetric breathing mode of boroxol rings, disappears. Upon decompression, the modulus decreases at a smaller rate, following a different path than on compression. The modulus decreases smoothly until a sharp transition occurs at about 2·7 GPa, at which point it returns to a value close to that during compression. At the same time, the 808 cm⁻¹ peak reappears in the Raman spectrum, along with a new peak at 881 cm⁻¹, which has not yet been assigned. The abrupt drop in modulus, which always occurs at the same pressure, regardless of what maximum pressure the sample was compressed to, marks a second order transformation (Figure 2). Note that for this transformation to appear, the glass has to be compressed beyond a certain threshold, which we observed to be around 14 GPa. Also the pressure release has to be very gradual in order to allow for adequate relaxation of the glass structure; typically we waited a day or longer between state points. From simulation, we know that the hysteresis in the modulus versus pressure data is due to the reversible transformation from 3- to 4-coordinate boron units with increasing pressure, and *vice versa*. We also know that all tetrahedrally coordinated boron disappears upon releasing the pressure.[11,31–33] After the first pressure cycle, the sample has a higher density than the quenched glass, but in subsequent cycles the constitutive behaviour of the glass is entirely reproducible. Importantly, for a glass with trigonal building blocks, permanent densification can be

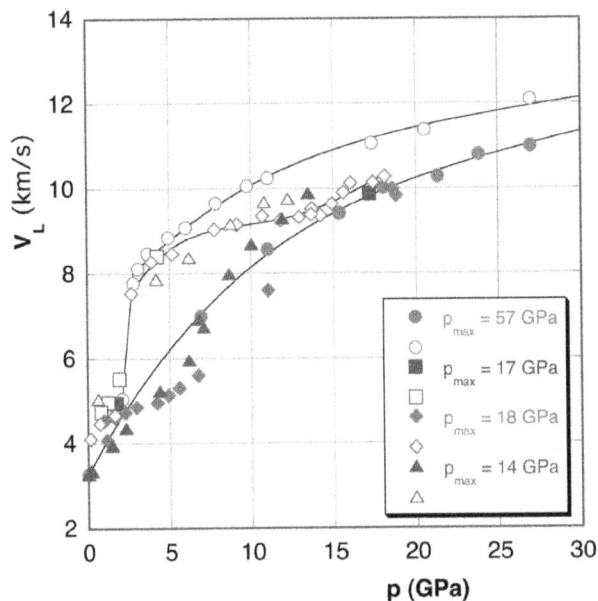

Figure 2. Longitudinal sound velocity of B_2O_3 glass, measured by Brillouin light scattering, as a function of pressure. Solid symbols represent data measured upon compression and open symbols upon decompression. Data sets are shown for four samples that differ in the maximum compression pressure: 14 GPa, 17 GPa, 18 GPa, and 57 GPa. Lines serve as guides to the eye

achieved without change in coordination. Finally, we recall the mysterious 'crystallisation anomaly' of B_2O_3: the compound has never been observed to crystallise from a dry melt at ambient pressure.[34,35] Even if the melt is seeded with crystals, and the crystals are melted back a bit, no crystal growth is observed at any imposed undercooling without pressure.

3. Simulation procedure

3.1 Interaction potential

A reliable interaction model for the simulation of substances such as B_2O_3 should be capable of simulating the network cation in both three- and fourfold coordination states, dynamically, and without preconception as to the prevalent state in a given physical or chemical condition. We have developed such a reactive force field, specifically designed to model mixed ionic–covalent bonding in oxides.[5,11,36,37] The force field includes three body terms to describe the directionality of bonds, capable of adapting to a variable coordination environment, and a charge transfer term. This latter feature, which is achieved by a distribution function designed to mimic s- and p-electronic orbital overlaps, provides a realistic description of the electron density redistribution upon formation or breaking of bonds. Both charge transfer and multiple coordination features are particularly important when simulating systems that undergo reactions and significant structural reorganisation. Parameterisation of the potential initially involves

matching the density, bond lengths and bond angles to experimental values of known crystalline polymorphs, such as α-cristobalite or B_2O_3-I. Parameters are further adjusted to achieve the best possible agreement between simulated and experimental data for elastic moduli, vibrational density of states, and infrared spectra. The charge transfer function is optimised based on results from density functional theory calculations. The parameters for the coordination-dependent part of the potential are adjusted so as to yield the experimentally observed pressures at which structural transitions occur that involve coordination changes, e.g. cristobalite to stishovite, or B_2O_3-I to B_2O_3-II.[37,38] The potential exhibits remarkably good transferability; a large number of silica polymorphs can be simulated with a single parameterisation.[38] Details concerning the functional form and values of potential parameters can be found in earlier publications.[8,11]

3.2 Generating glass structures

While the creation of silica glass structures can be achieved straightforwardly, e.g. by melting and cooling cristobalite (structural changes in the glass transition range are expectedly minor, given the minute thermal signature of this process), generating realistic structures of B_2O_3 glass represents a particular challenge due to the presence of boroxol rings. Experiments show that the fraction of boroxol rings increases with decreasing temperature.[39] Cooling rates in our simulations, which are of the order of 0·5 K/ps, do however not allow sufficient time for these ring structures to develop naturally. Instead, we started from caesium enneaborate ($Cs_2O.9B_2O_3$) crystal,[40] which has both boroxol groups and triborate groups. After extracting Cs_2O and some BO_3 triplets, a structure is obtained that has 75% of its boron atoms in boroxol rings.[41,42] These systems were equilibrated at 2500 K, under which condition the boroxol rings gradually dissolve. Quenching these systems after various amounts of annealing time allowed us to control the boroxol ring concentration in the resulting glass to range between 10% and 75%. The corresponding densities at ambient conditions are 1·75 to 1·81 g/cm^3, respectively, which is very close to the experimental value of 1·80 g/cm^3 for B_2O_3 glass.[43]

MD simulations were carried out for systems containing between 2500 and 3000 atoms, which proved to be free of system size effects. Long range Coulomb interactions were computed using the Ewald summation method. The hydrostatic compression–decompression was carried out at 300 K, between –20 GPa and 50 GPa, at 1 GPa intervals. The rate of compression was 0·5 GPa/ps. Temperature ramping was achieved by velocity scaling, and the density adjusts according to the Anderson constant pressure algorithm.[44] The bulk modulus of the glass

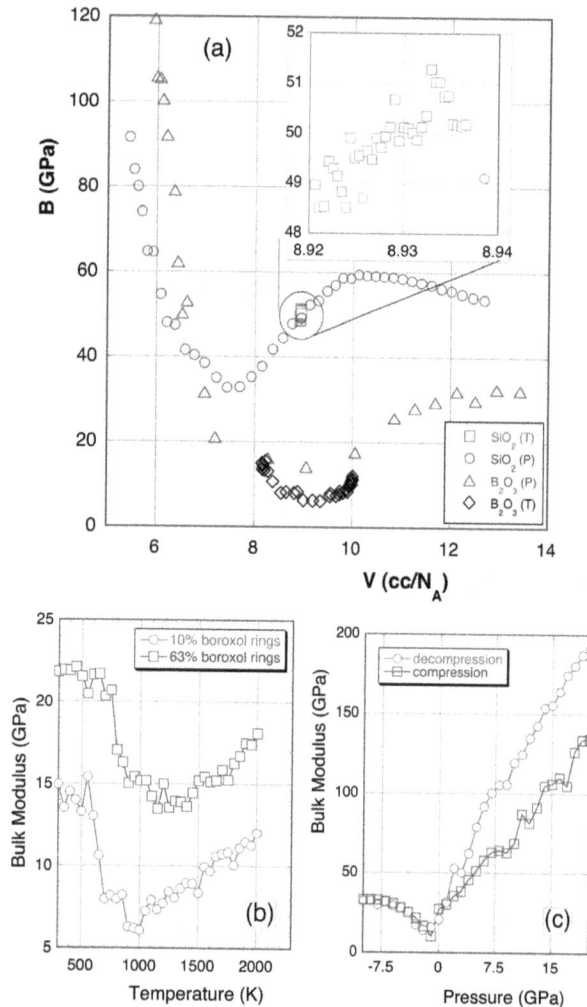

Figure 3. Bulk elastic moduli for a-SiO₂ and a-B₂O₃, generated using MD simulations: (a) as a function of the specific volume per Avogadro number of atoms. Symbols labelled with (P) or (T) indicate respectively that the volume change was induced mechanically or thermally. (b) Modulus as a function of temperature for two a-B₂O₃ systems with different concentration of boroxol rings. (c) Modulus of a-B₂O₃ containing 10% boroxol rings as a function of pressure upon compression and decompression

was calculated directly from the equation of state according to $B=\rho(\partial P/\partial \rho)$.[11]

4. Results

In this paper we focus on the anomalous increase of the elastic modulus of a-B₂O₃ with temperature and the underlying processes. This is of particular interest as it reveals a remarkable parallel to the behaviour in a-SiO₂. First, since both substances exhibit positive thermal expansion within the temperature ranges under consideration, the effect that either temperature or pressure has on the elastic modulus of these substances can be equivalently mapped onto the corresponding changes in the specific volume of the material. This is illustrated in Figure 3(a), where

we have plotted the bulk modulus versus specific volume for both a-SiO₂ and a-B₂O₃, with data obtained by varying temperature at ambient pressure, and by varying pressure at room temperature using MD simulations. In this graph, thermomechanical anomalous behaviour is evident as an increase in the elastic modulus with increasing specific volume. Note that the overlap between thermally and mechanically induced modulus changes is not perfect; typically thermal expansion brings about a more rapid increase in elastic modulus than pressure induced volume changes. This may be due to the added benefit that thermal activation provides for structural relaxation processes. It may also be the case that better overlap can be achieved when considering isothermal pressure and isobaric temperature changes at a different temperature and pressure intercept than that corresponding to ambient conditions.

From Figure 3(b) it is evident that our simulations reproduce the experimentally observed anomaly in the temperature dependence of the elastic modulus of B₂O₃. A more detailed comparison between simulation and experiments is given in Ref. 11. We find that the anomalous behaviour is independent of boroxol ring concentration; the modulus increases towards high temperature by the same amount, whether there are 10% or 65% of boron atoms in boroxol rings (although the experimental bulk modus data, $B=M-4G/3$, is more closely reproduced in simulations for the system with the lower concentration of boroxol rings). We can also exclude four-coordinated boron as playing a role in the anomaly. In Ref. 19 we show that boron transitions from three- to four-coordination upon compression of B₂O₃, and that this transformation is completely reversible upon decompression. Although a permanently densified glass results from such a compression–decompression cycle, and although the changeover to four-coordination is indeed necessary for the irreversible densification, the glass at ambient pressure and below is always devoid of any tetrahedral boron.

Note that, unlike a-SiO₂, which exhibits its thermomechanical anomaly between room temperature and at least 1600°C, for a-B₂O₃ the anomaly only appears at elevated temperatures when the material is in the liquid state.[1] A crucial test for whether this phenomenon is unequivocally related to the specific volume of a substance, was therefore to find the pressure range at which the anomaly might resurface at room temperature. Since the specific volume of molten B₂O₃ is significantly reduced from that of the room temperature glass, we evidently needed to look for this occurrence at negative pressures. Indeed, when isotropically expanding B₂O₃ glass in our simulations, a minimum in the bulk modulus was found at about −1·5 GPa, and below that pressure the modulus increases anomalously with the specific volume of the system (Figure 3(c)).

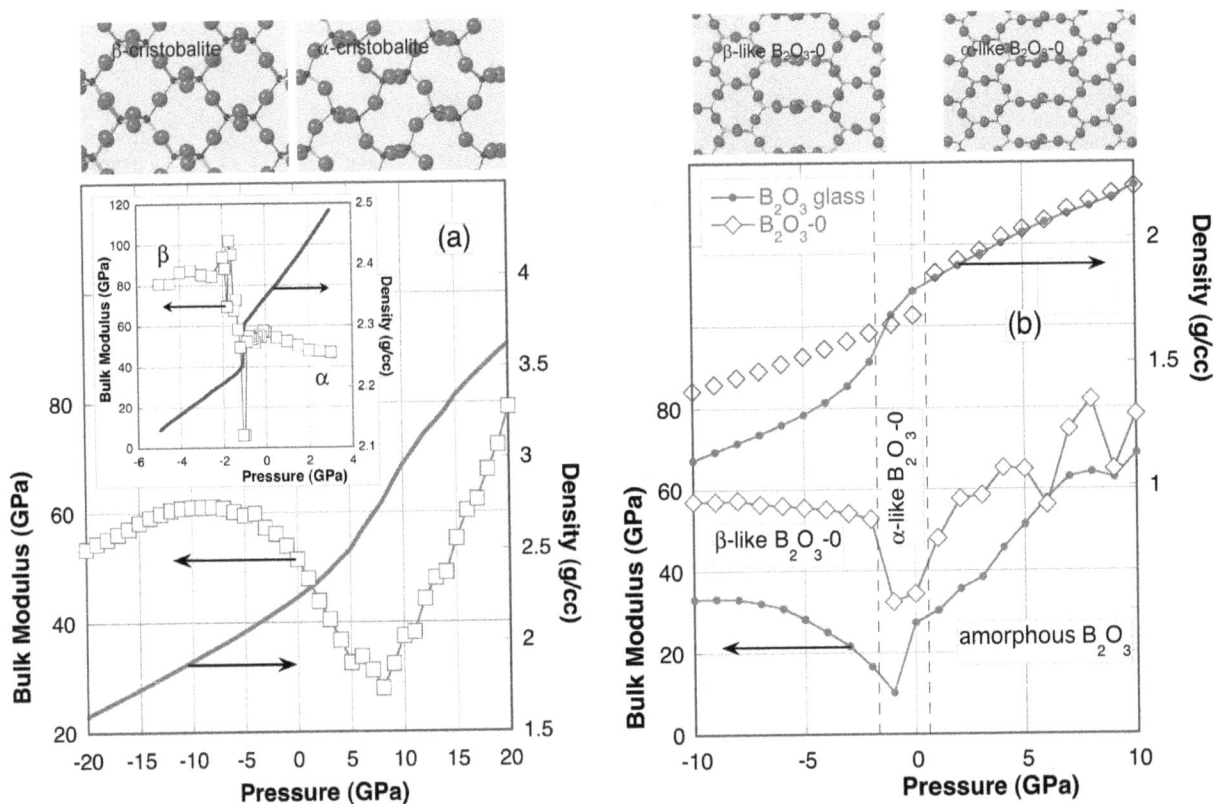

Figure 4. (a) Bulk elastic modulus and density of a-SiO$_2$ as a function of pressure, determined using MD simulations. Inset: Bulk elastic modulus and density of cristobalite SiO$_2$ as a function of pressure through the α-to-β transformation range. The ring structures of α- and β-cristobalite are depicted on top of the graph. (b) Bulk elastic modulus and density of crystalline and amorphous B$_2$O$_3$ as a function of pressure, determined using MD simulations

5. Discussion

Thus far, the observed behaviour has been consistent with expectations. The final step towards establishing thermomechanical anomalies as a universal phenomenon is to determine the mechanism that underlies the thermomechanical anomalies in B$_2$O$_3$. Recall that for a-SiO$_2$ the anomalous behaviour echoes the mechanical properties of α- and β-cristobalite, and can be understood in terms of the geometry of network interstices. In cristobalite there is only one type of ring delineating network interstices, and it contains six silicon atoms. This is also the prevalent ring size in a-SiO$_2$. In β-cristobalite, the low density modification, rings are bulged out into a highly symmetric shape, maximising the occupied volume; in α-cristobalite, the high density modification, the rings are partially twisted into a puckered shape that provides for denser packing (Figure 4(a)). The deformation of the symmetric interstices in any direction is difficult because it requires bond bending. Conversely, the puckered rings are more compliant; they continue to pivot lightly about Si–O bonds and fold onto themselves. As a result the elastic modulus of β-cristobalite is higher than that of α-cristobalite (Figure 4(a)). This was known from experiment, and we reproduced the behaviour in our simulations. Our simulations further showed that compressed silica glass contains more α-like structural motifs, and

expanded glass contains more β-like motifs. Upon expanding the glass, either thermally or mechanically, transitions between these structural motifs are abrupt, sporadic, and localised. They occur by invoking the same Si–O–Si bond plane rotation mechanism that underlies the α-to-β cristobalite transformation, and thus the elastic modulus gradually increases.

Using simulations, we were recently able to show that the same mechanism explains the anomalous thermomechanical properties in networks consisting of trigonal building blocks. This required us to search for and discover previously unknown low density crystalline phases of B$_2$O$_3$.[11,37] B$_2$O$_3$-I is the known crystalline phase that is stable at ambient conditions. By expanding this structure isotropically at room temperature, a new low density phase emerges through displacive transformation, which we called B$_2$O$_3$-0 to acknowledge its place in the density sequence following the existing naming scheme. B$_2$O$_3$-0 has an open network structure with two types of rings, one large and oblong, the other one small and symmetric. Upon reducing the isotropic tensile stress, this structure transitions into yet another structure, in which the small ring puckers. Hence, as with cristobalite, we have a low density β-like B$_2$O$_3$-0 and a higher density α-like B$_2$O$_3$-0 (these structures are depicted in Figure 4(b)), and indeed, the elastic modulus of β-like B$_2$O$_3$-0 is higher than that of α-like B$_2$O$_3$-0. This strongly sup-

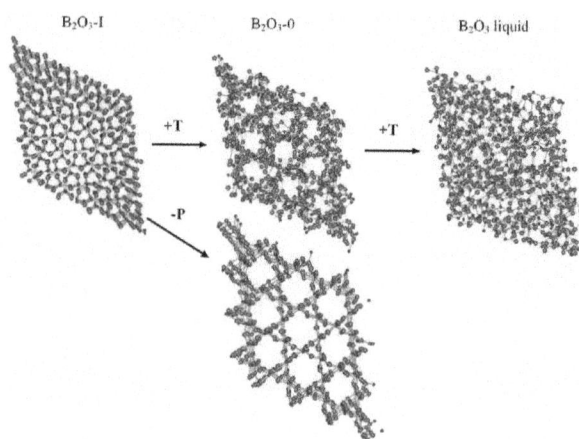

Figure 5. Thermally and mechanically induced transformation pathways in B_2O_3

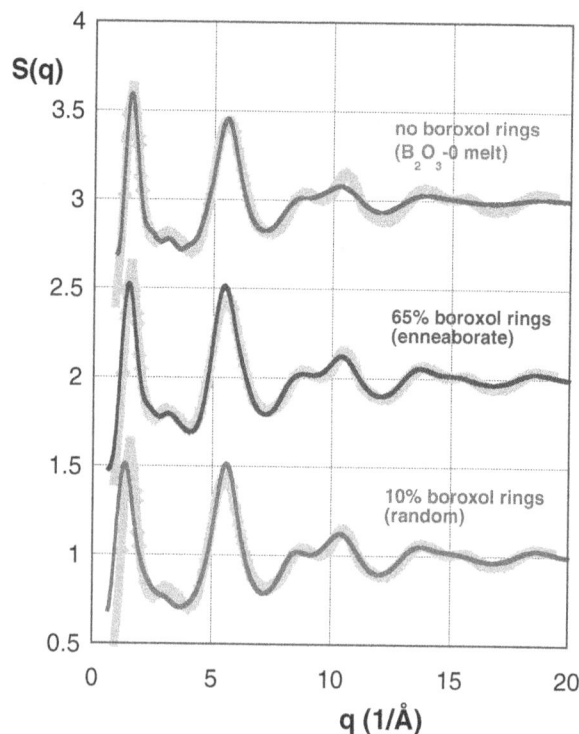

Figure 6. Comparison of the total neutron scattering structure factors obtained from three different simulated structures (thin coloured lines) and experiment (underlying gray broad lines)

ports our explanation that thermomechanical anomalies are based on structural motifs that take root in the crystalline counterparts of network glasses. Our findings also indicate that this phenomenon may be universal for all major network formers.[11] The thermomechanical anomalies discussed here play an important role in the context of glass stability and, in particular, the ability of the glass to undergo permanent densification.[45] Note that an anomalous glass becomes more compliant the more strongly it is compressed! This means that, the closer atoms in the structure are pushed together, the less resistance they offer towards further approach. The progressive ease with which atoms impinge upon each other creates configurations that facilitate the exchange of chemical bonds and lasting reorganisation. This is exactly what happens when a-SiO_2 is compressed.[6] Once the glass is compressed beyond the modulus minimum, transitions become irreversible and the structure permanently densifies. In silica glass the corresponding bond reconfigurations lead to a larger overall ring size, which can be understood when considering that larger rings possess more flexibility to fold up onto themselves for efficient packing. If we argue that thermomechanical anomalous behaviour promotes permanent densification, then the absence thereof would imply more structural stability. The case of a-B_2O_3 supports this argument. Recall that on a relative pressure axis, at ambient conditions, a-B_2O_3 is above the modulus minimum, i.e. the structure no longer exhibits accelerated deformation that invites bond reconfiguration. Accordingly, as-quenched a-B_2O_3 could be considered inherently more stable than as-quenched a-SiO_2.

The fact that B_2O_3 is a very good glass former and very stable is apparent from the crystallisation anomaly of this substance.[11] In closing, we offer a line of reasoning that connects this resistance towards crystallisation to the existence of B_2O_3-0. Liquid B_2O_3 has a large thermal expansion coefficient around the melting point, and a density that is commensurate

with that of B_2O_3-0 (Figure 4(b)). Indeed, when heating B_2O_3-I in simulations, it first transforms into B_2O_3-0 before melting, as illustrated in Figure 5. Hence, when the liquid is supercooled, the reverse process may be taking place, and it may be natural for the system to form low density B_2O_3-0 nuclei. Upon further cooling, the volume surrounding the nuclei contracts, compressing the nuclei. The B_2O_3-0 structure is no longer stable, but by now the temperature is too low and thermal activation is insufficient to allow transition into B_2O_3-I, which is the stable structure at ambient pressure and temperature. Hence, the system becomes frustrated in an amorphous structure with some remnant B_2O_3-0 character. An indication that the latter is the case may be gleaned from Figure 6, where we compare the experimental neutron scattering structure factor, $S(Q)$,[33,46,47] with those of different simulated B_2O_3 glasses: one with 10% boroxol rings, another with 65% boroxol rings, both prepared from caesium enneaborate as described above, and a third with no boroxol rings prepared by melting and quenching B_2O_3-0. Note that, while the structure factors of all simulated systems provide good qualitative agreement with the experimental data, regardless of the boroxol ring concentration, only in the structure factor for the glass derived from B_2O_3-0 does the first peak (at ~1·6 Å^{-1}) have a larger magnitude than the second (at ~5·8 Å^{-1}), as for the experimental one. At larger wavevector magnitude, the simulated glass with about 2/3 of borons in boroxol rings shows the best agreement with the experimental data. Accordingly, we conclude that the extended range network connectivity is derived from the open structure of B_2O_3-0, and the near to intermediate range order bears the signature of boroxol rings.

6. Conclusions

Using MD simulations based on a multi-coordination reactive interaction potential, we were able to reproduce the experimentally observed thermomechanical anomalies in both a-SiO_2 and a-B_2O_3. Whether induced by temperature or pressure changes, this behaviour can be unequivocally mapped onto a change in specific volume of the substances. For network glasses based on either trigonal or tetrahedral building blocks, the mechanism that underlies the anomalies are buckling-like, localised transitions, involving the sudden rotation of X–O–X linkages about the X–X axis, where X represents the network cation, similar to those between the α- and β-modifications of the crystalline counterparts of these materials. High density structures are more compliant than low density ones, because the topology of the former allows for folding of network rings by pivoting about X–O bonds, while the latter require bending of X–O–X linkages. The process responsible for thermo-mechanical anomalous behaviour appears to be universal for all major network formers. This insight has led to the discovery of two previously unknown crystalline structures, i.e. α- and β-B_2O_3-0.

Acknowledgments

The authors acknowledge the support of the National Science Foundation under Grant No. DMR-0605905 and of Saint-Gobain Recherche. Simulations were done on super-computers in the Center for Advanced Computing (CAC) at the University of Michigan.

References

1. Youngman, R. E., Kieffer, J., Bass, J. D. & Duffrène, L. *J. Non-Cryst. Solids*, 1997, **222**, 190.
2. Green, P. F., Brown, E. F. & Brow, R. K. *J. Non-Cryst. Solids*, 1999, **255**, 87.
3. Kim, Y. K., Bruckner, R. & Murach, J. *Glastech. Ber. Glass Sci. Technol.*, 1998, **71**, 67.
4. Lee, J. W., Sigel, G. H. & Li, J. *J. Non-Cryst. Solids*, 1998, **239**, 57.
5. Anderson, D. C., Kieffer, J. & Klarsfeld, S. *J. Chem. Phys.*, 1993, **98**, 8978.
6. Huang, L. & Kieffer, J. *Phys. Rev. B*, 2004, **69**, 224204.
7. Stanley, H. E., Kumar, P., Franzese, G., Xu, L., Yan, Z., Mazza, M. G., Buldyrev, S. V., Chen, S. H. & Mallamace, F. *Eur. Phys. J.-Special Topics*, 2008, **161**, 1.
8. Huang, L. & Kieffer, J. *Phys. Rev. B*, 2004, **69**, 224203.
9. Shen, G. Y., Liermann, H. P., Sinogeikin, S., Yang, W. G., Hong, X. G., Yoo, C. S. & Cynn, H. C. *Proc. Natl. Acad. Sci. USA*, 2007, **104**, 14576.
10. Takada, A., Richet, P., Catlow, C. R. A. & Price, G. D. *J. Non-Cryst. Solids*, 2007, **353**, 1892.
11. Huang, L. & Kieffer, J. *Phys. Rev. B*, 2006, **74**, 224107.
12. Deriano, S., Rouxel, T., LeFloch, M. & Beuneu, B. *Phys. Chem. Glasses*, 2004, **45**, 37.
13. Sehgal, J. & Ito, S. *J. Non-Cryst. Solids*, 1999, **253**, 126.
14. Babcock, C. L., Barber, S. W. & Fajans, K. *Ind. Eng. Chem.*, 1954, **46**, 161.
15. Vukcevich, M. R. *J. Non-Cryst. Solids*, 1972, **11**, 25.
16. Grimsditch, M. *Phys. Rev. Lett.*, 1984, **52**, 2379.
17. Grimsditch, M. *Phys. Rev. B*, 1986, **34**, 4372.
18. Grimsditch, M., Bhadra, R. & Meng, Y. *Phys. Rev. B*, 1988, **38**, 7836.
19. Huang, L., Nicholas, J. D., Kieffer, J. & Bass, J. D. *J. Phys.: Condens. Matter*, 2008, **20**, 075107.
20. Meade, C., Hemley, R. & Mao, H. *Phys. Rev. Lett.*, 1992, **69**, 1387.
21. Mishima, O., Calvert, L. D. & Whalley, E. *Nature*, 1984, **310**, 393.
22. Nicholas, J. D., Sinogeikin, S. V., Kieffer, J. & Bass, J. D. *Phys. Rev. Lett.*, 2004, **92**, 215701.
23. Poole, P. H., Grande, T., Sciortino, F., Stanley, H. E. & Angell, C. A. *Comp. Mater. Sci.*, 1995, **4**, 373.
24. Trachenko, K. & Dove, M. T. *J. Phys.: Condens. Matter*, 2002, **14**, 1143.
25. Trachenko, K. & Dove, M. T. *Phys. Rev. B*, 2003, **67**, 064107.
26. Tse, J. S., Klug, D. D. & Lepage, Y. *Phys. Rev. B*, 1992, **46**, 5933.
27. Walker, A. M., Sullivan, L. A., Trachenko, K., Bruin, R. P., White, T. O. H., Dove, M. T., Tyer, R. P., Todorov, I. T. & Wells, S. A. *J. Phys.: Condens. Matter*, 2007, **19**, 275210.
28. Grimsditch, M., Polian, A. & Wright, A. C. *Phys. Rev.*, 1996, *B* **54**, 152.
29. Nicholas, J. D., Youngman, R. E., Sinogeikin, S. V., Bass, J. D. & Kieffer, J. *Phys. Chem. Glasses*, 2003, **44**, 249.
30. Nicholas, J. D., Sinogeikin, S. V., Kieffer, J. & Bass, J. D. *J. Non-Cryst. Solids*, 2004, **349**, 30.
31. Lee, S. K., Eng, P. J., Mao, H. K., Meng, Y., Newville, M., Hu, M. Y. & Shu, J. F. *Nature Materials*, 2005, **4**, 851.
32. Lee, S. K., Mibe, K., Fei, Y. W., Cody, G. D. & Mysen, B. O. *Phys. Rev. Lett.*, 2005, **94**, 165507.
33. Wright, A. C., Stone, C. E., Sinclair, R. N., Umesaki, N., Kitamura, N., Ura, K., Ohtori, N. & Hannon, A. C. *Phys. Chem. Glasses*, 2000, **41**, 296.
34. Aziz, M. J., Nygren, E., Hays, J. F. & Turnbull, D. *J. Appl. Phys.*, 1985, **57**, 2233.
35. Uhlmann, D. R., Hays, J. F. & Turnbull, D. *Phys. Chem. Glasses*, 1967, **8**, 1.
36. Huang, L. & Kieffer, J. *J. Chem. Phys.*, 2003, **118**, 1487.
37. Huang, L., Durandurdu, M. & Kieffer, J. *J. Phys. Chem. C*, 2007, **111**, 13712.
38. Huang, L., Durandurdu, M. & Kieffer, J. *Nature Materials*, 2006, **5**, 977.
39. Hassan, A. K., Torell, L. M., Borjesson, L. & Doweidar, H. *Phys. Rev. B*, 1992, **45**, 12797.
40. Krogh-Moe, J. & Ihara, M. *Acta Cryst.*, 1967, **23**, 427.
41. Takada, A. *Phys. Chem. Glasses*, 2004, **45**, 156.
42. Takada, A., Catlow, C. R. A. & Price, G. D. *Phys. Chem. Glasses*, 2003, **44**, 147.
43. Johnson, P. A. V., Wright, A. C. & Sinclair, R. N. *J. Non-Cryst. Solids*, 1982, **50**, 281.
44. Andersen, H. C. *J. Chem. Phys.*, 1980, **72**, 2384.
45. Huang, L. & Kieffer, J. *Appl. Phys. Lett.*, 2006, **89**, 141915.
46. Mozzi, R. L. & Warren, B. E. *J. Appl. Cryst.*, 1970, **3**, 251.
47. Suzuya, K., Yoneda, Y., Kohara, S. & Umesaki, N. *Phys. Chem. Glasses*, 2000, **41**, 282.

Proc. VI Int. Conf. Borate Glasses, Himeji, Japan, 18–22 August 2008   *Phys. Chem. Glasses: Eur. J. Glass Sci. Technol. B*, June 2009, **50** (3), 229–235

Thermal poling induced structural changes in sodium borosilicate glasses

D. Möncke,* M. Dussauze, E. I. Kamitsos, C. P. E. Varsamis

National Hellenic Research Foundation, 48 Vassileos Constantinou Avenue, 11635 Athens, Greece

D. Ehrt

Otto-Schott-Institut für Glaschemie, Friedrich-Schiller-Universität, Fraunhoferstr. 6,
07743 Jena, Germany

Manuscript received 31 October 2008
Revised version received 4 December 2008
Accepted 4 December 2008

Inorganic glassy materials can exhibit large second order nonlinear optical (NLO) coefficients, $\chi^{(2)}$, after thermal poling. Sodium borosilicate glasses with various SiO_2 contents and constant $Na_2O:B_2O_3$ ratio of 0·2, which characterises the boric oxide anomaly, were thermally poled under argon. The second harmonic generation signal was measured for different poling parameters. Structural changes between poled and unpoled glasses were observed by infrared reflectance and Raman spectroscopy. It was found that under poling Na^+ ions migrate from the anode to the cathode side. This displacement of cations from the anode side is compensated by structural rearrangements of the glassy network, which lead eventually to the release of oxygen anions from the anode side. Both effects are facilitated by the conversion of borate tetrahedra into neutral trigonal borate units. The reverse transformation occurs at the cathode side, where borate tetrahedra are formed together with electron centres, also known as oxygen vacancies. The presence of vacancies was revealed by optical and EPR spectroscopy.

1. Introduction

The polarisation of materials in the presence of external electric fields is usually represented by the following equation

$$P=\varepsilon_0[\chi^{(1)}E(\omega)+\chi^{(2)}E(\omega_1)E(\omega_2)+\chi^{(3)}E(\omega_1)E(\omega_2)E(\omega_3)+...] \tag{1}$$

where P is the polarisation, $\chi^{(1)}$ is the linear susceptibility, $\chi^{(2)}$ and $\chi^{(3)}$ are higher order nonlinear susceptibilities, and $E(\omega_i)$ denotes electric fields with an angular frequency ω_i. Glasses show as a rule no nonlinear optical (NLO) properties. The reason is that the even order terms of Equation (1) are zero for materials with macroscopic inversion symmetry, and the third order term is in most cases negligible. However, an induced internal electric field, E_{int}, may generate second order optical response, $\chi^{(2)}$, through the following relationship

$$\chi^{(2)}=3\chi^{(3)}E_{int}+N\beta LE_{loc} \tag{2}$$

where N denotes the number of permanent dipoles per unit volume, L is the orientation factor, and β is the microscopic hyperpolarisability induced by the local field E_{loc} during poling.[1–4]

According to Equation (2) a strong NLO response in glasses can be associated with compositions exhibiting high $\chi^{(3)}$ values. Such compositions include elements which show maximal p–d-orbital overlap,

like the d^0-ions, or ions with lone electron pairs, such as the chalcogenides or post transition elements. Besides high $\chi^{(3)}$ values, the occurrence of highly mobile cations in glasses is just as important, because the migration of cations from the anode to the cathode during poling induces the formation of a space charge layer and the creation of the internal electric field.[1–5] In addition, the high NLO activity of borate crystals such as β-BaB_2O_4 and LiB_3O_5, has been attributed to the presence of delocalised electrons in boroxol or borate rings.[6,7]

In the present work, we report on the thermal poling induced structural changes in sodium borosilicate (NBS) glasses of composition $KSiO_2.0·2Na_2O.B_2O_3$, where K is the $SiO_2:B_2O_3$ ratio. The compositions and some properties of the investigated glasses are reported in Table 1. The properties of these NBS glasses, like their high melting temperature, low thermal expansion coefficient and high transparency values from the ultraviolet to the visible region, resemble those of vitreous silica, owing to the absence of nonbridging oxygen atoms.[8–14] The $Na_2O:B_2O_3$ ratio of $R=0·2$ characterises the boric oxide anomaly, that is the composition of modified borate glasses at which numerous of their properties exhibit a distinct extremum.[8] For R<0·2, all added Na_2O is used for the conversion of trigonal $BØ_3$ groups into charged $BØ_4^-$ tetrahedra (Ø denotes an oxygen atom bridging two boron centres). Nonbridging oxygen atoms will form only at higher Na_2O contents.

* Corresponding author. Email moenke@eie.gr or dorismoenke@web.de

Table 1. Molar compositions and selected properties of NBS glasses $KSiO_2.RNa_2O.B_2O_3$

mol%	NBSA	NBSD	NBS2	NBDB[b]
Na_2O	12·5	6·5	4·3	3·0
B_2O_3	62·5	33·5	20·7	48·5
SiO_2	25	60	74	48·5
Al_2O_3	-	-	1	-
K (SiO_2:B_2O_3)	0·40	1·79	3·57	1
R (Na_2O:B_2O_3)	0·2	0·2	0·2	0·06
T_g (°C) ±2°C	415	445	442	380
TEC[a] (ppm/K) ±0·3	8·5	5·3	3·5	6·4
ρ (g/cm³) ±0·01	2·18	2·50	2·18	2·04
n_e ±0·04 %	1·496	1·476	1·471	1·472
v_e ±0·3	62·5	64·2	65±2	62·4

[a]TEC is the thermal expansion coefficient measured in the temperature range 100–300°C.

[b]NBSB is included because the composition with its low R value simulates the alkali oxide depleted composition of the anodic side of the poled samples (Figure 5(a))

Previous NMR studies[10] showed that for NBS2, which is a silica rich glass ($K=3·57$) and a model for Duran® type glasses, 18% of the borate units form $BØ_4^-$ tetrahedra, while the remaining 82% are found in $BØ_3$ triangular units in various arrangements. In particular, more than half of the borate triangles are linked together in planar boroxol rings, while the other half is found outside such ring structures.[10] Numerous investigations of NBS2 glasses were conducted over the last decade and the most important structural effects concern the strong dependence of the medium range order structure on the thermal history.[9–11] For example, a combined NMR and vibrational spectroscopic investigation of NBS2 glasses with different thermal histories showed that structural variations concern primarily the connectivity of $BØ_3$ groups and the number of mixed B–O–Si bonds.[11]

Thermal poling induced structural changes of sodium borosilicate glasses were studied by infrared (IR) and Raman, as well as by optical and electron paramagnetic resonance (EPR) spectroscopic techniques in order to explain the underlying mechanisms which are responsible for the generation of NLO response, evaluated by measurements of the second harmonic generation (SHG) signal in poled glasses.

2. Experimental

All glasses were prepared from high purity optical grade materials SiO_2, $CaCO_3$, Na_2CO_3, H_3BO_3 and $Al(OH)_3$ in Pt-crucibles using a resistant heated furnace. Al_2O_3 was added to NBS2 ($R=0·2$, $K=3·57$) in order to avoid phase separation and to circumvent the metastable immiscibility gap of the ternary composition. NBSA ($R=0·2$, $K=0·40$) and NBSD ($R=0·2$, $K=1·79$) and NBSB ($R=0·06$, $K=1$) compositions were obtained by remelting semi-technical preforms. The polished samples had dimensions of 10×20 mm and were either 1 or 2 mm thick.

Homemade systems were used for thermal poling experiments and for SHG measurements. The poling

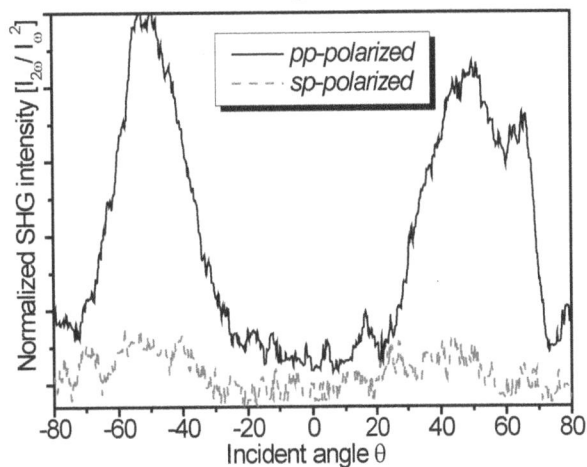

Figure 1. Typical normalised SHG signal measured in pp- and sp-polarisation on a 2 mm thick NBS2 glass after thermal poling (300°C, 3·0 kV)

voltage was typically between 1 to 3·5 kV, and it was applied for 1 h at temperatures between 190 to 300°C, which are considerably below the T_g value of NBS glasses (415–445°C). The maximum applied voltage was limited to about 0·2 kV below the breakthrough current, which in turn depends primarily on the sample thickness and the applied poling temperature. Poling was sustained during the cooling period until the samples reached room temperature. Silicon wafers acted as electrodes and were placed with their coarse sides towards the sample in order to avoid anodic sealing between the glasses and the electrodes. All poling experiments were conducted in argon, in order to eliminate possible reactions of atmospheric gases with the glass during poling.

A Nd:YAG laser beam (1064 nm) was used as source frequency, I_w, for SHG measurements. The SHG signal at 532 nm, I_{2w}, was analysed for incident angles of −80° to +80°, and it was measured in both sp and pp polarisations.

The poling induced structural changes were probed by infrared reflectance and Raman spectroscopy. Spectral mapping by an infrared microscope was conducted along an 8 to 10 mm long transect on the sample surface, including a poled section. Spectra were taken every 50 to 200 μm in the frequency range from 600–7500 cm⁻¹ in the backscattering reflectance mode. Besides spectral mapping, specular reflectance spectra were also measured in the 500–7500 cm⁻¹ frequency range on both poled and unpoled glasses using an infrared beam of 3 mm spot size at the sample. Confocal Raman microscopy allowed a 'depth scan' in order to study the poling induced structural changes along a cross section of the sample. Thus, micro-Raman spectra were collected every 0·5 μm along a break of a poled sample, starting near the surface and ending at ~100 μm inside the glass.

Optical spectra were recorded from 190 to 1000 nm. The induced absorbance spectra, that is, the dif-

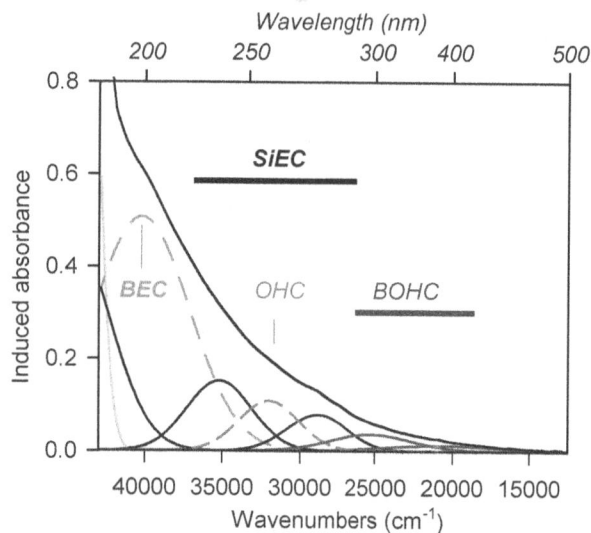

Figure 2. Poling induced absorbance spectrum of NBSD glass (300°C, 3·0 kV, d=2 mm), including band deconvolution (dots denote the experimental and the solid line the simulated spectrum). SiEC and BEC stand for electron centre connected to the silicate and borate network, respectively; OHC and BOHC indicate oxygen hole and boron related oxygen hole centres, respectively

ference spectra between poled and unpoled samples, were fitted with Gaussian component bands in accordance with literature data for irradiation induced defects.[12–15] EPR measurements, using a frequency band of v~9·78 GHz, were used to verify the existence of these defects.

3. Results

3.1 Second harmonic generation signals

Poling of all NBS samples resulted in NLO response, as shown by the SHG signal of a thermally poled NBS2 glass reported in Figure 1. In general, the SHG intensity

increases for samples that show enhanced structural rearrangements in their vibrational characteristics. In sodium rich NBSA glasses, poling resulted in profound surface degradation and structural inhomogeneities which were observed by optical microscopy and by infrared spectral mapping of the poled samples, respectively. These effects caused the attenuation of the SHG signal due to intense light scattering.

3.2 Optical spectra

All unpoled NBS samples were colourless, while NBS2 samples showed a yellow colouration after poling at 300°C. The colour centres are solely localised in a thin layer of the samples beneath and in the outlines of the cathodic electrode. As evidenced in Figure 2, optical spectroscopy revealed the development of substantial absorbance below 300 nm. Poled NBSD samples exhibited the same coloration, while no poling induced absorbance was observed for NBSA samples. Substantial bleaching occurred in the days following the poling experiments.

All induced absorbance spectra can be fitted with defect bands of colour centres that typically appear in glasses after UV irradiation. These bands arise from different positively charged hole centres (HC), which often absorb in the visible, and negatively charged electron centres (EC) which absorb mainly in the UV region.[13–15]

3.3 EPR spectra

Figure 3(a) and (b) show EPR reference signals of intrinsic paramagnetic defects in NBS glasses after UV irradiation.[13,14,16,17] EPR spectroscopy is much more sensitive to hole centres than to electron centres. Therefore, the signals of EC are often hidden by the intense HC signals of Figure 3(a). However, non-

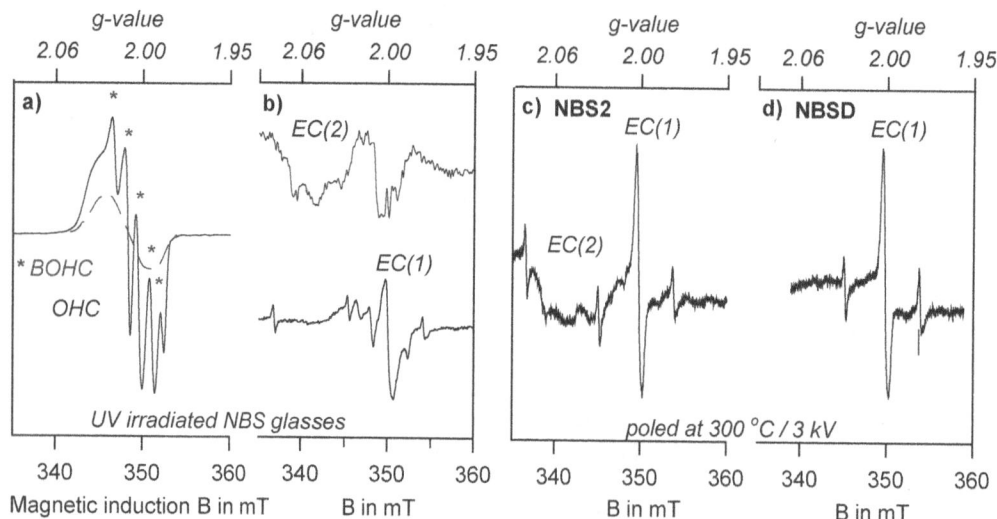

Figure 3. EPR reference spectra of UV-laser irradiated NBS glasses showing: (a) intrinsic hole centres (HC) and (b) intrinsic electron centres (EC) [see Refs 12,13]. EPR spectra of poled glasses: (c) NBS2 (300°C, 3·0 kV, d=2 mm), and (d) NBSD (300°C, 3·5 kV, d=2 mm)

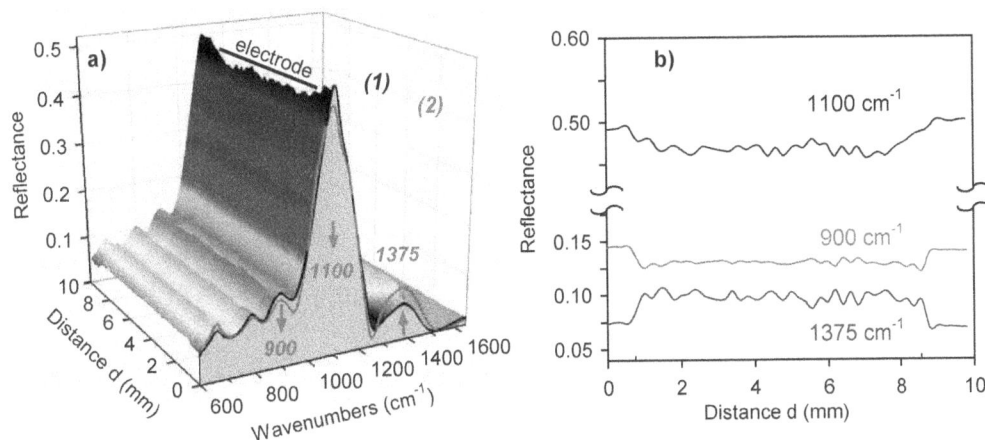

Figure 4. (a) Spectral mapping by micro-infrared reflectance spectroscopy of a thermally poled NBS2 sample (190°C, 3·0 kV, d=1 mm). Spectra were recorded along a line of the sample, including a section in contact with the anodic electrode (see label). For comparison, the figure includes the spectra: (1) with grey filling, is the spectrum measured outside the poling area, and (2) in red, is the spectrum measured at d=5 mm. (b) Values of measured reflectance from spectra shown in (a) plotted as a function of distance for bands at ~1100 cm^{-1}, 900 cm^{-1}, and 1375 cm^{-1}. The section of the sample probed under the anodic electrode is confined within the vertical broken lines

paramagnetic HC can arise from the photo-oxidation of dopants and consequently intrinsic EC signals may be observed unimpeded, like those displayed in Figure 3(b). The EPR signals of thermally poled NBS samples reported in Figure 3(c) and (d) are in good agreement with those of the irradiation induced intrinsic ECs in Figure 3(b).

3.4 Infrared spectra

Micro-infrared spectral mapping showed that spectra taken outside the poled area are generally comparable to spectra of unpoled samples, while spectra measured inside the poled area reveal substantial differences. A typical spectral mapping for the anode side of NBS2 poled glass is reported in Figure 4(a). The most intense bands of spectra are found in the 950–1250 cm^{-1} range and originate mostly from the asymmetric stretching vibrations of fully polymerised silicate groups, Q^4, without excluding the contribution of the asymmetric stretching vibrational modes of B$_4$–O–Si, at 1050 cm^{-1}, and of B$_3$–O–B$_4$ (the suffix denotes the boron coordination) bridges around 1080 and 1250 cm^{-1}.[7,11,18–22] At higher frequencies, the weak profile in the 1250–1500 cm^{-1} range can be attributed to the asymmetric stretching vibrations of BØ$_3$ groups in different ring and non-ring configurations.[21] The 900–1000 cm^{-1} feature may arise from the asymmetric stretching vibration of mixed Si–O–B$_3$ and B$_4$–O–B$_3$ bridges and from the asymmetric stretching vibration of BØ$_4^-$ tetrahedra.[11] Peaks at frequencies lower than 900 cm^{-1} are associated with the bending modes of various structural entities and of B–O–B, B–O–Si and Si–O–Si bridges. Deconvolution of infrared spectra is rather complicated since the structural units of the network may be coupled in many different configurations resulting in strongly overlapping bands.

The contour of the reflectance spectra in Figure 4(a) shows significant changes between the poled and unpoled regions of the anode side of the sample. In particular, there is an increase in the intensity of the 1250–1500 cm^{-1} profile, and a decrease in the intensity of the bands below 1000 cm^{-1}. Moreover, the intensity of the main band around 1100 cm^{-1} decreases, suggesting a progressive decrease of the relative population of B$_4$–O–Si and B$_3$–O–B$_4$ bridges since structural units with nonbridging oxygen atoms were not observed in poled NBS samples. These spectral changes are highlighted in Figure 4(b) where the reflectance measured inside and outside the pole area is shown for the three characteristic bands at ~1100 cm^{-1}, 900 cm^{-1} and 1375 cm^{-1}.

Infrared specular reflectance spectra for poled and unpoled NBS glasses are depicted in Figure 5. As the borate content increases from NBS2 to NBSA, structural changes become more pronounced. Such changes are primarily connected with the borate rather than with the silicate part of the glass network. It is worth noting that significant changes were found between the spectra of NBSA glass measured on the anode side directly after poling and that measured four days later. It was also found that structural changes increase with increasing poling temperature for samples of the same composition. For low poling temperatures up to 250°C no significant changes were discerned on the cathode side of poled NBS2 samples, which is in agreement with previous studies.[2–4] However, all NBS samples poled at 300°C showed distinct changes for both the anode and the cathode side. With respect to the unpoled glasses, the relative intensity of the peak of BØ$_3$ units at ~1400 cm^{-1} increases at the anode side and decreases at the cathode side. At the same time, the characteristic profile of borate tetrahedra centred around 900 cm^{-1} is strengthened in

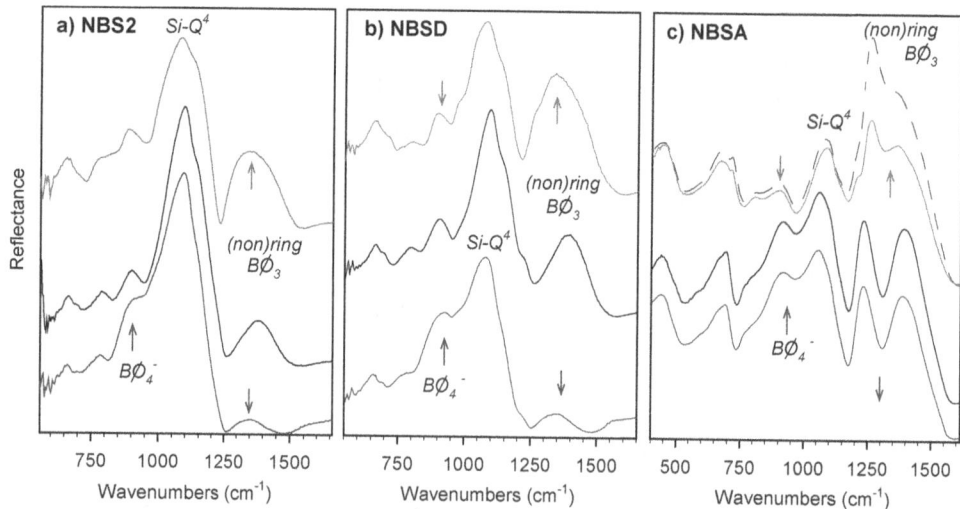

Figure 5. Infrared reflectance spectra of NBS glasses before poling (mid-black) and after poling (top-red: anode side, bottom-blue: cathode side): (a) NBS2 (300°C, 2·2 kV, d=1 mm), including for comparison the spectrum of NBSB simulating the alkali and oxygen depleted composition of the anodic side of poled NBS2 (R=0·06/dark red-dots); (b) NBSD (300°C, 3·5 kV, d=2 mm); (c) NBSA (300°C, 3·0 kV, d=2 mm), where the spectrum on the anode side (dark red-dash-dots) transforms within 4 days to the one marked with the solid-line

the spectrum of the cathode side and is weakened on the anode side. Finally, it should be emphasised that structural units having nonbridging oxygen atoms were not observed in any of the poled NBS samples.

3.5 Raman spectra

Figure 6(a) shows the Raman spectra of poled and unpoled NBS2 glass. The sample was broken in half in order to study the depth profile of the structural changes. The top six spectra were taken along this cut with increasing distance from the anode surface. The main features of the Raman spectra are a sharp peak at 805 cm^{-1} due to six-membered boroxol rings, a feature at 770 cm^{-1} due to six-membered borate rings with one BØ$_4^-$, and a feature at 750 cm^{-1} due to six-membered rings with two BØ$_4^-$ units.[11,20,21] The broad absorption around 450 cm^{-1} is caused by

various silicate and borate network vibrations, while the origin of the well defined strong peak at 260 cm^{-1}, which appears only on the anode side of poled samples, is still unclear. A plausible origin of this band could be the formation of 8-membered silicate rings.[23] The intensity of this peak correlates well with the intensities of the six-membered borate rings. The Raman spectra reveal that the poling induced structural changes on the anode side are limited to the top 1–2 μm of the glass. The spectrum of the cathode side is characterised by a higher intensity of the peak around 1075 cm^{-1} which is attributed to the asymmetric stretching vibrational modes of Si–O–Si bridges connecting Q^4 tetrahedra.[11,22]

The Raman spectra of unpoled and poled NBSD glass are depicted in Figure 6(b). As observed for NBS2 glass, the strong 260 cm^{-1} peak develops at the anode side in a similar way to the boroxol peak at 805

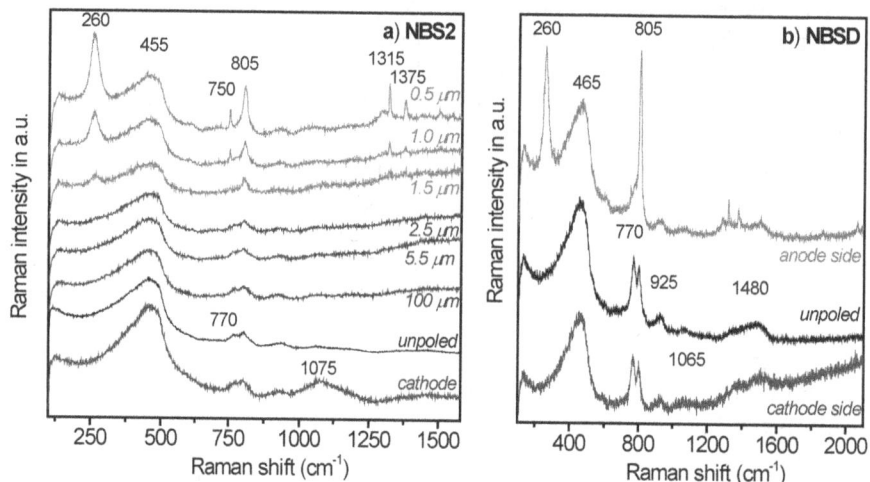

Figure 6. Effect of thermal poling on the micro-Raman spectra of: (a) NBS2 (300°C, 3·0 kV, d=2 mm) measured at various distances under the surface of the anode side, and (b) of NBSD glass (300°C, 3·5 kV, d=2 mm)

cm^{-1}. Unlike the 260 and 805 cm^{-1} features, the band at 770 cm^{-1} (borate rings with one $BØ_4^-$ tetrahedral unit) has almost disappeared at the anode side. Weaker peaks at 925 and around 1450 cm^{-1}, due to vibrational modes of tetrahedral and trigonal borate units, respectively,[11] show no significant changes at both the anode and cathode side of the poled NBSD glass.

4. Discussion

The structural changes observed in the infrared spectra measured after poling at the anode side can be described by the following transformation reaction of borate tetrahedra into triangular units

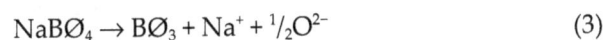

$$NaBØ_4 \rightarrow BØ_3 + Na^+ + \tfrac{1}{2}O^{2-} \tag{3}$$

In the presence of the external electric field the highly mobile sodium ions migrate through the glassy network towards the cathode, while oxygen anions tend to leave the glass at the anode side through conversion to oxygen gas. Migration of Na^+ ions from the anode is well established in the literature,[1–5] and it is supported also by preliminary results of far infrared measurements on poled NBSD glasses undertaken in our laboratory. Oxygen depletion at the anode of poled glasses has been described before as a key mechanism to reduce the excessive negative charge in the cation depleted anodic layer.[1–4] As shown in previous studies of poled sodium niobium borophosphate glasses, cation and oxygen migration triggers structural rearrangements of the glass matrix, which leads to conversion of nonbridging to bridging oxygen atoms in phosphate units and to transformation of corner to edge sharing niobate octahedra.[2,3] The remaining net space charge provides the basis for the NLO response of the poled glass.

A similar structural rearrangement mechanism due to thermal poling is supported also by the spectra reported in Figure 5(a), where the spectrum of the anode side NBS2 glass compares well with that of the unpoled NBSB glass (Figure 5(a)). The NBSB glass with its much lower Na_2O to B_2O_3 ratio ($R=0·06$) than the NBS2 glass ($R=0·2$) simulates the alkali and oxygen depleted composition of the anodic side of poled NBS2 sample well. The infrared spectra show accordingly more $BØ_3$ and less $BØ_4^-$ groups than seen in unpoled NBS2.

Whereas the migration of Na^+ ions in the presence of an external field is well understood, the mechanism associated with the creation of oxygen ions has to be elaborated further. As a first step, and in the absence of NBO atoms in the original glass, breakage of B_4–O–Si or B_3–O–B_4 bridges would lead to creation of NBOs according to the following schemes

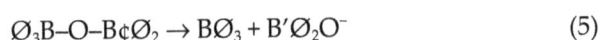

$$Ø_3B–O–SiØ_3 \rightarrow BØ_3 + SiØ_3O^- \tag{4}$$

$$Ø_3B–O–B¢Ø_2 \rightarrow BØ_3 + B'Ø_2O^- \tag{5}$$

A subsequent condensation reaction between two NBOs, by combination of transient $SiØ_3O^-$ and $BØ_2O^-$ units, would create bridging oxygen and release an oxygen ion as follows

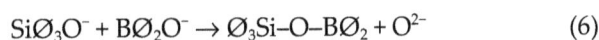

$$SiØ_3O^- + BØ_2O^- \rightarrow Ø_3Si–O–BØ_2 + O^{2-} \tag{6}$$

The overall effect of the mechanism described by Equations (4)–(6) is to transform two borate tetrahedra into two neutral triangular borate units and an O^{2-} anion. This mechanism is in agreement with the observed spectral changes in the infrared spectra.

It is noted that the above mechanism cannot be applied to B_3–O–B_3 or B_3–O–Si bridges. In fact, breakage of such bridges would result in positively charged $[BØ_2]^+$ or $[SiØ_3]^+$ entities, which cannot be stabilised by the release of surplus oxygen. Also, B_4–O–B_4 bridges are unlikely to break for two main reasons. Firstly, these bridges are encountered in borate rings which are very stable and, in fact, both infrared and Raman spectra show that their population increases on the anode side of poled glasses (see Figure 6(a)). Secondly, the existence of B_4–O–B_4 bridges outside ring structures can be excluded on the basis of the intertetrahedral link avoidance rule.[24]

At the cathode side, the infrared spectra show the conversion of $BØ_3$ into $NaBØ_4$ groups, which is the reverse transformation of Equation (3). The Na^+ ions involved in this reaction are those released from the anode side of the sample. On the other hand, the origin of the extra oxygen ions on the cathode side required for the reverse transformation of Equation (3) needs further explanation. It is noted that all poling experiments were performed under Ar, and this excludes the capture of oxygen atoms from the atmosphere.

As mentioned in Section 3.2, the formation of colour centres in poled glasses is localised in a thin layer on the cathode side. Moreover, deconvolution of the induced absorbance spectra in Figure 2 showed that the most prominent bands are due to intrinsic electron centres, as confirmed later by EPR spectroscopy. These findings provide a plausible explanation for the origin of extra oxygen ions on the cathode side. In fact, an electron centre is actually a form of oxygen deficient vacancy, which is formed during poling according to the following equations

$$Ø_2B–O–SiØ_3 + e^- \rightarrow Ø_2BO^- + \dot{S}iØ_3 \; (SiEC) \tag{7}$$

$$Ø_2B–O–BØ_3 + e^- \rightarrow Ø_2BO^- + \dot{B}Ø_3 \; (BEC) \tag{8}$$

An electron from the cathode may be trapped by an oxygen bond of the glassy network. The breakage of that bond can result in the formation of intrinsic EC and $BØ_2O^-$ units. Next, $BØ_2O^-$ units are transformed into the isomeric borate tetrahedra which in turn trap the migrating Na^+ ions released from the anode side of the samples. The remaining SiEC and BEC are both radicals, therefore not very stable, and can combine with adjacent EC or atmospheric components when located near to the surface.

5. Conclusions

Thermal poling of sodium borosilicate glasses with a $Na_2O:B_2O_3$ ratio of $R=0.2$ resulted in the creation of nonlinear optical properties. These were manifested by the detection of second harmonic generation signals with strength that is correlated to the induced structural variations. Infrared reflectance and Raman spectroscopic techniques showed that such variations involve mainly the borate part of the glassy network. On the anode side, the conversion of borate tetrahedra, $B\emptyset_4^-$, into neutral triangular units, $B\emptyset_3$, was observed for all samples. This conversion is then accompanied by the migration of sodium ions to the cathode and the release of oxygen from the glass on the anode side.

On the cathode side, the reverse transformation of $B\emptyset_3$ into $B\emptyset_4^-$ units occurs. EPR and optical spectroscopy attest to the presence of electron centres, or oxygen vacancies, which provide the required extra oxygen ions, although all samples were poled under argon. Borate tetrahedra trap sodium ions which reach the cathode through migration from the anode side. The relatively short lifetime of the electron defects can be explained by their recombination or by subsequent reactions with atmospheric components, as samples were kept in ambient atmosphere after poling.

Acknowledgments

Financial support of this work by the EU through the Marie Curie Actions – NANONLO project (MTKD-CT-2006-042301) is gratefully acknowledged. The authors thank M. Friedrich and B. Rambach for EPR measurements, R. Atzrodt for help in sample preparation, and D. Palles for help with SHG measurements.

References

1. Borelli, N. F. & Hall, D. W. Ch.3 in *Optical properties of glass*, Eds. D. R. Uhlmann & N. J. Kreidl, 1991, The American Ceramic Society, Westerville, Ohio, P 87–124.
2. (a) Dussauze, M., Fargin, E., Lahaye, M., Rodriguez, V. & Adamietz, F. *Opt. Express*, 2005, **13**, 4064–9. (b) Malakho A., Dussauze, M., Fargin, E., Lazoryak, B., Rodriguez, V. & Adamietz, F. *J. Solid State Chem.*, 2005, **178**, 1888–97.
3. Dussauze, M., Kamitsos, E. I., Fargin, E. & Rodriguez, V. *J. Phys. Chem. C*, 2007, **111**, 14560–6.
4. Franchina, E., Corbari, C., Kazansky, P., G., Chiodini, N., Lauria, A. & Palear, A. *Solid State Commun.*, 2005, **136**, 300–3.
5. (a) An, H. & Fleming, S. *Appl. Phys. Lett.* 2006, **89**, 181111. (b) An, H. & Fleming, S. *Opt. Commun.*, 2008, **281**, 1263–7. (c) An, H. & Fleming, S. *J. Opt. Soc. Am. B*, 2006, **23**, 2303–9.
6. Chen, C., Wu, Y. & Li, R. *J. Crystal Growth*, 1990, **99**, 790–8.
7. Moryc, U. & Ptak, W. S. *J. Molecular Structure*, 1999, **511&512**, 241–9.
8. (a) Vogel, W. *Glass Chemistry*, 2nd Edition, Springer-Verlag, 1992, p. 138. (b) Ehrt, D., Reiss, H. & Vogel, W. *Silikattechnik*, 1977, **28**, 359–364. (In German.)
9. Möncke, D. & Ehrt, D. *Glass Sci. Technol.*, 2002, **75**, 163–73.
10. Möncke, D., Ehrt, D., Eckert, H. & Mertens, V. *Phys. Chem. Glasses*, 2003, **44**, 113–6.
11. Möncke, D., Ehrt, D., Varsamis, C. P., Kamitsos, E. I. & Kalampounias, A. G. *Glass Technol.: Eur. J. Glass Sci. Technol. A*, 2006, **47**, 133–7.
12. Möncke, D., Dussauze, M. & Kamitsos, E. I. *82nd Conf. of the German Glass Technical Society and Glass Trend Seminary*, CD-ROM. 2008. Hameln, Germany. S4-1130.
13. (a) Möncke, D. & Ehrt, D. *Glass Sci. Technol.*, 2002, **75**, 243–53. (b) Möncke, D. & Ehrt, D. *Opt. Mater.*, 2004, **25**, 425–37. (c) Möncke, D. & Ehrt, D. *Phys. Chem. Glasses*, 2007, **48** (5), 317–23.
14. (a) Ehrt, D. *C. R. Chimie*, 2002, **5**, 679–92. (b) Ehrt, D. & Ebeling, P. *Glass Technol.*, 2003, **44**, 46–9. (c) Natura, U. & Ehrt, D. *Glastech. Ber. Glass Sci Technol.*, 1999, **72**, 295–301.
15. Carvalho, I. C. S., Fokine, M., Cordeiro, M. B., Carvalho, H. & Kashyap, R. *Opt. Mater.*, 2008, **12**, 1816–21.
16. Skuja, L. *J. Non-Cryst. Solids*, 1998, **239**, 16–48.
17. Shkrob, I. A., Tadjikov, B. M. & Trifunac, A. D. *J. Non-Cryst. Solids*, 2000, **262**, 6–34 and 35–65.
18. Jellyman, P. E. & Procter, J. P. *J. Soc. Glass Technol.*, 1959, **39**, T173–T192.
19. Tenney, A. S. & Wong, J. *J. Chem. Phys.*, 1972, **56**, 5516–23.
20. Bell, R., J., Carnevale, A., Kurkjian, C. R. & Peterson, G. E. *J. Non-Cryst. Solids*, 1980, **35&36**, 1185–90.
21. (a) Kamitsos, E. I., Karakassides, M. A. & Chryssikos, G. D. *J. Phys. Chem.*, 1987, **91**, 1073–79. (b) Kamitsos, E. I., Patsis, A. P., Karakassides, M. A. & Chryssikos, G. D. *J. Non-Cryst. Solids*, 1990, **126**, 52–67. (c) Varsamis, C. P., Kamitsos, E. I. & Chryssikos, G. D. *Phys. Rev. B*, 1999, **60**, 3885–98.
22. (a) Kamitsos, E. I. *Phys. Rev. B*, 1996, **53**, 14659–62. (b) Kamitsos, E. I., Kapoutsis, J. A., Jain, H. & Hsieh, C. H. *J. Non-Cryst. Solids*, 1994, **171**, 31–54.
23. Mihailova, B., Zotov, N., Marinov, M., Nikolov, J. & Konstantinov, L. *J. Non-Cryst. Solids*, 1994, **168**, 265–74.
24. Chan J. C. C., Bertmer M. & Eckert H. *J. Am. Chem. Soc.*, 1999, **121**, 5238–5248.

Proc. VI Int. Conf. Borate Glasses, Himeji, Japan, 18–22 August 2008   *Phys. Chem. Glasses: Eur. J. Glass Sci. Technol. B*, April 2009, **50** (2), 71–78

Recent studies of polyborate anions

David M. Schubert*

US Borax Inc., Rio Tinto Minerals, 8051 E. Maplewood Avenue, Greenwood Village, Colorado 80111, USA

Carolyn B. Knobler

McCullough Crystallography Laboratory, Department of Chemistry and Biochemistry, University of California, Los Angeles, 90095, USA

Manuscript received 19 September 2008
Revised manuscript received 5 November 2008
Accepted on 6 November 2008

Recent work in our laboratory has resulted in the synthesis and structural characterisation of crystalline compounds containing a number of unusual isolated borate anions associated with non-metal cations. These include a new example of the triborate monoanion, $B_3O_3(OH)_4^-$, a series of compounds containing the pentaborate monoanion, $B_5O_6(OH)_4^-$, and several larger polyborate anions. A new heptaborate, $B_7O_9(OH)_5^{2-}$, is an isomer of another recently reported anion. An unusual octaborate, $B_8O_{10}(OH)_6^{2-}$, was characterised, as well as two examples of a nonaborate anion, $B_9O_{12}(OH)_6^{3-}$. Several new borates containing coordinative covalent B–N bonds were also characterised. These include the organoborate $C_8H_{16}(NH_2)_2B_{10}O_{12}(OH)_8^{2-}$, composed of two pentaborate units covalently linked by a central diamino-octane moiety, and related B–N bond containing zwitterions, $B_5O_6(OH)_4NH_2C_nH_{2n}NH_3$ (n=5, 6).

Introduction

Borate compounds containing boron that is bonded only to oxygen serve important roles in a wide range of industrial applications.[1,2] Use as raw materials in the manufacture of vitreous products, including glasses, ceramic glazes, and enamels, accounts for more than half of the global consumption of borate minerals and refined borate chemicals. Crystalline borates, the subject of this paper, exhibit a complex and diverse structural chemistry, the study of which can provide useful insights into the structures of related vitreous borate systems.

A wide range of polyborate anions can be found in crystalline minerals and synthetic borates. These anions contain BO_3 and/or BO_4 units that can link together by sharing oxygen to form isolated rings and cages or extended 1D chains, 2D sheets, and 3D networks. Most borate compounds are iono-covalent, with the negative charges associated with BO_4 groups in covalently bound borate or polyborate components balanced by the net positive charge of interstitial cations.[3,4]

Polyborate anion structures are the result of formal condensation reactions between B–OH groups. In general, formation of more highly condensed extended chain, sheet, and network borate structures involves more extreme reaction conditions, often requiring prolonged hydrothermal treatments (or, as with many borate minerals, geological conditions). However, notable exceptions exist. For example, the triborate chain containing zinc polytriborate, $ZnB_3O_4(OH)_3$ (known as an article of commerce referred to as $ZnO.3B_2O_3.3.5H_2O$) forms in the reac-

tion of zinc oxide with excess boric acid under mild aqueous conditions.[5]

Although a number of groups have recently described fruitful investigations of crystalline borates formed under more extreme hydrothermal conditions, our studies have primarily focused on the synthesis of borates under mild, non-hydrothermal conditions, more suitable for larger scale industrial manufacturing processes. These conditions typically produce compounds containing less condensed isolated ring or cage borate anions. Our work has also made extensive use of non-metal cations, resulting in the characterisation of several unusual new borate anions.[6]

Most crystalline borate compounds formed in aqueous media, even under prolonged hydrothermal conditions, are not fully condensed and retain multiple B–OH groups. As a rule, these groups always form hydrogen bonds with neighbouring oxygen atoms in the system. For this reason, H-bonding plays an important role in these compounds and may be a key factor in directing the structures of these materials.

The spontaneous formation of several polyborate anions in concentrated aqueous borate solutions is well known.[2,7,8] These anions include the tetrahydroxyborate monoanion, $B(OH)_4^-$ (**1**), triborate dianion, $B_3O_3(OH)_5^{2-}$ (**2**), triborate monoanion, $B_3O_3(OH)_4^-$ (**3**), tetraborate dianion, $B_4O_5(OH)_4^{2-}$ (**4**), and pentaborate monoanion, $B_5O_6(OH)_4^-$ (**5**), shown in Figure 1. Anions **1**, **4**, and **5**, are found in commercial borate products produced on an industrial scale. These anions can be ordered according to decreasing basicity of their associated aqueous solutions as **1**>**2**>**4**>**3**>**5**.

A number of other isolated borate anions are found in solid state mineral and synthetic borates. For exam-

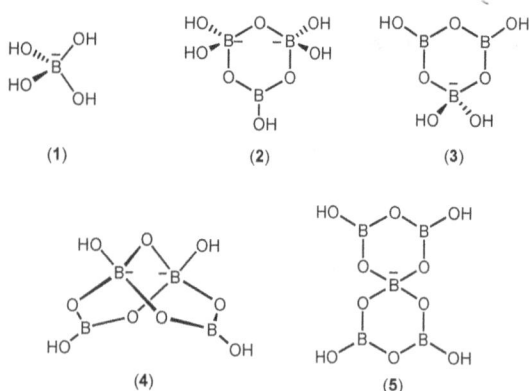

Figure 1. Isolated borate in concentrated aqueous solution

ple, the isolated hexaborate anion, $B_6O_7(OH)_6^{2-}$ **(6)**, is found in several mineral and synthetic borates, including the synthetic compound $Mg[B_6O_7(OH)_6].5H_2O$,[9] and the minerals aksaite ($Mg[B_6O_7(OH)_6].2H_2O$),[10] mcallisterite ($Mg_2[B_6O_7(OH)_6]_2.9H_2O$),[11] and rivadavite ($Na_6Mg[B_6O_7(OH)_6]_4.10H_2O$).[12]

The majority of isolated borate anions reported in the literature contain from one to six boron atoms, and larger isolated borate anions are relatively rare. In recent years, we have prepared and structurally characterised a number of compounds containing isolated borate anions having from three to ten boron atoms. Our work has focused on cation–anion interactions and the role played by cations in controlling borate structure. Here we will review recent work in our laboratory resulting in the characterisation of several new larger isolated polyborate anions, including $B_7O_9(OH)_5^{2-}$ **(7)**, $B_8O_{10}(OH)_7^{2-}$ **(8)**, $B_9O_{12}(OH)_6^{3-}$ **(9)**, and $C_8H_{16}(NH_2)_2B_{10}O_{12}(OH)_8^{2-}$ **(10)**, and related B–N coordinative covalent bond containing zwitterions. We will also discuss recent studies involving the triborate, **3**, and pentaborate, **5**, monoanions. Although these studies involve hydroxy-hydrated borates that crystallise from aqueous media, there are clear relationships between the borate structures observed in these crystalline compounds and those occurring in borate glasses.

Triborates

Although the triborate monoanion **3** is believed to be abundantly present in concentrated aqueous borate solutions in the intermediate pH range, a relatively small number of crystalline compounds are known to contain this anion, and structurally characterised examples are rare. We recently described the structure and some properties of an unusual crystalline alcohol-amine borate containing the triborate monoanion, **3**.[13]

Alcoholamine borate solutions have many commercial uses, especially in industrial fluids and agricultural products. As important components of lubricants and metal working and hydraulic fluids, they serve as buffers, corrosion inhibitors, lubrication

enhancers, and biostats.[1] As a necessary element for plant life, boron is an important agricultural micro-nutrient.[14–16] For this reason, concentrated solutions of borate in alcoholamine-water solvents are used as liquid fertilizer concentrates.[17] In the latter case, alcoholamines primarily serve to solubilise boric acid, which has limited solubility in water. Alcoholamines and glycols are among the most effective solubilisers for borates. Concentrated borate solutions are also used as fire retardant treatments for cellulosic materials and as wood preservatives.[1]

The alcoholamine–boric acid–water system is complex. Due to the basicity of alcoholamines and the weak acidity of boric acid, aqueous mixtures of these reagents contain equilibrium mixtures of both protonated and free base alcoholamine together with the various polyborate anions appropriate for the pH and concentration, where pH is largely a function of basicity of the alcoholamine and the boric acid–alcoholamine molar ratio. Boric acid also forms esters with alcoholamine hydroxyl groups to an increasing extent with decreasing water concentration. These borate esters are labile in the presence of water and exist in rapid equilibrium with the non-esterified alcoholamine and borate species.

Alcoholamine–boric acid–water solutions are often very stable, even when highly supersaturated. We found that certain alcoholamine-boric acid solutions containing a large excess of boric acid may be induced to phase separate, yielding crystalline alcohol ammonium borate salts. In the case where the alcoholamine is 2-amino-2-methylpropanol (AMP) with boric acid in a 1·00:3·33 molar ratio, the crystalline product was found to be the triborate, $[HOCH_2C(CH_3)_2NH_3][B_3O_3(OH)_4]$, formed according to Equation (1)

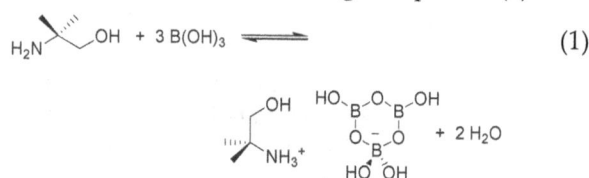

$$\tag{1}$$

The analogous reaction of excess boric acid with 2-aminoethanol, which lacks the two methyl substituents of AMP, yielded a crystalline pentaborate of composition $[HOCH_2CH_2NH_3][B_5O_6(OH)_4].H_2O$ instead of a triborate.[13]

The structure of the triborate monoanion **3** found in $[HOCH_2C(CH_3)_2NH_3][3].H_2O$ is shown in Figure 2. As usual for anion **3**, the B_3O_3 ring is not flat. This anion exhibits a twisted (reverse chair) conformation, with the four-coordinate boron atom and an adjacent ring oxygen atom displaced 0·096 Å and 0·097 Å, respectively, from the plane passing through the six ring atoms, and lying on opposite sides of this plane. The four-coordinate boron atom lies 0·206 Å from the plane passing through the three ring oxygen atoms. Bond angles about this boron atom are in the 108·3–111·2°

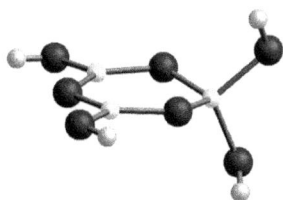

Figure 2. Structure of the triborate monoanion, $B_3O_3(OH)_4^-$ (3) found in $[HOCH_2C(CH_3)_2NH_3][B_3O_3(OH)_4]$

range, and angles about the two three-coordinate boron atoms are in the 116·2–122·6° range.[13]

We are aware of only two other structurally characterised examples of anion **3**, and of no previous examples containing non-metal cations. These examples are found in the mineral ameghinite, $NaB_3O_3(OH)_4$,[18] and the synthetic borate $KB_3O_3(OH)_4$.[19] The structure of a related anion, $B_3O_5(OH)_2^{3-}$, formed by formal deprotonation of **3**, was also reported.[20] The fundamental building block of anion **3**, denoted 3:□2Δ in the notation of Burns,[21] is found in some anhydrous borates, and notably as part of extended networks in so-called enneaborates, including CsB_9O_{14} ($Cs_2O.9B_2O_3$).[22]

The previously published structures of the anion **3** show a range of B_3O_3 ring geometries. The mineral ameghinite exhibits a triborate B_3O_3 ring having an envelope shape with the four-coordinate boron lying 0·315 Å from the plane of the three oxygen atoms and 0·146 Å from the least squares plane of the six ring atoms. The B_3O_3 ring of anion **3** in $KB_3O_3(OH)_4.H_2O$ is also envelope shaped, but is much more nearly planar than the ring in ameghinite, with the four-coordinate boron atom lying only 0·062 Å from the plane of the three ring oxygen atoms and only 0·036 Å from the least squares plane of the six ring atoms. Anion **3** in $[HOCH_2C(CH_3)_2NH_3][3]$ presents an intermediate ring geometry that that is more planar than that found in Na[**3**] (ameghinite) and less planar than in K[**3**].

Many borate compounds exhibit incongruent solubility behaviour, wherein the composition of the solid borate is different from the mother liquid with which it is in equilibrium. Thus, even when a solution has the composition of a particular crystalline borate salt, the borate that crystallises from it may have a different composition. This is usually the case for solutions having the cation-to-boron mole ratio corresponding to the triborate monoanion. Such solutions generally yield salts of either the pentaborate monoanion, **5**, or the tetraborate dianion, $B_4O_5(OH)_4^{2-}$, **4**, instead of triborate **3**. As suggested by Salentine, this may occur because salts of anion **3** are generally more soluble than those of **4** or **5**.[19] A solid state method was used to prepare $KB_3O_3(OH)_4$, allowing a mixture of solid potassium pentaborate and tetraborate to react slowly under high humidity conditions. It was reported that this compound could not be crystallised directly from aqueous solution. This suggests that $KB_3O_3(OH)_4$ may be the thermodynamically favoured product, but it

is more soluble in water than either $K_2[4]$ or $K[5]$, which crystallise first from concentrated solution.[19]

Pentaborates

A large number of compounds containing the pentaborate monoanion, **5**, have been described in the chemical literature. This anion is also the isolated polyborate anion most frequently encountered in non-metal borates. This is attributed in part to the low Lewis acidity of this anion, which is more closely matched with the Lewis acidity of larger organic cations.[23]

A notable feature of all hydroxy-hydrated borate anions is their propensits to self assemble via hydrogen bonding to form a wide range of supramolecular frameworks that host interstitial cations and other guest species. The term tecton is used in crystal engineering in reference to rigid molecular building blocks that can be used in the construction of extended framework structures.[24] The $[B_5O_6(OH)_4]^-$ anion, **5**, shown in Figure 3, is a tecton having a highly regular geometry that assembles with adjacent anions via H-bonding in a variety of extended structure arrangements. Variations in these arrangements arise from the fact that this anion donates four H-bonds to network formation and accepts H-bonds at any of ten oxygen atom positions of three distinct types. Pentaborate frameworks must contain cations and may also host other chemical species that reside in interstitial spaces that take the form of cavities, channels, or layers.

These oxygen-rich pentaborate H-bonded frameworks can further accept H-bonds from host cations and other interstitial species, including water molecules, where the specific framework structure adopted is strongly influenced by the steric and H-bonding requirements of interstitial components. In some cases, the system is non-centrosymmetric. For example, we have structurally characterised the pentaborate compound $[N(CH_2CH_2)_3NH][B_5O_6(OH)_4]$, which crystallises from the aqueous reaction of 1,4-diazabicyclo[2.2.2]octane with excess boric acid. This

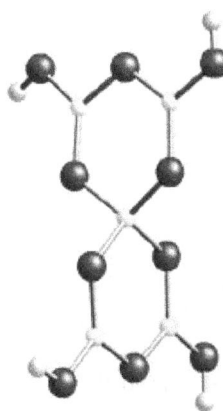

Figure 3. Structure of a typical pentaborate monoanion, $B_5O_6(OH)_4^-$ (5)

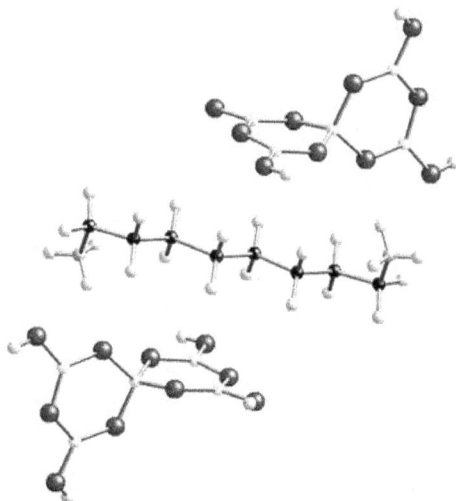

Figure 4. The structure of one formula unit of $[H_3N(CH_2)_8NH_3][B_5O_6(OH)_4]_2$

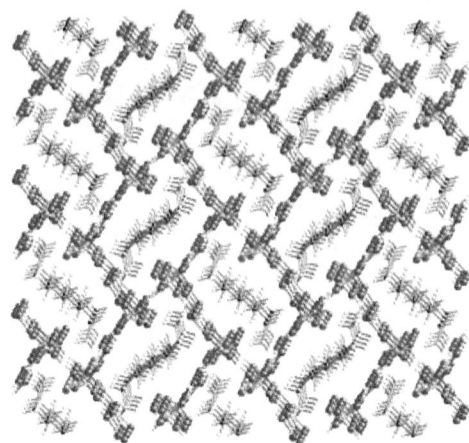

Figure 5. A view of the extended structure of $[H_3N(CH_2)_8NH_3][B_5O_6(OH)_4]_2$, showing $H_3N(CH_2)_8NH_3^{2+}$ cations arranged in channels within the pentaborate H-bonded supramolecular framework

compound contains directional H-bonds and crystallizes in the non-centrosymmetric monoclinic space group Cc (a=10·1970(16), b=14·136(2), c=10·9963(17) Å, β=113·949(3)°; V=1448·6(4) Å^3, Z=4).[25,26]

A series of crystalline nonmetal borate compounds was prepared by the reaction of linear a,w-diaminoalkanes, $H_2N(CH_2)_nNH_2$ (n=5–12), with boric acid in aqueous solution. When boric acid is present in ten-fold or greater excess these reagents react in water to form complex ordered structures. The products were characterised by elemental and thermal analyses, as well as single crystal x-ray diffraction studies.[26] In each case, except for n=7 and n=12, a crystalline bis(pentaborate) product of general formula $[H_3N(CH_2)_nNH_3][B_5O_6(OH)_4]_2$ was obtained in good yield. The structure of one formula unit of the compound obtained when n=8, $[H_3N(CH_2)_8NH_3][B_5O_6(OH)_4]_2$, is shown in Figure 4, as an example of compounds in this series. Individual members of this series may differ in the conformation of the cation and crystallographic details of symmetry, but all are based upon one diammonium dication paired with two pentaborate monoanions. When n=12, the product, $[H_3N(CH_2)_{12}NH_3][B_5O_6(OH)_4]_2.4H_2O$, is similar, but contains interstitial water. The case involving the n=7 diamine is anomalous and is discussed below under the headings of heptaborate and octaborate.[26]

In the series of alkyldiammonium bis(pentaborates) ($[H_3N(CH_2)_nNH_3][5]_2.xH_2O$, n=5, 6, 8, 9, 10, 11, 12), anion 5 forms three different types of supramolecular H-bonded frameworks that host alkyldiammonium dications (and water in the case of n=12). These cations contribute to H-bonding and arrange either in linearly extended or in folded conformations within channels in the frameworks. This study revealed an intriguing interplay between the anionic borate host framework and the cationic guest in which the characteristics of the cation control the borate framework structure, which in turn influences cation conformation.[26]

Figure 5 provides a view of a portion of the extended framework structure of $[H_3N(CH_2)_8NH_3]$ $[B_5O_6(OH)_4]_2$, showing the $H_3N(CH_2)_8NH_3^{2+}$ cations arranged in channels within the pentaborate H-bonded framework. In this case, the eight carbon atoms of the alkyldiammonium cation are arranged in a linear conformation, with the terminal NH_3 groups turned out from the carbon chain. All six N–H hydrogen atoms of this cation are involved in H-bonding with oxygen atoms of the 5 framework, as is the case for all the compounds in this series.

Figure 6 shows a view of a portion of the extended framework structure of $[H_3N(CH_2)_{12}NH_3]$ $[B_5O_6(OH)_4]_2.4H_2O$. In this compound, pentaborate anions assemble into a different type of supramolecular framework that contains much larger channels in which $H_3N(CH_2)_{12}NH_3^{2+}$ cations are arranged in a linear conformation with the terminal NH_3 groups in line with the carbon chain. All six hydrogen atoms of each $H_3N(CH_2)_{12}NH_3^{2+}$ cation are involved in H-bonding. Water molecules also lie within the framework and participate in H-bonding.[26]

Examination of the entire diamine pentaborate series reveals three different pentaborate framework types that serve as hosts for $H_3N(CH_2)_nNH_3^{2+}$ cations. One type of framework is formed when n=5 and 6. When n=7, pentaborate frameworks were not found, as discussed below. When n=8–11, a second type of pentaborate framework is formed. In the case of n=8, the diamine carbon chains adopt a linear conformation. As n increases from 9 to 11, a progressive folding of diamine carbon chains is observed as the framework channels become more crowded. Finally, when n=12, a third type of pentaborate framework, that can accommodate a longer chain length, was observed and the diamine cation is again found in linear conformation.[26]

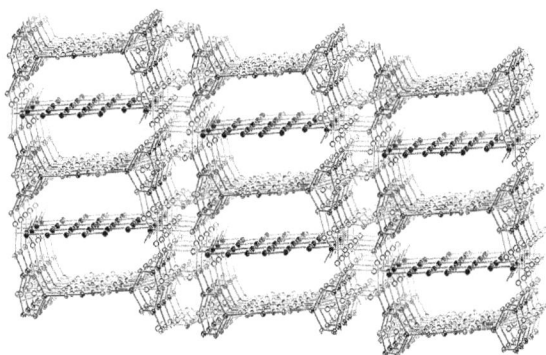

Figure 6. A view of the extended structure of $[H_3N(CH_2)_{12}NH_3]$ $[B_5O_6(OH)_4]_2.4H_2O$, showing the $H_3N(CH_2)_{12}NH_3^{2+}$ cations arranged linearly across channels within the pentaborate H-bonded supramolecular framework

Relevant previously described borates derived from diaminoalkanes include $[H_3N(CH_2)_2NH_3][B_4O_5(OH)_4]$ from 1,2-diaminoethane[27] and two "organically templated" borates, $[H_3N(C_6H_{10})NH_3][B_4O_5(OH)_4]$ and $[H_3N(C_6H_{10})NH_3][B_5O_8(OH)]$, from trans-1,4-diaminocyclohexane.[28] The first two compounds contain isolated tetraborate anions, whereas the third has a covalent borate sheet structure similar to that found in the mineral nasinite, $NaB_8O_{10}(OH)$, also known as Auger's borate.[29] It should also be noted that diamine and long chain polyamine borates have been exploited as templates in the synthesis of porous tectosilicates.[30]

Heptaborate

As mentioned above, the reaction of 1,7-diaminoheptane with excess boric acid in aqueous media is anomalous, producing an octaborate anion containing product when carried out in aqueous media. In another anomaly, reaction of the same reagents under limiting water conditions provides a different crystalline compound, $[H_3N(CH_2)_7NH_3][B_7O_9(OH)_5]$. H_2O, which contains a novel heptaborate anion (7), This reaction is expressed as Equation (2)[31]

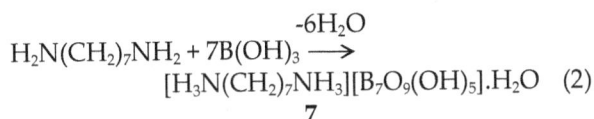

$$H_2N(CH_2)_7NH_2 + 7B(OH)_3 \xrightarrow{-6H_2O}$$
$$[H_3N(CH_2)_7NH_3][B_7O_9(OH)_5].H_2O \quad (2)$$
$$\mathbf{7}$$

The asymmetric unit for $[H_3N(CH_2)_7NH_3][\mathbf{7}].H_2O$ contains one $H_3N(CH_2)_7NH_3^{2+}$ dication paired with one 7 dianion and one water molecule. The heptaborate anion, 7, is shown in Figure 7.

Heptaborate anion 7 contains four fused B_3O_3 rings having one three coordinate oxygen atom and two four coordinate boron atoms in common. Two of these rings share an additional four coordinate boron atom. This anion contains both the tetraborate anion 4, as found in tincal (borax decahydrate)[32] and tincalconite (borax pentahydrate),[33,34] and the hexaborate anion 6.[18–20]

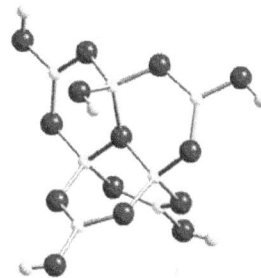

Figure 7. Structure of the heptaborate dianion, $B_7O_9(OH)_5^{2-}$ (7)

Two crystalline metal borate compounds were recently reported, both formed under hydrothermal conditions, that contain topologically the same heptaborate anion, $B_7O_9(OH)_5^{2-}$, which is a structural isomer of heptaborate 7, as shown in Figure 8.[35,36]

A related crystalline diamine heptaborate, $[H_3N(CH_2)_6NH_3][B_7O_{10}(OH)_3]$, was recently reported.[37] This compound consists of extended 2D layers of heptaborate fundamental building blocks (FBBs)[38] composed of hexaborate moieties elaborated with pendant $B(OH)_2$ groups, with interstitial $H_3N(CH_2)_6NH_3^{2+}$ cations. This compound is reported to form in the reaction of 1,6-diaminohexane with 3·8 molar equivalents of boric acid under solvothermal conditions (5 days at 150°C) in ethanol.[37] These conditions would be expected to favour the formation of more condensed borate phases having 2D or 3D extended structures. This reaction is similar to that which produces $[H_3N(CH_2)_7NH_3][\mathbf{7}].H_2O$, but the use of 1,6-diamine and different conditions results in inter-anion rather than intra-anion condensation, making the bridgeable hydroxy groups unavailable to participate in further cyclisation.

Octaborate

As mentioned above, the reaction of 1,7-diaminoheptane with a ten-fold excess of boric acid in aqueous solution produces a product that is anomalous in the series of products obtained under the same conditions for the series of a,w-diaminoalkanes ($[H_2N(CH_2)_nNH_2$, n=5–12). When n=7, and the reaction is carried out in the presence of added water, the crystalline product obtained has the composition $[H_3N(CH_2)_7NH_3][B_8O_{10}(OH)_6].2B(OH)_3$, and contains

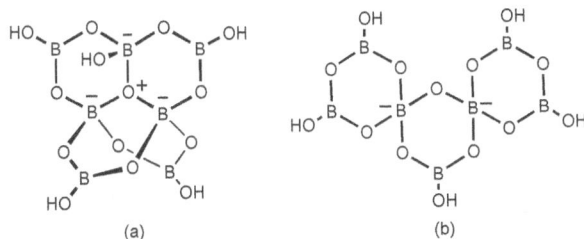

Figure 8. Structural isomers of the $B_7O_9(OH)_5^{2-}$ heptaborate dianion. Figure 8(a) is heptaborate anion 7

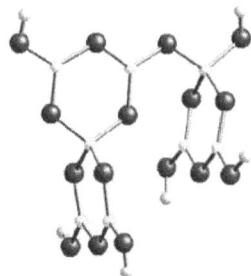

Figure 9. Structure of the octaborate dianion, $B_8O_{10}(OH)_6^{2-}$ (8)

Figure 10. Formal condensation of $[B_5O_6(OH)_4]^-$ (1) and $[B_3O_3(OH)_4]^-$ (3) anions to form the $[B_8O_{10}(OH)_6]^{2-}$ (8)

an unusual octaborate anion, $B_8O_{10}(OH)_6^{2-}$ (8), together with the $H_3N(CH_2)_7NH_3^{2+}$ cation and interstitial boric acid molecules.[26]

This compound contains the $[H_3N(CH_2)_7NH_3]^{2+}$ cation in fully extended linear conformation, As shown in Figure 9, this octaborate anion consists of three linked B_3O_3 rings, featuring both the pentaborate, 5, and triborate, 3, moieties. The octaborate anion, 8, can be thought of as formally produced by the condensation of the more familiar pentaborate (5) and triborate (3) anions, as shown in Figure 10.

The reaction of 1,7-diaminoheptane with a 10-fold or greater excess of boric acid repeatedly provided a means, using different concentrations, reaction times, and temperatures at or below 100°C, to see whether a bis(pentaborate) analogous to the other compounds in the diamine series described above would form. However, the boric acid included octaborate compound was obtained in each case. Occurring at a transition between the two types of pentaborate supramolecular framework structures, the phase containing octaborate 8 is apparently a more stable phase than the alternative bis(pentaborate). Condensation reactions involving borate polyanions are generally promoted by reaction conditions, including elevated concentrations and temperatures. However, formation of the more highly condensed octaborate anion 8, rather than the pentaborate that is found with other diamines under similar conditions, suggests that condensation reactions involving borates can be promoted by the steric or H-bonding demands of interstitial cations in addition to the reaction conditions. Other examples of borate condensation under mild conditions have been observed with certain metals, such as Zn^{2+}.[5]

Nonaborate

We previously described syntheses and single crystal structures of compounds containing the first examples of a novel isolated nonaborate trianion. $B_9O_{12}(OH)_6^{3-}$ (9), shown in Figure 11.[39,40] This anion was found in two compounds, guanidinium nonaborate, $[C(NH_2)_3]_3[B_9O_{12}(OH)_6]$, and imidazolium nonaborate, $[C_3H_5N_2]_3[B_9O_{12}(OH)_6]$.[39,40]

As seen in Figure 11, the $B_9O_{12}(OH)_6^{3-}$ anion, 9, consists of four B_3O_3 rings that share three tetrahedral

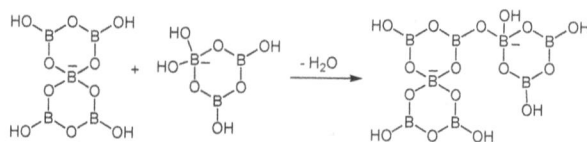

boron centres in a linear arrangement. The remaining six boron centres are trigonal with one hydroxyl group attached to each. The two inner B_3O_3 rings each contain one three-coordinate and two four-coordinate boron atoms and the two outer rings each contain two three-coordinate and one four-coordinate boron atoms. The nonaborate anions in guanidinium and imidazolium nonaborate are crystallographically unique, but are topologically equivalent.

In both $[C(NH_2)_3]_3[9]$, and $[C_3H_5N_2]_3[9]$, extensive $O–H\cdots O$ and $N–H\cdots O$ H-bonding occurs, involving all hydroxyl and all guanidinium or imidazolium hydrogen atoms, which link the anions with adjacent anions and anions with cations. Notably, $[C(NH_2)_3]_3[9]$ exhibits a remarkable 24 independent H-bonds per formula unit, with each anion oxygen atom participating in an average of three H-bonds.[39]

At least three binary guanidinium borates exist. The guanidinium tetraborate, $[C(NH_2)_3]_2[B_4O_5(OH)_4]$ $.2H_2O$, was described in 1921,[41] and a single crystal x-ray structure of this compound was reported in 1985.[42] Several literature references have been made to guanidinium pentaborates, including $[C(NH_2)_3]$ $[B_5O_6(OH)_4].2H_2O$.[39,43] In our study of this system, we found that guanidinium nonaborate crystallises from aqueous solution at temperatures above ~45°C at $B_2O_3:[C(NH_2)_3]_2O$ mole ratios greater than ~2·5, over a range of concentrations. At lower temperatures or lower mole ratios, the tetraborate crystallises. At $B_2O_3:[C(NH_2)_3]_2O$ ratios >~5, either boric acid or guanidinium pentaborate crystallises, depending on concentration. Guanidinium nonaborate can be prepared in aqueous media by stoichiometric, or near stoichiometric, reaction of guanidinium carbonate with boric acid, as shown in Equation (3). Alternatively

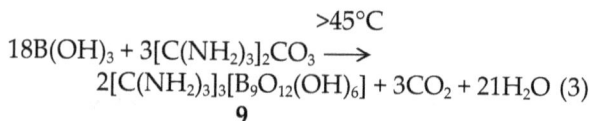

$$\xrightarrow{>45°C}$$
$$18B(OH)_3 + 3[C(NH_2)_3]_2CO_3 \longrightarrow$$
$$2[C(NH_2)_3]_3[B_9O_{12}(OH)_6] + 3CO_2 + 21H_2O \quad (3)$$
$$9$$

this compound can be prepared by the reaction of guanidinium salts with borax and boric acid under appropriate conditions.[39,43] Guanidinium nonaborate exhibits an interesting temperature dependent stability behavior in contact with water. At room temperature, an aqueous slurry of this compound disproportionates to guanidinium tetraborate and boric acid in about one day. At moderately elevated temperatures, e.g. 90°C, aqueous slurries of guanidin-

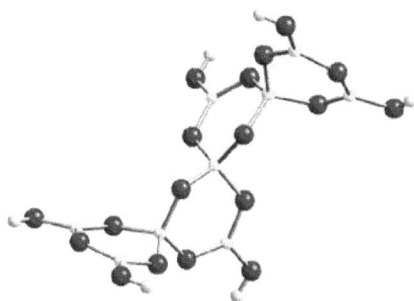

Figure 11. Structure of the nonaborate trianion, $B_9O_{12}(OH)_6^{3-}$ *(9)*

Figure 12. The structure of $C_8H_{16}(NH_2)_2B_{10}O_{12}(OH)_8^{2-}$ *(10)*

of spherical BN particles.[45]

ium nonaborate appear to be stable, and this phase can be recrystallised from hot water.[39]

Prior to our study of the imidazolium borates, little previous work on this system had been reported. We found that imidazolium nonaborate crystallises from an aqueous stoichiometric mixture of boric acid and imidazole between room temperature and 100°C (Equation (4)). In contrast to guanidinium nonaborate, imidazolium nonaborate is stable in contact with water at room temperature[39]

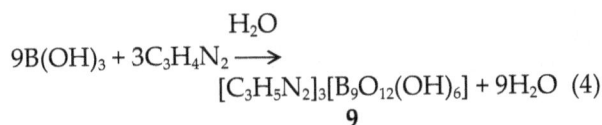

$$9B(OH)_3 + 3C_3H_4N_2 \xrightarrow{\text{H}_2\text{O}}$$
$$[C_3H_5N_2]_3[B_9O_{12}(OH)_6] + 9H_2O \quad (4)$$
$$9$$

This compound can be prepared easily by stirring together a 1:3 mole ratio mixture of imidazole and boric acid, respectively, in water at room temperature (although the compound obtained from hot water is more highly crystalline). The guanidinium and imidazolium borate systems differ from the extensively studied ammonia-borate system, where only ammonium tetraborate and pentaborate occur under comparable conditions.[43] The enthalpy of formation of imidazolium nonaborate has been measured.[44]

We noted that guanidinium nonaborate has a B:N mole ratio of 1, suggesting that it might serve as a precursor to boron nitride (BN). We found that this compound does not melt below 1000°C, but decomposes to a refractory material. GC-MS analysis showed that [C(NH₂)₃]₃[9] begins to lose weight at ~220°C. Initial weight loss is primarily due to elimination of water associated with the condensation of B–OH groups. Upon further heating, some NH_3 and CO_2 is lost, resulting in a ceramic material having B/N>1.

Following our initial experiments, we collaborated with Professor R. T. Paine, of the University of New Mexico, whose group studied the pyrolysis of guanidinium nonaborate, tetraborate, and pentaborate in an ammonia atmosphere, in both bulk samples and aqueous aerosols. These studies showed that guanidinium borates can serve as precursors to BN having relatively low oxygen content and specific particle morphologies. In particular, suitable processing of guanidinium nonaborate resulted in formation

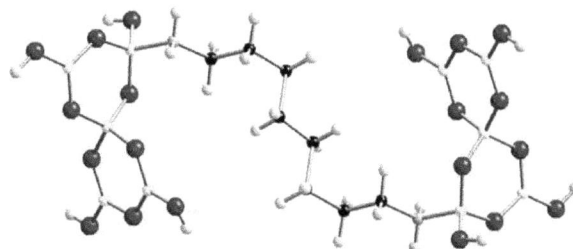

Borates with coordinative covalent B–N bonds

Reactions of diaminoalkanes with excess boric acid may lead to the pentaborate and octaborate salts described above. The specific case involving the reaction of 1,7-diaminoheptane with boric acid under limiting water conditions, resulting in an unusual heptaborate anion, **7**, is also described. However, when analogous reactions were carried out between diaminoalkanes and boric acid under limiting water conditions, a number of crystalline pentaborate-based compounds were obtained that contain coordinative covalent B–N bonds. Examples of these include [H₃N(CH₂)₈NH₃] [B₅O₆(OH)₄NH₂(CH₂)₈NH₂–B₅O₆(OH)₄] (**10**, triclinic, P-1; $a=8·9871(5)$, $b=10·1041(6)$, $c=13·6052(8)$ Å; $\alpha=84·6160(10)$, $\beta=72·3950(10)$, $\gamma=65·3050(10)°$; $V=1069·13(11)$ Å³; $Z=2$), the anionic portion of which is composed of two pentaborate units linked via B–N coordinative covalent bonds by a central diaminooctane moiety, as shown in Figure 12. This dianion is paired with a H₃N(CH₂)₈NH₃²⁺ cation. Each pentaborate unit in this compound contains two four coordinate boron atoms, one of which is bonded to the nitrogen atom of a diamine moiety.[26,46,47]

Also structurally characterised were related B–N coordinative covalent bond containing zwitterions of the general type, B₅O₆(OH)₄NH₂CₙH₂ₙNH₃, where $n=5$, 6 ($n=5$: triclinic, P-1; $a=7·4366(17)$, $b=9·585(2)$, $c=11·116(3)$ Å; $\alpha=71·046(2)$, $\beta=75·347(2)$, $\gamma=77·236(2)°$; $V=716·6(3)$ Å³, $Z=2$). These compounds contain similar pentaborate moieties having two four-coordinate boron atoms, one bonded to a diamine nitrogen atom. However, the diamine moiety in these compounds

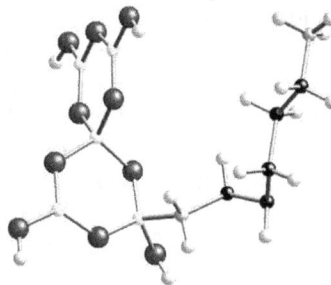

Figure 13. Structure of coordinative covalent B–N bond containing pentaborate-based zwitterion, $B_5O_6(OH)_4$–$H_2N(CH_2)_5NH_3$

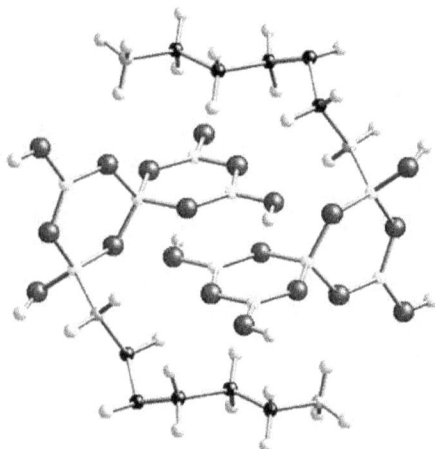

Figure 14. View of a pair of $B_5O_6(OH)_4H_2N(CH_2)_5NH_3$ zwitterions found in the crystal structure, showing the relatively close approach of terminal $-NH_3^+$ groups and anionic 4-coordinate boron centres

contains a terminal NH_3^+ group, the cationic charge of which balances the anionic charge associated with the central four-coordinate boron atom of the pentaborate moiety, giving the overall molecule a net neutral charge. The structure of the zwitterion having $n=5$ is shown in Figure 13.

These zwitterions show a large charge separation between the central anionic boron atom and the terminal ammonium group for each individual zwitterion. Examination of the crystal structures of these compounds shows that the zwitterions pairs up in such a way that the terminal $-NH_3^+$ groups approach closely to the cationic boron centres of adjacent zwitterions, as shown in Figure 14. The resulting zwitterion pairs also form extensive H-bonds with adjacent zwitterion pairs.[46,47]

Conclusions

The formation of compounds containing polyborate anions resulting from reactions of boric acid with Brønsted bases is a general reaction in borate chemistry. The specific polyborate anion formed and the degree of condensation is often a function of temperature, solvation, and the relative concentrations of boric acid and base. In addition, it is apparent that the intrinsic basicity, size, shape, and H-bond donating capabilities of the conjugate acid cation can play important roles. The use of non-metal bases permits the control of these latter parameters. For example, formation of the heptaborate anion 7, instead of other potential polyborate anions or B–N bond-containing products, found with other diamines under similar conditions, apparently results from the specific characteristics of the $H_3N(CH_2)_7NH_3^{2+}$ cation. Study of the structural chemistry of borate anions in crystalline compounds can provide useful insights into other areas of boron chemistry, including the structures of borate-based glasses and ceramics.

References

1. Schubert, D. M. *Group 13 Chemistry III, Industrial Applications*, Eds H. W. Roesky & D. A. Atwood, Structure and Bonding Series, Vol 105, Ch. 1, p 1–40. Spinger-Verlag, Berlin, 2003.
2. Schubert, D. M. *Encyclopedia of Inorganic Chemistry*, 2nd ed, Ed. R. B. King, Wiley, 2005, p 499–524.
3. Hawthorne, F. C. *Am. Miner.*, 1985, **70**, 455–473.
4. Grice, J. D., Burns, P. C. & Hawthorne, F. C. *Can. Miner.*, 1999, **37**, 731–762.
5. Schubert, D. M., Alam, F., Visi, M. Z. & Knobler, C. B. *Chem. Mater.*, 2003, **15**, 866–871.
6. Schubert, D. M., Smith, R. A. & Visi, M. Z. *Glass Technol.*, 2003, **44**, 63–70.
7. Ingri, N. *Sven. Chem. Tidskr.*, 1963, **75**, 199–230.
8. Salentine, C. G. *Inorg. Chem.*, 1983, **22**, 3920–39.
9. Hanic, F., Lindqvist, O., Nyborg J. & Zedler, J. *Coll. Czech Chem. Commun.*, 1971, **36**, 3678–3701.
10. Dal Negro, A., Ungaretti, L. & Sabelli, C. *Am. Miner.*, 1971, **56**, 1553–1566.
11. Dal Negro, A., Sabelli, C. & Ungaretti, L. *Acc. Noz. Lincei*, 1969, **47**, 353–364.
12. Dal Negro, A. & Ungaretti, L. *Naturwissen.* 1973, **60**, 350.
13. Schubert, D. M., Visi, M. Z. & Knobler, C. B. *Inorg. Chem.*, 2008, **47**, 2017–2023.
14. O'Neill, M. A., Eberhard, S., Albersheim, P. & Darvill, A. G. *Science*, 2001, **284**, 847–849.
15. Gutpa, U. C. *Handbook of Plant Nutrition: 117*, Eds A. V. Barker & D. J. Pilbeam, CRC Press, Taylor & Francis Group, 2006, Ch. 8 , p. 241–277.
16. Blevins, D. G. & Lukaszewski, K. M. *Ann. Rev. Plant Physiol. Mol Biol.*, 1998, **49**, 481–500.
17. Ott, L. E. US Patent 4332,609, 1982.
18. Dal Negro, A., Martin Pozaz, J. M. & Ungarretti, L. *Am. Miner.*, 1975, **60**, 897–883.
19. Salentine, C. G. *Inorg. Chem.*, 1987, **26**, 128–132.
20. Corazza, E., Menchetti, S. & Sabelli, C. *Acta Cryst.*, 1975, **B31**, 1993–1997.
21. Burns, P. C., Grice, J. D. & Hawthorne, F. C. *Can. Miner.*, 1995, **33**, 1131–1151.
22. Wright, A. C., Sinclair, R. N., Stone, C. E., Knight, K. S., Polyakova, I.G., Vedishcheva, N. M. & Shakhmatkin, B. A. *Phys. Chem. Glasses*, 2003, **44**, 197–202.
23. Schindler, M. & Hawthorne, F. C. *Can Miner.*, 2001, **39**, 1225–1242.
24. Simard, M., Su, D. & Wuest, J. D. *J. Am. Chem. Soc.*, 1991, **113**, 4696–4698.
25. Schubert, D. M., Visi, M. Z. & Knobler, C. B. *Natl Meeting Am. Chem. Soc.*, San Diego, USA, March 2005.
26. Visi, M. Z., Knobler, C. B., Owen, J. J., Khan, M. I. & Schubert, D. M. *Cryst. Growth Des.*, 2006, **6**, 538–545.
27. Silina, E,, Bel'skii, V. K., Tetere, I. & Ozolins, G. *Latv. PSR Zinat. Akad. Vestis, Kim. Ser.*, 1985, **4**, 399–985.
28. Wang, G.-M., Sun, Y.-Q. & Yang, G.-Y. *Solid State Chem.*, 2004, **177**, 4648–4654.
29. Corazza, E., Menchetti, S. & Sabelli, C. *Acta Cryst. B*, 1975, **31**, 2405–2410.
30. Gunawardane, R. P., Gies, H. & Marker, B. *Zeolite* 1988, **8**, 127–131.
31. Schubert, D. M., Visi, M. Z., Khan, S. & Knobler, C. B. *Inorg. Chem.*, 2008, **47**, 4740–4745.
32. Morimoto, N. *Miner. J. (Jpn.)*, 1956, **2**, 1–18.
33. Levy, H. A. & Lisensky, G. C. *Acta Cryst.*, 1978, **B34**, 3502–3510.
34. Powell, D. R., Gaines, D. F., Zerella, P. J. & Smith, R. A. *Acta Cryst.*, 1991, **C47**, 2279–2282.
35. Liu, Z.-H., Li, L.-Q. & Zhang, W.-J. *Inorg. Chem.*, 2006, **45**, 1430–1432.
36. Liu, Z.-H. & Li, L.-Q. *Cryst. Growth Des.*, 2006, **6**, 1247–1249.
37. Yang, S., Li, G., Tian, S., Liao, F., Xiong, M. & Lin, J. *J. Solid State Chem.*, 2007, **180**, 2225–2232.
38. Burns, P. C. *Can. Miner.*, 1995, **2**, 59–87.
39. Schubert, D. M., Visi, M. Z. & Knobler, C. B. *Inorg. Chem.*, 2000, **39**, 2250–2251.
40. Schubert, D. M. 2005, US Pat. 6,919,036, and related international patents.
41. Rosenheim, A. & Leyser, F. *Z. Anorg. U. Allg. Chem.*, 1921, **119**, 1–38.
42. Weakley, T. J. R. *Acta. Cryst.*, 1985, **C41**, 377–379.
43. Bowden, G. H. in *Supplement to Mellor's Comprehensive Treatise on Inorganic and Theoretical Chemistry*, Vol. V, Boron, Part A: Boron–Oxygen Compounds, Longman Group Ltd., London, 1980, p 685–693.
44. Liu, Z.-H. & Zhang, W. *Thermochim. Acta*, 2005, **436**, 156–158.
45. Wood, G. L., Janik, J. F., Visi, M. Z., Schubert, D. M. & Paine, R. T. *Chem. Mater.*, 2005, **17**, 1855–1859.
46. Schubert, D. M., Visi, M. Z., Knobler, C. B., Khan, S. I. & Hardcastle, K. I. *234th Natl. Meeting Am. Chem. Soc.*, Boston, MA, Aug 19, 2007, Abstr. INOR 246.
47. Schubert, D. M., Visi, M. Z. & Knobler, C. B. *Rocky Mountain Regional ACS Meeting*, Aug 29, 2007, Abstr. 35.

Proc. VI Int. Conf. Borate Glasses, Himeji, Japan, 18–22 August 2008   *Phys. Chem. Glasses: Eur. J. Glass Sci. Technol. B, June 2009, **50** (3), 212–218*

Boromolybdate glasses containing rare earth oxides

Yanko Dimitriev

University of Chemical Technology and Metallurgy, "Kl. Ohridski" bl. 8, 1756 Sofia, Bulgaria

Reni Iordanova,* Lyubomir Aleksandrov & Krassimir L. Kostov

Institute of General and Inorganic Chemistry, Bulgarian Academy of Sciences, "Acad. G. Bonchev" bl. 11, 1113 Sofia, Bulgaria

Manuscript received 18 August 2008
Revised version received 12 December 2008
Accepted 30 January 2009

Glasses in the MoO_3–La_2O_3–B_2O_3 and MoO_3–Nd_2O_3–B_2O_3 systems were obtained between 20 and 40 mol% Ln_2O_3 (Ln=La, Nd). A liquid phase separation region was observed near the MoO_3–B_2O_3 side for compositions containing below 20 mol% Ln_2O_3. New original glasses containing between 45 and 70 mol% ZnO were prepared in the MoO_3–ZnO–B_2O_3 system. The amorphous phases were characterised by x-ray diffraction (XRD), differential thermal analysis (DTA), UV-VIS, infrared spectroscopy (IR), x-ray photoelectron spectroscopy (XPS) and scanning electron microscopy (SEM). According to the DTA data, the thermal stability drastically decreased in glasses with a high MoO_3 content. Most of the glasses were transparent in the visible region. Structural models of the glass networks are suggested on the basis of IR and XPS spectroscopic studies. It was established that BO_3 (1380 cm^{-1}) and BO_4 (1100–950 cm^{-1}) units and isolated MoO_4 (870–840 cm^{-1}) groups build up the borate glass network, while MoO_6 units (band at 880 cm^{-1}) form the molybdate glass network for compositions with a high MoO_3 content (80–90 mol%). Different types of microheterogeneities in the range of stable liquid phase separation were determined. The reason for the immiscibility was explained by the low tendency to generate mixed Mo–O–B bonds at the expense of B–O–B and Mo–O–Mo bridges.

Introduction

Boromolybdate glasses have been a subject of investigation from different points of view. Most of the glass formation regions in the complex boromolybdate systems B_2O_3–MoO_3–Me_2O (Me=Li, Na, K), B_2O_3–MoO_3–MeO (Me=Ca, Sr, Ba, Co, Mn) were summarised in the book of Mazurin *et al*[1] and in some other papers published more recently.[2–9] For most of these systems, the glass compositions had below 30 mol% MoO_3[2–4] and liquid phase separation was detected.[4,5] The presence of Mo^{5+} was proven by EPR.[2,3,5] Its content decreases with increasing total molybdenum content. New fast ion conducting silver boromolybdate glasses[6] and dielectric materials in the system PbO–B_2O_3–MoO_3[7] were obtained. The group of Komatsu developed a technique for laser induced crystallisation in Ln_2O_3–MoO_3–B_2O_3 (Ln=lanthanide) glasses and formation of a homogeneous crystal line with $Ln_2(MoO_4)_3$ ferroelectrics.[8,9] On the other hand, rare earth borate glasses are very promising due to their good mechanical strength, laser effect and nonlinear optical properties.[10] The main difficulty with the preparation of such glasses is the stable liquid phase separation in the binary Ln_2O_3–B_2O_3 systems.[11,12] The other problem is the increasing melting temperature of compositions containing above 30 mol% rare earth oxides and the high crystallisation tendency. Some glass for-

mation improvement was achieved by introducing two Ln_2O_3 oxides into the borate system.[13] In our previous studies we examined the tendency to glass formation and immiscibility in the three component boromolybdate systems containing transition metal oxides[14,15] and rare earth oxides.[16–18] Complex microheterogeneous structures were detected due to metastable immiscibility in the MoO_3–B_2O_3 system[19] and stable liquid phase separation in the binary system B_2O_3–Ln_2O_3.[11,12] It was found that MoO_3 could be a suitable component for decreasing the melting temperature and modifying the properties. From a structural point of view, it is a challenge to combine one traditional glass former (B_2O_3) with a conditional one (MoO_3) because they are characterised by different short range order and different connectivity between the polyhedra in the amorphous structure.

The objects of the present study are glasses from these systems: MoO_3–La_2O_3–B_2O_3, MoO_3–Nd_2O_3–B_2O_3 and MoO_3–ZnO–B_2O_3. The aim is to analyse the structural transformation depending on the modifier content (Ln_2O_3) and to clarify the possibility to generate new amorphous structures with ZnO.

2. Experimental

All batches (10 g) were prepared using reagent grade MoO_3 (Merck, p.a), La_2O_3 (Fluka, puriss), Nd_2O_3 (Fluka, puriss), ZnO (Fluka, puriss) and H_3BO_3 (Reachim, chem. pure) as starting materials. The batches were

* Corresponding author. Email reni@svr.igic.bas.bg

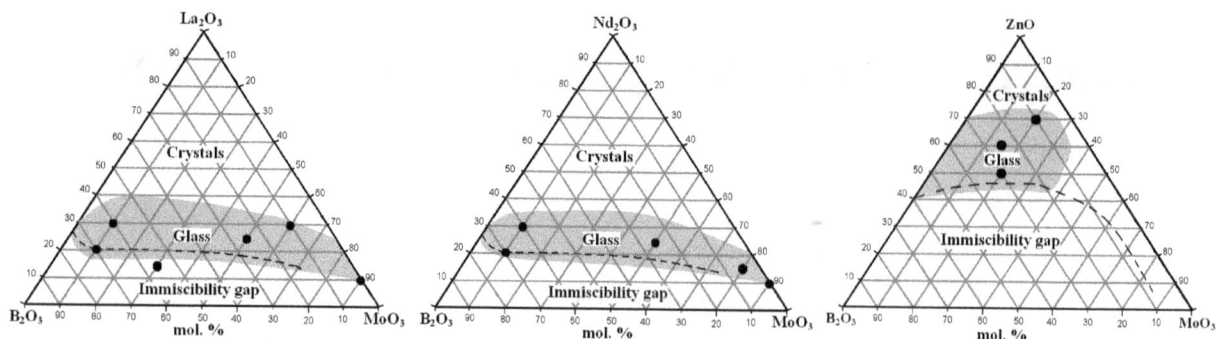

Figure 1. Glass formation region and immiscibility gap in the systems MoO_3–La_2O_3–B_2O_3, MoO_3–Nd_2O_3–B_2O_3 and MoO_3–ZnO–B_2O_3. ● - compositions for structural studies

melted for 20 min in air atmosphere in alumina crucibles for MoO_3–Ln_2O_3–B_2O_3 and in a platinum crucible for MoO_3–ZnO–B_2O_3. The melting temperature was limited to 1300°C in order to decrease the volatility of the components. The glasses rich in B_2O_3 were obtained by press quenching between two copper plates (cooling rate ~10^2 K/s), while for the compositions rich in MoO_3, glasses were only obtained at higher cooling rates (10^4–10^5 K/s) using a roller quenching technique. The liquid phase separation was observed visually by the appearance of two layers (lower and upper) in the crucibles after free cooling of the melts outside the furnace. The boundaries of vitrified compositions were outlined on the base on visual observations and x-ray phase analysis (Bruker D8 Advance diffractometer, Cu K_α radiation) changing the nominal composition through 5 mol% steps. The thermal stability of the selected glasses was determined by differential thermal analysis (DTA Stanton Redcroft) at a heating rate of 10 K/min. The optical transmission spectra in the range 300–900 nm were obtained using a UV-VIS spectrophotometer (Cary 100 Scan, Varian). The IR spectra were measured using the KBr pellet technique on a Nicolet-320 FTIR spectrometer with a resolution of ±1 cm^{-1}, by collecting 64 scans in the range 1500–400 cm^{-1}. X-ray photoelectron spectroscopy (XPS) was carried out on an ESCALAB Mk II (VG Scientific) electron spectrometer. The spectra were recorded with an Al K_α excitation source with a photon energy of 1486·6 eV. Energy correction of the XPS spectra was performed to account for the charging of the samples due to the photoelectron emission. This correction was done by comparing the C 1s peak position to 285 eV. Scanning electron microscopy (JEOL, JEM-200, CX) and electron microprobe analysis EMRA (JEOL, Superprobe 733) were performed on polished samples.

3. Results

Figure 1 presents the glass formation regions in the MoO_3–La_2O_3–B_2O_3 and MoO_3–Nd_2O_3–B_2O_3 systems the study of which started recently.[16–18] The glasses were obtained between 20 and 40 mol% Ln_2O_3 (Ln=La,

Nd). A liquid phase separation region was observed near the MoO_3–B_2O_3 side for compositions containing below 20 mol% Ln_2O_3 (Ln=La, Nd). For the first time data on glass formation in the MoO_3–ZnO–B_2O_3 system are presented. The glass formation region is situated between 45 and 70 mol% ZnO. A liquid phase separation region was registered in many compositions containing below 45 mol% ZnO. For the three investigated systems it is difficult to determine exactly the boundaries of immiscibility due to the overlap with the region of glass formation. Depending on the cooling rate, different degrees of microheterogeneity or transparent glass may be obtained from one composition located near the boundary. Selected compositions for structural studies are marked on the phase diagrams (Figure 1). More complex glass compositions containing 1, 10 and 25 mol% Nd_2O_3 were also investigated. DTA curves of the representa-

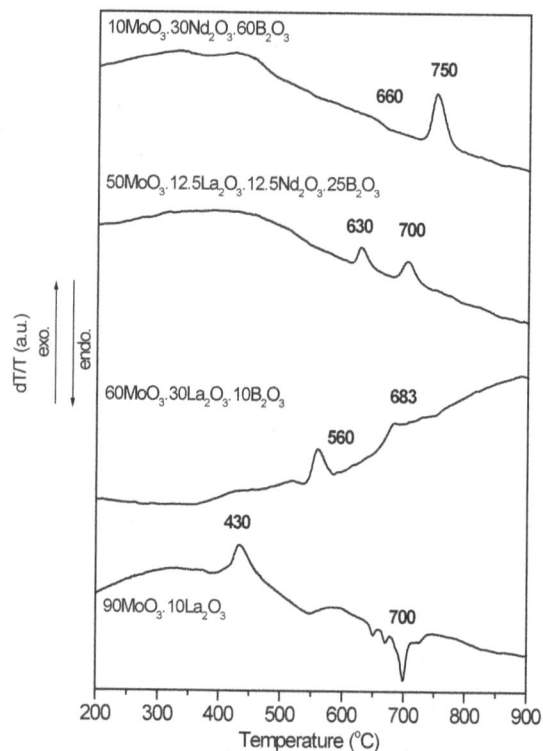

Figure 2. DTA curves of representative glasses from the MoO_3–Ln_2O_3–B_2O_3 system

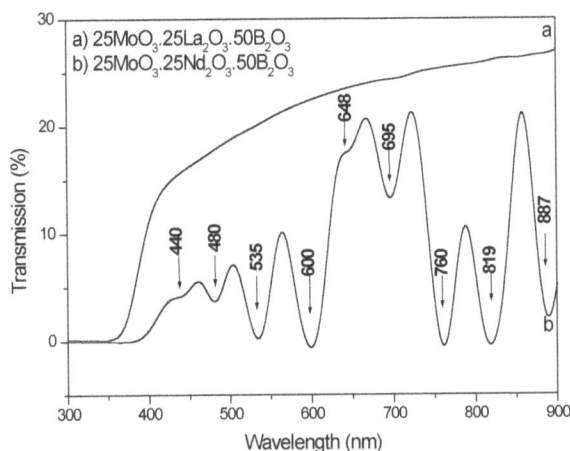

Figure 3. *UV-VIS spectra of glasses from the MoO_3–$La_2O_3(Nd_2O_3)$–B_2O_3 system*

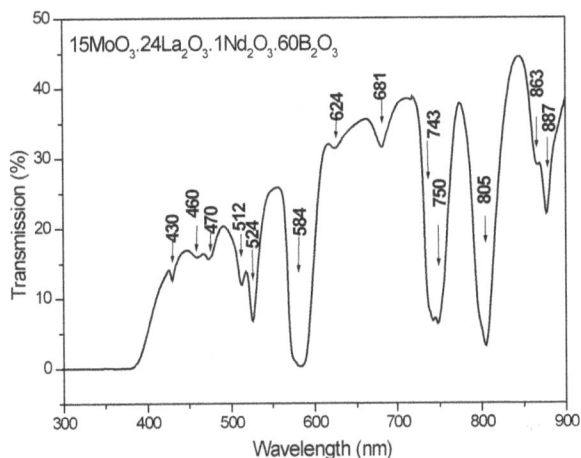

Figure 4. *UV-VIS spectra of glasses from the MoO_3–La_2O_3–Nd_2O_3–B_2O_3 system*

tive glasses are shown in Figure 2. The glass transition temperature, T_g, of $10MoO_3.30Nd_2O_3.60B_2O_3$ glass is $660\pm5°C$, which is close to the value obtained for $Nd_2O_3.3B_2O_3$ glass.[13] The exothermic effect due to crystallisation appears at $750\pm5°C$. The increase in MoO_3 content, at the expense of B_2O_3, with constant Ln_2O_3 ($60MoO_3.30La_2O_3.10B_2O_3$) decreases the thermal stability of the glasses, and the crystallisation temperature, T_x, shifts to $560\pm5°C$. Further increase in MoO_3 content results in a low melting glass. For example, the endothermic effect at $700\pm5°C$ for $90MoO_3.10La_2O_3$ glass is due to the appearance of a liquid phase, and the crystallisation temperature is decreased to $430\pm5°C$. As has been established for most of the molybdate glasses, T_g and T_x are in the same temperature region.[20,21]

The optical transmission spectra of glasses with and without Nd_2O_3 are shown in Figures 3–5. The absorption band edge is shifted towards longer wavelength (350–400 nm) as compared to lanthanum borate glasses[10,13,22] due to the presence of MoO_3. Sharp absorption peaks, typical of Nd^{3+} ions and due to 4f electron transitions, are observed. The obtained spectra are similar to the spectra of other borate glasses containing neodymium.[22–28] The intensity of the transition at 584 nm is higher in all glasses as compared to the other absorption transitions. Stark splitting of 4f electron levels is evident. It was observed in glasses containing 1 and 10 mol% Nd_2O_3, but for glasses containing 25 mol% Nd_2O_3 the bands were not split. More precise spectrum analyses are in progress and will be published separately.

The IR spectra of representative glasses are shown in Figures 6 and 7. The spectra are generally divided into three absorption regions: between 1400 and 1100 cm^{-1}, between 1100 and 950 cm^{-1} and between 950 and 700 cm^{-1}. X-ray photoelectron spectra were obtained in order to reveal the formation of different bonds in amorphous samples depending on the compositions. The O 1s spectra of the glasses were fitted to several peaks, which is an indication of the presence of dif-

ferent types of oxide ions in the glass network, with peak energies as follows: 532·8–533·0 eV, 531·6–531·8 eV, 531·0 eV and 529·8–530·1 eV (Figure 8).

The samples situated outside the glass formation region exhibit a complex heterogeneous structure. For example, a sample with the nominal composition $30MoO_3.15La_2O_3.55B_2O_3$, slowly cooled after melting, was separated into two macrophases as a result of stable liquid phase separation. The upper layer (Figure 9(a)) was milky-greenish, and the lower layer (Figure 9(b)) was dark violet or black. By SEM observation, it was shown additionally that in each separate macro volume a complex microstructure had developed as a result of the metastable phase separation. The microstructure of the upper layer (Figure 9(a)) was an amorphous borate matrix containing small light droplets rich in MoO_3 and La_2O_3. The upper layer is a borate glass with average composition $96.1B_2O_3.3.8MoO_3.0.1La_2O_3$ (in mol%) according to microprobe analysis. The lower macrolayer (Figure 9(b)) was also amorphous, with a composition of $38·5B_2O_3.55·2MoO_3.6·3La_2O_3$ (in mol%). The sample with higher B_2O_3 content (nominal composition $10MoO_3.20Nd_2O_3.70B_2O_3$), (Figure 10) was microheterogeneous and most of the volume was

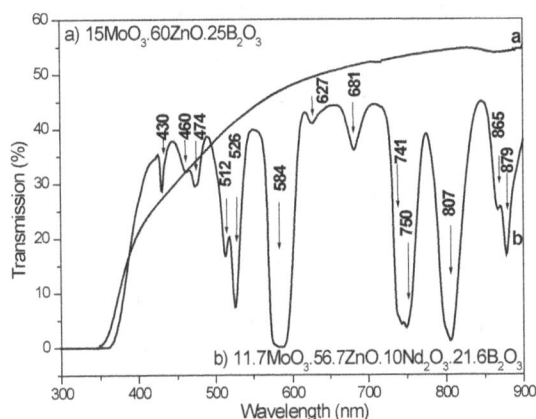

Figure 5. *UV-VIS spectra of glasses from the MoO_3–ZnO–Nd_2O_3–B_2O_3 system*

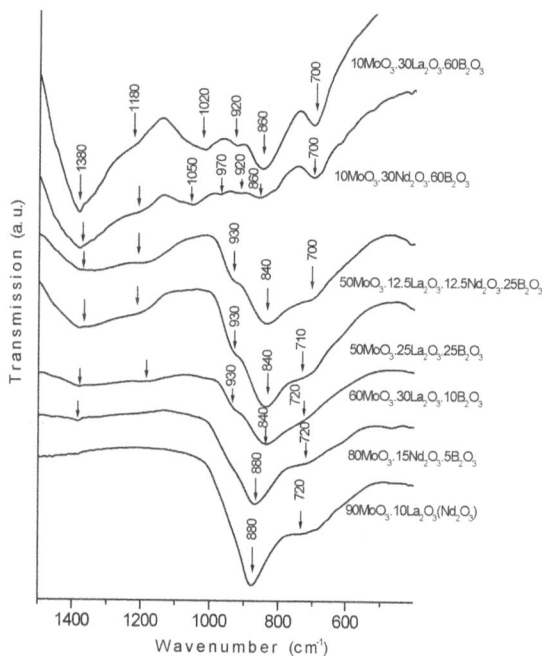

Figure 6. IR spectra of glasses from the MoO_3–Ln_2O_3–B_2O_3 system

crystalline. According to XRD data and microprobe analysis, two phases with compositions close to the compounds $NdMoBO_6$ and NdB_3O_6 were identified.

4. Discussion

The discussion of the glass structure is made on the basis of the IR and XPS spectra obtained in this work. Our previous assignment of the spectral data on molybdate glasses,[20,29] the infrared analysis of borate glasses performed by the team of Kamitsos,[30] as well as the data on the crystal structure of LnB_3O_6,[31,32] $LnMoBO_6$,[33,34] $ZnMoO_4$,[35] and $4ZnO.3B_2O_3$[36] were taken into account. We assumed that the network of the glasses in the MoO_3–La_2O_3–B_2O_3 system situated in the metaborate composition range is built up by BO_3 triangles associated with BO_4 tetrahedra. Evidence for this statement is the broad infrared band (Figure 6) centred at 1380 cm^{-1} (asymmetric stretch of BO_3 triangles), the band at 700 cm^{-1} (deformation mode of borate network) and the bands in the region 1100–950 cm^{-1} (stretching vibration of BO_4 tetrahedra). The structural model indicates that Ln_2O_3 acts as a modifier producing BO_4 tetrahedra in a narrow concentration range which is in agreement with the data on other borate glasses.[11,12,37–39] The influence of oxides on the $BO_3 \Leftrightarrow BO_4$ conversion is a very old problem that has been described intensely.[40,41] Recently, this problem arose again with borate glasses containing heavy metal oxides (PbO, Bi_2O_3) and rare earth oxides.[42–45] Small amounts of PbO (modifier) in the borate matrix convert BO_3 to BO_4 units. With high PbO concentrations, a back conversion occurs. The local order changes around the boron atoms in the presence of Bi_2O_3 are more complicated, and BO_4

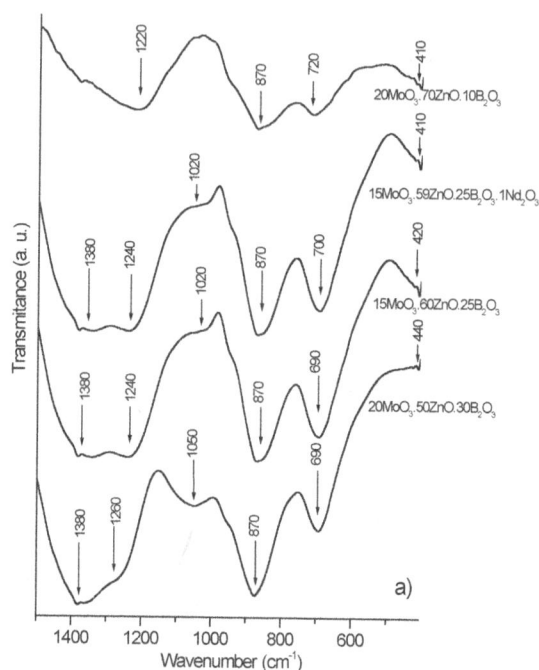

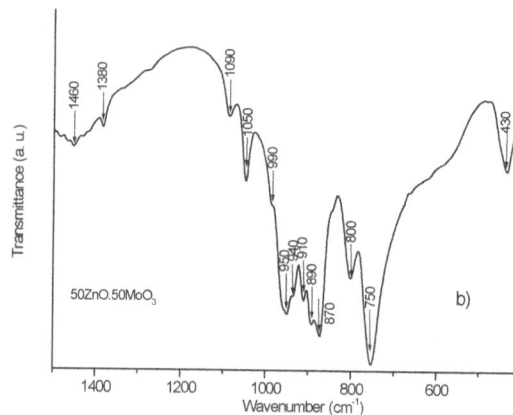

Figure 7 (a) IR spectra of glasses from the MoO_3–ZnO–B_2O_3 system; (b) IR spectra of crystalline $ZnMoO_4$

remains over a wider concentration range. In our case the replacement of B_2O_3 by MoO_3 (50 and 60 mol%) leads to the disappearance of the bands between 1100 and 950 cm^{-1}. This is an indication that BO_4 to BO_3 conversion takes place, which is accompanied by gradual depolymerisation of the borate network. In a wide concentration range (10–60 mol% MoO_3) an absorption band at 860–840 cm^{-1} with shoulders at 930–920 and 720–710 cm^{-1} shows the presence of distorted, isolated MoO_4 tetrahedra.[20,29] We compared the spectral results obtained with those on crystalline $LaMoBO_6$ ($50MoO_3.25B_2O_3.25La_2O_3$) whose composition is located in the glass formation area. Its crystal structure consists of chains of BO_3 triangles, where isolated MoO_4 units are bonded with BO_3 through lanthanum and there are no Mo–O–B bonds in the structure.[34] The similarity between the infrared spectra of crystalline and glassy $LaMoBO_6$ is a reason to suggest that in the amorphous network the Mo–O–B linkage is not favoured. The distorted MoO_4 units enter the metaborate glass network and form

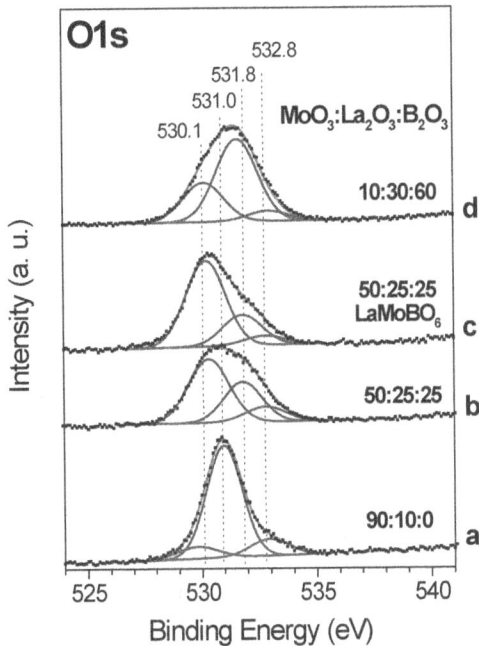

Figure 8. O1s spectra of different MoO_3:La_2O_3:B_2O_3 glass compositions (a, b, d) and crystalline $LaMoBO_6$ (c). The deconvolutions are based on the contributions of different metal–oxygen bonds

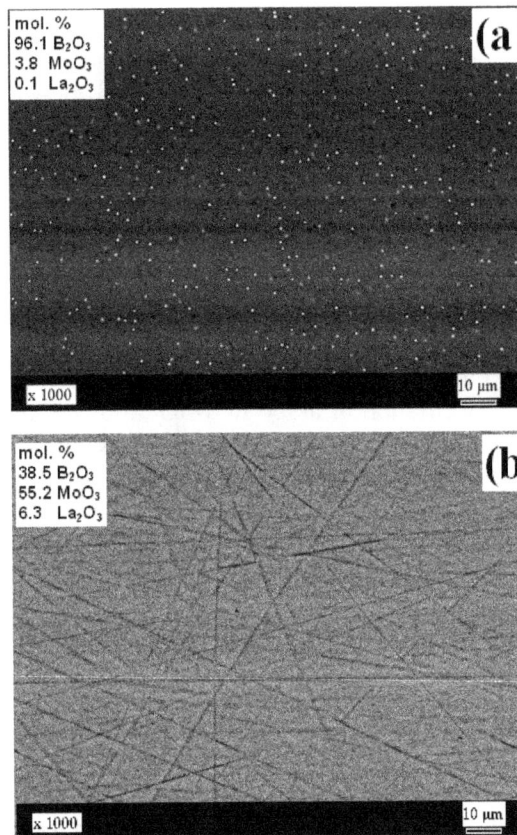

Figure 9. SEM micrograph of the slowly cooled sample with nominal composition $30MoO_3.15La_2O_3.55B_2O_3$ (a) upper layer; (b) lower layer

Mo–O–La linkages only. That is why lanthanum plays an important role in the formation of a homogeneous boromolybdate network. With further increase in MoO_3 content, a network transformation occurs due to MoO_4 to MoO_6 transition. In the spectra of glasses containing above 60 mol% MoO_3, the strong band at 880 cm^{-1} is dominant, which may be assigned to the vibration of Mo–O–Mo bridging bonds between corner shared MoO_6 octahedra.[29,46]

The replacement of La_2O_3 by ZnO contributed to the formation of a new type of amorphous network in which there is no typical modifier (Figure 7). In the spectra of the selected glasses with 50–60 mol% ZnO content, the envelope at 1380–1240 cm^{-1} along with the band at 1050–1020 cm^{-1} indicate the presence of both BO_3 and BO_4 units in the amorphous network. These data are similar with the results obtained by Efimov et al[47] for melt quenched zinc borate glasses and our previous IR spectra for a $50B_2O_3.50ZnO$ "gel glass".[48] In the corresponding crystallised product only BO_4 units were detected, as in the structure of the compound $4ZnO.3B_2O_3$, and there are no bands above 1100 cm^{-1}.[36] The increase in the intensity of the band near 1220 cm^{-1} at the expense of the band at 1380 cm^{-1} and the absence of the band at 1050–1020 cm^{-1} ($20MoO_3.70ZnO.10B_2O_3$) are a result of a predominance of isolated BO_3 in the amorphous network. The other part of the spectra below 1000 cm^{-1} is typical of a molybdate network containing distorted MoO_4 units.[20,29,46] Comparison with the spectrum of crystalline $ZnMoO_4$ (Figure 7(b)) which is composed of MoO_4 units[35] gives a reason to claim that the bands centred at 870 cm^{-1} and

700 cm^{-1} correspond to the vibrations of distorted MoO_4 units. In the spectrum of crystalline $ZnMoO_4$ the band at 440 cm^{-1} should be assigned to the vibration of ZnO_6 (ZnO_5) units present in this crystal structure.[35] In the glass spectra the absorption in this region is not well resolved. It can be connected with transformation of the oxygen environment of Zn atoms in the glass network.

Additional information on the transformation of the glass structure was obtained by XPS analysis. The XPS data on MoO_3–La_2O_3–B_2O_3 glasses (Figure 8) are compatible with the IR analysis. The high intensity peak of the O 1s spectrum at 531·0 eV for the $90MoO_3.10La_2O_3$ glass can be attributed to Mo–O–Mo bridging bonds between MoO_6 octahedra.[49] Having in mind the $LaMoBO_6$[34] crystal structure, the low binding energy O 1s component (at about 530·0 eV) in the XPS spectrum of this phase can be assigned mainly to the oxygen environment of tetra-coordinated Mo atoms participating in Mo–O–La bonds (Figure 8(c)). Therefore, this low binding energy component (529·8–530·1 eV) present in all glass spectra was attributed to Mo–O–La and B–O–La bonds (nonbridging oxygen, NBO). A similar analysis has been made for other molybdate glasses: MoO_3–CuO, MoO_3–CuO–Bi_2O_3, MoO_3–CuO–PbO.[49–51] The O 1s component of La–O–La bonds is in the same spectral region.[52,53] The assignment made above is in accordance with the XPS data on binary glasses in which the addition of a modifier oxide to network formers SiO_2, P_2O_5 or B_2O_3 breaks up the M–O–M (M=Si, P, B) network, converting bridging oxygen (BO) to NBO.[54–59] In the glass compositions with a high B_2O_3 content, for example $10MoO_3.30La_2O_3.60B_2O_3$, one should expect an increase in the fraction of BO (B–O–B), despite some formation of NBO (Mo–O–La, B–O–La). The dominant higher binding energy component at 531·8 eV of this glass is a result of the formation of B–O–B bridging bonds.[15] This conclusion is in agreement with the IR spectra (Figure 6), showing evidence for the formation of associated BO_3 and BO_4 units through a B–O–B linkage, as in the LnB_3O_6 metaborate crystal structure. The low intensity component at 532·8–533·0 eV is most probably attributable to adsorbed OH groups.[14]

The above results raise some basic questions concerning glass formation in non-traditional molybdate glasses and their comparison with classical glass forming systems. It is well known that pure orthorhombic MoO_3 is impossible to vitrify. The question is why the addition of a second component stabilises glass formation. We assume that the creation of nonbridging bonds in molybdate systems is crucial for glass formation. Actually for the investigated glasses the weak mixed bonds $Mo-_{strong}O_{weak}$–Ln and $B-_{strong}O_{weak}$–Ln (Ln=La, Nd) are nonbridging bonds. By means of XANES investigations, it has been determined that Ln–O distances in the binary molybdate glasses are too long, above 2·50 Å.[17] Our hypothesis for the molybdate glasses is that a small amount of modifier (Ln_2O_3) added to MoO_3 hinders the formation of edge shared MoO_6 octahedra, typical of the layered crystal structure of orthorhombic MoO_3, confirmed by the absence of a band at 590 cm^{-1}.[49] Thus, the generation of corner shared MoO_6 (Mo–O–Mo bridging bonds) and some amount of nonbridging Mo–O–Ln linkages facilitate the disorder in the system and the easier vitrification of the supercooled melt. This process explains the existence of the upper boundary of glass composition. On the other hand, the accumulation of a critical number of MoO_4 tetrahedra participating in weak Mo–O–Ln bonds in the network is the reason for the deterioration of the glass formation tendency in the corresponding binary MoO_3–Ln_2O_3 compositions (lower boundary). The consequence is an increase in the number of small mobile units in the melt (isolated MoO_4 tetrahedra). When the introduction of a second component does not stimulate formation of mixed bonds, which has been established in the systems B_2O_3–MoO_3 and B_2O_3–V_2O_5, liquid phase separation is observed.[14,15]

According to microprobe analysis of three component samples with immiscibility, MoO_3 and Ln_2O_3 had a stronger trend to be in the same network, while B_2O_3 dissolved too small an amount of these oxides. This

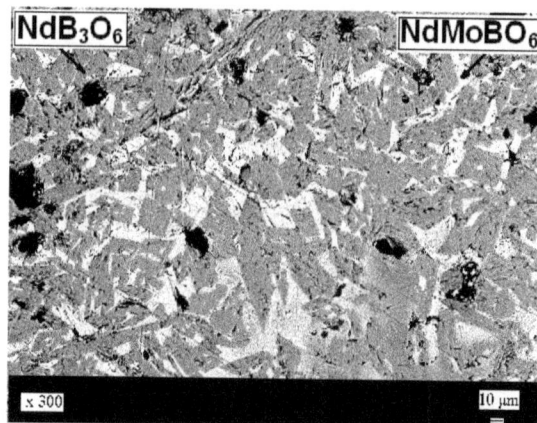

Figure 10. SEM micrograph of the slowly cooled sample with nominal composition $10MoO_3.20Nd_2O_3.70B_2O_3$, in which microheterogeneous structure had developed. Two crystalline phase were detected: NdB_3O_6 (grey area) and $NdMoBO_6$ (white area)

fact confirmed Zarzycki's statement[60] that it is not definitively established if mixed chains are formed, or if there are microdomains when two network formers were mixed in the melt. That is one of the reasons to verify experimentally which tendency is dominant. In the ternary MoO_3–La_2O_3–B_2O_3 system, according our hypothesis, LaO_n polyhedra have a crucial role for the formation of a homogeneous amorphous network because they connect incompatible borate and molybdate units.

In the present data as well in other studies,[61,62] glasses with high ZnO content up to 80 mol% were obtained, showing that the replacement of Ln_2O_3 by ZnO has changed the vitrification conditions. That is why it is not correct to regard ZnO as a modifier. According to the sub-solidus phase relations in the MoO_3–ZnO–B_2O_3 system[63] our glass compositions were situated in several 3-phase regions connected mainly with $4ZnO.3B_2O_3$, $5ZnO.2B_2O_3$ and $ZnMoO_4$ compounds. In these crystal phases the content of classical network former, B_2O_3, and the conditional one, MoO_3, were below 40 mol%. Obviously ZnO contributed to the development of the amorphous network which is built mainly of ZnO_n polyhedra. More investigations are needed to elucidate the coordination of Zn atoms by oxygen in a complex amorphous network.

Conclusions

The glass network in the three component metaborate compositions containing B_2O_3, MoO_3 and Ln_2O_3 is build up by chains consisting of BO_3 and BO_4 units. Molybdenum participates as isolated MoO_4 groups linked by lanthanum polyhedra (nonbridging bonds). The increase in MoO_3 content leads to a $BO_4 \rightarrow BO_3$ transformation, and BO_3 and MoO_4 units remain as building polyhedra in the glass network containing up to 60 mol% MoO_3. MoO_6 units form the network for glasses with the highest MoO_3 content (80–90 mol%) through Mo–O–Mo bridging bonds. Glasses in the MoO_3–ZnO–B_2O_3 system with a high ZnO content were obtained. In these glasses associated BO_3 and BO_4 units were detected up to 30 mol% B_2O_3. Complex microheterogeneous structure may develop inside the liquid phase separation region. The layer rich in B_2O_3 (upper layer) contains only a small amount of MoO_3 and Ln_2O_3. The lower layer is rich in MoO_3 and Ln_2O_3 and crystallises more easily. The reason for the immiscibility is the low ability for the generation of mixed B–O–Mo bonds.

Acknowledgements

The study was performed with the financial support of The Ministry of Education and Science of Bulgaria, The National Science Fund of Bulgaria, Contracts: TK-X-1718/07 and TK-X-1702/07.

References

1. Mazurin, O., Strelcina, M. & Shvaiko-Shvaikovska, T. *Glass Properties and Glassforming Melts*, Vol. 3, 1979, Nauka, Leningrad. (In Russian.)
2. Simon, S. & Nicula, Al. *J. Non-Cryst. Solids*, 1983, **57**, 23.
3. Bobkova, N., Rakov, I. & Solovei, N. *J. Non-Cryst. Solids*, 1989, **111**, 98.
4. Bobkova, N. *Bol. Soc. Esp. Ceram. Vidr.*, 1992, **31-C-3**, 257.
5. Berger, R., Beziade, P., Levasseur, A. & Servant, J. *Phys. Chem Glasses*, 1990, **31**, 231.
6. Agrawal, R., Verma, M., Gupta, R. & Kumar, R. *J. Phys. D: Appl. Phys.*, 2002, **35**, 810.
7. Prasad, P. Srinivasa, S., Kumar V. & Veeraiah, R. *Phys. Pagasine*, 2007, **87**, 5763
8. Abe, M, Benino, Y., Fujiwara, T., Komatsu, T. & Sato, R. *J. Appl. Phys.*, 2005, **97**, 123516
9. Nakajima, R. Abe, M., Benino Y., Fujiwara T. & Komatsu T., *J. Non-Cryst. Solids*, 2007, **353**, 85
10. Terashima, K. Tamura, S., Kim, S. & Yoko T., *J. Am. Ceram Soc.*, 1997, **80**, 2903
11. Chakraborty, I., Day, D., Lapp J. & Shelby, J. *J. Am. Ceram. Soc.*, 1985, **68**, 368.
12. Chakraborty, I., Shelby, J. & Condrate, R. *J. Am. Ceram. Soc.* 1984, **67**, 782.
13. Vinogradova, N., Dmitruk, L. & Petrova, O. *Glass Phys. Chem.*, 2004, **30**, 3. (In Russian.)
14. Dimitriev, Y., Kashchieva E., Iordanova, R. & Tyuliev, G. *Phys. Chem. Glasses*, 2003, **44**, 155.
15. Dimitriev, Y., Kashchieva E. & Iordanova, R. *Phys. Chem. Glasses Eur. J. Glass Sci. Technol. B*, 2006, **47**, 435.
16. Aleksandrov, L., Iordanova, R. & Dimitriev, Y. *Phys. Chem. Glasses Eur. J. Glass Sci. Technol. B*, 2007, **48**, 242
17. Aleksandrov, L., Iordanova, R., Dimitriev, Y., Handa, K. Ide, J. & Milanova, M. In: *Glass – The Challenge for the 21st Century*, Trans Tech, Zurich, 2008, **39–40**, 37.
18. Aleksandrov, L., Iordanova, R. & Dimitriev, Y., *Proceedings of the Sixth International Symposium on Non-Oxide and New Optical Glasses*, Montpellier, France, 2008 (in press). NOTE to author: Please update if possible
19. Malzev, V., Chobanan, N. & Volkov, V., *J. Inorg. Chem.*, 1973, **19** 2010. (In Russian.)
20. Iordanova, R., Dimitriev, Y., Dimitrov, V., Kassabov, S. & Klissurski, D. *J. Non-Cryst. Solids*, 1998, **231**, 227.
21. R. Iordanova, E. Lefterova, I. Uzunov,Y. Dimitriev, D. Klissurski, *J. Therm. Anal. Calorim.*, 2002, **70**, 393.
22. Das, M., Annapurna, K., Kundo, P. Dwivedi, R. & Buddhudu, S. *Mater. Lett.*, 2006, **60**, 222.
23. Bamfort, C. *Control of Colour and Generation in Glass*, Elsevier-North Holland Publication, Amsterdam, 1977.
24. Ehrt, D. *Glass Tech.*, 2000, **41**, 182.
25. Gattere, K., Pucker, G., Fritzer, H. & Arafa S. *J. Non-Cryst. Solids*, 1994, **176**, 237.
26. Balda, R., Fernandez, J., Sanz, M., Oleaga, A. & Pablos, A. *J. Non-Cryst. Solids*, 1999, **256&257**, 271.
27. Chen, Y., Huang, Y., Chen H. & Luo Z. *J. Am. Ceram. Soc.*, 2005, **88**, 19.
28. Li, H., Vienna, J., Qian, M., Wang Z. & Peeler D. *J. Non-Cryst. Solids*, 2000, **278**, 35.
29. Iordanova, R., Aleksandrov, L., Stoyanova, A. & Dimitriev, Y. In: *Glass – The Challenge for the 21st Century*, Trans Tech, Zurich, 2008, **39–40**, 73
30. Varsamis, C., Kamitsos, E. & Chrissikos, G. *Solid State Ionics*, 2000, **136&137**, 1031.
31. Ysker, J. & Hoffman W. *Naturwissenschaften*, 1970, **10**, 129.
32. Pahomov, V., Silinickaya, G., Medvedev, A. & Djurinskii, B. *Inorg. Mater.*, 1972, **8**, 1259. (In Russian.)
33. Lisanova, G., Djurinskii, B., Komova, M., & Tananaev, I. *J. Inorg. Chem.*, 1983, **28**, 2369. (In Russian.)
34. Palkina, K., Saifuddinov, V., Kuznecov, V., Djurinskii, B., Lisanova, G., & Reznik, E. *J. Inorg. Chem.*, 1979, **24**, 1193. (In Russian.)
35. Abrahams, S. *J. Chem. Phys.* 1967, **46**, 2052.
36. Smith, P., Garcia-Blanco S. & Rivoir, L. *Z. Kristallogr.*, 1961, **115**, 460.
37. Burns, A., Winslow, D., Clarida, W., Affatigato, M., Feller, S. & Brow, R. *J. Non-Cryst. Solids*, 2006, **352**, 2364.
38. Pisarski, W., Goryczka, T., Plonska, M. & Pisarska, J. *Mater. Sci. Eng. B*, 2005, **122**, 94.
39. Pisarski, W., Pisarska, J. & Romanovski, W. *J. Mol. Struct.*, 2005, **744**, 515.
40. Varshneya, A. *Fundamentals of Inorganic Glasses*, 1993. Harcourt Brace & Company.
41. Kapoutsis, J., Kamitsos, E. & Chryssikos, G. In: *Borate glasses, crystals and melts*, Eds. A. C. Wright, S. A. Feller & A. C. Hannon, Society of Glass Technology, Sheffield, 1997, p. 303.
42. Fragoso, W., Donega, C. & Longo, R. *J. Non-Cryst. Solids*, 2005, **352**, 3121.
43. Baia, L., Stefan, R., Popp, J., Simon, S. & Kiefer W. *J. Non-Cryst. Solids*, 2003, **324**, 109.
44. Bale, Sh., Rahman, S., Awashti A. & Sathe, V. *J. Alloys Compounds*, 2008, **460**, 699.
45. Cheng Y., Xiao, H. & Guo, W. *Mater. Sci. Eng. A*, 2008, **480**, 56.
46. Seguin, L., Figlarz, M., Cavagant, J. & Lassegues, C. *Spectrochim. Acta*, 1995, **51A**, 1323.
47. Efimov, A., Mihailov, B. & Arkatova, T. *Glass Phys. Chem.*, 1979, **5**, 692. (In Russian.)
48. Ivanova, Y., Kashchieva, E. & Dimitriev, Y. *Phys. Chem. Glasses*, 2000, **41**, 349.
49. Iordanova, R., Milanova, M. & Kostov, K. L. *Phys. Chem. Glasses Eur. J. Glass Sci. Technol. B*, 2006, **47**, 631–637.
50. Milanova, M., Iordanova, R., Kostov, K. & Dimitriev, Y. *Phys. Chem. Glasses Eur. J. Glass Sci. Technol. B*, 2007, **48**, 255
51. Iordanova, R., Milanova, M. & Kostov, K. L. *J. Non-Cryst. Solids*, in press. NOTE to author: Please update if possible
52. Talik, E. Kruczek, M., Sakowska, H. & Szyrski, W. *J. Alloys Compounds*, 2003, **361**, 282.
53. Honma, T., Benino, Y., Fujiwara, T. & Komatsu, T., *J. Appl. Phys.*, 2002, **91**, 2942.
54. Jain, H., Miller, A., Kamitsos, E & Kapoutsis, J. In *Borate glasses, crystals and melts*, Eds. A. C. Wright, S. A. Feller & A. C. Hannon, Society of Glass Technology, Sheffield, 1997, p.287.
55. Matsumoto, S., Miura Y., Murakami, C. & Nanba T. In: *Borate glasses, crystals and melts*, Eds. A. C. Wright, S. A. Feller & A. C. Hannon, Society of Glass Technology, Sheffield, 1997, p. 173.
56. Dimitrov, V., Komatsu, T. & Sato, R. *J. Ceram. Soc. Jpn.*, 1999, **107**, 21.
57. Honma, T., Benino Y., Komatsu, T., Sato, R. & Dimitrov, V. *Phys. Chem. Glasses*, 2002, **43**, 32.
58. Mekki, A., Khattak, G. & Wenger L. *J. Non-Cryst. Solids*, 2005, **351**, 2493.
59. Chowdari, B., Tan, K., Chia, W. & Gopalakrishnan, R. *J. Non-Cryst. Solids*, 1991, **128**, 18.
60. Zarzycki, J. *Glasses and the Vitreous State*, Cambridge Solid State Science Series, Cambridge University Press, Cambridge, 1991, p.229.
61. Hoppe, U., Dimitriev, Y. & Jovari, P. Z. *Naturforsch.*, 2055, **60**, 517.
62. Bersani, D., Antonioli, G., Dimitriev, Y., Dimitrov, V. & Kobourova, P. *J. Non-Cryst. Solids*, 1998, **232–234**, 293.
63. Xue, L., Lin, Z., Chen D. Huan, F. & Liang J. *J. Alloys Compounds*, 2008, **458**, 144.

Proc. VI Int. Conf. Borate Glasses, Himeji, Japan, 18–22 August 2008   *Glass Tech.: Eur. J. Glass Sci. Technol. A, August 2009, 50 (4), 221–226*

Glass formation and crystallisation behaviour of yttrium aluminium tetraborate glass-ceramic

Rafael M. Hovhannisyan,* Bella V. Grigoryan, Hovakim A. Alexanyan, Hovsep G. Shirinyan, Manuk A. Poghosyan, Berta V. Petrosyan,

Scientific-Production Enterprise of Materials Science, 17 Charents st., 0025 Yerevan, Armenia

Vahan P. Toroyan & Zhanna M. Abramyan

Institute of General and Inorganic Chemistry of NAS RA, Argutyan st., district 2, 10, 0051 Yerevan, Armenia

Manuscript received 21 August 2008
Revised version received 27 October 2008
Accepted 27 March 2009

The phase diagram of the binary Y_2O_3–B_2O_3 system has been revised, the phase diagram in the pseudo-binary $YAlO_3$–YBO_3 system has been constructed and two binary eutectics have been revealed. A new ternary eutectic between $YAlO_3$, YBO_3 and an unknown ternary yttrium aluminium compound has been revealed. Glass forming ability in the ternary Y_2O_3–Al_2O_3–B_2O_3 system has been investigated and a glass forming diagram dependant on melt casting procedure has been determined. The expanded glass formation area occupies fields of composition with low liquidus, formed between binary yttrium aluminate, yttrium borate and yttrium aluminium borate eutectics. The crystalline phase formation in yttrium aluminium tetraborate have been determined by solid state sintering and by devitrification of the corresponding glass composition in powder, bulk and tape form. The binary aluminium and yttrium borates are formed as the first step (750–850°C) of solid state sintering and all forms of glass devitrification. Differential thermal analysis and x-ray diffraction showed the dominant ternary yttrium aluminium tetraborate phase formation at 1000–1135°C in solid sintering batches and glass powder crystallisation. This phase is formed in the range 800–950°C for bulk glass and tape crystallisation. The glass ceramics based on yttrium aluminium tetraborate are ferroelectric, with remanent polarization equal to 0·08–0·3 $\mu C/cm^2$ at external field 200–240 kV/cm.

Yttrium aluminium tetraborate is a representative of the extensive group of single crystals with general formula $RM_3(BO_3)_4$ (where R=Y or Ln and M=Al, Sc, Cr, Fe and Ga) which are promising multifunctional solids for nonlinear optics, acousto-electronics, laser techniques and others. They have high thermal and chemical stability and good mechanical strength.[1] According to Leonuk,[1] all these crystals show incongruent melting at 1020–1290°C, except the $RSc_3(BO_3)_4$ group of crystals with congruent melting temperatures around 1500°C.[2] Incongruent melting prohibits the growth of large single crystals.

Glass-ceramic technology combines the flexibility, simplicity and cheapness of glass technology with the possibility to form crystal structures of the desired size and compositions through controlled crystallisation and properties approximate to single crystals.[3]

$YAl_3(BO_3)_4$ shows the largest nonlinear optical coefficient among the $RM_3(BO_3)_4$ group single crystals.[1] From this point of view, the glass formation and crystallisation behaviour of the $12·5Y_2O_3·37·5Al_2O_3·50B_2O_3$ (mol%) composition corresponds to stoichiometric $YAl_3(BO_3)_4$ and is the focus of our work. It is the only known ternary compound in the Y_2O_3–Al_2O_3–B_2O_3 (YAB) system at present. The absence of a phase diagram for the ternary YAB system hinders effec-

tive studies.

The binary Y_2O_3–Al_2O_3 system has been studied carefully by various groups. There are three stable congruently melting compounds, $Y_3Al_5O_{12}$, $YAlO_3$ and $Y_4Al_5O_9$, and four simple eutectics in the Y_2O_3–Al_2O_3 system.[4] All characteristic points are at high temperatures (more than 1800–1900°C) and hence there is a wide scatter of the melting and eutectic point temperatures given by various authors.[4]

Three compounds are known in the binary Y_2O_3–B_2O_3 system: YBO_3 congruently melting at 1650°C, with two irreversible phase transitions at 1024°C and 600°C;[5] incongruently melting Y_3BO_6[6] and $Y(BO_3)_2$, which decomposes at 959°C to YBO_3 and B_2O_3.[7] However, Y_3BO_6 and $Y(BO_3)_2$ compounds are not shown on the Y_2O_3–B_2O_3 phase diagram constructed by Levin,[8] which shows only the simple eutectic $30Y_2O_3·70B_2O_3$ (mol%) with melting point 1373°C between YBO_3 and a stable liquation area containing 70–98 mol% B_2O_3.[8]

Two compounds: $9Al_2O_3·2B_2O_3$ congruently melting at 1965°C, and $2Al_2O_3·B_2O_3$ incongruently melting at 1035°C, are known in the Al_2O_3–B_2O_3 binary system.[9,10] The presence of immiscibility in the liquid phase is not excluded by Gielisse *et al.*[10]

The limited area of glass formation has been determined in the YAB system by Chakraborty *et al.*[11] The $12·5Y_2O_3·37·5Al_2O_3·50B_2O_3$ (mol%) glass composition lies inside the area of clear glasses. The

---

Corresponding author. Email hovhannisyan@netsys.am

field of stable liquation composition extends to pure boron oxide. The properties of aluminoborate glasses formed in the YAB system were investigated by Rutz *et al.*[12] Nine compositions, containing from 10 to 25 mol% Y_2O_3, were investigated. The properties measured were density, refractive index, microhardness, chemical durability, thermal expansion and dc electrical resistivity.[12] Recently there has been increased interest in transmitting $YAl_3(BO_3)_4$ glass ceramics obtained both from bulk glass samples and from the sol-gel method.[13,14] Glass formation from yttrium aluminate melts using various super cooling techniques,[15,16] has expanded the glass formation region in the YAB system.

Experimental

Sample preparation

About sixty samples of various binary and ternary compositions in the YAB system have been synthesised and tested. Compositions were prepared from "chemically pure" Y_2O_3, Al_2O_3 and H_3BO_3 at 2·5–5·0 mol% intervals. Most samples were obtained as glasses using various cooling methods depending on melt crystallisation tendency: casting between two metallic plates (bulk samples) or super cooling techniques constructed by our group (tape samples with thickness 30–400 μm). Glasses were melted (50–70 g batch, 50 ml uncovered Pt crucible, air atmosphere, electric furnace, 1 h) at 1550°C. The chemical composition of some glasses was controlled and corrected according to the results of traditional chemical analysis. The final analysis results indicate a good correspondence of calculated and analytical values of B_2O_3, Y_2O_3 and Al_2O_3.

Compositions lying outside of the region of glass of formation or having high melting temperature have been synthesised by solid phase sintering. Mixes (15–20 g) were carefully mixed in an agate mortar, pressed as tablets, located on platinum plates and heated in electric muffles. After regrinding, powders were tested by DTA (differential thermal analysis) and x-ray methods. Tablets of the pressed samples were given repeated thermal treatment at 800–1200°C for 24 h on the basis of obtained analysis data.

Thermal analysis

DTA of the glasses was conducted on a Q-1500 in platinum crucibles, with powder glass samples (weight 0·5–0·6 g, heating rate 15 K/min, sensitivity 250 μV) and monolithic glass samples (weight 1·25–1·8 g, heating rate 3·75 K/min, sensitivity 100 μV). DTA of monolithic samples required a sensitivity 2·5 times higher than for powder samples. Monolithic glass samples for DTA were obtained by fusing powders in DTA platinum crucibles in an electric furnace at 1550°C. The accuracy of temperature measurement is ±5 K.

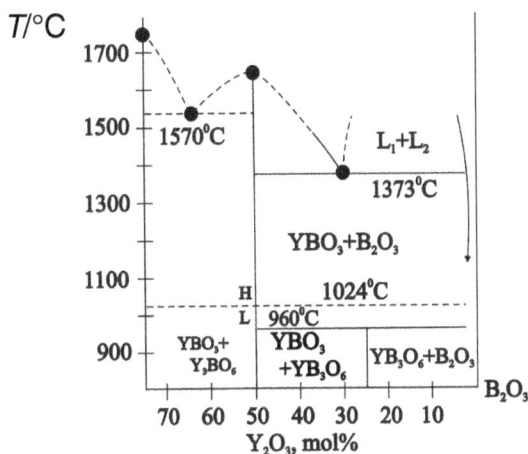

Figure 1. New version of the Y_2O_3–B_2O_3 system phase diagram

Phase identification

X-ray diffraction patterns were obtained on a DRON-3 type diffractometer (powder method, Cu K_α radiation, Ni-filter). Samples for glass crystallisation were prepared with glass powder pressed in the form of rods or tablets. Crystallisation was a single stage heat treatment for 6–24 h around the temperature at which the maximum exothermic effects had been observed by DTA on glasses. Crystalline phases of binary and ternary compounds have been identified by using the JCPDS-ICDD database.

Polarisation measurements

A computerised system was created for ferroelectric hysteresis tests and ferroelectric property measurements. This technique is based on the well known Sawyer–Tower modified scheme, which measures compensating phase shifts concerned with dielectric losses and conductivity. The two channel ultrahigh input impedance amplifiers (for signal conditioning) reduce the influence of measuring equipment on the results of experiments. The desired frequency signal from a waveform generator is amplified by a high voltage amplifier and applied to the sample. The signals, from the measuring circuit output, proportional to applied field and spontaneous polarisation (P_s) are passed through high impedance conditioning amplifiers, converted by an ADC, and controlled and analysed by a computer. Remanant polarisation on field rising P_{r-} and reducing P_{r+} are not equal so the characterisation for ferroelectric properties is given by $2P_r = |P_{r+}| + |P_{r-}|$.

Results and discussion

Phase diagram and phase identification (powders)

It is well known that the combination of eutectic compositions with rapid melt cooling promotes glass formation in non-traditional inorganic systems.[17] Eutectic depth and cooling speed are parameters limiting the borders of glass formation.

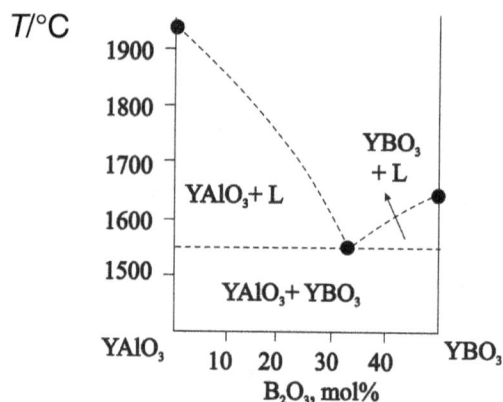

Figure 2. YAlO₃–YBO₃ system phase diagram

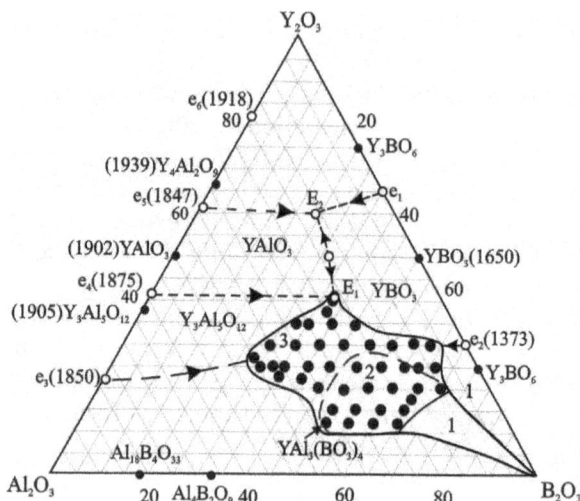

Figure 3. Glass forming diagram of the Y_2O_3–Al_2O_3–B_2O_3 system
1 stable liquation area; 2 area of clear glasses revealed by Chakraborty et al[11]; 3 area of clear glasses obtained by supercooling method

Y_2O_3–B_2O_3

First of all we have continued the construction of the binary Y_2O_3–B_2O_3 system phase diagram. The areas of compositions containing 30–45 and 70–90 mol% B_2O_3 have been tested in this system. The corrected version is shown in Figure 1. The eutectic area between YBO_3 and Y_3BO_3 has been determined. Eutectic e_1 (64–66) Y_2O_3.(34–36)B_2O_3 (mol%) composition has a melting point more than 1500°C (the temperature limit of our DTA). This melting point is in the range 1560–1580°C, as revealed by heat treatment of pre-sintered powders in an electric furnace. Two clear endothermic effects, at 960 and 1373°C, have been observed in the DTA curves of compositions containing 70–90 mol% B_2O_3. According to Tananaev et al[7] the first effect at 960°C is due to decomposition of YB_3O_6, with the formation of YBO_3 and B_2O_3 (Figure 1). The second effect at 1373°C, according to Levin et al,[8] is connected with eutectic composition 30Y_2O_3.70B_2O_3 melting.

$YAlO_3$–YBO_3

The pseudo-binary $YAlO_3$–YBO_3 system has been studied and a pseudo-binary eutectic (16–18)Al_2O_3.(34–33)B_2O_3.50Y_2O_3 (mol%) composition, with melting point 1540–1560°C, has been determined (Figure 2). The presence of both binary Y_2O_3–B_2O_3 and pseudo-binary $YAlO_3$–YBO_3 phase diagrams, together with the known diagram of the Y_2O_3–Al_2O_3 system, provides an understanding of the glass formation processes taking place in the investigated ternary YAB system. It is possible to observe a clear correlation between the glass forming area and the eutectic fields. From the binary high temperature yttrium aluminate and yttrium borate eutectic, the system goes to two ternary deeper YAB eutectics – E_1 and E_2. The melting point of the E_1 eutectic is equal to 1320°C (Figure 3).

$YAl_3(BO_3)_4$

The 12·5Y_2O_3.37·5Al_2O_3.50B_2O_3 (mol%) glass composition, corresponding to the stoichiometric yttrium aluminium tetraborate $YAl_3(BO_3)_4$ compound, is close to the 50Al_2O_3.50B_2O_3 (mol%) composition in the Al_2O_3–B_2O_3 binary system, which has a liquidus temperature around 1800–1900°C.[9,10] $YAl_3(BO_3)_4$ sharply reduces the liquidus temperature and promotes glass formation, which starts from the 11Y_2O_3.39Al_2O_3.50B_2O_3 (mol%) composition. The glass forming area determined occupies the field with low liquidus temperature in the YAB system (Figure 3). Glass formation by supercooling is an effective method for phase diagram construction in systems with high liquidus temperatures and having poor glass forming ability by traditional methods. As shown on Figure 3, the transparent glasses are good indicators for determining fields of each crystalline phase together with eutectic points. $YAl_3(BO_3)_4$ crystal phase formation using batch sintering and thermal treatment of powder, bulk and tape glass samples has been studied. DTA of batch pretreated at 850°C for 12 h shows two endothermic effects at 960 and 1100°C, and a strong endothermic effect in the interval 1230–1300°C (minimum at 1285°C) together with a strong exothermic effect at 1135°C (Figure 4, curve 1).

YBO_3 is the main crystalline phase (75%) in the initial batch (850°C, 12 h) according to data from x-ray analysis (Figure 5, curve 1).[18] However, the endothermic effect at 960°C indicates the presence of YB_3O_6, but this is not observed in the x-ray patterns. It is logical to suppose that the solid state reactions in ternary aluminium borates take place via initial high boron content and low melting point phase formation but x-ray analysis does not confirm this. The second crystalline phase is $Al_4B_2O_9$ (25%).[19,20] These crystalline phases remain in the same proportions in batch products after heat treatment at 850°C for 12 h and 1000°C for 1 h. An exothermic effect at 1135°C (Figure 4, curve 1) indicates the formation of the $YAl_3(BO_3)_4$

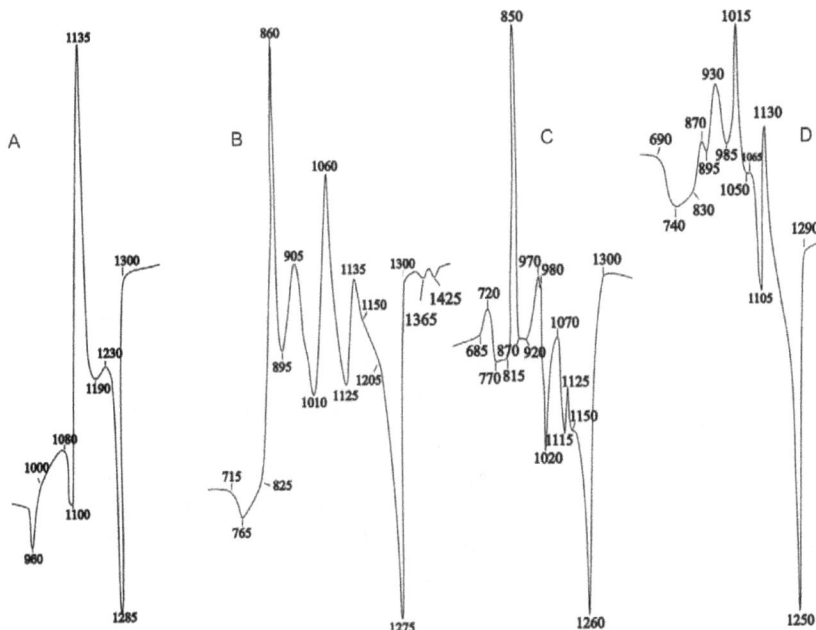

Figure 4. DTA curves of YAl₃(BO₃)₄ samples (heating rate 15 K/min, sensitivity 250 mV)
A solid state sintering sample (850°C, 24 h); B glass powder sample; C bulk sample (heating rate 3·75 K/min, sensitivity 100 mV); D sample thermal treated at 1300°C for 0·5 h and cold water quenched

phase in the interval 1000–1135°C (Figure 5, curve 2).

In the DTA curve of $YAl_3(BO_3)_4$ glass powder, we observe four exothermic effects at 860, 905, 1060 and 1135°C, a strong endothermic effect in the interval 1205–1305°C (minimum at 1275°C), and two weak endothermic effects at 1365 and 1425°C (Figure 4, curve B). The first strong exothermic effects at 850–860°C in both curves are connected with $Al_4B_2O_9$ and YBO_3 crystallisation (Figure 5, curve 3). $Al_4B_2O_9$ and YBO_3 are observed in glass powder samples crystallised at 900–930°C. However, the $YAl_3(BO_3)_4$ phase appears at 900–930°C (Figure 5, curve 4). Its amount increases in glass powder samples crystallised at 1000°C, and $YAl_3(BO_3)_4$ becomes practically single phase at 1065°C (Figure 5, curve 5). Two weak endothermic effects, at 1365 and 1425°C (Figure 4, curve B), are associated with incongruent melting of the $YAl_3(BO_3)_4$ compound and dissolution of the crystalline phase in the melt, which is formed during the incongruent melting. X-ray characterisation of $YAl_3(BO_3)_4$ glass and crystallised glass samples thermally processed at 1300°C (25°C above the melting point) and 1400°C (125°C above the melting point) for 0·5 h and cold water quenched, shows $9Al_2O_3.2B_2O_3$ (9A2B)[22] together with an amorphous phase. The DTA curve confirms the presence of glassy phase in the water quenched product, and there is a sharp decrease of the first

exothermic effect at 850–860°C (Figure 4, curve D) in contrast to the initial glass (Figure 4, curve A). This reflects the reduction in the alumina content of the glass phase and the decrease of $Al_4B_2O_9$ crystallisation

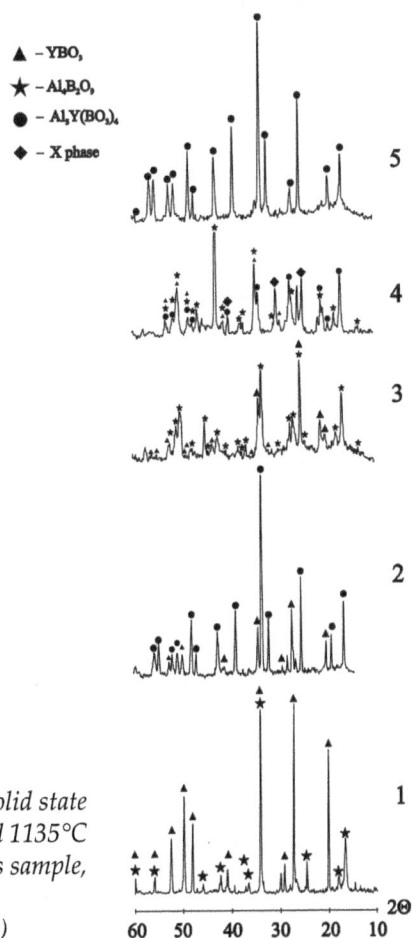

Figure 5. XRD patterns of the YAl₃(BO₃)₄ composition obtained by: curve 1 solid state sintering, at 850°C for 24 h; curve 2 solid state sintering, at 850°C for 24 h and 1135°C for 12 h; curve 3 powder glass sample, at 850°C for 12 h; curve 4 powder glass sample, at 930°C for 12 h; curve 5 powder glass sample, at 1065°C for 12 h.
(Crystalline phases: ▲ YBO₃;[18] ★ Al₄B₂O₉;[19,20] ● YAl₃(BO₃)₄;[21] ◆ X-phase)

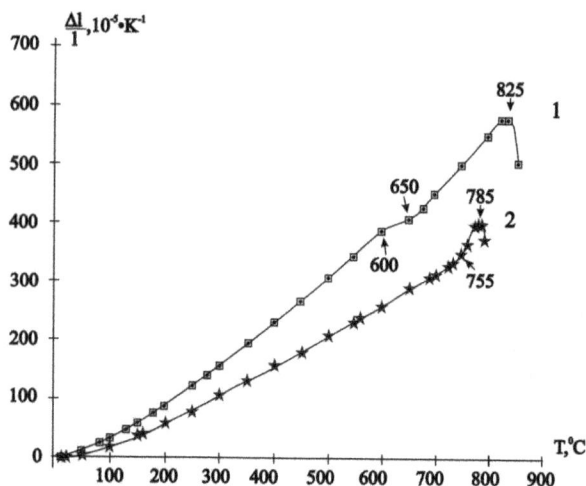

Figure 6. Thermal expansion curves of $YAl_3(BO_3)_4$ monolith glass samples. 2 initial glass (quartz vertical dilatometer); 1 initial glass thermal treated at 750°C for 48 h (alumina vertical dilatometer)

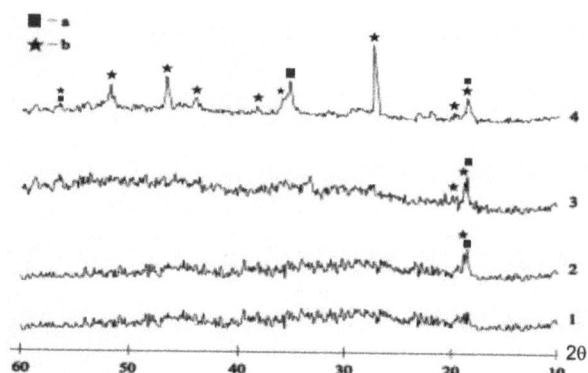

Figure 7. XRD patterns of the $YAl_3(BO_3)_4$ crystallised monolith glass samples.
curve 1 (regime 1) 720°C 24 h – transparent, x-ray amorphous; curve 2 (regime 2) regime 1+760°C 24 h – transparent; curve 3 (regime 3) regime 1+770°C 24 h – transparent; curve 4 (regime 4) regime 1+790°C 24 h – not transparent (Crystalline phases: ■ a $Al_{18}B_4O_{33}$;[22] ★ b $Al_4B_2O_9$[19,20])

in the interval 825–890°C (Figure 4, curve A).

Phase identification of monolithic samples

DTA of the monolithic glass samples gives a similar picture to that of the powder samples (Figure 4, curve C): four exothermic effects at 850, 970, 1070 and 1125°C, and a strong endothermic effect in the interval 1150–1300°C (minimum at 1260°C). The occurrence of an exothermic effect in the interval 685–770°C (maximum at 720°C) in the monolithic glass DTA curve (Figure 4, curve C) seems very interesting. This effect is absent from the DTA curve of the powder sample (Figure 4, curve B) and can be connected to nucleation processes taking place in the monolithic glass sample.

The initial glass, of $YAl_3(BO_3)_4$ composition, has thermal expansion coefficient TEC(20–300°C)=44·8×10^{-7} K^{-1}, T_g=755°C, and T_{def}=785°C (Figure 6, curve 2). After a one step thermal treatment at 750°C for 48 h, clear opalescence was visually observed as result of metastable phase separation, which is accompanied by displacement and increase of T_{def} (825°C) and the thermal expansion curve bending at 600°C and 650°C. TEC increases to 57×10^{-7} K^{-1} (Figure 6, curve 1).

Taking into account the DTA data for $YAl_3(BO_3)_4$ monolithic glasses (Figure 4, curve C), the crystallisation sequence of these glasses has been investigated in the temperature interval 680–790°C. Samples heat treated for 24 h at 680, 700 and 720°C were x-ray amorphous. Samples crystallised in the two step regimes 2 and 3 were visually transparent. However, the presence of crystalline material is readily observed by x-ray analysis: it is a mixture of crystalline $Al_{18}B_4O_{33}$ and $Al_4B_2O_9$ (Figure 7, curves 2–4).

The thermal treated (at 720°C 24 h) $YAl_3(BO_3)_4$ glass tape fabricated by supercooling is transparent

and x-ray amorphous (Figure 8, curve 1). A glass tape sample crystallised in two step regime 2 (720°C 24 h+820°C 12 h) is visually transparent, and the products of its crystallisation are $YAl_3(BO_3)_4$ together with $Al_4B_2O_9$[19,20] and YBO_3[18] (Figure 8, curve 2). The same crystal phases are present after regime 3 (720°C 12 h+910°C 12 h): $YAl_3(BO_3)_4$[21] (main phase>70%), together with $Al_4B_2O_9$[19,20] and YBO_3[18] (Figure 8, curve 3), however the sample became visually non-transparent. $Al_4B_2O_9$[19,20] and YBO_3[18] are the

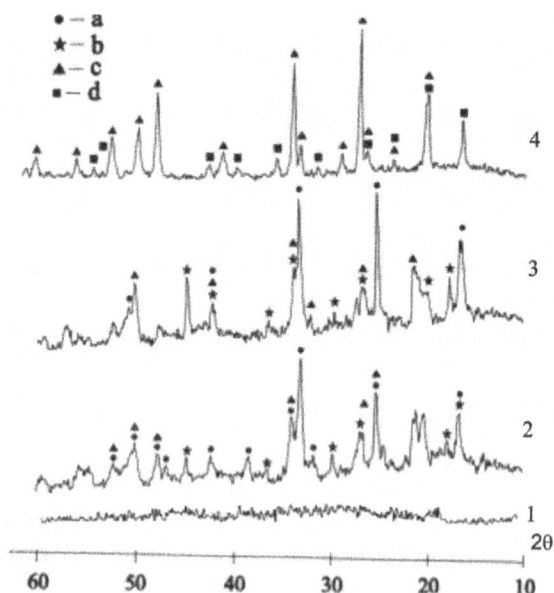

Figure 8. XRD patterns of the crystallised $YAl_3(BO_3)_4$ glass tape samples obtained by supercooling technique
curve 1 (regime 1) 720°C 24 h – transparent, x-ray amorphous; curve 2 (regime 2) regime 1+820°C 12 h – transparent; curve 3 (regime 3) regime 1+910°C 12 h – not transparent; curve 4 (regime 4) regime 1+1040°C 24 h) – not transparent; (Crystalline phases: ● a $YAl_3(BO_3)_4$;[21] ★ b $Al_4B_2O_9$;[19,20] ▲ c YBO_3;[18] ■ d $Al_{18}B_4O_{33}$[22])

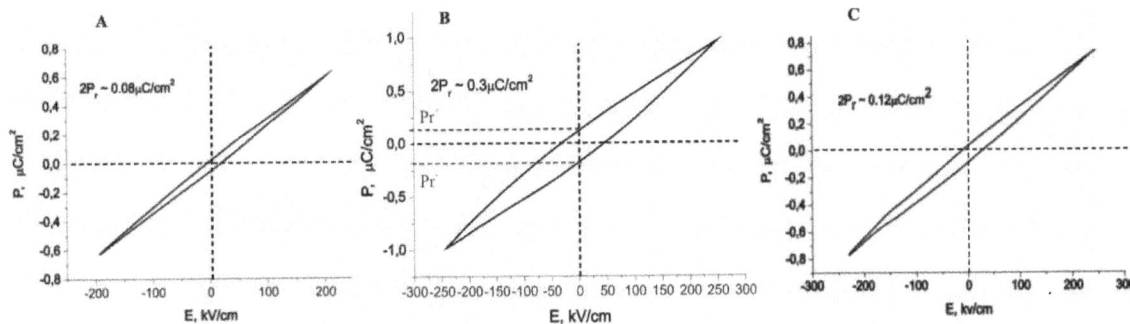

Figure 9. Dependence of polarisation, P, on electric field, E, for YAl₃(BO₃)₄ crystallised glass tape samples: A 720°C 24 h+820°C 12 h glass tape of thickness 0·08 mm; B 720°C 24 h+910°C 12 h glass tape of thickness 0·075 mm; C 720°C 24 h+1040°C 12 h glass tape of thickness 0·055 mm

main crystalline phases at increasing crystallisation temperature (Figure 8, curve 4).

Polarisation behaviour

Electric field induced polarisation measurement was used for ferroelectric characterisation. Polarisation, P, and remanent polarisation, P_r, were measured at room temperature for YAl$_3$(BO$_3$)$_4$ glass tape samples crystallised using various regimes (Figure 9). Linear P–E curves were observed up to fields of 100–120 kV/cm for all measured samples with thickness 0·055–0·08 mm. The polarisation increases with electric field. P values of 0·6, 1·0 and 0·8 μC, and P_r values of 0·08, 0·3 and 0·12 μC/cm^2 were found under fields of 200–240 kV/cm for samples A, B and C, respectively (Figure 9 A, B, C). It is possible to say that samples crystallised under regime B are better ferroelectrics than samples C and A. YAl$_3$(BO$_3$)$_4$ is ferroelectric and increasing its proportion of crystallised material led to increasing polarisation and remanent polarisation values. The conclusion is that YAl$_3$(BO$_3$)$_4$ is a better ferroelectric than other crystalline phases: aluminium and yttrium binary borates.

Conclusions

The phase diagram of the binary Y$_2$O$_3$–B$_2$O$_3$ system has been revised, and also the phase diagram in the pseudo-binary YAlO$_3$–YBO$_3$ system has been constructed and two binary eutectics have been revealed. New ternary eutectics between AlBO$_3$, YBO$_3$ and an unknown ternary yttrium aluminium compound have been revealed.

The glass forming ability in the ternary Y$_2$O$_3$–Al$_2$O$_3$–B$_2$O$_3$ system has been investigated and a glass forming diagram dependent on melt casting procedure has been determined. The expanded glass formation area occupies compositional regions with low liquidus, formed between binary yttrium aluminate, yttrium borate and yttrium aluminium borate eutectics.

Crystal phase formation in yttrium aluminium tetraborate samples made by solid-state sintering and the equivalent glasses as powder, bulk and tape samples has been determined. Binary aluminium and yttrium borates are formed at the first stage (750–850°C) of solid state sintering, and for all routes of glass crystallisation. DTA and x-ray analyses showed extensive formation of a ternary yttrium aluminium tetraborate phase at 1000–1135°C in solid-state sintered and glass powder samples. This phase is formed in the interval 800–950°C in bulk glass and tape samples.

The glass ceramics based on yttrium aluminium tetraborate are ferroelectric with remanent polarisation of 0·08–0·3 μC/cm^2 at external field 200–240 kV/cm.

Acknowledgement

This work was supported by the International Science and Technology Center (Projects # A-952 and A-1486).

References

1. Leonyuk, N. I. & Leonyuk, L. I. *Prog. Cryst. Growth Charact. Mater.,* 1995, **31**, 179.
2. Ivonina, N. P., Kutovoy, S. A., Laptev, V. V. & Simonova, I. N. *Zh. Neorg. Mater.,* 1991, **27**, 64 (In Russian).
3. Jain, H. *Ferroelectrics,* 2004, **306**, 111.
4. Phase Equilibria Diagrams. CD-ROM Database, Version 3.0, A CerS-NITS, 2004, Fig. # 00311, 02344, 04370, 09264, 09265, 10402, 10403.
5. Levin, E. M., Roth, R. S. & Martin, J. B. *Am. Miner.,* 1961, **46**, 1030.
6. JCPDS-ICDD, PDF Data Base, 1997, USA, File # 34-0291.
7. Tananaev, I. V., Dzhurinskiy, B. F. & Chistova, V. I. *Zh. Neorg. Mater.,* 1975, **11**, 86 (In Russian).
8. Levin., E. M. In: *Phase Diagrams,* Ed A. M. Alper, Academic Press, New York, 1970.
9. Dorner, P., Gauckler, L. J., Krieg, H., Lukas, H. L., Petzow, G. & Weiss, J. *CALPHAD,* 1979, **3**, 241.
10. Gielisse., P. J. M. & Foster., W. R. *Nature,* 1962, **195**, 69.
11. Chakraborty, I. N., Rutz, H. L. & Day, D. E. *J. Non-Cryst. Solids,* 1986, **84**, 86.
12. Rutz, H. L., Day, D. E. & Spencer Jr., C. F. *J. Am. Ceram. Soc.,* 1990, **73**, 1788.
13. Yamamoto, Y., Hashimoto, T., Nasu, H. & Kamiya, K. *Jpn. J. Appl. Phys.,* 2003, **42**, 5043.
14. Maia, L. J. Q., Mastelaro, V. R., Fick, J., Hernandes, A. C. & Ibanez, A. *Glass Technol.: Eur. J. Glass Sci. Technol. A,* 2006, **47**, 141.
15. Wilding, M. C., McMillan, P. F. & Navrotsky, A. *Physica A,* 2002, **314**, 379.
16. Skinner, L. B., Barnes, A. C., Salmon, P. S. & Crichton, W. A. *J. Phys.: Condens. Matter,* 2008, **20**, 205103.
17. Rawson, H. *Inorganic glass-forming systems.* Academic Press, London, 1967.
18. JCPDS-ICDD, PDF Data Base, 1997, USA, File # 16-0277.
19. JCPDS-ICDD, PDF Data Base, 1997, USA, File # 09-0158.
20. JCPDS-ICDD, PDF Data Base, 1997, USA, File # 29-0010.
21. JCPDS-ICDD, PDF Data Base, 1997, USA, File # 15-0117.
22. JCPDS-ICDD, PDF Data Base, 1997, USA, File # 32-0003.

Proc. VI Int. Conf. Borate Glasses, Himeji, Japan, 18–22 August 2008   *Phys. Chem. Glasses: Eur. J. Glass Sci. Technol. B,* June 2009, **50** (3), 175–182

Structure–property studies of SrBr$_2$–SrO–B$_2$O$_3$ glasses

Randall E. Youngman,* Lauren K. Cornelius, Shari E. Koval, Carrie L. Hogue & Adam J. G. Ellison

Science & Technology, Corning Incorporated, Corning, NY 14831, USA

Manuscript received 10 November 2008
Revised version received 17 April 2009
Accepted 23 April 2009

Strontium borate glasses containing significant quantities of SrBr$_2$ have been made and studied to ascertain their local network structure and the impact of high halide content on their physical properties. Two regions of glass formation have been mapped for the SrBr$_2$–SrO–B$_2$O$_3$ ternary; one at very low SrBr$_2$ levels and another along the 20 mol% B$_2$O$_3$ join. The structure of these glasses, as determined by ^{11}B MAS and 3QMAS NMR spectroscopies, is consistent with modification of the borate network in the form of tetrahedral boron upon addition of Sr, but with substantial depolymerisation at the highest SrBr$_2$ levels due to formation of a significant fraction of asymmetric BO$_3^-$ units.

1. Introduction

Unlike the alkali borates, binary and multicomponent alkaline earth borate glasses have been the subject of few structural studies. Many of the structure/property aspects of alkali borate glasses appear to hold for the alkaline earth borates, although a few differences have been discussed in light of recent experimental studies. For example, addition of alkaline earth oxides to B$_2$O$_3$ results in formation of tetrahedral boron until reaching a maximum value at around 30 to 40 mol% modifier, similar to the alkali borates.[1–9] But unlike the alkali borates, the fraction of boron in tetrahedral coordination, N_4, increases as one moves to larger alkaline earth cations, presumably in response to relative field strength of the cations and their ability to function in the role of glass formers.[8–10]

The glass forming regions within the alkaline earth borate binaries appear to be smaller than the corresponding alkali borates. The latter glasses can be made using conventional methods from 0 to ~45 mol% alkali oxide, with some glasses found at higher Na, K and Rb contents.[9] Most of the published work on alkaline earth borates suggests a much narrower glass forming region, constrained between approximately 20 and 45 mol% alkaline earth oxide.[9,11,12]

Another interesting difference between alkali and alkaline earth borate glasses is the ability to incorporate halides into the former. Many well characterised glassy electrolytes are based on lithium or sodium borates containing LiX (or NaX), where X=F, Cl, Br.[13–17] These glasses typically contain a large fraction of glass former (B$_2$O$_3$), but also are capable of having tens of mol% alkali halide. Another system which has been of great interest over the years is the AgI–Ag$_2$O–B$_2$O$_3$ ternary, which exhibits high ionic conductivity, and thus has been examined as a glassy electrolyte.[18–21] While halide containing borate glasses are generally known and show great promise for technological application, there is little evidence concerning alkaline earth borate glasses doped with halides. In particular, there is only one previously described example in the strontium borates, where glasses containing large quantities of SrCl$_2$ were successfully fabricated.[22]

As part of our on-going research into halide containing borate materials for fining of glass melts, glass-formation was discovered in the CaBr$_2$–CaO–B$_2$O$_3$, BaBr$_2$–BaO–B$_2$O$_3$ and SrBr$_2$–SrO–B$_2$O$_3$ ternaries. The latter system, which yielded the largest glass forming region, is the focus of the present study. We describe two distinct regions in which ternary SrBr$_2$–SrO–B$_2$O$_3$ glasses can be obtained using traditional melt-quench techniques. ^{11}B nuclear magnetic resonance (NMR) and physical property measurements were made on these glasses to better understand their microstructure and its impact on the glass transition temperature, T_g. Ternary glasses near the established SrO–B$_2$O$_3$ glass forming region can be made with a few mol% SrBr$_2$, and their structure and properties closely resemble the halide free strontium borates. The second compositional regime, where up to ~55 mol% SrBr$_2$ can be incorporated, displays a strong dependence of T_g on the SrBr$_2$/SrO ratio, and ^{11}B NMR examination of the borate structure reveals significant network depolymerisation at these higher SrBr$_2$ levels.

2. Experimental

Appropriate mixtures of high purity B$_2$O$_3$, SrO and SrBr$_2$ were melted in Pt crucibles for 30 min at temperatures ranging from 1000 to 1300°C. 25 g melts

---

Corresponding author. Email youngmanre@corning.com

Table 1. Composition, physical properties and fraction of 4-fold coordinated B for glasses in the $SrBr_2$–SrO–B_2O_3 ternary. Dashes denote properties not measured, as discussed in the text

Sample	Analysed glass compositions (mol%) (±0·02)			T_g (°C) (±5)	Density ($g\,cm^{-3}$)	Refractive index (±0·01)	N_4 (±2%)
	B_2O_3	SrO	$SrBr_2$				
A	71·3	27·4	1·4	630	2·88	1·577	29
B	70·6	26·4	3·0	630	2·85	1·581	28
C	68·0	32·0	0	625	2·99	1·592	34
D	65·5	32·6	1·9	635	3·06	1·601	34
E	64·4	34	1·7	630	3·10	1·607	34
F	21·4	45·5	33·1	440	3·96	—	1
G	20·4	44·4	35·2	430	3·95	—	2
H	21·2	36·9	41·9	—	3·97	—	3·4
I	21·6	36·1	42·3	420	3·95	—	5
J	21·6	33·2	45·2	410	3·96	—	5·6
K	20·7	32·9	46·5	410	3·96	—	7
L	20·6	31·2	48·2	400	3·95	—	9·4
M	20·9	26·4	52·7	390	3·98	—	14·5
N	19·7	24·4	55·9	380	3·97	—	15·4

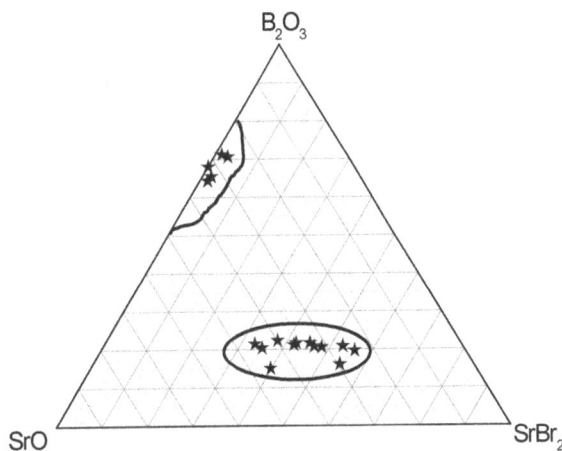

Figure 1. Ternary composition plot, showing two regions of glass formation in the $SrBr_2$–SrO–B_2O_3 system. The solid curves denote approximate glass forming regions based on this study

were quenched by pouring onto a graphite plate and cooling to room temperature. Due to potential losses of highly volatile components, the resulting glasses were analysed by inductively coupled plasma mass spectrometry and titration methods (for Br) to determine the actual compositions, which are reported in Table 1. Melts outside of the described glass forming regions typically crystallised upon cooling or exhibited substantial opaqueness, and were therefore not included in the study.

T_g measurements were made using a Perkin Elmer Pyris Diamond TG/DTA at a 20°C/min heating rate and with a N_2 purge. Density measurements were conducted using He pycnometry, where the volume of the sample was determined by the displacement of helium in a gas pycnometer. Refractive indices were determined for a limited number of samples using the Becke line method at 589·3 nm.[23] Ground glass samples were immersed in oils of known index and viewed through a microscope. Many of the samples were too hygroscopic to survive grinding, and thus could not be measured with the Becke technique.

[11]B magic angle spinning (MAS) NMR experiments were carried out at 11·7 T (160·3 MHz resonance frequency), using a commercial spectrometer and probes (Varian/Chemagnetics). Powdered samples were packed into 3·2 mm zirconia rotors under a dry nitrogen atmosphere, and sample spinning in a 3·2 mm MAS NMR probe was typically controlled to 20 kHz. ^{11}B MAS NMR data were obtained using 0·6 μs pulse widths (<π/12 tip angle) and 2 s recycle delays to ensure quantitative excitation of all resonances. These data were processed without any additional line broadening. External referencing was performed with aqueous boric acid (19·6 ppm).

Triple quantum MAS (3QMAS) NMR data were obtained for [11]B in order to provide additional resolution of overlapping MAS lineshapes. The data were recorded using the three-pulse z-filtered method,[24] using excitation and conversion pulse widths of

2·4 and 0·8 μs, respectively. A z-filter pulse width of 20 μs followed a delay of 0·2 ms. 48 acquisitions at each of 80 t_1 values, with a dwell time of 15 μs, were typically sufficient for these experiments. The resulting two-dimensional correlation plots were analysed in terms of isotropic chemical shift (δ_{iso}) and second order quadrupolar product ($P_Q=C_Q(1+h/2)^{1/2}$) by comparing the centres of gravity of the projection along the MAS and isotropic dimensions.[25] Isotropic projections of these 3QMAS NMR spectra were obtained by summing the two-dimensional data onto the isotropic frequency axis, resulting in a one-dimensional spectrum free from second order quadrupolar broadening.

3. Results

3.1 Compositions and physical properties

Glasses in the $SrBr_2$–SrO–B_2O_3 ternary system were obtained in two composition regions, as shown in the ternary composition plot of Figure 1. Samples were initially batched to explore a much larger range of compositions, but all analysed glassy samples appear to fall into two general categories: strontium borates with a small amount of $SrBr_2$ (<3 mol%), which are stable at relatively high B_2O_3 levels, or as highly modified borate glasses containing only around 20 mol% B_2O_3. This latter glass forming region is apparently distinct from the one at high B_2O_3 (i.e. not obviously connected), but sufficiently large to produce a number of distinct glass compositions for the purpose of this study. Analysed compositions for all glasses in this study are provided in Table 1.

Density, T_g and limited refractive index measurements were made for these glasses and are included in Table 1. Density data are plotted against total Sr content in Figure 2(a), where a linear dependence is shown to exist, although a considerable compositional gap does exist between the high and low

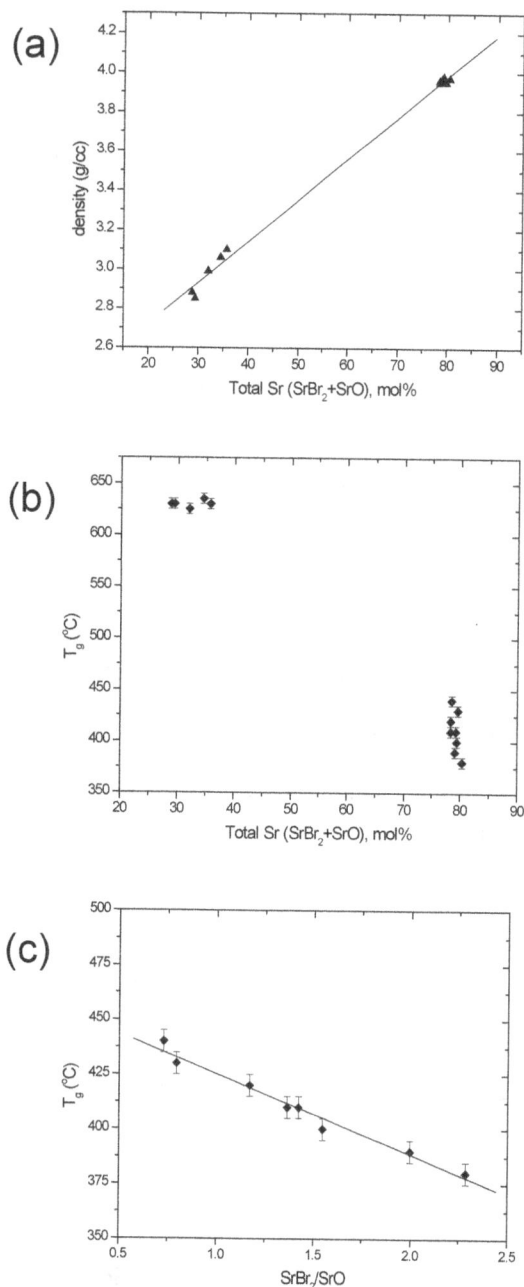

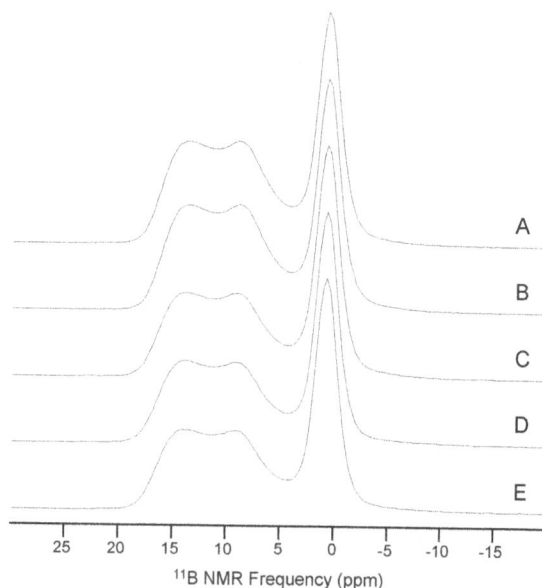

Figure 3. ^{11}B MAS NMR spectra for glasses in the high B_2O_3 regime, where labels to the right correspond to compositions in Table 1

Figure 2. Plots of several physical properties obtained for glasses in the $SrBr_2$–SrO–B_2O_3 ternary: (a) density as a function of total Sr, with a linear fit drawn as a solid line; (b) T_g as a function of total Sr; and (c) T_g plotted against the $SrBr_2/SrO$ ratio, along with a linear fit to these data. Error bars for composition and density values are smaller than the symbol sizes in these plots

modifier levels. T_g data are similarly plotted in Figure 2(b), where the data are clustered in two groups. The first of these, at low Sr, shows a fairly constant T_g with increasing alkaline earth, at least over a range of approximately 10 mol% Sr. The second grouping in Figure 2(b) is essentially fixed in Sr content, but shows a strong variation in T_g, ranging from 375 to ~450°C (see Table 1). By plotting T_g in a slightly different manner, against the $SrBr_2/SrO$ ratio as in Figure 2(c), a linear relationship with composition is

apparent. Here, T_g decreases with increasing $SrBr_2$. Due to incorporation of large and polarisable halide anions in these glasses, refractive index data were collected, but owing to poor durability, only those in the high B_2O_3 regime were actually measured. These limited data are listed in Table 1 and are consistent with other measurements of alkaline earth borates,[26] with no apparent change on addition of $SrBr_2$ in small quantities.

3.2 Nuclear magnetic resonance

^{11}B MAS NMR spectra for the high B_2O_3 glasses are plotted in Figure 3. The spectra all contain multiple signals, as demonstrated by the broad, asymmetric resonance between ~5 and 20 ppm, and a relatively narrow peak around 0 ppm. The characteristics of these resonances, including relative populations, do not appear to vary significantly with composition in the glasses with B_2O_3 contents between 64 and 75 mol%. ^{11}B MAS NMR spectra from glasses near 20 mol% B_2O_3 are shown in Figure 4. Distinct changes in relative peak intensities and shape of the broader spectral feature are apparent as a function of glass composition.

Representative ^{11}B 3QMAS NMR spectra for glasses in both regimes are given in Figure 5. The spectrum in Figure 5(a) is comprised of two distinct sets of contours which, when projected onto the isotropic shift axis (shown to the right), result in clear separation of two boron resonances at approximately 0 and 20 ppm. These are the same two features detected in the one-dimensional MAS NMR spectrum, but resolution is greatly enhanced in the 3QMAS NMR data. The other two spectra in Figure 5, obtained for glasses at

168

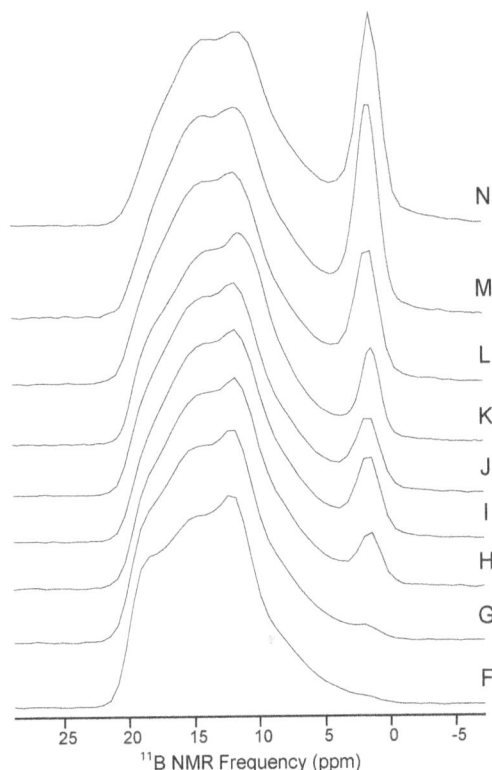

Figure 4. ^{11}B MAS NMR spectra for glasses in the low B_2O_3 regime, where labels correspond to compositions in Table 1. The spectra are plotted from bottom to top as a function of increasing $SrBr_2:SrO$

low B_2O_3 content, are quite different. These contain multiple sets of contours, especially the spectrum of sample M in Figure 5(c), corresponding to the broad and narrow peaks in the MAS NMR data. Each of these 3QMAS NMR spectra also appear to contain multiple overlapping resonances within the contours centred at 20 ppm in the isotropic dimension, with apparent differences in their relative intensities as a function of glass composition (i.e. $SrBr_2$ content).

Figure 6 shows reconstructed MAS NMR spectra, obtained as slices through the ^{11}B 3QMAS NMR spectrum of sample M. The slices in Figure 6(a) and (b) correspond to isotropic peak maxima at 26 and 22 ppm respectively.

Figure 7 contains a series of isotropic spectra obtained by projecting ^{11}B 3QMAS NMR data onto the isotropic shift axes, where only the 18 to 30 ppm frequency region is plotted. The data show pronounced changes in the relative population of two overlapping resonances, with nearly equal intensities at low $SrBr_2$ content, and a significantly more intense peak at 24 ppm for the highest Sr content.

4. Discussion

Glass formation in the $SrBr_2$–SrO–B_2O_3 ternary is apparently constrained to two regions: a high B_2O_3 region where only small amounts of $SrBr_2$ can be added, and a high modifier region where significant

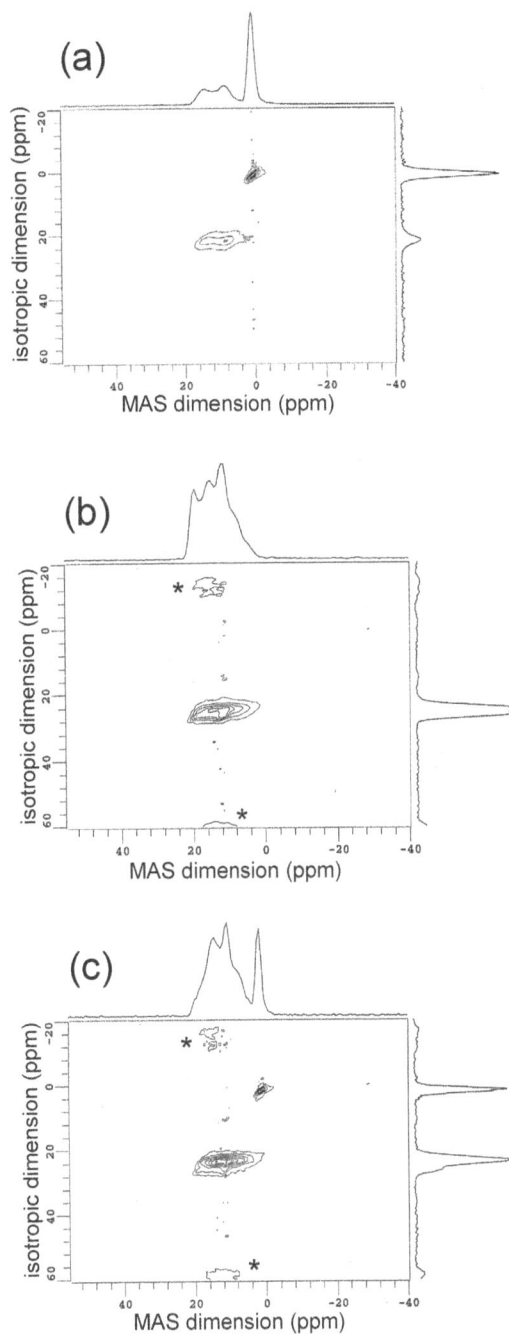

Figure 5. Representative ^{11}B 3QMAS NMR spectra for (a) sample B, (b) sample F and (c) sample M. MAS and isotropic projections are shown to the top and right of each spectrum. Asterisks denote spinning sidebands

levels of halide are retained. The ternary composition plot in Figure 1 shows the estimated glass forming regions based on analysed compositions from this study. The high B_2O_3 region is closely related to the known binary strontium borates, where glasses have been made with SrO contents ranging from 20 to 47 mol%.[9,11,12] The second regime, where the B_2O_3 content is low and the halide content remains high, is a seemingly new discovery for Br-containing borate glasses. One example of Cl-containing strontium borates is known,[22] and appears to be similar to the

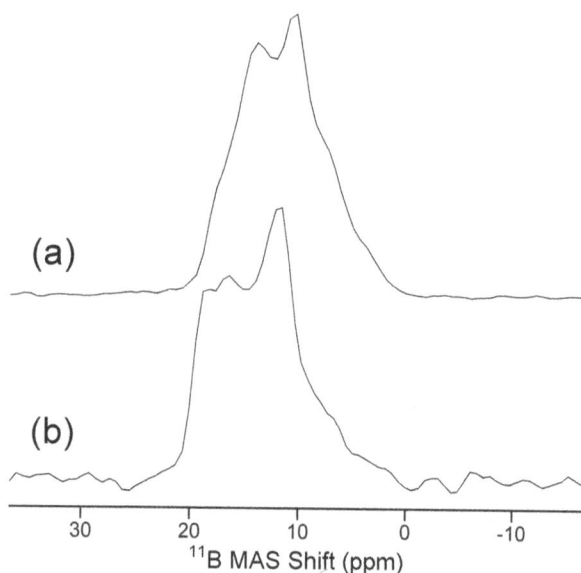

Figure 6. Slices through the ^{11}B 3QMAS NMR spectrum of sample M at isotropic frequencies of (a) 22 and (b) 26 ppm, corresponding to two resolved trigonal resonances

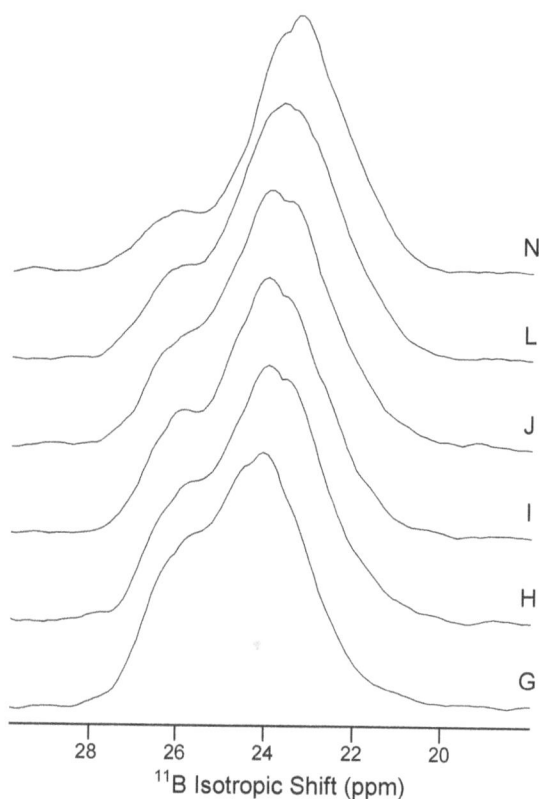

Figure 7. Isotropic projections of the ^{11}B 3QMAS NMR spectra of low B_2O_3 glasses, where labels to the right of each spectrum correspond to the glass compositions in Table 1

Br-containing glasses in this study. Minaev *et al*, in their calculations of glass formation in alkaline earth borates, suggested that glasses near the $60SrO.40B_2O_3$ eutectic might be possible with proper quenching,[27] which would describe glass formation at lower B_2O_3 levels, but not at 20 mol% B_2O_3 or less.

Attempts to make other glasses in the $SrBr_2$–SrO–B_2O_3 ternary were compromised by substantial volatilisation of Br, resulting in analysed compositions similar or identical to those shown. Some of the melts also crystallised during cooling, resulting in non-glassy samples outside of these two compositional regions. The two areas of glass formation are based on standard melt-quench approaches, which probably means that they can be extended by other methods, including very rapid quenching of some melts to avoid crystallisation, as demonstrated for a variety of borate glasses,[9,28] or more controlled melting conditions to aid in retention of the Br.

These glasses were analysed for some of their physical properties, including density and T_g. The results of these measurements (Table 1) are plotted in Figure 2. The density increases linearly with total Sr content, which is not necessarily surprising due to replacement of the lighter B and O atoms with heavier Sr and Br atoms. These data are clustered into two groupings, due to compositional constraints, but nonetheless the density appears to increase substantially with Sr content. The glass densities at the highest Sr contents are approaching the values for pure SrO and $SrBr_2$, 4·7 and 4·216 g/cc, respectively.[29] T_g values were similarly plotted against total Sr content (Figure 2(b)), showing the two groupings for high and low B_2O_3 compositions. The high B_2O_3 regime shows a nearly independent T_g over a narrow range of compositions, with values near 630°C for samples with 30 to 35 mol% Sr-containing

modifier. These compare favourably with T_g values for the binary strontium borates.[30] The other T_g data in Figure 2(b) clearly behave differently with composition, showing a highly varying T_g with little to no change in total Sr content. T_g ranges from 375 to 450°C; for these compositions the glasses are essentially fixed in their B_2O_3 content (Figure 1), but vary in $SrBr_2/SrO$ ratio. Plotting the same T_g data versus this ratio, as in Figure 2(c), shows a linear decrease with increasing halide content. Thus, at a fixed B_2O_3 content, T_g is sensitive to changes in modifier type (i.e. oxide versus halide), with significantly lower values at the highest $SrBr_2$ contents.

In order to further understand these unique glasses and rationalise the dependence of T_g on composition, a detailed ^{11}B NMR study was performed. ^{11}B MAS NMR spectroscopy has proven invaluable in identifying and quantifying the short range atomic order in simple and complex borate glasses.[31,32] These NMR spectra typically provide resolution of trigonal (BO_3) and tetrahedral (BO_4) units, allowing one to quantify N_4 as a function of composition or thermal history.[31–35] ^{11}B MAS NMR spectra for the ternary strontium borate glasses are shown in Figure 3 for glasses with high B_2O_3 levels. These data contain the typical broad, asymmetric resonance from BO_3 units, which is found between 0 and 20 ppm. The relatively sharp spectral feature near 0 ppm indicates the presence of BO_4 tetrahedra. These ^{11}B MAS NMR

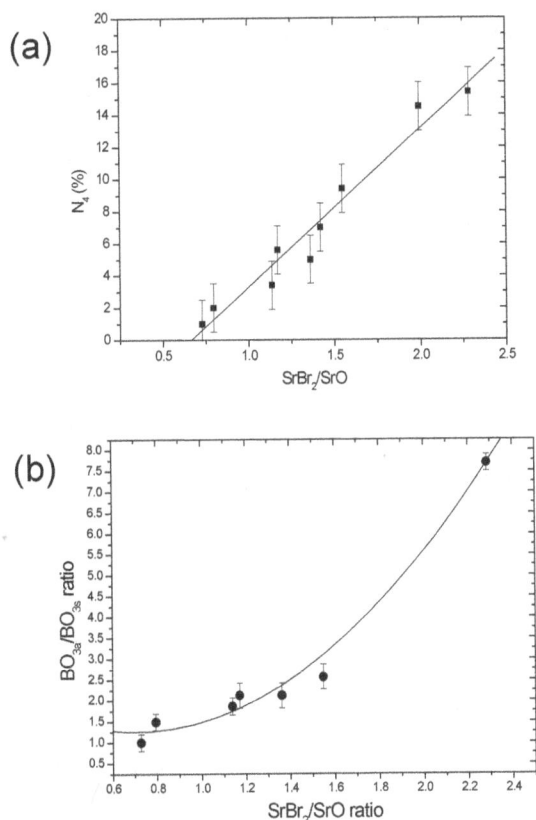

Figure 8. Plots of (a) N_4 and (b) ratio of asymmetric BO_3 to symmetric BO_3 units, for low B_2O_3 glasses, as determined from ^{11}B MAS and 3QMAS NMR data. Solid curves are provided as a guide to the eye. Uncertainties in composition yield error bars which are smaller than the symbol sizes

spectra do not appear to change appreciably with small compositional changes in the high B_2O_3 glasses. The shape of the BO_3 resonance is similar throughout the series, as is the shift and width of the BO_4 peak.

Based on glass compositions and the MAS NMR spectra in Figure 3, assignment of the ^{11}B NMR resonances is relatively straightforward. The trigonal lineshape for the peak centred around 12 ppm is consistent with symmetric BO_3 triangles having all bridging oxygen atoms, as one might expect for a modifier-poor glass. The separation of the two features at approximately 9 and 15 ppm reflects a relatively small quadrupolar asymmetry parameter, η, which is characteristic of these symmetric BO_3 units.[36] Similar lineshapes have been observed in a variety of alkali and alkaline earth borates with lower modifier contents.[37–39] Thus, these ternary glasses at high B_2O_3 content appear to be structurally similar to many of the binary borate glasses studied over the years. The boron speciation is essentially divided between symmetric BO_3 units and a large fraction of BO_4 tetrahedra. The latter appears to vary in concentration over the limited composition range of these particular glasses, and N_4 can be determined by integration of the two resonances, as MAS NMR measurements were made using conditions in which

the two boron coordination environments are equally excited and detected. N_4 values obtained from ^{11}B MAS NMR data are listed in Table 1, where a small variation is seen. The primary influence on N_4 in these high B_2O_3 glasses appears to be the total Sr:B ratio, i.e. essentially the modifier to glass-former ratio which controls boron speciation in binary alkali and alkaline earth glasses, consistent with other studies of Sr borates.[7,8] Thus, T_g and boron speciation in these high B_2O_3 glasses are consistent with other moderately modified borate glasses; the addition of SrO and $SrBr_2$ results in modification of the borate network, converting a fraction of the trigonal boron into tetrahedral units, without obvious formation of nonbridging oxygen.

The latter observation, based on the MAS NMR lineshapes of Figure 3, is corroborated by high resolution 3QMAS NMR spectra. The spectrum of sample B (Figure 5(a)) shows well separated resonances from BO_3 and BO_4 units. Only one type of trigonal boron is detected, as indicated by the single set of contours at an isotropic shift of 21·8 ppm. The resonance from 3-fold coordinated boron is assigned to symmetric BO_3 units without NBOs (nonbridging oxygens), as slices through this resonance show a relatively large span between the singularities, reflecting small h values, as in the above discussion of ^{11}B MAS NMR data.[39] This confirms that, for ternary glasses at high B_2O_3 content, the amount of modifier is insufficient to move N_4 beyond its maximum and begin stabilization of NBO, in general agreement with the position of N_4 maxima in binary alkaline earth borate glasses.[2,9]

The situation appears to be substantially different in the second compositional regime, where the B_2O_3 content is around 20 mol%. Here, T_g values extend over a larger range, with decreasing values at higher halide levels. The ^{11}B MAS NMR spectra of these glasses are shown in Figure 4, where the data contain at least two different types of boron resonance. The broad BO_3 signal is distinctly different from that in the high B_2O_3 glasses (cf. Figure 3), showing less apparent resolution with respect to the quadrupolar broadened lineshape, but also indications of more than one trigonal boron resonance. A tetrahedral boron resonance, present at a shift of ~2 ppm, is most prevalent at higher $SrBr_2$ content. N_4 was extracted from these data in the manner described above, resulting in the graph in Figure 8(a) (also listed in Table 1); N_4 ranges from 1 to 16%, which is both lower and more strongly varying than observed in the high B_2O_3 glasses.

As the B_2O_3 content is fixed, the total Sr content is fixed too, and this variation in N_4 would not be expected from a simple consideration of the modifier to glass former ratio, but the modifier is changing in the sense that the amounts of SrO and $SrBr_2$ vary throughout this composition range. The replacement of oxide with halide apparently leads to strong variations in both T_g and N_4. However, if one assumes that higher N_4 leads to a more polymerised glass network, then a decrease in T_g as SrBr$_2$/SrO increases (Figure 2(c)), should correlate with a decrease in N_4, which is clearly opposite to the trend in Figure 8(a). This correlation between N_4 and T_g is only valid at low modifier levels, where a simple relation exists between BO$_3$ and BO$_4$ units. In this situation, conversion of trigonal BO$_3$ units with all bridging oxygen atoms to BO$_4$ units does increase the overall connectivity of the network, resulting in pronounced increases in T_g. Eventually the BO$_4$ content reaches a maximum and additional modifier is incorporated via formation of NBOs so that N_4 decreases, serving as the basis of the so-called 'borate anomaly.' Thus at higher modifier levels, NBO on trigonal boron units (i.e. BO$_3^-$) are formed at the expense of N_4. This leads to a depolymerised network and a lowering in T_g. This description of highly modified borate glasses is probably applicable to the low B$_2$O$_3$ glasses in this study, for which the total modifier content is very high. The ^{11}B MAS NMR spectra in Figure 4 provide evidence for NBO formation, because the trigonal boron lineshapes contain spectral features which are much closer together, possibly reflecting a larger h value and thus the presence of asymmetric BO$_3$ units (i.e. BO$_3$ with at least one NBO). This is consistent with the NMR spectra of highly modified binary borate glasses for which the NBO content is large.[35]

Changes in trigonal boron geometry due to formation of NBO were further examined using the high resolution 3QMAS NMR technique. The spectrum in Figure 5(b) is from a low B$_2$O$_3$ glass with a relatively low SrBr$_2$/SrO ratio (sample F), where only one distinct set of contours is detected in the 2D correlation plot. Closer examination of these contours, as well as their projection onto the isotropic shift axis, shows evidence for multiple trigonal boron peaks having similar intensities. At higher SrBr$_2$/SrO ratios, as for sample M in Figure 5(c), the ^{11}B 3QMAS NMR spectrum changes considerably. In addition to two distinct trigonal boron resonances near 26 and 22 ppm in the isotropic dimension, there is a contribution from tetrahedral boron, giving rise to the peak near 0 ppm. The BO$_3$ and BO$_4$ peak intensities in these data are consistent with the ^{11}B MAS NMR spectra in Figure 4, showing higher N_4 at higher SrBr$_2$ contents. In addition, the higher resolution afforded by the 3QMAS NMR technique provides strong evidence for two types of trigonal boron species, which accounts for the added complexity of the ^{11}B MAS NMR spectra.

Based on their corresponding MAS dimension lineshapes, obtained by taking slices through the ^{11}B 3QMAS NMR data at each isotropic peak position, the two trigonal boron resonances can be assigned to symmetric and asymmetric units. Representative slices are plotted in Figure 6 for sample M. The 22 ppm isotropic resonance possesses a MAS lineshape with substantial quadrupolar asymmetry, as reflected

by the small separation in its singularities (10·5 and 14 ppm). The MAS lineshape for the less intense trigonal boron resonance, with an isotropic shift of approximately 26 ppm, appears very different. As can be seen in Figure 6(b), the separation between features in this lineshape is larger, and overall the spectrum looks somewhat similar to those from the high B$_2$O$_3$ glasses (Figure 3), confirming the presence of symmetric BO$_3$ (BO$_{3s}$) units in these low B$_2$O$_3$ glasses, where BO$_{3s}$ could arise from either all bridging O or all NBO on the three-fold coordinated B. The feature at 22 ppm in the isotropic projections is assigned to asymmetric BO$_3$ units (BO$_{3a}$), having one or two NBO. The presence of this BO$_{3a}$ resonance accounts for the marked changes in the MAS NMR lineshapes of Figure 4. It should be noted that another possible source of asymmetric trigonal boron could involve the formation of BO$_2$Br units, where terminal Br groups disrupt the local symmetry around the boron. Unfortunately, local symmetry and its impact on MAS NMR lineshapes cannot be used to distinguish between these asymmetric trigonal boron types.

Although assignment of the two trigonal boron resonances to BO$_{3a}$ and BO$_{3s}$ units is clear, the isotropic shifts of these two peaks (Figure 5(b) and (c)) are not entirely consistent with the BO$_{3s}$ peak in the high B$_2$O$_3$ glasses (cf. Figure 5(a)). In the latter, a single isotropic peak is detected near 22 ppm, which is the same frequency as the BO$_{3a}$ resonance in the low B$_2$O$_3$ glasses. In the low B$_2$O$_3$ glasses, the resonance assigned to BO$_{3s}$ units is found at an isotropic shift of around 26 ppm, distinctly different from the same apparent coordination environment in the high B$_2$O$_3$ glasses. This inconsistency must be related to how the trigonal BO$_3$ groups are arranged in the network, including the nature of neighbouring boron–oxygen polyhedra. For example, the formation of modified ring structures in alkali borate glasses has been shown to give rise to a resonance which is very similar to that for boron in boroxol rings, even though these species are not truly boroxol rings.[40] Thus one possible explanation for the above inconsistency is the presence of a significant number of tetrahedral boron in the high B$_2$O$_3$ glasses, which as second neighbours might change the shift of BO$_{3s}$ groups relative to their shift in glasses with lower N_4 values. Another possible explanation is that BO$_{3s}$ in low B$_2$O$_3$ glasses are comprised of three-fold coordinated B with all NBO neighbours. The symmetry would be approximately the same as for trigonal boron with all bridging oxygen, but electron densities might be sufficiently different to impact the chemical shielding of ^{11}B. In addition, the impact of Br on boron chemical shielding in the low B$_2$O$_3$ glasses is not completely understood and may play a role. Further studies of these glasses, as well as related alkaline earth borates, are underway to more accurately determine the short range structure and address these possibilities.

The isotropic spectra for the group of low B_2O_3 glasses are shown in Figure 7, where the compositional dependence of the two trigonal boron peaks is apparent. By fitting these projections as a sum of two Gaussian peaks, the relative populations of BO_{3a} and BO_{3s} can be determined. The resulting fractions are plotted in Figure 8(b), showing a large increase in BO_{3a} with increasing $SrBr_2/SrO$. By the highest halide levels ($SrBr_2$ ~56 mol%), the borate network is mostly a loosely connected assemblage of trigonal boron with a high population of NBO (or possibly terminal B–Br bonds), which is apparently held together by only a modest number of tetrahedral boron, and possibly strong Sr–O bonding, as suggested by Chryssikos et al.[8] The increasing number of depolymerised BO_{3a} units at higher halide content overwhelms the modest increase in N_4, accounting for the significant decrease in T_g. Additional factors, including the O/B ratio, which plays a significant role in boron speciation and T_g in AgI-doped borate glasses,[41] probably also contribute to structural changes in these low B_2O_3, high $SrBr_2$ glasses.

5. Conclusions

The $SrBr_2$–SrO–B_2O_3 ternary system has been examined and two regions of glass formation discovered: one at high B_2O_3, where only a small amount of halide can be added to binary strontium borates, and a second, larger region in which the B_2O_3 content is constrained to about 20 mol%, while large variations in $SrBr_2$ and SrO are possible. The properties of these glasses, notably T_g, depend significantly on the borate network, which is strongly interconnected at high B_2O_3, due to a large fraction of tetrahedral boron. However, the low B_2O_3 ternary glasses are comprised mostly of tetrahedral boron and asymmetric trigonal boron units. The latter structural feature, reflecting depolymerisation of the glass network, leads to relatively low T_g values in these novel halide rich alkaline earth borate glasses.

References

1. Bishop, S. G. & Bray, P. J. *Phys. Chem. Glasses*, 1966, **7**, 73.
2. Greenblatt, S. & Bray, P. J. *Phys. Chem. Glasses*, 1967, **8**, 190.
3. Park, M. J. & Bray, P. J. *Phys. Chem. Glasses*, 1972, **13**, 50.
4. Kim, K. S. & Bray, P. J. *Phys. Chem. Glasses*, 1974, **15**, 47.
5. Dell, W. J. & Bray, P. J. *Phys. Chem. Glasses*, 1982, **23**, 98.
6. Tang, Y., Jiang, Z. & Song, X. *J. Non-Cryst. Solids*, 1989, **112**, 131.
7. Moon, S.-J., Kim, M.-S., Chung, S.-J. & Kim, H.-T. *J. Korean Phys. Soc.*, 1996, **29**, 213.
8. Chryssikos, G. D., Kamitsos, E. I. & Yiannopoulos, Y. D. *J. Non-Cryst. Solids*, 1996, **196**, 244.
9. Yiannopoulos, Y. D., Chryssikos, G. D. & Kamitsos, E. I. *Phys. Chem. Glasses*, 2001, **42**, 164.
10. Suzuki, Y., Ohtori, N., Takase, K., Handa, K., Itoh, K., Fukunaga, T. & Umesaki, N. *Phys. Chem. Glasses*, 2003, **44**, 150.
11. Imaoka, M. Glass Formation Range and Glass Structure, in *Adv. Glass Technol., Techn. Papers of the VI International Congress on Glass*, New York: Plenum, 1962, part 1, pp. 149–164.
12. Polyakova, I. G. & Litovchik, E. O. *Glass Phys. Chem.*, 2008, **34**, 369.
13. Kamitsos, E. I. & Karakassides, M. A. *Solid State Ionics*, 1988, **28-30**, 783.
14. Chryssikos, G. D., Kamitsos, E. I. & Karakassides, M. A. *Phys. Chem. Glasses*, 1989, **30**, 243.
15. Trunnell, M., Torgeson, D. R., Martin, S. W. & Borsa, F. *J. Non-Cryst. Solids*, 1992, **139**, 257.
16. Tatsumisago, M., Angell, C. A. & Martin, S. W. *J. Chem. Phys.*, 1992, **97**, 6968.
17. Ngai, K. L. *J. Chem. Phys.*, 1993, **98**, 6424.
18. Schiraldi, A., Pezzati, E. & Caramella Crespi, V., *Solid State Ionics*, 1981, **5**, 675.
19. Chiodelli, G., Magistris, A., Villa, M. & Bjorkstam, J. L. *J. Non-Cryst. Solids*, 1982, **51**, 143.
20. Dalba, G., Fornasini, P., Fontana, A., Rocca, F. & Burattini, E. *Solid State Ionics*, 1988, **28-30**, 713.
21. Varsamis, C. P. E., Kamitsos, E. I., Tatsumisago, M. & Minami, T. *J. Non-Cryst. Solids*, 2004, **346**, 93.
22. Rahman, M. H., Dwivedi, B. P., Kumar, Y. & Khanna, B. N. *Pramana – J. Phys.*, 1992, **39**, 597.
23. Stroiber, R. E. & Morse, S. A. *Crystal Identification with the Polarising Microscope*, Chapman & Hall, 1994.
24. Amoureaux, J.-P., Fernandez, C. & Steuernagel, S. *J. Magn. Reson. A*, 1996, **123**, 116.
25. Amoureaux, J.-P., Huguenard, C., Engelke, F. & Taulelle, F. *Chem. Phys. Lett.* 2002, **356**, 497.
26. Mazurin, O. V., Strel'tsina, M. V. & Shvaiko-Shvaikovskaya, T. P. *A Handook on Properties of Glasses and Glass-Forming Melts, vol. 2*, Leningrad: Nauka, 1975.
27. Minaev, V. S., Petrova, V. Z., Timoshenkov, S. P., Khafizov, R. R. & Sharagov, V. A. *Glass Phys. Chem.*, 2004, **30**, 215.
28. Lower, N. P., McRae, J. L., Feller, H. A., Betzen, A. R., Kapoor, S., Affatigato, M. & Feller, S. A. *J. Non-Cryst. Solids*, 2001, **293-295**, 669.
29. *CRC Handbook of Chemistry and Physics*, Boca Raton, Fl: CRC Press, 1988.
30. Klyuev, V. P., Pevzner, B. Z. & Polyakova, I. G. *Phys. Chem. Glasses*, 2006, **47**, 524.
31. Turner, G., Kirkpatrick, R. J., Risbud, S. & Oldfield, E. *Am. Ceram. Soc. Bull.*, 1987, **66**, 656.
32. Aitken, B. G. & Youngman, R. E. *Phys. Chem. Glasses*, 2006, **47**, 381.
33. Sen, S., Xu, Z. & Stebbins, J. F. *J. Non-Cryst. Solids*, 1998, **226**, 29.
34. Möncke, D., Ehrt, D., Eckert, H. & Mertens, V. *Phys. Chem. Glasses*, 2003, **44**, 113.
35. Kroeker, S. & Stebbins, J. F., *Inorg. Chem.*, 2001, **40**, 6239.
36. Youngman, R. E. & Zwanziger, J. W. *J. Non-Cryst. Solids*, 1994, **168**, 293.
37. Stebbins, J. F., Zhao, P. & Kroeker, S. *Solid State Nucl. Mag.*, 2000, **16**, 9.
38. Aguiar, P. M. & Kroeker, S. *Solid State Nucl. Mag.*, 2005, **27**, 10.
39. Aguiar, P. M. & Kroeker, S. *J. Non-Cryst. Solids*, 2007, **353**, 1834.
40. Youngman, R. E. & Zwanziger, J. W. *J. Phys. Chem.*, 1996, **100**, 16720.
41. Varsamis, C. P., Kamitsos, E. I. & Chryssikos, G. D. *Phys. Rev. B*, 1999, **60**, 3885.

Proc. VI Int. Conf. Borate Glasses, Himeji, Japan, 18–22 August 2008, xxx–yyy

Fabrication of low temperature foaming glass by hydrothermal glass chemistry

Toshihiro Tanaka* & Takeshi Yoshikawa

Division of Materials and Manufacturing Science, Graduate School of Engineering,
Osaka University, 2-1 Yamadaoka, Suita, Osaka 565-0871, Japan

Manuscript received 23 August 2008
Revised version received 16 June 2009
Accepted 15 November 2009

The authors have attempted to use hydrothermal reactions to make low temperature foaming glass. Under hydrothermal conditions, with the presence of H_2O and with temperatures in the range 200–300°C and pressures in the range 30–40 MPa, powders made of glass can be sintered to become solidified glass materials containing about 10 mass% H_2O. When a glass containing H_2O is heated again under normal pressure, the glass expands, releasing H_2O to make a porous microstructure. H_2O starts to be emitted just above the glass transition temperature. Therefore, if a glass has a low glass transition temperature, it can be used to make a low temperature foaming glass. The SiO_2–Na_2O–B_2O_3 glass is a candidate to be a low temperature foaming glass. In this paper, we describe the fabrication of a low temperature foaming glass by using a hydrothermal reaction.

Introduction

A large amount of slag is discharged from iron and steel making industries, and is mainly recycled for use in construction materials, such as road bed materials. In addition, there is also a need to promote recycling of waste glass by developing a reasonable recycling processes. Since slag and glass consists mainly of SiO_2, CaO, Al_2O_3, MgO, FeO, etc, which are general components of ceramic materials, general sintering processes can be applied to produce some solidified materials from slag or glass. We must, however, keep in mind that heating processes are accompanied by energy consumption, and sometimes CO_2 emission. Therefore, the authors have focused on the application of hydrothermal reactions to solidify the slag or glass powders to produce functional ceramic materials, such as architectural materials, etc, as shown in Figure 1.[1,2] A hydrothermal reaction occurs when liquid H_2O is present under high pressure, that is to say, water at 120–350°C, as shown in Figure 2. This temperature can be obtained from exhaust heat coming out of iron and steel making processes, or waste melting furnaces, etc. The application of hydrothermal reactions is an ideal environmentally friendly process to cope with recycling issues for slag, as Jung *et al* have pointed out[3,4] – these authors have already applied hydrothermal reactions to solidify slag powder with some additives. However, in order to add more useful functions to those hydrothermally solidified materials, we have tried to make porous materials from slag or waste glass by applying hydrothermal reactions as described below.

The microstructure of molten slag, which mainly

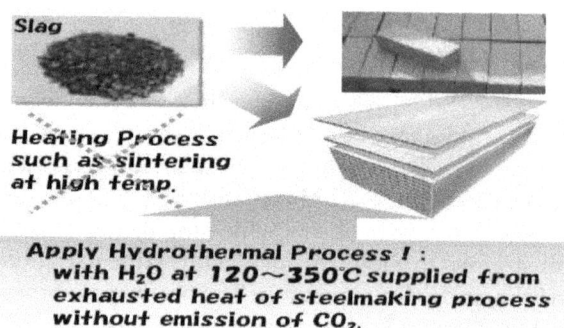

Figure 1. Application of the hydrothermal process for recycling of slag

consists of SiO_2 based network material, is controlled by the addition of alkaline or alkaline earth oxides to change the chemical and physical properties of molten slag, such as basicity, viscosity, etc, in iron and steel making processes at high temperature, as shown in Figure 3. Under hydrothermal conditions, the microstructure of slag may be controlled by H_2O,

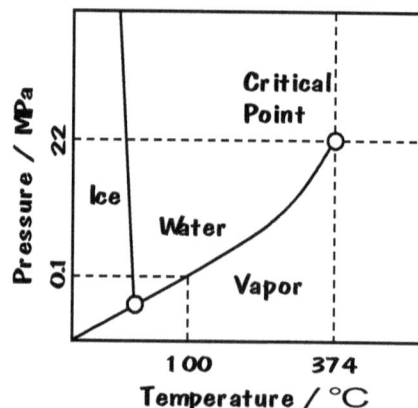

Figure 2. Equilibrium phase diagram of H_2O

---

* Corresponding author. Email tanaka@mat.eng.osaka-u.ac.jp

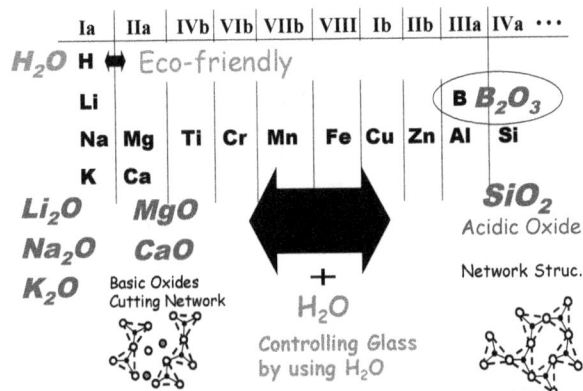

Figure 3. Hydrothermal slag/glass chemistry

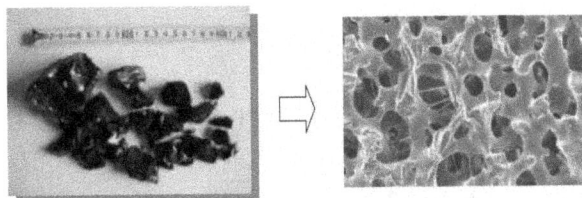

Figure 4. Making porous material from glass or slag with low exergy value

and this new approach can be called "Hydrothermal Slag/Glass Chemistry".

Even when we are able to solidify glass or slag powder to make materials, it is necessary to add some additional functions to these solidified glass or slag materials, because glass or slag is regarded as having a low value of "Exergy", which indicates how valuable a material is. The exergy is defined in the following equation

$$Exergy = \Delta H - T_0 \Delta S \tag{1}$$

where ΔH is the enthalpy, ΔS is the entropy of a material and T_0 is room temperature.

When a material has large ΔH and small ΔH, the exergy of the material is large. This means that the material is regarded as highly valuable and it can generate useful work.

Glass and slag are generally stable multicomponent oxide materials. Oxides are more stable than metals, which means that the enthalpy of oxides is smaller than that of metals. In addition, glass and slag are mixtures of many cations and anions, this means that the mixing entropy is very large. Thus, glass and slag have small exergy values. This suggests that it is very difficult to create value-added materials from the glass and slag in any recycling processing.

We assume in the above discussion that materials have infinite size and uniform structure. If some "interfaces" can be imported into the structure, we can create some additional valuable functions in the materials. Porous structure is one of these valuable interfacial structures. For example, when we have porous glass or slag materials, as shown in Figure 4, these materials can be used as an insulator to control heat transfer, as a filter to remove impurities from polluted air or water, and furthermore as water reserve materials which can be applied to decrease the temperature of a wall or pavement (sidewalk) by vaporizing H_2O absorbed in the porous structure of the materials on sunny days, which has previously been absorbed and kept during rainy days. These materials are expected to contribute largely to ecology.

Thus, if porous materials can be made from waste glass or slag by hydrothermal reactions, we may

propose an environmentally friendly processing to create value added materials. The authors have been attempting to make various kinds of porous materials based on slag and glass. In this paper, we describe an example of fabrication of low temperature foaming glass by using hydrothermal glass chemistry.

Fabrication of low temperature foaming glass

Matamoros-Veloza et al[5,6] have previously imported H_2O into waste glass mainly composed of SiO_2, Na_2O and CaO by hydrothermal synthesis, and obtained a hydrated glass compact. When the prepared glass was heated, it started to expand and foam around 650°C, with the release of water, and became a porous material. This expansion of glass is a suitable approach for making porous materials, but foaming at a lower temperature would be beneficial. Since glass containing H_2O can be made at around 200°C by hydrothermal reactions, it is desirable that the foaming or expansion occurs at around the same temperature.

When the sodium silicate based glass was hydrothermally reacted, the constituents of the glass considerably influenced the water content of the glass under identical hydrothermal conditions. Moreover, it was found[7,8,9] that an increase in water content lowers the glass transition temperature, as shown in Figure 5. For the fabrication of low temperature foaming glass, a glass sample of composition $63SiO_2.27Na_2O.10B_2O_3$ in mass%, which exhibited a suitable low glass transition temperature around 150°C after hydrothermal treatment, as reported in

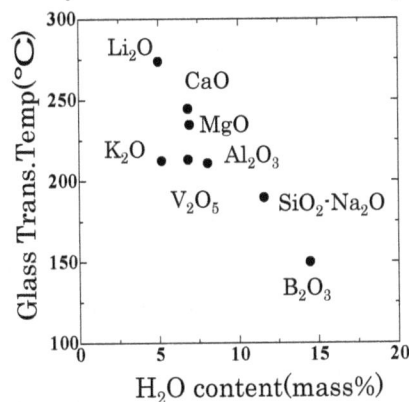

Figure 5. Relationship between glass transition temperature and H_2O content in $63SiO_2.27Na_2O.10MO$ (mass%) glass

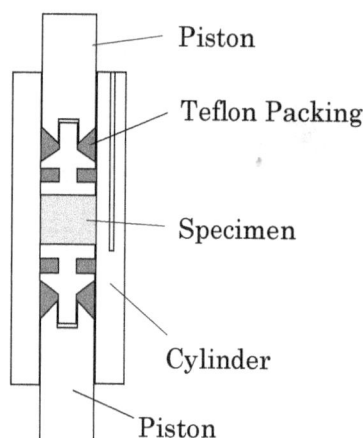

Figure 6. Hydrothermal hot pressing machine

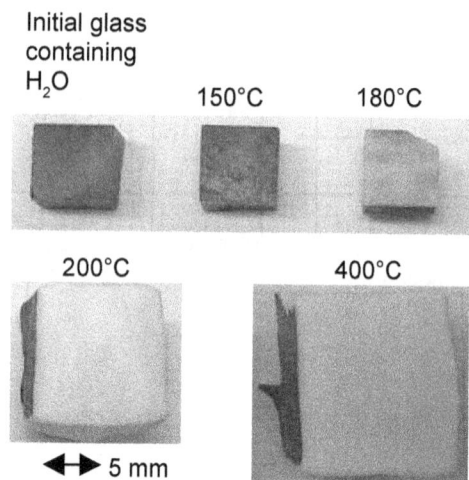

Figure 7. Change in macroscopic shapes of the hydrothermal hot pressed $63SiO_2.27Na_2O.10B_2O_3$ (mass%) glass after firing at 150–400°C

the previous study,[7–9] was selected for the present work. The glass was subjected to hydrothermal treatment at 250°C, and its water releasing and foaming behaviour with heat treatment at 150–400°C was investigated.[10]

Experimental

The original glass sample was prepared from reagent grade quartz, Na_2CO_3 and H_3BO_3. Mixed powders, corresponding to a composition of $63SiO_2.27Na_2O.10B_2O_3$ in mass%, were melted in a Pt–20%Rh crucible at 1200°C for 3 h in air, and then rapidly cooled on a copper cooling block. The glass was subjected to hydrothermal hot pressing (HHP), using the experimental apparatus shown in Figure 6. Pre-melted glass was ground and sieved at 63 μm. Glass powders were well-mixed with a small amount of purified water and charged into the autoclave. After the mixture was pressed at 40 MPa, the autoclave was heated to 250°C within 20 min and then immediately cooled to room temperature within the same time.

The HHP treated SiO_2–Na_2O–B_2O_3 glass prepared above was cut into cubic blocks of almost 6 mm. Each block was placed on a platinum pan, and heated in a horizontal electric resistance furnace, controlled to 150–400°C, in air for 5 min. The weight change of the sample was measured before and after the heat treatment. The apparent density of the sample after firing was determined by measuring its shape.

Results

To investigate the macroscopic change in the HHP treated SiO_2–Na_2O–B_2O_3 glass, a block of the glass was heat treated at 150–350°C in air. The samples after heat treatment are shown in Figure 7. Although no apparent change was observed when heated at 150°C, the sample heated at 180°C appeared to be whitened. Macroscopic expansion, namely foaming, of hydrated glass was observed for the samples

heated over 200°C, which corresponds to the starting temperature for water release observed by the TG-DTA analysis. A higher heating temperature resulted in a larger expansion of the glass materials. Here, a lower temperature foaming was successfully obtained,[10] compared to the temperature of around 650°C for soda–lime–silica glass, previously reported by Matamoros-Veloza et al.[5,6] It was confirmed by XRD analysis that the material foamed at any temperature possesses a glassy structure.

Discussion

Apparent densities of the foamed glasses were determined from the sample shapes, under the assumption of isotropic expansion, and are summarized in Figure 8, with the weight losses during heat treatment. The apparent density decreases drastically at 200°C due to foaming, and continues to gradually decrease as the heating temperature increases. The lowest apparent density of 0·25 g/cm³ was obtained when the

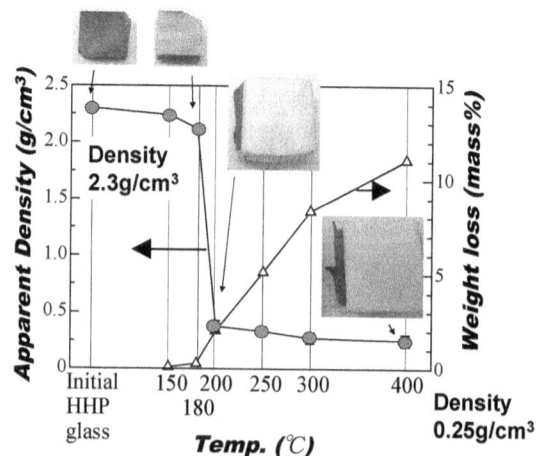

Figure 8. Apparent densities and weight loss after firing of the hydrothermal hot pressed $63SiO_2.27Na_2O.10B_2O_3$ (mass%) glass

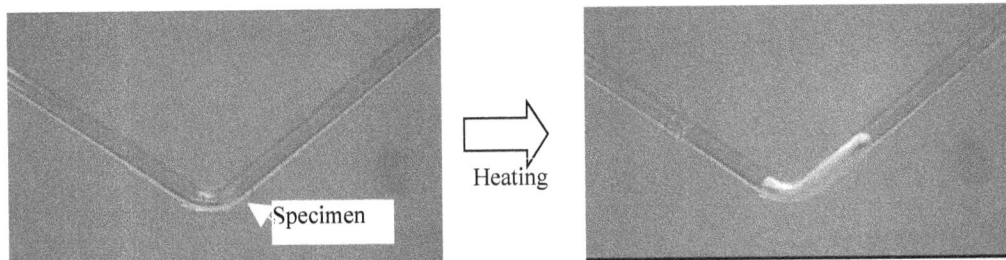

Figure 9. Expansion in a tube of a glass containing H_2O

heat treatment was conducted at 400°C. The weight loss was larger at a higher heating temperature, and the value of 11% at 400°C was almost in accordance with the water content of the HHP treated glass. The relationship between the weight loss and heating temperature indicates the possibility that the porosity of the foamed glass can be controlled by changing the heating process. Those porous materials can be used as a filter to remove impurities in water or air. As shown in Figure 9, when we heat up a piece of glass containing H_2O which is set in a tube, the glass expands to fit the inner shape of the tube. This means that the above glass can be applied as a filter to remove impurities from water or air flowing in a pipe.

Conclusion

The fabrication of low temperature foaming glass, by using a hydrothermal reaction, has been described as an example of the application of Hydrothermal Glass Chemistry. Hydrothermal hot pressed SiO_2–Na_2O–B_2O_3 glass was heated at 150–400°C in air to determine its macroscopic change by heating. Foaming behaviour was observed even at 200°C, and this foaming temperature is much lower than that of hydrated glass reported previously.

Acknowledgement

This study was partially supported by Priority Assistance for the Formation of Worldwide Renowned Centers of Research - The Global COE Program (Project: Center of Excellence for Advanced Structural and Functional Materials Design) from the Ministry of Education, Culture, Sports, Science and Technology (MEXT), Japan.

References

1. Tanaka, T., Hirai, N., Maeda, S., Nakamoto, M. & Hosokawa, M. In: *Proceedings of the Fourth International Congress of the Science and Technology of Ironmaking* The Iron and Steel Institute of Japan, Tokyo, 2006, p. 704.
2. Tanaka, T., Maeda, S., Takahira, N., Hirai, N. & Lee, J. *Mater. Sci. Forum*, 2006, **512**, 305.
3. Jing, Z., Jin, F., Hashida, T., Yamasaki, N. & Ishida, E. H. *J. Mater. Sci.*, 2007, **42**, 8236 .
4. Jing, Z., Ishida, E. H., Jin, F., Hashida, T. & Yamasaki, N. *Ind. Eng. Chem. Res.*, 2006, **45**, 7470.
5. Matamoros-Veloza, Z., Yanagisawa, K., Rendon-Angeles, J. C. & Yamasaki, N. *J. Mater. Sci. Lett.*, 2002, **21**, 1855.
6. Matamoros-Veloza, Z., Yanagisawa, K., Rendon-Angeles, J. C. & Oishi, S. *J. Phys.: Condens. Matter*, 2004, **16**, S1361.
7. Nakamoto, M., Lee, J., Tanaka, T., Ikeda, J. & Inagaki, S. *ISIJ Int.*, 2005, **45**, 1567.
8. Nakamoto, M., Lee J. & Tanaka, T. *Mater. Sci. Forum*, 2006, **512**, 319.
9. Sato, S., Yoshikawa, T., Nakamoto, M., Tanaka, T. & Ikeda, J. *ISIJ Int.*, 2008, **48**, 245.
10. Yoshikawa, T., Sato S. & Tanaka, T. *ISIJ Int.*, 2008, **48**, 130.

Proc. VI Int. Conf. Borate Glasses, Himeji, Japan, 18–22 August 2008   *Phys. Chem. Glasses: Eur. J. Glass Sci. Technol. B, June 2009, **50** (3), 172–174*

Effects of rare earth oxides (La$_2$O$_3$, Gd$_2$O$_3$) on optical and thermal properties in B$_2$O$_3$–La$_2$O$_3$ based glasses

Satoru Tomeno,* Jun Sasai & Yuki Kondo

Research Center, Asahi Glass Co., 1150 Hazawacho, Kanagawa-ku, Yokohama 221-8755, Japan

Manuscript received 20 August 2008
Revised version received 23 January 2009
Accepted 17 February 2009

The optical and thermal properties of B$_2$O$_3$–La$_2$O$_3$ based glasses, designed for digital imaging lenses, were investigated as a function of the Gd/(La+Gd) ratio in the glasses. The refractive index decreased gradually with increasing Gd/(La+Gd) ratio, although the change of refractive index was small. The liquidus temperature (LT) showed a minimum of 970°C for the glass with a Gd/(La+Gd) ratio of 0·5. The viscosity at the LT reached a maximum for the glass with a ratio of 0·5, although the viscosity determined at 1040°C for the glasses was found to increase gradually with increasing Gd/(La+Gd) ratio. X-ray diffraction measurements indicated that the crystalline phase in the glasses below the LT depended strongly on the Gd/(La+Gd) ratio. The effect of the Gd/(La+Gd) ratio on the refractive index, dispersion, density, LT and viscosity is discussed in terms of glass structure.

Introduction

A combination of high performance and compact size is strongly required for digital imaging devices. The development of lens material and lens processing technology is essential to make such imaging devices.

B$_2$O$_3$–La$_2$O$_3$ based glasses are known as lens glasses with high refractive index and low dispersion.[1] Since high refractive index leads to short focal length, this leads to a compact lens unit. Since the amount of chromatic aberration depends on dispersion, aspheric lenses with low dispersion can reduce the number of lenses in a lens unit.

The structure and properties of B$_2$O$_3$–La$_2$O$_3$ based glasses were reported in the 1980s.[2–4] The present authors also reported that the high packing fraction of the glasses in this system is an important factor for achieving high refractive index and low dispersion.[5]

On the other hand, precision moulding technology for producing lens preforms has recently attracted considerable attention. Development of the technology is the key to the mass production of aspheric lenses. In this technology, the viscosity of melt which goes through a small flow pipe for preparing the preform, should range from 8 to 15 dPa.s at the liquidus temperature (LT) to prevent crystallisation in the flow pipe. In general, the LT of B$_2$O$_3$–La$_2$O$_3$ based glasses is relatively high, whereas the viscosity is low. To our knowledge, a few studies have been reported on B$_2$O$_3$–La$_2$O$_3$ based glasses in relation to precision moulding. Hayashi *et al*, found that the viscosity at LT depends on the Gd/(La+Gd) ratio in glasses which they developed for precision moulding.[6] However, they did not explain the reason why the viscosity at LT changed with this ratio. The

purpose of this work is to investigate the influence of rare earth oxides on the optical and thermal properties of B$_2$O$_3$–La$_2$O$_3$ based glasses.

Experiments

Glasses in the 47(B$_2$O$_3$+SiO$_2$)–34(Li$_2$O+ZnO)–15(La$_2$O$_3$+Gd$_2$O$_3$)–4(ZrO$_2$+Ta$_2$O$_5$) (mol%) system were prepared. The total amount of rare earth oxides was kept at 15 mol%. The ratio of Gd$_2$O$_3$ to La$_2$O$_3$+Gd$_2$O$_3$ was varied from 0·3 to 0·7. Batches of 200 g were melted at 1250°C for 2 h in a platinum crucible. The glass melts were poured into a heated carbon mould to obtain glass blocks of thickness about 15 mm. The glasses obtained were annealed at 10°C above the glass transition temperature, T_g, for 4 h and cooled to room temperature at a rate of 60°C/h.

The refractive index at 486·1 nm (H$_2$ F-line), n_F, 587·6 nm (He d-line), n_d, and 656·3 nm (H$_2$ C-line), n_C, were measured by the V-block method using a Kalnew KPR-2 refractometer. The Abbe number, n_d, is defined by

$$v_d = \frac{n_d - 1}{n_F - n_C} \tag{1}$$

The density was measured by Archimedes' method using a Shimadzu SGM300P. T_g was measured by a thermomechanical analyser (TMA) at a heating rate of 5°C/min (BRUKER AXS TD5000SA). A glass rod of 5 mm in diameter and 20 mm in length was used for the TMA measurement. The viscosity (10^0~10^2 dPa s) of the glass melt above the LT was measured by using a rotation viscometer in the temperature range from 900 to 1250°C. A glass melt of volume 85 cm^3 was used in a platinum crucible with diameter of 40 mm. A rotor was sunk in the glass melt and the

*Corresponding author. Email satoru-tomeno@agc.co.jp

Proceedings of the VI International Conference on Borate Glasses, Himeji, Japan, 18–22 August 2008   **179**

Table 1. The properties of the prepared glasses

	Gd/(La+Gd)	Density (g/cm³)	n_d	v_d	T_g (°C)	R_m	V_m
sample0·3	0·3	4·60	1·77270	47·1	575	10·65	25·56
sample0·4	0·4	4·64	1·77155	47·1	575	10·60	25·47
sample0·5	0·5	4·67	1·77024	47·1	575	10·56	25·40
sample0·6	0·6	4·71	1·76914	47·2	579	10·51	25·32

torque value of the rotor was measured. The LT was defined as the lowest temperature where no crystal is precipitated after keeping for 1 h at that temperature. An optical microscope (×100) was used to observe the existence of crystal in the quenched glasses. X-ray diffraction (XRD) was used to identify the crystalline phases present in the glasses.

Results

Density, refractive index, and dispersion

The density, refractive index and dispersion of the B_2O_3–La_2O_3 based glasses are shown in Table 1. The refractive index, n_d, was found to decrease gradually with increasing Gd/(La+Gd) ratio. The Abbe number, n_d, almost remained constant with increasing Gd/(La+Gd) ratio. The molar refraction, R_m, is calculated using the Lorentz–Lorenz equation, given by

$$\frac{R_m}{V_m} = \frac{n^2 - 1}{n^2 + 2} \qquad (2)$$

where n is the refractive index and $V_m=M/d$ is the molar volume (where M is the molar weight of the glass, which is calculated from the composition, and d is the density). The refractive index increases with increasing R_m/V_m ratio, according to Equation (2). Both R_m and V_m values for the glasses in this study decreased simultaneously with increasing Gd/(La+Gd) ratio, as shown in Table 1.

Liquidus temperature

When La_2O_3 was substituted for Gd_2O_3, the LT changed with the Gd/(La+Gd) ratio as shown in Figure 1. The LT showed a minimum of 970°C, when the ratio was approximately 0·5. X-ray diffraction analysis was used to evaluate the precipitated crystalline phases. As shown in Figure 2, the crystalline phase consisted mainly of $LaBO_3$ for the glasses with a ratio of less than 0·5, whereas the main phase

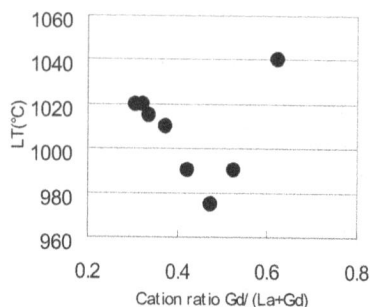

Figure 2. XRD patterns of samples with the Gd/(La+Gd) ratio ranging from 0·3 to 0·6 after heating for 1 h just below the liquidus temperature (▲)$LaBO_3$ (●)$GdBO_3$

was $GdBO_3$ for glasses with a ratio of more than 0·5. Consequently, it was found that the crystalline phase formed at the LT depended strongly on the ratio of the content of rare earth oxides.

Viscosity

The viscosity of the glass melts determined at 1040°C increased gradually with increasing Gd/(La+Ga) ratio, as shown in Figure 3. The viscosity at each LT is plotted as a function of the Gd/(La+Gd) ratio in Figure 4. The viscosity at LT has a maximum when the ratio is approximately 0·5.

Discussion

We discuss the following items: (1) density and optical properties, and (2) viscosity at the LT.

Density and molar volume are related to glass structure. Table 1 shows that the density increases and the molar volume decreases simultaneously with increasing Gd/(La+Gd) ratio. The increase of density, or the decrease of molar volume, is attributed to the radius of rare earth ions. However, since R_m showed stronger dependence on the ratio than V_m, as shown in Table 1, the decrease of the refractive index is mainly due to the decrease of R_m with increasing Gd/(La+Gd) ratio.

The control of viscosity at LT is very important for obtaining a glass preform. Generally, there are two techniques to increase the viscosity at LT. One is to decrease the LT, and the other is to increase the

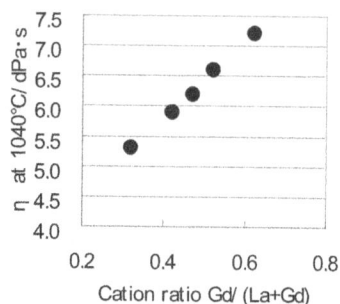

Figure 1. The liquidus temperature (LT) as a function of the Gd/(La+Gd) ratio

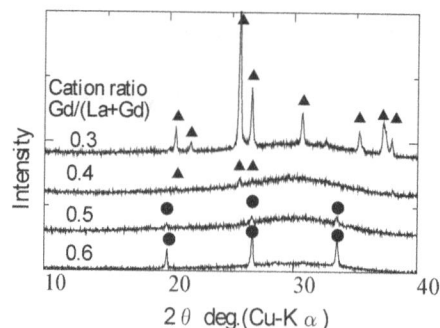

Figure 3. Change in viscosity at 1040°C as a function of the Gd/(La+Gd) ratio

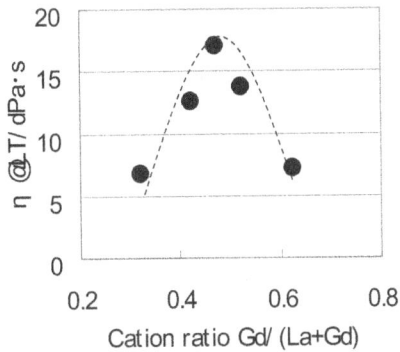

Figure 4. Viscosity at each liquidus temperature as a function of the Gd/(La+Gd) ratio

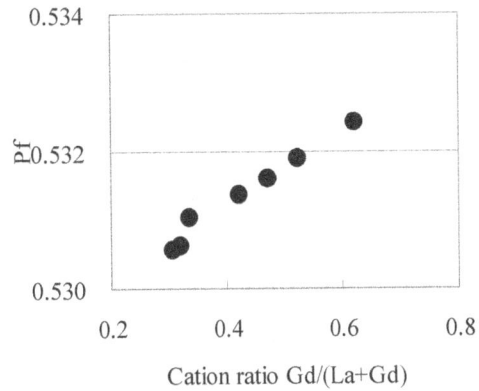

Figure 5. Packing fraction of glasses as a function of the Gd/(La+Gd) ratio

content of network formers such as SiO_2. We have investigated the effect of rare earth oxides on the LT and viscosity. The LT in this glass system showed a minimum, as shown in Figure 1. According to the phase diagram of the $LaBO_3$–$GdBO_3$ system,[7] the LT decreases by mixing $LaBO_3$ and $GdBO_3$, and shows a minimum at the eutectic point where the ratio of $GdBO_3/(LaBO_3+GdBO_3)$ is approximately 0·8. This may be the reason why the LT in this glass system shows a minimum.

On the other hand, the viscosity increased gradually with increasing Gd/(La+Gd) ratio, as shown in Figure 3. The increase of the viscosity can be explained by using the field strength[9] ($=Z/a^2$, where Z is the valence of the ion, and a is a typical ion–oxygen distance) of the rare earth ion. Since the field strength of the Gd^{3+} ion is larger than that of the La^{3+} ion, as shown in Table 2, the ability for Gd^{3+} to attract an oxygen ion is larger than that of La^{3+}. The ionic radii reported by Shannon[10] were used. Although the main network structure of the glasses in this study is considered not to change, the local structure around Gd^{3+} and La^{3+} ions can change, depending on their field strength. Consequently, the bond strength of Gd–O becomes larger than that of La–O, and hence the viscosity increases with increasing Gd/(La+Gd) ratio.

The packing fraction (PF) is given by

$$pf = \sum_i \frac{4\pi r_i^3 n_i}{3V_m} \tag{3}$$

where r_i is the radius of the ith atom, and n_i is the number of i atoms per molar volume.[11] The coordination numbers of cations for the glasses were estimated from their crystalline structure. It is said that the glass structure is similar to the structure of the crystalline phase precipitated firstly in glass.

The coordination numbers of La^{3+} in $LaBO_3$ and of Gd^{3+} in $GdBO_3$, respectively, are 9 and 6.[8] This suggests that the coordination numbers of La^{3+} and

Gd^{3+} in glass are also different. The packing fractions calculated from Equation (3) are shown in Figure 5. The larger packing fraction arising from the smaller radius of Gd^{3+} ions compared to La^{3+} ions may explain the viscosity increase of the glasses.

Conclusions

The influence of Gd/(La+Gd) ratio on the optical and thermal properties of B_2O_3–La_2O_3 based glasses has been investigated. A reason for the viscosity at the liquidus temperature (LT) changing with the ratio has been given. The LT showed a minimum of 970°C, and the viscosity at the LT showed a maximum, for the glass with a ratio of 0·5, although the viscosity determined at 1040°C was found to increase gradually with increasing Gd/(La+Gd) ratio. X-ray diffraction measurements indicated that the crystalline phase in the glasses below the LT depended strongly on the Gd/(La+Gd) ratio, although the change of refractive index is small. The Gd/(La+Gd) ratio is an important factor for controlling the viscosity at the LT in B_2O_3–La_2O_3 based glasses without change of refractive index and dispersion. A new glass composition with high viscosity at the LT, as required for precision moulding, has been obtained.

References

1. Martienssen, W. & Warlimont, H. (Eds.) *Springer Handbook of Condensed Matter and Materials Data*, Springer, Berlin, 2005.
2. Chakraborty, I. N., Shelby, J. E. & Condrate Sr., R. A. *J. Am. Ceram. Soc.*, 1984, **67**, 782.
3. Chakraborty, I. N., Day, D. E., Lapp, J. C. & Shelby, J. E. *J. Am. Ceram. Soc.*, 1985, **68**, 368.
4. Chakraborty, I. N. & Day, D. E. *J. Am. Ceram. Soc.*, 1985, **68**, 641.
5. Sasai, J. & Sugimoto, N. presented at the meeting of the Glass & Optical Materials Division of the American Ceramics Society, Spring, 2006. Note to author: Please provide a proper reference.
6. Hayashi, K., Zou, X. & Hirota, S. *Proceedings of the XX International Congress on Glass*, 2004, O-04-023.
7. Roth, R. S., Waring, J. L. & Levin, E. M. *Proceedings of the Third Rare Earth Conference*, Clearwater, Florida, 1963.
8. Levin, E. M., Roth, R. S. & Martin, J. B. *Am. Miner.*, 1961, **46**, 1030.
9. Dietzel, A. *Z. Elektrochem.*, 1942, **48**, 9.
10. Shannon, R. D. *Acta Cryst. A*, 1976, **32**, 751.
11. Giri, S., Gaebler, C., Helmus, J., Affatigato, M., Feller, S. & Kodama, M. *J. Non-Cryst. Solids*, 2004, **347**, 87.

Table 2. Difference in field strength of cations

Element	Valence Z	Coordination number	Ionic radius	Field strength
La	3	9	1·22	2·02
Gd	3	6	0·94	3·4

Proc. VI Int. Conf. Borate Glasses, Himeji, Japan, 18–22 August 2008 *Glass Tech.: Eur. J. Glass Sci. Technol. A,* October 2009, **50** (5), 277–279

Sintering behaviour of $BaO-P_2O_5-B_2O_3$ glass powders

Hiromichi Takebe,[1]* *Wataru Nonaka*[2] *& Makoto Kuwabara*[2]

[1] *Graduate School of Science and Engineering, Ehime University, 3, Bunkyo-cho, Matsuyama, Ehime 790-8577, Japan*
[2] *Graduate School of Engineering Sciences, Kyushu University, 6-1, Kasugakouen, Kasuga, Fukuoka 816-8580, Japan*

Manuscript received 31 October 2008
Revised version received 21 February 2009
Accepted 25 April 2009

The effect of B_2O_3 additions on the sintering behaviour of $BaO-P_2O_5-B_2O_3$ (BPB) glass powders has been studied. The sintering temperature and time were optimised for the densification of BPB powder compacts. The factors affecting the optical transmittance in the visible region for BPB sintered compacts are discussed in terms of microstructure.

Introduction

Boric oxide (B_2O_3) is a typical network former, and the addition of B_2O_3 extends the glass formation region in various multicomponent glasses, e.g. phosphate,[1] tellurite[2] and antimonate[3] glasses, in addition to archetypal borosilicate glasses.

Previous studies reveal that B_2O_3 addition improves the thermal stability of $BaO-P_2O_5$ metaphosphate glasses against surface crystallisation at ≥3 mol% B_2O_3 for the bulk form[1] and at 7·5–12 mol% B_2O_3 for the powdered form.[4] This study reports the effect of B_2O_3 additions on the sintering behaviour of $BaO-P_2O_5-B_2O_3$ (BPB) powders, which are used for sealing and transmitting applications. The factors affecting the transmittance of BPB sintered compacts are also discussed in terms of microstructure.

Experimental

The glass compositions studied are from the series $(50-x/2)BaO.(50-x/2)P_2O_5.xB_2O_3$, where x=3, 6, 8, 12 mol% B_2O_3. BPB bulk glasses were prepared by a conventional melt quenching method using commercial 99·99% $Ba(PO_3)_2$ and 99·9% B_2O_3 powders.[1] The bulk samples were ground using an alumina mortar and pestle, and the particle size distribution was adjusted to ≤32 μm, or the range 32–45 μm by sifting using two types of stainless sieves.

Powder pellets with 10 mm diameter were pressed using a stainless die with pressures of 186 or 19 MPa. The samples were heated at a continuous rate of 10°C min⁻¹ and were then pressureless sintered at 480–610°C for 1–10 h in air. After sintering, x-ray diffraction analysis confirmed that no crystalline phases were present for all BPB sintered samples.

The density, ρ, was evaluated by Archimedes' method, using kerosene as an immersion liquid.

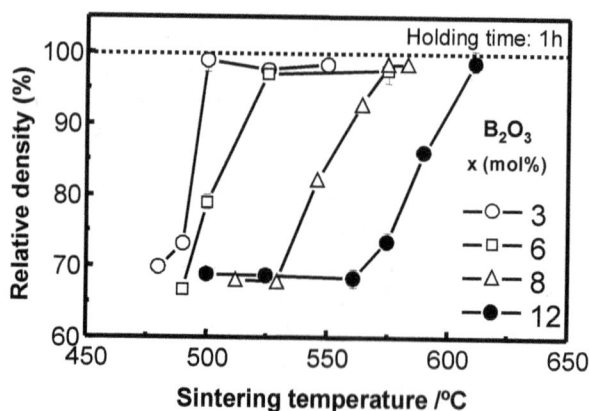

Figure 1. Variation of relative density with sintering temperature for BPB powdered compacts. The samples were sintered for 1 h at each temperature

Relative density was calculated as the ratio of ρ of the sintered compact and ρ of the corresponding bulk glass for each composition. The density was evaluated at least three times for each sintering condition. The microstructure was observed by FE-SEM (S-4500, Hitachi). Optical transmittance in the region of 200–900 nm was measured by a spectrophotomer (V-550, JASCO) for BPB sintered samples with 0·45 mm thickness.

Results

Figure 1 shows the variation of relative density with sintering temperature for various B_2O_3 concentrations. The samples made using BPB powders ground to less than 32 μm as starting materials were pressed at 186 MPa and were sintered for 1 h at each temperature. The minimum temperature required for densification with ≥98% relative density (RD) increases with increasing B_2O_3 concentration, with values of 500, 525, 570 and 615°C for 3, 6, 8 and 12 mol% B_2O_3, respectively.

*Corresponding author. Email takebe@eng.ehime-u.ac.jp

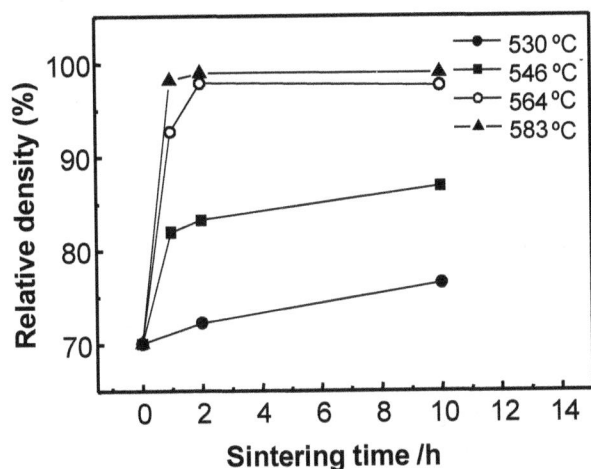

Figure 2. Variation of relative density with sintering time for BPB powder compacts with a composition of $46BaO.46P_2O_5.8B_2O_3$

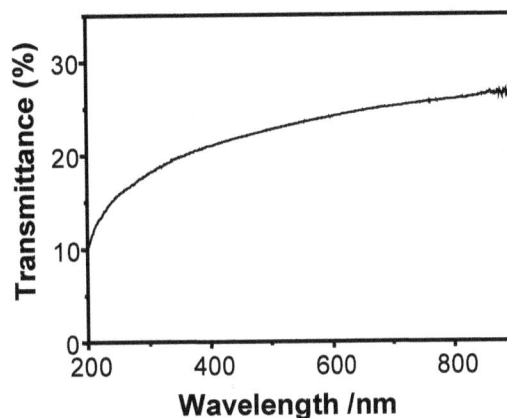

Figure 4. Transmission spectrum of a 0·45 mm thick BPB sintered sample with a composition of $46BaO.46P_2O_5.8B_2O_3$. The particle size range of the starting powder is 32–45 μm

The effects of sintering temperature and time were studied for an optimised composition of $46BaO.46P_2O_5.8B_2O_3$. These samples were also prepared from BPB powder ground to less than 32 μm and were pressed at 186 MPa. Figure 2 shows the variation of RD with sintering time at 530, 546, 564 and 583°C. The RD increases with increasing sintering time at all temperatures. Full densification (≥98% RD) occurs after 2 h at 564 and 583°C, but takes longer for lower temperatures.

Figures 3 and 4 show the appearance and transmission spectrum of a sintered BPB plate sample with ≥98% RD as a representative sample. The sample was sintered at 583°C for 4 h using the ground powder with particle size in the range 32–45 μm as the starting material. The BPB power compacts were formed with a relatively low pressure of 19 MPa for transmittance experiments to minimise the change of particle size distribution during the forming process. Table 1 summarises the transmittance, T, at 700 nm for representative BPB samples sintered at 583°C with a composition of $46BaO.46P_2O_5.8B_2O_3$. A narrower particle size distribution (32–45 μm) of the BPB start-

ing material results in higher transmittance in the visible region, compared to the result for particle size ≤32 μm. The longer sintering times also give higher transmittance for the 32–45 μm samples.

Figure 5 shows a typical example of the microstructure of the BPB sintered samples with ≥98% RD. The microstructure includes micro-size spherical pores with large size in comparison to visible light wavelengths. All the other sintered samples exhibited similar microstructures. The effect of microstructure on visible transmittance is considered in the discussion.

Discussion

As shown in Figure 1, the minimum temperature required for densification with ≥98% RD increases with increasing B_2O_3 concentration. A previous study of BPB glasses[5] has used a penetration method to measure the viscosity in the range from 10^7 to 10^{11} Pa s, showing that the viscosity increases gradually with increasing B_2O_3 content. The viscosity was also related to modified glass networks of PO_4 Q^2 chains with diborate groups and PO_4–BO_4 tetrahedra. A comparison between the results shown in Figure 1 and the viscosity data[5] shows that the viscosities required for densification (≥98% RD) are 10·8, 10·0, 8·3 and 8·0 Pa s for 3, 6, 8, 12 mol% B_2O_3, respectively.

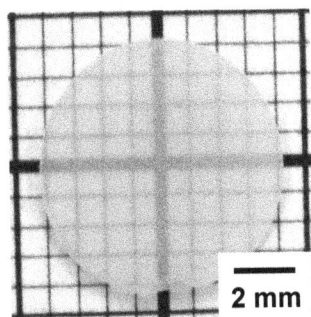

Figure 3. Appearance of a BPB sintered compact with a composition of $46BaO.46P_2O_5.8B_2O_3$. The sample was sintered at 583°C for 4 h. The thickness of the sample is 0·45 mm. The surface was optically polished. The particle size range of the starting powder is 32–45 μm

Figure 5. A SEM micrograph of the fracture surface of a sintered glass sample with a composition of $46BaO.46P_2O_5.8B_2O_3$

Table 1. *The values of transmittance, T, at 700 nm, the number density, n, and the diameter, a, of the micro pores and the distance, d, between nearest neighbouring pores for sintered BPB glasses with a composition of $46BaO.46P_2O_5.8B_2O_3$*

Particle size and sintering time	T (%) (at 700 nm)	n (mm⁻²)	a (μm)	d (μm)
≤32 μm, 2 h	0·9	1033	5	20
32–45 μm, 2 h	2·1	303	16	34
32–45 μm, 2·6 h	15·9	198	14	40
32–45 μm, 4 h	25·1	198	15	42

If the sintering of BPB ground powders occurs by viscous flow, the viscosity required for densification may show a similar value for all B_2O_3 concentrations. The ground surface of low B_2O_3 samples (3 and 6 mol% B_2O_3) mainly consists of P–O–P and P–O⁻ bonds and that of high B_2O_3 samples (≥8 mol% B_2O_3) contains P–O–B bonds in addition to P–O–P and P–O⁻ bonds.[4] The surface bonding state relates to surface tension and may affect the sintering behaviour of the BPB powder compacts.

In order to understand the factors affecting the microstructure, parameters were evaluated for the fracture surfaces of sintered BPB glasses. The average number density, *n*, and diameter, *a*, of the micro pores and distance, *d*, between pores were evaluated from SEM micrographs with 430 μm × 220 μm area. As shown in Table 1, the transmittance increases with decreasing number density of the micropores and with increasing distance between the pores. Our results (Table 1) show qualitatively that the scattering centres due to the micropores affect the transmittance of the BPB sintered glasses.

Conclusions

The sintering behaviour of $(50–x/2)BaO.(50–x/2)P_2O_5.xB_2O_3$ (BPB) glass powder compacts has been studied. The sintering temperature required for densification to ≥98% relative density increases from 500°C to 615°C as the B_2O_3 content increases from 3 to 12 mol% B_2O_3. The microstructure reveals that the transmittance in the visible region for BPB sintered glasses with ≥98% relative density increases with decreasing number density of micropores and with increasing distance between the pores.

References

1. Harada, T., In, H., Takebe, H. & Morinaga, K., *J. Am. Ceram. Soc.*, 2004, **87**, 408.
2. Cho, D. H., Choi, Y. G. & Kim, K. H., *ETRI J.*, 2001, **23**, 151.
3. Holland, D., Hannon, A. C., Smith, M. E., Johnson, C. E., Thomas, M. F. & Beesley, A. M., *Solid State Nucl. Mag.*, 2004, **26**, 172.
4. Harada, T., Takebe, H. & Kuwabara, M., *J. Am. Ceram. Soc.*, 2006, **89**, 247.
5. Takebe, H., Nonaka, W., Cha, J. & Kuwabara, M., *Phys. Chem. Glasses: Eur. J. Glass Sci. Technol. B*, 2007, **48**, 113.

Proc. VI Int. Conf. Borate Glasses, Himeji, Japan, 18–22 August 2008   *Phys. Chem. Glasses: Eur. J. Glass Sci. Technol. B*, December 2009, **50** (6), 335–342

Studies of borate glass structure using laser ionization time of flight mass spectrometry

Mario Affatigato,* Steve Feller, Rob Kramer, Amy Marquardt, Kristin Mertel, Sarah Blair, Dale Stentz & David Crist

Physics Department, Coe College, 1220 First Av. NE, Cedar Rapids, Iowa 52402, USA

Manuscript received 18 August 2008
Revised version received 15 July 2009
Accepted 30 September 2009

We report on our investigations on the structure of borate glasses using laser ionization time of flight mass spectrometry (LITOF-MS). This technique has allowed for new insights on the characteristics of borate networks at the short and intermediate range order, with the most recent work aimed at quantifying the structural changes. LITOF-MS measures the mass of negative and positive fragments gently desorbed from the sample at low laser powers. In particular, this report discusses binary borates containing lead, tellurium, and bismuth oxides; ternary systems (sodium-doped lead borates, europium-doped lead borates); and contrasts with other glass systems. Comparisons of the time of flight data to the results of Raman and other spectroscopies are also made.

1. Introduction

Borate glasses have proven to be a useful playground for structural investigation, and for the testing of new characterisation techniques. Among the most productive such techniques we can include optical techniques such as FTIR and Raman,[1] neutron and x-ray scattering, and solid state nuclear magnetic resonance (NMR). Amongst their more important contributions is a clear understanding of the short range order of the glass.

One of the most important remaining questions on the structure of borates pertains to the presence or absence of superstructural units in the glass network. These units establish an intermediate or medium range order in the oxide structure, affecting a variety of the properties of the glass, and they are typically constituted by 7–20 or more atoms. Information on the possible arrangement of these meso-units can be derived from the study of crystal structures isocompositional with the glasses. Figure 1 shows some[2] of these larger groupings obtained from alkali borate crystals. The infrared or Raman signatures of the meso-units can be obtained from the crystals, and the samples can be checked for the presence or absence of these signatures, but quantification of the abundance of the superstructural units is difficult. Though the techniques mentioned above have the advantage of decades of development, they do have some disadvantages. Infrared and Raman spectroscopies do not yield quantitative information as a matter of course; neutron and x-ray scattering often require expensive isotope enrichment and large scale instrumental facilities; and solid state NMR can be

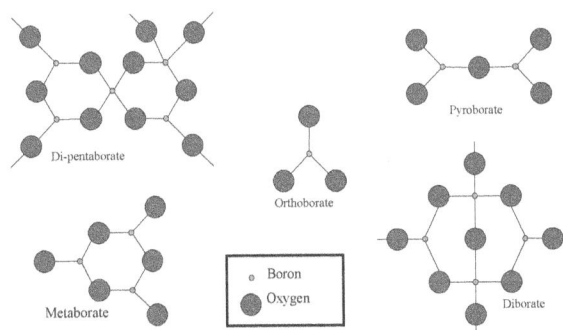

Figure 1. Intermediate range borate groupings derived from alkali borate crystals

difficult to interpret.

Recent advances in neutron[3] and x-ray scattering[4] as well as NMR[5–8] have allowed for further scrutiny of the intermediate range order. Inelastic neutron scattering provides a better look[9,10] at the vibrational density of states than either infrared or Raman spectroscopy, and it can attain quantitative results.

Laser induced ionization time of flight mass spectrometry (LITOF-MS) is a technique that has only been applied to the field of glass science for a little over a decade.[11–15] The method works by gently desorbing fragments of the glass network using a low power laser, avoiding any surface damage. The mass of the units thus removed is then determined with sub-isotopic accuracy using a time of flight mass spectrometer. Computer simulation based on the natural (or enriched) isotopic abundances can then yield the exact chemical composition of the fragments. This information can be used to obtain insights about the borate network, and to try to reconstruct the glass structure. The technique has several advantages, including ease of measurement, relatively low cost,

* Corresponding author. Email maffatig@coe.edu

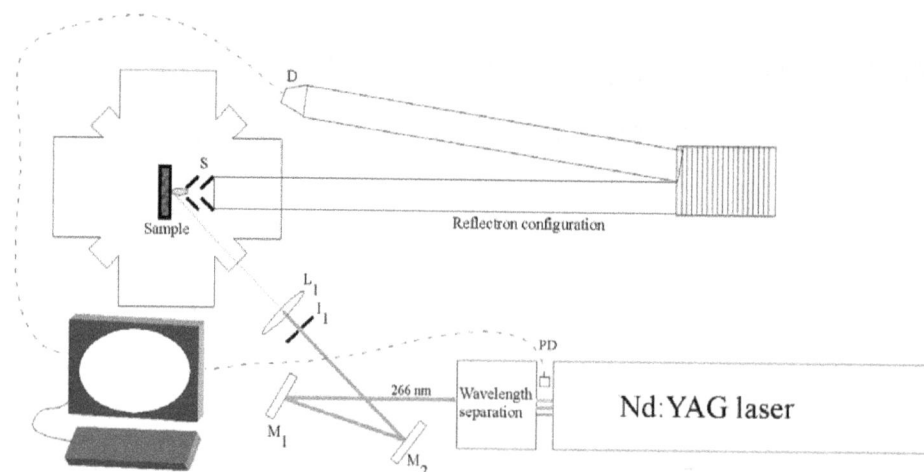

Figure 2. Instrumental diagram of the laser ionization time of flight mass spectrometer. M_1 and M_2 are mirrors, L_1 is the long focus lens, PD is the photodetector, D is the MCP detector, S is the source region, and I is the iris

no need for special sample preparation, and applicability to all glass systems that absorb at the laser wavelength.

In this paper we review our past LITOF-MS work on borate-containing systems.

2. Experimental procedure

2.1 Glassmaking

All glasses were made from high purity (>99·99%) starting materials to reduce impurities in the final mass spectra. The powders were mixed in a platinum crucible and heated in an electric muffle furnace. Furnace times varied, but were typically in the range 10–20 min. A weight loss measurement was taken after 5–10 min, and the result was compared to our expectation from stoichiometric calculations. All glasses had losses within 0·2 g of expected, a variation attributed to fusion of the powders. After the glass was remelted, it was rapidly cooled using a roller quencher, which attains a cooling rate of 100 000°C/s. This ensured wide ranges of glass formation, as well as a uniform thermal history for all the samples.

2.2 Setup of LITOF-MS instrument

Figure 2 illustrates the instrumental setup of the spectrometer. Both a low power nitrogen laser (337·1 nm, <100 μJ per pulse, 10 Hz) and, more recently, a frequency quadrupled Nd:YAG laser (266 nm, <100 μJ/pulse, 10 Hz) have been used to irradiate the sample which is inside a vacuum chamber at 10^{-7} Torr. All the laser pulses have a duration of 4 ns, yielding a peak power of only 25 kW (on sample). The fragments that are removed are extracted from the source region using a pulsed, time delayed electric field, and then fly down a drift tube. An ion mirror reverses their direction, attaining an even better mass resolution, and the ionic fragments then arrive at a microchannel plate detector. The instrument can measure positive or negative ions separately, but cannot detect neutral atoms or molecules.

The instrument was previously calibrated for mass arrival times using known standards, such as NaCl, KCl, or, if higher masses are needed, PbO. The computer displays a mass spectrum, in which the mass/unit charge (in amu/e) is on the x-axis, and the signal intensity, proportional to the number of ions of that particular mass, is shown on the y-axis. The charge on a particular group cannot be determined from its location on the mass spectrum, but is easily measured from the separation of the individual isotopic components.

The resolution of the instrument is sub-isotopic, with a typical value $M/\Delta M=1500$. Considering both the peak width, and the shot-to-shot and sample-to-sample variability, we estimate the net mass error to be 0·4 amu. We also note that each spectrum reported here is an average of 100 laser shots and their subsequent mass spectra.

All the borate glasses reported herein were analysed for evidence of laser damage. This technique can only be used for structural characterisation if each laser shot does not alter the surface significantly. Otherwise the next laser shot might analyse a chemically different glass after every laser pulse. To check for this we obtained scanning electron micrographs (SEMs) for multiple glass samples after several thousand shots at full power (no attenuation). The surfaces are unscathed after this treatment, virtually unrecognisable from the unirradiated control samples. They also yield similar TOF spectra.

2.3. Spectral analysis and relative quantification

Figure 3(a) displays a typical negative ion time of flight spectrum from a lead borate glass. The resolution of the data is high enough that mass multiplets appear condensed in the figure, and thus Figure

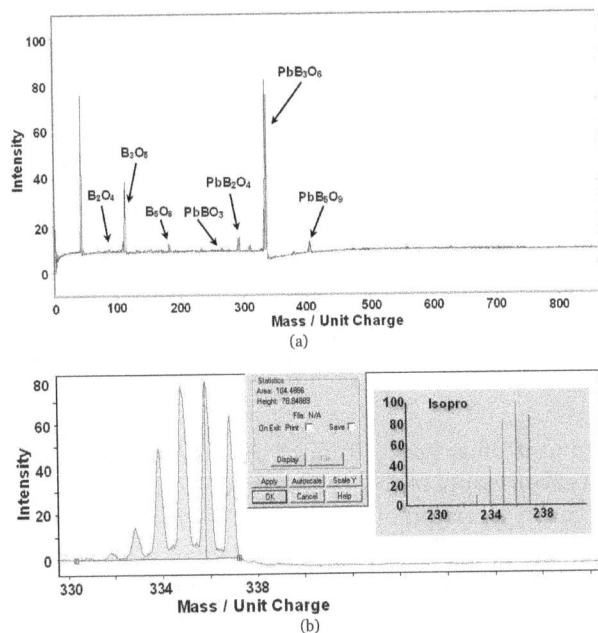

Figure 3. (a) Typical negative ion spectrum of $58 \cdot 3PbO.41 \cdot 7B_2O_3$ glass with larger peaks marked; (b) Enlarged portion of the spectrum of $58 \cdot 3PbO.41 \cdot 7B_2O_3$ glass, highlighting the $(PbB_3O_6)^-$ mass multiplet and its area integration. Also shown on the right is the IsoPro® simulation of the same multiplet with $M/\Delta M = 1000$

3(b) is an enlargement of the $(PbB_3O_6)^-$ multiplet. Each individual peak corresponds to a variation of the $(PbB_3O_6)^-$ group, each composed of a different combination of isotopes of lead, boron, and oxygen. It is this signature pattern that allows for the unambiguous identification of the elements that compose each mass grouping. IsoPro® is modelling software that simulates the appearance of the spectral feature given a guess of the atomic components and the natural abundance of their isotopes. Figure 3(b) also shows the result of one such simulation for the $(PbB_3O_6)^-$ group. The clear match between the simulation and the data, down to the smallest features, is unambiguous evidence that this mass multiplet is indeed $(PbB_3O_6)^-$. Further confirmation can be obtained by using isotopically enriched starting compounds.

For clarity, we will refer to the largest mass peak in the multiplet when discussing the individual units, although the multiplets are often made up of several peaks. Also, all of the units reported herein have a charge of ±1, regardless of the valences of the ions that constitute them. This is quite common in laser desorption.

For the quantification calculations, we first acquired all spectra under the same laser power and gain conditions. We then integrated the area under each mass multiplet using GRAMS32® for each composition. By dividing the calculated area by the net area (e.g. the sum of the integrations of all the multiplets) we were able to quantify the relative abundance of each unit.

3. Results

The LITOF-MS technique can provide a new approach to give qualitative insights into the structure of borates glasses. Some examples are given below:

3.1. Binary borates

3.1.1 Lead borates

The lead borate system $xPbO.(100-x)B_2O_3$, yields very stable and repeatable data over a wide range of compositions. It was one of the original systems investigated with LITOF-MS because of its strong optical absorption at the available laser wavelengths, 337·1 and 266 nm. Both wavelengths yielded the same mass spectra for all compositions.

Figure 3(a) shows the negative ion spectrum for $58 \cdot 3PbO.41 \cdot 7B_2O_3$ glass. The dominant feature is the $(PbB_3O_6)^-$ multiplet, though the $(B_3O_5)^-$ grouping also shows some strength. We have previously assigned[11] the $(PbB_3O_6)^-$ unit to a boroxol/metaborate (B/MB) ring, noting that it is impossible for the LITOF-MS technique to describe the geometry of a unit.

A comparison of negative ion spectra as a function of composition is shown in Figure 4. It is especially notable that the $(PbB_3O_6)^-$ unit not only survives at high lead oxide contents, but actually appears to increase in relative abundance. Some smaller groupings $(PbB_2O_4^-, PbBO_3^-)$, which we associated with the formation of ortho- and pyroborate units, also show some growth. Also of interest is the behaviour of the $(PbB_5O_9)^-$ unit, which we attribute to the diborate ring structure. We will discuss the relative abundance of the intermediate range units in a later section.

The positive ion spectrum displayed in Figure 5 highlights the presence of lead and lead oxide units in a glass with 58 mol% PbO. By this composition we also see the formation of many $(Pb_2BO_3)^+$ units, indicating the fragmentation of the borate network by the lead, and the formation of more ionic groupings. In general, as the lead content is increased from 58·3 mol% PbO, we see a decrease in single lead ions, and a growth in the abundance of groups with two or three lead ions, as we would expect.

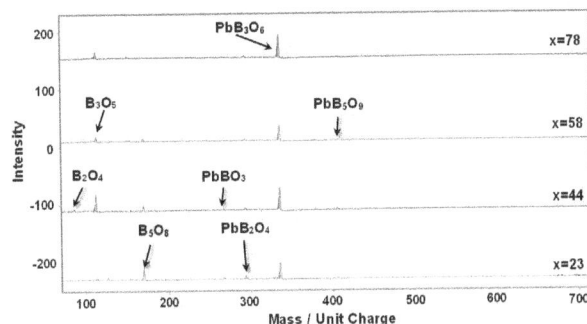

Figure 4. Comparison of the mass spectra of lead borate glasses of varying compositions. "x" denotes the molar percent of PbO in the glass

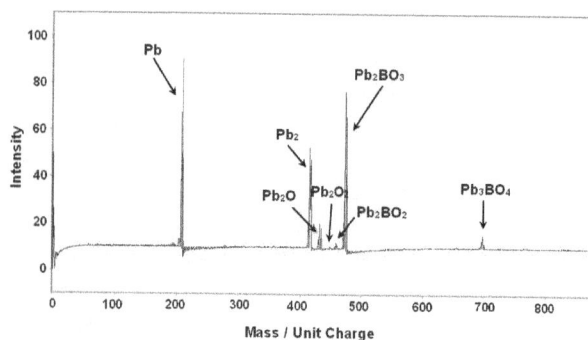

Figure 5. Typical positive ion spectrum of 58·3PbO.41·7B$_2$O$_3$ glass

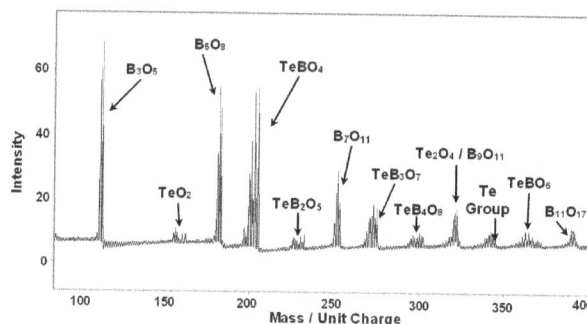

Figure 7. Typical negative ion spectrum of 70TeO$_2$.30B$_2$O$_3$ glass

3.1.2 Bismuth borates

The bismuth borate system shares some similarities with the lead borates. In the negative ion spectra of 15Bi$_2$O$_3$.85B$_2$O$_3$ glass, shown in Figure 6, the $(B_3O_5)^-$ and $(B_5O_8)^-$ groupings are very abundant, indicating a strong borate network. The mixed units include $(BiB_2O_5)^-$, $(BiB_2O_4)^-$, and $(BiB_4O_8)^-$. In general, the bismuth borate network does not show the same ring dominance that we observe in the lead borates, exhibiting greater fragmentation for similar compositions. This might be due to a weaker bismuth oxide subnetwork, not as strong or covalently bonded as the lead oxides. It has been observed[16] that the bismuth ions have a (3+3) coordination and an ill defined coordination sphere. This may lead to a larger variety of environments and more fragmentation of the network.

In the positive spectra, many similarities are noted with the lead borates: at high Bi$_2$O$_3$ contents, the LITOF-MS technique sees an increase in the formation of $(Bi_x)^+$ and $(Bi_xO_y)^+$ groupings. Even though bismuth may not form a strong subnetwork, there is plenty of evidence of bismuth oxide in the glass, and perhaps even small metallic clusters of a few atoms.

3.1.3 Tellurium borates

The case of tellurium borates is interesting as it combines a traditional glass former (B$_2$O$_3$) and a conditional glass former (TeO$_2$). To attain good spectra it was necessary to have a tellurium oxide

content of at least 33 mol% as this produced the necessary optical absorption. Figure 7 shows a negative ion spectrum of 70TeO$_2$.30B$_2$O$_3$ glass, showing a variety of single glassformer and mixed glassformer multiplets. Some of the more dominant features are purely borate in nature, including $(B_5O_8)^-$ and $(B_7O_{11})^-$, but the spectrum is also rich in mixed units, including $(TeB_2O_5)^-$, $(TeB_3O_7)^-$, and $(TeBO_6)^-$. The absence of a pure boroxol/metaborate ring group is notable, and it is likely to be linked to the high tellurium content and the mixing of the two networks. Ring structures may well be present with tellurium cations attached, as in the multiplets $(TeB_3O_7)^-$ and $(TeB_4O_8)^-$, but in lower concentrations than in the lead borate system. The pure boron oxide groups appear to grow by the sequential addition of a unit of B$_2$O$_3$ ($B_3O_5 \rightarrow B_5O_8 \rightarrow B_7O_{11} \rightarrow B_9O_{14}$), suggesting a chain-like construction, quite different from the lead borate case.

Figure 8 shows the changes in the different mass spectra as a function of the glass composition. It is not surprising to observe the growth of the tellurium containing units as the TeO$_2$ content increases. It is worthy of note that the largest increase occurs for the TeBO$_4$ unit, indicating the formation of smaller ionic units, and the diminishing role of intermediate range units. From an integration of the multiplets, we note that the relative abundance of the TeBO$_4$ unit grows from approximately 18% at x=50 mol% TeO$_2$ to 30% at x=78 mol% TeO$_2$, with an error of ±10%. In the glass with x=70 mol% TeO$_2$, shown in Figure 7, only 7% of the boron atoms are found to be in the possible ring

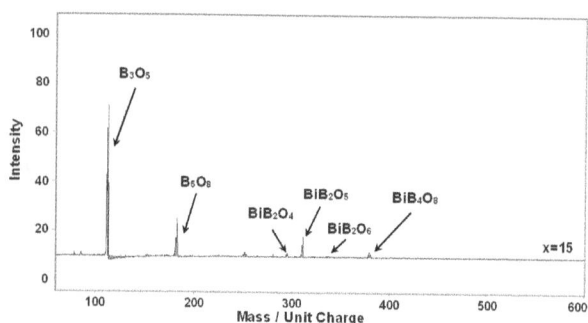

Figure 6. Typical negative ion spectrum of 15Bi$_2$O$_3$.85B$_2$O$_3$ glass

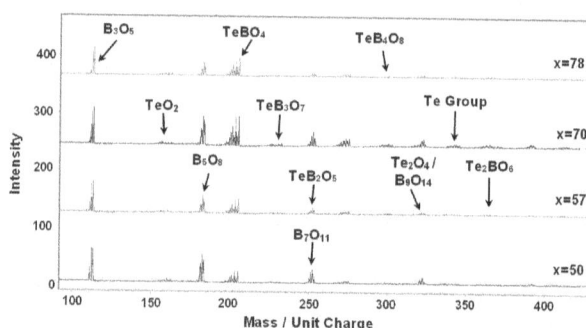

Figure 8. Comparison of the mass spectra of tellurium borate glasses of varying compositions. "x" denotes the molar percent of TeO$_2$ in the glass

(a)

(b)

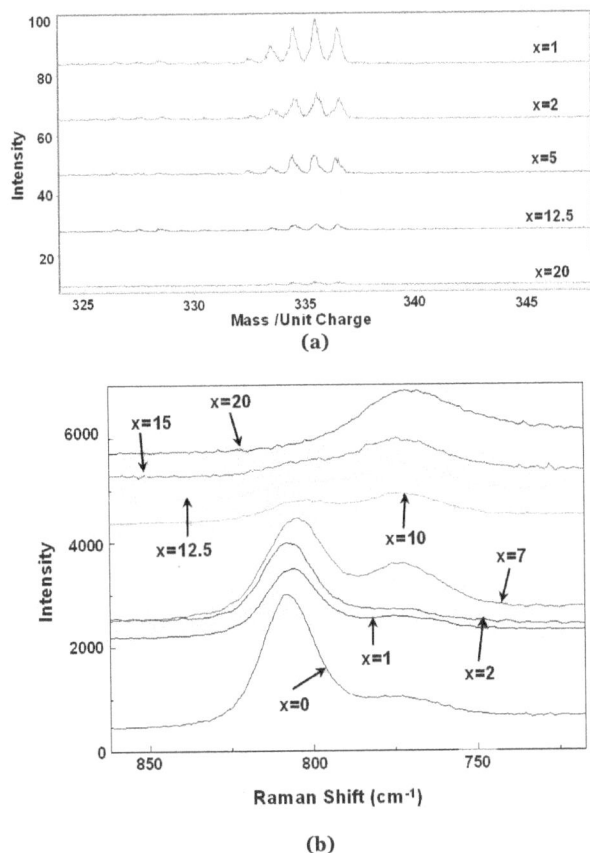

Figure 9. (a) Comparison of the mass spectra, highlighting the region around the $(PbB_3O_6)^-$ mass multiplet, as a function of sodium content in the glasses $xNa_2O.(100-x)(10PbO.90B_2O_3)$. All spectra are shown with the same scale. (b) Comparison of the Raman spectra of the glasses $xNa_2O.(100-x)(10PbO.90B_2O_3)$, near the 808 cm^{-1} boroxol ring signature

structures, $(TeB_3O_7)^-$ and $(TeB_4O_8)^-$.

From the analysis of the tellurium borate negative spectra, we are led to believe that the tellurium borates may form some intermediate range units, but these groupings are not as stable as the ones in lead or bismuth borates. It is likely that the glassforming tellurite units mix well with the boron network and lead to more inhomogeneity over our reported range.

3.2. Ternary borates

3.2.1 Na-doped lead borates

Sodium-doped lead borate glasses, made according to the stoichiometry $xNa_2O.(100-x)(10PbO.90B_2O_3)$, show the impact of alkali doping on the larger structural units. Figure 9(a) illustrates the clear decay in the abundance of the $(PbB_3O_6)^{-1}$ unit as the amount of alkali is increased, eventually leading to its disappearance by $x=20$ mol%. This is in accord with Figure 9(b), in which the Raman spectra of the same glasses show fading of the well known 808 cm^{-1} boroxol ring resonance, though there is growth in the triborate region (~780 cm^{-1}). This suggests a

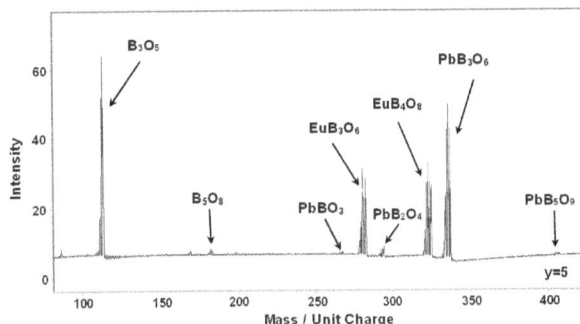

Figure 10. Typical negative ion spectrum of $5Eu_2O_3$. $95(50PbO.50B_2O_3)$ glass, with relevant peaks marked

very different behaviour from the lead borates: the sodium ions appear to fragment the borate network far more efficiently, leading to a quick decrease in the B/MB ring population.

3.2.2 Eu-doped lead borates

Figure 10 shows the negative ion mass spectrum for the sample $5Eu_2O_3.95(50PbO.50B_2O_3)$, over a relevant mass range. The spectrum shows pure borate ionic groups $[(B_3O_5)^{-1}, (B_5O_8)^{-1}]$, some mixed lead borate groups $[(PbBO_3)^-, (PbB_2O_4)^-, (PbB_3O_6)^-,$ and $(PbB_5O_9)^-]$, and two mixed europium borates, $[(EuB_3O_6)^-, (EuB_4O_8)^-]$. Given our previous comments on the lead borate LITOF-MS data, we tentatively identify the $(EuB_3O_6)^-$ unit with a metaborate ring containing a europium ion in substitution for a lead ion. There appears to be a competition between the lead and europium cations for occupation of sites next to the rings. The $(EuB_4O_8)^-$ multiplet, on the other hand, is slightly less abundant, and could easily be formed by the addition of a BO_2 unit to the $(EuB_3O_6)^-$ ionic group. It is worthy of note that there are no *negative* pure europium oxide (Eu_xO_y) groups.

In contrast with many glass families we have studied, the positive spectra are structurally interesting, and a typical one is shown in Figure 11. The figure displays the relevant mass spectrum for the $5Eu_2O_3.95(50PbO.50B_2O_3)$ sample. Other than the expected lead ions, Pb^+, there are several lead borate

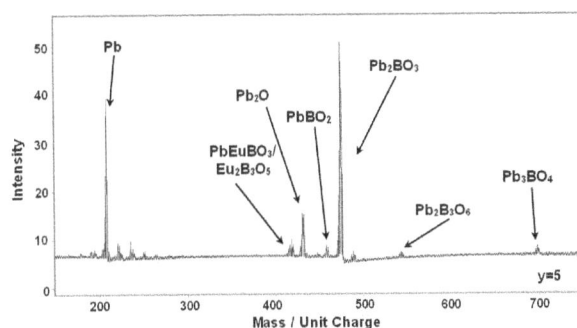

Figure 11. Typical positive ion spectrum of $5Eu_2O_3.95(50PbO.50B_2O_3)$ glass, with relevant peaks marked

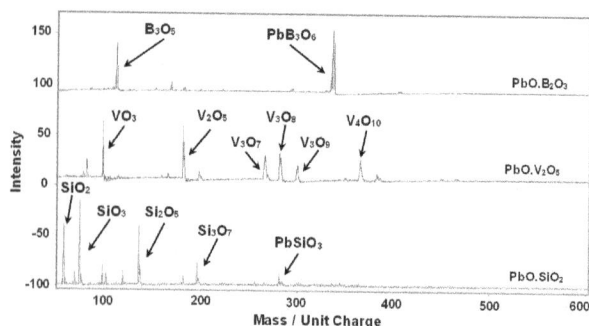

Figure 12. Comparison of the time of flight mass spectra of lead borate, vanadate, and silicate glasses. All glasses had 50 mol% PbO

species $[(Pb_2BO_2)^+, (Pb_2BO_3)^+, (Pb_2B_3O_6)^+, (Pb_3BO_4)^+]$, one europium borate group, $(Eu_2B_3O_5)^+$, in small abundance, and, interestingly, one mixed europium lead borate species, $(PbEuBO_3)^+$. Our calculations show that the strength of this feature is more than expected from stoichiometry and random substitution alone, pointing to preferential occupation of borate sites by europium ions and a consequent displacement of the lead ions.

4. Discussion

4.1. Comparison with other glass systems

Figure 12 illustrates the differences between a lead borate LITOF-MS spectrum and ones from the lead vanadate and lead silicate systems, all with 50 mol% PbO. The greatest difference is that the spectra in the case of the borates point to stable, distinct super-structural units. This is very different from the other two cases. The spectra for the vanadates and the silicates show a larger number of different fragments, molecular groupings that are often simply additions of smaller units. In the vanadates, for instance, most of the peaks in this negative spectrum can be constructed by adding VO_2 or VO_3 units. Thus, for the most intense mass peaks in the lead vanadates

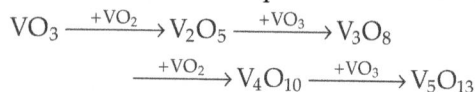

$$VO_3 \xrightarrow{+VO_2} V_2O_5 \xrightarrow{+VO_3} V_3O_8$$
$$\xrightarrow{+VO_2} V_4O_{10} \xrightarrow{+VO_3} V_5O_{13}$$

The negative lead silicates can be similarly reconstructed yielding

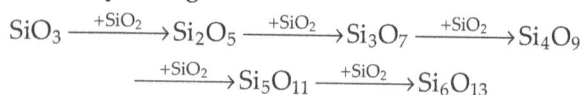

$$SiO_3 \xrightarrow{+SiO_2} Si_2O_5 \xrightarrow{+SiO_2} Si_3O_7 \xrightarrow{+SiO_2} Si_4O_9$$
$$\xrightarrow{+SiO_2} Si_5O_{11} \xrightarrow{+SiO_2} Si_6O_{13}$$

and

$$SiO_2 \xrightarrow{+SiO_2} Si_2O_4 \xrightarrow{+SiO_2} Si_3O_6 \xrightarrow{+SiO_2} Si_4O_8$$

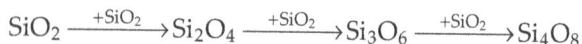

The other silicate units can be obtained by adding a lead cation to the anionic units, as in $PbSiO_3$, $PbSi_2O_5$ and $PbSi_3O_7$. Hence, the larger structures of the vanadates and silicates appear to be more chain-like, made up of small links (fragments) and do not have any added stability. This can be seen by the diminishing number of larger molecular groupings: the $(Si_6O_{13})^-$

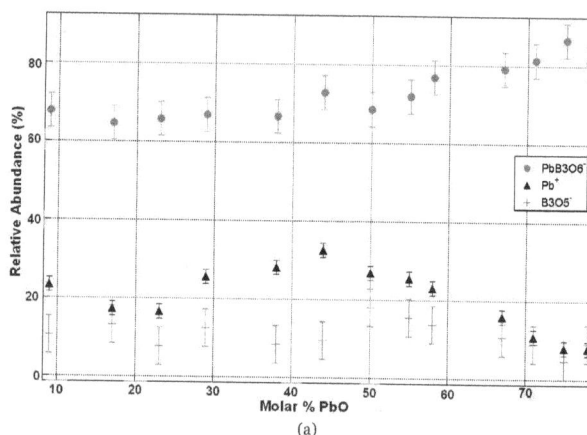

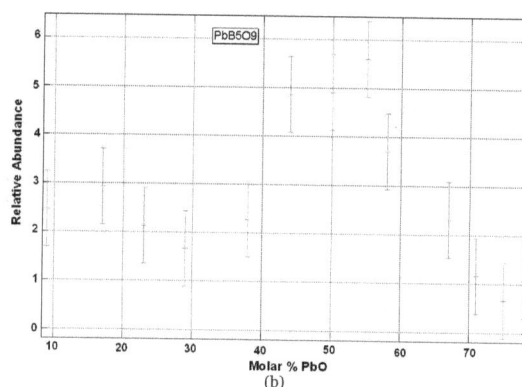

Figure 13. (a) Relative abundance of three mass multiplets for lead borate glasses: $(Pb)^+$; $(B_3O_5)^-$; and $(PbB_3O_6)^-$. The abundances were calculated from area integration of the multiplets. (b) Relative abundance of the $(PbB_5O_9)^-$ multiplet, identified as the diborate unit, as a function of PbO content in lead borate glasses

unit is less abundant than the $(Si_2O_5)^-$ unit, as would be expected. The lead borates differ significantly: the $(PbB_3O_6)^-$ unit is stable, present even at higher lead contents, and more abundant than some of the smaller units.

4.2 Quantification: lead borates

Our most recent work has aimed to quantify the abundance of intermediate range order units present in borate glasses. A relative measure can be obtained by integrating the area under a multiplet, and then dividing the result by the net area under all the multiplets in the spectrum. This calculation can be especially useful when comparing unit trends across compositions, and it has allowed us to create more coherent models.

Figure 13(a) shows the relative abundance of three different units present in a lead borate glass: the single, positive lead cation, Pb^+; the negative $(B_3O_5)^-$ pure borate unit; and the mixed $(PbB_3O_6)^-$ unit. The lead cation shows a general increase that roughly follows its concentration in the glass, starting from 0% concentration, but begins to level off at approximately 30 mol% PbO. After 45 mol% PbO there is a steady decrease in its abundance. This behaviour can be explained by the formation of Pb_2^+ and $(Pb_xO_y)^+$ units, which grow to dominate the positive spectrum as the lead oxide concentration is increased. We do note, however, that single lead ions are present throughout the compositional range, suggesting that some of the lead continues to fulfil the role of an ionic modifier.

The $(B_3O_5)^-$ unit does not vary in abundance greatly as the lead content is increased. Our interpretation is that a portion of the glass network retains a pure borate character (with single lead ions as modifiers), and so borate units survive in the entire range. This unit is unlikely to arise from fragmentation of the $(PbB_3O_6)^-$ unit, because the complementary PbO fragment is not observed in the positive or negative spectra.

The $(PbB_3O_6)^-$ units show an interesting behaviour, as shown in Figure 13(a). As mentioned previously, we attribute this unit to the boroxol/metaborate ring. It is a dominant feature of the negative spectrum, with a relative abundance nearing 70%, and, starting at 55 mol% PbO, it shows an actual increase in abundance. The continued presence of ring structures in lead borates up to 75 mol% PbO has been noted in neutron scattering,[17] in accord with these results. Similar behaviour has also been observed in alkali borates through NMR[18] measurements. The high abundance may also be due to a particular stability (energy of formation) associated with this superstructural unit, which would allow it to fragment in its entirety.

Figure 13(b) displays the abundance of the $(PbB_5O_9)^-$ unit as a function of the lead content. Previously we have identified this multiplet with the diborate unit, and the abundance trend appears to support this assignment. The curve peaks at roughly 55 mol% PbO, and declines quickly afterwards. This is in line with the observations of Meera & Ramakrishna[1] on the range in composition over which it is expected to occur. Krogh-Moe & Wold-Hansen[19] also point to the existence of a larger superstructural unit, composed of two diborate rings joined by a short chain of two (BO_3) triangles. This unit was detected in crystalline lead borates at 54·5 mol% PbO.

Finally, in Figure 14 we show the relative percentage of boron atoms that belong to boroxol/metaborate ring structures. We note that, in the case of pure B_2O_3 glass, the percentage has been the subject of much discussion. The most recent experimental data points to a relative abundance of 78–82%. The LITOF-MS measurements show a percentage value that averages to approximately 72% for the compositional range 10 to 60 mol% PbO, and then increases. We interpret this as evidence that the ring structures are quite stable, and most of the borons are in B/MB rings up to high lead contents. The physical picture that emerges is one in which the lead oxide begins to form a subnetwork relatively early, and slowly engulfs the borate matrix as the lead content is increased. Though the

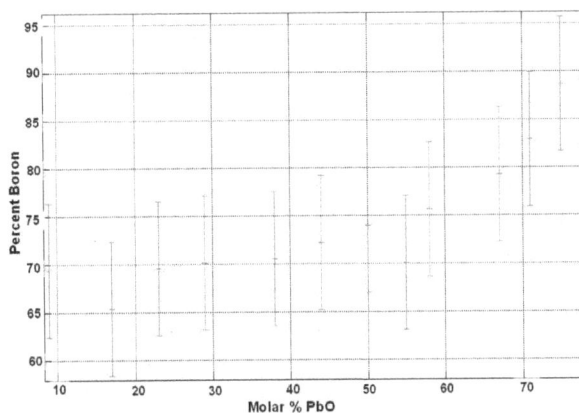

Figure 14. Percentage of boron atoms in boroxol/metaborate rings as a function of composition in lead borate glasses

borate mesh becomes fractured, the boron remain in their most stable oxide units, namely the boroxol/metaborate rings. The addition of sodium oxide, as noted previously, has a much greater effect on the B/MB rings, leading to their disappearance at much lower net modifier contents.

5. Conclusions

Laser ionization time of flight mass spectrometry measurements of borate glasses have yielded interesting, novel insights into possible structural changes. Generally, borate glasses show the presence of superstructural units that are highly stable. This is in contrast to other glassformer networks, such as the vanadates and the silicates, which show more chain-like, fragmentary structures with no evidence of added stability for larger, superstructural units. We have also reported on the quantification of LITOF-MS results for lead borate glasses. In this particular case, we note that:

- The spectra show a stable mass multiplet, $(PbB_3O_6)^-$, that survives up to 75 mol% PbO. This has been identified with a boroxol/metaborate ring superstructural unit, also based on Raman measurements of sodium-doped lead borates. Previous neutron scattering work also supports the long survivability of this unit. The addition of soda leads to much faster decrease in the abundance of the $(PbB_3O_6)^-$ multiplet as a function of net modifier content.

- The relative abundance of the $(PbB_5O_9)^-$ mass multiplet, identified with the diborate superstructural unit, reaches a maximum at 55 mol% PbO, but never exceeds 6% in value. We thus conclude that these units are present in far lesser numbers than the B/MB rings. It is suggested that units containing a four-coordinated boron may be less stable than the planar boroxol/metaborate rings.

- The fraction of borons that belong to the $(PbB_3O_6)^-$ unit is approximately 72% over most of the compositional range. This compares reasonably well with values obtained from NMR measurements in pure B_2O_3 glass.

6. Acknowledgements

We would like to acknowledge the financial support of the US National Science Foundation under grants DMR-PECASE 9733724, DMR-0502051, and MRI-0320861, as well as grant PHY-0649007, funded by the US Department of Defense. The authors would also like to thank Coe College for its unwavering help, and Dr Norimasa Umesaki and the other organisers of the Sixth International Conference on Borate Glasses, Crystals and Melts.

6. References

1. Meera, B. N. & Ramakrishna, J. *J. Non-Cryst. Solids*, 1993, **159**, 1.
2. Kreidl, N. J. *Glass Science and Technology. I. Glass-Forming Systems*, Academic Press, New York, 1983.
3. Wright, A. C., Vedishcheva, N. M. & Shakhmatkin, B. A. In: *Advances in X-ray Analysis*, Eds. J. V. Gilfrich, I. C. Noyan, R. Jenkins, T. C. Huang, R. L. Snyder, D. K. Smith, M. A. Zaitz & P. K. Predecki, Plenum, New York, 1997, p.535.
4. Hoppe, U., Kranold, R., Weber, H. J., Neuefeind, J. & Hannon, A. C. *J. Non-Cryst. Solids*, 2000, **278**, 99.
5. Joo, C., Werner-Zwanziger, U. & Zwanziger, J. W. *J. Non-Cryst. Solids*, 2000, **261**, 282.
6. Youngman, R. E., Werner-Zwanziger, U. & Zwanziger, J. W. *Z. Naturforsh.*, 1996, **51a**, 321.
7. Ratai, E.-M., Janssen, M., Epping, J. D., Chan, J. C. C. & Eckert, H. *Phys. Chem. Glasses*, 2002, **44**, 45.
8. Du, L. S. & Stebbins, J. *J. Non-Cryst. Solids*, 2003, **315**, 239.
9. Wright, A. C. In: *Experimental Techniques in Glass Science*, Eds. C. J. Simmons & O. H. El-Bayoumi, The American Ceramic Society, Westerville, 1993, p.205.
10. Sinclair, R., Stone, C. E., Wright, A. C., Polyakova, I. G., Vedishcheva, N. M., Shakhmatkin, B. A., Feller, S., Johanson, B. C., Venhuizen, P., Williams, R. B. & Hannon, A. C. *Phys. Chem. Glasses*, 2000, **41**, 286.
11. Stentz, D., Blair, S., Goater, C., Feller, S. & Affatigato, M. *Appl. Phys. Lett.*, 2000, **76**, 1.
12. Stentz, D., Blair, S., Goater, C., Feller, S. & Affatigato, M. *Phys. Chem. Glasses*, 2000, **41**, 259.
13. Blair, S., Stentz, D., Goater, C., Feller, S. & Affatigato, M. *J. Non-Cryst. Solids*, 2001, **293–295**, 416.
14. Affatigato, M., Feller, S., Schue, A., Blair, S., Stentz, D., Smith, G. B., Liss, D., Kelley, M. J., Goater, C. & Leelesagar, R. *J. Phys.: Condens. Matter*, 2003, **15**, S2323.
15. Nieuwendaal, R., Schue, A., Blair, S., Stentz, D., Feller, S., Affatigato, M. & Singleton, S. *Phys. Chem. Glasses: Eur. J. Glass Sci. Technol. B*, 2005, **46**, 302.
16. Stone, C. E., Wright, A. C., Sinclair, R., Feller, S., Affatigato, M., Hogan, D., Nelson, N. D., Vira, C., Dimitriev, Y. B., Gattef, E. and Ehrt, D. *Phys. Chem. Glasses*, 2000, **41**, 409.
17. Vedischeva, N. M., Shakhmatkin, B. A., Wright, A. C., Grimley, D. I., Etherington, G. & Sinclair, R. N. In: *Fundamentals of Glass Science and Technology (ESG Conf., Venice)* 1993, p.459.
18. Zwanziger, J. W., Youngman, R. E. & Braun, M. In: *Borate glasses, crystals and melts*, Eds. A. C. Wright, S. A. Feller & A. C. Hannon, The Society of Glass Technology, Sheffield, 1997, p. 21.
19. Krogh-Moe, J. & Wold-Hansen, P. S. *Acta Cryst. B*, 1973, **29**, 2242.

Proc. VI Int. Conf. Borate Glasses, Himeji, Japan, 18–22 August 2008   *Glass Tech.: Eur. J. Glass Sci. Technol. A, August 2009,* **50** (4), 233–235

Luminescence properties of YBO$_3$:Eu^{3+} crystallised from borosilicate glass

Etsuko Fujinaka, Yusuke Daiko, Atsushi Mineshige, Masafumi Kobune, Tetsuo Yazawa*

Department of Materials Science & Chemistry, Graduate School of Engineering, University of Hyogo Himeji Hyogo 671-2201, Japan

Tetsuro Jin

National Institute of Advanced Industrial Science and Technology (AIST), 1-8-31 Midorigaoka Ikeda Osaka 563-8577, Japan

Toshihiro Okajima

Kyusyu Syncrotron Light Reseach Center, 8-7 Yayoigaoka Tosu Saga 841-0005, Japan

Manuscript received 21 November 2008
Revised version received 26 December 2008
Accepted 9 January 2009

Nanocrystalline YBO$_3$:Eu^{3+} dispersed glasses were prepared utilising the spinodal-type phase separation of Na$_2$O–B$_2$O$_3$–SiO$_2$ glass. The YBO$_3$:Eu^{3+} crystals were selectively precipitated inside the Na$_2$O–B$_2$O$_3$ phase after heat treatment of the obtained glasses above 800°C, and the size of which was estimated to be around 15 nm by use of the Scherrer equation. The local structure around Eu^{3+} ions was estimated from extended x-ray absorption fine structure results using synchrotron radiation. The formation of YBO$_3$:Eu^{3+} in the glasses, which is strongly related to the emission properties, was found to be affected by the addition of a small amount of Al$_2$O$_3$.

1. Introduction

Rare earth ion doped fluorescent glasses are expected to be useful for various applications, such as fluorescent lamps, backlights of LCDs and display devices, because of their good shapability and unique optical properties.[1,2] Crucial aspects of these glasses for practical applications are their brightness and energy conversion efficiency, which are strongly related to the glass/phosphor-crystalline structures. Usually, phosphors are coated on a glass substrate. However, for this arrangement, electrons or ions directly hit the phosphors due to discharge phenomenon, and the phosphors are damaged, resulting in a decrease of the emission intensity.

We have studied the preparation of functionally advanced glasses focusing on the spinodal-type phase separation of Na$_2$O–B$_2$O$_3$–SiO$_2$ glass.[3–5] Sodium borosilicate (Na$_2$O–B$_2$O$_3$–SiO$_2$) glasses show a spinodal-type phase separation into Na$_2$O–B$_2$O$_3$ and SiO$_2$ phases when they are heated above ~550°C. Since the Na$_2$O–B$_2$O$_3$ phase is much more polar than the SiO$_2$ matrix phase, various ions are infiltrated into the Na$_2$O–B$_2$O$_3$ phase. We anticipate, as shown in Figure 1, that nanocrystalline phosphor-dispersed glasses will be obtained by utilising the spinodal-type phase separation of Na$_2$O–B$_2$O$_3$–SiO$_2$ glasses.

In this study, a typical phosphor of Y$_2$O$_3$:Eu^{3+} doped sodium borosilicate glasses was prepared, and the structure and its effect on the fluorescent properties are discussed, based on the results of x-ray diffraction (XRD), x-ray absorption fine structure (XAFS) and emission spectroscopies. Since the Al$_2$O$_3$ component affects the glass structure significantly, including the phase separation phenomena, various Al$_2$O$_3$-doped glasses were also investigated.

2. Experimental procedure

A parent glass of composition 9·4Na$_2$O.25·4B$_2$O$_3$.65·2SiO$_2$ (mol%), which can exhibit spinodal phase separation after heat treatment, was prepared by the conventional melting method. To investigate the effect of Al$_2$O$_3$ on the structure of the glass, various amounts of Al$_2$O$_3$ in the range from 0 to 1·9 wt% were added to the parent glass. A typical phosphor crystal of Y$_2$O$_3$:Eu^{3+} was also added to the parent glass. The preparation method for Y$_2$O$_3$:Eu^{3+} crystals was as follows: reagent grade Y$_2$O$_3$ (5 wt%) and Eu$_2$O$_3$ (0·35 wt%) were mixed in concentrated HNO$_3$ solution at 50°C. After the solution became transparent, a large amount of H$_2$C$_2$O$_4$ was mixed into the solution. The resultant powders were filtrated and washed. After drying at 120°C for 12 h, the obtained powders were calcined at 900°C for 10 h in air atmosphere. The formation of Y$_2$O$_3$:Eu^{3+} crystals was confirmed by XRD.

Mixtures of appropriate amounts of Na$_2$CO$_3$, H$_3$BO$_3$, SiO$_2$, Al$_2$O$_3$ and Y$_2$O$_3$:Eu^{3+} powders were placed in a platinum crucible and melted at 1500°C for 1 h in air. The melt was poured onto carbon moulds, and then the quenched glass was slowly cooled from 600°C in an electric furnace. The resultant glass was annealed at 800–900°C for 48 h in air.

* Corresponding author. Email yazawa@eng.u-hyogo.ac.jp

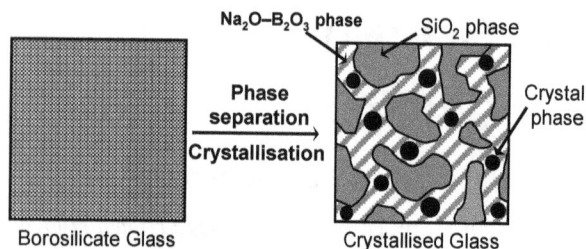

Figure 1. *Schematic illustration of the concept of nanocrystal-dispersed glasses prepared by utilising spinodal-type phase separation*

XRD patterns and fluorescence spectra of the obtained samples were measured using a Rigaku Mini-Flex and a Hitachi F-2500 spectrophotometer equipped with a 150 W Xe lamp, respectively. XAFS spectra at the Eu L_3 edge and Y K edge were measured.[6–8]

3. Results

3.1 Structures and emission properties

Figure 2 shows the XRD patterns of the obtained glasses doped with 1·9 wt% Al_2O_3 and annealed at 800 to 900°C. At annealing temperatures lower than 800°C, no crystalline phases were observed. On the other hand, it is clear that YBO_3 crystals were observed at annealing temperatures 875 and 900°C. The YBO_3 crystal sizes, estimated by using the Scherrer equation, are around 15 nm. Note that Y_2O_3 crystals (raw material) were not observed.

Since the $Na_2O–B_2O_3$ phase is dissolved by acid solution, characteristic continuous pores can be obtained from the sodium borosilicate glasses. The samples shown in Figure 2 were treated with 3 mol/l of HNO_3 solution at 98°C for 72 h to dissolve the $B_2O_3–Na_2O$ phase, and then XRD patterns were measured. Figure 3 shows the XRD patterns of the samples shown in Figure 2 after the acid treatment. Note that no crystalline phases were observed.

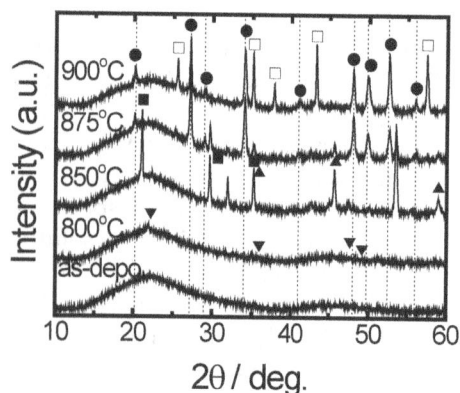

Figure 2. *XRD patterns of the samples doped with 1·9 wt% Al_2O_3 and annealed at 800 to 900°C for 48 h in air. The XRD pattern of the sample before annealing (labelled "as-depo") is also shown as a comparison. The precipitated crystals were YBO_3 (●), Al_2O_3 (□), $NaAlO_4$ (■), SiO_2 (▲), and cristobalite SiO_2 (▼)*

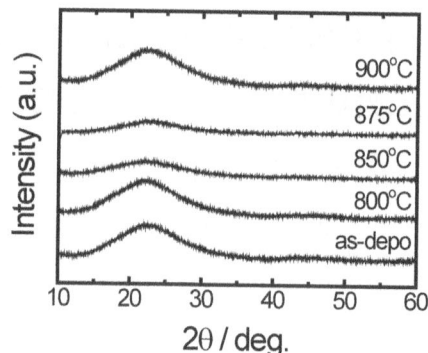

Figure 3. *XRD patterns of the samples shown in Figure 2 after acid treatment*

The samples shown in Figure 2 were excited at a wavelength of 254 nm, and the emission spectra were measured at room temperature. Figure 4 shows the emission spectra of the samples. Three peaks were clearly observed for the samples annealed above 875°C. The peak observed around 590 nm corresponds to the $^5D_0 \rightarrow {}^7F_1$ (orange emission) transition, and two peaks around 620 nm correspond to the $^5D_0 \rightarrow {}^7F_2$ (red emission) transitions, respectively. Furthermore, photoluminescence intensities increased with increasing heat treatment temperature.

3.2 Effect of Al_2O_3

YBO_3:Eu^{3+} nanocrystals were selectively precipitated inside the $Na_2O–B_2O_3$ phase, and the emission intensity was found to improve as a result of heat treatment. Since Al_2O_3 affects the structure and phase separation behaviour of the glass, samples doped with various amounts of Al_2O_3 were also investigated. Figure 5 shows the XRD patterns of glasses with Al_2O_3 doping from 0 to 1·9 wt%. The glasses were annealed at 900°C. It is clear that YBO_3 crystals were observed for all of the samples. Figure 6(a) shows the emission spectra of the samples, and Figure 6(b) shows the relationship between the emission intensities of the $^5D_0 \rightarrow {}^7F_1$ and $^5D_0 \rightarrow {}^7F_2$ transitions and the amount of doped Al_2O_3. Note that, in the range from 0 to 1·9 wt%, both the emission intensities increased with increasing Al_2O_3 content.

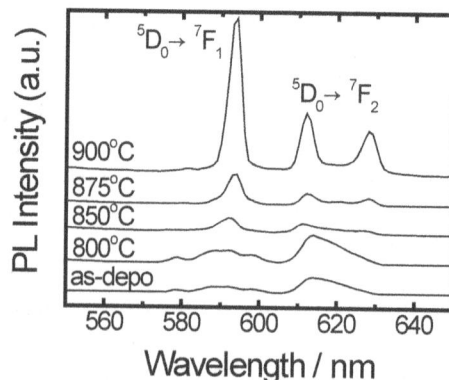

Figure 4. *Emission spectra of the samples annealed at 800 to 900°C for 48 h in air. The excitation wavelength was 254 nm*

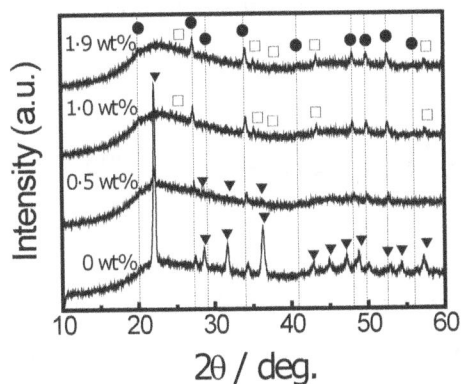

Figure 5. XRD patterns of the samples doped with Al_2O_3 and annealed at 900°C for 48 h in air. The precipitated crystals were YBO_3 (●), Al_2O_3 (□), and cristobalite SiO_2 (▼)

4. Discussion

XRD results (shown in Figures 1 and 2) strongly suggest that the YBO_3 nanocrystals were selectively precipitated inside the Na_2O–B_2O_3 phase as expected (Figure 1). The temperature of 875°C, at which YBO_3 crystals were precipitated, and the characteristic three peaks of the emission spectra in Figure 4, strongly suggest that a fraction of the Y^{3+} ions of the YBO_3 crystal were substituted by Eu^{3+} ions.

The coordination number of the Eu^{3+} ions was estimated by extended x-ray absorption fine structure (EXAFS) spectroscopy. The coordination number of Eu^{3+} changed from 6 (for the unannealed "as-depo" glass) to about 8 after heating above 875°C. The coordination number of the Y^{3+} ion in Y_2O_3 is 6, and that of the Y^{3+} ion in YBO_3 is 8. Since the coordination number of Y^{3+} in YBO_3 and Eu^{3+} in the glass annealed after 900°C is

same, EXAFS analysis also suggests that the Y^{3+} ions of YBO_3 were partially substituted by Eu^{3+} ions. Thus, nanocrystalline YBO_3:Eu^{3+} can be selectively precipitated inside the Na_2O–B_2O_3 glass phase by utilising the spinodal-type phase separation of sodium borosilicate glass, resulting in high red emission intensities.

It is very interesting to note the behaviour of the three characteristic emission peaks shown in Figures 4 and 6(a), which shows that the substitution of Y^{3+} ions in YBO_3 with Eu^{3+} ions, can be observed only for samples with more than 1·0 wt% of Al_2O_3. This suggests no YBO_3:Eu^{3+} crystal was precipitated for samples with 0 and 0·5 wt% Al_2O_3. The XRD and emission results strongly suggest that a small amount of Al_2O_3 is indispensable to the precipitation of the red phosphor YBO_3:Eu^{3+} crystal.

YBO_3:Eu^{3+} is widely used as a red phosphor, which yields high vacuum ultraviolet (VUV) transparency and exceptional optical damage threshold.[9,10] It has been reported that the red emission transition of $^5D_0 \rightarrow {}^7F_2$ is hypersensitive to the symmetry of the crystal field and will be relatively strong if the symmetry of the crystal field is relatively low.[11] Further investigations, including the effect of Al_2O_3 on the precipitation of YBO_3:Eu^{3+}, as well as the local strains which are related to the nano-ordered spinodal type phase separation, and their relationship to the photoluminescence properties are in progress.

5. Conclusions

Nanocrystalline YBO_3 dispersed glasses were prepared utilising the spinodal-type phase separation of the $9·4Na_2O.25·4B_2O_3.65·2SiO_2$ (mol%) glass. The YBO_3 nanocrystals were selectively precipitated in the Na_2O–B_2O_3 phase. The emission spectroscopy and EXAFS measurements revealed that a fraction of the Y^{3+} ions in the YBO_3 nanocrystals were was substituted with Eu^{3+} ions during the annealing treatment. Al_2O_3 doping was found to be very important for the precipitation of red phosphor YBO_3:Eu^{3+} nanocrystals.

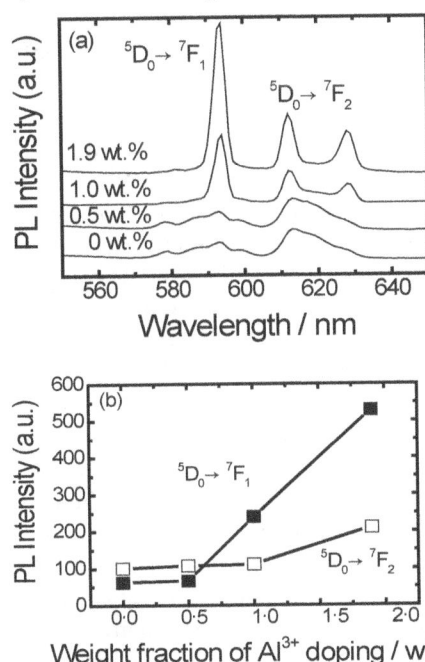

Figure 6. (a) Emission spectra of the samples with various amounts of Al_2O_3, and (b) the relationships between the intensities of the $^5D_0 \rightarrow {}^7F_1$ (at 593nm) and $^5D_0 \rightarrow {}^7F_2$ (at 611 nm) transitions and the amount of doped Al_2O_3

References

1. Ronda, C. R., Jüstel, T. & Nikol, H. *J. Alloys Comp.*, 1998, **275–277**, 669.
2. Kim, C. H., Kwon, I. E., Park, C. H., Hwang, Y. J., Bae, H. S., Yu, B. Y., Pyun, C. H. & Hong, G. Y. *J. Alloys Comp.*, 2000, **311**, 33.
3. Yazawa, T., Kuraoka, K. & Du, W.-F. *J. Phys. Chem. B*, 1999, **103**, 9841.
4. Du, W.-F., Kuraoka, K., Akai, T. & Yazawa, T. *J. Phys. Chem. B*, 2001, **105**, 11949.
5. Yazawa, T., Kuraoka, K., Akai, T., Umesaki, N. & Du, W.-F. *J. Phys. Chem. B*, 2000, **104**, 2109.
6. Okajima, T., Umesaki, N., Konishi, A., Jin T., Tanaka. I. & Yazawa. T., *Phys. Chem. Glasses: Eur. J. Glass Sci. Technol. B*, 2006, **47**, 558.
7. Qi, Z., Shi, C., Zhang. W., Zhang, W. & Hu, T., *Appl. Phys. Lett.*, 2002, **81**, 2857.
8. Wei, Z.-G., Sun, L.-D., Jiang, X.-C., Liao, C.-S. & Yan, C.-H., *Chem. Mater.*, 2003, **15**, 3011.
9. Ren, M., Lin, J. H., Dong, Y., Yang, L. Q., Su, M. Z. & You, L. P. *Chem. Mater.*, 1999, **11**, 1576.
10. Chadeyron, G., El-Ghozzi, M., Mahiou, R., Arbus, A & Cousseins, J. C., *J. Solid State Chem.*, 1997, **128**, 261
11. Wei, Z., Sun, L., Liao, C., Yin, Jialu, Jiang, X., Yan, C. & Lü, S., *J. Phys. Chem. B*, 2002, **106**, 10610.

Proc. VI Int. Conf. Borate Glasses, Himeji, Japan, 18–22 August 2008   *Phys. Chem. Glasses: Eur. J. Glass Sci. Technol. B*, February 2009, **50** (1), 63–70

Indentation induced densification of sodium borate glasses

Satoshi Yoshida,* Yasuhiro Hayashi, Akiko Konno, Toru Sugawara, Yoshinari Miura & Jun Matsuoka

Center for Glass Science and Technology, The University of Shiga Prefecture, 2500 Hassaka, Hikone, Shiga 522-8533, Japan

Manuscript received 20 August 2008
Revised version received 20 September 2008
Accepted 24 November 2008

Glass is densified under a high compressive stress, because of its structural flexibility and high free volume. Densification of glass also occurs during an indentation test using a sharp diamond indenter, such as a Vickers indenter. The purpose of this study is to evaluate the densification contribution to total indentation deformation beneath a Vickers indenter for binary sodium borate glasses, xNa$_2$O.(100–x)B$_2$O$_3$ (x=10, 15, 20, 25, 30, and 35 mol%). The densification contribution is estimated from the ratio of the densified volume to the total volume displaced by the indenter. By using an atomic force microscope, 3D-images of Vickers imprints on sodium borate glasses were obtained before and after annealing. Only the densified region under the indenter can be recovered by annealing at around the glass transition temperature. The volume recovered on annealing corresponds to the densified volume under the indenter. A large volume recovery (more than 65%) of Vickers indentation was observed for every borate glass. The large contribution of densification for sodium borate glass stems from the peculiar mechanism of densification under stress. Sodium borate glasses have various types of ring structures involving 3- and 4-coordinated borons. It is found that a rearrangement of the B–O ring structures takes place during indentation, and that this structural change is responsible for indentation induced densification of sodium borate glasses.

1. Introduction

Under a high hydrostatic compressive stress, glass experiences a permanent densification.[1] The densified structure of glass is retained even after unloading. Bridgman & Simon[1] showed that silica glass, boric oxide glass, and three kinds of Na$_2$O–SiO$_2$ glasses underwent a significant increase in density after application of 20 GPa uniaxial pressure. In their paper, boric oxide glass exhibited approximately 6% increase in density. After this discovery of permanent densification of glass, the permanent densification of various kinds of inorganic glasses has now been reported.[2–10] Permanent densification of glass is a general property for oxide, fluoride, and chalcogenide glasses which have a large free volume in their open random network.

On the other hand, permanent densification of glass also occurs under a sharp diamond indenter, such as a Vickers indenter. Ernsberger[11] first reported a change in refractive index beneath indentation imprints and scratch grooves on silica and soda–lime–silica glasses. This change in refractive index results from an increase in the density of glass. During an indentation test, shear flow of glass also occurs even at room temperature. Peter[12] summarised densifica-

tion and flow phenomena of oxide glass through indentation tests. He concluded that most silicate glasses, which include silica and window glasses, underwent densification beneath an indenter, and that shear flow required a minimum percentage of network modifiers in glass. In Peter's report,[12] piling up around indentation imprints and slip lines beneath an indenter can be observed for plate glass and ternary silicate glass (14Na$_2$O.10CaO.76SiO$_2$), but an occurrence of these phenomena was not found for silica and 15Na$_2$O.85SiO$_2$ glasses. The compositional variation of flow behaviours in glass comes from the diversity of glass connectivity, or the degree of polymerisation of the glass network.[13] The permanent change in density by pressure treatment depends on free volume in the glass network.

Using a classical definition, glasses can be divided into two categories, 'anomalous' and 'normal'. For example, anomalous behaviours may involve positive temperature coefficient of shear or bulk modulus, or negative pressure coefficient of shear or bulk modulus.[14] Arora *et al*[15] reported that the indentation deformation for anomalous glass (silica or Pyrex), which has relatively low atomic packing density, is governed by densification, whilst the deformation for normal glass, such as soda–lime–silica or lead silicate glasses, is governed by shear plasticity. However,

* Corresponding author. Email yoshida@mat.usp.ac.jp

the normal and anomalous behaviours must be a continuous function of glass composition, as shown in a previous report,[16] not classified into only two groups.

One of the present authors (SY) reported that the contribution of indentation induced densification to total indentation deformation varied with glass composition, and showed a relationship between Poisson's ratio of glass and the densification contribution beneath an indenter.[17] The contribution of densification to total deformation can be estimated from the change in volume of indentation imprint by annealing. The densified volume can be relaxed by annealing at around the glass transition temperature.[18] Mackenzie[18] showed that more than 99% of the original volume of hydrostatically densified silica glass was recovered after annealing at 1000°C for 1 h. The annealing recovery in volume also occurs for the indentation imprints.[19,20] The indentation imprint shrinks by annealing.[17,19,20] This means that measuring the volumes of indentation imprint before and after annealing allows discrimination of the densified volume from the total indentation volume. In the previous report,[17] various kinds of glasses including bulk metallic glass were investigated in order to assess the densification contribution to the indentation deformation. As far as the authors know, however, there has been no report on the indentation induced densification of borate glasses. In this study, the densification contributions to indentation deformation for binary sodium borate glasses are investigated, and the mechanism of indentation induced densification is discussed in terms of structural change by pressure treatment.

2. Experimental procedure

The glasses employed in this study were binary sodium borate glasses (xNa$_2$O.(100−x)B$_2$O$_3$, x=10, 15, 20, 25, 30, 35 mol%). The glass samples were prepared from reagent grade Na$_2$CO$_3$ and anhydrous B$_2$O$_3$ powders. The appropriately weighed powders were mixed thoroughly and melted in Pt crucibles at 1200°C for 20 min in an electric furnace. Batches were prepared to obtain about 15 g of the glasses. The glass transition temperature (T_g) of each glass was determined by using a thermal dilatometer (SEIKO TMA/SS6600, Japan). The obtained glasses were annealed at T_g+10°C for 1 h, cooled to room temperature in the furnace and then polished to the required size for the indentation test (10 mm × 10 mm × 2 mm). The density was measured using Archimedes' technique with kerosene as the immersion liquid.

Vickers indentations were performed in air (20°C, 60% r.h.) using a hardness tester (Akashi MVK-H2, Japan). Indentation load and dwell time at maximum load were 245 mN and 15 s, respectively. The indenta-

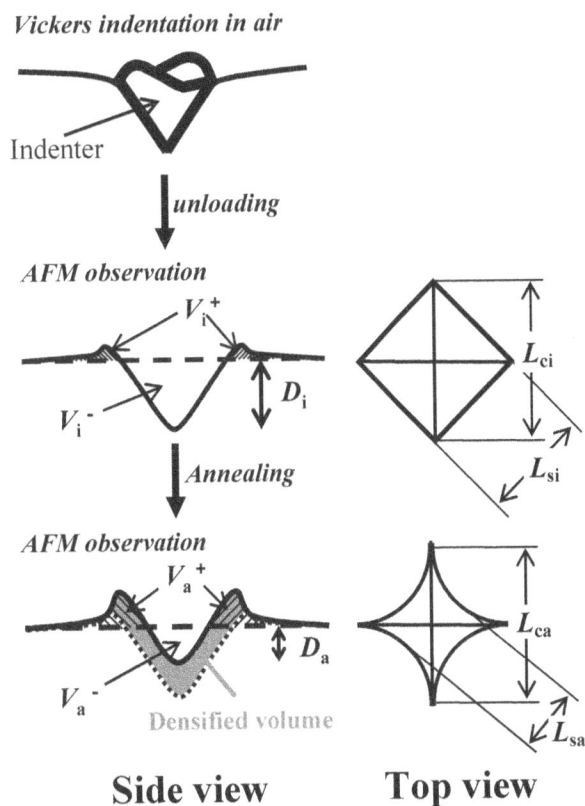

Figure 1. Schematic procedure for measuring the indentation geometries and volume before and after annealing

tion geometries, including the indentation volume, were measured before and after annealing through the procedure described below and schematically illustrated in Figure 1. Annealing temperature and time were T_g×0·9 (K) and 2 h, respectively.

An atomic force microscope (AFM, Veeco, Nanoscope E, USA) was used in contact mode in order to measure the indentation geometries and volumes.[21] The indentation geometries measured were the diagonal (corner-to-corner) length, L_c, the opposite-side (face-to-face) length, L_s, and the indentation depth, D. All the parameters obtained before annealing and those after annealing have the additional subscript "i" and "a", respectively. The volume measurement was also performed before and after annealing. The superscripts "+" and "−" correspond to piling up and inside volumes of indentation imprint, respectively. The piling up and inside volumes of indentation were defined as the volumes above and below the zero-level surface, respectively. Determination of the zero-level surface is critical for measuring the indentation volumes. After capturing the AFM image, flattening modification included in the AFM software was carried out for eliminating bow and slope of the surface in order to determine the zero-level surface. The bearing analysis included in the AFM software allowed us to analyse how much of the volume exists below or above the original surface.

The indentation imprints shrank by annealing.

Table 1. Thermal and physical properties of binary sodium borate glasses

	T_g (°C)	Density (g/cm³)	Vickers hardness (GPa)
$10Na_2O.90B_2O_3$	354	2·038	2·99
$15Na_2O.85B_2O_3$	405	2·116	3·28
$20Na_2O.80B_2O_3$	456	2·182	3·83
$25Na_2O.75B_2O_3$	475	2·266	4·20
$30Na_2O.70B_2O_3$	475	2·340	4·44
$35Na_2O.65B_2O3$	460	2·384	4·20
Experimental uncertainty	±2	±0·005	±0·05

The recovery ratio, or the shrinkage ratio, of each indentation geometry is defined as the ratio of the change in each indentation geometry to the initial value. For example, L_{cR} means the recovery ratio of the diagonal length by annealing. On the other hand, the volume ratio of annealing recovery, V_R, is defined as the recovery ratio calculated by the following equation[17]

$$V_R = \frac{\left(V_i^- - V_a^-\right) + \left(V_a^+ - V_i^+\right)}{V^-} \quad (1)$$

The pile-up volume before annealing is always larger than that obtained after annealing ($V_a^+ > V_i^+$). Assuming that the increase in pile up volume, V^+, is induced by the recovery of densified volume under the Vickers indenter (see Figure 1), the sum of the changes in the pile up volume, $(V_a^+ - V_i^+)$, and in the inside volume, $(V_i^- - V_a^-)$, is defined as the densified volume.[17,21,22]

Vickers hardness, H_V, was obtained by using the following equation

$$H_V = 1·8544 \frac{P}{L_{ci}^2} \quad (2)$$

where P is the peak indentation load, and L_{ci} is the diagonal length obtained by AFM before annealing.

In order to investigate the structural change of borate glass beneath the indenter, a Raman scattering measurement was performed on each sodium borate glass. Raman spectra from the centre and from the outside of the indentation imprint were acquired using a Raman spectrometer (JASCO NRS-3100, Japan) with 1800 lines/mm grating and with a dichroic mirror. The laser beam (532 nm, 65 mW) was focused on the surface at the centre and the outside of the indentation imprint. The laser spot diameter was estimated to be about 1 μm using a ×100 objective.

The acquisition time was 2 s per run, and ten runs were performed for each point.

3. Results and discussion

Table 1 shows the measured physical and thermal properties of sodium borate glasses. The experimental uncertainty given in Table 1 is the average of standard deviations. The glass compositions shown in this work are batch compositions. With increasing Na_2O content, T_g, density, and hardness increase, but T_g and hardness show a maximum at a composition of around $30Na_2O.70B_2O_3$. It is known that further addition of Na_2O in glass produces nonbridging oxygens,[23–25] although the onset composition of nonbridging oxygen formation in this glass system also depends on glass preparation conditions.[26] An increase in nonbridging oxygen is one of the reasons for the decreases in T_g and hardness.

Figure 2 shows the change on annealing of the top-views of indentation imprints. All indentation imprints shrink on annealing. It is found that the annealing recovery of the diagonal length of indentation, L_c, is very limited, and that the recovery of the opposite-side length, L_s, is remarkable. It is noted that no crack is observed at the corners and edges of indentations. Figure 3 shows AFM cross-section profiles of indentation imprints before and after annealing. The cross-section planes include the centre of indentation, and are parallel to the diagonal direction (a) or the face-to-face direction (b). Lower profiles in Figures 3(a) and (b) represent the cross-section profiles before annealing, upper ones the profiles after annealing. These figures also confirm the shrinkage of the indentation imprint on annealing, and show the significant recoveries of indentation depth (centre of indentation) and of opposite-side (face-to-face) distance on annealing. In Figures 3(a) and (b), with increasing Na_2O content, the depth of indentation decreases, but the shallowest depth is obtained from the profile of $30Na_2O.70B_2O_3$, not of $35Na_2O.65B_2O_3$. This compositional change in indentation depth corresponds to the compositional variation in hardness shown in Table 1.

In Table 2, the indentation geometries and volumes for every sodium borate glass before and after anneal-

Table 2. Indentation geometries and volume for binary sodium borate glasses before and after annealing

	L_c (mm)		L_s (mm)		D (mm)		V^+ (mm³)		V^- (mm³)	
	Before annealing	After annealing	Before annealing	After annealing	Before annealing	After annealing	Before annealing	After annealing	Before annealing	After annealing
$10Na_2O.90B_2O_3$	12·34	11·77	7·27	4·83	0·963	0·647	0·19	0·33	11·16	3·52
$15Na_2O.85B_2O_3$	11·77	11·28	6·81	4·30	0·913	0·603	0·16	0·30	9·34	2·72
$20Na_2O.80B_2O_3$	10·89	10·13	6·20	3·83	0·895	0·574	0·27	0·53	8·30	2·26
$25Na_2O.75B_2O_3$	10·41	10·03	6·00	3·71	0·848	0·560	0·16	0·42	7·17	1·98
$30Na_2O.70B_2O_3$	10·12	9·66	5·57	3·53	0·828	0·547	0·21	0·49	6·39	1·82
$35Na_2O.65B_2O_3$	10·40	10·13	6·02	4·17	0·848	0·587	0·16	0·33	7·19	2·56
Experimental uncertainty	±0·07	±0·07	±0·07	±0·07	±0·009	±0·004	±0·02	±0·04	±0·02	±0·04

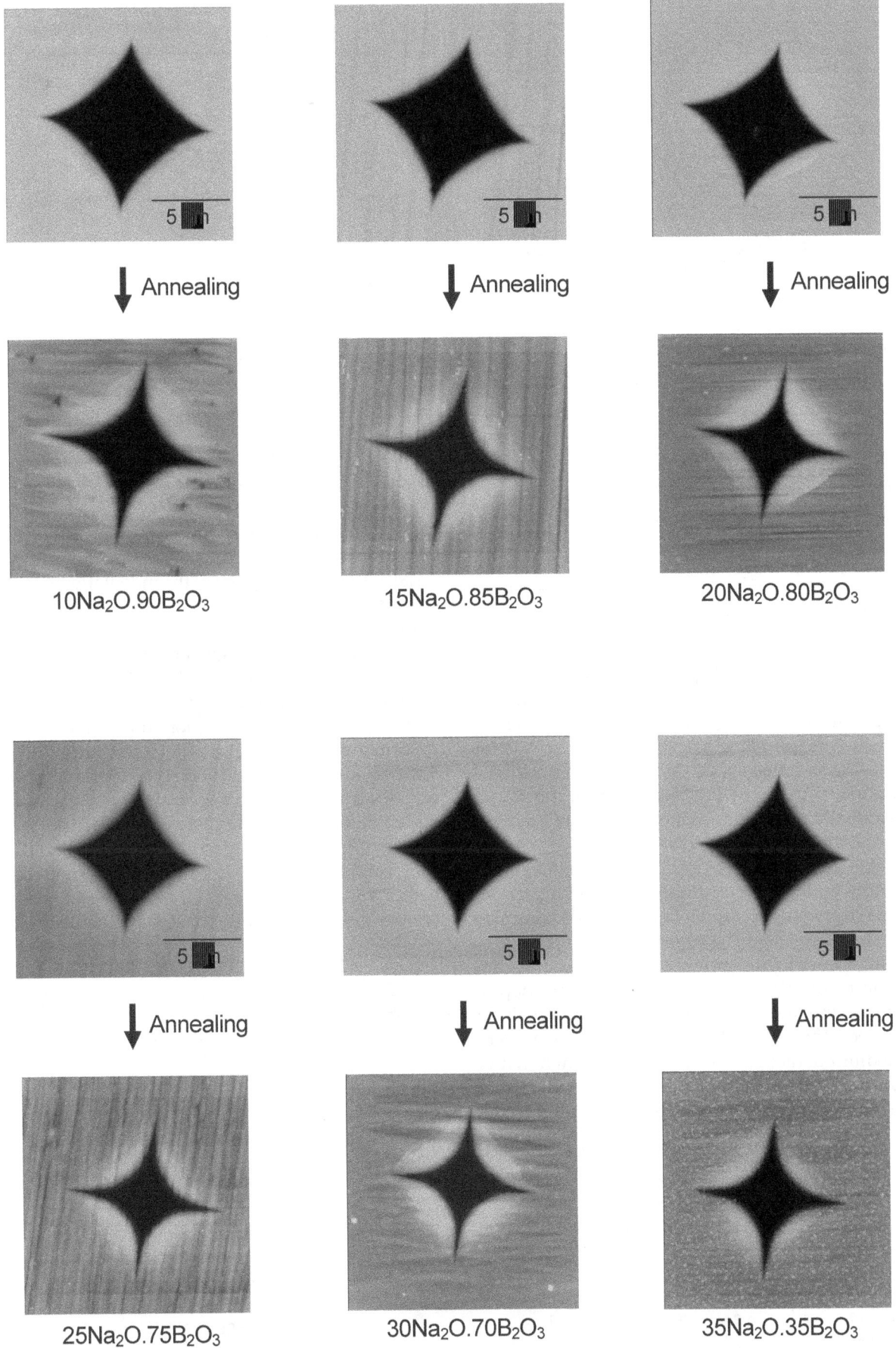

Figure 2. Change in AFM top-views of Vickers indentation imprints by annealing at $T_g \times 0.9$ for 2 h

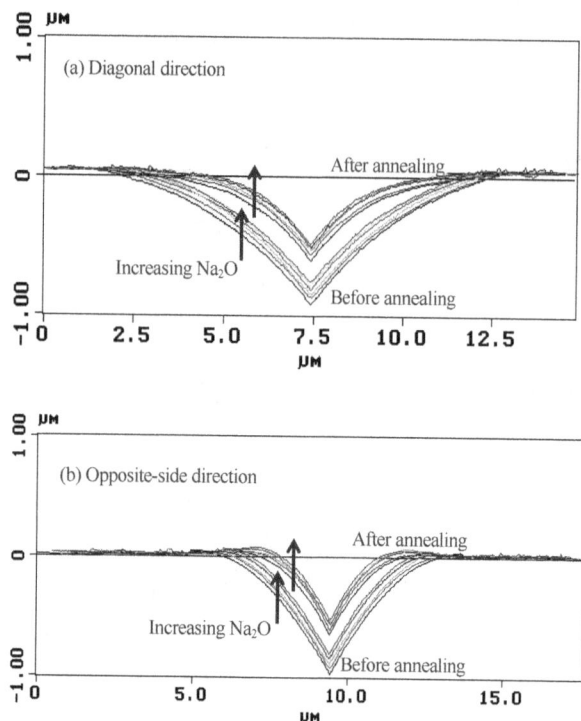

Figure 3. Change in AFM cross-section profiles of Vickers indentation imprints by annealing at $T_g \times 0.9$ for 2 h. (a) diagonal direction, (b) opposite-side direction

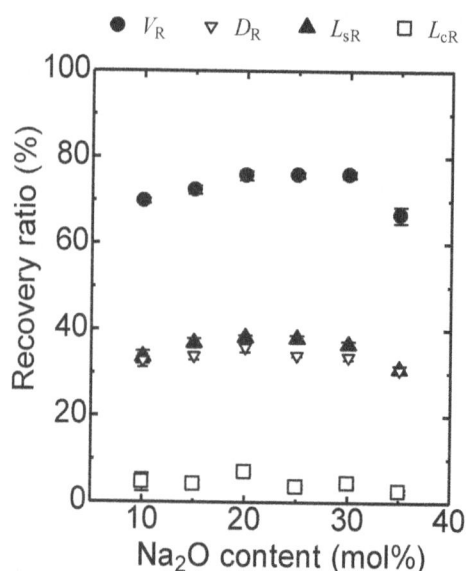

Figure 4. Variation of recovery ratios with Na_2O content in glass

ing are summarised. The experimental uncertainty here is again the average of standard deviations. All the parameters except V^+ decrease after annealing. V^+, which is the piling up volume, increases after annealing as stated above. As for the compositional variation, with increasing Na_2O content, the size of the indentation imprint reduces, and shows a minimum at the composition $30Na_2O.70B_2O_3$.

Table 3 shows the recovery ratios of indentation geometries and volume on annealing. For comparison, the recovery ratios for soda–lime–silica glass are also shown in this table. The recovery ratios for sodium borate glasses in this table are also plotted in Figure 4. The recoveries of opposite-side length, depth, and volume show a maximum at a composition between 20 and 30 mol% Na_2O. The volume recovery of sodium borate glass is a little larger than that of soda–lime–silca glass, and this comes from the larger recovery of opposite-side length. The region recov-

ered on annealing corresponds to the densified region beneath the indenter. A considerable part (more than 65%) of the inside volume of the indentation imprint for sodium borate glasses is densified under a high compressive stress.

In the previous report,[17] however, the densification contribution to total indentation deformation decreases with increasing Poisson's ratio of the glass. The ratio of densification to total indentation deformation of silica glass (Poisson's ratio 0.17) is 92%, whereas the densification contribution for bulk metallic glass (Poisson's ratio 0.40) is only 5%. In Figure 5, the relationship between the densification contribution (the volume ratio of recovery) and Poisson's ratio from the previous report[17] is plotted together with the present data for sodium borate glasses (values for Poisson's ratio for sodium borate glasses

Table 3. Recovery ratios of indentation geometries and volume for binary sodium borate glasses

	L_{cR} (%)	L_{sR} (%)	D_R (%)	V_R (%)
$10Na_2O.90B_2O_3$	4.6	33.6	32.8	69.8
$15Na_2O.85B_2O_3$	4.2	36.9	33.9	72.4
$20Na_2O.80B_2O_3$	7.0	38.2	35.9	75.8
$25Na_2O.75B_2O_3$	3.6	38.0	34.1	75.9
$30Na_2O.70B_2O_3$	4.5	36.6	33.8	76.0
$35Na_2O.65B_2O_3$	2.6	30.9	30.8	66.7
Soda–lime–silica*	1.1	25.3	35.4	63.6
Experimental uncertainty	±0.9	±0.9	±0.8	±1.0

*: Matsunami S-0050, Japan, Indentation load=300 mN

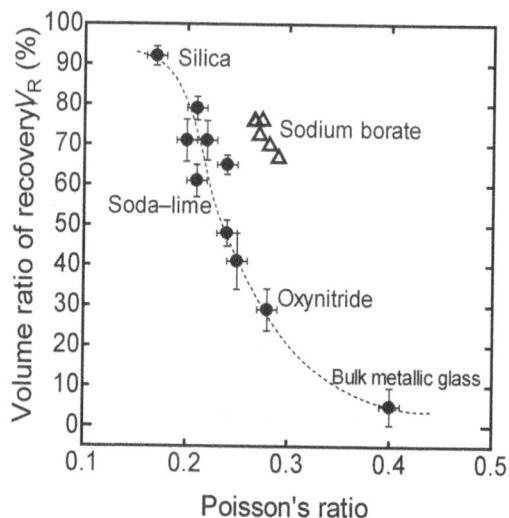

Figure 5. Relationship between the volume ratio of recovery and Poisson's ratio

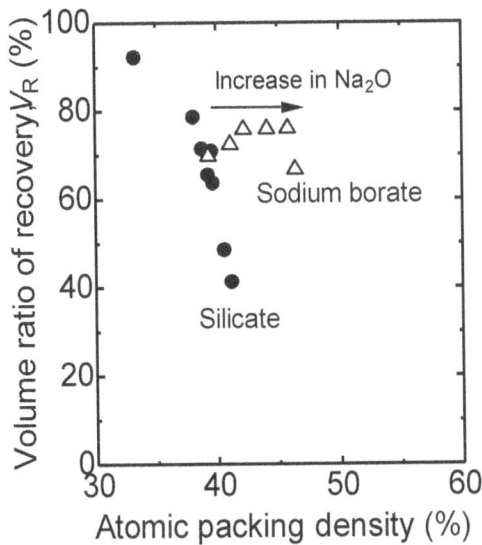

Figure 6. Relationship between the volume ratio of recovery and the atomic packing density

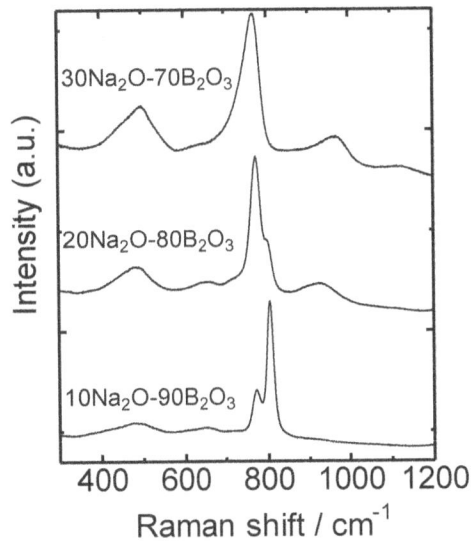

Figure 7. Raman spectra of 10Na$_2$O.90B$_2$O$_3$, 20Na$_2$O.80B$_2$O$_3$, and 30Na$_2$O.70B$_2$O$_3$ glasses without indentation imprint

were estimated from Ref. 27). The large contribution of densification to total indentation deformation for sodium borate glass is not expected from the reported relationship between the densification contribution and Poisson's ratio. Poisson's ratio for glass is directly related to the atomic packing density.[28] The atomic packing density is the ratio of the volume occupied by ions to the total volume of glass.[29] Using Shannon's ionic radii,[30] the atomic packing densities for sodium borate glasses were calculated and plotted in Figure 6. The atomic packing fraction, C_g, is calculated using the following equation

$$C_g = \frac{\rho}{M} \sum_i x_i V_i \qquad (3)$$

where ρ, M, x_i and V_i are respectively density, mean molar weight, the molar fraction of component i and the ionic packing factor obtained from Equation (4) for an oxide A$_X$O$_Y$

$$V_i = \frac{4\pi}{3} N_A \left(X R_A{}^3 + Y R_O{}^3 \right) \qquad (4)$$

where R_A and R_O are the ionic radii of the cation and the oxygen ion, and N_A is Avogadro's number. The ionic radii of sodium, 3-fold coordinated boron, 4-fold coordinated boron and oxygen are 0·114 nm, 0·015 nm, 0·025 nm and 0·121 nm, respectively.[30] The fraction of 4-fold coordinated boron in xNa$_2$O.(100−x)B$_2$O$_3$ glass is assumed to be $x/(1−x)$ for $x<0·33$ and $(300−4x)/(500−5x)$ for $x>0·33$.[31] The atomic packing densities obtained are those for undensified glasses. In Figure 6, solid circles denote silicate glasses, including silica and soda–lime–silica glasses, and open triangles represent sodium borate glasses. In Figures 5 and 6, it is clear that indentation induced densification for sodium borate glass is quite different from that for other types of glass. For sodium borate glasses

the densification contribution beneath the indenter does not decrease with increasing packing density, as shown in Figure 6. This suggests the existence of a peculiar mechanism of densification for sodium borate glass.

Figure 7 shows the Raman spectra of 10Na$_2$O.90B$_2$O$_3$, 20Na$_2$O.80B$_2$O$_3$, and 30Na$_2$O.70B$_2$O$_3$ glasses prior to indentation test. With the addition of Na$_2$O to the glass, the 805 cm^{-1} band reduces in intensity and the 770 cm^{-1} band appears. The 770 cm^{-1} band shifts to lower frequency and becomes asymmetric with increasing Na$_2$O content. As already reported,[24] the 805 cm^{-1} band is assigned to the breathing mode of boroxol rings. The assignment of the 770 cm^{-1} band is a little controversial,[24] but it is presumably assigned to the vibration of rings having BO$_4$ tetrahedron(s) (i.e. the pentaborate or tetraborate group). With increasing Na$_2$O content, the 770 cm^{-1} peak shifts to 765 cm^{-1} and a band appears at around 1100 cm^{-1}, which is assigned to diborate groups.[24]

Figures 8(a)–(f) show close-up Raman spectra in the range from 600 cm^{-1} to 900 cm^{-1} of sodium borate glasses both from the centre and from the outside of the indentation imprint. The indentation induced structural change for every sodium borate glass can be deduced from comparison of Raman spectra before and after indentation. As far as the present authors know, only one paper has been reported on the structural change for permanent densification of binary sodium borate glasses produced by applying hydrostatic pressure.[32] Although *in situ* inelastic x-ray scattering spectra of lithium diborate glass clearly show a coordination change of boron from three to four under a high compressive stress, no evidence of permanent densification after unloading was observed.[33] In Figures 8(a)–(c), the 805 cm^{-1} band of boroxol rings shifts a little (~1 cm^{-1}) toward higher frequency and the 770

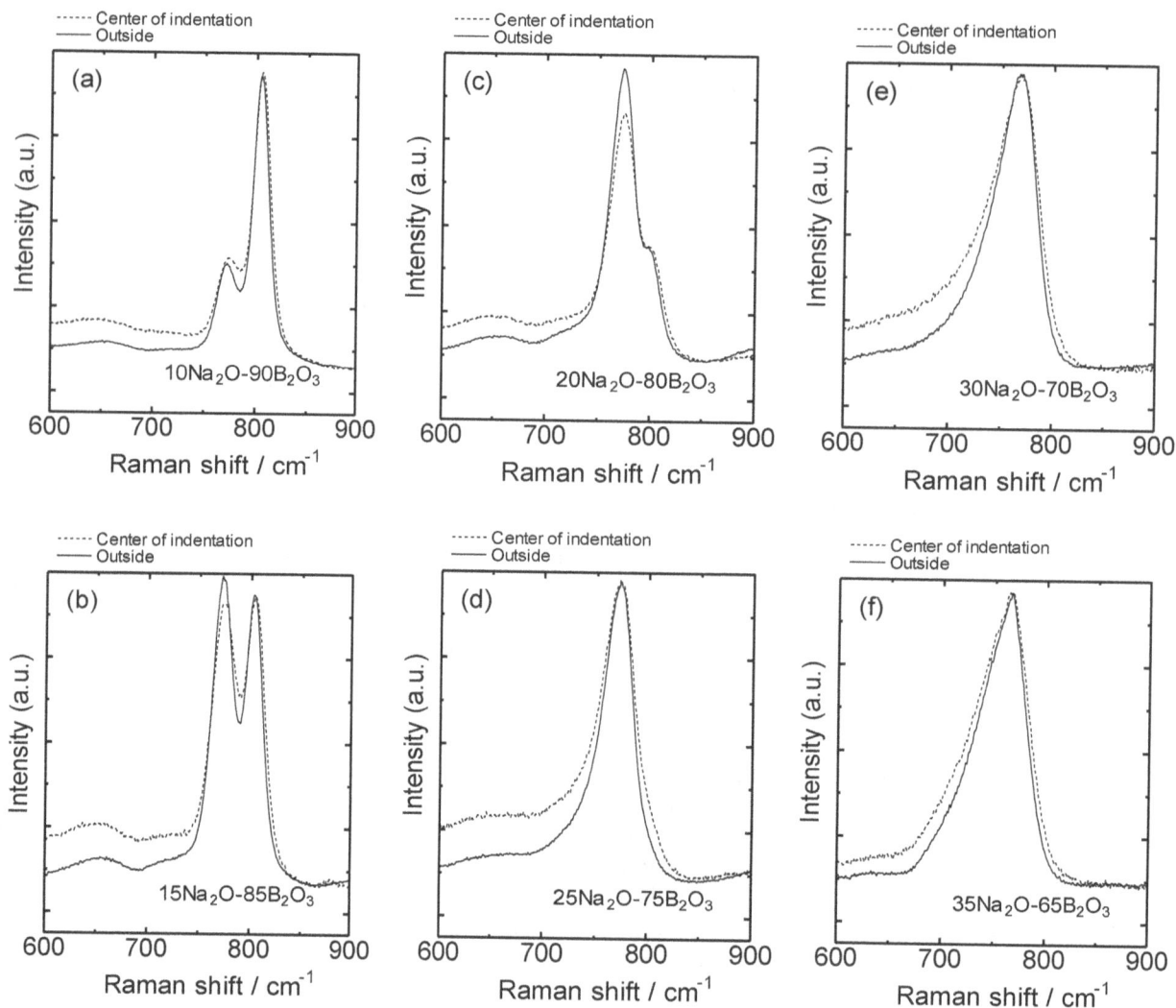

Figure 8. Raman spectra of (a) 10Na₂O.90B₂O₃, (b) 15Na₂O.85B₂O₃, (c) 20Na₂O.80B₂O₃, (d) 25Na₂O.75B₂O₃, (e) 30Na₂O.70B₂O₃, (f) 35Na₂O.65B₂O₃ glasses from the centre and from the outside of the indentation imprint

cm^{-1} band is broadening by applying indentation induced pressure. This indentation induced broadening is more remarkable in the spectrum from the centre of indentation for $25Na_2O.75B_2O_3$ glass in Figure 8(d). The 766 cm^{-1} band assigned to diborate groups and shown in Figures 8(e) and (f) is also broadening after indentation. In all Raman spectra shown in Figures 8(a)–(f), the relative Raman intensity in the range from 600 cm^{-1} to 700 cm^{-1} increases after indentation. The B–Ô–B bending vibration falls in this frequency region. Therefore, an increase in polarisability caused by a distortion of the glass network is one of the reasons for the increase in Raman intensity in the range from 600 cm^{-1} to 700 cm^{-1}.[32]

This is the first report on the indentation induced structural change of sodium borate glasses. The previous reports on the structural change of B_2O_3 glass by application of hydrostatic pressure[6,34–36] show that high compressive pressure causes 3-membered B–O rings (boroxol rings) to break up and increases the numbers of four-coordinated boron and three-

coordinated oxygen. After unloading, however, 3-membered B–O rings are regenerated, and a little decrease in boroxol ring fraction is also confirmed in permanently densified boric oxide glass.[34,36]

On the other hand, the densification mechanism of SiO_2 glass is quite different from that of B_2O_3 glass. Hydrostatic pressure on silica glass makes the 430–470 cm^{-1} Raman band become narrower and shift to higher frequency.[37–41] This corresponds to a narrower distribution of Si–Ô–Si angles in densified silica glass. The high compressive pressure also changes the ring statistics towards smaller rings (the numbers of three- and four- membered rings increase).[39] In the case of densified B_2O_3 glass, the change in mean bond angle is not so remarkable, and the change in ring statistics (i.e. the formation of 4- to 7-membered rings) is responsible for permanent densification.[35]

As for sodium borate glasses, Zhang & Soga[32] concluded that BO_4 units were destroyed to form boroxol rings, because the relative intensity of the 780

202

cm^{-1} band decreased in comparison with that of the 808 cm^{-1} band. This relative decrease in the intensity of the 780 cm^{-1} band (the 770 cm^{-1} band in our study and in Ref. 24) is also observed in Figures 8(b) and (c) in the present study. However, the small broadening of the 770 cm^{-1} band in our study supports another explanation involving the rearrangement among different groups containing BO_4 units, as proposed by Zhang & Soga.[32] Indentation induced high compressive pressure would cause the reduction in inter ring distances (but with boron remaining 3-coordinated). This reduction in inter-ring distances results in the distortion of B–Ô–B bridges involving BO_4 units. This is the reason why the Raman band (the 770 cm^{-1} band) assigned to rings containing BO_4 units is broadening by application of pressure. The change in ring statistics among different groups containing at least one BO_4 unit (pentaborate, tetraborate, triborate, diborate, etc.) by application of pressure is responsible for indentation induced densification of sodium borate glasses.

In Figure 4, the recovery ratio of indentation volume shows a maximum at a composition between 20 and 30 mol% Na_2O. In other words, the densification contributions to total indentation deformation for $10Na_2O.90B_2O_3$ and $35Na_2O.65B_2O_3$ glasses are smaller than those of other glasses. It is suggested that a slip motion between planar boroxol rings in $10Na_2O.90B_2O_3$ glass or a redistribution of nonbridging oxygens in $35Na_2O.65B_2O_3$ glass becomes the origin of irreversible shear flow and reduces the densification contribution of these glasses.

4. Conclusion

The densification contribution to total indentation deformation for binary sodium borate glass has been estimated from the ratio of the densified volume to the total volume displaced by the indenter. By using an atomic force microscope, 3D-images of Vickers indentations on sodium borate glasses were obtained before and after annealing. Only the densified region under the indenter can be recovered on annealing at $T_g \times 0.9$. The recovered volume by annealing corresponds to the densified volume under the indenter. A large recovery (more than 65%) of Vickers indentation volume was observed for every borate glass. The large contribution of densification for sodium borate glass stems from the peculiar mechanism of densification under a stress. It is elucidated that the change in ring statistics among different B–O rings containing BO_4 unit(s) beneath the indenter is responsible for indentation induced densification of sodium borate glasses. As for the compositional variation of densification contribution, it is found that the densification contributions of $10Na_2O.90B_2O_3$ and $35Na_2O.65B_2O_3$ glasses are smaller than those of

other glasses. It is suggested that a decrease in densification contribution of these glasses results from the irreversible shear flow originating from the slip motion of planar structures or the redistribution of nonbridging oxygens in the glass.

Acknowledgement

The authors would like to thank to Mr T. Fukuda and Mr Y. Morishita at Mitsue Mold Engineering Co. Ltd., Japan for their help in measuring Raman spectra.

References

1. Bridgman, P. W. & Simon, I. *J. Appl. Phys.*, 1953, **24**, 405.
2. Cohen, H. M. & Roy, R. *J. Am. Ceram. Soc.*, 1961, **44**, 523.
3. Cohen, H. M. & Roy, R. *Phys. Chem. Glasses*, 1965, **6**, 149.
4. Uhlmann, D. R. *J. Non-Cryst. Solids*, 1973/74, **13**, 89.
5. Hirao, K., Yoshimoto, M., Soga, N. & Tanaka, K. *J. Non-Cryst. Solids*, 1991, **130**, 78.
6. Wright, A. C., Stone, C. E., Sinclair, R. N., Umesaki, N., Kitamura, N., Ura, K., Ohtori, N. & Hannon, A. C. *Phys. Chem. Glasses*, 2000, **41**, 296.
7. Sakida, S., Miyauchi, K., Kawamoto, Y. & Kitamura, N. *J. Non-Cryst. Solids*, 2000, **271**, 64.
8. Miyauchi, K., Qiu, J., Shojiya, M., Kawamoto, Y. & Kitamura, N. *J. Non-Cryst. Solids*, 2001, **279**, 186.
9. Kawamoto, Y., Miyauchi, K., Shojiya, M., Sakida, S. & Kitamura, N. *J. Non-Cryst. Solids*, 2001, **284**, 128.
10. Stone, C. E., Hannon, A. C., Ishihara, T., Kitamura, N., Shirakawa, Y., Sinclair, R. N., Umesaki, N. & Wright, A. C. *J. Non-Cryst. Solids*, 2001, **293–295**, 769.
11. Ernsberger, F. M. *J. Am. Ceram. Soc.*, 1968, **51**, 545.
12. Peter, K. W. *J. Non-Cryst. Solids*, 1979, **5**, 103.
13. Rouxel, T., Ji, H., Hammouda, T. & Moréac, A. *Phys.Rev. Lett.*, 2008, **100**, 225501.
14. Krause, J. T. & Kurkjian, C. R. *J. Am. Ceram. Soc.*, 1968, **51**, 226.
15. Arora, A., Marshall, D. B. & Lawn, B. R. *J. Non-Cryst. Solids*, 1979, **31**, 415.
16. Bertoldi, M. & Sglavo, V. M. *J. Non-Cryst. Solids*, 2004, **344**, 51.
17. Yoshida, S., Sanglebouef, J.-C. & Rouxel, T. *J. Mater. Res.*, 2005, **20**, 3404.
18. Mackenzie, J. D. *J. Am. Ceram. Soc.*, 1963, **46**, 470.
19. Neely, J. E. & Mackenzie, J. D. *J. Mater. Sci.*, 1968, **3**, 603.
20. Yoshida, S., Isono, S., Matsuoka, J. & Soga, N. *J. Am. Ceram. Soc.*, 2001, **84**, 2141.
21. Sawasato, H., Yoshida, S., Sugawara, T., Miura, Y. & Matsuoka, J. *J. Ceram. Soc. Jpn.*, 2008, **116**, 864.
22. Yoshida, S., Sanglebouef, J.-C. & Rouxel, T. *Int. J. Mater. Res.*, 2007, **98**, 360.
23. Bray, P. J. & O'Keefe, J. G. *Phys. Chem. Glasses*, 1963, **4**, 37.
24. Meera, B. N. & Ramakrishna, J. *J. Non-Cryst. Solids*, 1993, **159**, 1.
25. Ratai, E.-M., Janssen, M., Epping, J. D., Chan, J. C. C. & Eckert, H. *Phys. Chem. Glasses*, 2003, **44**, 45.
26. Machowski, P. M., Varsamis, C. P. E. & Kamitsos, E. I. *J. Non-Cryst. Solids*, 2004, **345&346**, 213.
27. Namilov, S. V. *Zh. Prikl. Khim.*, 1972, **45**, 256.
28. Rouxel, T. *J. Am. Ceram. Soc.*, 2007, **90**, 3019.
29. Makishima, A. & Mackenzie, J. D. *J. Non-Cryst. Solids*, 1975, **17**, 147.
30. Shannon, R. D. *Acta. Cryst.*, 1976, **A32**, 751.
31. Shelby, J. E. *Introduction to Glass Science and Technology*, 2005, Royal Society of Chemistry, P 98.
32. Zhang, Z. & Soga, N. *Phys. Chem. Glasses*, 1991, **32**, 839.
33. Lee, S. K., Eng, P. J., Mao, H.-K., Meng, Y. & Shu, J. *Phys. Rev. Lett.*, 2007, **98**, 105502.
34. Grimsditch, M., Polian, A. & Wright, A. C. *Phys. Rev. B*, 1996, **54**, 152.
35. Takada, A., *Phys. Chem. Glasses*, 2004, **45**, 156.
36. Lee, S. K., Eng, P. J., Mao, H.-K., Meng, Y., Newville, M., Hu, M. Y. & Shu, J. *Nature Mater.*, 2005, **4**, 851.
37. McMillan, P., Pirou, B. & Couty, R. *J. Chem. Phys.*, 1984, **81**, 4234.
38. Grimsditch, M. *Phys. Rev. Lett.*, 1984, **52**, 2379.
39. Hemley, R. J., Mao, H. K., Bell, P. M. & Mysen, B. O. *Phys. Rev. Lett.*, 1986, **57**, 747.
40. Sugai, S. & Onodera, A., *Phys. Rev. Lett.*, 1996, **77**, 4210.
41. Poe, B. T., Romano, C. & Henderson, G. *J. Non-Cryst. Solids*, 2004, **341**, 162.

Proc. VI Int. Conf. Borate Glasses, Himeji, Japan, 18–22 August 2008   *Glass Technol.: Eur. J. Glass Sci. Technol. A, October 2009, 50 (5), 288–292*

Pulsed laser induced nanostructure and thermal properties of bismuth borate glass

Hirotoshi Yasunaga, Hirokazu Masai, Yoshihiro Takahashi, Takumi Fujiwara*

Department of Applied Physics, Tohoku University, 6-6-05, Aoba, Sendai, 980-8579, Japan

Takayuki Komatsu

Department of Materials Science and Technology, Nagaoka University of Technology, 1603-1 Kamitomioka, Nagaoka, 940-2188, Japan

Manuscript received 18 August 2008
Revised version received 30 October 2008
Accepted 7 January 2009

We have investigated the formation of nanostructure at the surface of $50Bi_2O_3.50B_2O_3$ glass using XeCl pulsed laser irradiation with a heated support. The growth of a protuberant nanostructure was enhanced by either increasing the support temperature or the laser pulse repetition rate. It was found that the surface protuberances were amorphous with the same composition as the original glass. In order to discuss the relation between the temperature and the structural change, the Urbach rule was applied to analyse the optical absorption. As a result, we found that the effect of temperature was insufficient for crystallisation in the $50Bi_2O_3.50B_2O_3$ glass. Considering the characteristics of the glass, it appears that dynamic randomness is independent of the morphology and only affects the phase change.

Introduction

Recently, nanoscale local structural modification of a glass matrix, induced by laser irradiation, has attracted much interest because of the possibility of fabricating new functional optical materials.[1-12] In particular, crystallisation is a promising method for giving lightwave controllability (a characteristic of crystalline materials) to the glassy material, because the precipitated crystallites provide optical functions not achieved in conventional optical glass materials. One of the methods used to induce nanoscale structural change is pulsed laser irradiation. By controlling the laser irradiation conditions, it is possible to control the time for which the sample is heated above both the glass transition temperature, T_g, and the crystallisation onset temperature, T_x. Therefore, it is expected that pulsed laser irradiation can enable us to control the morphology of the surface nanostructure by controlling the time and area over which structural change is induced in the glass.

Our group have succeeded in fabricating nanoparticles at the surface of $K_2O–Nb_2O_5–TeO_2$ and $CaO–Bi_2O_3–B_2O_3–Al_2O_3–TiO_2$ glasses, in which transparent nanocrystallisation occurs by heat treatment, by means of ultraviolet pulsed laser irradiation.[13,14] Under pulsed laser irradiation, due to the short term instantaneous temperature increase and relaxation, there is a possibility of behaviour which is different from that for conventional heat treatment. However, the mechanism for pulsed laser irradiation is not yet fully clarified because the glass composition is not

the same as the crystal composition. In the present study, we selected $50Bi_2O_3.50B_2O_3$ glass as a virgin sample for irradiation by an excimer laser because the glass transition temperature in this glass is 395°C, which is suitable for a thermally or optically induced structural change. In addition, since the glass is fragile, it is also expected that rearrangement of atoms occurs in a short time for temperatures above T_g. Furthermore, this glass composition corresponds to the stoichiometry of crystalline $BiBO_3$. Since there has been no report on transparent nanocrystallisation of the glass with heat treatment, there is a possibility of exhibiting transparent nanocrystallisation by the laser irradiation process. Moreover, the present glass is interesting from the point of view of the high optical nonlinearity of $BiBO_3$.[15]

In this study, $50Bi_2O_3.50B_2O_3$ glass has been fabricated and irradiated by a XeCl pulsed laser with different conditions for creating nanostructures at the glass surface. The fabricated protuberant nanostructure was analysed using several measurement techniques. In addition, we have examined and discussed the optical properties on the basis of the Urbach rule[16] in order to study the relation between the heat and the structural change of the glass. From these results, we have found a guide of how to control the fabrication of surface crystallised nanostructures.

Experimental

To prepare $50Bi_2O_3.50B_2O_3$ glass, Bi_2O_3 (50 mol%) and B_2O_3 (50 mol%) were melted in a gold crucible at 850°C for 40 min. The glass melt was quenched on

* Corresponding author. Email fujiwara@laser.apph.tohoku.ac.jp

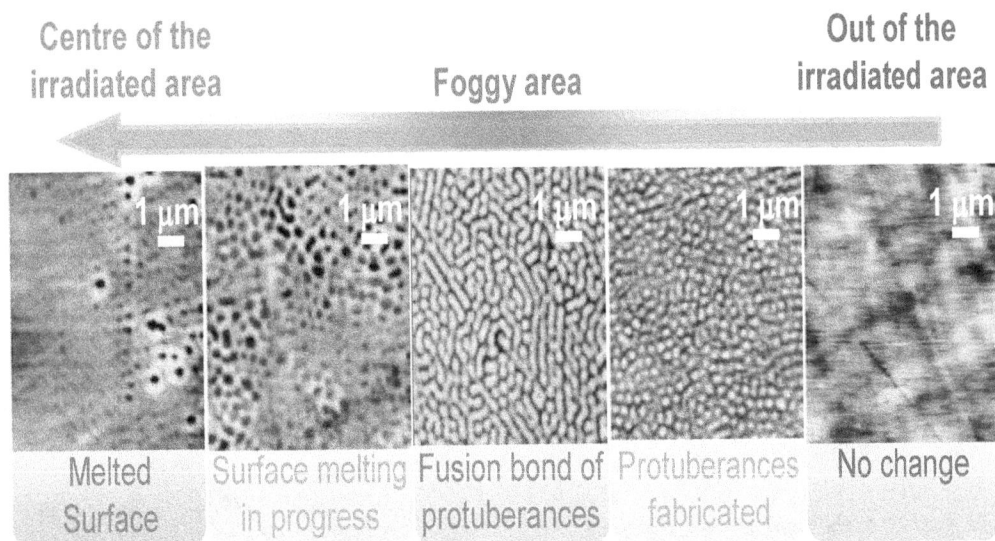

Figure 1. AFM images of different surface regions of the $50Bi_2O_3.50B_2O_3$ glass after XeCl laser irradiation. The repetition rate and the support heating temperature were 1 Hz and 250°C, respectively

a steel plate at 150°C, annealed at T_g for 30 min, and then polished to obtain a glass with a mirror surface. We used a XeCl pulsed laser (λ=308 nm) with a full width at half maximum (FWHM) pulse width of 8 ns, a laser intensity of 8·0 mJ/cm^2, and a laser spot size of 4×8 mm^2. For the laser irradiation, the number of pulses was fixed at 1000 shots, and repetition rates of either 1 Hz or 10 Hz were selected. A sample holder with a heat controller was used for heat supported laser irradiation. On the other hand, optical absorption spectra were measured using a metallic sample holder with a ceramic heater. The temperature for both the support heating of the laser irradiation and the optical absorption was changed from room temperature up to 300°C by steps of 25°C. Atomic force microscopy (AFM), transmission electron microscopy (TEM), x-ray photoelectron spectroscopy (XPS), and x-ray diffraction (XRD) measurements were used as evaluation methods.

Results and discussion

T_g, T_x (crystallisation onset temperature) and T_p (peak crystallisation temperature) of $50Bi_2O_3.50B_2O_3$ glass were estimated[14] to be 395°C, 451°C, and 469°C, respectively. The irradiated part of the glass surface became foggy without decrease of transparency. The foggy area became deeper in colour and larger as the support temperature increased, whereas the centre of the irradiated area became more transparent with increasing support temperature over 225°C. Figure 1 shows AFM images of $50Bi_2O_3.50B_2O_3$ glass after XeCl laser irradiation at 250°C with a repetition rate of 1 Hz. From the outside to the centre of the irradiated area, we observed the following surface changes: flat surface with no change (outside the irradiated area), a protuberant nanostructure and its development and then bonding together to form connected lines

(foggy area), and melted flat surface (the centre of the irradiated area). The observed structural differences in the irradiated area are thought to correspond to the distribution of the energy density of the laser, and it is supposed that the area closer to the centre of the irradiated area became higher in temperature. We can speculate that Figure 1 shows a morphological conversion from a protuberant nanostructure to a flat smooth surface as the centre of the irradiated area is approached. This morphological conversion was also observed by increasing the support temperature, each time observing the same region of the irradiated area.

Surface structures at the centre of the irradiated area of the $50Bi_2O_3.50B_2O_3$ glasses at each temperature were observed by AFM. Increasing the support temperature resulted in morphological conversion from a protuberant nanostructure to a flat smooth surface; the protuberances developed (up to 200°C), fused with each other (from 200°C to 250°C), and finally a flat smooth surface was formed (from 250°C to 300°C).

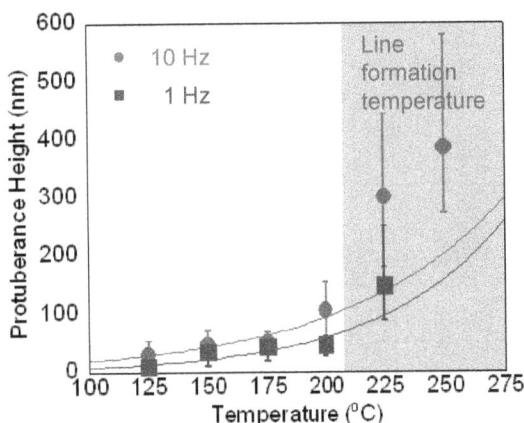

Figure 2. Temperature dependence of the protuberance height at the centre of the irradiated spot. The lines indicate fits which were performed excluding data in the temperature range where line formation occurs

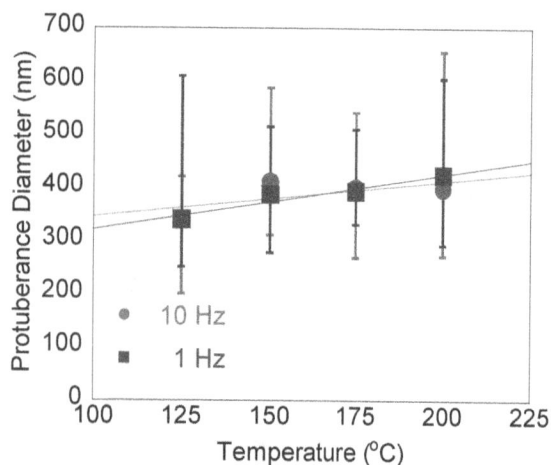

Figure 3. Temperature dependence of the protuberance diameter at the surface of the samples with different laser repetition rates

Figure 2 shows the correlation between the sample temperature and the protuberance height of the nanostructure at the surface, in the temperature range before the smoothing of the surface starts. The protuberance height increased with increasing sample temperature or repetition rate. The protuberance height of the sample irradiated with a laser repetition rate of 10 Hz showed a higher rate of increase than for a repetition rate of 1 Hz. In the temperature range where line formation occurs, the observed protuberance heights exceeded the trend which was followed at lower temperatures (i.e. the fit lines in Figure 2). Figure 3 shows the correlation between the support temperature and the protuberance diameter. The diameter increased at a small rate as the sample tem-

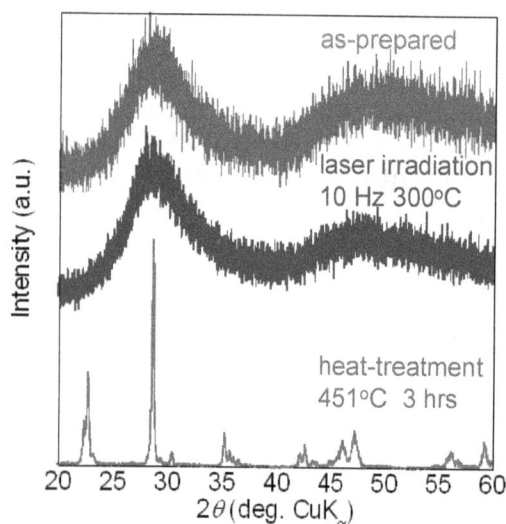

Figure 4. XRD patterns of the as-prepared glass, and the glass with protuberant nanostructure fabricated by laser irradiation. The repetition rate and support heating temperature of the laser condition were 10 Hz and 300°C, respectively. The XRD pattern of the sample heat treated at T$_x$ (=451°C) for 3 h in an electric furnace is also shown for comparison

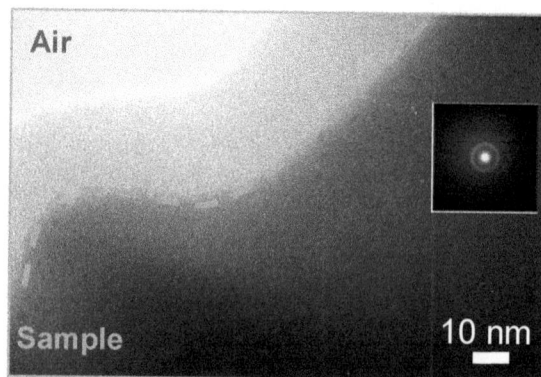

Figure 5. TEM image of protuberant nanostructure at the surface of the 50Bi$_2$O$_3$.50B$_2$O$_3$ glass after laser irradiation. The repetition rate and support heating temperature for the laser irradiation were 10 Hz and 300°C, respectively

perature increased. Therefore, it can be said that the protuberance size increases with increasing sample temperature or laser repetition rate.

The crystalline phases in the protuberances were examined by XRD. Figure 4 shows the XRD patterns of the as-prepared glass and the glass with the protuberant nanostructure fabricated by laser irradiation with repetition rate of 10 Hz at 300°C. The XRD pattern of the 50Bi$_2$O$_3$.50B$_2$O$_3$ glass heat treated at 451°C (T$_x$) for 3 h is also shown in the figure for comparison. Although the BiBO$_3$ crystalline phase was observed in the heat treated glass, no crystalline peaks were observed for the laser irradiated sample. This result suggests that the protuberances are not crystals, or the diffraction peaks are too weak to confirm the crystalline phase. We, therefore, utilised TEM and XPS measurements to examine the protuberant nanostructure at the surface. Figure 5 shows the TEM image of the sample irradiated at 300°C and 10 Hz. Neither precipitation of crystal nor clusters were observed from the TEM image which showed a uniform structure. Electron beam diffraction also showed no clear satellites. Figure 6 shows the Bi 4f XPS spectra of a laser irradiated sample at different surface regions; the foggy region, the centre of the irradiated area, and the non-irradiated area (see Figure 1). There is little difference of electron states between these three regions. Since other electron states also showed little difference in the XPS spectra between the observed surface regions, considered together with the electron beam diffraction result, it is concluded that the composition of the surface shows little change in composition after laser irradiation. From both the TEM and XPS measurements, we concluded that the protuberant nanostructure is almost the same as the base glass, exhibiting neither crystallisation nor compositional change. Although the protuberant nanostructure shows a similar morphology to phase separation, we think that phase separation has not taken place because of the following results. First, no contrast difference can be seen in the TEM image

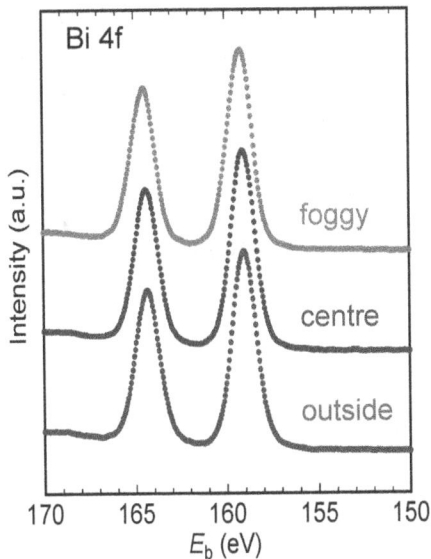

Figure 6. Bi 4f XPS spectra of a laser irradiated sample at different surface regions of the $50Bi_2O_3.50B_2O_3$ glass; protuberances existed in the foggy region, the centre region showed melting behaviour, and the outside is the non-irradiated region

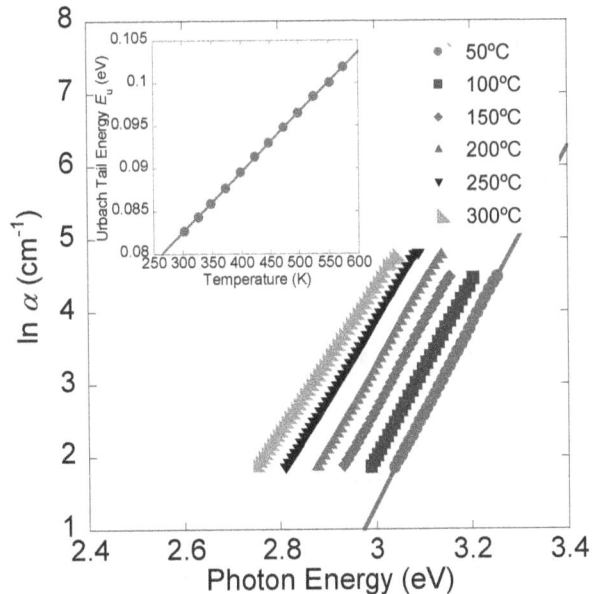

Figure 7. Absorption coefficient of as-prepared $50Bi_2O_3.50B_2O_3$ glass at different temperatures. Inset shows the Urbach tail energy at different temperatures. The Urbach tail energy is estimated from the gradient of the absorption coefficient plot

of the protuberance. Second, no clear difference can be seen in the XPS spectra between protuberance and non-irradiated area. Third, no crystalline peaks were observed in the XRD pattern despite the phase diagram,[17] which shows that $Bi_3B_5O_{12}$ could be precipitated. We, therefore, conclude that the observed structure is not caused by phase separation.

In order to accomplish the fabrication of crystal nanoparticles, we attempted to estimate the optical randomness of the glass based on the Urbach rule, which is often used for chalcogenide glasses and semiconductors.[16] The Urbach rule is written as follows

$$\log a = (E-E_0)/E_U + \log a_0 \qquad (1)$$

where a is absorption coefficient, E_U is the Urbach tail energy, E is the phonon energy, E_0 is an optical band gap energy, and a_0 is the value of the absorption coefficient at E_0. By fitting the linear correlation part of the measured absorption coefficient at each temperature, the optical band gap and the Urbach tail energy can be determined. In this study, we focused on the Urbach tail energy which is the reciprocal of the gradient of the fitted line and reflects the randomness inside the material.[16] The value of E_U at 0 K shows the static randomness, which originates in the glass structure. On the other hand, dE_U/dt, the gradient of the plot of E_U versus temperature, is due to dynamic randomness, which arises from phonon oscillations. Randomness is a parameter for the degree of disorder in the glass structure, that is to say, how much the structure is different from the crystal structure. For example, a small value of randomness means the structure is rather crystal like. Therefore,

it is clear that the randomness of the structure links directly to a phase transition, such as crystallisation.

Figure 7 shows the absorption coefficient of $50Bi_2O_3.50B_2O_3$ glass, which we modelled by the Urbach rule. By linear fitting of the Urbach tail energy, E_U, versus temperature, the dynamic randomness dE_U/dT and the static randomness E_U (T=0 K) can be determined. Table 1 shows the values of the static randomness and dE_U/dT for the $50Bi_2O_3.50B_2O_3$ glass and various other materials for comparison. From the table, it is clear that $50Bi_2O_3.50B_2O_3$ glass shows smaller dE_U/dT than the other glasses. Although a small value for the dynamic randomness indicates that the structure is hardly affected by changes in temperature, this seems to contradict the fact that the present glass is fragile.

We thought that consideration of viscosity may solve this contradiction. Viscosity is a parameter that is related to the morphology whereas dynamic randomness may be a parameter that is related to a phase transition, i.e. crystallisation. We presume that these parameters are almost independent during laser irradiation to simplify the explanation. The viscosity has little effect on the phase transition and similarly the dynamic randomness has little effect on the morphology. Since viscosity and dynamic randomness of the $50Bi_2O_3.50B_2O_3$ glass are low, it is easy to induce morphological change, but difficult to induce a phase transition. If the contribution of the heat to the phase transition, in other words the thermodynamic driving force, is insufficient, no crystallised nanostructure will be fabricated. On the other hand, applying a higher temperature to the sample during laser irradiation

Table. 1. The determined values of E_U and dE_U/dT for $50Bi_2O_3.50B_2O_3$ glass and values for various other materials for comparison

Composition	E_U (0 K) (eV)	dE_U/dT ($\times 10^{-5}$ eV K^{-1})
$50Bi_2O_3.50B_2O_3$ glass	0·061	7·0
$5CaO.10Bi_2O_3.65B_2O_3.20TiO_2$ glass[18]	0·15	118·8
$15K_2O.15Nb_2O_5.70TeO_2.0·5Er_2O_3$ glass[13]	0·066	89·9
TeO_2 (001) single crystal[13]	0·001	13·5
Silica glass[19]	0·073	11·2

can produce a crystallised nanostructure. Although the centre of the irradiated area has shown melting behaviour with high temperature, no crystallisation took place. It is considered that the contribution of the support temperature during laser irradiation to the phase transition is insufficient. Therefore, it is proposed that applying a higher temperature over a short range of time, i.e. higher laser fluence, would be effective for fabrication of crystalline nanostructure.

Conclusion

Protuberant nanostructures were formed on the surface of $50Bi_2O_3.50B_2O_3$ glass by means of pulsed XeCl laser irradiation. The size of protuberances increased with increasing support temperature or laser pulse repetition rate. At high temperatures, protuberances bonded together and connected to form lines, and eventually, changed to a smooth surface. It is suggested that these protuberances consist of amorphous material with a composition almost identical to that of the original glass. Optical analysis using absorption spectra and the Urbach rule showed that the dynamic randomness (which governs the temperature dependence) is small. In order to resolve the paradox between the fragility and dynamic randomness, we assumed that the dynamic randomness does not

affect the morphological change and only affects the phase change. Considering the result that the surface is melting at high support temperature, increase of the laser fluence is thought to be effective to fabricate crystallised nanostructure. Although there is still a lot of unclear behaviour about the structural changes in the glass material, we have showed the possibility of gaining a guide to the fabrication of surface nanostructure by applying the Urbach rule.

References

1. Davis, K. M., Miura, K., Sugimoto, N. & Hirao, K. *Opt. Lett.*, 1996, **21**, 1729–1731.
2. Maruo, S., Nakamura, O. & Kawata, S. *Opt. Lett.*, 1997, **22**, 132–134.
3. Watanabe, M., Juodkazis, S., Sun, H.-B., Matsuo, S. & Misawa, H. *Appl. Phys. Lett.*, 2000, **77**, 13–15.
4. Watanabe, W., Asano, T., Yamada, K., Itoh, K. & Nishii, J. *Opt. Lett.*, 2003, **28**, 2491–2493.
5. Nogami, M., Ohno, A. & You, H. *Phys. Rev. B*, 2003, **68**, 104204/1–104204/7.
6. Shimotsuma, Y., Hirao, K., Kazansky, P. G. & Qiu, J. *Jpn. J. Appl. Phys.*, 2005, **44**, 4735–4748.
7. Fujiwara, T., Ogawa, R., Takahashi, Y., Benino, Y. & Komatsu, T. *Phys. Chem. Glasses*, 2002, **43C**, 213–216.
8. Nishiyama, H., Miyamoto, I., Matsumoto, S., Saito, M., Fukumi, K., Kintaka, K. & Nishii, J. *Appl. Phys. Lett.* 2004, **85**, 3734–3736.
9. Nishiyama, H., Miyamoto, I., Hirata, Y. & Nishii, J. *Opt. Express*, 2007, **15**, 2047–2054.
10. Masai, H., Mizuno, S., Fujiwara, T., Benino, Y., Komatsu, T. & Mori, H. *J. Mater. Res.*, 2007, **22**, 1270–1274.
11. Sato, R., Benino, Y., Fujiwara, T. & Komatsu, T. *J. Non-Cryst. Solids*, 2001, **289**, 228–232.
12. Honma, T., Benino, Y., Fujiwara, T., Komatsu, T. & Sato, R. *Appl. Phys. Lett.*, 2003, **83**, 2796–2798.
13. Mizuno, S., *Doctor Thesis, Tohoku University*, 2008.
14. Masai, H., Mizuno, S., Fujiwara, T., Mori, H. & Komatsu, T. *Opt. Express*, 2008, **16**, 2614–2620.
15. Ihara, R., Honma, T., Benino, Y., Fujiwara, T. & Komatsu, T. *Opt. Mater.*, 2004, **27**, 403–408.
16. Urbach, F. *Phys. Rev.*, 1953, **92**, 1324.
17. Levin, E. M. & McDaniel, C. *J. Am. Ceram. Soc.*, 1962, **45**, 356.
18. Tachibana, N., Masai, H., Takahashi, Y. & Fujiwara, T. *Annual Meeting of the Ceramic Society of Japan, Abstract Book*, 2008, **1E29**, 30.
19. Saito, K. & Ikushima, A. J. *Phys. Rev. B*, 2000, **62**, 8584–8587.

Proc. VI Int. Conf. Borate Glasses, Himeji, Japan, 18–22 August 2008   *Phys. Chem. Glasses: Eur. J. Glass Sci. Technol. B,* October 2009, **50** (5), 271–283

The structure of tin borate based glasses

Alex C. Hannon,[1] *Emma R. Barney*

ISIS Facility, Rutherford Appleton Laboratory, Chilton, Didcot, Oxon OX11 0QX, UK

Diane Holland

Department of Physics, University of Warwick, Coventry CV4 7AL, UK

Manuscript received 17 August 2009
Revised version received 4 September 2009
Accepted 4 September 2009

A series of tin borate glass samples was prepared by melting in alumina crucibles, and these were studied by means of neutron diffraction, ^{27}Al and ^{11}B MAS (magic angle spinning) NMR, density, XRF and EDX measurements. The results show that the samples contain about 6·5 mol% Al$_2$O$_3$ from the crucibles, and when this was taken into account the B–O coordination numbers from neutron diffraction showed reasonable agreement with ^{11}B MAS NMR results. These results serve as a case study for showing the importance of sample composition for a correct analysis of diffraction results for glasses. The presence of Al$_2$O$_3$ has only a small effect on the number of 4-coordinated borons because few of the aluminium are on 4-coordinated sites. For low SnO content, both the fraction of 4-coordinated borons and the mean B–O bond length increase as SnO is added. However, for greater than 50 mol% SnO the fraction of 4-coordinated borons decreases, but the mean B–O bond length remains constant, and this behaviour of the bond length is as yet unexplained. For the Sn–O coordination, the results may be interpreted as showing that for high SnO content the Sn^{2+} ions are predominantly on asymmetric, trigonal pyramid SnO$_3$ sites, but that, for low SnO content, the Sn^{2+} ions are on a mixture of asymmetric SnO$_3$ sites and symmetric, high coordination sites.

1. Introduction

From a technological point of view there has been considerable recent interest in tin borate based glasses due to their potential advantages for use as anode materials for lithium ion batteries.[1–12] The structural role of lone pair ions such as Sn^{2+} in glass is also of interest due to the associated nonlinear optical properties.[13,14]

From a fundamental point of view, the structure of tin borate glasses is of interest primarily due to the composition dependence of N_4, the fraction of boron atoms which are 4-coordinated, which is associated with the *borate anomaly* in the thermophysical properties. In a brief, early study, Bray[15] used continuous wave ^{11}B NMR to show that N_4 increases up to a value ~0·18 at about 25 mol% SnO, and then falls as further SnO is added to the glass. More recently, Hayashi *et al*[4] and Holland *et al*[16] have both used ^{11}B MAS (magic angle spinning) NMR to confirm that N_4 exhibits a maximum, although these workers found that the maximum value of N_4 is ~0·3, and that it occurs at about 50 mol% SnO. The behaviour of N_4 in alkali borate glasses is understood, at least to some extent,[17,18] but as Holland *et al*[19] have discussed, there is greater variability in the behaviour of N_4 for glass systems with modifier cation valence greater than one. Hence the behaviour of N_4 in these systems provides a greater challenge to understanding, and an opportunity to gain greater insight into the mechanisms which govern this behaviour.

In an NMR study of GeO$_2$–B$_2$O$_3$ glasses, Baugher & Bray[20] have shown that if a network former is added to B$_2$O$_3$ then N_4 remains at a value of zero, as for pure B$_2$O$_3$, and hence the non-zero values of N_4 for tin borates indicate that SnO is not behaving as a network former. However, in a recent neutron diffraction (ND) and Reverse Monte Carlo (RMC) study of a glass with 50 mol% SnO it was concluded that SnO acts more like a network former.[6]

In order to further investigate these issues, we have followed up Holland *et al*'s NMR study of tin borate glasses[16] with a study involving ND and further NMR measurements on an almost identical set of samples.

2. Outline of neutron diffraction theory

The quantity measured in a neutron diffraction experiment[21–24] is the differential cross-section

$$\frac{\mathrm{d}\sigma}{\mathrm{d}\Omega} = I^S(Q) + i(Q) \tag{1}$$

where $\hbar Q$ is the magnitude of the momentum transfer, $I^S(Q)$ is the self scattering and $i(Q)$ is the distinct scattering. The self scattering, which can be calculated approximately, is subtracted from the data to give the distinct scattering. Structural information may then be obtained by a Fourier transformation of $i(Q)$, yielding the total correlation function

$$T'(r) = T^0(r) + \frac{2}{\pi} \int_0^\infty Q i(Q) M(Q) \sin(rQ) \, \mathrm{d}Q \tag{2}$$

Corresponding author. Email alex.hannon@stfc.ac.uk

where $M(Q)$ is a modification function introduced to take into account the maximum experimentally attainable momentum transfer, Q_{max}, and the prime on $T'(r)$ indicates the real space broadening which results from the finite value of Q_{max}. The average density contribution to the correlation function is

$$T^0(r) = 4\pi r g^0 \left(\sum_l c_l \bar{b}_l \right) \tag{3}$$

where g^0 is the average atomic number density and the l summation is over elements. c_l and $\bar{b}_l$ are, respectively, the atomic fraction and coherent neutron scattering length for element l. The correlation function is a weighted sum of partial correlation functions, $t_{ll'}(r)$

$$T(r) = \sum_{ll'}^{l \geq l'} C_{ll'} t_{ll'}(r) \tag{4}$$

where the l,l' summations are over all unique pairs of elements in the sample, and the coefficients, $C_{ll'}$, are given by

$$C_{ll'} = c_l (2 - \delta_{ll'}) \bar{b}_l \bar{b}_{l'} \tag{5}$$

where $\delta_{ll'}$ is the Kronecker delta. If a peak at a distance $r_{ll'}$ in the total correlation function is identified as arising from a particular pair of elements, l and l', then the relevant coordination number may be calculated from the area, $A_{ll'}$, under the peak according to

$$n_{ll'} = \frac{n_{ll'} A_{ll'}}{C_{ll'}} \tag{6}$$

3. Experimental

3.1. Sample preparation and characterisation

Tin borate, $x\text{SnO}.(100-x)\text{B}_2\text{O}_3$, glass samples were prepared with nominal compositions, x, of 20, 30, 40, 50, 60 and 70 mol% SnO, as well as pure B_2O_3. The pure B_2O_3 sample was made in a platinum crucible, whilst the tin borate glasses were made using the preparation method described by Holland et al,[16] which involves the use of an alumina crucible. The only difference from the previous preparation is that B_2O_3 containing boron enriched in ^{11}B was used, in order to avoid the extremely high neutron absorption of ^{10}B. The $^{11}\text{B}_2\text{O}_3$ was supplied by Eagle-Picher Inc. with a quoted isotopic purity of 99·62% ^{11}B. The isotopic purity of a similar batch of $^{11}\text{B}_2\text{O}_3$ was checked by secondary ion mass spectroscopy (SIMS), and found to be correct within experimental error. Alumina crucibles were used for the tin borate glasses because disproportionation of Sn^{2+} ($2\text{SnO} \rightarrow \text{Sn} + \text{SnO}_2$) can occur at the surface of the melt,[16,25] and metallic Sn^0 attacks platinum. An attempt was made to prepare a sample with 80 mol% SnO, but crystallisation occurred, and laboratory XRD (x-ray diffraction) revealed Bragg peaks which could be ascribed to crystalline SnO_2.

Two samples of nominal composition 20 mol%

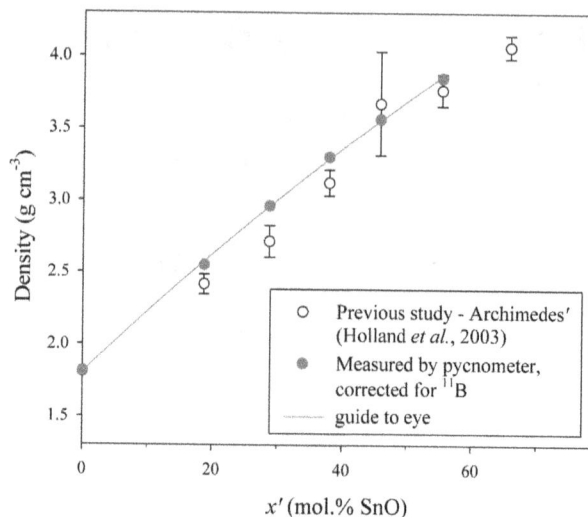

Figure 1. The density of tin borate glass samples as a function of revised SnO content, x'. The current data (solid circles and guide to the eye) were measured using a pycnometer with helium as the working fluid and have been corrected to show the density for a sample containing natural boron. The previous data (open circles and error bars) were measured using Archimedes' method with distilled water

SnO were prepared separately, but near identical results were obtained from them, and the neutron diffraction figures (2, 5 and 6) show the results for only one of these samples. Nevertheless, the derived parameters for both samples are presented, since this gives some indication of the reliability of the results.

The glass densities were measured using a Quantachrome Micropycnometer with helium gas as the working fluid, since this gives more accurate results than Archimedes' method with distilled water, as reported previously.[16] For each glass composition, Table 1 gives the measured densities for the isotopic $^{11}\text{B}_2\text{O}_3$-containing samples, and also a 'corrected density' value, which is the density that would be measured for a sample with the same atomic number density, but containing natural boron. As shown in Figure 1, the current results are consistent with the previous results, but the data points show much less scatter due to the superior accuracy of the current method. This consistency of density data indicates that the current samples are closely similar to those on which the previous ^{11}B and ^{119}Sn NMR measurements were made,[16] as is expected since the same preparation conditions were used. The close agreement of the measured density of the $^{11}\text{B}_2\text{O}_3$ sample (after correction) with a literature value[26] is further confirmation that the quoted isotopic composition is reasonable. (Because boron is a light element, the effect of isotopic substitution on the density is more significant than is the case for heavier elements.)

The samples were also examined by x-ray fluorescence (XRF) using a Panalytical MiniPal4 spectrometer, and by energy dispersive x-ray (EDX) analysis using a Jeol 6100 scanning electron microscope with an EDAX Genesis analytical system.

3.2. Neutron diffraction

Time-of-flight (TOF) ND measurements were performed using the GEneral Materials diffractometer, GEM,[27] at the ISIS spallation neutron source of the Rutherford Appleton Laboratory, UK. The samples were broken into pieces with maximum dimension 5 mm and placed in 8·3 mm diameter containers made of thin 25 μm vanadium foil so as to minimise the container background and absorption. The absolute calibration for momentum transfer, Q, was achieved with reference to diffraction data measured on crystalline $Y_3Al_5O_{12}$.[28] The uncertainty in the absolute values of the interatomic distances arising from this calibration is approximately 0·1%. The diffraction data were normalised with reference to the measured scattering from a 8·34 mm diameter vanadium rod.

TOF ND data are very sensitive to the presence of hydrogen in a sample, which would be manifest as both a strong rise in the scattering at low Q, and as a reduction of the scattering level with increasing scattering angle, 2θ, so that the data from different detector banks are inconsistent. Borate glasses can easily absorb water from the environment,[29] and hence care was taken with the preparation (in particular, this is why the neutron diffraction samples were in the form of pieces, rather than ground to a powder) and storage of the samples to avoid the presence of water. No evidence for the presence of hydrogen was found in the diffraction data for any of the samples.

Data from detector banks 1, 2, 3, 4 and 5 (covering a 2θ-range from 5·32° to 107·07°) were reduced and corrected for attenuation and multiple scattering using the standard ATLAS software,[24] by treating the sample as a homogenous material, on the assumption that the effect of the variations in path length due to the coarse nature of the sample is negligible. Oscillations were apparent in the data up to a maximum momentum transfer, Q_{max}, of 40 Å^{-1} or greater, and Figure 2 shows the distinct scattering, $i(Q)$, measured for each of the samples. The data were extrapolated to zero Q by means of fitting a quadratic of the form $A+BQ^2$, so that a Fourier transform could be performed using data over the Q-range from zero to Q_{max}.

The experimental ND data, in both reciprocal- and real-space, are available from the ISIS Disordered Materials Database.[30]

3.3. MAS NMR

^{27}Al MAS NMR measurements were performed on the samples with nominal compositions of x=20, 30, 40 and 60 mol% SnO. These measurements were performed using a Bruker Avance II+ 600 spectrom-

eter, operating at 156·37 MHz and 14·1 T, and with a Bruker 3·2 mm probe with spinning speed 16 kHz. A single pulse programme was used with a 0·5 ms pulse width and 30 s pulse delay. Chemical shifts were referenced using yttrium aluminium garnet (0·7 ppm with respect to the primary reference $Al(H_2O)_6^{3+}$). ^{11}B NMR measurements were also performed, in exactly the same way as described in the previous report.[16]

4. Results

4.1. Neutron diffraction

TOF ND is very sensitive to the presence of a small quantity of crystalline material in a glass sample, due to both the highly penetrating nature of a neutron beam, and the high reciprocal-space resolution which is normally achieved with the use of pulsed neutrons. The samples with x less than 60 mol% SnO showed

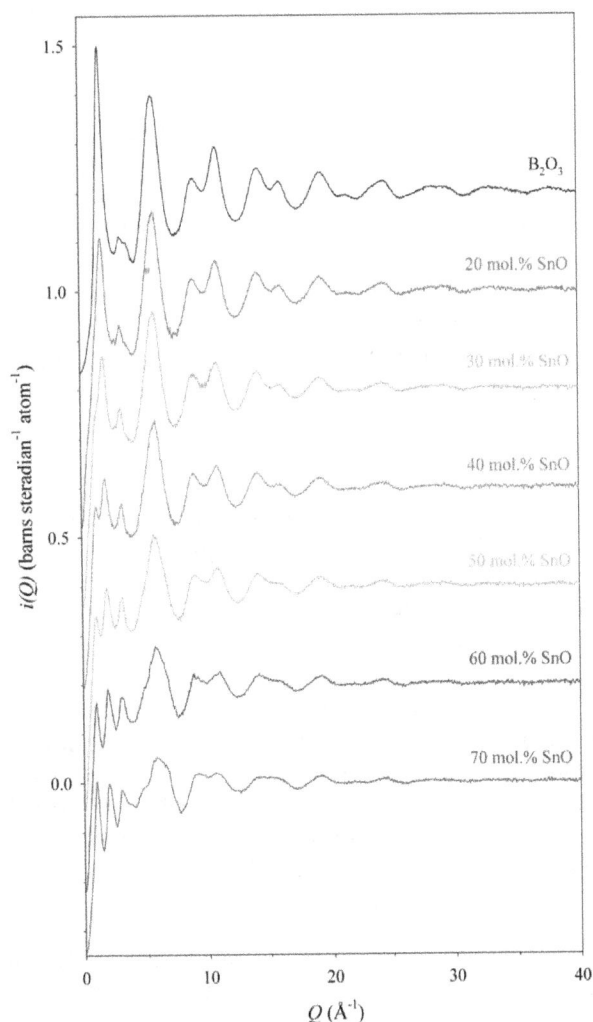

Figure 2. The distinct scattering, i(Q), for the tin borate glass samples, with successive offsets of 0·2 for clarity. The number by each curve indicates the nominal composition, x, of the glass sample in units of mol% SnO. Note that the data for the samples with x=60 and 70 mol% SnO are shown after the removal of Bragg peaks, as described in the text

no Bragg peaks, indicating that they were fully non-crystalline. Small Bragg peaks were observed for the x=60 mol% SnO sample, and larger Bragg peaks with the same d-spacings were observed for the sample with x=70 mol% SnO. The Bragg peaks were removed from the diffraction patterns for these two samples by the simple expedient of a linear interpolation underneath them, so that only the glass contribution to the diffraction pattern remained. The main contribution to the Bragg peaks was identified as arising from crystalline SnO_2,[31] although there were a few additional, small Bragg peaks which were not consistent with this crystal phase. The d-spacings of these additional peaks were also not consistent with either crystalline SnO,[32] or α-Al_2O_3,[33] and their origin was not identified. The data shown in Figure 2 are after the removal of Bragg peaks. The density of the sample with x=70 mol% SnO is not reported in Table 1, since this sample was not a single phase sample, to a sufficient extent that it would be misleading to give its density as though it were completely glassy. It must be acknowledged that the presence of crystalline SnO_2 material in the two samples with highest SnO content means that their true SnO content is likely to be less than quoted, but this has not been taken into account. The final normalisation of the ND results was achieved so that the low r region of the Fourier transform of Equation (2) was consistent with the average density contribution, $T^0(r)$, given by Equation (3), i.e. the data were normalised so that the total correlation function, $T'(r)$, oscillates about zero at low r.

4.2. MAS NMR

The ^{27}Al MAS NMR spectra are relatively weak, and they are shown in Figure 3. The spectra show three clear peaks at chemical shifts of about 3, 30 and 50 ppm. Attempts to fit the spectra with quadrupole lineshapes for three sites were not successful. The S/N (signal to noise ratio) is insufficient and, being a glass, there is a distribution of quadrupole parameters. Instead, an estimate of the relative contributions from the three sites was obtained by fitting the peak at ~50 ppm with a Gaussian, the peak at ~30 ppm with a bi-Gaussian and the area of the third peak was obtained by difference. The fitting procedure was carried out using Origin software (OriginLab

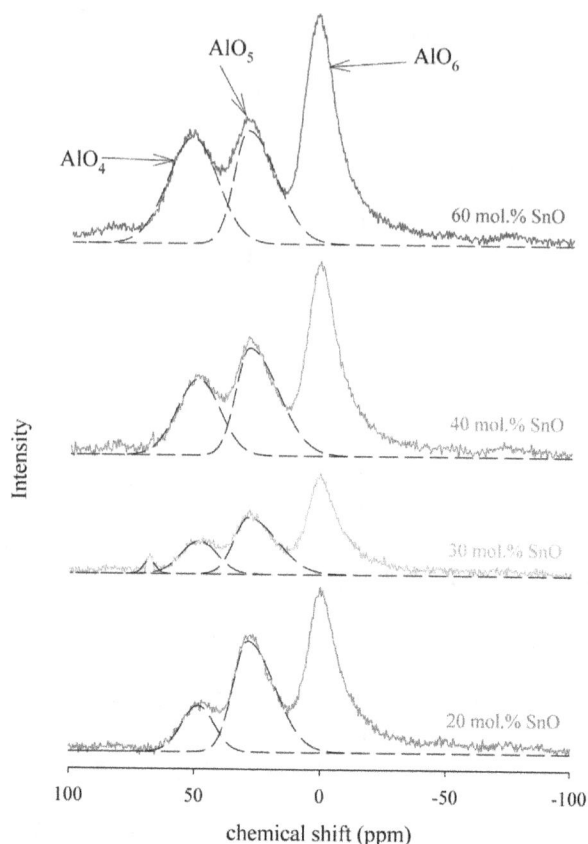

Figure 3. The ^{27}Al NMR spectra of tin borate glass samples with nominal compositions x=20, 30, 40 and 60 mol% SnO, together with the individual fit components (dashed lines)

Corporation). The fitted areas of the three peaks are given in Table 2, and plotted in Figure 4(a). N_4 results from the ^{11}B MAS NMR measurements are given later, in Table 4.

5. Discussion

5.1. Initial analysis of correlation functions

The ND data shown in Figure 2 were Fourier transformed according to Equation (2), using the Lorch modification function[34] with a value of 40 Å^{-1} for Q_{max}, to obtain the total correlation functions, $T'(r)$, shown in Figure 5. The first physical peak at about 1·4 Å is due to B–O bonds,[35] and as SnO is added to the glass, this peak becomes asymmetric and extends

Table 1. Densities and compositions of tin borate glass samples. (The revised composition parameters, x' and y', were determined from the ND and NMR results, as discussed later)

Nominal composition, x (mol% SnO)	Revised SnO content, x' (mol%)	Al₂O₃ content, y' (mol%)	Density of isotopic sample (g cm⁻³)	Corrected density (g cm⁻³)	Previous density[16] (g cm⁻³)
0 (pure ¹¹B₂O₃)	-	-	1·823	1·813	1·810(3)[26]
20	18·7 & 19·4	6·6 & 3·2	2·562	2·552	2·42(7)
30	28·7	4·4	2·971	2·962	2·72(11)
40	37·9	5·4	3·309	3·301	3·12(9)
50	45·7	8·7	3·573	3·566	3·7(4)
60	55·3	7·8	3·865	3·849	3·77(11)
70	65·7	6·1			4·07(8)

Table 2. Relative areas (as a percentage) of the peaks fitted to the ^{27}Al NMR spectra. (The errors on the percentages are estimated as ±4, leading to errors on n_{AlO} and r_{AlO} of ±0·16 and ±0·0015 Å, respectively)

Nominal composition, x (mol% SnO)	AlO$_4$ peak (~50 ppm)	AlO$_5$ peak (~30 ppm)	AlO$_6$ peak (~3 ppm)	Average coordination number, n_{AlO}	Predicted average bond length, r_{AlO} (Å)
20	12·5	38·5	49	5·365	1·869
30	17	34	49	5·32	1·867
40	22	31	47	5·25	1·862
60	29	26	45	5·16	1·857

to higher r, due to the formation of four-coordinated boron. For pure B$_2$O$_3$ the second peak at about 2·4 Å is due to both O···O distances in BO$_3$ triangles, and also B···B distances in B$_3$O$_6$ boroxol groups. The sharp peak at about 3·6 Å is also characteristic of the boroxol group.[35] As SnO is added to the tin borate glasses, another peak at about 2·2 Å grows, due to Sn–O bonds.

A fit was performed to the first B–O peak in the B$_2$O$_3$ correlation function, $T'(r)$, yielding the parameters given in Table 3 (see Hannon *et al*[35] for a description of the fitted function and parameters).

Table 3. Parameters from fitting the first B–O peak in the B$_2$O$_3$ correlation function (statistical errors from the fit are given in brackets)

B–O bond length, r_{BO} (Å)	RMS variation in bond length, $\langle u_{BO}^2 \rangle$ (Å)	Coordination number, n_{BO}
1·3655(2)	0·0420(3)	2·934(8)

The best coordination number accuracy that can be achieved with pulsed ND is probably about 1%,[36] and the measured B–O coordination number in Table 3 differs from the ideal value of three by about 2%. Therefore the measured value is consistent with a three-coordinated boron environment. In addition, the peak position and width, r_{BO} and $\langle u_{BO}^2 \rangle^{1/2}$, are very close to the values found in a previous ND study of B$_2$O$_3$.[35]

In order to gain an initial estimate of n_{BO} for the tin borate samples, a similar approach was adopted

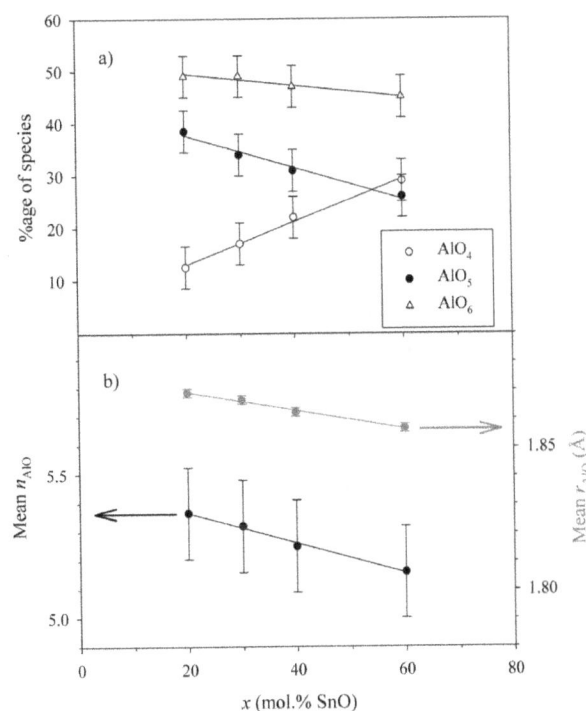

Figure 4. (a) The x-dependence of the relative occurrence of the aluminium species (as a percentage) in the tin borate glass samples, determined from fitting the ^{27}Al MAS NMR spectra (open circle, AlO$_4$; solid circle, AlO$_5$; open triangle, AlO$_6$), and (b) the x-dependence of the predicted mean Al–O coordination number (open squares, left axis) and bond length (solid circles, right axis). The lines are guides to the eye

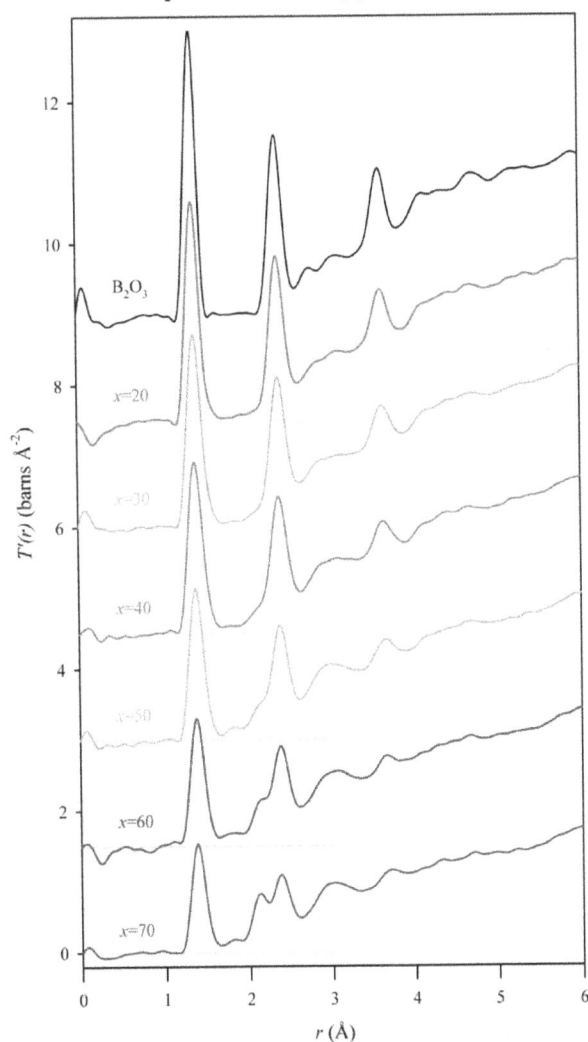

Figure 5. The total correlation function, T'(r), for the tin borate glass samples, obtained by Fourier transformation of the data shown in Figure 2. The curves are shown with successive offsets of 1·5 for clarity, and a dashed line indicates the zero level for each curve. The nominal composition, x, is indicated against each curve

to their correlation functions, fitting a single peak to the B–O region, and this led to values in the range 2·6–3·0. (For example, the sample with x=50 mol% SnO yielded a value n_{BO}=2·92, whereas ^{11}B NMR[16] shows that the B–O coordination number is at a maximum of 3·32 for this composition.) This behaviour is not consistent with the results of ^{11}B NMR,[16] which show that n_{BO} is greater than three for all non-zero values of x. Furthermore, B–O coordination numbers less than three cannot be supported by any accepted model for the structure of borates. The result for n_{BO} for pure B_2O_3 shows that reasonable coordination numbers can be obtained from ND data with a full knowledge of the experimental conditions, and suggests that further information is needed for the samples containing SnO. Although the use of a single symmetric peak to fit the asymmetric B–O peak in the correlation function leads to an inexact estimate of n_{BO}, this effect does not account for the discrepancies noted here.

Figure 6 shows the region of the correlation functions from 1·1 to 2·7 Å in detail. For the pure B_2O_3 sample the correlation function is approximately zero between the B–O and O···O peaks, and this is as expected because the B_2O_3 network structure has no interatomic distances in this range. However, for the tin borate glasses the correlation function is not zero between the B–O and Sn–O peaks. For a pure tin borate glass there are no expected interatomic distances in this range. However, the area under the correlation functions in this region of r can be interpreted as arising from Al–O distances due to an impurity Al_2O_3 component in the glass composition.

5.2. ^{27}Al MAS NMR

The ^{27}Al MAS NMR spectra shown in Figure 3 give clear evidence that the samples contain aluminium. Furthermore, the XRF and EDX spectra of these samples both show clear signals for aluminium. Thus it may be concluded that some of the crucible material became dissolved in the melt and remained in the glass. Therefore the samples studied here, and previously by Holland et al,[16] are actually tin aluminoborate glasses. It may thus be concluded that alumina crucibles are unsuitable for the preparation of pure tin borate glasses. Instead, it appears from the literature[2–12] that a more profitable approach may be to use carbon crucibles (either graphite or vitreous carbon) and a controlled dry, low oxygen atmosphere, or possibly mechanical milling.

There is widespread support in the literature for the assignment of the peak at ~3 ppm in the ^{27}Al NMR spectra to Al in an octahedral AlO_6 environment.[37] There is also little controversy regarding the assignment of the peak at ~50 ppm to Al in a tetrahedral AlO_4 environment. However, the assignment of the intermediate peak is more controversial,

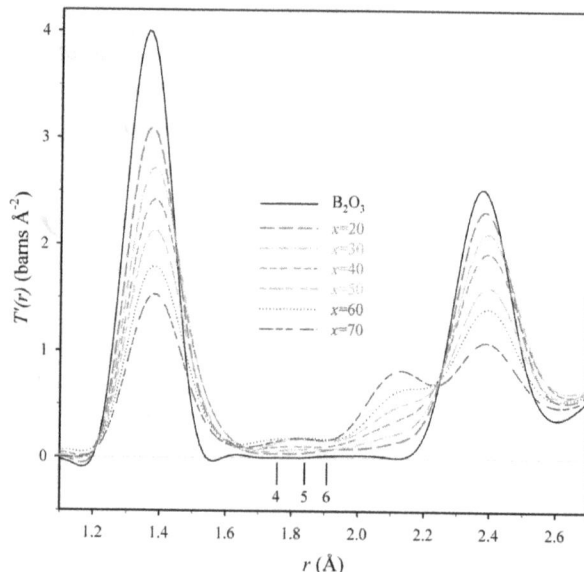

Figure 6. The total correlation function, T'(r), for the tin borate glass samples, showing the distance region from 1·1 to 2·7 Å in detail. The vertical lines indicate the Al–O bond lengths predicted by a simple bond–valence calculation for 4-, 5- and 6-coordinated aluminium[49,50]

and there are three possible interpretations which may be considered. Perhaps the most common interpretation of the ~30 ppm peak is as arising from five-coordinated AlO_5 sites. However, Schmücker et al[38–41] have presented a body of work arguing that for aluminosilicates this peak arises instead from Al in a tetrahedral site which is distorted due to being part of a tricluster. (A tricluster is formed when three cation–oxygen tetrahedra are linked together by one common oxygen atom.[42]) However, we have rejected the tricluster interpretation for our tin borate glass samples, because they have too low a concentration of tetrahedral sites (either AlO_4 or BO_4) for clusters of three tetrahedra to be common. Dupree et al,[43] in a study of magnesium aluminoborate glasses, have argued that the peak at ~30 ppm is due to Al on sites, $Al(OB)_4$, which are tetrahedrally coordinated by oxygen bridges to boron atoms. However, if the ~30 ppm peak observed for our tin borate samples is assigned to such sites then this leaves the ~50 ppm peak without an explanation, since the concentration of Al_2O_3 is too low for $Al(OAl)_4$ sites to be common. Thus, we also rejected this interpretation, and instead we assigned the intermediate peak to AlO_5 sites – an assignment which has also been adopted in most studies of aluminoborate glasses.[44–46]

The relative areas of the three NMR peaks show that approximately half the aluminium are on AlO_6 sites for all compositions, but that as SnO is added the number of AlO_4 increases at the expense of a loss of AlO_5. The average Al–O coordination number, n_{AlO}, was calculated from the relative peak areas (see Table 2 and Figure 4(b)), and this is a little larger than five with only a slight composition dependence.

214

5.3. Detailed analysis of correlation functions

5.3.1. B–O coordination

For an NMR experiment, correct results may be obtained without a full knowledge of the composition of the sample. For example, the values of N_4 reported previously from ^{11}B NMR measurements on glasses in this system[16] are a valid determination of the proportion of four-coordinated boron, even though the present results show that the samples were not pure tin borates as previously reported, but actually contain some aluminium. However, for a structural description of a glass it is necessary to know the composition reliably, and hence valid NMR results combined with incorrect sample composition data can lead to an incorrect structural model.

Contrastingly, for ND it is essential to have an accurate knowledge of the sample composition and density for correct results to be obtained. If the density and composition are not known accurately, then the average density contribution, $T^0(r)$ (see Equation (3)), to the correlation function cannot be calculated correctly, with the result that the area under a peak in the correlation function is not determined correctly, and incorrect coordination numbers are reported. Furthermore, incorrect compositional information leads to errors in the partial coefficients, $C_{ll'}$ (see Equations (4) and (5)), with the result that incorrect coordination numbers are calculated from the areas under correlation function peaks (see Equation (6)). Thus a ND experiment on a glass can provide a very useful comparison for NMR results, so that a correct structural interpretation can be developed. For example, see the recent study of tellurium borate glasses by Barney et al.[47]

In the present case, we use a combination of the results from ^{27}Al NMR and ND to investigate the glass composition, and then show that this leads to consistent results from ND and ^{11}B NMR. The length of Al–O bonds as a function of coordination number, n_{AlO}, can be estimated by using bond–valence parameters,[48] as has been discussed by Hannon & Parker.[49,50] The bond lengths predicted for 4-, 5- and 6-coordination are 1·757, 1·840 and 1·907 Å, respectively (assuming that all Al–O bonds in a unit have the same length), and these are indicated as vertical lines in Figure 6. Clearly these bond lengths occur at suitable values for explaining the extra area in the neutron correlation functions for the tin borate glass samples.

The correlation function, $T'(r)$, for each of the tin borate glasses was fitted over the range 0·75–2·48 Å by five peaks. The first peak was fixed at a position of 1·3655 Å, the same as the B–O distance, r_{BO}, in BO_3 triangles found for pure B_2O_3 (see Table 3). The second peak, at about 1·45 Å, was introduced to allow for the asymmetric broadening of the B–O peak, arising from the presence of longer B–O bonds, which is observed when BO_4 units are present in the glass.

(To our knowledge, the structure of only one tin borate crystal has ever been reported – the high pressure phase β-SnB_4O_7 (33·3 mol% SnO) in which the borons are all 4-coordinated, with B–O bond lengths varying from 1·408 to 1·561 Å.[51]) The third peak was introduced to allow for Al–O bonds, and its position and width, r_{AlO} and $\langle u^2_{BO}\rangle^{1/2}$, were fixed as follows: The predicted average Al–O bond length for each glass was calculated by a combined use of the frequencies of the three Al-centred units (see Table 2), and the bond–valence predictions of the Al–O bond lengths for each of these units (see Hannon & Parker[49,50]). The standard deviation of the predicted Al–O bond lengths of the three Al-centred units is 0·06 Å, and also the root mean square (RMS) variation in Al–O bond lengths in tetrahedral aluminate glasses[49,52,53] (due to thermal and static disorder) is commonly found to be 0·06 Å. These two values were added in quadrature to obtain a value of 0·085 Å for the width of the Al–O peak in the current glasses. The fourth peak at ~2·11 Å was introduced to account for Sn–O bonds, whilst the fifth peak at ~2·39 Å was introduced to account mainly for O···O and B···B distances.

An example of the results of this fitting is shown in Figure 7. The fitted area, A_{AlO}, of each Al–O peak was used together with the average coordination number, n_{AlO}, from ^{27}Al NMR (see Table 2) and Equation (6) to determine the Al–O partial coefficient, C_{AlO}. Then, by use of Equation (5), the aluminium atomic fraction, c_{Al}, was determined. On the assumption that the aluminium enters the glass network in the form of Al_2O_3, and that SnO and B_2O_3 are in the relative proportions given by the nominal composition of the sample, the value of c_{Al} was then used to calculate the Al_2O_3 content, y', and revised SnO content, x', as given in Table 1 (i.e. the revised glass compositions are of the form x'SnO.$y'Al_2O_3$.$(100-x'-y')B_2O_3$). On average, the samples were found to contain 6·5 mol% Al_2O_3, with a small spread of values and no obvious dependence on SnO content.

The ND data were renormalised to take account of the revised composition (which includes Al_2O_3) reported in Table 1, and the data (in both reciprocal- and real-space) shown in the figures are after the renormalisation has been applied, with the sole exception of Figure 7. This renormalisation was achieved with reference to the calculated average density term given in Equation (3), which is affected by both the change in atomic number density, and the change in average neutron scattering cross-section, that results from the change in composition. Table 4 gives the B–O coordination number, n_{BO}, calculated from the fits to the correlation function, $T'(r)$, based on both the nominal composition and the revised (Al_2O_3-containing) composition. These coordination numbers, and also the mean B–O bond length, r_{BO}, are derived from the sum of the contributions from the first two fitted peaks. Table 4 also gives the B–O

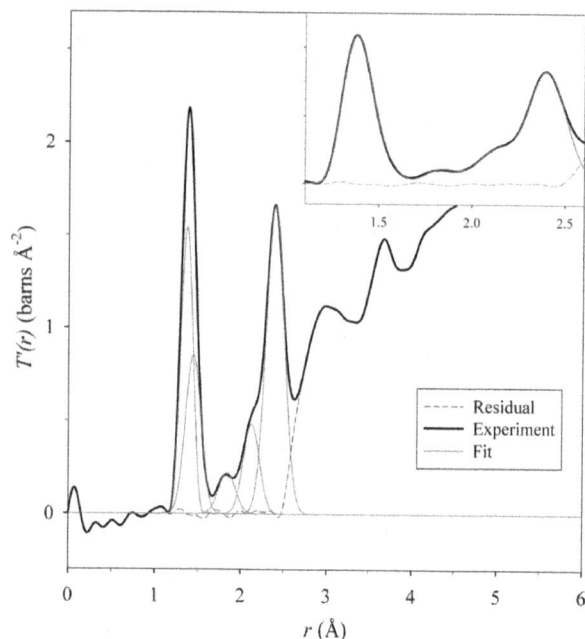

Figure 7. An example of fitting the correlation function, for the sample with a nominal composition x=50 mol% SnO (thick line, experimental data; thin line, fit; dashed line, residual). The main figure shows the components of the fit, whilst the inset shows the full fitted function

coordination number calculated from the N_4 values reported in the previous ^{11}B NMR study (it is useful to note that $n_{BO}=3+N_4$).[16] It is apparent that the B–O coordination numbers from ND based on the revised composition are much more consistent with the NMR results than is the case for the ND results based on the nominal composition (see Figure 8(a)). This demonstrates that it is essential to have an accurate knowledge of the sample composition to obtain reliable coordination numbers from ND, and as such this report serves as a useful ND case study. In general, the effect of the composition on both the calculated number density, g^0, and the scattering cross-section and coefficients (see Equations (3) and (5)) have significant effects on the derived coordination number.

In summary, the area of the Al–O region of the correlation function has been used in combination with the average coordination number from ^{27}Al NMR to determine the likely Al_2O_3 content of the samples. With this knowledge, it has then been pos-

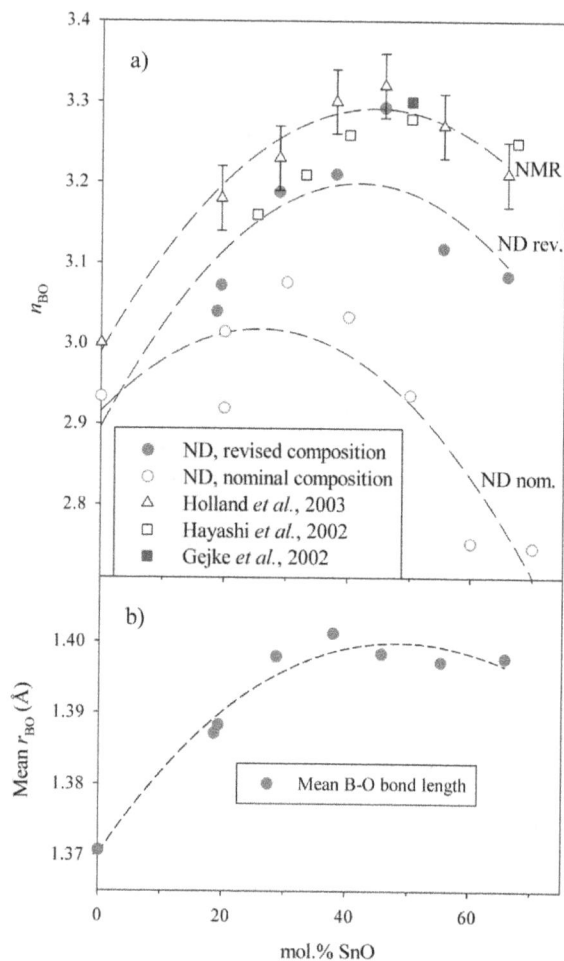

Figure 8. (a) The B–O coordination number, n_{BO}, of tin borate based glasses as a function of SnO content, as determined by ND, both with and without correction for Al_2O_3 content (solid and open circles, respectively), and as previously reported in the literature (open triangles, Holland et al;[16] open squares, Hayashi et al;[4]; closed square, Gejke et al[6]). The dashed lines are guides to the eye for the two sets of ND data (ND nom. and ND rev.) and for the data reported by Holland et al[16] (NMR). (b) The mean B–O bond length determined by ND. The dashed line is a guide to the eye

sible to extract B–O coordination numbers which are consistent with ^{11}B NMR. The B–O coordination numbers are not new knowledge, but the fact that we have achieved consistency with NMR demonstrates

Table 4. Coordination numbers and mean bond lengths of tin borate based samples (note that $n_{BO}=3+N_4$). For n_{BO} from ND (revised composition), the numbers in brackets indicate the errors arising solely from the uncertainty in n_{AlO} from NMR, as given in Table 2. (*The NMR n_{BO} values in brackets were measured by ^{11}B MAS NMR on the current samples)

Nominal composition, x (mol% SnO)	Revised SnO content, x' (mol%)	n_{BO} from ND (nominal composition)	n_{BO} from ND (revised composition)	n_{BO} from NMR*[16]	Mean r_{BO} (Å)	r_{SnO} (Å)	n_{SnO}	Sn valence
0 (pure ^{11}B$_2$O$_3$)	-	2·934	-	3	1·3655	-	-	-
20	18·7	2·920	3·040(4)	3·18 (3·05)	1·3872	2·109	1·56	1·11
20	19·4	3·015	3·072(2)	3·18	1·3883	2·104	1·75	1·26
30	28·7	3·076	3·188(4)	3·23 (3·26)	1·3980	2·111	1·65	1·17
40	37·9	3·033	3·210(6)	3·3 (3·28)	1·4012	2·119	2·03	1·41
50	45·7	2·936	3·293(12)	3·32 (3·27)	1·3984	2·125	2·62	1·79
60	55·3	2·752	3·118(13)	3·27 (3·26)	1·3972	2·125	2·13	1·46
70	65·7	2·746	3·084(12)	3·21 (3·22)	1·3977	2·114	2·27	1·60

that a reliable knowledge of the sample composition, and normalisation of $T'(r)$, has been achieved. This then enables results about the B–O bond length and the Sn–O coordination to be derived with a greater degree of confidence. Nevertheless, in view of the complex procedure used to determine the sample composition and data normalisation, and the possibility that the samples may be inhomogeneous, with different Al_2O_3 contents in different regions, the final coordination numbers should be regarded as indicating a probable behaviour, rather than being of the highest achievable numerical accuracy.

Table 4 gives the coordination numbers, n_{BO}, determined by ^{11}B MAS NMR for both the previous study,[16] and the current samples (in brackets). There is mostly good agreement between the two sets of results, giving further evidence that the current samples are closely similar to those for the previous study.[16] The largest difference between the two sets of data is for a nominal composition $x=20$ mol% SnO, and this is the reason why two samples were made at this nominal composition. However, there is a high level of consistency between the B–O coordination numbers derived from both ND and NMR for the current samples with this composition.

An important question that arises in this study is the effect which the presence of Al_2O_3 in the glasses may have on the presence of 4-coordinated boron in the structure. The most relevant study in the literature is Bunker et al's[44] ^{11}B and ^{27}Al MAS NMR investigation of the structure of alkaline earth aluminoborate glasses. In this study, it was found that the addition of Al_2O_3 causes a reduction in the value of N_4 (see Bunker et al's[44] Figure 4). However, the ^{27}Al NMR results showed that this reduction is mostly due to the replacement of BO_4 units with AlO_4 units, and it was found that a large proportion (typically of order 50%) of aluminium atoms are on AlO_4 sites. Contrastingly, the current results show a relatively low frequency of occurrence for AlO_4 sites (of order 20% – see Table 2 and Figure 4(a)). Hence it may be concluded that the results on N_4 for the samples studied by Holland et al[16] are a reasonable guide to the behaviour for pure tin borates, despite the presence of Al_2O_3. Some support for this conjecture is given by Figure 8(a) which shows that the N_4 values reported by Holland et al[16] are actually slightly larger than those reported by Hayashi et al[4] for samples which are presumably pure tin borates because they were made using carbon crucibles and a dry N_2 atmosphere.

Table 4 gives values for the mean B–O bond length, r_{BO}, calculated from the sum of the contributions from the first two peaks fitted to the correlation function, and shown in Figure 8(b). It is notable that r_{BO} increases with x for smaller values of x, and then becomes approximately constant. This behaviour is a challenge to current understanding of the structure of borate glasses. A simple model would be to assume

that all B–O bonds in BO_3 units are of one shorter length, say 1.3655 Å as found for pure B_2O_3, and that all B–O bonds in BO_4 units are of one longer length, of order 1.45 Å. However, in this model the observed growth and decline in the values of N_4 would be accompanied by a growth and decline in the mean B–O bond length. Our results, shown in Figure 8(b), do not exhibit a significant decline in the value of r_{BO} for high SnO contents, contradicting the simple model. For example, since the sample with $x=70$ mol% SnO has a similar n_{BO} value to the sample with $x=20$ mol% SnO (see Table 4), it is reasonable to expect that these two samples would have similar r_{BO} values; in fact the mean B–O bond length for the high SnO sample is about 0.01 Å higher, only slightly less than the maximum observed bond length, for the sample with $x=40$ mol% SnO. A development of the simple model mentioned above would be to take into account the effect of nonbridging oxygens (NBOs), which occur in the glass for high SnO content. However, according to current understanding, the bond lengths for NBOs are shorter than for bridging oxygens, and thus the inclusion of NBOs in the model does not immediately lead to an explanation for the observed behaviour of r_{BO}. A similar problem is found for germanate glasses. For example, for caesium germanate glasses it has been found[36] that the mean Ge–O bond length increases and then becomes constant, as Cs_2O is added to the glass, even though the Ge–O coordination number increases and then declines for high Cs_2O content. Henderson[54] has pointed out that this behaviour is also observed in EXAFS and XRD results for germanate glasses, and is as yet unexplained. Thus the failure of the observed mean bond length to decrease for high modifier content in variable coordination number systems, such as borate and germanate glasses, is a current challenge in glass science. It may be that the solution to this problem will involve a revised understanding of the behaviour of NBOs for high modifier content. To our knowledge, this is the first time that the anomalous behaviour of the bond length in relation to a coordination anomaly has been noted for borate glasses.

The coordination numbers, n_{BO}, in Table 4 have been obtained by a sum of the contributions from the first two peaks which were fitted to the experimental correlation functions, $T'(r)$. Although the results obtained from this sum are shown to be reasonable by comparison with ^{11}B NMR results, the individual coordination numbers from each of the two peaks are probably not physically meaningful alone, perhaps due to the effects discussed in the previous paragraph. Thus this two-peak fit is a convenient way of parameterising the B–O contribution to the correlation function, rather than having a simple interpretation. For example, it is doubtful that the distribution of B–O distances involves two well defined distances, one each for BO_3 and BO_4 units.

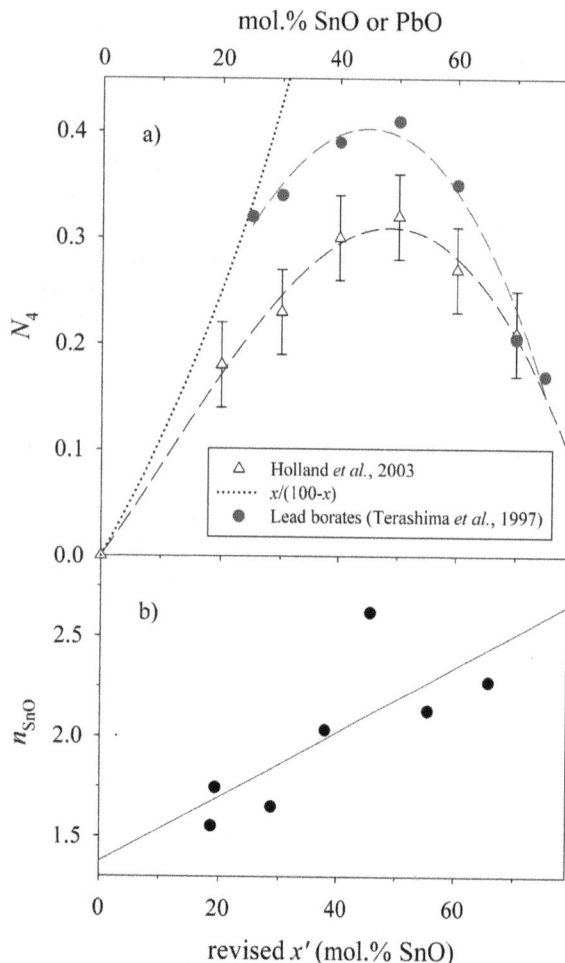

Figure 9. (a) The composition dependence of N_4 for tin borate based glasses[16] (open triangles) and for lead borate glasses[55] (solid circles). The dashed lines are guides to the eye for the two datasets, whilst the dotted line is the theoretical $x/(100-x)$ curve. (b) The Sn–O coordination number, n_{SiO}, for tin borate based glasses, calculated from the area of the fourth fitted correlation function peak

Other fitting strategies were attempted, but did not prove any more successful in yielding bond lengths and coordination numbers which were amenable to such a simple interpretation.

Figure 9(a) shows a comparison of the N_4 values from ^{11}B NMR measurements on tin borate glasses[16] (against revised composition) compared with recent measurements of N_4 for lead borate glasses.[55] The values of N_4 for tin borates are significantly smaller than for lead borates. Figure 9(a) also shows the theoretical curve (dotted line), $x/(100-x)$, which N_4 follows if the addition of modifier results solely in the formation of BO$_4$ units, and no NBOs.[17,18] For lead borates it appears that the experimental results may agree with this curve for less than ~25 mol% PbO, and this composition is similar to the prediction of Wright et al's[17] model that NBOs must start to form by a composition of 30 mol% in order to avoid the formation of bridges between two BO$_4$ units. However, the behaviour for tin borate glasses is different, and there

is no clear evidence to show that at any composition the values of N_4 are as large as predicted by the $x/(100-x)$ curve. Hence it appears that, even for low SnO content, the addition of SnO results in the formation of both BO$_4$ units and NBOs. Some support for this interpretation is given by XPS (x-ray photoelectron spectroscopy) results[4] which show evidence for the presence of NBOs in a glass with 25 mol% SnO. (This implies at least three distinct types of B–O correlation in the first peak.)

5.3.2. Sn–O coordination

Figure 9(b) shows the Sn–O coordination numbers, n_{SnO}, calculated (after correction for the Al$_2$O$_3$ content) from the area of the fourth peak fitted to the correlation function, $T'(r)$ (see Table 4). As shown by the guide to the eye, the overall trend is for the values to increase as the SnO content increases. However, the true average Sn–O coordination number probably has the opposite trend, but this is obscured in the correlation functions because of the overlap between the contribution due to longer Sn–O distances with other contributions (mainly O$\cdots$O and B$\cdots$B) which give rise to the fifth peak. For each of the fitted peaks, the Sn^{2+} valence was calculated from the values of n_{SnO} and r_{SnO} using the bond–valence parameter for Sn^{2+}–O,[48–50] and this value is given in Table 4. If all of the coordination to the Sn^{2+} has been correctly identified, then the calculated valence is expected to be close to the formal valence of two. However, with the possible exception of the value determined for x=50 mol% SnO, the calculated Sn^{2+} valence is significantly less than two, indicating that it is likely that the fitted correlation function peak has not included all of the Sn–O coordination, i.e. it is likely that there are additional Sn–O bonds whose contribution has been included in the fifth fitted peak.

Figure 10 shows the oxygen environments of the three Pb sites in crystalline Pb$_9$Al$_8$O$_{21}$.[53] These environments are typical of the various types of co-ordination that can be found around a lone pair ion, such as Pb^{2+} or Sn^{2+}, in an oxide network. Lone pair ions may either be found on sites such as Pb1 and Pb3 in Figure 10, with a low coordination number (in the range 2–5) and a highly asymmetric coordination shell, or sites such as Pb2, with a high coordination number (6 or more) and a highly symmetric coordination shell.[56] For the asymmetric sites, the lone pair ion may either be found with a smaller coordination number (~3) and all oxygens at about the same short distance (as for Pb3), or with a larger coordination number (~5) and some oxygens at similar short distances, whilst other oxygens are considerably more distant (as for Pb1). The Sn–O distances for several coordination numbers were calculated from the Sn^{2+}–O bond–valence parameter,[48] assuming that all bonds have the same length,[49,50] leading to the

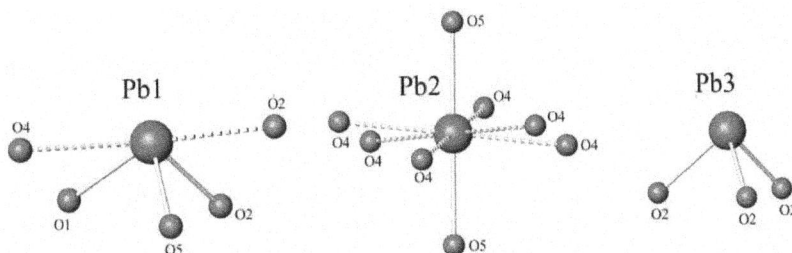

Figure 10. The oxygen environments of the three Pb sites in crystalline $Pb_9Al_8O_{21}$ (longer bonds are shown dashed)[53]

values given in Table 5. Clearly the position of the fitted Sn–O peak (see Table 4) is consistent with the bond length expected for 3-coordination. It is worth mentioning that the most common crystalline form of SnO involves SnO_4 square pyramids with a Sn—O bond length of 2·224 Å,[57] but this distance coincides with a minimum in the correlation functions (see Figure 6) and hence there is no direct evidence in the experimental data in favour of the presence of such units in the glass. There is also a high pressure form of SnO with a crystal structure composed of SnO_3 trigonal pyramids with a bond length of 2·123 Å, and this appears to be more relevant to the glass structure. Although the observed Sn–O bond length in the glass is consistent with 3-coordination, the observed coordination number, n_{SnO}, is not consistent because it is mostly significantly too low (see Table 4). However, these observations can be reconciled by application of the traditional model for the role of lone pair ions, according to which an oxide such as PbO acts as a network modifier[58] for low PbO content, but acts as a network former (or intermediate) for high PbO content.[59,60] In structural terms, we interpret this to mean that the predominant Pb coordination is symmetric for low PbO content, but that it becomes asymmetric for high PbO content. Applying this model to tin borate glasses implies that for high SnO content the predominant coordination is trigonal pyramidal SnO_3, as observed. However, for low SnO content there is an increasing number of Sn^{2+} ions on symmetric sites with a high coordination number of 6 or more, and with a mean Sn–O bond length of order 2·39 Å (see Table 5). The contribution to $T'(r)$ due to such symmetric Sn^{2+} sites is coincident with the fifth fitted peak at ~2·39 Å (due mainly O···O and B···B correlations, see Figure 7) so that it is not resolved by the peak fitting, with the result that as the SnO content decreases there is a steady reduction in the apparent coordination number, n_{SnO}, calculated from the fitted peak at ~2·11 Å. Thus the fit results may be

interpreted as showing that, for low SnO content, the Sn^{2+} ions are on a mixture of asymmetric SnO_3 sites and symmetric, high coordination sites. This interpretation could be further tested in the future by atomistic modelling, such as RMC or molecular dynamics, although we note that the simulation of the asymmetric type of lone pair site is especially challenging for these kinds of modelling.

This interpretation is consistent with the Sn–O environment observed in SnO–GeO_2 glasses by neutron diffraction.[61] Here the O···O peak in the correlation function due to GeO_4 tetrahedra is centred at ~2·8 Å, which is significantly longer than O···O distances in borate glasses (~2·39 Å), with the result that a longer distance Sn–O component was clearly observed. The Sn–O correlations in germanate glasses were interpreted in terms of 3-coordinated Sn^{2+} at ~2·12 Å, and 6-coordinated Sn^{2+} at ~2·35 Å. The 6-coordinated site was also found to be more abundant for low SnO concentrations. The ^{119}Sn MAS static NMR measurements reported in the previous study of tin borates[16] showed the tin environment to be close to axial at high SnO concentrations – as expected for a trigonal pyramidal SnO_3 unit – but becoming more asymmetric at lower SnO contents. The peak narrows slightly but there is no indication of a highly symmetric SnO_6 environment and something akin to a 3+3 unit is more likely.

It is of note that, although the traditional model for the role of PbO was originally developed to explain the wide glass formation range for lead silicate glasses,[59,60] recent structural studies suggest that in fact PbO may be on asymmetric PbO_3 sites for all lead silicate compositions.[62,63] However, a recent structural study[64] also shows that in lead borate glasses the Pb–O coordination number is about 6 for low PbO content, falling to about 3 for high PbO content, consistent with the traditional model for the role of PbO. This raises the possibility that the traditional model may be more appropriate for glasses (e.g. borates) in which the network forming cation itself (e.g. boron) undergoes coordination number change, and it is interesting to note that the values of n_{SnO} reported in Table 4 are at a maximum at the same composition as for the maximum of N_4, suggesting an association between the Sn–O and B–O coordination numbers. Most considerations of the factors causing the B–O coordination number to change in borate glasses have

Table 5. Sn^{2+}–O bond lengths calculated from the bond–valence parameter, assuming all bonds have the same lengths

Coordination number, n_{SnO}	Bond length, r_{SnO} (Å)
3	2·134
4	2·240
5	2·323
6	2·390

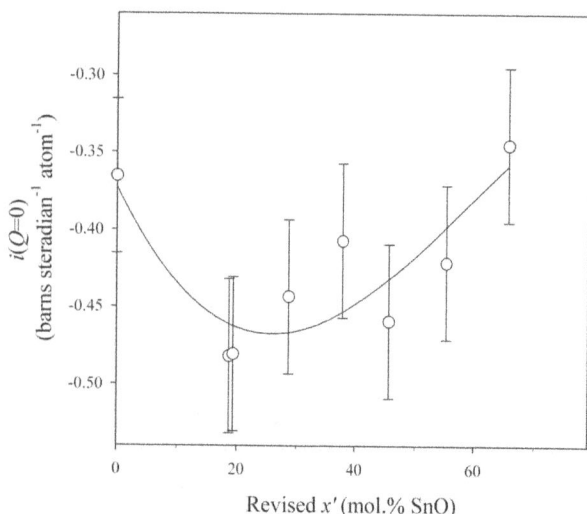

Figure 11. The composition dependence of the Q=0 limit of the distinct scattering, i(Q). The line is a guide to the eye

focused on alkali borate glasses. For alkali borates, it is likely that the alkali–oxygen coordination number changes little as the glass composition is changed, and thus the alkali–oxygen coordination is an essentially unchanging context in which the factors governing the B–O coordination operate. This study shows that the situation is more complex for borate glasses containing a modifier cation (such as a lone pair cation) whose own coordination number is subject to change, and thus such glasses provide a greater challenge for the understanding of the borate anomaly.

5.4. Longer range order

The *borate anomaly* is commonly observed in the thermophysical properties of borate glasses, and these are long range, bulk properties. The region of the diffraction pattern which is related to long range properties is the $Q=0$ limit. Figure 11 shows the composition dependence of the extrapolated zero Q limit of the distinct scattering, $i(Q)$. The addition of a small amount of SnO causes a rapid decrease in this value, followed by a minimum at ~20–30 mol% SnO and then a slight increase. This behaviour is not so similar to the behaviour of N_4 (as shown in Figure 8(a)), but shows greater similarity to the composition dependence of the glass transition temperature, T_g, which increases rapidly when a small amount of SnO is added to the melt, followed by a maximum at ~40 mol% SnO and then a slight decline.[16] In a careful and exhaustive small angle neutron scattering study of SiO_2 glass, Wright *et al*[65] have demonstrated that the zero Q limit of the diffraction data depends directly on the isothermal compressibility of the glass melt at the fictive temperature. Therefore the composition dependence shown in Figure 11 is apparently dominated by the behaviour of T_g, rather than any other factor, and does not have any close relation to the borate anomaly.

6. Conclusions

It is concluded that alumina crucibles are not suitable for making pure tin borate glasses, because typically about 6·5 mol% Al_2O_3 enters the glass melt from the crucible. The measurements described here are a case study for showing the importance of an accurate knowledge of the sample composition for obtaining reliable results from neutron diffraction on a glass. When the Al_2O_3 content is taken into account, the B–O coordination number, n_{BO}, from neutron diffraction shows reasonable agreement with ^{11}B NMR results. However, it appears that the effect of Al_2O_3 on the fraction, N_4, of borons which are 4-coordinated is small, because most of the aluminiums are on 5- or 6-coordinated sites. As SnO is added to the glass, N_4 increases up to a maximum at ~50 mol% SnO, and then decreases again. The values of N_4 show, however, that even for low SnO content, the addition of SnO results in the formation of both 4-coordinated boron and nonbridging oxygens. Thus tin borates are an unusual borate glass system, worthy of further investigation. The mean B–O bond length, r_{BO}, also increases up to ~50 mol% SnO, in accordance with the increase in N_4. However, for more than 50 mol% SnO, r_{BO} is constant, despite the fall in the values of N_4; this behaviour is as yet unexplained. For the Sn–O coordination, the results may be interpreted as showing that, for high SnO content, the Sn^{2+} ions are predominantly on asymmetric, trigonal pyramid SnO_3 sites, but that, for low SnO content, the Sn^{2+} ions are on a mixture of asymmetric SnO_3 sites and symmetric, high coordination sites, in accordance with the traditional model for the structural role of lone pair ions in glasses.

Acknowledgements

The sample preparation, NMR and EDX measurements were carried out at the University of Warwick Physics Department. The authors would like to acknowledge the assistance of Dr Richard Morris for carrying out the SIMS for this study, and Dr Andrew Howes and Dr Tom Kemp for their assistance in collecting the NMR data. EPSRC are thanked for partial funding of the NMR equipment at Warwick University. The neutron diffraction work was funded by EPSRC and the Centre for Materials Physics and Chemistry, grant number CMPC04108.

References

1. Idota, Y., Kubota, T., Matsufuji, A., Maekawa, Y. & Miyasaka, T. *Science*, 1997, **276**, 1395.
2. Nakai, M., Hayashi, A., Morimoto, H., Tatsumisago, M. & Minami, T. *J. Ceram. Soc. Japan*, 2001, **109**, 1010.
3. Hayashi, A., Nakai, M., Tatsumisago, M. & Minami, T. *Comptes Rendus Chimie*, 2002, **5**, 751.
4. Hayashi, A., Nakai, M., Tatsumisago, M., Minami, T., Himei, Y., Miura, Y. & Katada, M. *J. Non-Cryst. Solids*, 2002, **306**, 227.
5. Gejke, C., Nordstrom, E., Fransson, L., Edstrom, K., Haggstrom, L. &

Borjesson, L. *J. Mater. Chem.*, 2002, **12**, 2965.

6. Gejke, C., Swenson, J., Delaplane, R. G. & Borjesson, L. *Phys. Rev. B*, 2002, **65**, 212201.

7. Hayashi, A., Nakai, M., Tatsumisago, M., Minami, T. & Katada, M. *J. Electrochem. Soc.*, 2003, **150**, A582.

8. Gejke, C., Borjesson, L. & Edstrom, K. *Electrochem. Commun.*, 2003, **5**, 27.

9. Gejke, C., Zanghellini, E., Swenson, J. & Borjesson, L. *J. Power Sources*, 2003, **119**, 576.

10. Hayashi, A., Konishi, T., Nakai, M., Morimoto, H., Tadanaga, K., Minami, T. & Tatsumisago, M. *J. Ceram. Soc. Japan*, 2004, **112**, S713.

11. Hayashi, A., Nakai, M., Morimoto, H., Minami, T. & Tatsumisago, M. *J. Mat. Sci.*, 2004, **39**, 5361.

12. Tatsumisago, M. & Hayashi, A. *Glass Technol.: Eur. J. Glass Sci. Technol. A*, 2007, **48**, 6.

13. Friberg, S. R. & Smith, P. W. *IEEE J. Quantum Electron.*, 1987, **23**, 2089.

14. Hall, D. W., Newhouse, M. A., Borrelli, N. F., Dumbaugh, W. H. & Weidman, D. L. *Appl. Phys. Lett.*, 1989, **54**, 1293.

15. Bray, P. J. In: *Tenth International Congress on Glass*, vol. 13, The Ceramic Society of Japan, Kyoto, 1974, p. 1.

16. Holland, D., Smith, M. E., Howes, A. P., Davies, T. & Barrett, L. *Phys. Chem. Glasses*, 2003, **44**, 59.

17. Wright, A. C., Vedishcheva, N. M. & Shakhmatkin, B. A. *Mater. Res. Soc. Symp. Proc.*, 1997, **455**, 381.

18. Hannon, A. C. & Holland, D. *Phys. Chem. Glasses: Eur. J. Glass Sci. Technol. B*, 2006, **47**, 449.

19. Holland, D., Hannon, A. C., Smith, M. E., Johnson, C. E., Thomas, M. F. & Beesley, A. M. *Solid State Nucl. Mag.*, 2004, **26**, 172.

20. Baugher, J. F. & Bray, P. J. *Phys. Chem. Glasses*, 1972, **13**, 63.

21. Wright, A. C. *Adv. Struc. Res. Diffr. Meth.*, 1974, **5**, 1.

22. Hannon, A. C. In: *Encyclopedia of Spectroscopy and Spectrometry*, vol. 2, Eds. J. Lindon, G. Tranter & J. Holmes, Academic Press, London, 2000, p. 1479.

23. Hannon, A. C. In: *Encyclopedia of Spectroscopy and Spectrometry*, vol. 2, Eds. J. Lindon, G. Tranter & J. Holmes, Academic Press, London, 2000, p. 1493.

24. Hannon, A. C., Howells, W. S. & Soper, A. K. *Inst. Phys. Conf. Ser.*, 1990, **107**, 193.

25. Paul, A., Donaldson, J. D., Donoghue, M. T. & Thomas, M. J. K. *Phys. Chem. Glasses*, 1977, **18**, 125.

26. Wright, A. C., Stone, C. E., Sinclair, R. N., Umesaki, N., Kitamura, N., Ura, K., Ohtori, N. & Hannon, A. C. *Phys. Chem. Glasses*, 2000, **41**, 296.

27. Hannon, A. C. *Nucl. Instrum. Meth. A*, 2005, **551**, 88.

28. Nakatsuka, A., Yoshiasa, A. & Yamanaka, T. *Acta Cryst. B*, 1999, **55**, 266.

29. Shelby, J. E. *Phys. Chem. Glasses*, 2003, **44**, 106.

30. Hannon, A. C. ISIS Disordered Materials Database, http://www.isis.rl.ac.uk/disordered/Database

31. Yamanaka, T., Kurashima, R. & Mimaki, J. *Z. Kristallogr.*, 2000, **215**, 424.

32. Izumi, F. *J. Solid State Chem.*, 1981, **38**, 381.

33. Graafsma, H., Souhassou, M., Harkema, S., Kvick, A. & Lecomte, C. *Acta Cryst. B*, 1998, **54**, 193.

34. Lorch, E. *J. Phys. C*, 1969, **2**, 229.

35. Hannon, A. C., Grimley, D. I., Hulme, R. A., Wright, A. C. & Sinclair, R. N. *J. Non-Cryst. Solids*, 1994, **177**, 299.

36. Hannon, A. C., Di Martino, D., Santos, L. F. & Almeida, R. M. *J. Phys. Chem. B*, 2007, **111**, 3342.

37. MacKenzie, K. J. D. & Smith, M. E. *Multinuclear solid-state nuclear magnetic resonance of inorganic materials.* Pergamon, Oxford, 2002 p. 740.

38. Schmücker, M. & Schneider, H. *Ber. Bunsen Phys. Chem.*, 1996, **100**, 1550.

39. Schmücker, M., MacKenzie, K. J. D., Schneider, H. & Meinhold, R. *J. Non-Cryst. Solids*, 1997, **217**, 99.

40. Schmücker, M., Schneider, H. & MacKenzie, K. J. D. *J. Non-Cryst. Solids*, 1998, **226**, 99.

41. Schmücker, M. & Schneider, H. *J. Non-Cryst. Solids*, 2002, **311**, 211.

42. Lacy, E. D. *Phys. Chem. Glasses*, 1963, **4**, 234.

43. Dupree, R., Holland, D. & Williams, D. S. *Phys. Chem. Glasses*, 1985, **26**, 50.

44. Bunker, B. C., Kirkpatrick, R. J., Brow, R. K., Turner, G. L. & Nelson, C. *J. Am. Ceram. Soc.*, 1991, **74**, 1430.

45. Brow, R. K., Tallant, D. R. & Turner, G. L. *J. Am. Ceram. Soc.*, 1997, **80**, 1239.

46. Züchner, L., Chan, J. C. C., Müller-Warmuth, W. & Eckert, H. *J. Phys. Chem. B*, 1998, **102**, 4495.

47. Barney, E. R., Hannon, A. C. & Holland, D. *Phys. Chem. Glasses Eur. J. Glass Sci. Technol. B*, 2009, **50**, 156, these proceedings.

48. Brese, N. E. & O'Keeffe, M. *Acta Cryst. B*, 1991, **47**, 192.

49. Hannon, A. C. & Parker, J. M. *J. Non-Cryst. Solids*, 2000, **274**, 102.

50. Hannon, A. C. & Parker, J. M. *Phys. Chem. Glasses*, 2002, **43C**, 6.

51. Knyrim, J. S., Schappacher, F. M., Pöttgen, R., Gunne, J., Johrendt, D. & Huppertz, H. *Chem. Mater.*, 2007, **19**, 254.

52. Barney, E. R., Hannon, A. C., Holland, D., Feller, S. A., Winslow, D. & Biswas, R. *J. Non-Cryst. Solids*, 2007, **353**, 1741.

53. Hannon, A. C., Barney, E. R., Holland, D. & Knight, K. S. *J. Solid State Chem.*, 2008, **181**, 1087.

54. Henderson, G. S. *J. Non-Cryst. Solids*, 2007, **353**, 1695.

55. Terashima, K., Shimoto, T. H. & Yoko, T. *Phys. Chem. Glasses*, 1997, **38**, 211.

56. Shimoni-Livny, L., Glusker, J. P. & Bock, C. W. *Inorg. Chem.*, 1998, **37**, 1853.

57. Pannetier, J. & Denes, G. *Acta Cryst. B*, 1980, **36**, 2763.

58. Sun, K. H. *J. Am. Ceram. Soc.*, 1947, **30**, 277.

59. Fajans, K. & Kreidl, N. J. *J. Am. Ceram. Soc.*, 1948, **31**, 105.

60. Stanworth, J. E. *J. Soc. Glass Technol.*, 1948, **32**, 154.

61. Holland, D., Smith, M. E., Poplett, I. J. F., Johnson, J. A., Thomas, M. F. & Bland, J. *J. Non-Cryst. Solids*, 2001, **293–295**, 175.

62. Hoppe, U., Kranold, R., Ghosh, A., Landron, C., Neuefeind, J. & Jóvári, P. *J. Non-Cryst. Solids*, 2003, **328**, 146.

63. Takaishi, T., Takahashi, M., Jin, J., Uchino, T. & Yoko, T. *J. Am. Ceram. Soc.*, 2005, **88**, 1591.

64. Takaishi, T., Jin, J. S., Uchino, T. & Yoko, T. *J. Am. Ceram. Soc.*, 2000, **83**, 2543.

65. Wright, A. C., Hulme, R. A. & Sinclair, R. N. *Phys. Chem. Glasses*, 2005, **46**, 59.

Proc. VI Int. Conf. Borate Glasses, Himeji, Japan, 18–22 August 2008   *Phys. Chem. Glasses: Eur. J. Glass Sci. Technol. B,* June 2009, **50** (3), 205–211

Structure and the mechanism of rapid phase change in amorphous Ge$_2$Sb$_2$Te$_5$

M. Takata,[1] *Y. Tanaka, K. Kato, F. Yoshida*

RIKEN SPring-8 Center, 1-1-1 Kouto, Sayo-cho, Sayo-gun, Hyogo 679-5148, Japan

Y. Fukuyama, N. Yasuda, S. Kohara, H. Osawa, T. Nakagawa, J. Kim, H. Murayama,[2] *S. Kimura*

Japan Synchrotron Radiation Research Institute/SPring-8, 1-1-1 Kouto, Sayo-cho, Sayo-gun, Hyogo 679-5198, Japan

H. Kamioka, Y. Moritomo

The Graduate School of Pure and Applied Sciences, University of Tsukuba, Tennodai, Ttsukuba, Ibaraki 305-8571, Japan

T. Matsunaga, R. Kojima, N. Yamada

Panasonic Corporation, 3-1-1 Yagumo-Nakamachi, Moriguchi, Osaka 570-8501, Japan

K. Toriumi

Graduate School of Material Science, University of Hyogo, 3-2-1 Kouto, Kamigori-cho, Ako-gun, Hyogo 678-1297, Japan

T. Ohshima & H. Tanaka

XFEL Project Head Office/RIKEN, 1-1-1 Kouto, Sayo-cho, Sayo-gun, Hyogo 679-5148, Japan

Manuscript received 22 October 2008
Revised version received 26 January 2009
Accepted 2 February 2009

Understanding the mechanism of rapid phase change process in DVD (digital versatile disc) materials is one of the important topics in materials science, and hence numerous studies investigating the phase change process as well as structural analysis of the crystal and amorphous phases have been reported. Nevertheless, the mechanism of rapid phase change is still unclear, owing to the lack of detailed structure analysis, in particular on the amorphous phase and its crystallisation process. We have studied the amorphous structure of Ge$_2$Sb$_2$Te$_5$ and the crystallisation process by high energy synchrotron x-ray diffraction with the aid of structure modelling and time-resolved synchrotron x-ray diffraction, in order to obtain key information for revealing the rapid phase change mechanism. We found a large fraction of 4-fold and 6-fold rings in the atomic configuration of the amorphous phase. Intriguingly the bond angle distributions of rings exhibit a peak at approximately 90°, which corresponds to that in the crystal phase. Therefore, it is suggested that the 4-fold and 6-fold rings are nuclei for crystallisation. Furthermore, time-resolved x-ray diffraction measurements combined with photoreflectivity measurements during the crystallisation of the amorphous phase suggest that the crystallisation of Ge$_2$Sb$_2$Te$_5$ can be explained by a nucleation driven process. We conclude that a large amount of nuclei is the reason for rapid crystallisation in Ge$_2$Sb$_2$Te$_5$ with a nucleation driven process.

Introduction

Phase change materials, which utilise the melt quenching (amorphisation for record) and annealing (crystallisation for erase) processes of chalcogenides, are essential for the development of information technologies. The idea of applying an amorphous–crystal reversible phase change phenomenon for memory devices was proposed by Ovshinsky in the 1960s, in which a memory switch can be generated on the basis of changes in the electrical properties of both phases in chalcogenide materials.[1] The development of the Ge–Sb–Te[2] and Ag–In–Sb–Te[3] systems has allowed

us to produce rewritable compact discs (CDs), DVDs and Blu-ray discs. Thus, phase change materials are well established media; however, the rapid phase change mechanism as well as the 3-dimensional atomic configuration in the amorphous phase is still unclear.

The structure of amorphous Ge$_2$Sb$_2$Te$_5$ (*a*-Ge$_2$Sb$_2$Te$_5$) has been studied by diffraction,[4–7] spectroscopy[8–10] and theoretical simulations,[11–15] particularly after the landmark study reported by Kolobov *et al,*[8] but the detailed structure is still controversial. We have studied the structure of *a*-Ge$_2$Sb$_2$Te$_5$ and amorphous GeTe (*a*-GeTe) by a combination of high energy x-ray diffraction and reverse Monte Carlo (RMC) modelling[16,17] and found the formation of a large amount

[1] Corresponding author. Email takatama@spring8.or.jp
[2] Present address: Applied Chemistry, Science and Engineering, Chuo University, 1-13-27 Kasuga, Bunkyo-ku, Tokyo 112-8551, Japan

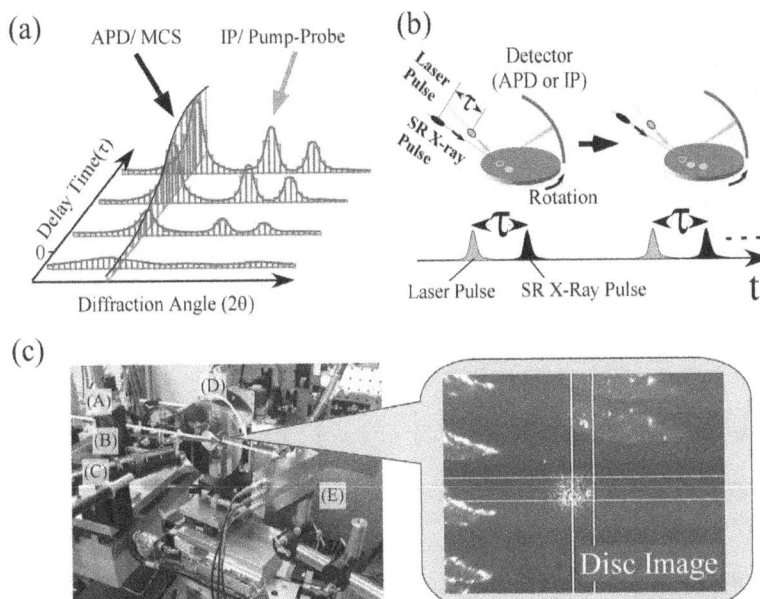

Figure 1. (a) Schematic diagram of APD/MCS and IP/pump-probe measurement. (b) Scheme of DVD rotating system and the time chart of pump-probe measurement. (c) Photograph of the time-resolved experimental apparatus for APD/ MCS measurement. (A) X-ray beam, (B) femtosecond laser, (C) monitor camera for sample, (D) DVD rotating system, (E) APD detector. A DVD disc is rotated during the measurement to supply the virgin amorphous surface. The dark area in the disc image is the amorphous phase and the bright area is the area crystallised by laser irradiation

of even numbered rings in a-Ge$_2$Sb$_2$Te$_5$, which can be nuclei for the crystal phase.[6] This model was modified by taking into account the formation of Ge–Ge homopolar bonds suggested by Baker *et al*,[9] in which the fraction of even numbered rings in a-Ge$_2$Sb$_2$Te$_5$ is significantly larger than that in a-GeTe.[18] This trend agrees well with the results obtained by a large scale combined density functional theory (DFT)/molecular dynamics (MD) simulation reported by Akola & Jones.[14] Furthermore, we recently performed time-resolved x-ray diffraction measurements combined with photoreflectivity measurements, and found that Ge$_2$Sb$_2$Te$_5$ exhibits a two stage crystallisation process,[19] whereas Ag$_{3.5}$In$_{3.8}$Sb$_{75.0}$Te$_{17.7}$ does not exhibit such a complicated process.[20] In this article, the structure of a-Ge$_2$Sb$_2$Te$_5$ and the structural origin of the rapid crystallisation process in Ge$_2$Sb$_2$Te$_5$ are discussed in detail on the basis of the amorphous structure and crystallisation behaviour.

Experimental

The specimens for x-ray diffraction experiments were made by laminating organic film sheet on a 120 mm diameter glass disc and sputtering to form the recording film with a thickness of 200–500 nm. The organic film sheet was removed from the glass disc and the specimen was manually removed from the glass substrate using a spatula. The composition of the sample was examined by inductively coupled plasma atomic emission spectrometry.

The high energy x-ray diffraction experiments were carried out at the beamline BL04B2.[21,22] The diffraction patterns of a powder sample in a thin walled (10 mm) tube of 1 mm diameter (supplier: GLAS Müller, D-13503 Berlin) and an empty tube were measured in the transmission geometry. The diffraction patterns of a liquid sample in a silica glass tube of 2·0 mm diameter and 0·5 mm wall thickness and an empty tube were measured at 953 K for comparison. The collected data were corrected using a standard program.[23] The fully corrected data were normalised to obtain the Faber–Ziman[24] total structure factor $S(Q)$.

The time resolved x-ray diffraction measurement was performed on a-Ge$_2$Sb$_2$Te$_5$ at the SPring-8 40XU beamline, in order to reveal the crystallisation process for the amorphous phase. We measured (i) the time constants of both crystallisation and optical reflectivity changes, and (ii) crystallisation behaviour. We employed avalanche photodiode (APD)/ multichannel scaling (MCS) measurement with a time resolution of 3·2 ns, coupled with photoreflectivity measurement for (i), and imaging plate (IP)/ pump–probe measurement of time resolution 40 ps for (ii) as schematically shown in Figure 1(a). Figure 1(b) shows the experimental method and the time chart for pump probe measurement using 40 ps x-ray pulses and synchronous femtosecond laser pulses. Figure 1(c) shows the time resolved experimental apparatus for APD/MCS measurement installed at the SPring-8 40XU beamline.[25] The details of the time resolved measurement are described in Ref. 20.

The RMC simulation was performed with an ensemble of 3686 particles for a-Ge$_2$Sb$_2$Te$_5$, starting with a random configuration. Throughout the RMC

(a)

(b)

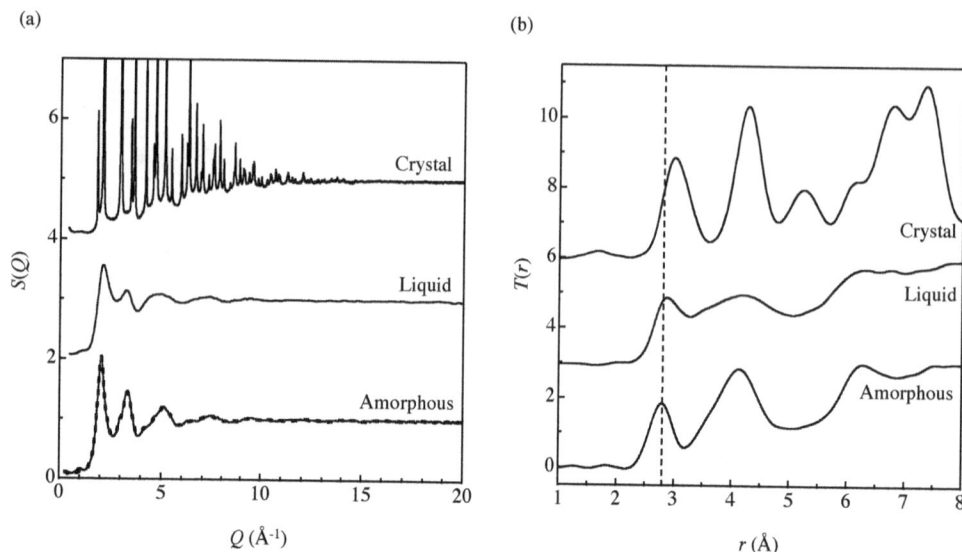

Figure 2. (a) Total structure factors, S(Q), and (b) total correlation functions, T(r), for crystal (300 K), liquid (953 K), and amorphous $Ge_2Sb_2Te_5$ (300 K). Solid lines and dashed lines represent the experimental data and the result of the RMC simulation, respectively

simulation, the constraint of closest atom–atom approach was applied so as to avoid the appearance of unphysical spikes in the partial pair distribution functions. The atomic number density of a-$Ge_2Sb_2Te_5$ was 0·0315 $Å^{-3}$ which is consistent with the reported data.[26]

Results

High energy x-ray diffraction experiment and reverse Monte Carlo simulation for amorphous $Ge_2Sb_2Te_5$

Figure 2(a) shows the measured structure factors, $S(Q)$, for $Ge_2Sb_2Te_5$. The diffraction pattern of $Ge_2Sb_2Te_5$ crystal consists of sharp Bragg reflections,

indicating long range periodicity in the atomic arrangement. It is well known that $Ge_2Sb_2Te_5$ crystal exhibit a NaCl-type structure with a lattice parameter of 6·02 Å.[27] The 4(a) sites are wholly occupied only by Te atoms, while the 4(b) sites are randomly occupied by Ge and Sb atoms. There is 20% vacancy at the 4(b) site, which may be related with the large atomic displacement factor of the Ge atom.[5] On the other hand, the diffraction pattern of $Ge_2Sb_2Te_5$ liquid shows a typical halo pattern which is different from that of a-$Ge_2Sb_2Te_5$, suggesting $Ge_2Sb_2Te_5$ is in a highly disordered state during the recording process (crystal–amorphous phase change) by a laser heated melt–quench process.

Figure 2(b) shows the total correlation functions,

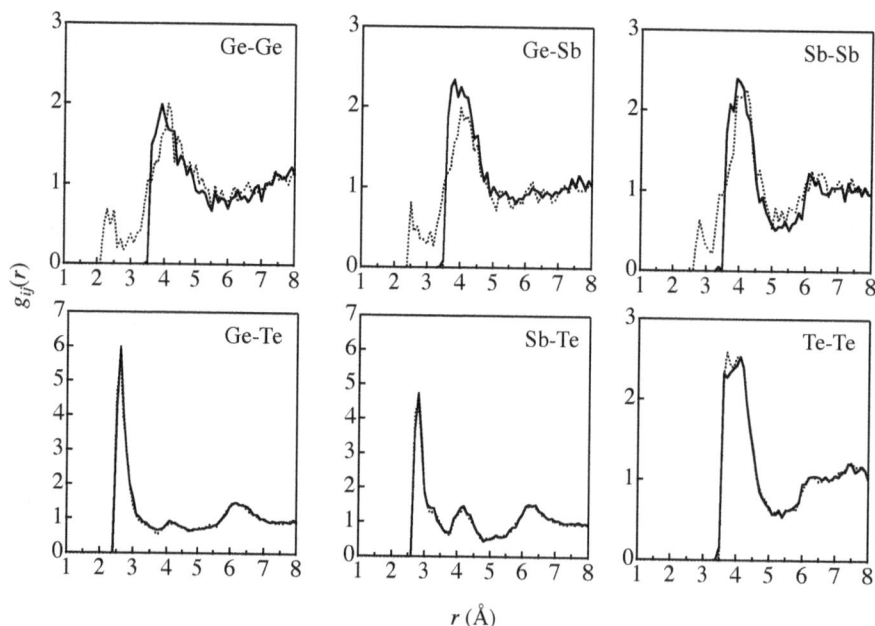

Figure 3. The partial pair distribution functions, $g_{ij}(r)$, of a-$Ge_2Sb_2Te_5$. Solid lines: non-homopolar bond model), Dotted lines: homopolar bond model

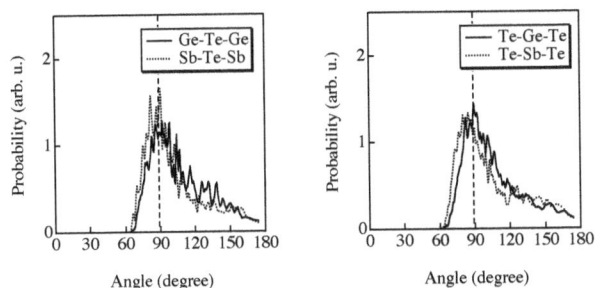

Figure 4. Bond angle distributions in a-$Ge_2Sb_2Te_5$

$T(r)$, for $Ge_2Sb_2Te_5$. The first peak, observed at approximately 3·0 Å in $T(r)$ for the crystal phase, is assigned to the Ge(Sb)–Te correlations. It is notable that the bond length becomes shorter (2·78 Å) and the peak intensity becomes smaller in a-$Ge_2Sb_2Te_5$.

The total structure factor, $S(Q)$, for a-$Ge_2Sb_2Te_5$, derived from the RMC model is shown in Figure 2(a) as a dashed line. It is confirmed that the RMC model is sufficiently consistent with the experimental data. The partial pair distribution functions, $g_{ij}(r)$, for a-$Ge_2Sb_2Te_5$ are shown in Figure 3 as solid lines.

To characterise the structural units, we investigated the bond angle distributions within the first coordination shell of the $g_{ij}(r)$ as shown in Figure 4. Te–Ge(Sb)–Te and Ge(Sb)–Te–Ge(Sb) in a-$Ge_2Sb_2Te_5$ exhibit a peak at approximately 90°.

The total coordination number around the Ge(Sb) derived from the RMC model of a-$Ge_2Sb_2Te_5$ is estimated to be 3·7 (3·0), and hence it is suggested that the dominant short range structural units are $GeTe_3$, $GeTe_4$ and $SbTe_3$. The coordination number of the Ge(Sb) atom around the Te atom is estimated to be 1·5 (1·2). Thus the estimated coordination number (2·7) around the Te atom in a-$Ge_2Sb_2Te_5$ is consistent with that in Olson's argument,[28] which insists that the coordination number should be approximately three, taking into account only the Ge–Te and Sb–Te bonds.

To obtain insights into the relationship between the speed of phase change and the atomic configuration, the ring statistics[29] in both crystal and amorphous phase for $Ge_2Sb_2Te_5$ were calculated up to 12-fold

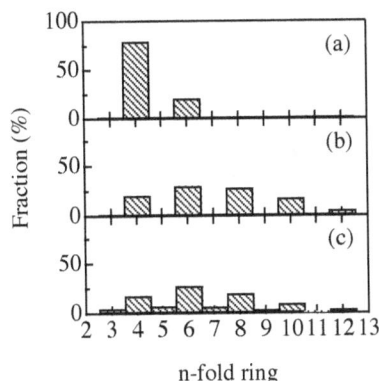

Figure 5. Ring size distribution in $Ge_2Sb_2Te_5$. (a) crystal (b) amorphous (non-homopolar bond model), (c) amorphous (Ge–Ge, Ge–Sb, Sb–Sb homopolar bonds model)

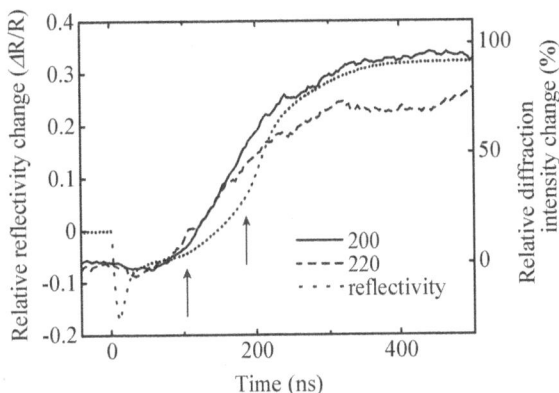

Figure 6. Photoreflectivity and time resolved x-ray diffraction profiles of $Ge_2Sb_2Te_5$ obtained by APD/MCS measurement. Photoreflectivity profiles were measured from the right and rear sides of the discs. The profile from the rear side is almost the same as that of the right side except for a negative peak at approximately 20 ns. The changes in relative diffraction intensity were normalised by the intensity at 5 μs. The grey arrows indicate the boundary of the three stages discussed in Ref. 19

rings, as shown in Figure 5(a) and (b), respectively. In the case of $Ge_2Sb_2Te_5$ crystal, the 20% of vacancies on the mixed Ge/Sb sites produce about 20% 6-fold rings.

Time resolved experiment on $Ge_2Sb_2Te_5$

To investigate the crystallisation behaviour of a-$Ge_2Sb_2Te_5$, it is necessary to understand the time evolution of the x-ray diffraction pattern. Accordingly, we performed time resolved x-ray diffraction measurements combined with photoreflectivity measurements on $Ge_2Sb_2Te_5$. The time resolved x-ray diffraction pattern and photoreflectivity profile of a 300 nm thick $Ge_2Sb_2Te_5$ sample are shown in Figure 6. The photoreflectivity profile exhibits a rapid increase between 100 and 200 ns.

To further investigate the phase change process, we analysed the time resolved x-ray diffraction profiles as shown in Figure 6. As can be observed in the figure, the x-ray diffraction intensity profiles of all the Bragg peaks show good accordance with the photoreflectivity profile. To estimate the time constant, the x-ray diffraction data for the 200 Bragg peak were fitted by a linear function, and the times at which the function exhibits 0 and 100% of diffraction intensity change, were defined as start and end time, respectively: the start and the end times are 90(1) and 273(1) ns, respectively.

From the intensity profile measured by the APD/MCS method, the delay times, τ, were determined for the IP/pump-probe method. Figure 7(a) shows the diffraction patterns obtained from the IP/pump-probe method for a 40 ps snapshot. Since the intensity of each diffraction peak has a uniform time dependent increase, there is no crystal–crystal phase transition during the crystal growth. However, the positions

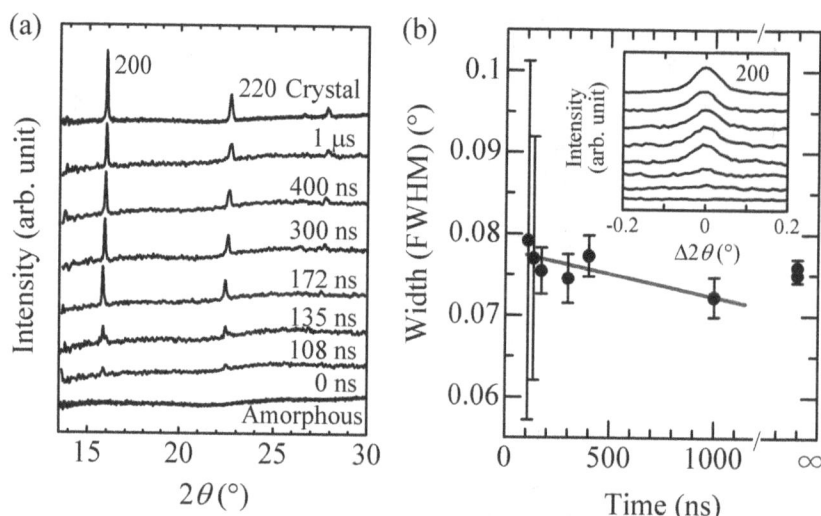

Figure 7. (a) Time dependent x-ray diffraction patterns of Ge$_2$Sb$_2$Te$_5$ obtained by the IP/pump-probe method. (b) Changes in peak width for the 200 reflection of Ge$_2$Sb$_2$Te$_5$. The inset shows close-ups of the 200 reflection peaks. The positions of the Bragg peaks calculated by curve fitting were offset to zero for clarity

of the diffraction peaks shifted to higher angle, corresponding to a lattice parameter shrinkage of about 1%, due to the time dependent temperature decrease. We estimated the grain sizes from the line width of the Bragg reflection, as shown in Figure 7(b); 69(1) nm at 300 ns to 71(1) nm at 1 μs.

Discussion

Structure of amorphous Ge$_2$Sb$_2$Te$_5$

As can be seen in Figure 2(b), the variations of the atomic distance and the peak intensity can be ascribed to the formation of Ge–Te covalent network and the reduction of the coordination number of the Te atom around the Ge(Sb) atom. It is worth mentioning that the peak corresponding to the Te–Te correlation distance observed at approximately 4·2 Å becomes broader in the liquid phase and sharper again in the amorphous phase. In addition, a peak observed at approximately 5 Å, which corresponds to a half of the diagonal length in a cubic cell, disappears in both liquid and amorphous phases. It is suggested from these features that local and long range atomic orders of the crystal phase are lost in the liquid phase, and that in the amorphous phase some specific atomic orders are regained. These observations seem to contradict the "umbrella model" proposed by Kolobov *et al*, in which a rigid Te sublattice is postulated.[8] Unfortunately, the comparison between the crystal, liquid, and amorphous Ge$_2$Sb$_2$Te$_5$ data cannot enable us to realise a remarkable structural resemblance between the amorphous and crystal phases for giving an insight into rapid phase change. However, there must be some hidden structural resemblances that cause the rapid phase change between the crystal and amorphous structures.

It is worth mentioning that the first peak in $g_{ij}(r)$ for Ge–Te and Sb–Te shown in Figure 3, is sharp but skewed towards the high-r side, indicating a distribution of Ge–Te and Sb–Te distances from 2·4 to 3·4 Å, due to highly distorted polyhedra. In the case of the Ge–Te correlation, this feature implies that regular GeTe$_4$ tetrahedra are not formed in the amorphous phase, which is consistent with the fact that there is no chalcogenide crystal with GeTe$_4$ tetrahedra.

As can be seen in Figure 4, the bond angle distributions Te–Ge(Sb)–Te and Ge(Sb)–Te–Ge(Sb) in a-Ge$_2$Sb$_2$Te$_5$ exhibit a peak at approximately 90°, which corresponds to that in the crystal phase, although the peak width is larger than that obtained by recent DFT/MD simulations.[14,15] Therefore it is suggested that a-Ge$_2$Sb$_2$Te$_5$ possesses crystal-like bond angle distributions. These features are confirmed by recent DFT/MD simulations,[14,15] and are consistent with the results of x-ray photoemission spectroscopy, where the spectrum of a-Ge$_2$Sb$_2$Te$_5$ resembles that of the crystal phase.[10]

It is of note that the structure of a-Ge$_2$Sb$_2$Te$_5$ can be regarded as an "even numbered ring structure",

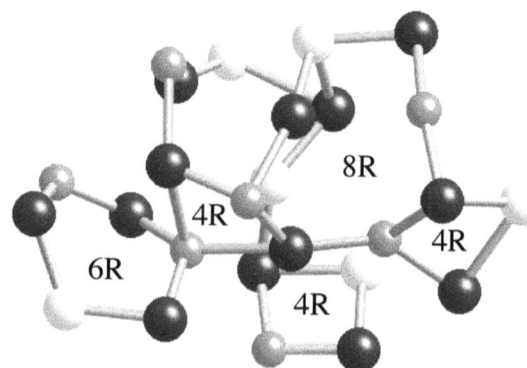

Figure 8. An enlarged atomic configuration framework of a-Ge$_2$Sb$_2$Te$_5$ from RMC simulation. The Ge(Sb)–Te bonds shorter than 3·4 Å are indicated by sticks. Grey: Ge, White: Sb, Black: Te

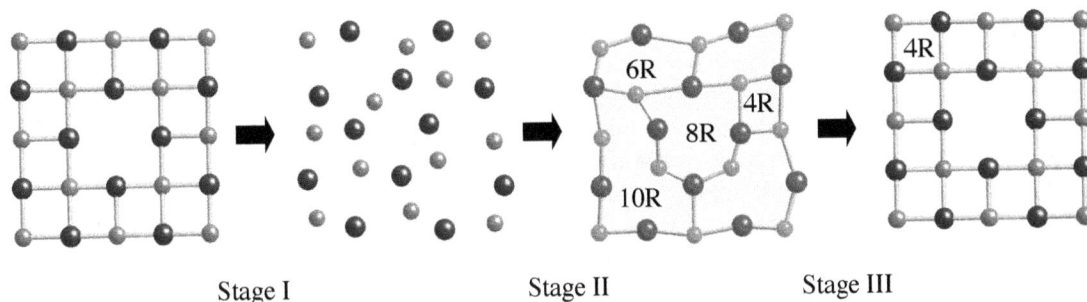

Stage I   Stage II   Stage III

Figure 9. A schematic representation for the possible ring size transformation in the crystal–liquid–amorphous phase change (record) and the amorphous–crystal phase change (erase) in $Ge_2Sb_2Te_5$. Stage I and II: recording process, stage III: erasing process

because the ring statistics are dominated by 4- and 6-fold rings, analogous to the crystal phase. This feature of the ring structures is clearly visible in the 3-dimensional atomic configurations obtained from the RMC simulation, as shown in Figure 8. Recently Lucovsky *et al* confirmed the formation of only the Ge–Ge homopolar bond in a-$Ge_2Sb_2Te_5$ on the basis of EXAFS data,[30] whereas Akola *et al* suggested the formation of Ge–Ge, Ge–Sb, Sb–Sb, and a very small amount of Te–Te homopolar bonds.[14] Although it is difficult to estimate the fraction of Ge–Ge homopolar bonds, owing to the small weighting factor for x-rays, RMC modelling with and without homopolar bonds (Ge–Ge, Ge–Sb, Sb–Sb) in a-$Ge_2Sb_2Te_5$ was performed. The RMC model with homopolar bonds exhibited good agreement with the experimental data within experimental errors. The calculated $g_{ij}(r)$ are shown as dotted lines in Figure 3, and the ring statistics are shown Figure 5(c). As can be seen in Figure 3, the small peaks corresponding to the homopolar bond are observed in $g_{ij}(r)$ for Ge–Ge, Ge–Sb, and Sb–Sb. Recent results of RMC modelling using a combination of diffraction and EXAFS data suggest the formation of homopolar bonds,[31,32] too. However, even numbered rings are still dominant in a-$Ge_2Sb_2Te_5$ with the formation of homopolar bonds, suggesting the addition of Sb_2Te_3 to GeTe can effectively prevent the formation of Ge–Ge bonds in a-$Ge_2Sb_2Te_5$. The ring statistics obtained from the homopolar bond model in our study is similar to that obtained by a recent large scale DFT/MD simulation.[14]

On the basis of the comparison of the ring statistics, the mechanisms of rapid crystal–liquid–amorphous (record) and amorphous–crystal (erase) phase changes in $Ge_2Sb_2Te_5$ are proposed, as shown by the schematic representation in Figure 9. In the crystal–liquid phase change process (stage I), the atomic configuration in the crystal phase is disarranged by laser heating and melting to a liquid. However, in the liquid–amorphous phase change process (stage II), mostly even numbered rings are formed in a-$Ge_2Sb_2Te_5$. In the amorphous–crystal phase change process (stage III), a-$Ge_2Sb_2Te_5$ transforms to the crystal phase via the transformation of only the large even numbered (8-, 10-, 12-fold) rings into the

crystal structure (4- and 6-fold rings). Of course, the formation of a small fraction of homopolar bond yields odd numbered rings in stage II, which requires the recombination of the various size rings in stage III. However, the contribution of recombination of rings is not significant in a-$Ge_2Sb_2Te_5$, and hence it is suggested that the presence of a large number of odd numbered rings in other phase change materials disturbs the rapid crystallisation of the amorphous phase. This topological order in a-$Ge_2Sb_2Te_5$, manifested by the ring statistics, is generally incompatible with the disordered structure in amorphous materials.[33] Thus a topological order in the form of a large fraction of 4-fold and 6-fold rings can provide nuclei for crystallisation.

Time resolved crystallisation of $Ge_2Sb_2Te_5$

The photoreflectivity profile of $Ge_2Sb_2Te_5$ shown in Figure 6 exhibits a rapid increase between 100 and 200 ns. A sharp negative peak can be observed at approximately 20 ns in the photoreflectivity profile. Although the origin of the negative peak is still unclear, we surmise that decrease in the photoreflectivity may be due to the roughness of the sample surface transiently induced by femtosecond laser irradiation, because such a decrease was not observed on the rear side of the samples.

Wei & Gan reported the change in photoreflectivity of a 30 nm thick $Ge_2Sb_2Te_5$ film deposited by dc magnetron sputtering and found three stages for the crystallisation: an onset stage (~40 ns), a nucleation stage (~120 ns), and a growth stage (~140 ns).[19] These stages can be observed in the photoreflectivity profile of $Ge_2Sb_2Te_5$ in Figure 6; an onset stage of ~100 ns, a nucleation stage of ~200 ns and a growth stage, as indicated by arrows. The diffraction intensity increases and stops at approximately 300 ns, which means that the crystallisation of a-$Ge_2Sb_2Te_5$ is almost finished within 300 ns. Consequently, although the thickness of the sample in our study is greater than that in commercially available devices, our results are evidence of a strong relationship between x-ray diffraction intensity and the photoreflectivity of the phase change materials, i.e. between the structure and the electronic properties.

From the experimental findings obtained from the time resolved measurements, we propose a model for the crystallisation of $Ge_2Sb_2Te_5$. Nucleation takes place in the whole area in the amorphous phase after laser irradiation, and the number of newly formed crystallites of 70 nm diameter increases during the cooling process until 300 ns. The crystal growth is then disturbed by the impingement of crystallites with each other. Our model is consistent with the nucleation driven crystallisation process discussed in Ref. 36. Furthermore we stress that the large fraction of 4-fold and 6-fold rings found in the RMC simulation of the amorphous structure can be regarded as nuclei for rapid crystallisation.

Conclusions

We performed a synchrotron x-ray diffraction experiment on amorphous $Ge_2Sb_2Te_5$, and a time resolved synchrotron x-ray diffraction on the crystallisation of amorphous $Ge_2Sb_2Te_5$, in order to obtain key information for revealing the rapid phase change mechanism. The structure model obtained by RMC modelling on the basis of diffraction data for the amorphous phase possesses a large fraction of 4-fold and 6-fold rings, and their bond angle distributions exhibit a peak at approximately 90°, which corresponds to that in the crystal phase. Accordingly, we surmise that they behave like nuclei in the rapid phase change process. A combination of time resolved x-ray diffraction and photoreflectivity measurements has revealed that crystallisation of amorphous $Ge_2Sb_2Te_5$ can be explained by a nucleation driven process. Therefore, we conclude that a large amount of nuclei, which can be formed in a thin amorphous film, is the reason for rapid crystallisation with a nucleation driven process.

Acknowledgment

This work was supported by the CREST "X-ray pinpoint structural measurement project – development of the spatial- and time-resolved structural study for nanomaterials and devices" from the Japan Science and Technology Agency.

References

1. Ovshinsky, S. R. *Phys. Rev. Lett.*, 1968, **21**, 1450.
2. Yamada, N., Ohno, E., Akahira, N., Nishiuchi, K., Nagata, K. & Takao, M. *Jpn. J. Appl. Phys.*, 1987, **26**, Suppl. 26-4, 61.
3. Iwasaki, H., Ide, Y., Harigaya, M., Kageyama, Y. & Fujimura, I. *Jpn. J. Appl. Phys.*, 1991, **31**, 461.
4. Naito, M., Ishimaru, M., Hirotsu, Y. & Takashima, M. *J. Appl. Phys.*, 2004, **95**, 8130.
5. Shamoto, S., Yamada, N., Matsunaga, T., Proffen, Th., Richadron, Jr., J. W., Chung, J.-H. & Egami, T. *Appl. Phys. Lett.*, 2005, **86**, 081904.
6. Kohara, S., Kato, K., Kimura, S., Tanaka, H., Usuki, T., Suzuya, K., Tanaka, H., Moritomo, Y., Matsunaga, T., Yamada, N., Tanaka Y., Suematsu, H. & Takata, M. *Appl. Phys. Lett.*, 2006, **89**, 201910.
7. Jóvári, P., Kaban, I., Steiner, J., Beuneu, B., Schöps, A. & Webb, M. A. *Phys. Rev. B*, 2007, **77**, 035202.
8. Kolobov, A. V., Fons, P., Frenkel, A. I., Ankudinov, A. L., Tominaga, J. & Uruga, T. *Nature Mater.*, 2004, **3**, 703.
9. Baker, D. A., Paesler, M. A., Lucovsky, G., Agarwal, S. C. & Taylor, P. C. *Phys. Rev. Lett.*, 2006, **96**, 255501.
10. Kim, J. J., Kobayashi, K., Ikenaga, E., Kobata, M., Ueda, S., Matsunaga, T., Kifune, K., Kojima, R. & Yamada, N. *Phys. Rev. B*, 2007, **76**, 115124.
11. Welnic, W., Pamungkas, A., Detemple, R., Steimer, C., Blügel, S. & Wuttig, M. 2006 *Nature Mater.*, 2006, **5**, 56.
12. Caravti, S., Bernasconi, Kühne, T. D., Krack, M. & Parrinello, M. *Appl. Phys. Lett.*, 2007, **89**, 171906.
13. Lang, C., Song, S. A., Manh, D. N. & Cockayne D. J. H. *Phys. Rev. B*, 2007, **76**, 054101.
14. Akola, J. & Jones, R. O. *Phys. Rev. B*, 2007, **76**, 235201; Akola, J. & Jones, R. O. *J. Phys.: Condens. Matter*, 2008, **20**, 465103.
15. Hegedüs, J. & Elliott, S. R. *Nature Mater.*, 2008, **7**, 399.
16. McGreevy, R. L. & Pusztai *Molec. Simul.*, 1988, **1**, 359.
17. McGreevy, R. L. *J. Phys.: Condens. Matter*, 2001, **13**, R877.
18. Kohara, S., Kimura, S., Tanaka, H, Yasuda, N., Fukuyama, Y., Murayama, H., Kim, J., Takata, M, Kato, K., Tanaka, Y., Usuki, T., Suzuya, K., Tanaka, H., Moritomo, Y., Matsunaga, T., Kojima, R. & Yamada, N. http://www.epcos.org/papers/pdf_2007/paper09_Kohara.pdf.
19. Wei, J. & Gan, F. *Thin Solid Films*, 2003, **441**, 292.
20. Fukuyama Y., Yasuda, N., Kim. J., Murayama, H., Tanaka, Y., Kimura, S., Kato, K., Kohara, S., Moritomo, Y., Matsunga, T., Kojima, R., Yamada, N., Tanaka, H., Ohshima, T. & Takata, M. *Appl. Phys. Exp.*, 2008, **1**, 045001.
21. Isshiki, M., Ohishi, Y., Goto, S., Takeshita, K. & Ishikawa, T. *Nucl. Instr. Meth. A*, 2001, **467–468**, 663.
22. Kohara, S., Suzuya, K., Kashihara, Y., Matsumoto, M., Umesaki, N. & Sakai, I. *Nucl. Instr. Meth. A*, 2001 **467-468**, 1030.
23. Kohara S., Itou, M., Suzuya, K., Inamura, Y., Sakurai, Y., Ohishi, Y. & Takata, M. *J. Phys.: Condens. Matter*, 2007, **19**, 506101.
24. Faber T. E. & Ziman J. M. *Phil. Mag.*, 1965, **11**, 153.
25. Kimura, S., Moritomo, Y., Tanaka, Y., Tanaka, H., Toriumi, K., Kato, K., Yasuda, N., Fukuyama, Y., Kim, J., Murayama, H. & Takata, M. *AIP Conf. Proc.*, 2007, **879**, 1238.
26. Njoroge, W. K., Wöltgens, H. & Wuttig M. *J. Vac. Sci. Technol. A*, 2002, **20**, 230.
27. Yamada, N. & Matsunaga, T. *J. Appl. Phys.*, 2000, **88**, 7020.
28. Olson, J.K., Li, H. & Taylor P. C. *J. Ovonic Res.*, 2005, **1**, 1.
29. The ring size distributions were calculated using shortest-path analysis, where we count the number of atoms from a starting atom, and returning to this atom through the shortest path length, in order to avoid the counting of a larger ring that can be divided into smaller rings.
30. Lucovsky, G. Baker, D. A., Paesler, M. A., Agrwal, S. C. & Phillips J. C. *J. Non-Cryst. Solids*, 2007, **353**, 1713.
31. Jóvári, P., Kaban, K., Steiner, J., Beuneu, B., Schöps, A. & Webb, A. *J. Phys.: Condens. Matter*, 2007, **19**, 335212.
32. Arai, T., Sato, M. & Umesaki, N. *J. Phys.: Condens. Matter*, 2007, **19**, 335213.
33. Disordered materials generally do not exhibit a characteristic topological order and show a broad ring size distribution due to the long range structural disorder.[34,35]
34. Cooper, A. R. *Phys. Chem. Glasses*, 1978, **19**, 60.
35. Gupta, P. K. & Cooper, A. R. *J. Non-Cryst. Solids*, 1990, **123**, 14.
36. Daly-Flynn, K. & Strand, D. A. *Proc. SPIE*, 2002, **4342**, 94.

Proc. VI Int. Conf. Borate Glasses, Himeji, Japan, 18–22 August 2008   *Phys. Chem. Glasses: Eur. J. Glass Sci. Technol. B, June 2009, **50** (3), 149–152*

Brillouin scattering study of elastic properties of sodium borate binary glasses

Yasuteru Fukawa,[1] Yu Matsuda,[1]* Mitsuru Kawashima,[1] Masao Kodama[2] & Seiji Kojima[1]

[1] *Graduate School of Pure and Applied Sciences, University of Tsukuba, Tsukuba, Ibaraki 305-8573, Japan*

[2] *Department of Applied Chemistry, Sojo University, Kumamoto 860-0082, Japan*

Manuscript received 18 August 2008
Revised version received 17 February 2009
Accepted 24 April 2009

The elastic properties of sodium borate glasses, xNa$_2$O.(100−x)B$_2$O$_3$, where x is the molar concentration of Na$_2$O, have been investigated by Brillouin scattering. The Brillouin spectra were measured using right angle (90A) scattering geometry. The advantage of 90A-scattering geometry is that the determination of the sound velocity is free from the refractive index of the sample. From the observed Brillouin spectra, the longitudinal sound velocity and the transverse sound velocity were determined. The elastic moduli, such as longitudinal modulus, shear modulus, bulk modulus, Young's modulus and Poisson's ratio were determined from the sound velocities. These elastic properties were shown to have strong composition dependences. It is concluded that various structural units in the glass network are related to these composition dependences.

Introduction

It is known that the physical properties of alkali metal borate glasses show anomalous behaviour with composition, such as exhibiting maxima or minima. This phenomenon is called the "borate anomaly".[1] This has attracted the attention of many investigators, and there are various studies on the structure of alkali borate glasses. For instance, the various kinds of intermediate structures in the borate network were discussed in reports of previous studies performed by NMR and Raman spectroscopy.[2–6] The random glass network of pure vitreous boron oxide is constructed from BO$_3$ triangular planar units. The addition of alkali oxide to B$_2$O$_3$ glass causes a change in the coordination number of some boron atoms from 3 to 4, resulting in the formation of BO$_4$ tetrahedral groups. These structural units combine to form various superstructural units, such as the boroxol ring, pentaborate group, tetraborate group, triborate group, diborate group, etc. In a previous study performed by modulated DSC,[7] the non-Debye nature of thermal relaxation in lithium borate glasses was investigated, and was discussed in terms of the structural units.

In this paper, we report the investigation of the elastic properties of sodium borate glasses, xNa$_2$O.(100−x)B$_2$O$_3$, where x denotes the molar concentration of Na$_2$O. We have already reported the composition dependences of the elastic properties of lithium borate glasses,[8] and those elastic properties have strong composition dependences. The composition dependences of the elastic properties of sodium borate glasses were previously reported by Kodama and

co-workers, using the ultrasonic pulse-echo overlap method.[9,10] However, the composition range studied was limited to 0≤x≤35 mol% Na$_2$O, which is the bulk glass formation range. Above 35 mol% Na$_2$O, the glass forming ability becomes poorer with increasing Na$_2$O content, and only thin and fragile glass samples can be prepared. Brillouin spectroscopy can solve this problem. Even if the glass sample is thin and fragile, the elastic properties can be obtained without any contact to the sample by using Brillouin spectroscopy, and its usefulness has been shown in our previous papers.[11,12] The purpose of this paper is to investigate the elastic properties of sodium borate glasses by Brillouin scattering with 90A-scattering geometry.

Experimental

All glasses were prepared with high homogeneity by the "solution method", in order to investigate the inherent nature of the binary system.[9,10] This method is composed of the following two processes. (1) The starting materials were analytical reagent grade NaOH.H$_2$O and H$_3$BO$_3$. To achieve high homogeneity of the glass forming materials, they were first made to react in an aqueous solution in a Teflon beaker. The beaker containing the solution was then placed in a drying oven at 140°C for 1–3 days. After complete evaporation of water, the chemically synthesised batch was obtained. (2) The batch was fused in a Pt crucible at about 1000°C for 1·5 h. Finally, the glasses were obtained by plate quenching.

In order to check for water contamination, we previously performed a weight loss measurement on the lithium borate glass system by comparison of the weight of the melt with that before melting,[13] and it

---

* Corresponding author. Email s-matsuda@ims.tsukuba.ac.jp

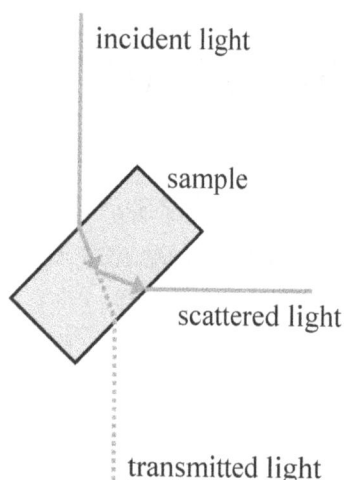

Figure 1. Schematic diagram of the 90A-scattering geometry

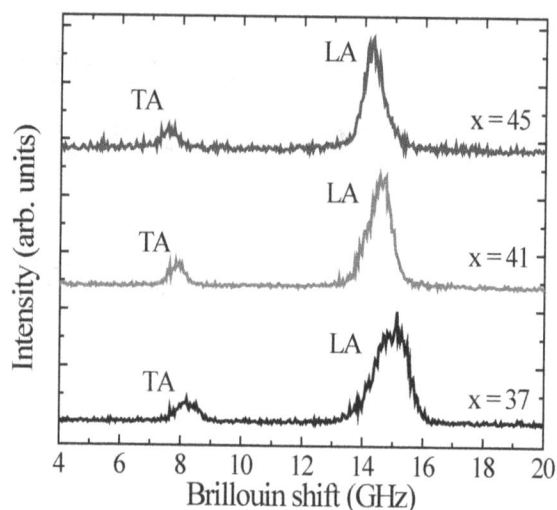

Figure 2. Observed Brillouin shift at the 90A-scattering geometry. TA and LA denote transverse and longitudinal acoustic modes, respectively

was shown that water contamination was negligible. From the result for lithium borate glasses, we conclude that water contamination is also negligible in the sodium borate glass system. As an extension of a previous study, the glass samples were newly prepared in the composition range $37 \leq x \leq 45$ mol% Na_2O.

The Brillouin scattering was measured by a Sandercock-type 3+3 pass tandem multipass Fabry-Perot interferometer (FPI), using a single frequency green YAG laser with wavelength 532 nm. In order to measure not only a longitudinal acoustic (LA) mode but a transverse acoustic (TA) mode, the Brillouin scattering spectra were measured using the right angle (90A) scattering geometry[14] for the composition range $37 \leq x \leq 45$ at room temperature. A schematic diagram of the 90A-scattering geometry is shown in Figure 1. A free spectral range of 30 GHz was applied for the measurements, and the scanning range was between ±25 GHz for the acquisition of the Brillouin spectra. A standard photon counting system and a multichannel analyser were used to accumulate the signals.

Results

Figure 2 shows the Brillouin spectra of sodium borate glasses (x=37, 41, 45 mol% Na_2O) measured with the 90A-scattering geometry. These spectra exhibit both the LA and TA modes. The longitudinal sound velocity, V_L, and the transverse sound velocity, V_T, of the sample can be determined from the Brillouin shifts, Δv_{90A}, by the equation

$$V_k = \frac{\Delta v_{90A}^k \lambda}{\sqrt{2}} \quad (k=L, T) \tag{1}$$

where λ is the wavelength of the incident beam (532 nm). Usually the refractive index is necessary to determine sound velocity from Brillouin shift. However, in the 90A-scattering geometry, the determination of sound velocity is free from the refractive index of

the sample.[13]

Figure 3 shows the composition dependence of the longitudinal sound velocity, V_L, and the transverse sound velocity, V_T, determined by Equation (1). The sound velocities in the composition range $0 \leq x \leq 35$ are results of the previous work by Kodama and co-workers using the ultrasonic pulse-echo overlap method.[9,10] The values of V_L and V_T increase with increasing Na_2O content up to about 30 mol% Na_2O, and then show maxima at about 35 mol% Na_2O. Comparing these results for V_L and V_T with our previous study of lithium borate glasses[8] shows that the form of the composition dependences is similar. However, the values of V_L and V_T for sodium borate glasses are lower than those for lithium borate glasses. This is caused by the difference in ionic radius between lithium and sodium ions.

The elastic moduli, such as longitudinal modulus, L, shear modulus, G, bulk modulus, K, Young's modulus, E, and Poisson's ratio, σ, can be calculated by the equations

$$L=\rho V_L^2 \tag{2}$$

$$G=\rho V_T^2 \tag{3}$$

$$K=L-(4/3)G \tag{4}$$

$$E=G(3L-4G)/(L-G) \tag{5}$$

$$\sigma=(L-2G)/2(L-G) \tag{6}$$

where ρ is the density of the sample. The ρ values of the samples are available in the literature as report by Feller's group.[15] The composition dependences of L, G, K and E are shown in Figure 4. The form of the composition dependences of these elastic moduli is similar to that for V_L and V_T. The elastic moduli increase with increasing Na_2O content up to about 30 mol% Na_2O, and then show a maximum at about 35 mol% Na_2O.

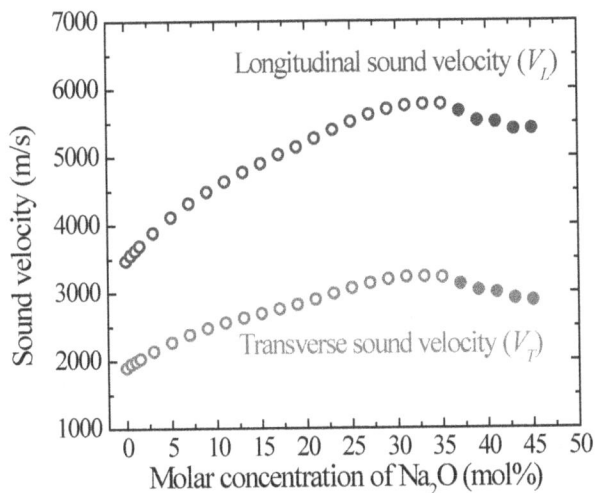

Figure 3. Composition dependences of longitudinal sound velocity, V_L, and transverse sound velocity, V_T. The sound velocities in the composition range $0 \leq x \leq 35$ are the results of previous work by Kodama and co-workers using the ultrasonic pulse-echo overlap method[9,10]

Figure 5 shows the composition dependence of Poisson's ratio, σ. The values of σ exhibit a minimum at around $x=7$ mol% Na_2O, a maximum at around $x=17$ mol% Na_2O, and another minimum at around $x=30$ mol% Na_2O. The behaviour of σ is more complicated than that of the other elastic moduli, L, G, K, and E.

Discussion

In the previous study of the lithium borate system,[16] the composition dependences of the elastic moduli were discussed in terms of three structural units, defined as $BØ_3$, $Li^+BØ_2O^-$ and $Li^+BØ_4^-$, where Ø represents a bridging oxygen and O^- a nonbridging oxygen, on the assumption that these have their respective elastic constants defined on the basis of the elastic internal energy due to deformation. A plot of M_sV^2 against the molar concentration ratio of Li_2O to B_2O_3, where M_s is the molar mass, reproduced well the behaviour of those composition dependences. Briefly, the increase of the fraction of the $BØ_4$ unit causes the increase in the value of the elastic moduli. In the sodium borate system, a detailed analysis, including a fitting procedure on the basis of the distribution of the different structural units, is in progress. According to the results of Raman scattering by Kamitsos et al,[17] the fraction of $BØ_4$ unit shows a maximum at around $x=30$ mol% Na_2O, and this composition is in agreement with the maximum of the elastic moduli. From this consistency, it can be understood that the fraction of $BØ_4$ units is the main influence on the elastic moduli in the composition range studied.

Poisson's ratio is defined as the ratio of transverse contracting strain to elongation strain. Hence, Poisson's ratio reflects the resistance of a material which opposes volume change with respect to shape change and is small for shear-resistant compressible materials, but tends toward 0.5 for incompressible bodies. In general, Poisson's ratio correlates with the atomic packing density and with the glass network dimensionality.[18] The values of Poisson's ratio decrease monotonically with increasing average coordination number, or with increasing number of bridging oxygens per glass forming cation. For instance, the value of Poisson's ratio of SiO_2 glass, which has a 3-dimensional tetrahedral unit, is low ($\sigma \approx 0.14$). On the other hand, the values of Poisson's ratio for $GeSe_4$ or Se glass, which have 2 or 1-dimensional structural units respectively, are large ($\sigma \approx 0.286$ or 0.323, respectively). In the case of sodium borate glasses, the value of Poisson's ratio decreases with increase of Na_2O content in the composition range $0 \leq x \leq 7$. This result indicates that the network connectivly or dimensionality of the borate network is increasing. In the composition range $0 \leq x \leq 7$, some of the structural units of the borate

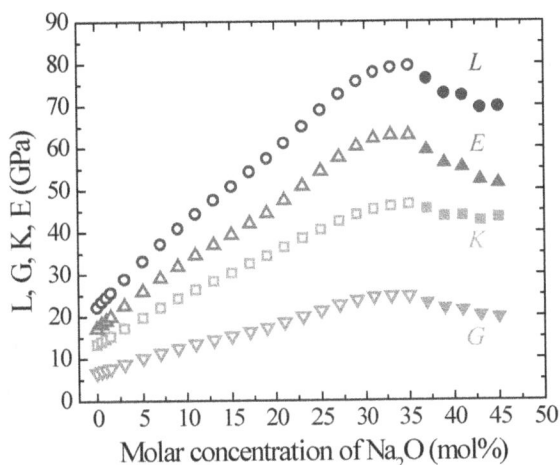

Figure 4. Composition dependences of longitudinal modulus, L, shear modulus, G, bulk modulus, K, and Young's modulus, E. Open symbols are values reported by Kodama and co-workers[9,10]

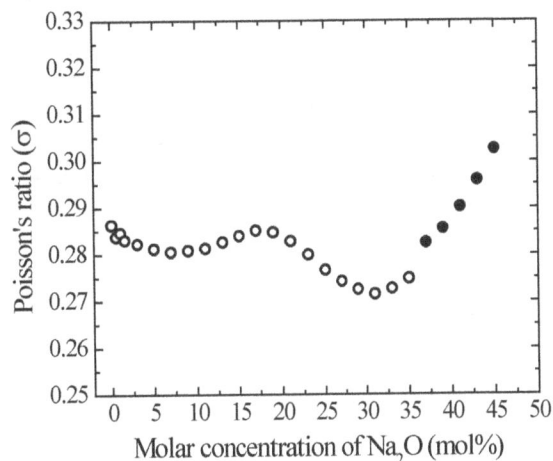

Figure 5. Composition dependence of Poisson's ratio, σ. Open circles are values reported by Kodama and co-workers[9,10]

network change from $BØ_3$ to $BØ_4$. This means that the network dimensionality increases with increasing Na_2O content, and the value of Poisson's ratio decreases. However, above $x=7$ mol% Na_2O, the value of Poisson's ratio increases, in spite of the increase of network dimensionality. It has been found that above $x=7$ mol% Na_2O, the general relationship between network dimensionality and Poisson's ratio does not hold for the sodium borate glass system. Although this breakdown may be due to the structures specific to the borate network, a father investigation will be necessary and the influence of the structural units on Poisson's ratio and their correlation remains to be solved.

The correlation between fragility, which is used to classify various glass forming liquids,[19] and the ratio of bulk modulus to shear modulus, K/G, was discussed by Novikov & Sokolov.[20] Fragility is based on the temperature dependence of the viscosity, η, of glass forming liquids. In the Angell plot,[19] the plot of $\log \eta$ or log scale of α-relaxation time versus T_g/T, the liquids exhibiting an Arrhenius behaviour are called "strong", and those exhibiting a Vogel–Fulcher behaviour are called "fragile". The deviation from an Arrhenius law is characterised by the fragility index m defined by the following equation

$$m = \lim_{T \to T_g} \left| \frac{d \log \tau}{d(T_g / T)} \right| \tag{7}$$

It was suggested by Novikov & Sokolov[20] that the correlation of fragility index, m, with K/G of glass can be described by the relationship

$$m-17=29(K/G-1) \text{ or } m=29(K/G-0.41) \tag{8}$$

This equation indicates that Poisson's ratio is directly related to the ratio K/G, and it also correlates with fragility. In the case of sodium borate glasses, According to our previous study[21] and Chryssikos et al,[22] the value of the fragility index, m, increases monotonically with increase of modifier content. However, the composition dependence of Poisson's ratio shows a complicated behaviour as shown here. Therefore, we have not found any correlation between fragility index, m, and Poisson's ratio. Although it is important to clarify the behaviour of Poisson's ratio of alkali borate glasses, the behaviour of Poisson's ratio of alkali borate glasses is more complicated than for other glasses.

Conclusions

The elastic properties of sodium borate glasses were investigated by Brillouin scattering with 90A-scattering geometry. From the observed Brillouin spectra, the longitudinal sound velocity and the transverse sound velocity were determined without the refractive index of the sample. Using the observed sound velocities and density of the sample, the elastic moduli such as longitudinal modulus, L, shear modulus, G, bulk modulus, K, Young's modulus, E, and Poisson's ratio, σ, were determined. The values of L, G, K and E increase with increasing Na_2O content up to about 30 mol% Na_2O, and then show a maximum at around 35 mol% Na_2O. These elastic moduli may relate to the structural units. On the other hand, the composition dependence of Poisson's ratio, σ, is more complicated than that of the other elastic moduli. It exhibits a minimum at around $x=7$ mol% Na_2O, a maximum at around $x=17$ mol% Na_2O, and another minimum at around $x=30$ mol% Na_2O. Any correlation between glass structure and Poisson's ratio is not clear at present.

Acknowledgements

One of the authors (Y.M.) is thankful for a JSPS Research Fellowship.

References

1. Shelby, J. E. *J. Am. Ceram. Soc.*, 1983, **66**, 225.
2. Bray, P. J. *Phys. Chem. Glasses*, 1963, **4**, 37.
3. Zhong, J. & Bray, P. J. *J. Non-Cryst. Solids*, 1989, **111**, 67.
4. Chryssikos, G. D. & Kamitsos, E. I. In: *Borate Glasses, Crystals and Melts*, Eds. A. C. Wright, S. A. Feller and A. C. Hannon, The Society of Glass Technology, Sheffield, 1997, p. 128.
5. Hassan, A. K., Torell, L. M., Börjesson, L. & Doweidar, H. *Phys. Rev. B*, 1992, **45**, 12797.
6. Fukumi, K., Ogawa, K. & Hayakawa, J. *J. Non-Cryst. Solids*, 1992, **151**, 217.
7. Matsuda, Y., Fukawa, Y., Ike, Y., Kodama, M. & Kojima, S. *J. Phys. Soc. Jpn.*, 2008, **77**, ??.
8. Fukawa, Y., Matsuda, Y., Ike, Y., Kodama, M. & Kojima, S. *Jpn. J. Appl. Phys.*, 2008, **47**, 3833.
9. Kodama, M. *J. Mater. Sci.*, 1991, **26**, 4048.
10. Kodama, M., Ono, A., Kojima, S., Feller, S. A. & Affatigato, M. *Phys. Chem. Glasses: Eur. J. Glass Sci. Technol. B*, 2006, **47**, 465.
11. Ike, Y., Matsuda, Y., Kojima, S. & Kodama, M. *Jpn. J. Appl. Phys.*, 2006, **45**, 4474.
12. Hushur, A., Kojima, S., Kodama, M., Whittington, B., Olesiak, M., Affatigato, M. & Feller, S. A. *Jpn. J. Appl. Phys.*, 2005, **44**, 6683.
13. Matsuda, Y., Fukawa, Y., Matsui, C., Ike, Y., Kodama, M. & Kojima, S. *Fluid Phase Equilibr.*, 2007, **256**, 127.
14. Krbecek, H. H., Kupisch, W. & Pietralla, M. *Polymer*, 1996, **37**, 3483.
15. Karki, A., Feller, S., Lim, H. P., Stark, J., Sanchez, C. & Shibata, M. *J. Non-Cryst. Solids*, 1987, **92**, 11.
16. Kodama, M., Matsushita, T. & Kojima, S. *Jpn. J. Appl. Phys.*, 1995, **34**, 2570.
17. Kamitsos, E. I., Patsis, A. P., Karakassides, M. A. & Chryssikos, G. D. *J. Non-Cryst. Solids*, 1990, **126**, 52.
18. Rouxel, T. *J. Am. Ceram. Soc.*, 2007, **90**, 3019.
19. Angell, C. A. *Science*, 1995, **267**, 1924.
20. Novikov, V. N. & Sokolov, A. P. *Nature*, 2004, **431**, 961.
21. Fukawa, Y., Matsuda, Y., Kawashima, M. & Kojima, S. *J. Therm. Anal. Cal.*, submitted. AUTHOR please update details if available
22. Chryssikos, G. D., Kamitsos, E. I. & Yiannopoulos, Y. D. *J. Non-Cryst. Solids*, 1996, **196** 244.

Proc. VI Int. Conf. Borate Glasses, Himeji, Japan, 18–22 August 2008   Phys. Chem. Glasses: Eur. J. Glass Sci. Technol. B, June 2009, **50** (3), 183–188

Network structure of
xB$_2$O$_3$.(22·5−x)Al$_2$O$_3$.7·5P$_2$O$_5$.70SiO$_2$ glasses

Randall E. Youngman* & Bruce G. Aitken

Science & Technology, Corning Incorporated, Corning, NY 14831, USA

Manuscript received 13 November 2008
Revised version received 20 March 2009
Accepted 24 March 2009

Quaternary glasses containing only the glass forming oxides B$_2$O$_3$, Al$_2$O$_3$, P$_2$O$_5$ and SiO$_2$, have been studied by multi-nuclear NMR as a function of B for Al substitution at 7·5 mol% P$_2$O$_5$ and 70 mol% SiO$_2$. Clear glasses along this join can be obtained for all B:Al ratios, except near the 1:1 composition, where liquidus temperatures are unusually high. As B+Al is always in excess of P in these glasses, P was found almost exclusively in symmetric Q^4 units with all bridging oxygen atoms and B or Al next nearest neighbours. ^{11}B and ^{27}Al NMR showed a clear preference for Al–P instead of B–P associations, regardless of the B:Al ratio, in contrast to the situation in AlBP ternary glasses where B has been determined to be more strongly associated with P.

1. Introduction

Glasses for silicon based technologies, including substrates for active matrix liquid crystal displays (LCDs), continue to be of commercial importance and scientific interest. Some of the key properties necessary for such applications include thermal stability, thermal expansion coefficients comparable to that of silicon, and high strain points (≥650°C). The latter, in particular, is critical for many of the additional glass processing steps required to successfully deposit thin film transistors onto glass substrates. Most of the currently available commercial display glass compositions are based on alkaline earth boroaluminosilicate compositions,[1] but these can be compromised by crystallisation of aluminosilicate phases (mullite) and may not have sufficiently high liquidus viscosities to be amenable to the fusion sheet manufacturing process.[2] One possible solution to these issues was described in an earlier study, where glasses in the B$_2$O$_3$–P$_2$O$_5$–SiO$_2$ (BPSi) ternary were proposed as alternative candidate materials for LCD applications.[3] These glasses were shown to have sufficiently high strain points, especially at high silica levels, as well as high liquidus viscosities.

The BPSi ternary was also of fundamental scientific interest in that all of the compositions were obtained using traditional glass forming oxides, without the addition of any modifiers. Therefore, it served as a simple system in which to examine the degree to which network forming cations tend to either associate or form random distributions. Through the use of multi-nuclear NMR, the network structure of BPSi glasses along the 70 mol% SiO$_2$ join were shown to consist of a large fraction of PO$_{4/2}$ tetrahedra with predominantly B next nearest neighbours (NNN),

hereinafter referred to as BPO$_4$-like groups, in which a strong association between B and P led to increased network connectivity and a resulting maximum in T_g near the 1:1 B$_2$O$_3$:P$_2$O$_5$ composition. More recent studies on the Al$_2$O$_3$–P$_2$O$_5$–SiO$_2$ (AlPSi) ternary system also established a close connection between network structure and physical properties.[4] In these glasses, ^{27}Al and ^{31}P NMR data showed a similarly strong association between Al and P, contributing to a large fraction of PO$_{4/2}$ tetrahedra with predominantly Al NNN (AlPO$_4$-like groups), and thus enhanced three dimensional polymerisation with mixtures of these and SiO$_{4/2}$ tetrahedra. The maximum in T_g and minimum in thermal expansion coefficient for glasses along the 70 mol% SiO$_2$ join were strongly correlated to the maximum in this level of network connectivity.

Both of the above ternary systems were studied in the context of high strain point glasses for commercial applications, with the added benefit of having relatively simple compositions for fundamental structure–property studies. In both systems, the network forming cations are amenable to characterisation by NMR, leading to a detailed understanding of the glass network. The other related ternary, B$_2$O$_3$–Al$_2$O$_3$–P$_2$O$_5$ (BAlP), was previously reported on by Buckermann *et al*, where the authors also used NMR methods to examine the coordination environments of the glass forming cations.[5] They described high P glasses, where B was entirely in 4-fold coordination and Al speciation was dependent on the remaining Al/P ratio, once formation of BPO$_4$ groups was complete. Their evidence indicated a preference for association of P with B rather than other structural entities, including AlPO$_4$-like groups.

In the present study, all three of the glass forming oxide ternaries described above have been combined to obtain a series of quaternary glasses based solely on

---

* Corresponding author. Email youngmanre@corning.com

Table 1. *Composition and structural properties for glasses in the series* $xB_2O_3.(22.5-x)Al_2O_3.7.5P_2O_5.70SiO_2$. N_4 *is the total fraction of boron which are* B^{IV}, *and* BPO_4 *values denote only those* B^{IV} *in* BPO_4-*like tetrahedra*

x (mol% B_2O_3)	N_4 (%) (±2)	BPO_4 (%) (±2)
0·0	-	0
2·5	0	0
5·0	0	0
10·0	0	0
17·5	13	11
20·0	18	13
22·5	24	19

SiO_2, B_2O_3, P_2O_5 and Al_2O_3. The competition between Al and B for association with P was studied in great detail for glasses at fixed 7·5 mol% P_2O_5 and 70 mol% SiO_2 content, using multi-nuclear NMR. Although Al and B each show strong preferences to associate with P in their respective ternaries with Si, the preference for BPO_4-like groups in the quaternary system was not as strong as that observed by Buckermann *et al* for the BAlP ternary.[5] Close association between Al and P in the form of $AlPO_4$-like groups was observed, and B was found to exist in mostly trigonal coordination.

2. Experimental

$xB_2O_3.(22.5-x)Al_2O_3.7.5P_2O_5.70SiO_2$, glasses were typically prepared from mixtures of Al_2O_3, $Al(PO_3)_3$, B_2O_3 and SiO_2 resulting in the compositions listed in Table 1. In the case of the x=10 sample, this glass was made from a mixture of previously synthesised $30P_2O_5.70SiO_2$, $30B_2O_3.70SiO_2$ and $22.5Al_2O_3.7.5P_2O_5.70SiO_2$ glasses. All samples were melted in covered Pt crucibles at 1650°C for ~20 h. Glasses were obtained by quenching the melts on stainless steel and annealing near T_g. Compositions were verified with ICP-MS (inductively coupled plasma mass spectroscopy) and found to be close to those in Table 1.

^{11}B MAS (magic angle spinning) NMR experiments were carried out at 11·7 T (160·3 MHz resonance frequency) using a 3·2 mm MAS NMR probe and sample spinning at 20 kHz. MAS NMR data were obtained using 0·6 μs pulse widths (<π/12 tip angle) and 2 s recycle delays to ensure quantitative excitation of all resonances. Similarly, ^{27}Al MAS NMR spectra were acquired at 11·7 T (130·2 MHz) with π/10 tip angles (0·6 μs) and 2 s pulse delays. ^{31}P wideline NMR spectra were collected using a 5 mm MAS NMR probe without sample spinning. π/2 pulse widths of 5 μs and recycle delays of 180 s were chosen in order to obtain quantitative spectra. The MAS and wideline data were processed without any additional line broadening. External referencing was performed with aqueous boric acid (19·6 ppm), aqueous aluminium nitrate and an 85% solution of H_3PO_4, for ^{11}B, ^{27}Al and ^{31}P, respectively.

Triple quantum MAS (3QMAS) NMR data were

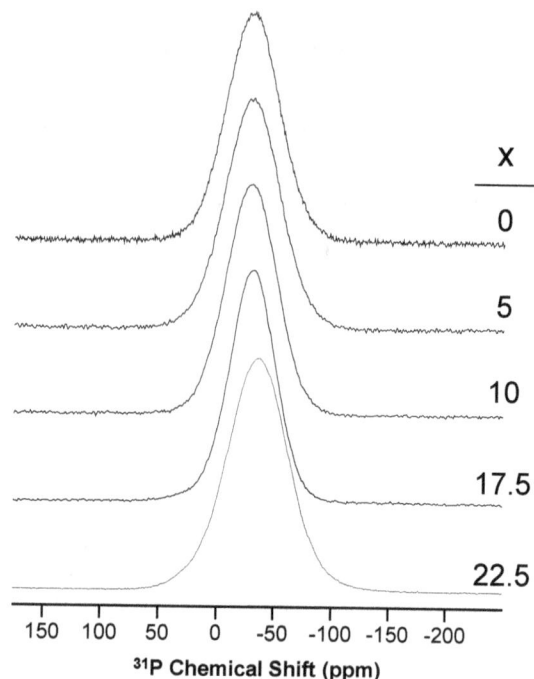

Figure 1. ^{31}P *wideline NMR spectra of glasses in the series* $xB_2O_3.(22.5-x)Al_2O_3.7.5P_2O_5.70SiO_2$ *with values for* B_2O_3 *content in mol% shown to the right*

obtained for both ^{11}B and ^{27}Al in these materials. The data were recorded using the three-pulse, z-filtered method.[6] Experimental conditions were similar to those used in the previous studies of glasses in the AlPSi and BPSi ternaries.[3,4] The resulting two-dimensional correlation plots for both ^{11}B and ^{27}Al were analysed in terms of the isotropic chemical shift (δ_{iso}) and second-order quadrupolar product ($P_Q=C_Q(1+\eta^2/3)^{1/2}$) by comparing the centres of gravity of the projection along the F_1 and F_2 dimensions. [7]

3. Results

(a) Glass formation

As shown in Table 1, glasses in the quaternary system at 70 mol% SiO_2 and 7·5 mol% P_2O_5 were successfully made for x=2·5, 5, 7·5, 10, 17·5 and 20 mol% B_2O_3. There is a pronounced gap at intermediate compositions between x values of 12·5 and 15 in which glasses were not readily obtained. For these compositions, the experimental conditions were insufficient to yield complete melts and, upon quenching, these samples were found to be partially crystallised. X-ray diffraction indicated that the crystalline phases were the cristobalite forms of $AlPO_4$ and BPO_4. Higher melting temperatures appear necessary in order to make glasses near the 1:1 B:Al composition in this specific series. Moreover, the B-rich samples at x=17·5 and 20 mol% B_2O_3 were found to exhibit a weak opalescence, indicating that these materials are sub-microscopically phase separated. In contrast, all Al-rich glasses were completely transparent.

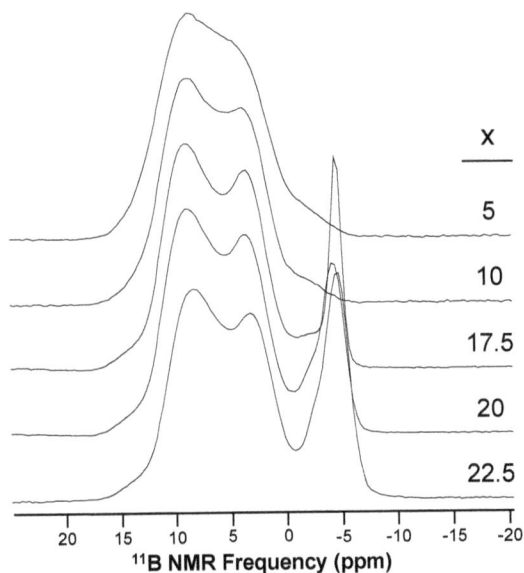

Figure 2. ^{11}B MAS NMR spectra of glasses in the series $xB_2O_3.(22\cdot5-x)Al_2O_3.7\cdot5P_2O_5.70SiO_2$ with values for B_2O_3 content in mol% shown to the right

(b) Nuclear magnetic resonance

^{31}P wideline NMR spectra are shown in Figure 1. These data are characterised by a single, symmetric lineshape, which does not appear to vary with glass composition. There are small changes in the ^{31}P chemical shift with increasing B, with the spectrum of the x=22·5 mol% B_2O_3 glass having the most shielded resonance.

The stack plot in Figure 2 contains ^{11}B MAS NMR spectra from B-containing glasses in the 7·5 mol% P_2O_5 series. These spectra contain one or two B resonances, including a complicated lineshape between 0 and 15 ppm, as well as a relatively narrow resonance near −4 ppm. The former is characteristic of trigonal boron (B^{III}) environments where all oxygen atoms are bridging to other cations.[8] The sharp feature at −4 ppm, which is found only in glasses with x≥17·5 mol% B_2O_3, indicates the presence of boron in 4-fold coordination.[9,10] The tetrahedral boron (B^{IV}) resonance increases in relative intensity with increasing x, and also develops considerable asymmetry due to the presence of a shoulder near −2 ppm. This indicates the formation of a second type of B^{IV} environment. The B^{III} resonances in Figure 2 also change with x, showing much higher definition in the B-rich compositions.

^{27}Al MAS NMR data are plotted in Figure 3. These spectra generally consist of a single Al resonance, with large changes in lineshape as a function of glass composition. At low x values, this lineshape is very broad and contains an asymmetric tail towards higher shielding. For x≥17·5, the resonances are significantly narrowed and additional peaks are observed for the x=20 sample. The majority of Al in these glasses appears to be four-coordinated (Al^{IV}), as reflected by the resonances between 40 and 50 ppm.[11] The well-resolved peaks at 40, 0 and −20 ppm in the x=20

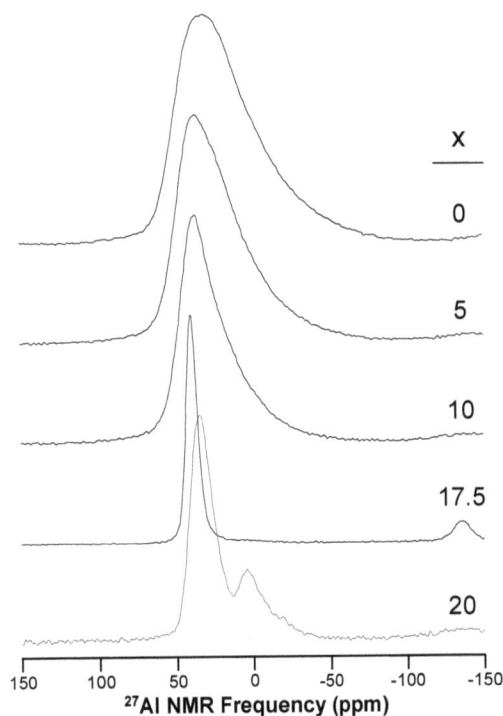

Figure 3. ^{27}Al MAS NMR spectra of glasses in the series $xB_2O_3.(22\cdot5-x)Al_2O_3.7\cdot5P_2O_5.70SiO_2$ with values for B_2O_3 content in mol% shown to the right. The small feature near −140 ppm in the x=17·5 spectrum is a spinning sideband

glass, are due to Al^{IV}, five-coordinated aluminium (Al^V) and six-coordinated aluminium (Al^{VI}) polyhedra, respectively.[12,13]

High resolution spectra from ^{11}B and ^{27}Al are shown in the contour plots of Figure 4. These 3QMAS NMR spectra provide additional resolution of multiple resonances which were obscured by broad peaks in the respective MAS NMR data. Figure 4(a) contains two ^{11}B 3QMAS NMR spectra, with the spectrum of the x=22·5 mol% B_2O_3 glass showing clear evidence for BO_4^- units. This B species is not detected in the x=7·5 mol% B_2O_3 glass. The ^{27}Al 3QMAS NMR spectra in Figure 4(b) provide additional resolution for glasses at low x, where the ^{27}Al MAS NMR spectra are characterised by very broad and asymmetric lineshapes. In these high resolution spectra, the contours show evidence for a second Al peak, appearing as a shoulder near 30 ppm in the isotropic dimension.

4. Discussion

Glass formation appears to be limited for the intermediate compositions, in particular for 11≤x≤16 where Al and B are in approximately equal concentrations. Similar behaviour was observed in the previously studied AlPSi ternary glasses with compositions close to the $AlPO_4$–SiO_2 join,[4] although in the present study all glasses contain 7·5% mol% P_2O_5, and are therefore on the P-poor side of these joins. Due to crystallisation of $AlPO_4$ in some of the intermediate

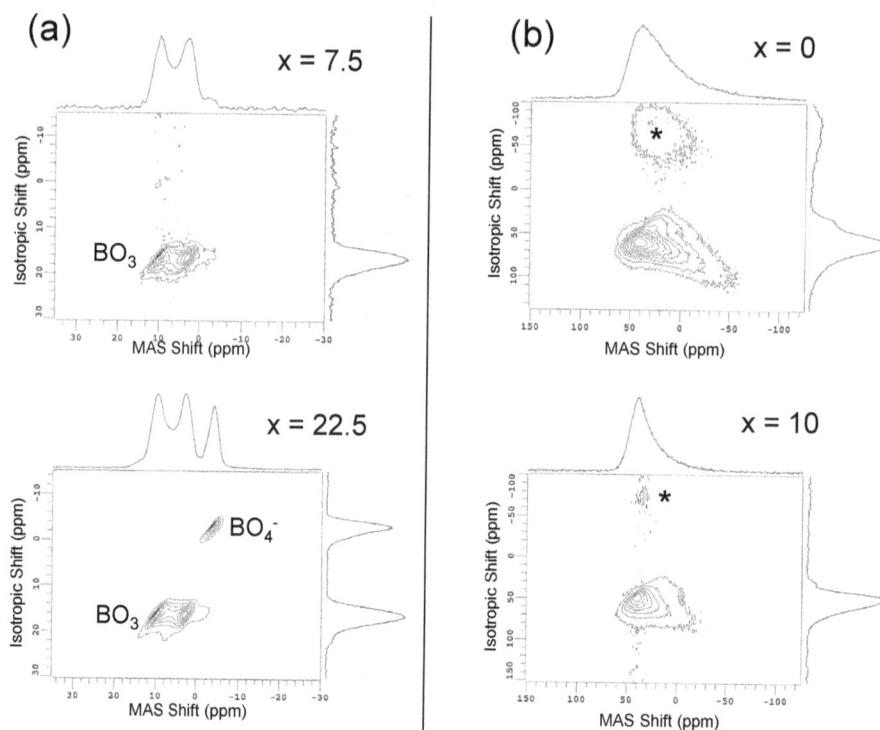

Figure 4. High resolution 3QMAS NMR spectra of (a) ^{11}B for x=7·5 and 22·5, and (b) ^{27}Al for x=0 and x=10 in the series $xB_2O_3.(22·5-x)Al_2O_3.7·5P_2O_5.70SiO_2$. MAS and isotropic projections are shown to the top and right of each plot, respectively. Spinning sidebands are marked with asterisks and the different B resonances are labelled

compositions, it appears that Al and P are strongly associated, which may be reflective of the situation in the closely related homogeneous glasses.

To better understand the limits for glass formation and structure of the $xB_2O_3.(22·5-x)Al_2O_3.7·5P_2O_5.70SiO_2$ series, the glass samples have been examined using a variety of NMR methods. ^{31}P wideline NMR has proven very useful in studying the local symmetry of P polyhedra, and especially the formation of symmetric Q^4 tetrahedra (i.e. $PO_{4/2}$ groups with all bridging oxygen) associated with $AlPO_4$- and BPO_4-like units.[3,4] The spectra in Figure 1 confirm that B for Al substitution does not change the local coordination environment of P in these glasses, and the symmetric lineshape observed in all spectra is simply a result of virtually all P being stabilised in Q^4 groups. The two end-member AlPSi and BPSi glasses were previously shown to have all P in such environments for P<B or Al, so the data in Figure 1 are consistent with prior studies.[3,4] Thus, as B+Al is always in excess of P in this series, Q^4 phosphate groups with B and/or Al neighbours appear to be the only phosphate groups present, at least with sufficient population to be detected in these data.

However, the symmetry arguments based on these wideline ^{31}P NMR spectra only confirm Q^4 polyhedra but do not specify the type of Q^4 groups, i.e. whether $AlPO_4$- or BPO_4-like. Some indication as to the nature of the next nearest neighbour cations can be obtained by close examination of the ^{31}P shielding. As clearly seen in Figure 1, the chemical shift of the Q^4 resonance

changes with composition, starting at −30 ppm for x=0 and decreasing to a value of −36·5 ppm for the BPSi glass (x=22·5). The change in shielding is not quite linear, as the largest change is seen on going from a glass containing only 2·5–5 mol% Al_2O_3 (x=17·5–20) to an Al-free composition (x=22·5). This pronounced increase in shielding upon complete removal of Al indicates that Al–P association is more important than B–P association in that (1) significant differences are not observed upon initial addition of B (x=2·5 or 5), and (2) the chemical shielding change is not linear between the two end member glasses. Furthermore, the ~6 ppm increase in shielding as x increases from 17·5–20 to 22·5 (Figure 1) indicates a shift in the next nearest neighbour cations around the Q^4 phosphate groups, from one mostly dominated by Al to one with predominantly B. The intermediate compositions, with mixed Al and B, are more similar to the B-free x=0 glass, suggesting a closer structural similarity to one in which $AlPO_4$ groups comprise the majority of P polyhedra. ^{31}P chemical shifts for crystalline $AlPO_4$ and BPO_4 are −24·5 and −29·5 ppm, respectively,[9] and are thus similar to the measured shielding values in these ternary and quaternary glasses. The measured chemical shift differences between the glasses of this study and the crystalline $AlPO_4$, BPO_4 analogues are likely due to the high SiO_2 content of these materials, in which some Si next nearest neighbours (NNNs) are probably contributing to the P chemical shielding and reducing the number of Al or B NNN to a value below four.[4]

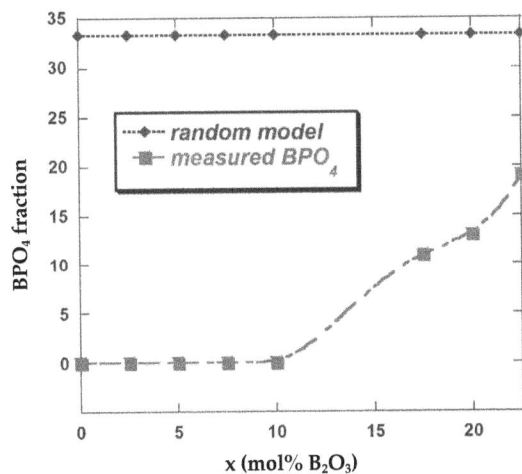

Figure 5. Plot of BPO$_4$ tetrahedra from ^{11}B MAS NMR (filled squares) and the fraction of BPO$_4$ groups predicted from the B:P ratio (filled diamonds)

Additional insight into the degree to which P Q^4 groups are associated with B is readily obtained from ^{11}B MAS NMR spectra. For low B content ($x \leq 10$) the ^{11}B MAS NMR spectra (Figure 2) show only a single resonance from B^{III} groups. It is only when B exceeds Al (i.e. $x \geq 17 \cdot 5$) that evidence for B^{IV} units is apparent in the data, indicating that formation of BPO$_4$-like units, which would require B^{IV}, does not occur for compositions with B<Al. For B:Al above 1:1, the presence and increasing relative intensity of the tetrahedral resonance at −4 ppm indicates the formation of BPO$_4$-like units, consistent with both previous results for BPSi ternary glasses and the ^{11}B chemical shift for crystalline BPO$_4$.[3,14,15] The total fraction of B^{IV} groups, N_4, was obtained by integrating the two resonances in the spectra of Figure 2 and is listed in Table 1. The results indicate an increase in N_4 with increasing B until a value of 24% is reached at the BPSi end member composition.

The direct quantification of B^{IV} with P NNN (i.e. the BPO$_4$-like resonance at −4·2 ppm) allows for comparison with the number of P Q^4 groups obtained from the ^{31}P wideline NMR data, which indicate that the entire quantity of P$_2$O$_5$ is present as Q^4 groups in all glasses. In order to satisfy this requirement for the Al-free glass ($x=22 \cdot 5$), there must be an equal number of B^{IV} groups. The value of N_4 for this composition is 24%, which is somewhat close but insufficient to completely balance the P Q^4 population, assuming that the entire 7·5% mol% P$_2$O$_5$ content requires a proportional number of B^{IV} groups (i.e. $N_4=33\%$). The correlation between B and P is even less evident for the Al-containing glasses, especially for low x values, where the spectra in Figure 2 show a complete absence of B^{IV}. To confirm the B speciation at low x, high resolution 3QMAS NMR spectra were collected on the $x=7 \cdot 5$ glass (see Figure 4(a)). The complete absence of any signal from B^{IV}, which is present as a set of narrow contours in the $x=22 \cdot 5$ glass spectrum of

Figure 4(a), verifies the N_4 determinations from MAS NMR spectra. Even when N_4 is non-zero ($x \geq 17 \cdot 5$), the amount of B^{IV} is insufficient to account for the entire Q^4 phosphate population, so it appears that all Al-containing glasses are comprised of some fraction of AlPO$_4$-like units.

An additional complication in the ^{11}B MAS NMR data is the appearance of a second B^{IV} resonance, which is seen as a downfield shoulder around −2 ppm. The 2 ppm chemical shift difference between the two B^{IV} peaks is consistent with a large fraction of BO$_4^-$ units with P NNN and a lesser number of BO$_4^-$ units having some other NNN cation, which is most likely to be Si.[10] The presence of B^{IV} not associated with BPO$_4$-like groups means that the entire N_4 fraction cannot be attributed to such structural units and that the actual BPO$_4$ content in high x glass compositions is even less than that suggested by the tabulated N_4 values. This discrepancy is further illustrated in Figure 5, where the measured BPO$_4$ content from ^{11}B MAS NMR is compared with the N_4 value one would obtain if P were equally associated with B and Al. Since the ratio of P to B+Al is a constant 33% in all the studied glasses, a random distribution of B and Al around Q^4 phosphate groups would require $N_4=33\%$, regardless of total B content. As can be seen in Figure 5, none of the compositions are close to this value, and in fact all of the Al-containing quaternary glasses ($x<22 \cdot 5$) are far from $N_4=33\%$, requiring much more interaction between Al and P than observed between B and P.

Although it is clearly shown that insufficient B^{IV} is present in these glasses to account for the measured quantity of Q^4 phosphate groups, it is only with closer examination of Al speciation that the structural description can be clarified. The ^{27}Al MAS NMR spectra in Figure 3 reflect changes in Al speciation with increasing B content. The Al-rich glasses appear to contain mostly Al^{IV} groups, as evidenced by the broad, asymmetric lineshape in the top three spectra of Figure 3. These resonances have frequencies consistent with 4-fold coordination, and are also consistent with the shifts observed for Al^{IV} groups with P NNN.[12,16] For example, the ^{27}Al NMR chemical shift of crystalline AlPO$_4$ is 39 ppm, similar to that of all of the Al^{IV} peaks in these glasses.[17] Al^{IV} groups in other local environments, for example those in aluminosilicate glasses, are typically shifted downfield from the present value of ~40 ppm to as high as 50–60 ppm.[18,19] Thus, the presence of AlPO$_4$-like units are confirmed by both the coordination number and NMR frequencies of Al.

As these glasses have Al+B in excess of P, the number of Al required for AlPO$_4$-like units is less than the total Al content for $x<15$. Therefore, some of the Al signal detected in these spectra must belong to another coordination environment. ^{27}Al MAS NMR spectra indicate that these glasses contain mostly Al^{IV}, but due to the large asymmetry of this Al^{IV} resonance, resonances from Al with higher coordination numbers may be unresolved. The 3QMAS NMR method allows for higher resolution spectra of ^{27}Al, and has been used for the glasses in this study. The spectra of two representative Al-rich glasses are shown in Figure 4(b). The two-dimensional correlation plot between MAS and isotropic frequencies contains a well defined set of contours from the Al^{IV} resonance. Upon closer inspection of the data, a second and very low intensity resonance is detected, appearing as a weak peak around 40 ppm in the isotropic projections to the right of each spectrum. This second resonance, by virtue of its position in the two dimensions, can be assigned to Al^V groups.[4,18] The chemical shift parameters are consistent with at least some P NNN atoms around the Al polyhedra, similar to what was previously demonstrated for Al-rich AlPSi ternary glasses.[4] Glasses with higher B content, as shown in Figure 3, provide ^{27}Al NMR spectra with much higher resolution and clear evidence for multiple Al resonances, at least for the x=20 sample. For these glasses, P is in excess of Al, and the spectra appear to be very similar to those of P-rich AlPSi ternary glasses.[4] The x=17·5 spectrum in Figure 3 contains only a single, narrow resonance from Al^{IV} groups and is very similar to that of crystalline AlPO$_4$.[17] This glass contains P in excess of Al, so if AlPO$_4$-like groups are the preferred structural unit, it is not surprising that all of the Al is found in this environment. However, a further increase in the P:Al ratio on going to larger x results in Al being present in environments other than AlPO$_4$-like units, as shown by the clearly resolved Al^V and Al^{VI} resonances. Thus the Al speciation in these Al-poor glasses is not dictated simply by formation of AlPO$_4$-like units, but higher coordinated Al species are formed, possibly to aid in charge-balancing various structural elements, including perhaps Al^{IV} that is not associated with Q^4 phosphate groups.[18]

The structural description for the x≤17·5 glasses is now fairly clear. Most of these glasses do not contain B^{IV}, so BPO$_4$-like units cannot occur. Therefore the Q^4 phosphate species in B-poor glasses are associated entirely with Al into AlPO$_4$-like groups, consistent with the Q^4 chemical shifts as well as the ^{27}Al MAS NMR spectra. Even the x=17·5 sample, which is the lowest B-containing glass with B^{IV}, appears to be comprised of mostly AlPO$_4$-like phosphate groups, as the entire Al population is found in this structural unit. In addition, the BPO$_4$-like B^{IV} content, as determined from the N$_4$ fraction with P NNN, is insufficient to fully compensate for the full complement of Q^4 P groups for all glasses in this series. This not only indicates that Al–P interactions are favoured over those involving B and P, but perhaps P is not entirely converted into Q^4 tetrahedra. This is almost certainly the case for the x=22·5 glass in which all Q^4 groups must be in BPO$_4$-like environments due to the complete absence of Al, and yet the N$_4$ fraction for this glass is much

lower than the necessary 33%. It is likely that some of the P are in other structural units, for example Q^3 groups found in P-rich examples of the AlPSi and BPSi ternary glasses,[3,4] and are simply too low in concentration to be detected by ^{31}P wideline NMR. It is also possible that the B:P ratio for some of these high x glasses is a little higher than nominal, reducing the N$_4$ value needed to correspond to all of the P in Q^4 tetrahedra. Further studies of glasses at other P$_2$O$_5$ contents are underway and may provide important details from which to better understand these and other questions pertaining to competition between Al and B in P-containing glasses.

5. Conclusions

Glasses can be made in the B$_2$O$_3$–Al$_2$O$_3$–P$_2$O$_5$–SiO$_2$ quaternary, where all components are well known glass forming, or conditional glass forming oxides. Some compositions at 7·5 mol% P$_2$O$_5$ and 70 mol% SiO$_2$ could not be formed into glasses, due to high liquidus temperatures and/or crystallisation of AlPO$_4$. There is clearly a strong association between Al and P, which dominates the formation of symmetric Q^4 phosphate groups, even in the x=17·5 glass where the B concentration is greater than that of Al. B also associates with P, as in the previously studied BAlP and BPSi ternary glasses, but is always found to interact with P only when Al≪P where insufficient Al is present to form only AlPO$_4$-like units.

References

1. Dumbaugh, W. H., Lapp, J. C. & Moffat, D. M. US Patent 5,374,595, 1994.
2. Dockerty, S. M. US Patent 3,338,696, 1967.
3. Aitken, B. G. & Youngman, R. E. *Phys. Chem. Glasses: Eur. J. Glass Sci. Technol. B*, 2006, **47**, 381.
4. Aitken, B. G., Youngman, R. E., Deshpande, R. R. & Eckert, H. *J. Phys. Chem. C*, 2009, **113**, 3322.
5. Buckermann, W. A., Müller-Warmuth, W. & Mundus, C. *J. Non-Cryst. Solids*, 1996, **208**, 217.
6. Amoureaux, J.-P., Fernandez, C., Steuernagel, S. *J. Magn. Reson. A*, 1996, **123**, 116.
7. Amoureaux, J.-P., Huguenard, C., Engelke, F., Taulelle, F. *Chem. Phys. Lett.* 2002, **356**, 497.
8. Turner, G. L., Smith, K. A., Kirkpatrick, R. J. & Oldfield, E. *J. Magn. Reson.*, 1986, **67**, 544.
9. Turner, G. L., Smith, K. A., Kirkpatrick, R. J. & Oldfield, E. *J. Magn. Reson.*, 1986, **70**, 408.
10. Gan, H., Hess, P. C. & Kirkpatrick, R. J. *Geochim. Cosmochim. Ac.*, 1994, **58**, 4633.
11. Turner, G. L., Kirkpatrick, R. J., Risbud, S. H. & Oldfield, E. *Am. Ceram. Soc. Bull.*, 1987, **66**, 656.
12. Brow, R. K., Kirkpatrick, R. J. & Turner, G. L. *J. Am. Ceram. Soc.*, 1990, **73**, 2293.
13. Zhang, L. & Eckert, H. *J. Phys. Chem. B.*, 2006, **110**, 8946.
14. Grimmer, A. R., Müller, D., Gözel, G. & Kniep, R. *Fresen. J. Anal. Chem.*, 1997, **357**, 485.
15. Youngman, R. E., Aitken, B. G. & Dickinson, J. E. *J. Non-Cryst. Solids*, 2000, **264&264**, 111.
16. Brow, R. K., Kirkpatrick, R. J. & Turner, G. L. *J. Am. Ceram. Soc.*, 1993, **76**, 919.
17. Müller, D., Jahn, E., Ladwig, G. & Haubenreisser, U. *Chem. Phys. Lett.*, 1984, **109**, 332.
18. Sen, S. & Youngman, R. E. *J. Phys. Chem. B*, 2004, **108**, 7557.
19. Weber, R., Sen, S., Youngman, R. E., Hart, R. T. & Benmore, C. J. *J. Phys. Chem. B*, 2008, **112**, 16726.

*Proc. VI Int. Conf. Borate Glasses, Himeji, Japan, 18–22 August 2008   Phys. Chem. Glasses: Eur. J. Glass Sci. Technol. B, April 2009, **50** (2), 89–94*

Structural characterisation of alkaline earth borosilicate glasses through density modelling

T. Mullenbach, M. Franke, A. Ramm, A. R. Betzen, S. Kapoor, N. Lower, T. Munhollon, M. Berman, M. Affatigato & S. A. Feller*

Physics Department, Coe College, Cedar Rapids, IA 52402, USA

Manuscript received 18 August 2008
Revised version received 21 December 2008
Accepted 9 January 2009

The densities of nearly 120 alkaline earth borosilicate glasses (MO–B$_2$O$_3$–SiO$_2$, where M is Ba or Ca) were determined using pycnometry. These results were used to make inferences about the underlying atomic level structure of alkaline earth borosilicate glasses. Previously Dell, Bray and Xiao produced structural models of alkali borosilicates where the glass is composed of two micro-phase separated borate and silicate networks, in which the sharing of the alkali oxide modifier between the two networks depends on the relative amounts of borate structural units. Density models were created that utilised both the Dell model and simple proportional sharing. The predictions of these models were compared with the experimental data, showing that the proportional sharing model works well, and suggesting that the Dell hypotheses overestimate the amount of alkaline earth oxide modifier associated with the borate network.

A. Introduction

In the early 1980s, both Dell & Bray[1] and Yun & Bray[2] produced semi-empirical structural models for alkali borosilicate glasses. These models use NMR data to describe the manner in which the alkali oxide modifier is shared. In our previous work on the alkali borosilicate systems, we analysed these glasses using the structural models of Dell, Bray, and Yun to predict density.[3–7] Barium and calcium borosilicate glasses were made as an extension of this research to determine the effects of alkaline earth oxide modifiers on the borosilicate network and the resulting properties. We report measurements of the density as a function of R (molar ratio of modifier oxide, MO, to borate, B$_2$O$_3$) for fixed K (molar ratio of silicate to borate) families where the glass composition is RMO. B$_2$O$_3$.KSiO$_2$. We also employed a proportional sharing model and one based on the Dell hypothesis to explain these observed density trends. Other properties of these glass systems, thermal and spectroscopic, were also measured and will be reported in a later paper.

B. Experimental procedures

Barium and calcium borosilicate glasses were made from reagent grade chemicals purchased from Sigma Aldrich, each with purities above 99%. The modifier oxide, MO (where M is Ba or Ca), was introduced to the mix via carbonates, BaCO$_3$ and CaCO$_3$.

The reagents were weighed in a platinum crucible, using electronic balances, to the nearest ten thousandth of a gram, to make 6–8 g batches. The samples were mixed for at least 5 min to ensure homogeneity. The batches were then placed in either a Thermoline 46100 or Carbolite 1400 high temperature furnace and heated for 15–20 min. The glasses were heated at 1200–1300°C, depending on the composition. The melts were then removed from the furnace and allowed to air cool to room temperature. Once at room temperature, the weight loss was recorded. Acceptable weight losses were within a tenth of a gram of the expected, and these weight losses ensured the sample had the correct composition.

After the mixture was shown to have an acceptable weight loss, it was placed back into the furnace to be melted again. This second heating allowed the sample to be quenched faster than air cooling. The mixture was heated for 10 minutes at the same temperature as the initial melt. Upon completion of the second heating, the melts were removed from the furnace and the bottom of the crucible was placed in a beaker of water and crushed ice. If this method of quenching failed to produce glass, the samples were roller quenched. At high modifier content the melts required roller quenching. High silica content required higher melting temperatures.

The densities were measured using 0·5 to 1·5 cm^3 of sample. Either a Quantachrome Micropycnometer or a Quantachrome Ultrapycnometer 1000 was used with helium gas. The density of each sample was determined 10 to 15 times and the average of the last five runs used as the actual datum. To ensure accuracy each run was calibrated against ultrapure aluminium. The density is accurate to ±1%.

---

* Corresponding author. Email sfeller@coe.edu

Table 1. Short range borate units, R=(molar% MO)/ (molar% B₂O₃)

F_i unit	Structure	R
F_1	trigonal boron with three bridging oxygens	0·0
F_2	tetrahedral boron with four bridging oxygens	1·0
F_3	trigonal boron with two bridging oxygens (one NBO)	1·0
F_4	trigonal boron with one bridging oxygen (two NBOs)	2·0
F_5	trigonal boron with no bridging oxygen (three NBOs)	3·0

C. Structural models

Structural ideas for alkaline earth borosilicate glasses were based on borate and silicate binary glass models. The alkali borates are comprised of five short range structural units. The five units are labelled F_1, F_2, F_3, F_4 and F_5 and are defined in Table 1. The units appear and disappear as a function of R (our model ignores the F_5 unit because glass formation with roller quenching does not reach high enough R values to need it). Pure B₂O₃ glass is completely F_1 units, and as R is increased each unit appears as numbered. As the first amounts of modifier are added the F_1 unit undergoes a coordination change from three-coordinated boron to four. This produces the F_2 unit with a nearby alkali ion for charge balance. Increased addition of alkali modifier causes a coordination change back to three-coordinated boron and forms increasing amounts of nonbridging oxygens (NBOs); the F_3, F_4, and F_5 groups form sequentially. This model has been examined by Bray and others.[8-10]

The silicate system works on a similar principle. The two main differences are notation and the lack of a coordination change. Binary alkali silicate glasses contain five tetrahedral units (Q^4, Q^3, Q^2, Q^1, and Q^0) with decreasing numbers of bridging oxygen atoms, and each NBO is paired with an alkali ion for charge balance,[11-13] see Table 2. Similar to the borate units, the silicate units appear and disappear depending on J, the ratio of modifier oxide to silicate.

A simple model of alkali borosilicate glasses describes the modifier oxide as being shared proportionally between the borate and silicate networks. This, however, is not the case according to Dell.[1] Dell stated that in these glass systems, the borate network initially uses the incoming alkali oxide to form F_2 units. This happens until $R=0.5$ when the alkali oxide begins being shared among the borate and silicate networks forming reedmergnerite groups (BSi₄O₁₀, a tetrahedral boron where each oxygen is bonded to a silica tetrahedron, Q^4). At $R_{max}=1/2+K/16$, the alkali oxide begins to form NBOs on the silica tetrahedra

Table 2. Short range silicate units, J=(molar% MO)/ (molar% SiO₂)

Q^i unit	Structure	J
Q^4	tetrahedral silica with four bridging oxygens	0·0
Q^3	tetrahedral silica with three bridging oxygens (one NBO)	0·5
Q^2	tetrahedral silica with two bridging oxygens (two NBOs)	1·0
Q^1	tetrahedral silica with one bridging oxygen (three NBOs)	1·5
Q^0	tetrahedral silica with no bridging oxygen (four NBOs)	2·0

of the reedmergnerite groups until $R_{d1}=1/2+K/4$. After R_{d1} the alkali oxide starts being shared between the borate and silicate networks.[1] The parameters for these processes are summarised in the following equations (in which F_i is the fraction of borons which are in the ith type of borate unit)

$$R_{max} = \frac{1}{2} + \frac{K}{16} \tag{1a}$$

$$R_{d1} = \frac{1}{2} + \frac{K}{4} \tag{1b}$$

$$R_{d2} = \frac{3}{2} + \frac{3K}{4} \tag{1c}$$

$$R_{d3} = 2 + K \tag{1d}$$

For $0 \leq R \leq R_{max}$

$$F_1 = 1 - R \tag{2a}$$

$$F_2 = R \tag{2b}$$

$$F_3 = 0 \tag{2c}$$

$$F_4 = 0 \tag{2d}$$

For $R_{max} \leq R \leq R_{d1}$

$$F_1 = 1 - R_{max} \tag{3a}$$

$$F_2 = R_{max} \tag{3b}$$

$$F_3 = 0 \tag{3c}$$

$$F_4 = 0 \tag{3d}$$

For $R_{d1} \leq R \leq R_{d2}$

$$F_1 = \left(1 - \frac{K}{8}\right)\left(\frac{3}{4} - \frac{R}{2+K}\right) \tag{4a}$$

$$F_2 = (8+K)\left(\frac{1}{12} - \frac{R}{24+12K}\right) \tag{4b}$$

$$F_3 = \frac{1}{3}(R - R_{d1})\left(\frac{2 - K/4}{2+K}\right) \tag{4c}$$

$$F_4 = \frac{1}{2}(R - R_{d1})\left(\frac{2 - K/4}{2+K}\right) + \frac{2}{3}\left(\frac{KR}{8+4K} - \frac{K}{16}\right) \tag{4d}$$

For $R_{d2} \leq R \leq R_{d3}$

$$F_1 = 0 \tag{5a}$$

$$F_2 = (8+K)\left(\frac{1}{12} - \frac{R}{24+12K}\right) \tag{5b}$$

$$F_3 = \left(\frac{4}{3} - \frac{K}{6}\right)\left(1 - \frac{R}{2+K}\right) \tag{5c}$$

$$F_4 = \frac{1}{2}\left(1 - \frac{K}{8}\right) + (R - R_{d2})\left(\frac{2 - K/4}{2+K}\right)$$
$$+ \frac{2}{15}\left(\frac{5KR}{8+4K} - \frac{5K}{16}\right) \tag{5d}$$

D. Density modelling

The density modelling was done in two different manners. One employed a proportional model that assumed uniform sharing of modifier oxide between the borate and silicate networks. This model assumes that two micro-phase separated glasses comprise the borosilicate. The following equation was used for the density according to this model

$$\rho(R,K) = \frac{1}{1+K}\left[\rho\left(\frac{R}{1+K}\text{MO.B}_2\text{O}_3\right)\right] \\ + \frac{1}{1+K}\left[\rho\left(\frac{R}{1+K}\text{MO.SiO}_2\right)\right]$$ (6)

where $\rho(R\text{MO.B}_2\text{O}_3)$ indicates the measured density of a glass with composition $R\text{MO.B}_2\text{O}_3$. The densities of the borate glasses were taken from our previous work[14] while the silicate densities were taken from literature.[15] Least squares fits to the binary density data were used to extrapolate the needed densities at compositions between those where actual data are available.[6,7]

The other model used the work of Dell et al[1] The F_i-unit abundances predicted by Dell were paired with the known atomic masses and the volumes determined from the least squares fitting procedure done earlier for the binary systems[14,15]

$$\rho(R,K) = \rho(0)\frac{F_1 + F_2M_2 + F_3M_3 + F_4M_4 + KM_{Si}/2}{F_1V_1 + F_2V_2 + F_3V_3 + F_4V_4 + KV_{Si}/2}$$ (7)

Equation (7) was used for this density model[6] where F_i is the fraction of borons which are in the ith type of borate unit, M_i is the mass of each borate unit relative to that of the F_1 unit, M_{Si} is the average relative mass of the silicate units, V_i is the relative volume of the ith borate structural unit, and V_{Si} is the average relative volume of the silicate units. $\rho(0)$ is the density of B_2O_3 glass,[4,6] $1\cdot823$ g/cm^3, and all relative volumes are referenced to V_1 in B_2O_3.

The values of M_i were determined from atomic masses using

$$M_1 = \frac{B+1\cdot5O}{B+1\cdot5O} = 1$$ (8a)

$$M_2 = M_3 = \frac{B+2O+0\cdot5M}{B+1\cdot5O}$$ (8b)

$$M_4 = \frac{B+2\cdot5O+M}{B+1\cdot5O}$$ (8c)

In Equations (8a)–(8c), Si, O, B, and M represent the atomic masses of silicon, oxygen, boron and the modifier (barium or calcium), respectively. The Dell fractions described earlier provide the values for F_i, but M_{Si} and V_{Si} are defined below

For $0{\le}R{\le}R_{max}$: $$M_{Si} = \frac{Si+2O}{B+1\cdot5O} = 1\cdot7261$$ (9a)

For $R_{max}{\le}R{\le}R_{d1}$:

$$M_{Si} = \frac{Si+(2+(R-R_{max})/K)O+(2(R-R_{max})/K)M}{B+1\cdot5O}$$ (9b)

For $R_{d1}{\le}R{\le}R_{d3}$:

$$M_{Si} = \frac{Si+\left(\frac{35}{15}+\frac{13A}{15}\right)O+\left(\frac{3}{8}+\frac{26A}{15}\right)M}{B+1\cdot5O}$$ (9c)

where

$$A = 5\left(\frac{R-R_{d1}}{8+4K}\right)$$ (9d)

M_{Si} is a function of R, as given by Equations (9a) to (9c).[6,7] For the ranges of R used in this model V_{Si} is a combination of two terms: V_6 and ΔV_6. V_6 represents the relative volume of the Q^4 unit. ΔV_6 is the additional relative volume compared to the Q^4 unit, due to the formation of Q^3 or Q^2 units. V_{Si} is thus defined as

$$V_{Si}=V_6+\Delta V_6$$ (10)

The term ΔV_6, is given by the following equations, in which ΔV_a is the added volume when a Q^4 unit changes to a Q^3, and ΔV_b is the added volume when a Q^3 unit changes to a Q^2

For $0{\le}N(R,K){\le}1$: $\Delta V_6=N(R,K)\Delta V_a$ (11a)

For $1{\le}N(R,K){\le}2$: $\Delta V_6=\Delta V_a+[N(R,K)-1]\Delta V_b$ (11b)

The parameter $N(R,K)$ is the number of added NBOs per silicon. This number can be determined from the Dell equations

For $0{\le}R{\le}R_{max}$: $N(R,K)=0$ (12a)

For $R_{max}{\le}R{\le}R_{d1}$: $N(R,K) = 2\left(\frac{R-R_{max}}{K}\right)$ (12b)

For $R_{d1}{\le}R{\le}R_{d3}$: $N(R,K) = \frac{3}{8}+\frac{26A}{15}$ (12c)

The values of V_i are unknown parameters of the model. These values were previously determined by least squares analysis of the density data for binary borate and silicate glasses using earlier models for the fractions of the borate and silicate units.[6,7] Another least squares analysis was performed using all the borosilicate density data to determine the values of V_i that produced the best fits. This least squares analysis allowed V_1 through V_4 and V_6, ΔV_a, and ΔV_b to vary against all the borosilicate density data. These values were then compared with the values determined from the binary glasses, see the Results and Discussion sections below.

D. Results

Table 3 lists the determined densities for each barium borosilicate sample. Table 4 lists the densities for the calcium borosilicates. Table 5 lists the borate and

Table 3. *Densities for barium borosilicate glass (g/cm³, ±1%)*

R	K=0·0	K=0·5	K=1·0	K=2·0
0·0	1·81	1·90	1·91	1·87
0·2	2·68	2·49		
0·2	2·66			
0·4	3·35			
0·4	3·29			
0·5		3·30	3·06	
0·5		3·38		
0·6	3·71	3·46		
0·6	3·68	3·54		
0·7		3·60	3·40	
0·8	3·95	3·71	3·54	
0·8	3·90	3·73		
0·9	4·09		3·61	
1·0		3·89	3·73	3·40
1·0		3·89		3·42
1·2	4·22	4·02	3·86	3·57
1·2				3·56
1·3	4·31			
1·4		4·13	3·97	3·71
1·4			3·99	
1·5	4·4	4·12		3·75
1·6		4·23	4·08	
1·6		4·21		
1·7	4·5			
1·8		4·31	4·15	3·91
2·0	4·53	4·37	4·23	3·97
2·0		4·38	4·23	
2·4		4·51	4·36	4·12
2·5			4·38	4·15
2·8			4·47	4·22
3·0			4·45	4·28
3·2			4·57	4·31
3·6				4·40
4·0				4·45

Table 4. *Densities for calcium borosilicate glass (g/cm³, ±1%)*

R	K=0·0	K=0·5	K=1·0	K=2·0
0·0	1·81	1·90	1·91	1·87
0·2		2·16		
0·35	2·45			
0·4	2·48			
0·5	2·5	2·59	2·25	2·17
0·6	2·68	2·61	2·43	
0·7		2·69	2·44	
0·8		2·68	2·54	
0·81	2·72			
1·0	2·79	2·74	2·63	2·47
1·2		2·77	2·68	
1·22	2·8			
1·4		2·80	2·80	2·60
1·6		2·81	2·79	
1·8		2·83	2·80	2·68
2·0		2·83	2·83	
2·4			2·86	2·80
2·6			2·87	
2·8			2·86	2·84
3·0			2·87	
3·2				2·86
3·6				2·85
4·0				2·88

silicate structural unit volumes determined by least squares fitting of the density data for the binary systems.[14,15] Figures 1 and 2 show respectively the predictions of the proportional sharing model for the density of barium and calcium borosilicates, calculated using these volumes. This simple model agrees reasonably well with the experimental data for both systems.

Using the binary volumes in Table 5 for the corresponding V_i values in the Dell model produces the poor agreement shown in Figures 3 and 4. Table 6 shows the results of the least squares analysis with the Dell model where all the volumes were varied against all the borosilicate density data. This produced values of V_i useful for comparison with those determined from the binary glasses.

E. Discussion

The values derived for V_1 through V_4 from the binary glasses and the borosilicates (Tables 5 and 6, respectively) show the volumes of the borate structural units to be similar in the binary borate and borosilicate systems. The barium borosilicate analysis suggests much larger values for ΔV_a and ΔV_b than in the binary silicate case (1·281 compared to 0·477 and, 1·388 compared to 0·617 for ΔV_a and ΔV_b,

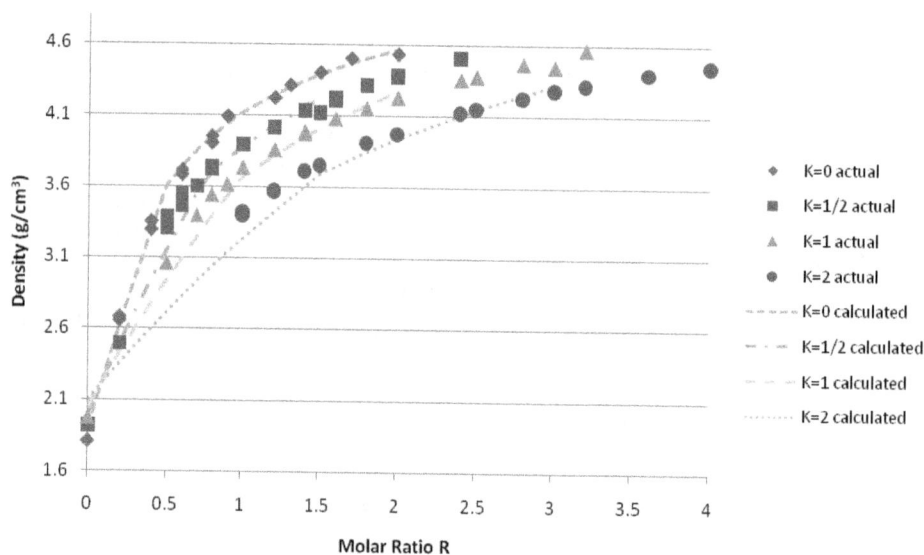

Figure 1. *The predictions of the proportional sharing density model (lines), compared with the experimental density data for barium borosilicate glasses (points)*

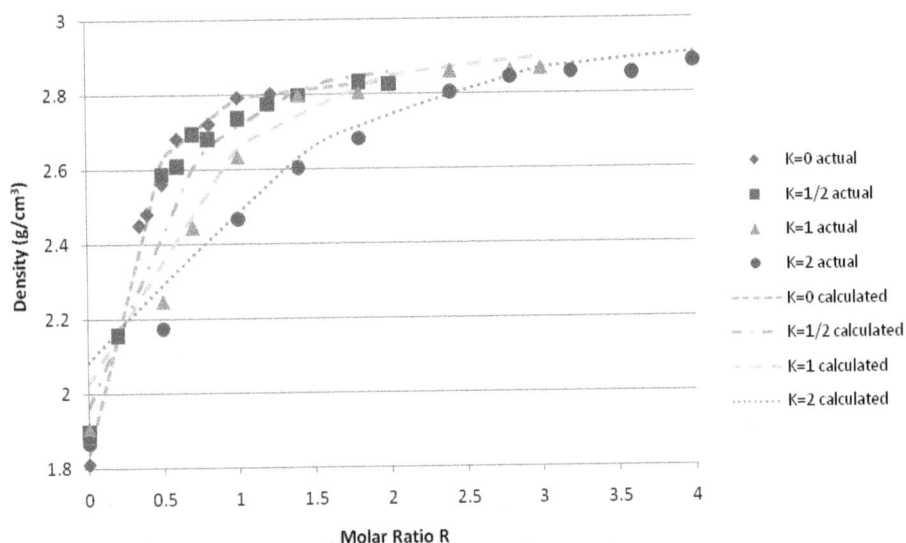

Figure 2. The predictions of the proportional sharing density model (lines), compared with the experimental density data for calcium borosilicate glasses (points)

respectively) and the calcium borosilicate analysis also suggests considerably higher values than in the binary silicate case (1·633 compared to 1·427, 0·419 compared to 0·290, and 0·785 compared to 0·375 for V_6, ΔV_a, and ΔV_b, respectively). With the exception of V_6 in barium borosilicates, this comparison suggests that, if the Dell model is correct, the volume of the silicate units increases drastically (ΔV_a and ΔV_b

Table 5. *Dimensionless least squares values for the relative volumes in binary borate and silicate glasses*

Barium		Calcium	
V_1	0·960	V_1	0·990
V_2	1·164	V_2	0·964
V_3	1·543	V_3	1·290
V_4	2·159	V_4	1·678
V_6	1·443	V_6	1·427
ΔV_a	0·477	ΔV_a	0·290
ΔV_b	0·617	ΔV_b	0·375

for barium containing glasses and ΔV_b for calcium containing glasses all increase by more than 200%) between the binary barium silicate glass and the barium borosilicate glass.

The poor fits of the Dell model using the binary volumes suggests two possibilities. One: the Dell model does not accurately describe alkaline earth borosilicate glasses, or two: the V_i values necessary to produce a good fit (see Table 6) are true. The downside to the second possibility is the discrepancy in values between the binary glass systems and those necessary to make the Dell model work. This drastic difference makes the first possibility more plausible. There is, however, the Stallworth & Bray model[16] to describe the barium borosilicates which could also be used for calcium borosilicates. This is an empirical model, relying on the NMR-determined experimental trends for

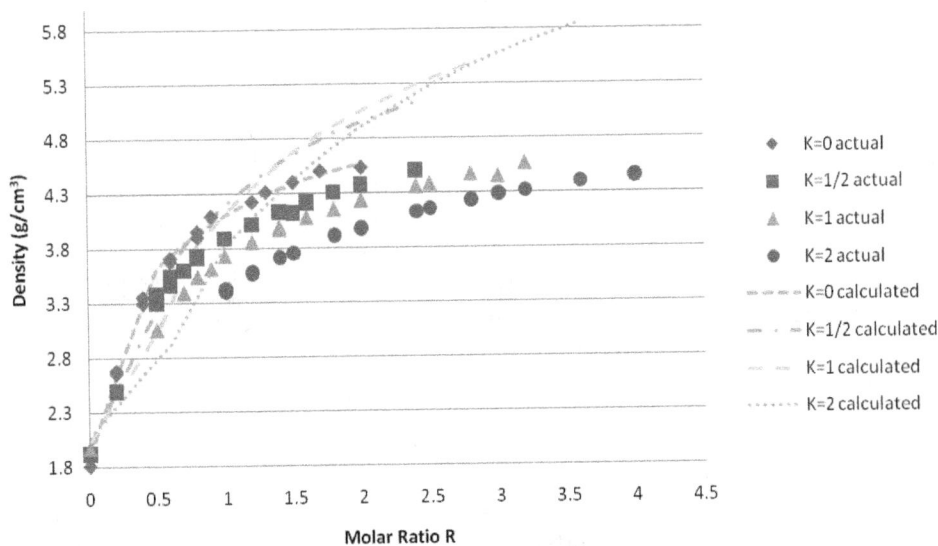

Figure 3. The predictions of the Dell model using volumes determined from density data for binary glasses (lines), compared with the experimental density data for barium borosilicate glasses (points)

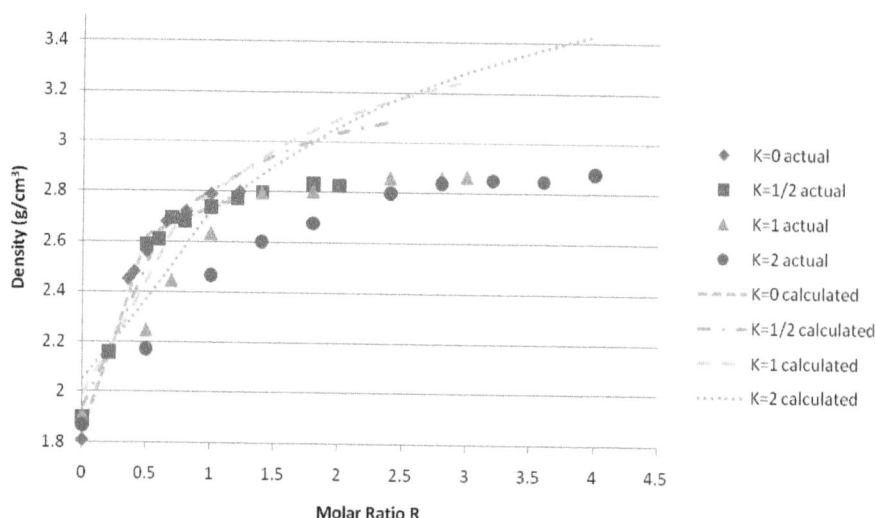

Figure 4. The predictions of the Dell model using volumes determined from density data for binary glasses (lines), compared with the experimental density data for calcium borosilicate glasses (points)

Table 6. Dimensionless Dell model forced fit values for relative volumes against all K families

Barium		Calcium	
V_1	0·990	V_1	0·983
V_2	1·110	V_2	0·931
V_3	1·589	V_3	1·277
V_4	2·160	V_4	1·672
V_6	1·485	V_6	1·633
ΔV_a	1·281	ΔV_a	0·419
ΔV_b	1·388	ΔV_b	0·785

the borate fractions F_1 through F_4, and on deduction for the Q^i units by applying charge conservation. Both the proportional sharing model and the Stallworth model predict density equally well using physically reasonable volumes, but the Stallworth model comes at the expense of extreme complexity. We will continue to investigate the short range structure of alkaline earth borosilicate glasses through their thermal and spectroscopic properties in later papers. In another paper in the proceedings of the *Sixth International Confrence on Borate Glasses, Crystals, and Melts*, the density data from this study were used to determine the packing fractions of these glasses.[17]

F. Conclusions

Based on our modelling of the densities of alkaline earth borosililcates, we have found the proportional sharing model to be both simple and accurate. This approach uses least squares fitting procedures to determine the volumes for the borate and silicate units in the binary barium and calcuim glasses. The success of this model suggests the hypothesis of micro-phase separated borate and silicate networks is a useful description of ternary alkaline earth borosilicates. The Dell model, originally derived for sodium borosilicates, fails to reproduce the observed density trends in the barium and calcium borosilicates, suggesting that the sharing of modifier oxide amongst the sub-networks

in alkaline earth borosilicate glasses is different than that found in the alkali borosilicate systems.

G. Aknowledgements

We would like to thank The National Science Foundation, for their support with Grant DMR 0502051; Coe College, for housing and student support; and the Carver Foundation of Eastern Iowa, for student support.

References

1. Dell, W. J., Bray, P. J. & Xiao, S.Z. *J. Non-Cryst. Solids*, 1983, **58**, 1.
2. Yun Y. H. & Bray P. J. *J. Non-Cryst. Solids*, 1978, **27**, 363.
3. Budhwani, K. Physical Properties of and Structures in Lithium, Sodium, and Potassium Borosilicate Glass System, *Coe College Honors Thesis*, 1993.
4. Boekenhauer, R. *The Density of Lithium Borosilicate Glasses Related to Atomic Arrangement*, Coe College Honors Thesis, 1991.
5. Kottke, J. *An Examination of the Physical Properties of Rubidium and Cesium Borosilicate Glasses and the Development of a Quantitative Density Model for the Alkali Borosilicate Glass System*, Coe College Honors Thesis, 1995.
6. Feil, D. & Feller, S. A. *J. Non-Cryst Solids*, 1990, **119**, 103.
7. Kapoor, S., George, H., Betzen, A., Affatigato, M. & Feller, S.A. *J. Non-Cryst. Solids*, 2000, **270**, 215.
8. Bray, P. J., Feller, S. A., Jellison Jr., G. E. & Yun, Y. H. *J. Non-Cryst. Solids*, 1980, **38/39**, 93.
9. Yun, Y. H. & Bray, P. J. *J. Non-Cryst. Solids*, 1981, **44**, 227.
10. Feller, S. A., Dell, W. J. & Bray, P. J. *J. Non-Cryst. Solids*, 1982, **51**, 21.
11. Emerson, J. F., Stallworth, P. E. & Bray, P. J. *J. Non-Cryst. Solids*, 1989, **111**, 253.
12. Dupree, R., Holland, D. & Martoza, M. G. *J. Non-Cryst. Solids*, 1990, **116**, 148.
13. Larson, C., Doerr, J., Affatigato, M., Feller, S. A., Holland, D. & Smith, M. E. *J. Phys.: Condens. Matter*, 2006, **18**, 11323.
14. Feller, S. A., Kottke, J., Welter, J., Nijhawan, S., Boekenhauer, R., Zhang, H., Feil, D., Paramswar, C., Budhwani, K., Affatigato, M., Bhasin, G., Bhowmik, S., Mackenzie, J., Royle, M., Kambeyanda, S., Pandikuthira, P. & Sharma, M. In *Borate Glasses, Crystals and Melts, 1997*, Edited by A. C. Wright, S. A. Feller, A. C. Hannon, Society of Glass Technology, Sheffield. p.246.
15. Bansal, N. & Doremus, R. *Handbook of Glass Properties*, Academic Press, New York, 1986.
16. Stallworth, P. *Nuclear Magnetic Resonance Structural Studies of Barium Borosilicate Glasses, Fluorine Doped Silicate Glasses, and Glasses Containing Selenium Oxide and Mercuric Oxide*, Brown University Doctoral Thesis, 1989.
17. Bista, S., O'Donovan-Zavada, A., Mullenbach, T., Franke, M., Affatigato, M., Feller, S. A. *Phys. Chem. Glasses Eur. J. Glass. Sci. Technol. B*, 2009, in press.

Proc. VI Int. Conf. Borate Glasses, Himeji, Japan, 18–22 August 2008   *Phys. Chem. Glasses: Eur. J. Glass Sci. Technol. B, June 2009, **50** (3), 153–155*

Viscosity of Bi$_2$O$_3$–B$_2$O$_3$–SiO$_2$ melts

Seiji Inaba,[1] Hirofumi Tokunaga,[2] Chawon Hwang[3] & Shigeru Fujino[1]*

[1] *Department of Chemical Engineering, Kyushu University, 744, Motooka, Nishi-ku, Fukuoka-shi, Fukuoka 819-0395, Japan*

[2] *Department of Materials Science and Engineering, Graduate School of Engineering, Kyushu University, 744, Motooka, Nishi-ku, Fukuoka-shi, Fukuoka 819-0395, Japan*

[3] *Department of Applied Sciences for Electronics and Materials, Kyushu University, 6-1, Kasugakouen, Kasuga-shi, Fukuoka 816-8580, Japan*

Manuscript received 23 August 2008
Revised version received 5 January 2009
Accepted 15 April 2009

The viscosity of melts from the Bi$_2$O$_3$–B$_2$O$_3$–SiO$_2$ system has been measured in the temperature range of 973–1473 K. The effect of composition on viscosity was investigated. The viscosity of the melt was found to decrease linearly with increasing Bi$_2$O$_3$ content. Comparisons with PbO–B$_2$O$_3$–SiO$_2$ ternary melts were carried out. Empirical equations to estimate viscosity from chemical composition were proposed.

1. Introduction

The importance of the viscosity of glass melts is widely recognised. For example, viscosity is the dominant parameter taken into consideration in the glass manufacturing process, because certain properties such as the working temperature, softening temperature, and annealing temperature are related to the viscosity of glass. In addition, viscosity data provide us with important information about the glass melt structure.[1–3]

PbO-containing glasses are characterised by a broad glass forming region, low thermal properties and a high refractive index.[4–6] Because of such characteristics of PbO-containing glasses, they have been used in many fields of industry.[7] However, the development of Pb-free glass has recently become necessary, in order to meet environmental demands resulting from concern over the toxicity of Pb.[8,9]

The Bi$_2$O$_3$–B$_2$O$_3$–SiO$_2$ system has attracted considerable attention due to its low characteristic temperatures, such as glass softening point and glass transition temperature.[4] However, there is no report on the viscosity of Bi$_2$O$_3$–B$_2$O$_3$–SiO$_2$ melts. In order to develop Pb-free glass suitable for use as sealing glass, it is desirable to understand the interfacial reactions between ceramics and glasses, and these interfacial reactions are closely connected to the melt properties of glass, such as density, surface tension, and viscosity.[10] Consequently, there is a need to measure density, surface tension, and viscosity of Bi$_2$O$_3$–B$_2$O$_3$–SiO$_2$ melts and to compare with that of PbO–B$_2$O$_3$–SiO$_2$ melts.

The purpose of this study is to measure the viscosity of Bi$_2$O$_3$–B$_2$O$_3$–SiO$_2$ melts and compare with that of PbO–B$_2$O$_3$–SiO$_2$ melts. Based on the measurement

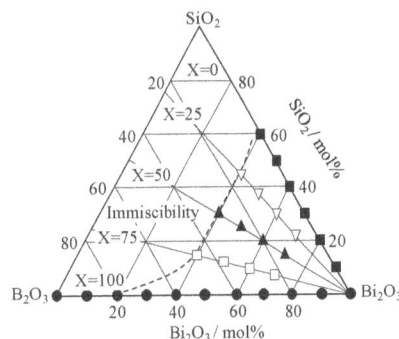

Figure 1. Compositions of the samples studied from the Bi$_2$O$_3$.XB$_2$O$_3$.(100–X)SiO$_2$ melt system

results, an empirical equation that can be used to estimate the melt viscosity from the chemical composition is derived.

2. Experimental procedure

The compositions of samples in this study are given by the symbols in Figure 1. Compositions were chosen based on the glass forming region of the Bi$_2$O$_3$–B$_2$O$_3$–SiO$_2$ system. There are five series of compositions; Bi$_2$O$_3$–SiO$_2$ and Bi$_2$O$_3$–B$_2$O$_3$ binary systems, and three ternary Bi$_2$O$_3$.XB$_2$O$_3$.(100–X)SiO$_2$ series for which the SiO$_2$/B$_2$O$_3$ ratio is 3/1 (X=25), 1/1 (X=50), and 1/3 (X=75). The batch of each sample was prepared by mixing appropriate amounts of reagent grade Bi$_2$O$_3$, B$_2$O$_3$ and SiO$_2$. The prepared batch was melted in a Pt crucible for 30 minutes in air at 1073–1473 K.

Viscosities of Bi$_2$O$_3$–B$_2$O$_3$–SiO$_2$ melts were measured systematically in the temperature range of 973 to 1573 K using the rotating crucible method. The details of the apparatus and the procedure for viscosity measurements are previously published.[11,12] The compositions of the samples after the measurement

*Corresponding author. Email fujino@chem-eng.kyushu-u.ac.jp

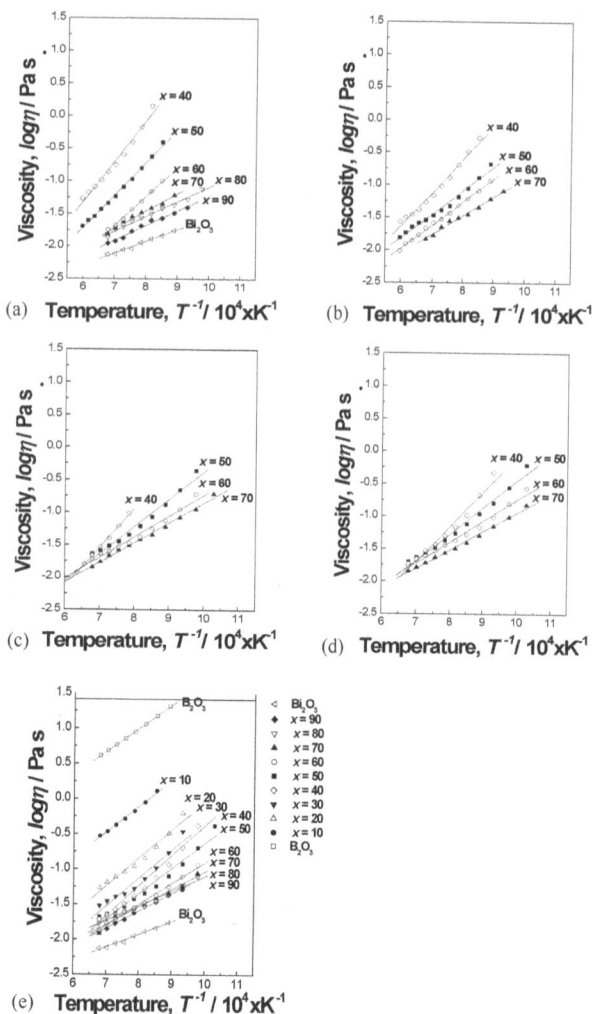

(a) **Temperature, T^{-1}/ 10^4xK^{-1}**

(b) **Temperature, T^{-1}/ 10^4xK^{-1}**

(c) **Temperature, T^{-1}/ 10^4xK^{-1}**

(d) **Temperature, T^{-1}/ 10^4xK^{-1}**

(e) **Temperature, T^{-1}/ 10^4xK^{-1}**

Figure 2. Viscosity of Bi_2O_3–B_2O_3–SiO_2 melts: (a) xBi_2O_3. (100–x)SiO_2 binary melts, (b) xBi_2O_3.(100–x)(B_2O_3.3SiO_2) melts, (c) xBi_2O_3.(100–x)(B_2O_3.SiO_2) melts, (d) xBi_2O_3. (100–x)(3B_2O_3.SiO_2) melts, (e) xBi_2O_3.(100–x)B_2O_3 binary melts (The symbol x gives the Bi_2O_3 content of each composition. The straight lines were fitted to data using the least squares method)

were analysed by the x-ray fluorescence method. The difference between the batched and analysed compositions was <1 mol%. Therefore, the compositions of the samples are given by the batched compositions in this paper.

3. Results

The viscosities, η (Pa s), of Bi_2O_3–B_2O_3–SiO_2 melts are shown in Figure 2(a–e) as a function of temperature, T(K). η decreased almost linearly with increasing T for all compositions.

Figure 3 shows the relationship between η at 1273 K and Bi_2O_3 content. η decreased steeply and continually with increasing Bi_2O_3 content. This tendency is attributed to the breakdown of the glass network, which results in the creation of nonbridging oxygens (NBOs). Since NBOs do not connect two network

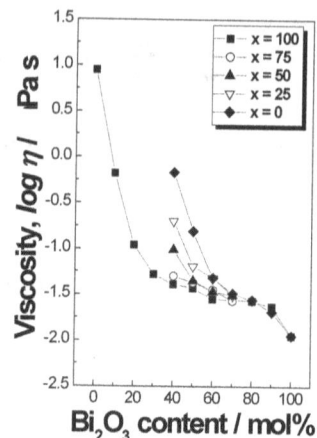

Figure 3. Relationship between viscosity η and Bi_2O_3 content in Bi_2O_3.XB_2O_3.(100–X) SiO_2 melts at 1273 K

forming cations, such as Si^{4+} and B^{3+}, the creation of NBOs leads to an increase in the mobility of the glass melt, and thus a decrease in η.

4. Discussion

4.1 Comparison of viscosity between Bi_2O_3–B_2O_3–SiO_2 and PbO–B_2O_3–SiO_2

The viscosities, η, for Bi_2O_3–B_2O_3–SiO_2 and PbO–B_2O_3–SiO_2 melts are compared here. As shown in Figure 3, the breakdown of the glass network, caused by the creation of NBOs, results in a reduction in η. One PbO introduces one oxide anion into the melt, while one Bi_2O_3 introduces three oxide anions into the melt. This difference results in a large structural change; almost three times more breaking of the glass network in Bi_2O_3–B_2O_3–SiO_2. Therefore, the compositional expressions $Bi_{2/3}O$–B_2O_3–SiO_2 and PbO–B_2O_3–SiO_2 are used to compare the role of Bi_2O_3 with PbO. Figure 4 shows a comparison of the viscosities of $Bi_{2/3}O$–B_2O_3–SiO_2 and PbO–B_2O_3–SiO_2 melts, where the SiO_2/B_2O_3 ratio is 3/1 (X=25) and 1/1 (X=50). The viscosity of the $Bi_{2/3}O$–B_2O_3–SiO_2 melt is higher than that of the PbO–B_2O_3–SiO_2 melt. This means that the viscosity lowering effect of PbO is larger than that of Bi_2O_3 in borosilicate melts. This may be attributed to the difference in field strength and polarisability of the cations.[13]

4.2 Estimatation of viscosity at 1273 K from chemical composition

From the viscosity data for Bi_2O_3–B_2O_3–SiO_2 and PbO–B_2O_3–SiO_2 melts, there is a tendency that the contribution of single component B_2O_3 to η at 1273 K is empirically one-half that of single component SiO_2. Thus, we introduced an empirical parameter P_η which was defined as follows

$$P_\eta = 0\cdot5\times[B_2O_3 \text{ mol%}]+[SiO_2 \text{ mol%}] \qquad (1)$$

As shown by Figure 5(a) and (b), a reasonable cor-

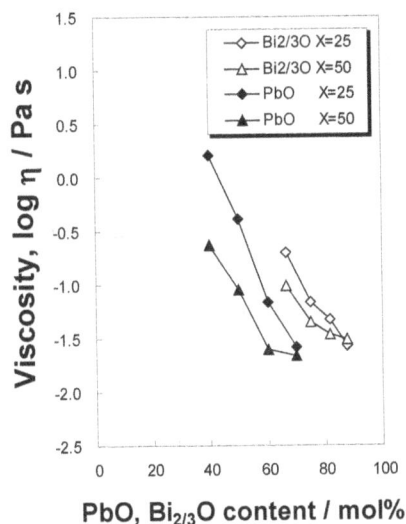

Figure 4. Comparison of viscosity between $Bi_{2/3}O.XB_2O_3.$ $(100–X)SiO_2$ and $PbO.XB_2O_3.(100–X)SiO_2$ melts at 1273 K

relation between P_η and η was found for different values of X for $Bi_2O_3–B_2O_3–SiO_2$ and $PbO–B_2O_3–SiO_2$ melts, and thus empirical equations can be obtained as follows; for the $Bi_2O_3–B_2O_3–SiO_2$ system

$$\log\eta_{Bi2O3}=1{\cdot}24–9{\cdot}26\times10^4\times P_\eta+4{\cdot}33\times10^4\times P_\eta^{\,2} \qquad (2)$$

for the $PbO–B_2O_3–SiO_2$ system

$$\log\eta_{PbO}=1{\cdot}36–3{\cdot}54\times10^2\times P_\eta+1{\cdot}4\times10^3\times P_\eta^{\,2} \qquad (3)$$

Moreover, based on the correlation between η_{Bi2O3} and η_{PbO} at 1273 K (see Figure 6), η_{Bi2O3} can be converted to

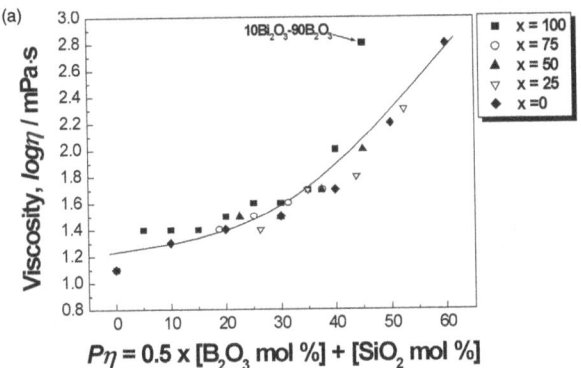

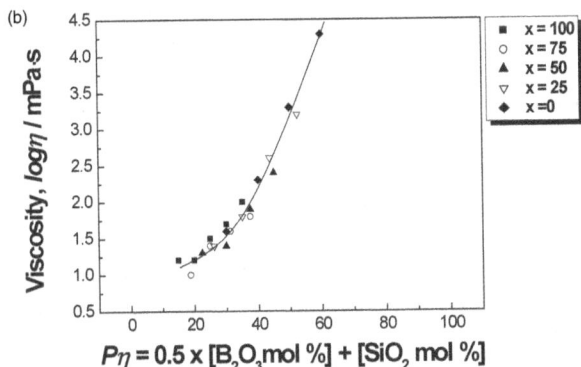

Figure 5. Relationships between the viscosity and the empirical parameter P_h in (a) $Bi_2O_3.XB_2O_3.(100–X)SiO_2$ and (b) $PbO.XB_2O_3.(100–X)SiO_2$ melts at 1273 K

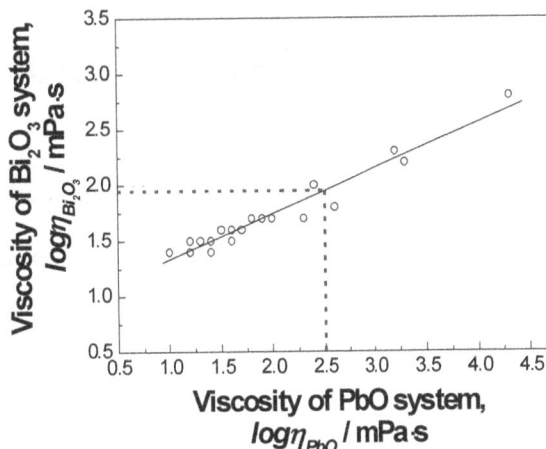

Figure 6. Correlation between the viscosity, η, of $Bi_2O_3–$ $B_2O_3–SiO_2$ and of $PbO–B_2O_3–SiO_2$ melts at 1273 K

η_{PbO} by the following equation when the composition is essentially the same

$$\log\eta_{Bi2O3}=0{\cdot}92+0{\cdot}41\times\log\eta_{PbO} \qquad (4)$$

The above equations (1)–(4) will be useful to explore the composition of $Bi_2O_3–B_2O_3–SiO_2$ melts with viscosities equivalent to those of $PbO–B_2O_3–SiO_2$ melts, and will be helpful in designing Pb-free glass from the $Bi_2O_3–B_2O_3–SiO_2$ system, attempting to solve problems that may be encountered in their practical applications.

5. Conclusions

The viscosities of $Bi_2O_3–B_2O_3–SiO_2$ melts were measured systematically in the temperature range 973 to 1573 K using the rotating crucible method. The viscosity decreased steeply and continually with increasing Bi_2O_3 content, and decreased with decreasing $SiO_2/$ B_2O_3 ratio. The viscosities of $Bi_2O_3–B_2O_3–SiO_2$ melts were compared with those of $PbO–B_2O_3–SiO_2$ melts, and it is found that the viscosity lowering effect of PbO is larger than that of Bi_2O_3 in borosilicate melts. Empirical equations to estimate the viscosity from chemical composition were derived for $Bi_2O_3–B_2O_3–$ SiO_2 and $PbO–B_2O_3–SiO_2$ melts.

References

1. Bockris, J. O'M., Mackenzie, J. D. & Kitchener, J. A. *T. Faraday Soc.,* 1955, **51**, 1734.
2. Ito, H. & Yanagase, T. *Trans. Jpn. Inst. Met.,* 1960, **1**, 115.
3. Ejima, T. & Kameda, M. *J. Jpn Inst. Met.,* 1967, **31**, 120.
4. Masuda, H., Nishida, H. & Morinaga, K. *J. Jpn Inst. Met.,* 1998, **62**, 444.
5. Kim, K. S. & Bray, P. J. *J. Chem. Phys.,* 1976, **11**, 4459.
6. Mazurin, O. V., Streltsina, M. V. & Shvaiko-Shvaikovskaya, T. P. *Handbook of glass data. Part B, Single-component and binary non-silicate oxide glasses.* Elsevier, Amsterdam, 1985, p.767.
7. Tera, R. *J. Mater. Integr. Jpn.,* 2004, **17**, 55.
8. Guadagnino, E. & Dall'Igna, R. *Glass Technol.,* 1996, **37**, 76.
9. Camphell, I. *Glass Technol,,* 1998, **39**, 38.
10. Kuromitsu, Y. *J. Am. Ceram. Soc.,* 1997, **80**, 1583.
11. Kawahara, M., Morinaga K. & Yanagase, T. *J. Jpn Inst. Met.,* 1977, **41**, 1047.
12. ASTM C 965-81, 1990.
13. Scholze, H. *Glass: Nature, Structure, and Properties,* Springer-Verlag, New York, 1991, p. 168.

Proc. VI Int. Conf. Borate Glasses, Himeji, Japan, 18–22 August 2008   *Phys. Chem. Glasses: Eur. J. Glass Sci. Technol. B*, October 2009, **50** (5), 284–288

First principle simulations of liquid and vitreous B₂S₃

G. Ferlat[1,]* *& M. Micoulaut*[2]

[1] *Institut de Minéralogie et de Physique des Milieux Condensés, UMR CNRS–IPGP–Universités Paris 6 et 7 no. 7590, 4 place Jussieu, 75252 Paris cedex 05, France*
[2] *Laboratoire de Physique Théorique de la Matière Condensée, Boîte 121, Université Paris 6, 4 place Jussieu, 75252 Paris cedex 05, France*

Manuscript received 27 August 2008
Revised version received 24 April 2009
Accepted 21 July 2009

This paper is part of an ongoing effort aimed at modelling the structure of liquid and vitreous B₂S₃ within a first principles framework. Using density functional theory, finite basis sets and pseudopotentials, we carried out molecular dynamics of the liquid phase. We report tests regarding some of the description parameters, such as the basis and the system size. Preliminary results obtained in the vitreous phase are also presented.

1. Introduction

Thioborate (i.e. B₂S₃ based) compounds have interesting technological applications, for instance as superionic conducting materials for batteries or fuel cells.[1] They are also of fundamental interest as prototypical III–VI network glasses, like B₂O₃ compounds for which they provide comparative insight concerning the structural chemistry.

From a microscopic point of view, it is well established that the molecular building block in both crystalline and glassy B₂S₃ is the planar BS₃ unit. In the B₂S₃-I crystal, these units connect into 4-fold (B₂S₄) and 6-fold (B₃S₆) rings, and there are no independent BS₃ triangles.[2] The 6-fold rings, called borsulphols or thioboroxols, are the equivalent of the boroxol rings in vitreous B₂O₃, while the 4-fold (edge sharing) rings are not observed in borates and are specific to chalcogenides. In the following, we shall refer to f_∞, f_4 and f_6, the relative proportions of boron atoms involved in independent BS₃ triangles, 4- and 6-fold rings, respectively. In B₂S₃-I, f_∞=0, f_4=25 and f_6=75%.

These quantities in the vitreous phase v-B₂S₃ are still uncertain and the information available in the literature is contradictory.[3–11] Support for the existence of both 4- and 6-fold rings, and thus for structural similarities between the glass and the crystal, was inferred from the first Raman studies[3,4] and from NMR data.[5] However, at least in the latter case, contamination of the samples was noticed.[5] More recently, Raman,[6] infrared[7] and diffraction[8,9] studies found no evidence for 4-fold rings, and instead show the glass to be predominantly made of BS₃ triangles and B₃S₆ thioboroxols. In this latter model, v-B₂S₃ would be isostructural to its oxide analogue, v-B₂O₃ for which f_∞~25 and f_6~75%.[12] In a theoretical aggregation model,[10] the B₃S₆ units were found to be energetically favoured over the B₂S₄ rings. Finally, a

previous molecular dynamics (MD) simulation using empirical force fields led to a model made almost entirely of chains of B₂S₄ rings (f_4~100%).[11] We note that at the time of this theoretical work[11] the experimental static structure factor of pure v-B₂S₃ was not available, since it was measured only recently.[8] The comparison of the calculated[11] and experimental[8] neutron structure factors clearly allows us to rule out the f_4~100% model (see Figure 5(c) later).

In order to clarify the situation, it is clear that a model which simultaneously reproduces all the mentioned information (diffraction, IR, Raman, NMR) is highly desirable. These considerations motivated us to undertake an extensive description of B₂S₃ disordered phases within a first principles scheme. The purpose of the present work is to assess the validity of our approach and to present preliminary results.

2. Method

2.1 Details of calculations

First principles molecular dynamics (FPMD) simulations were performed within the density functional theory (DFT) framework using generalised gradient approximations (GGA) to the exchange-correlation functional and norm-conserving Troullier–Martins pseudopotentials[13] in the Kleinman–Bylander[14] form. We tested two different GGA functionals, namely PBE[15] and BLYP.[15–17] The core radii used are summarised in Table 1. We used both the CPMD[18] and Siesta[19,20] codes. In the former case, an accurate

Table 1. *Core radii used in the Troullier–Martins pseudopotentials for the s, p, d channels depending on the exchange-correlation functional used (PBE[15] or BLYP[16,17])*

	r^B_{core} (a.u.)			r^S_{core} (a.u.)		
	s	p	d	s	p	d
PBE	1·7	1·7	1·7	1·4	1·4	1·7
BLYP	1·4	1·5	0·9	1·4	1·4	0·95

*Corresponding author. Email Guillaume.Ferlat@impmc.jussieu.fr

and easy-to-converge plane-wave (PW) basis set is used for the electronic wave functions expansion, rendering however, the simulations very time expensive for large systems. A cut off of 40 Ry was chosen for the PW kinetic energy. In the latter case, the basis set is a combination of localised pseudoatomic orbitals, allowing for a very efficient resolution of the Kohn–Sham equations. It is thus very adapted to simulations of disordered systems, for which large unit cells are required, but the size and shape of the basis set require to be tuned and the accuracy of the results to be checked. We used a double-ζ polarised (DZP) basis set, which amounts to two 2s, two 2p and one 3d shells for both sulphur and boron atoms. The shape and cutoff radii (r_c) of the support regions of the wave functions were obtained variationally[21,20] in a reference system, $B_3S_6H_3$, i.e. a borsulphol molecule with hydrogens terminating the nonbridging sulphurs. A similar procedure was used for our previous simulations of B_2O_3[22,12,23] and a more detailed description can be found in Ref. 22. This resulted in r_c values for the first-ζ of 6·6, 7·3 and 3·7 a.u., respectively, for the s, p, d orbitals of sulphur, and 4·7, 5·2, 4·8 a.u. for boron. In the following, this basis is referred to as DZP-opt. For Siesta calculations, the real-space integrations were carried out on a grid whose fineness is defined by a 250 Ry energy cutoff. This allowed the total force modulus to converge below 0·1 eV/Å at each MD step.

All the simulations were carried out within the Born–Oppenheimer approximation, using an ionic time step of 1 fs, except in simulation #5 (see Table 2) for which the Car–Parrinello method was used with a fictitious electron mass of μ_0=600 a.u. and a time step of 0·1 fs. All the trajectories were sampled at fixed density using NVT or NVE ensembles. The initial configurations were generated from snapshots taken from our previous B_2O_3 simulations.[12] The oxygens were substituted by sulphurs and the cell lattice rescaled so as to get the B_2S_3 glass density, $\rho{\sim}1{\cdot}7$ g cm^{-3}. We used two different initial configurations taken from either our boroxol-rich (BR) or boroxol-poor (BP) B_2O_3 glasses. The BR (f_6=75%) configuration contains 80 atoms in a monoclinic (almost hexagonal) box, while the BP (f_6=22%) configuration contains 100 atoms in a cubic box. We also generated a larger system of 320 atoms by using a 2×2×1 supercell (almost cubic) of the 80 atoms configuration. γ-point

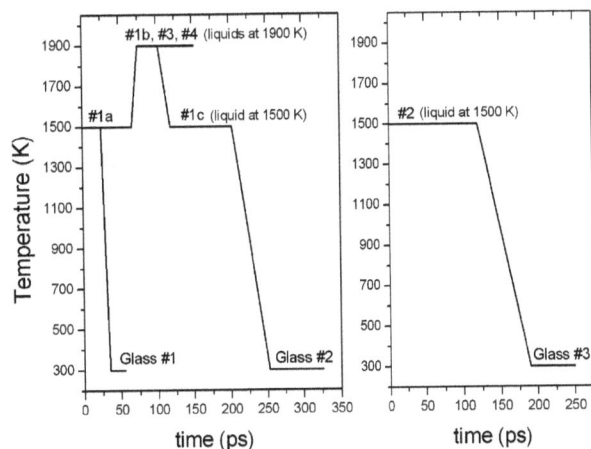

Figure 1. Schematic protocol of the liquid and glass simulations (see Tables 2 and 3)

sampling only was used for all the calculations.

Table 2 and 3 summarise the characteristics of the liquid and glassy simulations undertaken: basis set, exchange-correlation functional, initial configuration, system size, temperature and duration. Figure 1 shows schematically the protocol used for the thermal sampling, detailed in the next subsection.

2.2 Simulation protocol

We started simulations #1 and #2 (using PW and DZP-opt basis sets respectively) at 1500 K using the BR model as a starting configuration. This model appeared to be quite stable at this temperature; very long times (>50 ps) were required for the system to gradually melt. The melting was monitored from the evolution of the mean squared displacement (MSD). Only simulation #2 was run for a time long enough for the system to melt; in simulation #1a, the system is a hot glass, i.e. only local rearrangements have occurred due to the limited simulation time. Simulations using the BP model (#5) did not show such metastable behaviour; the diffusive regime was quickly reached. Thus, the temperature in simulation #1 was increased to a higher temperature, 1900 K and run for 75 ps (#1b), in order for the system to reach an equilibrium independent of the initial configuration. Simulations at 1900 K using the DZP-opt basis were started from the configuration obtained in #1b after 25 ps: one using the same number of atoms (#3) was run for 50 ps, another one using 320 atoms for 12 ps

Table 2. Parameters describing the high temperature simulations: PW stands for plane wave basis sets using a 40 Ry cutoff and the CPMD code, DZP-opt. stands for the optimised double-ζ polarised basis using Siesta, XC is the exchange-correlation functional (PBE[15] or BLYP[16,17]), and finally BR and BP stand for boroxol-rich and boroxol-poor respectively (see text)

	Basis set	XC	Initial configuration	T (K)	Duration (ps)
#1a	PW	PBE	BR (80 atoms)	1500	65
#1b	PW	PBE	continuation from #1a	1900	75
#1c	PW	PBE	branched from #1b	Quench -1500	16 + 84
#2	DZP-opt.	PBE	BR (80 atoms)	1500	120
#3	DZP-opt.	PBE	started from #1b, 80 atoms	1900	50
#4	DZP-opt.	PBE	started from #1b, 320 atoms	1900	12
#5	PW	BLYP	BP (100 atoms)	1500	35

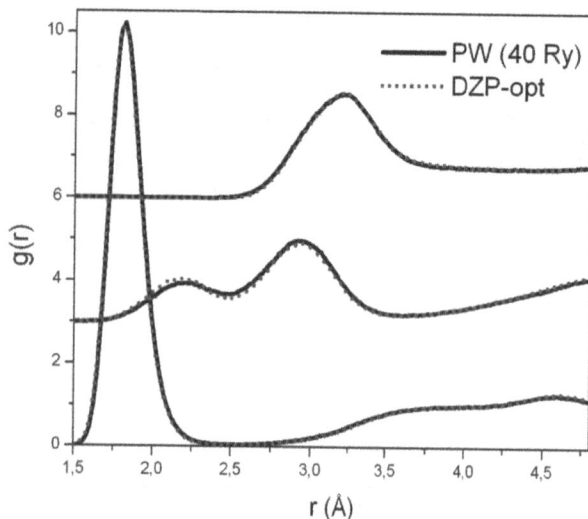

Figure 2. From bottom to top: B–S, B–B and S–S partial pair correlation functions obtained from simulations #1b and #3 (see Table 2)

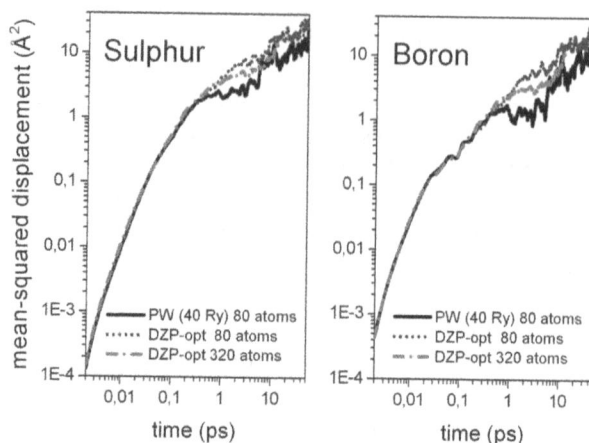

Figure 3. Mean squared displacements of sulphur (left) and boron (right) atoms at 1900 K as a function of time in simulations #1b, #3 and #4 (see Table 2)

(#4). Finally, a quench at 1500 K was branched from simulation #1b for a duration of 16 ps ($2 \cdot 5 \times 10^{13}$ K/s) and the liquid at 1500 K was sampled for 86 ps.

In the following section, the simulations carried out at 1900 K (#1b,#3, #4) are used to compare the calculation methodologies (basis and system sizes), since this temperature appeared to be high enough to reach the diffusive regime and thus an equilibrium independent from the initial configuration. On the contrary, simulations carried out at 1500 K (#1a, #1c, #2, #5) do show dependencies on the thermal history (due to the limited simulation times) and we shall use this dependency to generate several glassy models.

The first glass was obtained from a quench to 300 K branched from simulation #1a: this glass (G1) is fully isostructural to the starting configuration, i.e. the boroxol-rich model of B_2O_3. The second glass (G2) was obtained from a quench initiated from simulation #1c (quenched at $2 \cdot 5 \times 10^{13}$ K/s over 50 ps and then relaxed at 300 K for 70 ps); its thermal history thus includes the high temperature (1900 K) liquid state. Finally, the third glass (G3) was obtained from simulation #2 (quenched at $1 \cdot 7 \times 10^{13}$ K/s over 70 ps and then relaxed at 300 K for 60 ps).

3. Results

Figure 2 compares the partial pair correlation functions (PPCFs), calculated for the same duration (50 ps) at 1900 K from simulations #1b and #3, thus allowing to probe the effects of the basis size. The B–S and S–S curves are almost identical while very small differences are visible in the B–B PPCF, in particular

the height and shape of the first B–B peak at ~$2 \cdot 1$ Å. These differences are, however, marginal; the integration of the B–B PPCF up to the first minimum ($2 \cdot 45$ Å) gives n_{BB} (B–B coordination number) values of $0 \cdot 47$ and $0 \cdot 50$ for simulations #1b and #3, respectively. The same is true for the angular distributions (not shown); small or no differences at all were observed between both simulations. A more stringent test is provided by the dynamics; we monitored the MSD and the fraction of rings as a function of the simulation time. Both quantities showed a very similar behaviour in either type of simulation, see Figure 3.

Figure 4 compares again the PPCF obtained at 1900 K from simulations #3 and #4, thus allowing to probe system size effects. Although not fully negligible, the differences are rather small, the most important being found again in the first B–B peak (n_{BB} values of $0 \cdot 50$ and $0 \cdot 40$ were obtained). The dynamics were found to be unaffected by the system size within the statistical uncertainties (Figure 3). Longer simulations, however, are required for a more definitive assessment.

The comparison of the results from simulations #1c and #5 allows to probe primarily the effect of the XC functional; the obtained PPCFs (not shown) were found to be essentially the same with, however, a slight shift of the first peak position to higher r, by approximately $0 \cdot 01$–$0 \cdot 03$ Å in the BLYP case as compared to the PBE results. Since there are no experimental data available for the liquid, it is not possible yet to favour one or the other of these two functions. However, this choice does not seem to be critical.

The fraction of n-fold rings has been calculated for each liquid snapshot, and averaged over all snapshots. The results for the liquids at 1500 K are

Table 3. Parameters describing the obtained glassy models (using the same notation as in Table 2)

Glass	Basis set	XC	Initial configuration	T (K)	Duration (ps)
G1	PW	PBE	branched from #1a (after 25 ps)	Quench - 300	10 + 20
G2	PW	PBE	branched from #1c (after 200 ps)	Quench - 300	50 + 70
G3	DZP-opt.	PBE	branched from #2 (after 120 ps)	Quench - 300	70 + 60

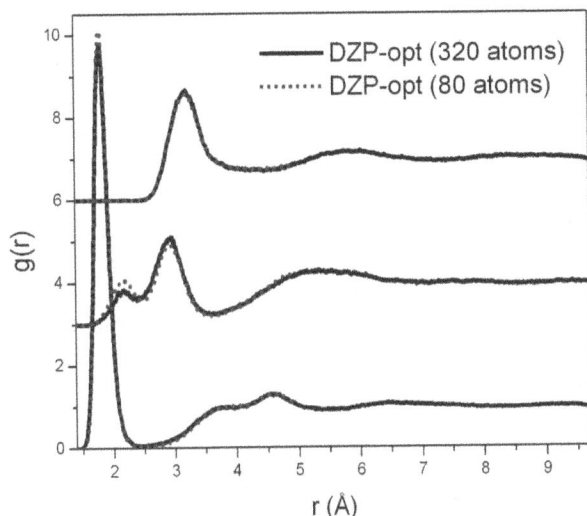

Figure 4. From bottom to top: B–S, B–B and S–S partial pair correlation functions obtained from simulations #3 and #4 (see Table 2)

Table 4. Average relative proportion (in %) of boron atoms in independent triangles (f_∞), 4-fold (f_4) and 6-fold (f_6) rings in the liquid samples at 1500 K

Simulation	f_∞	f_4	f_6
#1c	14	28	58
#2	19	16	65
#5	7	37	56

(Figures 3 and 4), at least in the range explored (from 80 to 320 atoms). It is quite reassuring that, despite the differences in methodologies and/or thermal histories used in the liquid simulations at 1500 K (#1c, #2 and #5), the average ring statistics (Table 4) is qualitatively the same in all three samples (Table 4); the liquid at 1500 K is predominantly made of 6-fold rings (~60%) with, however, a significant amount of 4-fold rings (~10–40%).

The presence of 4-fold rings in the liquid is likely a robust result; though they were absent from all of the initial configurations, their occurrence is a systematic output for long enough simulations, from values of ~10 to ~40%. The predominance of borsulphol rings at 1500 K is also strongly supported by the simulations; independently of the initial value (22% in simulation #5, 75% in the others), an equilibrium around 60±10% was observed. The determination of the ring statistics in the glassy state is more problematic due to the use of quenching rates which are much too fast; the obtained glass is in general strongly reminiscent of the liquid configuration from which the quench was initiated. Our strategy is thus to use the dependency of the glass structure upon its thermal history to generate several numerical glassy models. These models will then be compared with the available

presented in Table 4. Due to the limitations in both simulation times and system sizes, an error bar of at least 5 % should be assigned to these averaged values.

4. Discussion

The structural and dynamic quantities explored in our liquid simulations showed no or weak dependencies upon the methodology used. In particular, the system description is not degraded by the use of the localised basis set (in simulations #2, #3, #4) as compared to the more expensive PW basis set (Figures 2 and 3). Moreover, finite system size effects are minor

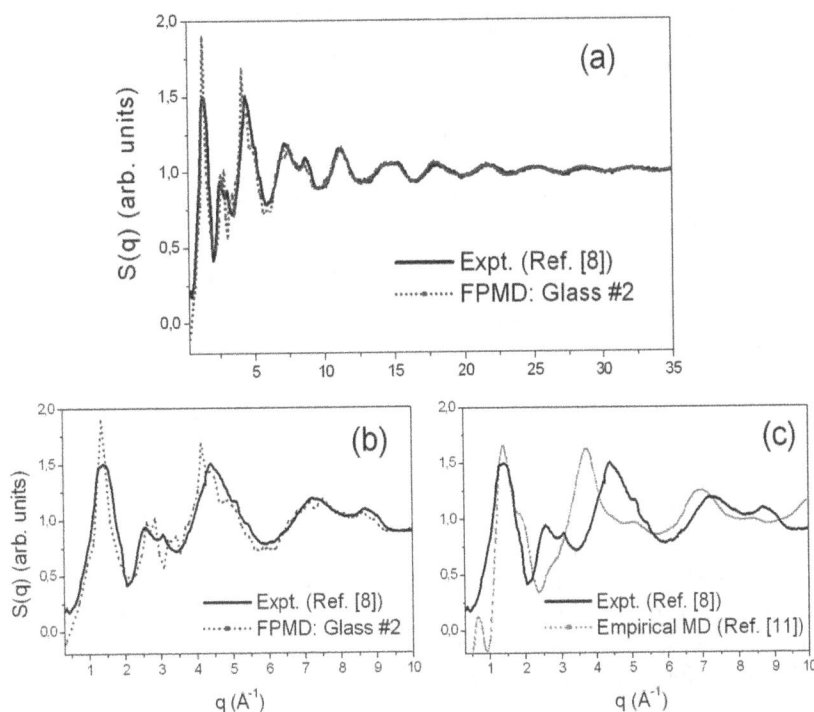

Figure 5. (a) calculated (dots) and experimental[8] (solid line) neutron weighted total static structure factor at 300 K, (b) the same data, expanded in the 0–10 $\mathring{A}^{-1}$ range, and (c) results from a previous empirical MD study[11]

Table 5. *Relative proportion (in %) of boron atoms in non-ring units (f_∞), 4-fold (f_4) and 6-fold (f_6) rings in the glassy samples*

Glass	f_∞	f_4	f_6
G1	25	0	75
G2	0	44	56
G3	22	12	66

experimental information (diffraction, infrared, Raman and NMR).

We generated three different glasses (see Table 3) for which the ring statistics noticeably differ (Table 5), thus allowing us to test different hypotheses proposed for v-B_2S_3. If the glass structure is close to that of the liquid obtained at 1500 K, then the G2 model should be the most realistic. We note that the numbers obtained for this model are pretty similar to those of the known crystal. On the contrary, if 4-fold rings do not survive in the glass, then the G1 model, isostructural to v-B_2O_3 should be closer to reality. The G3 model is somehow intermediate between G1 and G2.

As a first test, Figure 5(a) compares the neutron weighted total static structure factor calculated for the G2 glass to the experimental data.[8] The agreement is overall very good, even in the region of the first sharp diffraction peak (Figure 5(b)). Within the statistical uncertainty, a similar level of agreement was obtained for all of our three glasses (not shown), in spite of significant differences in the model B–B PPCFs; this is due to the fact that the total structure factor is dominated by the B–S partial contribution. Unfortunately, there are as yet no experimental partial structure factors available. However, the disagreement is obvious in the case of the chain of 4-fold rings model, formerly obtained from classical MD[11] (Figure 5(c)). This illustrates both the higher degree of realism of the first-principles models provided in this work, and the rather poor sensitivity of the total structure factor to the details of the ring statistics. As found in the case of B_2O_3,[12] the reproduction of the total structure factor is a necessary condition, but not a sufficient one.

To gain further insight, we plan to compute Raman, infrared and [11]B MQMAS (multiple quantum magic-angle spinning) spectra, which constitute much more stringent tests of the models.[12]

5. Conclusions

First-principle molecular dynamics simulations of liquid and vitreous B_2S_3 have been carried out. It has been shown that results of comparable accuracy to PW ones can be obtained with rather small basis (DZP like), making the extension of simulation times

and/or system sizes affordable. Simulations of the liquid at 1500 K show the presence of large amounts of both 4- and 6-fold rings. Several glassy models, differing by their amount of 4- and 6-fold rings, have been generated by varying the thermal history. The obtained models are all consistent with neutron diffraction experiments, which is not the case of the numerical model proposed formerly in the literature.[11] To quantify more precisely the amount of rings in the glass, we shall compare the predictions of our models with the results of spectroscopic probes (infrared, Raman and NMR), which may allow us to discriminate between them.

Acknowledgements

We thank A. P. Seitsonen and C. Massobrio for providing us with some of the pseudopotentials, and A. Takada for providing us with the initial B_2O_3 BR (boroxol-rich) configuration. The calculations have been performed at the French supercomputing centres IDRIS, CINES and CCR. G. Ferlat acknowledges financial support from IUCr2008 for attending the Borate2008 conference.

References

1. Kincs, J. & Martin, S. W. *Phys. Rev. Lett.*, 1996, **76**, 70.
2. Diercks, H. & Krebs, B. *Angew. Chem.*, 1977, **89**, 327.
3. Geissberger, A. E. & Galeener, F. L. In: *The structure of non-crystalline materials*, Eds. P. H. Gaskell, J. M. Parker & E. A. Davis, Taylor and Francis, London, 1983, p. 381.
4. Menetrier, M., Hojjaaji, A., Levasseur, A., Couzi, M. & Rao, K. J. *Phys. Chem. Glasses*, 1992, **33**, 222.
5. Hwang, S.-J., Fernandez, C., Amoureux, J., Han, J.-W., Cho, J., Martin, S. W. & Pruski, M. *J. Am. Chem. Soc.*, 1998, **120**, 7337.
6. Royle, M., Cho, J. & Martin, S. W. *J. Non-Cryst. Solids*, 2001, **279**, 97.
7. Cho, J. & Martin, S. W. *J. Non-Cryst. Solids*, 1995, **182**, 248.
8. Sinclair, R. N., Stone, C. E., Wright, A. C., Martin, S. W., Royle, M. L. & Hannon, A. C. *J. Non-Cryst. Solids*, 2001, **293**, 383.
9. Yao, W., Martin, S. W. & Petkov, V. *J. Non-Cryst. Solids*, 2005, **351**, 1995.
10. Micoulaut, M. In: *Current problems in condensed matter*, Ed. J. L. Morán-López, Plenum Press, 1998, p. 339.
11. Balasubramanian, S. & Rao, K. J. *Phys. Chem.*, 1994, **98**, 9216.
12. Ferlat, G., Charpentier, T., Seitsonen, A. P., Takada, A., Lazzeri, M., Cormier, L., Calas, G. & Mauri, F. *Phys. Rev. Lett.*, 2008, **101**, 065504.
13. Troullier, N. & Martins, J. L. *Phys. Rev. B*, 1991, **43**, 1993.
14. Kleinman, L. & Bylander, D. M. *Phys. Rev. Lett.*, 1982, **48**, 1425.
15. Perdew, J. P., Burke, K. & Ernzerhof, M. *Phys. Rev. Lett.*, 1996, **77**, 3865.
16. Becke, A. D. *Phys. Rev. A*, 1988, **38**, 3098.
17. Lee, C., Yang, W. & Parr, R. G. *Phys. Rev. B*, 1988, **37**, 785.
18. Copyright IBM Corp 1990–2006, Copyright MPI fuer Festkoerperforschung Stuttgart 1997–2001, CPMD V3.11.
19. Soler, J. M., Artacho, E., Gale, J. D., García, A., Junquera, J., Ordejón, P. & Sánchez-Portal, D. *J. Phys.: Condens. Matter*, 2002, **14**, 2745.
20. Artacho, E., Anglada, E., Diéguez, O., Gale, J. D., García, A., Junquera, J., Martin, R. M., Ordejón, P., Pruneda, J. M., Sánchez-Portal, D. & Soler, J. M. *J. Phys.: Condens. Matter*, 2008, **20**, 064208.
21. Anglada, E., Soler, J. M., Junquera, J. & Artacho, E. *Phys. Rev. B*, 2002, **66**, 205101.
22. Ferlat, G., Cormier, L., Mauri, F., Balan, E., Charpentier, T., Anglada, E. & Calas, G. *Phys. Chem. Glasses: Eur. J. Glass Sci. Technol. B*, 2006, **47**, 441.
23. Brazhkin, V. V., Katayama, Y., Trachenko, K., Tsiok, O. B., Lyapin, A. G., Artacho, E., Dove, M., Ferlat, G., Inamura, Y. & Saitoh, H. *Phys. Rev. Lett.*, 2008, **101**, 035702.

Proc. VI Int. Conf. Borate Glasses, Himeji, Japan, 18–22 August 2008   *Phys. Chem. Glasses: Eur. J. Glass Sci. Technol. B, December 2009, 50 (6), 372–377*

Elucidation of quadrupole parameters by simulation of ^{10}B NMR powder patterns

Jack Berkowitz,[1] *Michael R. McConnell,*[1] *Kevin Tholen,*[1] *Steve Feller,*[1,*] *Mario Affatigato,*[1]
Steve W. Martin,[2] *Diane Holland,*[3] *Mark E. Smith*[3] *& Tom F. Kemp*[3]

[1] *Coe College, Cedar Rapids, Iowa 52402, USA*
[2] *Iowa State University, Ames, Iowa 50011, USA*
[3] *University of Warwick, Coventry, England, CV4 7AL, UK*

Manuscript received 18 August 2008
Revised version received 16 July 2009
Accepted 19 August 2009

A method has been developed to analyse static ^{10}B NMR spectra and, using exhaustive simulation procedures, we have obtained values for quadrupole parameters and their distribution. Experimental data for vitreous boron oxide, vitreous and crystalline caesium triborate, vitreous caesium diborate, crystalline potassium diborate, and caesium enneaborate have been fitted to yield information on the multiple sites and their quadrupole parameters. These sites may result from differences in short range order, or intermediate range structure. The asymmetry parameter is particularly sensitive to differences in the environments of three-coordinated borons placed in differing intermediate range order positions.

1. Introduction

Nuclear magnetic resonance is one of the most frequently used spectroscopic methods. The measurement of chemical shifts in substances is one of the most effective methods of determining the presence and relative abundance of various structural components. The quadrupole effect provides yet more information and is extremely sensitive to short range structure. Defined as the interaction between the electric quadrupole moment of a nucleus and the electric field gradient of the electronic surroundings, this effect is highly sensitive to bond lengths and angles – the essence of chemical structure.[1]

Two parameters characterise the general line shape of the spectrum corresponding to a single, distinct atomic site: the asymmetry parameter, η, and the quadrupole coupling constant, Q_{cc}. Q_{cc} is defined as $e^2q_{zz}Q/h$ where e is the fundamental charge, eq_{zz} is the maximum size of the electric field gradient tensor in its principal axis system, h is Planck's constant divided by 2π, and Q is the electric quadrupole moment of the nucleus. Q_{cc} acts as a scaling factor of the overall strength of a quadupole interaction for a given atomic site. If eq_{yy} and eq_{xx} are defined as the other electric field gradient tensor components in the principal axis system, such that $q_{xx} < q_{yy}$, then η is defined to be $\eta = |q_{yy} - q_{xx}|/q_{zz}$. As a result, η ranges from 0 to 1.[2] In essence, η is a measure of the symmetry of the charge distribution around the nucleus, with 0 corresponding to perfect cylindrical symmetry, and 1 to full displacement of charge to one side. This value is very sensitive to bond angles and lengths.

It is also essential to take into account distributions of these parameters within a sample, which can be described by the width, σ, of a normalised distribution. A distribution in these structural parameters would correspond to a distribution in structural makeup of the sample. This is quite important within amorphous samples, as the structure shows local variations.[2]

The importance of the quadrupole interaction in an NMR study of borate glasses is paramount. For boron, chemical shift analysis is sensitive to changes in coordination number, but not nearly as sensitive to structural differences, as is the quadrupole interaction. Because of the high integer spin value of the ^{10}B nucleus ($I=3$), as well as its much higher quadrupole moment compared to ^{11}B, ^{10}B NMR is inherently more sensitive to differing quadrupole parameters than ^{11}B NMR. It is now possible to differentiate between three-coordinated boron atoms that are parts of different proposed superstructural units, such as boroxol or triborate rings.[2]

The goal of this research is to fit experimental spectra with the optimum combination of quadrupole parameters and distributions. In particular, we wish to examine distinct borate superstructural groups and their effect on the ^{10}B NMR spectra of 3 and 4 coordinated borons. This may come from a single site of three-coordinated boron atoms with three bridging oxygens (an F_1 unit), or a combination of several.

2. Procedure

2.1 Sample preparation

Samples for NMR were produced from alkali carbonates and either boric acid enriched with more than 97 at% ^{10}B (Aldrich Chemical Company) or from B_2O_3 enriched with 99·62% ^{10}B (Eagle Picher). Boron

*Corresponding author. Email sfeller@coe.edu

Table 1 Structural properties of samples examined

Sample	Short range structure	Superstructural group	Fraction of 4-coordinated borons, N_4 (F_2 abundance, from crystal studies)
v-B_2O_3 $R=0$	Symmetric [BO_3]	boroxol ring	0
v-CsB_3O_5 $R=1/3$	Trigonal borons with three bridging oxygens and tetrahedral borons with four bridging oxygens	triborate ring	0·33
c-CsB_3O_5 $R=1/3$	Same as v-CsB_3O_5	triborate ring	0·33
v-$Cs_2B_4O_7$ $R=1/2$	Trigonal borons with three bridging oxygens, trigonal borons with two bridging oxygens, and tetrahedral borons	diborate ring	0·50
v-CsB_9O_{14} $R=1/9$	Trigonal borons with three bridging oxygens and tetrahedral borons with four bridging oxygens	Mixture of boroxol rings and triborate rings	0·11

oxide glass was formed upon melting the sample in a platinum crucible at 1000°C for about 30 min and pouring the sample onto a metal plate. The sample was made in a furnace contained within a nitrogen glove box because of the hygroscopicity of boron oxide. Polycrystalline caesium triborate was prepared at Sojo University, Japan, in a 20 cm³ platinum crucible by cooling the melt from 1000°C to 500°C at a cooling rate of 200°C/h; then, the platinum crucible was transferred to an electric tube furnace and the crystallisation was further promoted in a high purity argon atmosphere by keeping the furnace temperature at 700°C for 24 h. X-ray powder diffraction of the polycrystal was in close agreement with the literature values for $Cs_2O.3B_2O_3$, though a trace of $CsBO_2$ was also present. The glass was prepared from some of the crystalline material by melt quenching.[2]

Glassy caesium enneaborate and diborate were prepared by melting stoichiometric quantities of caesium carbonate and [10]B enriched boric acid. The glasses were melted at 1000°C for approximately 20–25 min, with a weight loss measurement to confirm sample composition. The samples were plate quenched.

Crystalline potassium diborate was prepared by melting stoichiometric quantities of potassium carbonate and [10]B enriched boric acid. The batch was melted at 1000°C for approximately 20–25 min, with a weight loss measurement to confirm sample composition. The sample was slow cooled in the crucible to form the crystal.

2.2 Nuclear magnetic resonance

Standard static pulse echo ($\pi/2 \rightarrow \pi$) experiments were performed at a field of 7·0 T and an operating frequency of 31·49 MHz with the following conditions: τ=3 μs and $\tau/2$=1·5 μs (τ, $\tau/2$ are the pulse lengths used in the echo); 10–30 s pulse delay; 1 MHz spectral width. A novel aspect of the present work was the acquisition of spectra at incremented magnetic fields through the use of a field step unit. This was necessary because of the extreme width of the powder pattern. Typically a range of 1·44 MHz was covered in 57·6 kHz steps, with 1500 to 2500 acquisitions taken at each field to give an acceptable

signal to noise ratio. The final experimental powder pattern is the summation of the spectra obtained at each magnetic field. It was found that a significant number of steps were needed to avoid artefacts from the step size appearing in the spectrum.[2]

2.3 Simulation of [10]B spectra

Because no closed mathematical form exists for the shape of a quadrupole NMR powder pattern, experimental data must be fitted by repeatedly calculating theoretical powder patterns until the one that best reproduces the experimental line shape is found. This can be an arduous process, and brute force analytical techniques are not optimum for this kind of analysis. Physical intuition and attention to detail in the line shape are the most useful methods of comparison available.

Prior to fitting spectra of samples thought to contain multiple superstructural units, it was necessary to fit spectra of samples that were believed to contain only one superstructural group (see Table 1 for the structural properties of the samples). Once this was done, the other samples could be fitted with a combination of the spectra, adding them at

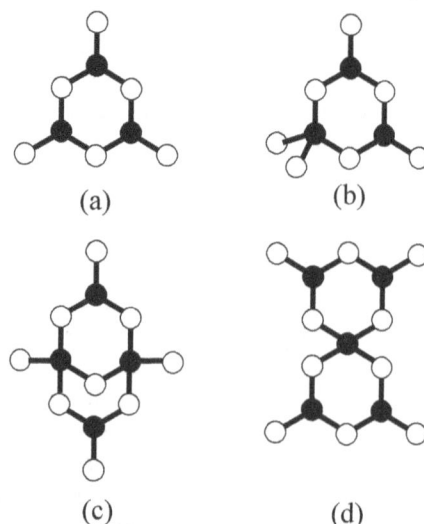

Figure 1. Examples of borate superstructural units: (a) boroxol, (b) triborate, (c) diborate, (d) di-pentaborate. Boron atoms are black, oxygens white. All oxygens shown are bridging

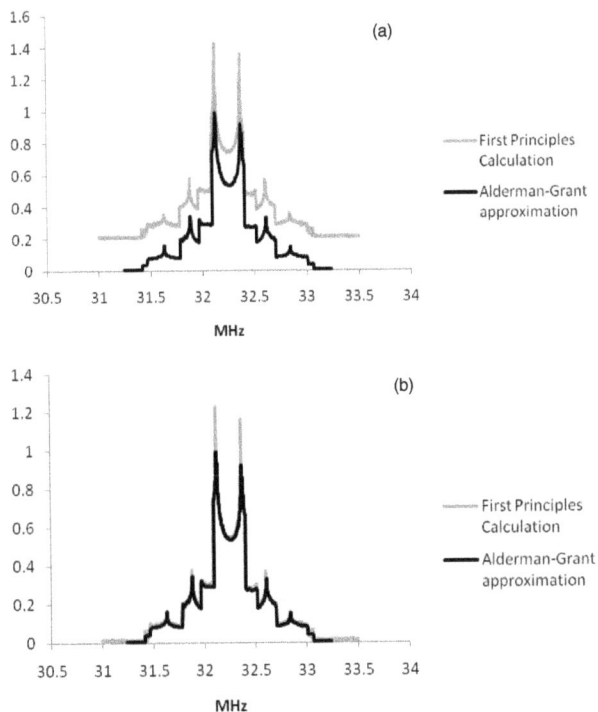

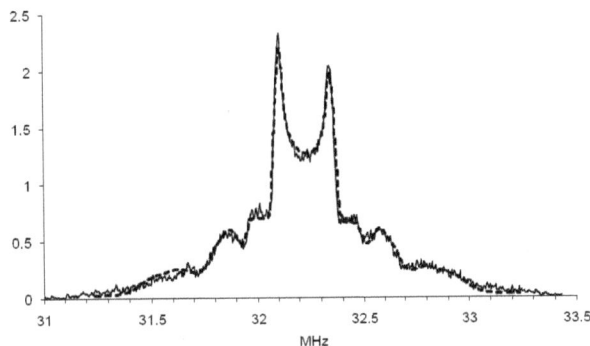

Figure 3. v-B$_2$O$_3$ fit. Solid line is experimental, dotted curve is the simulated powder pattern. NRP (normalised residual point, see Ref. 3) ~0·017

Figure 2. Comparison between ^{10}B NMR spectra calculated from first principles (gray) and using Alderman Grant approximation (black) for B$_2$O$_3$ without distributed quadrupole parameters (Q$_{cc}$=5·50 MHz, and h=0·13), with (a) and without (b) a vertical displacement of the first principles calculation

behind the development of Alderman–Grant interpolation), and thus was not useful for actually fitting the spectrum itself. Figure 2(a) and (b) displays a direct comparison between the two methods for the representative case of non-distributed B$_2$O$_3$. The agreement is close, as seen in Figure 2. For comparisons with spectra from the glasses, an isotropic broadening of 14·2 kHz was applied to the Alderman–Grant fits, to be consistent with earlier fitting.[2,3]

3. Results and discussion

The composition of the samples is written RM$_2$O. B$_2$O$_3$, where M is an alkali ion and R is the molar ratio of M$_2$O to B$_2$O$_3$. The values of Q$_{cc}$, σ(Q$_{cc}$), η, and σ(η) obtained from the best simulation(s) are collected in Table 2. In the discussions that follow, F$_1$ refers to a 3-coordinated boron with all bridging oxygens, F$_2$ refers to a 4- coordinated boron with all bridging oxygens, and F$_3$ refers to a 3-coordinated boron with one nonbridging oxygen.

3.1 Borate superstructural site fits

3.1.1 Vitreous boron oxide (R=0)

The simulated spectrum for a single F$_1$ site was fitted to the experimental data for v-B$_2$O$_3$ (Figure 3). The theoretical spectrum agrees closely with the experimental data and previous fits.[2] The F$_1$ site that is being fitted is due to the boroxol ring borons (Figure 1(a)), which are supposed to account for 65–75% of the boron atoms in boron oxide glass, as well as boron atoms outside the ring.[3] The trigonal boron atoms with bridging oxygen should have

various relative amplitudes until the sum accurately reflected the experimental spectra. From there, the relative area of each spectrum would represent the relative abundance of that type of 3 or 4 coordinated boron, which is often predicted by theory. Much of the analysis in this research made use of software written by one of the authors.[3]

To obtain the parameter sets of individual superstructural units, samples of vitreous boron oxide (boroxol rings, Figure 1(a)), vitreous and crystalline caesium triborate (triborate rings, Figure 1(b)), vitreous caesium diborate (diborate rings, Figure 1(c)) and crystalline potassium diborate were analysed. Since the software used in fitting these single sites made use of the approximate but efficient Alderman–Grant interpolation,[4] a first principles algorithm making use of sphere stepping was developed and used to confirm the resulting theoretical spectra. This algorithm was computationally intensive (the motivation

Table 2 Parameters of best fit

| Sample | F$_1$ | | | | | F$_2$ | | | | |
	Q$_{cc}$ (MHz)	σ (Q$_{cc}$)	η	σ(η)	% Abundance	Q$_{cc}$ (MHz)	σ(Q$_{cc}$)	η	σ(η)	% Abundance
v-B$_2$O$_3$ R=0	5·50	0·80	0·13	0·02	100					
v-CsB$_3$O$_5$ R=1/3	5·30	0·86	0·26	0·06	82·8	0·43	0·032	0·15	0·02	17·2
c-CsB$_3$O$_5$ R=1/3	5·60	1·1	0·25	0·02	86·3	0·35	0·096	0·90		13·7
v-Cs$_2$B$_4$O$_7$ R=1/2	5·30	0·30	0·30	0·20	61·0	1·20	0·60	0·54	0·08	39·0
c-K$_2$B$_4$O$_7$ R=1/2	5·40	0·60	0·11	0·06	74·6	0·74	0·16	0·54	0·096	25·4
v-CsB$_9$O$_{14}$ R=1/9 single site	5·50	1·0	0·19	0·06						

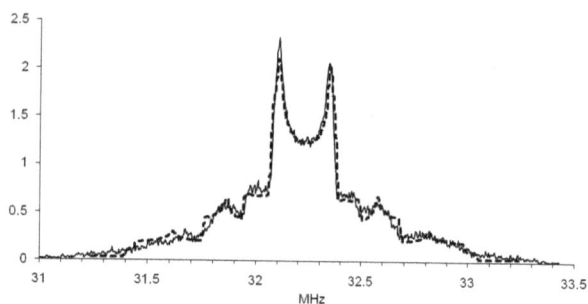

Figure 4. Non-distributed powder pattern fit for v-B_2O_3 (dotted curve) compared to experimental data. NRP of non-distributed pattern ~0·024

relatively low asymmetry (η), as well as $\sigma(\eta)$, as they are bonded to three other F_1s. The simulated values η=0·13 and $\sigma(\eta)$=0·02 agree with this prediction. Since only one simple fit was required, ^{10}B NMR is unable to differentiate boron atoms in the ring from those outside the ring.

Also, it is worth comparing a distributed powder pattern to a powder pattern with just one single combination of η and Q_{cc} (i.e. $\sigma(Q_{cc})$=$\sigma(\eta)$=0). For the case of vitreous boron oxide, these are displayed in Figures 3 and 4. The parameters used for the single combination powder pattern shown in Figure 4 are the peak parameters of the distributed powder pattern (see Table 2). Notice that, in the single combination powder pattern, there are very sharp features and peaks that are not actually realised in the experimental spectra.

3.1.2 Vitreous caesium triborate (R=1/3)

The simulated spectrum for a single F_1 site and a single F_2 site was fitted to the data for vitreous caesium triborate (Figure 5). One can observe the F_2 signal as the tall and narrow peak in the middle of the spectrum. This shape is due to its low Q_{cc} (0·43 MHz), resulting from its tetrahedral symmetry. The fitted spectrum agrees strongly with the experimental data.[2] These two sites most probably arise from the triborate superstructural unit (Figure 1(b)) that is

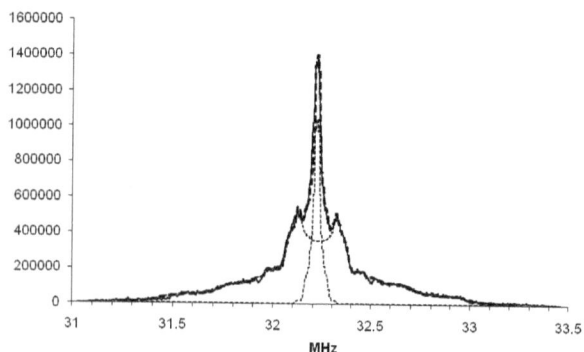

Figure 5. v-CsB_3O_5 ^{10}B NMR powder pattern fit. Solid curve is experimental, and dotted curve is overall calculated powder pattern. F_1 and F_2 contributions are shown as thin dotted lines. NRP ~0·0079

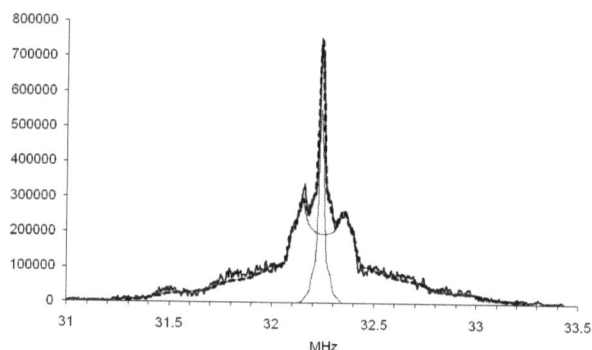

Figure 6. c-CsB_3O_5 ^{10}B NMR powder pattern fit. Solid curve is experimental, and dotted curve is overall calculated powder pattern. F_1 and F_2 contributions are shown as thin dotted lines. NRP ~0·012

thought to dominate borate glass at this composition. The increase in the F_1 value of h compared to vitreous boron oxide (from 0·13 to 0·26) is evidence of this, as each triborate F_1 is bonded within the structural unit to one F_1 and one F_2, while its exo-unit bond can be to a F_1 or F_2, as opposed to the boroxol structure where all F_1s are bonded to three other F_1s. The variability in the type of exo-unit bond is also evidenced in the increase of the F_1 value of $\sigma(\eta)$ from 0·02 to 0·06. The F_2 peak was more difficult to fit, as there is usually less detail for this kind of spectrum. Generally it was fitted by matching the height and width of the peak with experiment at several points. Their relative abundances (82·8% F_1, 17·2% F_2) also agree with previous ^{10}B fits, but not theory which predicts a two to one F_1 to F_2 ratio.[2] The reason for this is at present unclear.

3.1.3 Crystalline caesium triborate (R=1/3)

Again, the simulated spectrum for a single F_1 site and a single F_2 site was fitted to the experimental data for crystalline caesium triborate (Figure 6). The parameters agree with previous fits to experimental data.[2] The peak parameters, Q_{cc} and η, were very similar to the amorphous case. The relative abundances have changed to roughly 86·3% F_1 and 13·7% F_2, again in agreement with previous fits but not theory.

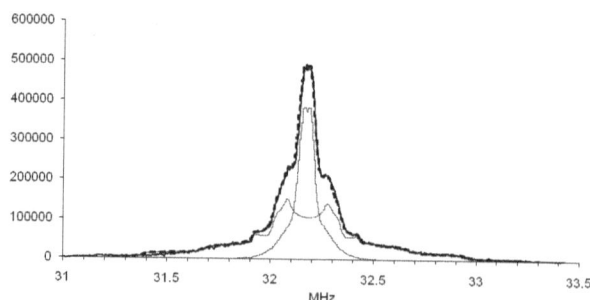

Figure 7. v-$Cs_2B_4O_7$ ^{10}B NMR powder pattern fit. Solid curve is experimental, and dotted curve is overall calculated powder pattern. F_1 and F_2 contributions are shown as thin dotted lines. NRP ~0·007

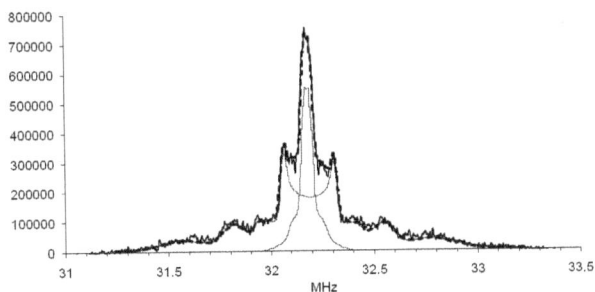

Figure 8. c-$K_2B_4O_7$ ^{10}B NMR powder pattern fit. Solid curve is experimental, and dotted curve is overall calculated powder pattern. F_1 and F_2 contributions are shown as thin dotted lines. NRP ~0·013

3.1.4 Vitreous caesium diborate (R=1/2)

No previous fits have been performed on data for this composition, but the simulated spectrum for two sites, F_2 and F_1, was fitted satisfactorily (Figure 7). The increase in the F_1 value of η compared to vitreous caesium triborate (from 0·26 to 0·30) is expected if the glass is composed of the diborate unit (Figure 1(c)), as the F_1s are now bonded to two F_2s in the unit, and either an F_1 or F_2 outside of the unit (note also the high F_2 value of $\sigma(\eta)$, 0·08). At this composition, the presence of F_3 units must be accounted for. Charge conservation states that the abundances of F_2 and F_3 must sum to R.[5] Without accounting for F_3, the fit shows ~60% F_1 and ~40% F_2. It is entirely possible that up to 1/6 of F_1 (10% of overall) are actually F_3, which would show up as a highly asymmetric signal, but this is likely too weak in intensity to detect. The data for crystalline lithium metaborate, which consists of F_3 units, has been fitted with Q_{cc}=5·14 MHz, and η=0·50.[2] ^{11}B NMR confirms the F_2 fraction for this glass.[6]

3.1.5 Crystalline potassium diborate (R=1/2)

The simulated spectrum for two sites (F_1 and F_2) was fitted to the experimental data for this sample (Figure 8). Unlike the triborate case, this crystalline form did not have peak parameters similar to the

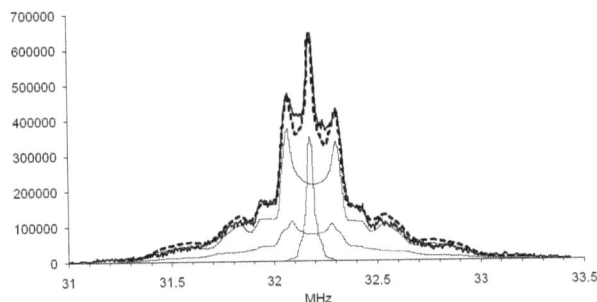

Figure 9. v-CsB_9O_{14} ^{10}B NMR powder pattern combination fit. Solid curve is experimental, and dotted curve is overall calculated powder pattern. Individual contributions are shown as thin dotted lines: 69·5% vitreous boron oxide F_1, 22·9% vitreous caesium triborate F_1, 7·6% F_2. NRP ~0·020

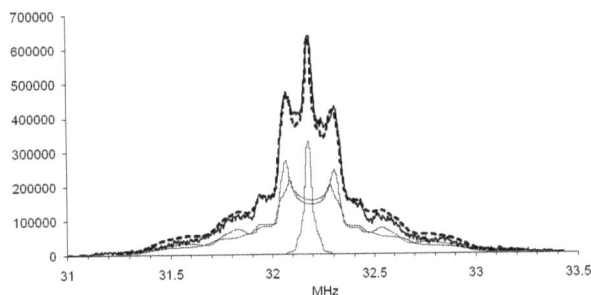

Figure 10. v-CsB_9O_{14} ^{10}B NMR powder pattern combination fit. Solid curve is experimental, and dotted curve is overall calculated powder pattern. Individual contributions are shown as thin dotted lines: 49·6% vitreous boron oxide F_1, 43·4% vitreous caesium triborate F_1, 7·0% F_2. NRP ~0·019

corresponding vitreous case. Since the modifying cation is different, the structure itself could be quite different from vitreous caesium diborate. According to crystallographic studies, crystalline potassium diborate should consist of equally abundant diborate rings (B_4O_7), di-triborate rings (B_6O_{11}, shown in Figure 1(d)), and free F_1 trigonal boron.[7] This is a much more complicated structure than the vitreous caesium diborate case. Before fitting this sample in a compound manner (as has been done for enneaborate, see below) lineshapes for di-triborate and free boron would be needed.

3.2 Multisite fits for vitreous caesium enneaborate (R=1/9)

In crystalline caesium enneaborate, the F_1 boron atoms exist in triborate rings and boroxol rings. Specifically the ratio is three boroxol F_1s to one triborate F_1. According to Krogh-Moe,[8] the glassy structure for enneaborate should resemble the crystalline structure. This means that the area of the theoretical powder pattern contributed by the boroxol boron should be three times that of the area contributed by triborate F_1. Using this theoretical prediction as a starting point, the individual F_1 sites fitted to vitreous boron oxide and vitreous caesium triborate were combined in the appropriate ratio. As can be seen in Figure 9, this rather accurately reproduces the lineshape of the experimental spectrum. However, as Figure 10 shows, it is quite possible to obtain another accurate fit with a different ratio, this case being 50% vitreous boron oxide and 40% vitreous caesium triborate. In both cases, the $F2$ peak parameters were taken from the triborate fit, and the starting value for the relative abundance was 10%. The F_2 peak in both cases does not appear to fit exactly and this is quite likely due to a difference in distribution values in both the F_1 and F_2 portions of the glass, resulting from the combination of sites or experimental issues. A single site fit was also attempted, with the result-

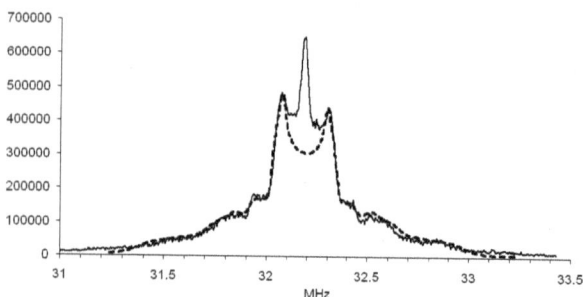

Figure 11. v-CsB$_9$O$_{14}$ ^{10}B NMR powder pattern single site fit. NRP ~0·017. Experiment is thin solid curve and fit is dotted curve

ing parameters seeming to arise from a combination of the parameters of vitreous caesium triborate and boron oxide (Figure 11); for this single site fit, the value of η is 0·19, intermediate between the boroxol value of 0·13 and the triborate value of 0·26. The Q_{cc} value matches that of boroxol at 5·50 MHz. The fit somewhat matches the character of the experimental lineshape, but fails to match some of the details, which indicates that a better fit is bound to come from a combination of two sites with independently spaced peaks in the signal.

4. Conclusions and future work

The success in fitting the data for vitreous caesium enneaborate with the theoretically predicted ratios of vitreous boron oxide and vitreous caesium triborate is a clear indicator of the value of this particular spectroscopic technique. The ability to differentiate among F$_1$ boron atoms that are parts of different superstructural units adds a whole new level of detail, both quantitative and qualitative, to our understanding of glass structure. The observation of increasing η among samples provides experimental corroboration to theoretically predicted trends. In addition to providing evidence for the existence of said superstructural units, it also enables analysis of the relative abundance of these units within samples of mixed intermediate range structure.

Certain work remains to be done towards improving analysis of these spectra:

- How much free trigonal boron is present in the samples? For instance, can adding a free, highly symmetrical F$_1$ group to the powder pattern for boron oxide improve the fit?
- How accurate are the assumptions surrounding the use of parameter distributions? Should a separate site be fitted for each exo-unit bond combination (see triborate and diborate discussion), or is it sufficient to assume a Gaussian distribution?
- How much F$_3$ will need to be fitted for compositions at or above $R=\frac{1}{2}$? Data exist for this unit in the form of lithium metaborate.[1] Will there be multiple sites of F$_1$, F$_2$, and F$_3$? If so, what details in the spectra can aid in fitting so many sites?

Acknowledgments

The authors would like to thank Jim Hoekstra of the Iowa State High Performance Computing group for his help. The Engineering and Physical Sciences Research Council of the United Kingdom is thanked for financial support. This work was funded in part by NSF grant No. DMR-0502051 and NSF grant No. DMR-0755231.

References

1. Jellison, Jr, G. E., Feller, S. A. & Bray, P. J. *J. Mag. Reson.*, 1977, **27**, 121.
2. Holland, D., Feller, S. A., Kemp, T. F., Smith, M. E., Howes, A. P., Winslow, D. & Kodama, M. *Phys. Chem. Glasses Eur. J. Glass Sci. Technol. B*, 2007, **48**, 1.
3. Kemp, T. F. & Smith, M. E. *Solid State Nucl. Magn. Reson.*, 2009, **34**, 243.
4. Alderman, D. W., Solum, M. S. & Grant, D. M. *J. Chem. Phys.* 1986, **84**, 3717.
5. Jellison, Jr, G. E., Feller, S. A. & Bray, P. J. *Phys. Chem. Glasses*, 1978, **19**, 52.
6. Berryman, J. R., Feller, S. A., Affatigato, M., Kodama, M., Kroeker, S., Meyer, B. M., Martin, S. W. & Borsa, F. *J. Non-Cryst. Solids* 2001, **293-295**, 489.
7. Wright, A. C., Vedishcheva, N. M. & Shakhmatkin, B. A. *Mater. Res. Soc. Symp. Proc.* 1997, **455**, 381.
8. Kreidl, N. J. In: *Borate Glasses: Structure, Properties, Applications*, Eds. L. D. Pye, V. D. Fréchette, & N. J. Kreidl, Plenum Press, New York, 1978, p. 1.

Proc. VI Int. Conf. Borate Glasses, Himeji, Japan, 18–22 August 2008   *Phys. Chem. Glasses: Eur. J. Glass Sci. Technol. B, June 2009, **50** (3), 144–148*

The mixed glass former effect on the thermal and volume properties of Na$_2$S–B$_2$S$_3$–P$_2$S$_5$ glasses

Michael J. Haynes, Christian Bischoff, Tobias Kaufmann & Steve W. Martin[1]

Department of Materials Science and Engineering, Iowa State University, Ames, IA 50011, USA

Manuscript received 15 October 2008
Revised version received 14 November 2008
Accepted 23 April 2009

The effect of mixing two glass formers at constant total modifier content has been studied on the thermal and volume properties of glasses in the series (Na$_2$S)$_y$.[(B$_2$S$_3$)$_x$.(P$_2$S$_5$)$_{(1-x)}$]$_{(1-y)}$, *for two modifying alkali sulphide contents,* y=0·5 *and* y=0·65, *with* 0≤x≤1 *in* 0·1 *steps. The glass transition temperature,* T$_g$, *shows a pronounced maximum at* x=0·5 *for both series, while the onset crystallisation temperature,* T$_c$, *shows a pronounced maximum at* x=0·4 *in both series of glasses. These two features gives rise to a maximum in the glass stability, as measured by the ratio* (T$_c$-T$_g$)/T$_g$, *in the region of* x=0·4 *for both series of glasses. In a similar manner, the density of the* x=0·5 *series of glasses shows a weak maximum at* x=0·5, *whereas the density of the* x=0·65 *series of glasses shows a more pronounced maximum at* x=0·3. *However, the calculated molar volumes of both series of glasses fail to show any significant deviation from essentially linear behaviour between the molar volumes of the two end member glasses in both of these series of glasses.*

1. Introduction and background

While much is known of the composition dependence of glass properties, especially of how the properties of a particular glass series change as a function of the amount of modifying oxide (typically an alkali oxide such as Na$_2$O) doped into a single base glass former (such as SiO$_2$ or B$_2$O$_3$), very much less is known about how these same properties change when the alkali modifier is held constant and the fraction of two different glass formers is changed from 0, one end member binary glassforming system, to 1, the other end member binary glass forming system. There are, nevertheless, a few very well known systems where such substitution of one glass former for another has yielded glasses of exceptional commercial and technical importance.

The so-called "Pyrex" sodium borosilicate glasses, discovered some years ago, are chief among the important mixed glass former glasses.[1–4] The substitution of SiO$_2$ by B$_2$O$_3$ at a constant amount of alkali oxide, typically Na$_2$O, was shown to significantly improve the glass forming ability of the binary soda–silica glass, dramatically improving its chemical durability, and perhaps as importantly, significantly lowering the thermal expansion coefficient and increasing the mechanical modulii of the ternary mixed glass former glasses (MGFGs), compared to the binary glasses.

These significant improvements in nearly all of the physical and chemical properties of these MGFGs, the so-called mixed glass former effect (MGFE), has led to the Pyrex class of glasses being among the most widely used of all glasses in residential, commercial,

and technical applications, second perhaps only to the most widely used glass, the common soda–lime–silica glass, which interestingly would be considered a mixed modifier glass.

Other improvements in other physical properties have been sought (and observed) in other MGFGs, and a few such glass series and their corresponding physical properties have been explored for these reasons. More recently, it has been observed that a particularly striking MGFE is observed in the alkali ion conductivity in MGFGs.[5–7] In these systems, the alkali ion conductivity has been shown to increase by as much as a few orders of magnitude over and above the linear mixing rule behaviour expected from the two end member binary glasses. For example, the MGFE for the Na$^+$ ion conductivity has been observed by us in the new Na$_2$S–B$_2$S$_3$–P$_2$S$_5$ glass system under study in this work, and is shown in Figure 1. Such alkali ion conducting glasses are gaining wider interest from the battery research community as new alkali ion, especially Li$^+$ ion, conducting electrolytes are being sought to replace liquid polymer electrolytes that suffer from low chemical durability and enable lithium dendrites to grow in these widely used Li-ion batteries. Such ion conducting glasses, with their high Li$^+$ ion conductivity, improved chemical durability, and solid structure offer potential solutions to these critical as yet unsolved problems in Li-ion batteries.[8–12]

For these reasons, we have begun an in-depth and detailed study of the MGFE in both oxide and, as we report here, sulphide glasses, to better understand and use this effect, in order to enable improved and optimised solid electrolytes for next generation Li batteries. In the present report, we have selected the Na$_2$S–B$_2$S$_3$–P$_2$S$_5$ system for a few very important rea-

---

[1] Corresponding author. Email swmartin@iastate.edu

sons. While Li⁺ ions are perhaps more appropriate for Li batteries, Na⁺ offers access to easily studied radio nuclides that can be used to measure the fundamentally important diffusion coefficient, and the ^{23}Na nuclide provides a common and easily studied NMR nucleus that will aid in probing the local chemistries in and around the mobile Na⁺ ion. In a similar way, B and P were chosen as the two glass forming cations as they also have easily measured NMR nuclei, ^{11}B and ^{31}P, respectively. They also provide strongly glass forming binary glass systems, and have the potential to yield strong glass forming ternary systems as well. Finally, the sulphide system was chosen for study as it has been widely found that sulphide glasses exhibit alkali ion conductivities as much as a million times higher than their equally modified oxide glass counterparts.[8] This higher ionic conductivity not only makes these sulphide glasses more interesting to study, but in many cases it makes them easier to study as well. A companion study of the all oxide system, Na_2O–B_2O_3–P_2O_5, is being studied in our group as well and is being reported by Christensen et al at this same conference.[13]

While the purpose of this study is to examine the effect of mixing the concentration of the two glass formers in the ternary glass forming series upon the ionic conductivity of the Na⁺ ions in these glasses at constant overall mole fraction of Na_2S, we do expect as seen in nearly all other such modified glasses,[8,14-17] higher ionic conductivity with greater Na_2S content in the glasses. There are several reports, however, including those of the author,[18] that such glasses can exhibit a conductivity maximum at the extremes of alkali content. We do not expect that y=0·65 will result in glasses which reach this point, and our preliminary results do not show such a conductivity maximum.

In this paper, we give the first of a series of reports on the properties, structures and transport properties the MGFG series $(Na_2S)_y \cdot [(B_2S_3)_x \cdot (P_2S_5)_{(1-x)}]_{(1-y)}$ by examining the effect on the volume and thermal properties of these glasses at two fixed (high alkali) values of y, 0·5 and 0·65, and for all values of x, $0 \leq x \leq 1$ in 0·1 steps. We report the glass transition temperature, T_g, the crystallisation temperature, T_c, the glass stability, as measured by the ratio $(T_c - T_g)/T_g$, the density, and the molar volumes as a first step in fully characterising these glasses. Future work will report the structures of the glasses as well as the ion conductivity with the ultimate goal being to develop detailed models of the MGFE so that these glasses and their lithium analogues may find use in next generation Li ion batteries.

2. Experimental methods

Glass samples were made in a high purity N_2 glove box with <0·1ppm O_2 and <5ppm H_2O. They were made by combining appropriate amounts of the

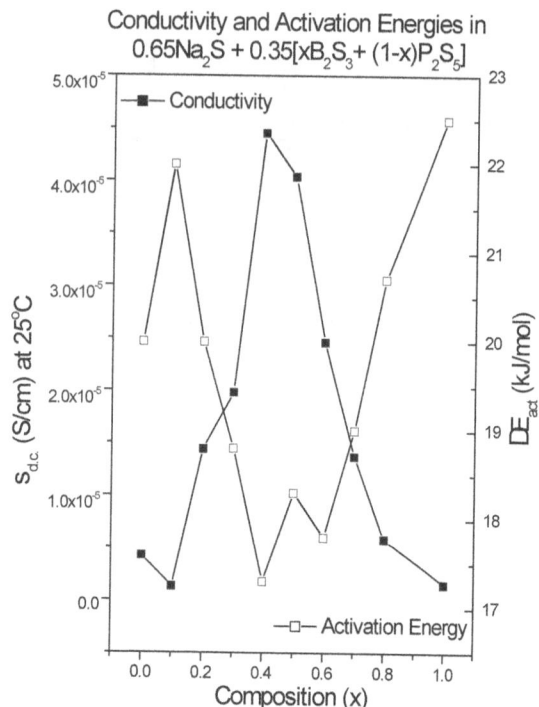

Figure 1. Composition dependence of the Na⁺ ion conductivity and the conductivity activation energy for the MGFG series $(Na_2S)_y \cdot [(B_2S_3)_x \cdot (P_2S_5)_{(1-x)}]_{(1-y)}$ for the high alkali series y=0·65. The conductivity shows a sharp maximum and the activation energy shows a sharp minimum in the middle of the composition range, at x~0·5

starting materials, Na_2S, B_2S_3, and P_2S_5, in a carbon crucible, mixing thoroughly to create a homogeneous powder. As high purity Na_2S is not available commercially, it was prepared in our laboratory by reacting NaOH with pressurised and liquefied H_2S at room temperature for extended periods of up to three weeks. The resulting NaHS was then calcined to Na_2S inside the glove box to quantitatively drive off H_2S, yielding stoichiometric Na_2S. IR spectroscopy showed that the calcination process removed all the –SH from the material and powder x-ray diffraction was used to confirm that high purity Na_2S was obtained. B_2S_3 was prepared in our laboratory by reacting appropriate amounts of B (99·99% Alfa-Aesar) and S (99·999% Alfa-Aesar) in previously vacuum flame dried and carbon coated silica ampoules, sealed under vacuum. The sealed charge was heated at 1°C/min to 850°C, held for 8 h, and then cooled to room temperature. IR spectroscopy was used to confirm the purity of the glassy B_2S_3 so prepared. P_2S_5 was used as received (99·9% Alfa-Aesar) and not otherwise purified, because IR and Raman spectroscopy showed that it was high purity. Appropriate amounts of the powders to create batch compositions of $(Na_2S)_y \cdot [(B_2S_3)_x \cdot (P_2S_5)_{(1-x)}]_{(1-y)}$ at two fixed values of y, 0·5 and 0·65, and for all values of x, $0 \leq x \leq 1$ in 0·1 steps, were melted inside the glove box furnace at 630–675°C for ~15 min inside covered vitreous carbon crucibles. Samples were splat quenched between

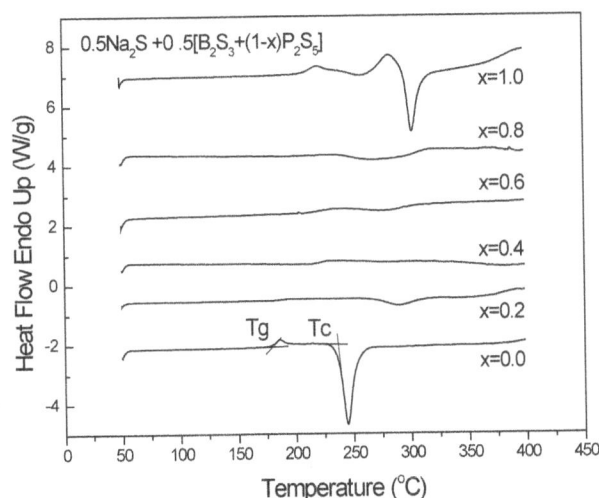

Figure 2. DSC scans for the $(Na_2S)_y.[(B_2S_3)_x.(P_2S_5)_{(1-x)}]_{(1-y)}$ series of glasses with y=0·5. The glass transition and onset crystallisation temperatures, T_g and T_c, for the different compositions were determined as shown in the DSC scan for the x=0 glass

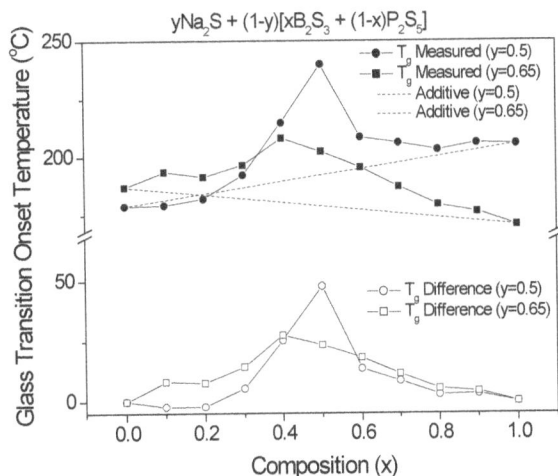

Figure 3. Composition dependence of T_g and its difference from simple additive behaviour for the y=0·5 and 0·65 series of glasses

copper plates and later crushed with a steel hammer to make powder specimens for DSC or annealed at a temperature 50°C below their T_g in a copper mould to create bulk samples for density measurements.

At this preliminary point, a full exploration of the glass forming region in this ternary system has not been completed. We have, rather, concentrated our efforts in examining the full composition range from one pure glass former, B_2S_3, to the other P_2S_5, at three levels of Na_2S content, y=0·35, 0·50, and 0·65, contents. In all three series, homogeneous, darkly coloured glasses can be prepared. We expect that glasses could be prepared at lower Na_2S contents (y<0·35), whereas we would expect significantly faster quenching would be required to yield glasses at higher Na_2S contents, (y>0·65). The glasses become less dark for higher Na_2S content, becoming a dark orange to yellow at the highest Na_2S contents.

DSC measurements were run on ~20 mg samples, hermetically sealed in Al sample pans, in a Perkin-Elmer Diamond DSC calorimeter, that had previously been calibrated on the indium melting point (±1°C). Four scans were run for each glass composition. First, a survey scan on the as-quenched glass was run at 20°C/min, heating the sample up through its T_g until it crystallised. This scan was used to determine T_c. A second sample was then heated at 20°C/min up to a temperature above T_g, but below T_c. The sample was then cooled at 20°C/min, and subsequently reheated two times to give two separate measurements of T_g for the glass under a -20/+20°C/min thermal history.

Density measurements were made on bulk plate quenched and annealed samples inside the glove box, using Archimedes' method with dried kerosene as the immersion fluid. The balance was calibrated to ±0·001 g, and the density of kerosene was determined using a density flask to ±0·01 g/ml. Repeated measurements

on three samples were used, and it is believed that the accuracy of the density measurements is ±0·01 g/ml.

3. Results

3.1 Glass transition and crystallisation temperatures

The survey glass transition scans are shown in Figure 2 for the y=0·5 series of glasses. On one of the scans, the onset T_g and T_c tangents are shown to indicate how these two quantities were determined. Since the sample mass varied slightly from composition to composition, the signal was divided by the sample mass so that the resulting signal was closer, but not exactly, to the heat capacity of the sample. As can be seen from the curves, the glass transition temperatures as well as the crystallisation temperatures show a marked increase and then decrease across the mixed glass former range, strongly indicative of an underlying glass structure MGFE. It is also significant to note that the crystallisation exotherms become more complicated in the mid-composition range, compared to those of the end member glasses, suggesting multiple crystallisation pathways, indicative of more complex underlying glass structures. At the very end of the thermal scan, in the range of 400 to 450°C, there is a sharp endotherm, suggestive of a melting transition. We believe, however, that because the onset temperature of this peak (not determined or reported) appears to be insensitive to glass composition, this peak is due to reaction of the now hot and molten sulphide with the inside surface of the Al sample pan. Lower temperature endothermic peaks, for the y=0·65 and x=1 and 0·8 compositions in particular, near 300°C we believe are due to melting of crystallised phases.

Figure 3 shows the onset glass transition temperatures and the difference glass transition temperatures (i.e. the difference between the value of T_g for the

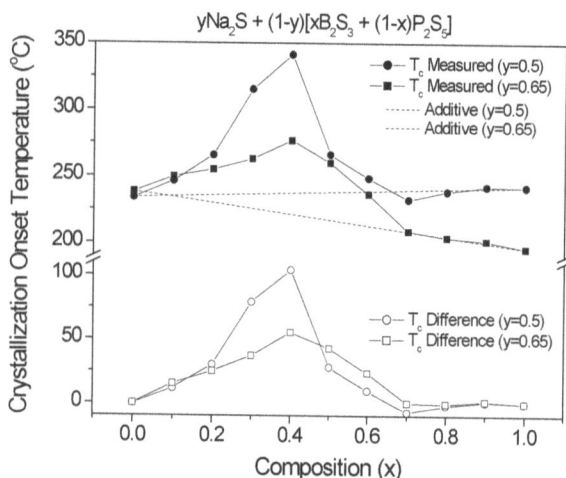

Figure 4. Composition dependence of T_c and its difference from simple additive behaviour for the $y=0.5$ and 0.65 series of glasses

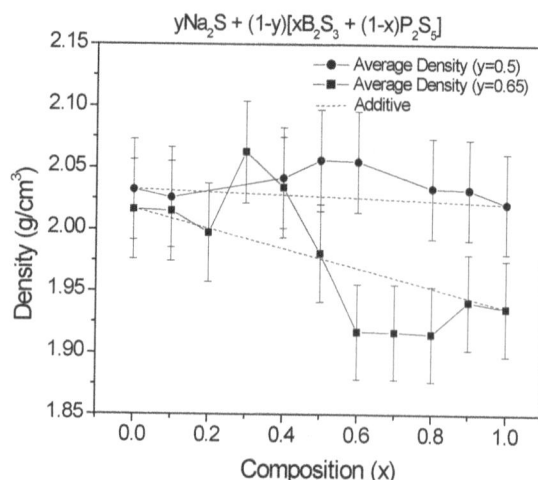

Figure 6. Composition dependence of the density for the $y=0.5$ and 0.65 series of glasses

MGFG and the linear interpolation of the values for the end member glasses) for the $y=0.5$ and $y=0.65$ series. The sharp and pronounced MGFE in both series of glasses is clearly evident. While the T_g maximum for the $y=0.5$ series of glass is centred near $x=0.5$, the T_g maximum for the $y=0.65$ series of glasses is centred near $x=0.4$. Interestingly, T_g for the end member NaPS$_3$ ($x=0$ and $y=0.5$) is slightly lower than T_g for NaBS$_2$ ($x=1$ and $y=0.5$), while T_g for the end member ~Na$_4$P$_2$S$_7$ ($x=0$ and $y=0.65$) is slightly higher than that of the end member ~Na$_4$B$_2$S$_5$ ($x=1$ and $y=0.65$).

Likewise, Figure 4 shows the onset crystallisation temperatures for the $y=0.5$ and $y=0.65$ series of glasses. Also shown at the bottom of the figures is the difference crystallisation temperature (i.e. the difference between the value of T_c for the MGFG and the linear interpolation of the values for the end member glasses). The increased glass stability, as evidenced by the increased crystallisation temperature of the MGFG, is clearly apparent. In both cases for these

glasses, the maximum in the crystallisation temperature is observed near the $x=0.4$ composition, i.e. for glasses richer in the P$_2$S$_5$ glass former.

Finally, Figure 5 shows the glass stability index, $(T_c-T_g)/T_g$, for the $y=0.5$ and $y=0.65$ series of glasses. While the trend is less systematic than either the trend for T_g or T_c alone, the clear increase in the thermal stability of the MGFG over that of the end member glasses is clearly apparent. Indeed, for the $y=0.5$ series, this stability increases nearly 60% above the end member NaPS$_3$, and to more than 70% above the end member NaBS$_2$. The increases in glass stability for the $y=0.65$ series of glasses are significantly less than the increase in glass stability for the $y=0.5$ series, and this is a signature of the overall lower glass stability of these series of glasses, due the higher fraction of alkali modifier in the glasses.

3.2 Density and molar volumes of the glasses

Figure 6 shows the densities of the $y=0.5$ and 0.65 series of glasses, respectively. While Figure 6 shows that there is a relatively weak MGFE in the density of the $y=0.5$ series of glasses, it also shows that the MGFE is present in the density of the $y=0.65$ series of glasses. Indeed, within experimental error, there may not even be a MGFE in the density of the $y=0.5$ series of glasses, and at best the MGFE is a weak phenomena in the density of the $y=0.65$ series of glasses. Figure 7 shows the calculated molar volumes for the $y=0.5$ and $y=0.65$ series of glasses. The molar volume does not show any MGFE and, within experimental error, the molar volumes appear to follow simple ideal mixing behaviour.

4. Discussion

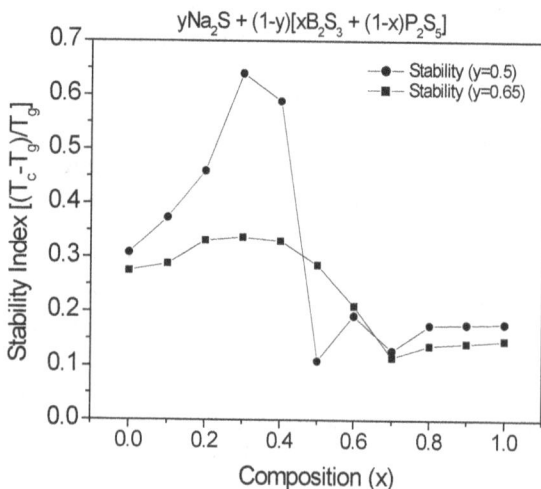

Figure 5. Composition dependence of the glass forming stability index, $(T_c-T_g)/T_g$, for the $y=0.5$ and $y=0.65$ series of glasses

From the data of the thermal and volume properties of the (Na$_2$S)$_y$·[(B$_2$S$_3$)$_x$·(P$_2$S$_5$)$_{(1-x)}$]$_{(1-y)}$ glasses, it is evident that while strong MGFEs are observed in the T_g and

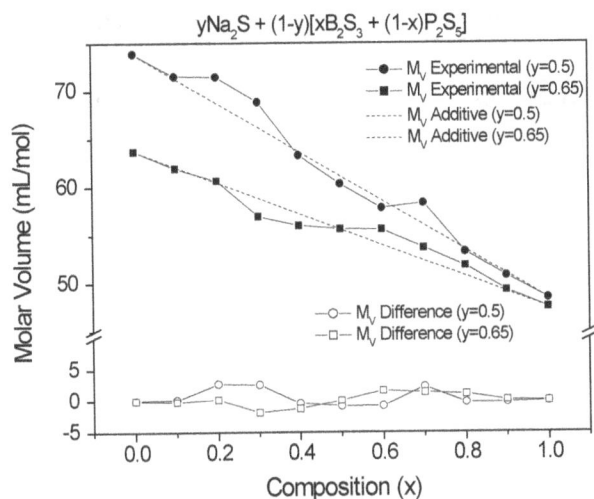

Figure 7. Composition dependence of the molar volume and the difference from simple linear mixing for the y=0·5 and 0·65 series of glasses

T_c data, there appears be no MGFE in the volume properties. This is an important finding in as much as T_g and T_c are most likely dominated by longer length scale structures, such as those beyond the first coordination shell, and are likely determined by the long range cohesiveness of the glassy network. The volume properties can be, on the other hand, dominated by the short range packing efficiency of the individual short range structures in the glass. So, while the behaviour of T_g is suggests that the mixing of the two thio-borate and thio-phosphate glass networks manifests itself in a more cohesive long range network, the density and molar volume compositional behaviour suggests that, whatever new MGF glass structures are forming, they apparently pack no more or no less efficiently than the parent end member binary thio-borate and thio-phosphate glass structures.

In our continuing work on the MGFE in these glasses, we will examine the exact nature of these new structures that form in the glasses using a variety of structurally sensitive techniques. IR and Raman vibrational spectroscopy will be used to characterise the variation of the short range structures of these glasses, while NMR, neutron diffraction, and x-ray diffraction, will be use to examine longer range correlations in the glass structure. We will also pursue both classical (such as reverse Monte Carlo, and molecular dynamics) and quantum (density functional theory) based simulation techniques.

5. Summary and conclusions

The thermal and volume properties of $(Na_2S)_y \cdot [(B_2S_3)_x \cdot (P_2S_5)_{(1-x)}]_{(1-y)}$ glasses have been measured at two values of y, 0·5 and 0·65, across the full range of x, $0 \leq x \leq 1$, to determine if the mixed glass former effect (MGFE) is observed in these two types of physical

properties of the glasses. Preliminary measurements show a strong MGFE in the Na^+ ion conductivity and in this work we have sought to determine if this behaviour is observed in other physical properties. While a strong and positive MGFE is observed in both the glass transition and onset crystallisation temperatures of the glasses, T_g and T_c, with the maximum in both T_g and T_c occurring in the mid-range of x, the density only showed a very weak MGFE and the molar volume showed no MGFE. These results suggest that the intermediate and long range structure of the glasses, which are more likely responsible for the compositional behaviour of T_g and T_c, are not simple mixtures of the intermediate and long range structures of the two end member parent sodium thio-borate and sodium thio-phosphate glasses. There was no evidence of the MGFE in the molar volume of the glasses, and hence the short range structures of the mixed glass former glasses, which may be more important for determining the volume properties of the glasses, appear to pack as efficiently as the short range structures of the two end member glass forming systems as there was no evidence of the MGFE in the molar volume of the glasses. Future work will be to determine the exact nature of the short and intermediate range structure of these glasses to further investigate these hypotheses.

Acknowledgements

This work was supported in part by a grant from the National Science Foundation, grant number DMR 0710564.

References

1. Kreidl, N. J. *Glass Science and Technology. I. Glass-Forming Systems.* Academic Press, New York, 1983.
2. Rawson, H. *Mater. Sci. Technol.,* 1991, **9**, 279.
3. Feller, S., Boekenhauer, R., Zhang, H., Bain, D., Feil, D., Parameswar, C., Budhwani, K., Ghosh, S. & Nijhawan, S. *Chim. Chron.,* 1994, **23**, 315.
4. Polyakova, I. G. & Tokareva, E. V. In: *Borate Glasses, Crystals and Melts,* Eds. A. C. Wright, S. A. Feller, & A. C. Hannon, The Society of Glass Technology, Sheffield, 1997, p. 223.
5. Pradel, A., Rau, C., Bittencourt, D., Armand, P., Philippot, E. & Ribes, M. *Chem. Mater.,* 1998, **10**, 2162.
6. Min, J. R., Wang, J., Chen, L. Q., Xue, R. J., Cui, W. Q. & Liang, J. K. *Phys. Status Solidi A,* 1995, **148**, 383.
7. Deshpande, V. K., Pradel, A. & Ribes, M. *Mater. Res. Bull.,* 1988, **23**, 379.
8. Pradel, A., Kuwata, N. & Ribes, M. *J. Phys.: Condens. Matter,* 2003, **15**, S1561.
9. Angell, C. A. *Solid State Ionics,* 1998, **105**, 15.
10. Rao, K. J. & Ganguli, M. In: *Handbook of Solid State Batteries & Capacitors,* Ed. M. Z. A. Munshi, World Scientific, Singapore, 1995, p. 189.
11. Burckhardt, W. & Rudolph, B. *Silikattechnik,* 1991, **42**, 275.
12. Martin, S. W. *J. Am. Ceram. Soc.,* 1991, **74**, 1767.
13. Christensen, R., Byer, J., Kaufmann, T. & Martin, S. W. *Phys. Chem. Glasses Eur. J. Glass Sci. Technol. B,* 2009, submitted. BORATE2008 proceedings – currently with author for minor modification.
14. Angell, C. A., Ngai, K. L., McKenna, G. B., McMillan, P. F. & Martin, S. W. *J. Appl. Phys.,* 2000, **88**, 3113.
15. Martin, S. W. *J. Am. Ceram. Soc.,* 1991, **74**, 1767.
16. Ingram, M. D. *Philos. Mag. B,* 1989, **60**, 729.
17. Tuller, H. L. *NATO ASI Ser., Ser. B,* 1989, **199**, 51.
18. Martin, S. W. & Angell, C. A. *J. Am. Ceram. Soc.,* 1984, **67**, C148.

Proc. VI Int. Conf. Borate Glasses, Himeji, Japan, 18–22 August 2008   *Phys. Chem. Glasses: Eur. J. Glass Sci. Technol. B, October 2009, 50 (5), 289–293*

Correlation between basicity and coordination structure in borosilicate glasses

Y. Tanaka, Y. Benino, T. Nanba[1]

Graduate School of Environmental Science, Okayama University, 3-1-1, Tsushima-Naka, Okayama 700-8530, Japan

S. Sakida

Environmental Management Center, Okayama University, 3-1-1, Tsushima-Naka, Okayama 700-8530, Japan

Y. Miura

University of Shiga Prefecture, 2500, Hassaka-Cho, Hikone-City, Shiga 522-8533, Japan

Manuscript received 19 August 2008
Revised version received 17 May 2009
Accepted 14 September 2009

Various quaternary borosilicate glasses in the general system $M_2O–M'O–Al_2O_3–B_2O_3–SiO_2$ (M=Li, Na or K; M'=Ca, Sr or Ba) were prepared, and the concentration of four-fold coordinated boron (B4) atoms was determined using ^{11}B magic angle spinning NMR. Regression analyses were performed to predict the B4 fraction, in which B4 fraction and glass basicity were used as dependent and independent variables, respectively. Except for Al_2O_3-containing glasses, regression formulae giving high correlation coefficients were successfully obtained by using B4/(B+Si×f) (where f is a tunable coefficient) as the dependent variable. As for the glasses containing Al_2O_3, however, the necessity of a different dependent variable was indicated.

The physical properties of glass change continuously with compositional change, and at the same time, the glass structure also changes, resulting in changes in properties. The additivity rule[1] is useful to predict physical properties of glasses. In a binary glass system, a property of a glass is estimated by a linear approximation connecting both ends of the glass composition. However, in some glasses, such as borate and germanate glasses, properties are not linearly related to composition. In binary alkali borate glasses, the coordination number of boron initially changes from 3 to 4 without forming nonbridging oxygen (NBO) as alkali is added, and then further alkali addition results in a decrease of the coordination number from 4 to 3, accompanied by NBO formation in trigonal BO_3 units. In multicomponent glasses containing B_2O_3 and GeO_2, prediction of glass structure and properties is quite difficult. Multiple regression analysis has been applied to multicomponent glasses to predict the glass properties.[2] As mentioned, the properties of a glass depend on its structure. Therefore, if the structure of a glass is predicted, accurate prediction of glass properties may be achieved, which will contribute to material design of glasses in practical use.

Changes in glass structure are associated with changes in chemical bonding character, that is, electronic states of the glass constituents. In the case of oxide glasses, the electronic states of oxide ions are predominant, because O2p is the main constituent of HOMO (highest occupied molecular orbital) levels.

The electronic population of oxide ions is associated with basicity, which is interpreted as an ability of electron donation of oxide ions.[3] It is therefore assumed that glass structure is influenced by basicity. Miura *et al*[4] discussed the structural changes in borosilicate glasses based on a basicity equalisation concept:[3] The borate groups of $BO_{4/2}$ and $BO_{2/2+1}$ units have the same group basicity, λ, of 0·60. As for (Si,B)–O–(Si,B) bridges, λ of bridging oxygens (BOs) is 0·42–0·57, and that of NBOs is 0·79–0·86. According to the basicity equalisation concept,[3] a structural unit having a group basicity λ which is close to a glass basicity Λ is preferentially formed in the glass. In acidic glasses at $\Lambda<0·5$, $BO_{4/2}$ units associated with BOs having small λs are stably present, and $BO_{2/2+1}$ units including NBO with large λ are not produced. In basic glasses at $\Lambda>0·6$, however, $BO_{2/2+1}$ units are stable, and unstable $BO_{4/2}$ units decrease in concentration with increasing Λ.

A number of numerical expressions for basicity have been proposed. Nanba *et al*[5] concluded that the numerical expression given by Duffy & Ingram[3] was best suited for reproducing the chemical shift of O1s binding energy of oxide glasses. Tateyama *et al*[6] have found good correlations between optical property and the basicity calculated from the equation given by Duffy & Ingram. In our previous study,[7] the correlation between glass structure and basicity was investigated in ternary borosilicate glasses in the general system $M_2O(M'O)–B_2O_3–SiO_2$ (M=Li, Na, K; M'=Ca, Sr, Ba), in which the basicity equation given by Duffy & Ingram was also used. As a result, the fraction of NBOs determined from

---

[1] Corresponding author. Email tokuro_n@cc.okayama-u.ac.jp

x-ray photoelectron spectroscopy was successfully expressed by a linear function of basicity, and also for the fraction of four-fold coordinated boron (B4) atoms, regression formulae giving high correlation coefficients were successfully obtained.

In the present study, quaternary borosilicate glasses were prepared, and the B4 fraction was determined by using ^{11}B MAS (magic angle spinning) NMR spectrometry. Regression analyses were performed to predict the B4 fraction, in which the B4 fraction and glass basicity calculated from the equation given by Duffy & Ingram[3] were used as dependent and independent variables, respectively. The third and fourth components added to B_2O_3 and SiO_2 were chosen from Al_2O_3, Na_2O, CaO, and BaO, and in total five kinds of quaternary borosilicate glasses were studied by this analysis.

Experimental

The glass compositions investigated are given in Table 1. All the glasses were prepared by a conventional melt quenching method. Batches of 10 g were melted for more than 30 min at 1100–1400°C in a Pt crucible with an alumina lid. Post annealing heat treatment was not done in order to avoid phase separation. Inductively coupled plasma analysis was performed for some of the prepared glasses, and the compositional deviation of the cations was less than 5 at%. Therefore, the nominal composition was used in the basicity calculation.

NMR measurements were carried out at 7·05 T on a Varian Unity Inova 300 spectrometer. ^{11}B MAS

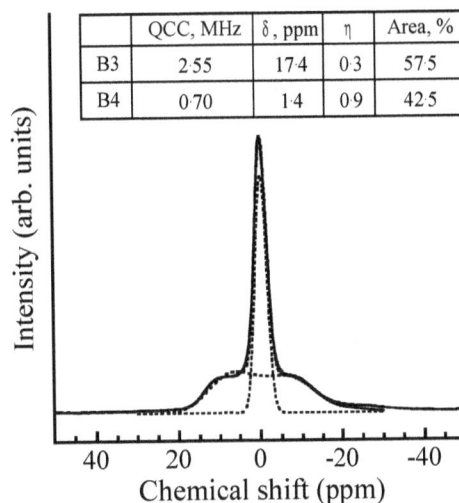

	QCC, MHz	δ, ppm	η	Area, %
B3	2·55	17·4	0·3	57·5
B4	0·70	1·4	0·9	42·5

Figure 1. An example of spectral deconvolution of ^{11}B MAS NMR spectrum for $0·75(0·5Na_2O.0·5CaO).B_2O_3.SiO_2$ glass (R=0·75, x=0·50, K=1·0). The simulated components and total curve are shown by the dotted lines. The fitting parameters are also shown (QCC: quadrupolar coupling constant; δ: isotropic chemical shift; η: asymmetry parameter)

NMR spectra were collected at 96·251 MHz with 0·6 μs pulses and 1·0 s recycle delays under the sample spinning speed of 6 kHz at the magic angle to the external field. BPO_4 was used as an external standard for ^{11}B MAS NMR, and the chemical shift of BPO_4 was −3·3 ppm from $BF_3(C_2H_5)_2O$ which was used as a standard reference.

The basicity, Λ, was calculated from the following equations[3]

$$\Lambda = \sum_i \frac{z_i \, r_i}{2 \, \gamma_i} \tag{1}$$

$$\gamma_i = 1·36(\chi_i - 0·26) \tag{2}$$

where z_i is the oxidation number of the cation i, and r_i is the ionic ratio with respect to the total number of oxides. γ_i is the basicity moderating parameter, which is given by the Pauling electronegativity χ_i.

Results

Compositional dependence of the B4 fraction

Figure 1 shows a typical ^{11}B MAS NMR spectrum for a quaternary borosilicate glass, where the symmetric sharp peak at around 0 ppm is attributed to B4 (i.e. it is attributed to BO_4 units), and the asymmetric broad peak at about 20 to −20 ppm is assigned to B3 (i.e. it is attributed to BO_3 units). Spinning side bands were not observed in this region. The spectra obtained were deconvoluted into the structural components for B3 and B4 species, obtaining the fraction of B4, ($N_4 \equiv B4/(B3+B4)$ in

Table 1. Compositions of the glasses prepared in this study

Glass system		Composition (molar ratio)
$R(xNa_2O.(1-x)CaO).B_2O_3.0SiO_2$	x=0·25	R=0·2, 0·3, 0·4, 0·5, 0·6, 0·7
(K=0)	x=0·50	R=0·1, 0·2, 0·3, 0·4, 0·5, 0·6, 0·7
	x=0·75	R=0·1, 0·2, 0·3, 0·4, 0·5, 0·6, 0·7
$R(xNa_2O.(1-x)CaO).B_2O_3.1SiO_2$	x=0·25	R=0·25, 0·5, 0·75, 1·0, 1·25, 1·5, 1·75
(K=1)	x=0·50	R=0·25, 0·5, 0·75, 1·0, 1·3, 1·6, 2·0
	x=0·75	R=0·25, 0·5, 0·75, 1·0, 1·3, 1·6, 2·0
$R(xNa_2O.(1-x)CaO).B_2O_3.3SiO_2$	x=0·25	R=1·5, 2·0, 2·5, 3·0, 3·5
(K=3)	x=0·50	R=0·5, 1·0, 1·5, 2·0, 2·5, 3·0, 3·5
	x=0·75	R=0·5, 1·0, 1·5, 2·0, 2·5, 3·0, 3·5
$R(xNa_2O.(1-x)BaO).B_2O_3.0SiO_2$	x=0·25	R=0·1, 0·25, 0·4, 0·55, 0·7, 0·85
(K=0)	x=0·50	R=0·1, 0·2, 0·3, 0·4, 0·5, 0·6, 0·7
	x=0·75	R=0·1, 0·2, 0·3, 0·4, 0·5, 0·6, 0·7
$R(xNa_2O.(1-x)BaO).B_2O_3.1SiO_2$	x=0·25	R=0·25, 0·5, 0·75, 1·0, 1·25, 1·6, 2·0
(K=1)	x=0·50	R=0·25, 0·5, 0·75, 1·0, 1·3, 1·6, 2·0
	x=0·75	R=0·25, 0·5, 0·75, 1·0, 1·3, 1·6, 2·0
$R(xNa_2O.(1-x)BaO).B_2O_3.3SiO_2$	x=0·25	R=1·0, 1·5, 2·0, 2·5, 3·0, 3·5
(K=3)	x=0·50	R=0·5, 1·0, 1·5, 2·0, 2·5, 3·0, 3·5
	x=0·75	R=0·5, 1·0, 1·5, 2·0, 2·5, 3·0, 3·5
$R(xCaO.(1-x)BaO).B_2O_3.0SiO_2$	x=0·25	R=0·3, 0·4, 0·5, 0·6, 0·7
(K=0)	x=0·50	R=0·3, 0·4, 0·5, 0·6, 0·7
	x=0·75	R=0·3, 0·4, 0·5, 0·6, 0·7
$R(xCaO.(1-x)BaO).B_2O_3.1SiO_2$	x=0·25	R=1·0, 1·3, 1·6, 2·0
(K=1)	x=0·50	R=1·0, 1·3, 1·6, 2·0
	x=0·75	R=1·3, 1·6, 2·0
$R(xCaO.(1-x)BaO).B_2O_3.3SiO_2$	x=0·25	R=2·0, 2·5, 3·0, 3·5
(K=3)	x=0·50	R=2·0, 2·5, 3·0, 3·5
	x=0·75	R=2·5, 3·0, 3·5
$RNa_2O.(yB_2O_3.(1-y)Al_2O_3).1SiO_2$	R=0·70	y=0·85, 0·70, 0·55, 0·40, 0·25
$RCaO.(yB_2O_3.(1-y)Al_2O_3).1SiO_2$	R=0·70	y=0·70, 0·55, 0·40, 0·25

the conventional notation). The N_4 values estimated in this way have margins of error less than ±0·03.

The fraction of B4 in the Na_2O–CaO–B_2O_3–SiO_2 system is shown in Figure 2. In the glasses with the same network modifier (NWM) content, $R=(M_2O+M'O)/B_2O_3$, the N_4 value increases with increasing SiO_2 content, $K=SiO_2/B_2O_3$, and the N_4 value also depends on the NWM mixing ratio, $x=M_2O/(M_2O+M'O)$. As for the series with constant K and x, the addition of NWMs (i.e. an increase in R), results in a transformation from B3 to B4 in the lower R region. With increasing R, the B4 fraction reaches a maximum and then decreases, due to the formation of nonbridging oxygen (NBO) associated with Si and B3. Moreover, the NWM content R at which N_4 shows a maximum is different depending on the SiO_2 content K and NWM mixing ratio, x, indicating that SiO_2 content and NWM species influence the N_4 value.

Regression analysis of the B4 fraction

It is obvious from Figure 2 that the B4 fraction depends on the SiO_2 content, K. In our previous study,[7] the concentration of B4 was expressed as B4/(B+Si×f), and, after optimising the coefficient f, correlations independent of SiO_2 content were successfully obtained. The conventional notation of B4 fraction, N_4=B4/(B3+B4) corresponds to the case with the coefficient f=0, and, in the case with f=1, the denominator, (B+Si×f), means the total amount of network formers (NWFs). In Na_2O or CaO ternary borosilicate glasses,[7] the optimal coefficients f were obtained as 0·226 for Na_2O and 0·399 for CaO, indicating that a fraction of the Si atoms contribute to the formation of tetrahedral BO_4 units.

Then, the B4 concentration expressed with the coefficient f was plotted against glass basicity, and the correlation between the B4 concentration and basicity was approximated by a biquadratic polynomial. Furthermore, the coefficient f was optimised by the quasi-Newton method so as to obtain the largest correlation coefficient, $|R|$. Figure 3 shows the results for Na_2O–CaO–B_2O_3–SiO_2 glasses. Low dispersion in the

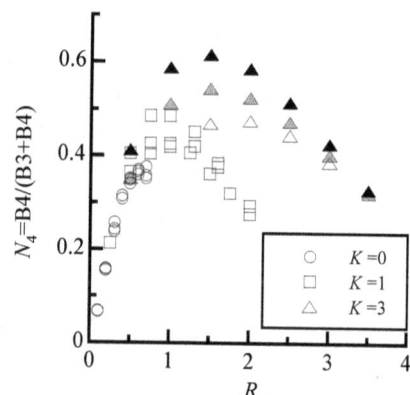

Figure 2. The fraction of four-fold coordinated boron atoms to total boron atoms, N_4 in $R(xNa_2O.(1-x)CaO).B_2O_3.KSiO_2$ glasses. The margins of error in N_4 are less than ±0·03. For K=3, the glasses with x=0·25, 0·50 and 0·75 are indicated by open, gray-coloured and solid markers, respectively

B4 fraction is successfully achieved by optimising the coefficient f. It is hence concluded that the notation for B4 concentration expressed using the coefficient f, B4/(B+Si×f), can be used as an dependent variable in the present regression analyses for the quaternary borosilicate glasses. The results of regression analyses for other quaternary borosilicate systems are shown in Figures 4 and 5, for Na_2O–BaO–B_2O_3–SiO_2 and CaO–BaO–B_2O_3–SiO_2 glasses, respectively, in which good correlations are also confirmed after optimising the coefficients, f.

Discussion

Estimation of coefficient f in quaternary borosilicate glasses

If the coefficients, f, for multicomponent glass systems were predictable without NMR measurements, it would greatly contribute to the prediction of glass structure. As shown in Figure 3, the optimal coefficient f for the quaternary Na_2O–CaO–B_2O_3–SiO_2 glass is 0·297, which is intermediate between the coefficients f for ternary Na_2O (0·226) and CaO (0·399) borosilicate glasses. Then, the coefficients f for the quaternary glasses were calculated from the follow-

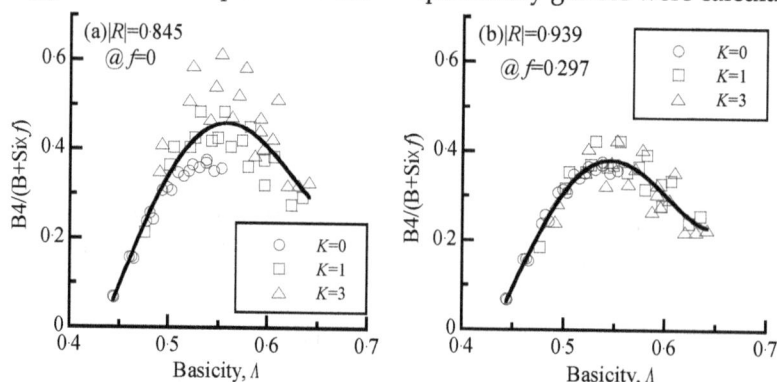

Figure 3. The fraction of four-fold coordinated boron atoms given by B4/(B+Si×f) for the Na_2O–CaO–B_2O_3–SiO_2 system, for f=0 and f=0·297. $|R|$ is the correlation coefficient for the relation between the B4 fraction and basicity, Λ, shown by the lines: (a) f=0, y=837·40Λ^4−1808·5Λ^3+1428·5Λ^2−486·82Λ+60·361; (b) f=0·297, y=996·17Λ^4−2096·7Λ^3+1619·8Λ^2−542·21Λ+66·298

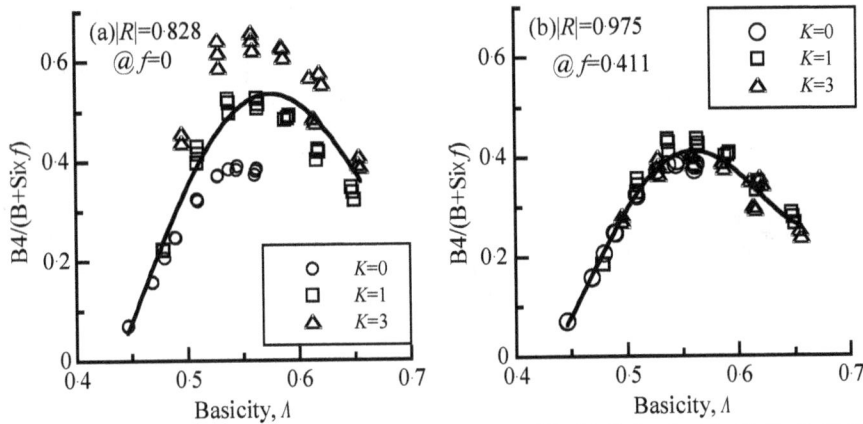

Figure 4. The fraction of four-fold coordinated boron atoms given by B4/(B+Si×f) for the Na_2O–BaO–B_2O_3–SiO_2 system, for f=0 and f=0·411. |R| is the correlation coefficient for the relation between the B4 fraction and basicity, Λ, shown by the lines: (a) f=0, y=493·93Λ^4–1113·4Λ^3+906·94Λ^2–314·21Λ+39·016; (b) f=0·411, y=965·78Λ^4–2098·8Λ^3+1677·1Λ^2–582·13Λ+74·084

ing equation

$$f(mix)=xf(Na_2O)+(1-x)f(CaO) \qquad (3)$$

where $f(Na_2O)$ and $f(CaO)$ are the coefficients obtained for the ternary borosilicate glasses. As shown in Figure 6, high correlation coefficients, |R|, are successfully obtained in the quaternary glasses. Compared to the |R| values shown in Figures 3 to 5, the |R| values shown in Figure 6 are slightly reduced, but the difference in |R| is quite small. For example, the |R| value changes from 0·939 to 0·914 in Na_2O–CaO–B_2O_3–SiO_2 glass. It is therefore concluded that the f values obtained from the ternary borosilicate glasses are applicable to the estimation of f values for multicomponent borosilicate glasses.

Then, is it possible to predict the coefficient f for alkali or alkaline earth ternary borosilicate glasses? The character of alkali and alkaline earth ions is often associated with field strength Z/r, where ionic charge Z is divided by ionic radius r. Hence, the correlation between the optimal coefficient f for ternary borosilicate glasses and field strength Z/r was examined. As shown in Figure 7, the coefficient f decreases with increasing Z/r, in both the alkali and alkaline earth series. However, the coefficients f for the alkaline

earth ions are much higher than those for the alkali ions, and a universal relationship is not observed. The chemical or physical meaning of the coefficient f is still obscure.

Dependence of the B4 fraction on basicity in aluminoborosilicate glasses

The correlation between B4 fraction and basicity, Λ, for Na_2O–Al_2O_3–B_2O_3–SiO_2 and CaO–Al_2O_3–B_2O_3–SiO_2 glasses is shown in Figure 8. In these glasses, NWM and SiO_2 contents are constant at R=0·7 and K=1×0, and B_2O_3 is substituted for Al_2O_3. Basicity increases with Al_2O_3 substitution, resulting in a decrease in N_4. As shown in Figures 3 to 6 for alkali and alkaline earth quaternary borosilicate glasses, the maximum B4 fractions are observed at around Λ=0·55. It is hence concluded that the change in coordination structure of boron atoms in aluminoborosilicate glasses is quite different from alkali and alkaline earth borosilicate glasses. In general, Al atoms take coordination numbers of 4 to 6 in glasses, but in the present glasses Al atoms are only in AlO_4 units, which is confirmed from ^{27}Al MAS NMR. As mentioned, the term (B+Si×f) in B4/(B+Si×f) represents the total amount of NWFs, and

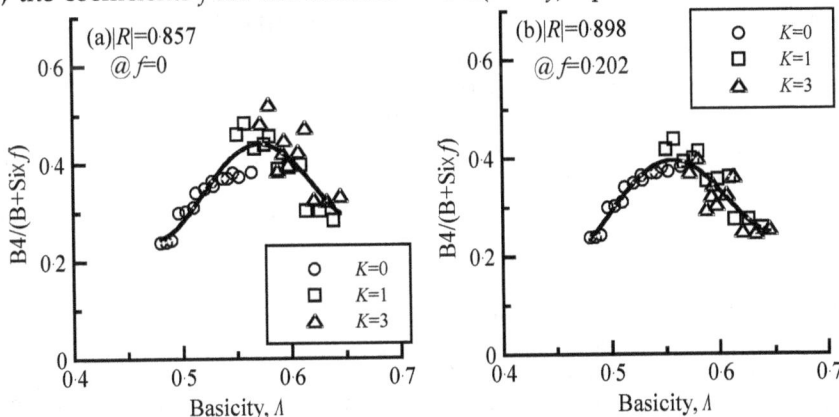

Figure 5. The fraction of four-fold coordinated boron atoms given by B4/(B+Si×f) for the CaO–BaO–B_2O_3–SiO_2 system, for f=0 and f=0·202. |R| is the correlation coefficient for the relation between the B4 fraction and basicity, Λ, shown by the lines: (a) f=0, y=2697·5Λ^4–6140·6Λ^3+5198·6Λ^2–1939·3Λ+269·26; (b) f=0·202, y=2112·9Λ^4–4678·5Λ^3+3847·2Λ^2–1391·8Λ+187·14

(a) Na_2O–CaO–B_2O_3–SiO_2 (b) Na_2O–BaO–B_2O_3–SiO_2 (c) CaO–BaO–B_2O_3–SiO_2

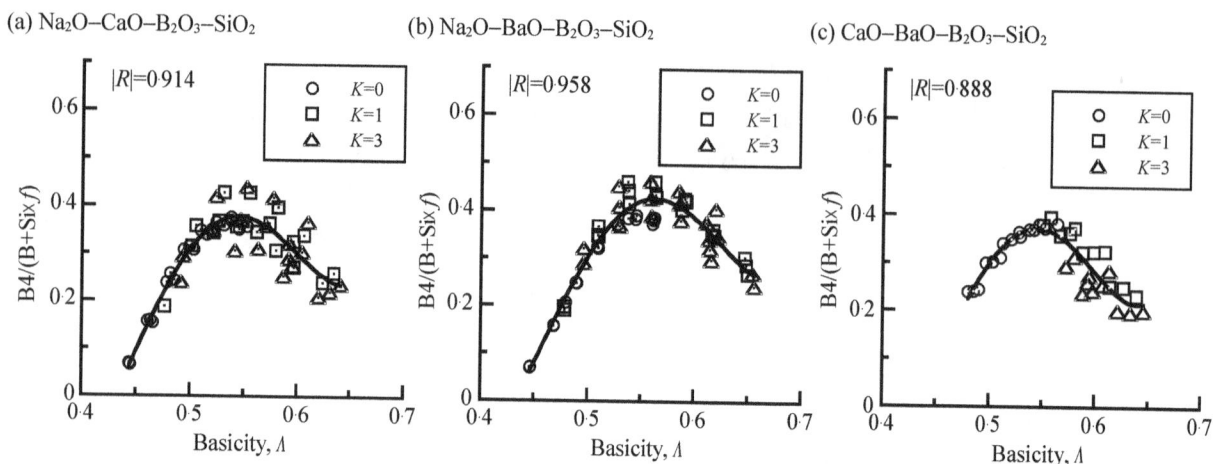

Figure 6. The fraction of four-fold coordinated boron atoms given by B4/(B+Si×f) in quaternary borosilicate glasses, where the coefficients f were obtained from Equation (3). |R| is the correlation coefficient for the relation between the B4 fraction and basicity, Λ, shown by the lines.
(a) $y=1014\cdot8\Lambda^4-2128\cdot0\Lambda^3+1638\cdot1\Lambda^2-546\cdot37\Lambda+66\cdot555$; (b) $y=827\cdot66\Lambda^4-1802\cdot0\Lambda^3+1438\cdot5\Lambda^2-497\cdot06\Lambda+62\cdot727$
(c) $y=1516\cdot0\Lambda^4-3231\cdot9\Lambda^3+2544\cdot9\Lambda^2-875\cdot73\Lambda+111\cdot19$

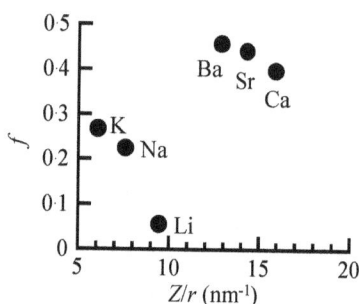

Figure 7. Field strength, Z/r and the optimal coefficient f in B4/(B+Si×f) for the ternary borosilicate glasses[7] (Z and r indicate ionic charge and radius, respectively)

hence B4/(B+Si×f+Al×g) is expected as the dependent variable in the present aluminoborosilicate glasses. However, a satisfactory result was not obtained, and hence a different expression for the B4 fraction is required in the case of aluminoborosilicate glasses.

Conclusions

Quaternary borosilicate glasses containing Na_2O, CaO, BaO and Al_2O_3 were prepared, and the concentration of four-fold coordinated boron (B4) atoms was determined by using ^{11}B MAS NMR. Regression analyses were performed to predict the B4 fraction, in which the B4 fraction and glass basicity were used as the dependent and independent variables, respectively. The B4 fraction, expressed as B4/(B+Si×f), was used as the dependent variable, obtaining high correlation coefficients after optimisation of the coefficients, f, in the quaternary glass systems containing Na_2O, CaO and BaO. High correlation coefficients were also obtained by averaging the coefficients f for ternary borosilicate glasses. As for the glasses containing Al_2O_3, however, the necessity of a different dependent variable was indicated.

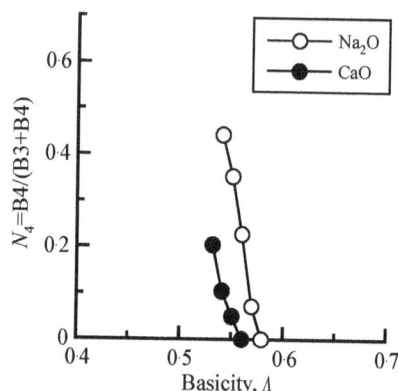

Figure 8. Fraction of four-fold coordinated boron atoms to total boron atoms, N_4 in the Na_2O–Al_2O_3–B_2O_3–SiO_2 and CaO–Al_2O_3–B_2O_3–SiO_2 systems

Acknowledgments

This work was financially supported by the project 'Research and Development to Promote the Creation and Utilization of an Intellectual Infrastructure/ Research and development for glass structure database construction FY2005 and FY2006' of the New Energy and Industrial Technology Development Organization (NEDO), Japan.

References

1. Appen, A. A. Chemie des Glases, Vol. 2, Verlag Chemie, Leningrad, 1974.
2. Utsuno, F., Inoue, H., Yasui, I., Tsuboi, S. & Iseda, T. Proceedings of the XX International Congress on Glass, 2004, Kyoto, P-07-028.
3. Duffy, J. A., Ingram, M. D. J. Non-Cryst. Solids, 1976, 21, 373.
4. Miura, Y., Kusano, H., Nanba, T. & Matsumoto, S. J. Non-Cryst. Solids, 2001, 290, 1.
5. Nanba, T., Miura, Y. & Sakida, S. J. Ceram. Soc. Jpn, 2005, 113, 44.
6. Tateyama, Y., Sakida, S., Nanba, T. & Miura, Y. In: Proceedings of Materials Science & Technology 2006, Materials and Systems, Vol. 1, MS&T Partner Societies, 2006, pp 555–565 (CD-ROM).
7. Nanba, T., Sakida, S. & Miura, Y. In: Proceedings of Materials Science & Technology 2006, Materials and Systems, Vol. 1, MS&T Partner Societies, 2006, pp 535–544 (CD-ROM).

Proc. VI Int. Conf. Borate Glasses, Himeji, Japan, 18–22 August 2008   *Phys. Chem. Glasses: Eur. J. Glass Sci. Technol. B*, April 2009, **50** (2), 95–97

Vibrational and elastic properties of potassium borate glasses

Mitsuru Kawashima,[1] *Yu Matsuda,*[1] *Yasuteru Fukawa,*[1] *Masao Kodama*[2] *& Seiji Kojima*[1]

[1] *Graduate School of Pure and Applied Sciences, University of Tsukuba, Tsukuba, Ibaraki 305-8573, Japan*

[2] *Department of Applied Chemistry, Sojo University, Kumamoto 860-0082, Japan*

Manuscript received 18 August 2008
Revised version received 24 October 2008
Accepted 23 December 2008

Potassium borate glasses, $xK_2O.(100-x)B_2O_3$ ($2 \leq x \leq 34$), have been investigated by Raman spectroscopy. We have analysed the three vibrational peaks at 805, 775 and 750 cm^{-1} in order to study the changes of structural groups. This shows that for low K_2O content, one 3-fold coordinated boron in each boroxol ring transforms into a 4-fold coordinated boron. Then from x=20 mol% K_2O, a second 3-fold coordinated boron in the ring transforms into a 4-fold coordinated boron. We have also investigated the low frequency boson peak by Raman scattering. The observed frequency of the boson peak increases as the K_2O content increases. From Raman scattering and sound velocity data, it is concluded that potassium borate glasses show a linear correlation between the boson peak frequency and the sound velocity.

1. Introduction

The structure and physical properties of potassium borate glasses, $xK_2O.(100-x)B_2O_3$, are of great interest, since their physical properties vary markedly with the addition of potassium oxide. For example, anomalous composition dependences of density, sound velocity and thermal expansion have been observed.[1,2] Such behaviour is called the "borate anomaly" and has been discussed on the basis of the variation of the intermediate range structural units.[3,4]

The structure of borate glasses involves a covalently bonded network. Pure B_2O_3 is a random three-dimensional network of BO_3 triangles with a large fraction of almost planar B_3O_6 boroxol rings.[5] As the alkali concentration increases, the network structure is modified in such a way that the fraction of boroxol rings decreases and there appear other structural units such as pentaborate, triborate, diborate, and metaborate groups. These structural groups have been studied extensively by Raman scattering,[3,4] but a sufficiently detailed understanding of the structural properties as a function composition is not yet available.

The boson peak (BP) is a broad band observed in the Raman spectrum of glasses at low frequency, and the origin of BP is still the subject of much debate. Various theoretical approaches have been proposed to explain the BP. In our previous paper,[6] the BP frequency of lithium borate glass was investigated. The results suggested that the BP is related to collective motions of transverse acoustic phonons within the intermediate range structure order of the glasses. However, the previous studies of the elastic and vibrational properties of alkali borate glass are

insufficient for a discussion of their correlation. The purpose of this paper is to investigate the structural units assigned to the Raman spectra in the region 750–810 cm^{-1} and the relation between the BP and sound velocity of potassium borate glass.

2. Experimental

The $xK_2O.(100-x)B_2O_3$ ($2 \leq x \leq 34$, every 2 mol%) glasses were prepared by the "solution method".[7] This method is very useful for obtaining very homogenous samples, which is necessary to investigate the inherent nature of the binary system. Analytical reagent grade KOH and H_3BO_3, were used as the starting materials, and were reacted in an aqueous solution to achieve high homogeneity. Then the mixed solution was transferred to a dry box and after the complete evaporation of water a chemically synthesised powder was obtained. The powder was melted in a furnace at temperatures between 900 and 1300°C. The melt was poured into a cylindrical graphite mould held at the glass transition temperature for 2 h, and then cooled at a rate of 1 K/min down to room temperature.[8] The glass obtained is free from strain and bubbles, which disturb the measurement. Potassium borate glass is hygroscopic; therefore the samples were immersed in silicone oil to prevent them from absorbing moisture from the air.

For the Raman measurements, light from a green YAG laser (532 nm) was incident on the sample. The scattered light was analysed by an additive dispersion triple monochromator (Jobin Yvon, T6400). All spectra were measured at room temperature with a resolution of 3 cm^{-1}. The output signal from the spectrometer was detected by a photon counting system.

*Corresponding author. Email m_kawa326@yahoo.co.jp

Proceedings of the VI International Conference on Borate Glasses, Himeji, Japan, 18–22 August 2008

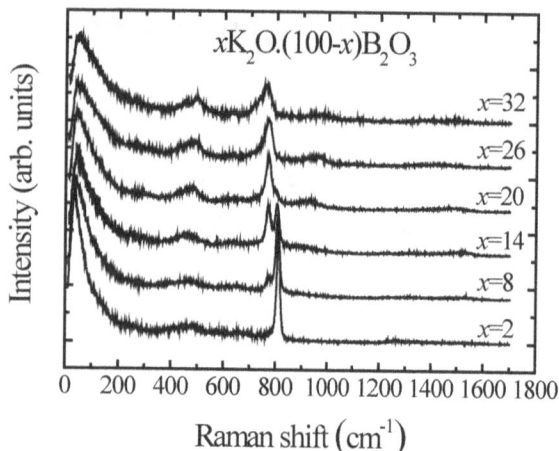

Figure 1. Raman spectra of selected $xK_2O.(100-x)B_2O_3$ glasses (x=2, 8, 14, 20, 26, 32 mol% K_2O)

3. Results

The measured Raman spectra of $xK_2O.(100-x)B_2O_3$ (x=2, 8, 14, 20, 26, 32) are shown in Figure 1. These spectra were normalised by

$$I_{normal}(\omega) = \frac{I(\omega)}{\int I d\omega} \tag{1}$$

where I_{normal} denotes the normalised intensity of the Raman spectrum and ω is the Raman frequency.

In order to discuss those composition dependences, it is important to consider the intermediate structural units present in the borate network. Therefore, the Raman spectrum in the region between 750 cm^{-1} and 810 cm^{-1} was deconvoluted into three Gaussian bands (denoted as 1, 2 and 3, respectively), and the composition dependence of the relative intensities of these three bands was studied. Figure 2 shows an example of a deconvoluted Raman spectrum. This is for x=16 mol% K_2O, and for this sample there are only two observable peaks, 1 and 2, in the region of interest. The three deconvoluted peaks frequencies are respectively 805, 775 and 750 cm^{-1}, and the areas of these peaks are denoted as A_1, A_2 and A_3. The peak

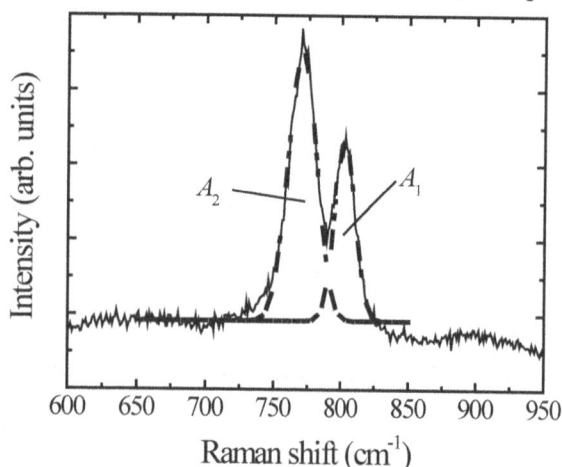

Figure 2. Deconvoluted bands of the Raman spectrum of $16K_2O.84B_2O_3$ glass

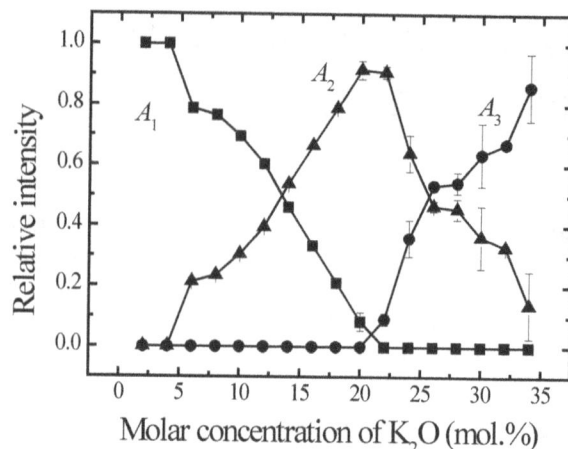

Figure 3. Relative band intensity ratio $A_i/(A_1+A_2+A_3)$ (i=1, 2, 3). The square, triangle and circle symbols denote the ratio for i=1, 2 and 3, respectively

area ratio $A_i/(A_1+A_2+A_3)$ (i=1, 2, 3) is shown in Figure 3.

Figure 4 shows the composition dependence of the BP of $xK_2O.(100-x)B_2O_3$ (x=2, 8, 14, 20, 26, 32 mol% K_2O). The frequency of the BP increases when the amount of network modifier, K_2O, increases. In addition, the longitudinal sound velocity (V_L) and the transverse sound velocity (V_T) were previously measured by the pulse echo overlap method (Figure 5).[8] Figure 6 shows the BP frequency versus the transverse sound velocity for potassium borate glasses.

4. Discussion

In the spectra of Figure 1, marked changes occur in the region between 750 cm^{-1} and 810 cm^{-1}. The main band for low K_2O content is around 805 cm^{-1}. For pure B_2O_3 glass, the band at 805 cm^{-1} is assigned to the breathing mode of the boroxol ring.[5] For low K_2O content, the intensity of the band at 775 cm^{-1} increases as the intensity of the band at 805 cm^{-1} decreases. The band at 775 cm^{-1} is assigned to the breathing mode of six membered rings with one BO$_4$ tetrahedron

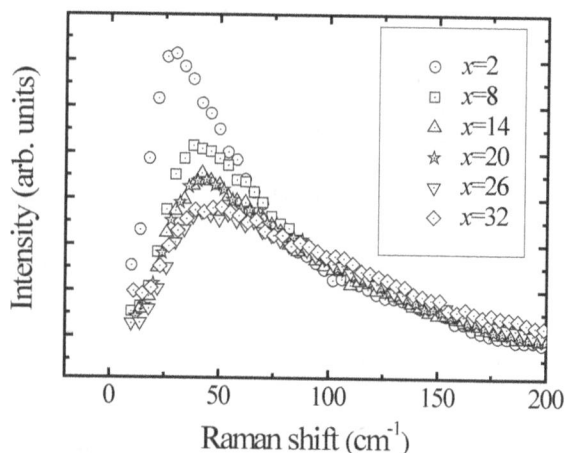

Figure 4. Raman spectra of selected $xK_2O.(100-x)B_2O_3$ glasses in the region of the boson peak

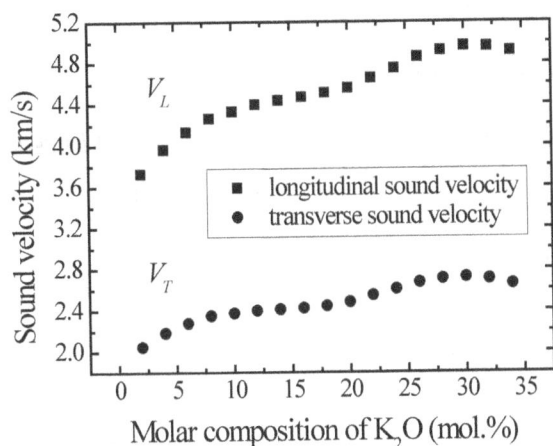

Figure 5. The longitudinal sound velocity (V_L) and the transverse sound velocity (V_T) of $xK_2O.(100-x)B_2O_3$ glasses. The square and circle symbols denote V_L and V_T, respectively

(i.e. pentaborate, tetraborate, and triborate rings).[3,4] Whereas the boroxol ring has three 3-fold coordinated borons, these six membered rings have one 4-fold coordinated boron. For $x \geq 22$ mol% K_2O, the intensity of the 750 cm^{-1} band increases while the intensity of the 775 cm^{-1} band decreases. The 750 cm^{-1} band is assigned to the breathing mode of six membered rings with two BO_4 tetrahedra (i.e. ditriborate and dipentaborate rings).[3,4] These six membered rings have two 4-fold coordinated borons.

Figure 3 clearly indicates the variation of these three peaks as a function of composition. For low K_2O content, peak 1 is dominant. However, peak 1 decreases as the K_2O content increases, and disappears at x=20 mol% K_2O. On the other hand, peak 2 grows up to x=20 mol% K_2O. Thus, it is shown that for low K_2O content, one 3-fold coordinated boron in each boroxol ring transforms into a 4-fold coordinated boron. The intensity of peak 2 has a maximum at x=20 mol% K_2O. Then it decreases, while at the same time peak 3 first appears at x=20 mol% K_2O, and then grows. Thus from x=20 mol% K_2O, a second 3-fold coordinated boron in the ring transforms into a 4-fold coordinated boron.

As shown in Figure 4, there is a general increase in the frequency of the maximum of the BP, up to 48 cm^{-1}, as the K_2O content increases. However, this increase is not monotonic with the increase of the K_2O content. In addition, the sound velocities (Figure 5) have a maximum at around x=30 mol% K_2O and do not change in a monotonic way with the K_2O content. Therefore, in a previous paper,[8] Kodama proposed that three structural units are present as primary entities in potassium borate glasses.

It has been proposed that the BP originates from the lowest torsional resonance mode of a localised medium range cluster.[6] In order to examine the validity of this theory, we show the relation between the transverse sound velocity and the peak frequency

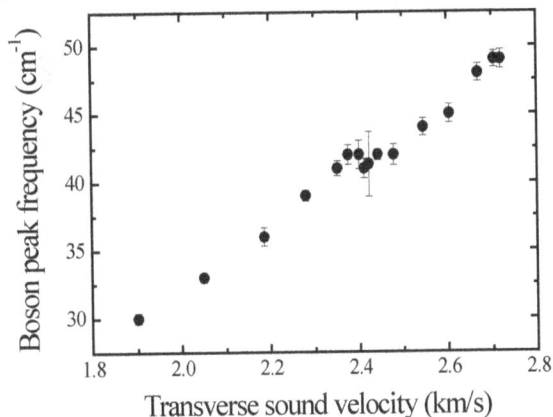

Figure 6. Boson peak frequency versus the transverse sound velocity of $xK_2O.(100-x)B_2O_3$ glasses

of the BP in Figure 6. The results show a clear linear relation. Therefore, it is suggested that the origin of the BP in potassium borate glass relates to transverse acoustic phonons.

5. Conclusions

The internal vibration bands and boson peak of potassium borate glass were investigated by Raman scattering. We analysed the changes of the peaks in the region between 750 cm^{-1} and 810 cm^{-1}. Three peaks at 805, 775 and 750 cm^{-1} were deconvoluted into Gaussian bands. These peaks were assigned to vibrations of intermediate structural units present in the borate glass network. The structural changes were estimated by analysing these peaks. This shows that for low K_2O content, one 3-fold coordinated boron in each boroxol ring transforms into a 4-fold coordinated boron. Then from x=20 mol% K_2O, a second 3-fold coordinated boron in the ring transforms into a 4-fold coordinated boron.

There is a general increase in the frequency of the boron peak as the K_2O content increases. We found a clear linear correlation between the boson peak frequency and the sound velocity.

Acknowledgements

One of the authors (YM) is grateful for the receipt of JSPS Research Fellowship 19.574.

References

1. Kodama, M., Kojima S., Feller, S. A. & Affatigato, M. *Ceram. Trans.*, 2006, **197**, 21.
2. Shelby, J. E. *J. Am. Ceram. Soc.*, 1983, **66**, 225.
3. Kamitsos, E. I., Karakassides M. A. & Chyssikos G. D. *Phys. Chem. Glasses*, 1989, **30**, 229–234.
4. Kamitsos, E. I. & Karakassides, M. A. *Phys. Chem. Glasses*, 1989, **30**, 19.
5. Krogh-Moe, J. *J. Non-Cryst. Solids*, 1969, **1**, 269.
6. Kojima, S. & Kodama, M. *Phys. B*, 1999, **263–264**, 336.
7. Kodama, M., Matsushita, T. & Kojima S. *Jpn. J. Appl. Phys.*, 1994, **34**, 2570.
8. Kodama, M. *J. Non-Cryst. Solids*, 1991, **27**, 65.

Proc. VI Int. Conf. Borate Glasses, Himeji, Japan, 18–22 August 2008   *Phys. Chem. Glasses: Eur. J. Glass Sci. Technol. B*, August 2009, **50** (4), 253–256

In situ XAFS measurement of transition metal in borate glass at high temperature

Akihiko Kajinami,[*1,3] *Haruaki Matsuura,*[2] *Shigehito Deki,*[3] *Shouhei Fujiwara*[3] *& Norimasa Umesaki*[4]

[1] *Center for Environmental Management, Kobe University, 1-1 Rokko-dai, Kobe, Hyogo, 657-8501, Japan*
[2] *Research Laboratory for Nuclear Reactors, Tokyo Institute of Technology,, Ookayama, Meguro-ku, Tokyo, 152-8550, Japan*
[3] *Graduate School of Engineering, Kobe University, 1-1 Rokko-dai, Kobe, Hyogo, 657-8501, Japan*
[4] *JAPAN Synchrotron Radiation Research Institute (JASRI/SPring-8), Kouto, Sayo-cho, Sayo-gun, Hyogo, 679-5198, Japan*

Manuscript received 31 October 2008
Revised version received 1 March 2009
Accepted 27 May 2009

The behaviour of transition metal ions in borate glass at high temperature was studied by in situ XAFS measurements. The XAFS measurements for manganese borate glass, $(MnO)_{0.21}(B_2O_3)_{0.79}$, and cobalt borate glass, $(CoO)_{0.25}(B_2O_3)_{0.75}$, from 298 to 1373 K were performed to examine the variations of short range structure and valence of the transition metal ions with temperature. It was found that cobalt ions in borate glass change to cobalt metal at 1273 K under helium, while an obvious valence change was not observed for manganese ions in borate glass at 1273 K. The effect of atmosphere on the valence of cobalt ions is also discussed.

1. Introduction

It is well known that alkali borate glass shows anomalous composition dependence of various physico-chemical properties, such as density and thermal expansion coefficient, with the metal oxide content, which is called the borate anomaly.[1,2] The glass structure has been extensively investigated by x-ray and neutron diffraction,[3,4] Raman spectroscopy,[5] ^{11}B NMR,[6] molecular dynamics simulation,[7] and so on. However, the structure of borate glasses containing transition metal ions has not been extensively investigated. It is important to examine the structure and physico-chemical properties of borate glasses containing transition metal ions because their presence has a great effect on the structure and properties. The valence of the transition metal ion varies with both composition and preparation condition, and also influences the physico-chemical properties. The authors have previously investigated the composition dependence of the short range structure in zinc borate glass by XAFS (x-ray absorption fine structure) and XRD (x-ray diffraction) measurements,[8] and manganese borate glass by HEXRD (high energy x-ray diffraction) and XAFS measurements,[9,10] respectively.

In the present study, the variations of short range structure and valence of manganese and cobalt ions in borate glass with temperature and atmosphere are examined by *in situ* XAFS measurements.

2. Experimental

2.1 Preparation of manganese and cobalt borate glasses

The manganese borate glass, $(MnO)_{0.21}(B_2O_3)_{0.79}$, was prepared from B_2O_3 and $MnCO_3$, which were mixed and then heated in aluminium crucibles at 1373 K for about 2 h under air. The cobalt borate glass, $(CoO)_{0.25}(B_2O_3)_{0.75}$, was prepared from B_2O_3 and CoO. They were mixed and heated in a platinum crucible at 1373 K for about 2 h under air. The melt was rapidly quenched to ambient temperature by the twin roller method.[11] The quenching rate was estimated to be 10^3 to 10^4 K/s. The amorphicity of the glasses was verified by an XRD measurement. The composition of the glasses was analysed by inductively coupled plasma atomic emission spectroscopy (ICP-AES) (Seiko Instruments Inc., SPS1500VR).

2.2 In situ *XAFS measurement*

The glass was well ground and mixed with hexagonal boron nitride (99%) powder to adjust the absorbance for XAFS measurement.[12] The mixture was pressed in a mechanical die press to form a tablet of 10 mm in diameter. The tablet was set in a sample holder (see Figure 1(a)), and heated in a high temperature furnace (Figure 1(b)) from 298 to 1373 K under helium with a flow rate of about 50 ml/min. The *in situ* XAFS measurements were carried out on beamlines BL-7C and BL-9C at the PF-Ring of the High Energy Accelerator Research Organization (KEK, Tsukuba). The K-edge XAFS spectra of manganese and cobalt were measured using Si (111) double monochromator crystals and two ionisation chambers in transmission mode around 6·5 and 7·7 keV, respectively. The XAFS spectra were measured by a step scanning technique, with fixed time of 1 s at every energy. Therefore, it took about 5 and 15 min to get each XANES (x-ray absorption near edge structure) and EXAFS (extended x-ray absorption fine structure) spectrum, respectively.

*Corresponding author. Email kajinami@kobe-u.ac.jp

(a)

(b)

X-ray beam

X-ray beam

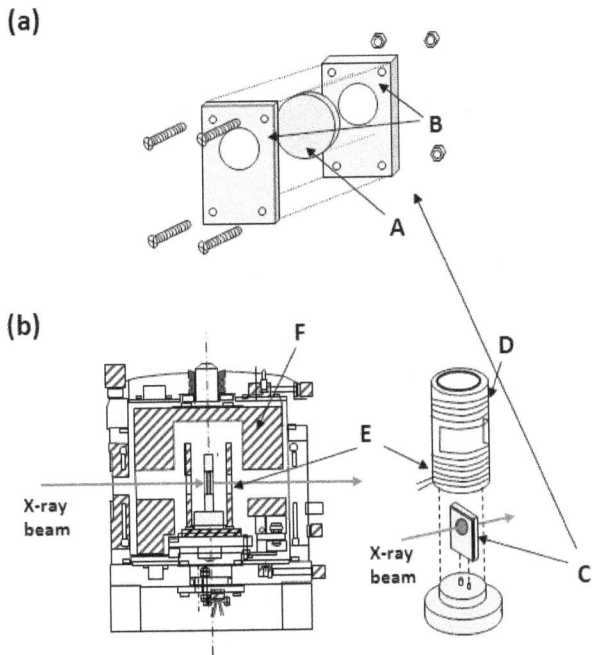

Figure 1. Sample holder (a) and high temperature furnace (b) for in situ XAFS measurement. A: sample tablet, B: alumina plates, C: sample holder, D: platinum heater, E: alumina bobbin, F: heat shield

The XANES and radial structural function, $\Phi(r)$, were calculated by commercial EXAFS software (Rigaku, REX2). The EXAFS interference functions, $k^3\chi(k)$, were extracted from the observed absorption spectra by a cubic spline method, and were Fourier transformed to evaluate the radial structural functions.[12]

2.3 Characterisation of mixture of borate glass and boron nitride

The cobalt borate glass and boron nitride were mixed and heated in a platinum vessel at various temperatures for 2 h under helium. The heated mixture of glass and boron nitride was examined by XRD (Rigaku, RINT-TTR) with Cu K_α radiation (50

kV, 300 mA). The microstructure of the mixture was observed by transmission electron microscopy (TEM, Hitachi, H-7500M).

3. Results

3.1 Variation of XAFS of manganese borate glass with temperature

Figure 2(a) shows the temperature variation of the XANES of $(MnO)_{0.21}(B_2O_3)_{0.79}$ glass under helium. The manganese borate glass at 298 K showed the manganese K-edge at around 6545 eV, which shows that the manganese ion is divalent (Mn^{2+}) in this glass.[9] The XANES spectrum at 1073 K shows fine structure in the range from 6545 to 6600 eV, which suggests some ordered structures in the glass. The fine structure disappears at 1273 K, which indicates fusion of the manganese borate, while the absorption edge does not shift from 6545 eV. It shows that the valence of the manganese ion does not change in the range from 298 to 1073 K. Figure 2(b) shows the temperature variation of the radial structural function, $\Phi(r)$, of the manganese borate glass, which has not been corrected for phase shift. The radial structural function at 298 K shows a broad peak at ~0·16 nm, which is assigned to Mn–O correlations.[9,10] The functions $\Phi(r)$ at 1073 and 1273 K show a peak at the same distance. Thus the Mn–O correlation shows no obvious variation with temperature.

3.2 Variation of XANES of cobalt borate glass with temperature

Figure 3(a) shows the temperature variation of the XANES of $(CoO)_{0.25}(B_2O_3)_{0.75}$ glass under helium. The cobalt borate glass at 298 K showed an absorption edge at 7715 eV, which shows that the cobalt ion is divalent (Co^{2+}) in the glass.[13] A fine structure is found in the range from 7730 to 7750 eV at 1073 K, which suggests some ordered structures in the glass. However, the fine structure disappears at 1273 K. Nevertheless, the pre-edge peak at 7708 eV gradu-

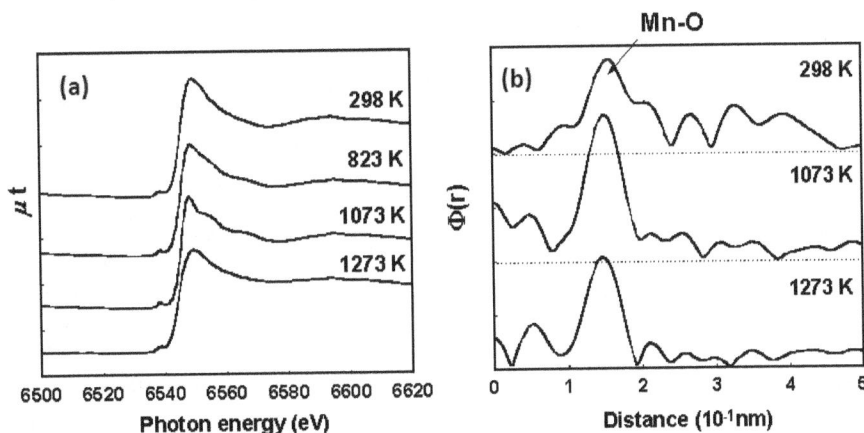

Mn–O

(a) 298 K / 823 K / 1073 K / 1273 K

μt

Photon energy (eV)

(b) 298 K / 1073 K / 1273 K

$\Phi(r)$

Distance (10^{-1} nm)

Figure 2. Temperature variation of XANES (a) and radial structural function, $\Phi(r)$, (b) for $(MnO)_{0.21}(B_2O_3)_{0.79}$ glass under helium

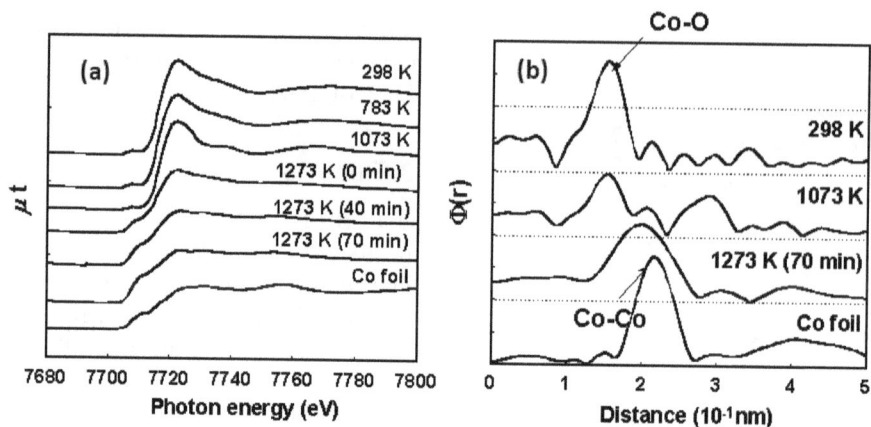

Figure 3. Temperature variation of XANES (a) and radial structural function, $\Phi(r)$, (b) for $(CoO)_{0.25}(B_2O_3)_{0.75}$ glass under helium

ally becomes larger, and the absorption edge shifts to lower energy. The XANES spectrum of the sample heated at 1273 K for 70 min shows a similar pattern to cobalt metal foil.[13]

Figure 3(b) shows the temperature variation of $\Phi(r)$ of the glass under helium. The function $\Phi(r)$ at 298 and 1073 K shows a peak around 0.15 nm, which is assigned to Co–O correlations. The function $\Phi(r)$, after heating at 1273 K for 70 min, shows a broad peak at around 0.20 nm. $\Phi(r)$ for cobalt foil, as shown in Figure 3(b), has a peak at 0.21nm which is assigned to Co–Co correlations.[14] Therefore, it is clear that the broad peak around 0.20 nm for the glass at 1273 K includes both Co–O and Co–Co correlations. Thus the XAFS spectra indicate that the cobalt ion changes to cobalt metal at 1273 K.

3.3 Effect of atmosphere on cobalt borate glass at high temperature

Figure 4 shows the temperature variation of the XANES of $(CoO)_{0.25}(B_2O_3)_{0.75}$ glass under air. A fine structure is observed from 7730 to 7750 eV at 1073 and 1273 K, which indicates some ordered structures. It suggests that the cobalt borate glass does not melt

at 1273 K under air. The absorption edge of cobalt shifts to lower energy at 1373 K. The XANES pattern at 1373 K is similar to that for cobalt metal foil (see Figure 3(a)), which shows that the reduction of cobalt ions occurs at 1373K under air. When the sample is cooled to 1053K, a broad peak appears at ~7722 eV, which is assigned to cobalt oxides.[14] It shows that cobalt metal is formed at 1373 K under air, and that the cobalt metal is oxidised to cobalt oxide at 1053 K.

4. Discussion

The in situ XAFS spectra show that the mixture of cobalt borate glass and hexagonal boron nitride reacted at high temperature, and that the cobalt ions (Co^{2+}) were reduced to Co metal. Figure 5 shows XRD patterns for a mixture of cobalt borate glass, $(CoO)_{0.1}(B_2O_3)_{0.9}$, and boron nitride, heated for 2 h under helium at various temperatures. There are XRD peaks of boron nitride, cobalt oxide and unknown materials in the mixtures below 1173 K. The peaks of unknown materials may be assigned to some crystal structure of cobalt borate. The unknown peaks and cobalt oxide peaks disappear above 1223 K, while

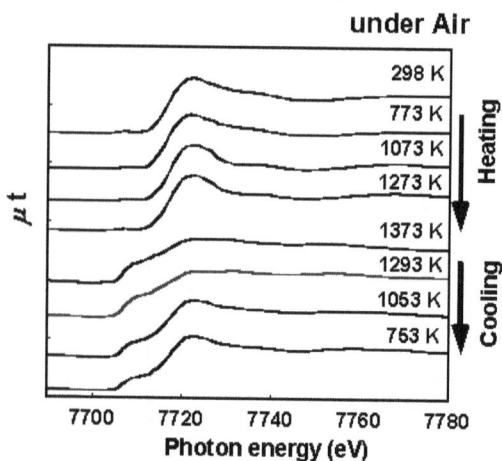

Figure 4. Temperature variation of XANES for $(CoO)_{0.25}(B_2O_3)_{0.75}$ glass under air

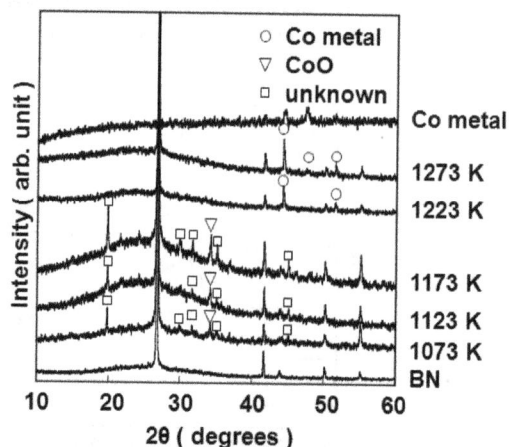

Figure 5. Temperature dependence of the XRD pattern for a mixture of $(CoO)_{0.1}(B_2O_3)_{0.9}$ glass and boron nitride heated for 2 h under helium

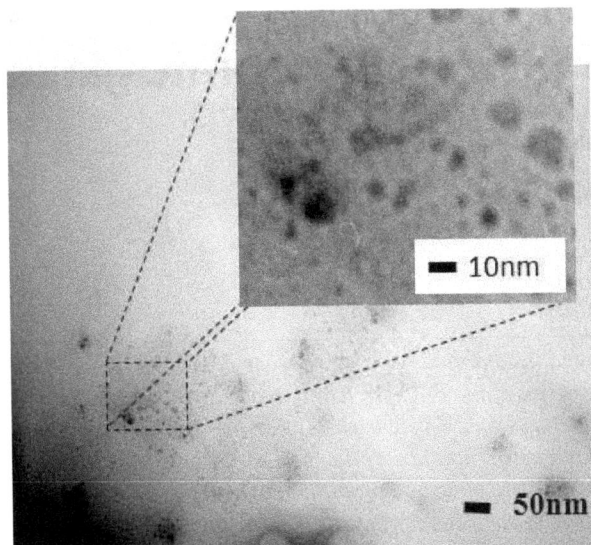

Figure 6. TEM image of the mixture of $(CoO)_{0.1}(B_2O_3)_{0.9}$ glass and boron nitride heated at 1273 K for 2 h under helium

Co metal peaks appear and their intensity increases as the temperature increases. Figure 6 shows a TEM image of the mixture of cobalt borate glass and boron nitride heated at 1273 K under helium for 2 h. It is clear that the sample includes fine particles of about 10 nm diameter. From the XRD and TEM results, it is concluded that fine particles of cobalt metal are formed by heating the mixture above 1273 K under helium.

Other XRD measurements reveal that cobalt metal is not formed by heating cobalt borate glass without boron nitride under helium at 1273 K, This shows that the formation of cobalt metal requires the presence of boron nitride and is closely related to any reactions of boron nitride. Though the reaction mechanism is unknown, the boron nitride reduces the divalent cobalt ion (Co^{2+}) in the glass to cobalt metal.

Contrastingly, a noticeable temperature dependence of the local structure and the valence of the manganese ions in manganese borate glass is not observed in the *in situ* XAFS spectra of manganese borate glass. It is found that the manganese borate glass is more stable against reduction than the cobalt borate glass.

It is well known that hexagonal boron nitride is chemically stable at high temperature under inert atmosphere, and it is used for crucibles, heat resistant containers or as a lubricant. Hence hexagonal boron nitride is generally used to dilute and hold glasses or melts for XAFS measurements at high temperature.[12] It is therefore notable that boron nitride is not suitable for cobalt borate glass at high temperature, even under inert gas.

5. Conclusions

The XAFS spectra of cobalt borate glass indicate that the local structure and the valence of the cobalt ions drastically change with temperature, because the glass is reduced by boron nitride at 1273 K under helium to form fine particles of cobalt metal. Reduction by boron nitride may occur for other metal oxides at high temperature, and therefore it is important to investigate the reactivity of hexagonal boron nitride with other metal oxides to examine the detailed reaction mechanism.

Acknowledgements

We would like to thank the *Center for Supports to Research and Education Activities* in Kobe University for use of the ICP-AES spectrometer.

References

1. Sakka, S., Kamiya, K. & Kato, K. *J. Non-Cryst. Solids*, 1982, **52**, 77.
2. Shelby, J. E. *J. Am. Ceram. Soc.*, 1983, **66**, 225.
3. Paschina, G., Piccaluga, G. & Magini, M. *J. Chem. Phys.*, 1984, **81**, 6201.
4. Herm, G., Derno, M. & Stiel, H. *J. Non-Cryst. Solids*, 1986, **88**, 381.
5. Meera, B. N. & Ramakrishna, J. *J. Non-Cryst. Solids*, 1993, **159**, 1.
6. Bray, P. J. *J. Non-Cryst. Solids*, 1987, **96**, 45.
7. Handa, K., Iwadate, Y. & Umesaki, N. *Jpn. J. Appl. Phys. Suppl.*, 1999, **38-1**, 140.
8. Kajinami, A., Harada, Y., Inoue, S., Deki, S. & Umesaki, N. *Jpn. J. Appl. Phys. Suppl.*, 1999, **38-1**, 132.
9. Kajinami, A., Deki, S. & Handa, K. *Phys. Scripta B*, 2005, **T115**, 477.
10. Kajinami, A., Kotake, T., Deki, S. & Kohara, S. *Nucl. Instrum. Meth. B*, 2003, **199**, 34.
11. Nassau, K., Wang, C. A. & Grasso, M. *J. Am. Ceram. Soc.*,1979, **62**, 74.
12. Filipponi, A. *J. Phys.: Condens. Matter*, 2001, **13**, R23.
13. Teo, B. K. *EXAFS: basic principles and data analysis*, Springer-Verlag, Berlin, 1986, p. 23.
14. Saib, A. M., Borgan, A., van de Loosdrecht, J., van Berge, P. M. & Niemantsverdriet, J. W. *Appl. Catal. A-Gen.*, 2006, **312**, 12.
15. Huffman, G. P., Shah, N., Zhao, J., Huggins, F. E., Hoost, E., Halvorsen, S. & Goodwin Jr, J. G. *J. Catal.*, 1995, **151**,17.

Proc. VI Int. Conf. Borate Glasses, Himeji, Japan, 18–22 August 2008   *Glass Tech.: Eur. J. Glass Sci. Technol. A, August 2009,* **50** (4), 230–232

Optical properties of Pr and Nd ions in a borate glass

J. A. Erwin Desa,* Wilson A. Vaz, Brenda L. De Souza & Manju Singh

Department of Physics, Goa University, Taleigao Plateau, Goa 403206, India

Manuscript received 31 October 2008
Revised version received 24 January 2009
Accepted 18 April 2009

A set of five borate glasses containing oxides of praseodymium and neodymium have been prepared with several relative molar ratios of Pr and Nd. FTIR measurements showed the presence of the B–O stretching mode in BO_4 tetrahedra at 900 cm^{-1}, as well as the B–Ô–B bending mode at 700 cm^{-1} in the glasses. UV-visible absorption spectra were measured and found to be due to the 4f electronic transitions of the Pr and Nd ions. These absorption peaks were only weakly linked to their structural neighbourhoods, as evidenced by the compositionally scaled patterns showing very good agreement, in peak wavelengths, with the measured data in these glasses as well as in the equivalent glasses with phosphate as the host matrix. However, the intensities of Nd absorption peaks are observed to be enhanced by the presence of Pr ions in these glasses. Excitation at 445 nm of the luminescent states of Nd and Pr yielded emission data in the range 800 nm to 950 nm and lifetimes between 1·094 μs and 1·225 μs, indicating that the local structural environment of the rare earth ions does not greatly affect these states.

1. Introduction

Rare earth oxide glasses often have high chemical durabilities and high refractive indices[1,2] and are known[3,4,5] to have important applications in optical filters, as optical waveguides, as active and passive multi-spectral components, optical fibres and as laser host materials. Some rare earth ions are known to have well defined optical absorption resonances in the visible region. The work described here was performed with the objective of containing such ions within borate glasses to yield potentially useful materials whose characterisation is reported here. The present study has used infrared, UV-visible and luminescence spectroscopy to examine the optical properties of these glasses, and the extent of the interaction of the outer electronic levels of the rare earth ions Pr and Nd with their borate host glass matrix and with each other.

2. Sample preparation

Five borate glasses containing oxides of La, Pr and Nd were prepared. The initial powder mixture compositions that were used are as shown in Table 1 (rare earth oxides (99·9%) from Indian Rare Earths Ltd, Al_2O_3 (99·8%) and B_2O_3 (99%) both from Aldrich Chemical Company Inc). All samples were melt quenched in air from temperatures between 1000°C and 1300°C to ambient in batch sizes of 15 g of the starting oxide mixtures using alumina crucibles.[6,7] Homogeneous, bubble free glasses were formed, having colours from purple for sample 1 to green for sample 5, and intermediate colours for the other glasses (samples 2, 3, 4).

Table 1. Initial powder mixture compositions of the glass samples

Sample No.	Components (mol%)				
	La_2O_3	Nd_2O_3	Pr_6O_{11}	Al_2O_3	B_2O_3
1	7	8	0	5	80
2	9·19	3·06	0·99	5·10	81·66
3	10·56	2·29	2·64	5·28	79·23
4	9·19	0·99	3·06	5·10	81·66
5	10·73	0	3·46	5·36	80·45

3. Results

X-ray diffraction patterns were taken with incident Cu K_α radiation in reflection mode in the 2θ angular range of 10° to 140° and are shown in Figure 1. Continuous lines in figures in this paper are drawn linking data points and are intended as guides to the eye. FTIR data (Shimadzu spectrometer) on the five glass samples in the range 400 to 1000 cm^{-1} are shown in Figure 2. UV-visible absorption spectra of the glasses were measured with a Shimadzu spectrophotometer from 300 nm to 900 nm and are plotted in a stack in

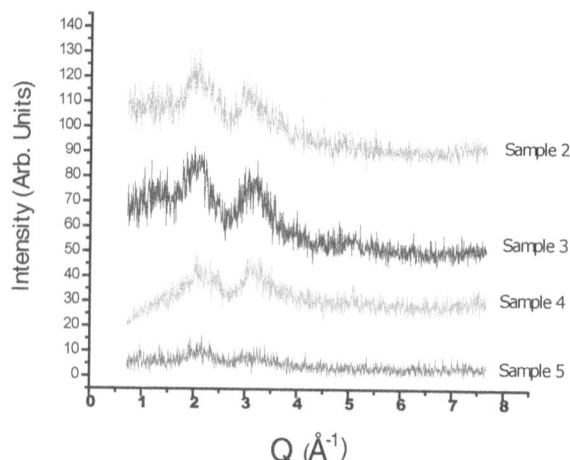

Figure 1. X-ray diffraction patterns of the borate glasses

*Corresponding author. Email edesa@unigoa.ac.in

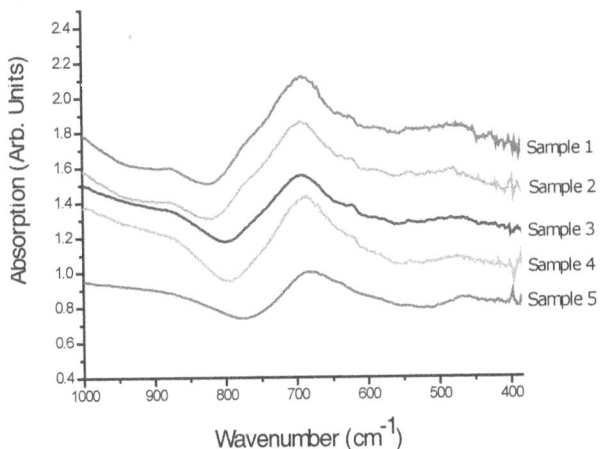

Figure 2. FTIR absorption spectra of the glasses

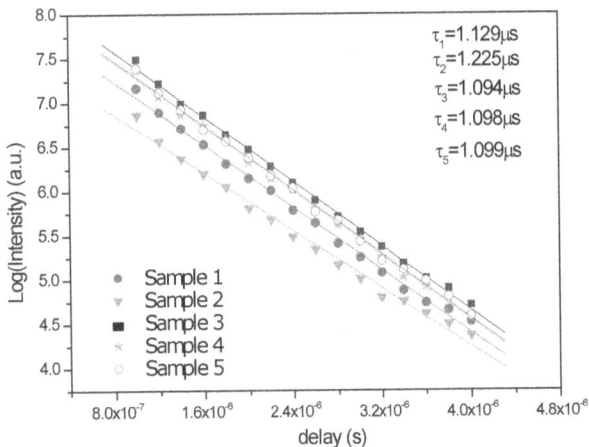

Figure 4. Lifetimes of luminescent emission from Nd and Pr in the borate glasses

Figure 3. For the luminescence measurements, the excitation wavelength for the rare earth ions was 445 nm. The resulting emission spectra for Nd and Pr transitions were measured in the range 800 to 950 nm (to be reported elsewhere). Lifetime measurements from the latter emission peaks (relating to $^4F_{3/2}$ to $^4I_{9/2}$ transitions[8]) yielded the purely exponential decay functions shown in Figure 4.

4. Discussion

The essentially vitreous nature of the samples was confirmed from the x-ray diffraction data of Figure 1. The B–O stretching mode due to BO_4 tetrahedra at 900 cm^{-1} and also the B–Ô–B bond bending mode at 700 cm^{-1} are observed in the FTIR spectra of the glasses, indicating the presence of BO_4 units and their linkages.[9]

The absorption peaks in both glass types were spectroscopically identified[4,5,8,10] as being from the 4f electronic levels of Nd and Pr, as indicated in Figure 3. The spectra of the two samples that had either Nd_2O_3 (sample 1) or Pr_6O_{11} (sample 5) but not both oxides were added together in the relative ratios

that these components occur in samples 2, 3 and 4 to yield three calculated spectra. The latter were then compared to the measured spectra from these glasses by multiplicative scaling of the calculated spectra until the peak heights due to the Pr transitions at 450 nm matched. The comparison is shown for sample 3 in Figure 5. This procedure was also tried by using the Nd transitions, but this resulted in an offset and apart from a particular chosen Nd peak (e.g. at 750 nm) the other peak heights in the spectra did not match (Figure 6). Scaling by using the Nd peaks was thus not pursued further. This scaling procedure (using the Pr peaks) resulted in agreement to within 2 nm of all peak positions and about 2% of peak height for the Pr peaks. Peak wavelengths relate to energies of electronic transitions and the fact that these wavelengths remain approximately unchanged strongly suggests that the 4f states of the rare earth ions (at these concentrations) in all these glasses interact only weakly with the host glass structure. The extent that this is true is emphasised when the absorption spectra for phosphate glasses having the

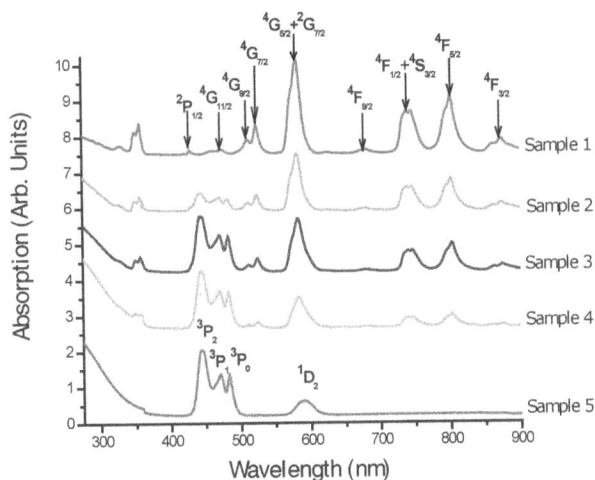

Figure 3. Comparison of the UV-visible absorption spectra of the borate glasses

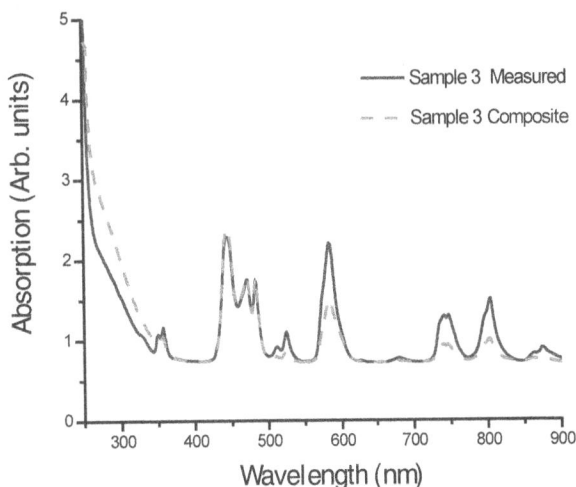

Figure 5. Comparison of measured and calculated composite UV-visible spectra of sample 3, scaled to the Pr peaks at 450 nm

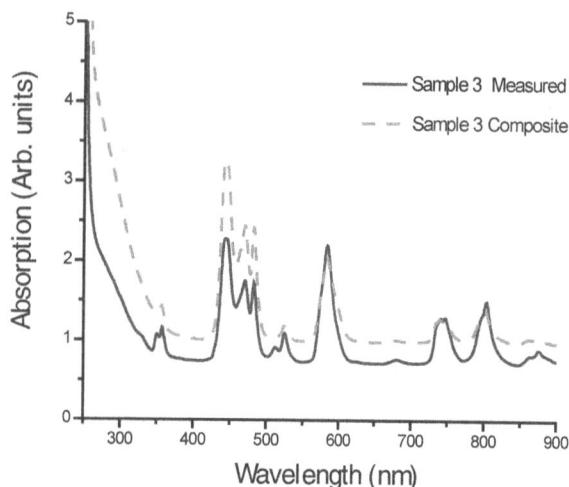

Figure 6. Comparison of measured and calculated composite UV-visible spectra of sample 3, scaled to the Nd peak at 750 nm

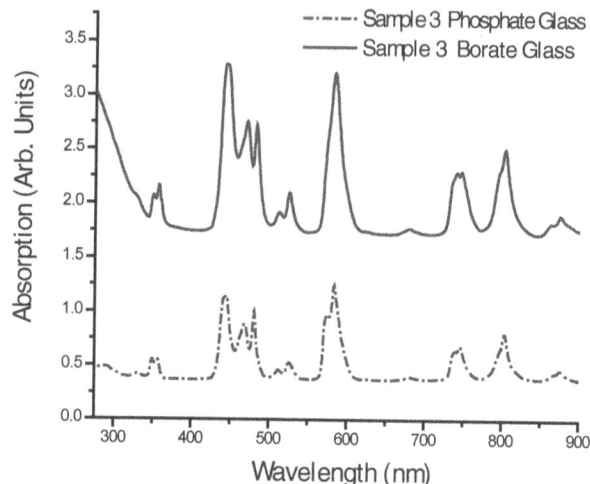

Figure 7. Comparison of the UV-visible spectra of a borate and equivalent phosphate glass with the same rare earth ion compositions

same rare earth concentrations as the borate glasses (as for example in sample 3) are compared (Figure 7). The main absorption features are present in both glass types but relative peak intensities which relate to transition probabilities are, as expected, slightly different. However, for the Nd absorption peaks the heights of the measured peaks were up to 60% higher than the calculated, indicating that the presence of Pr in the actual glasses (samples 2, 3, 4) enhances the intensities of these Nd absorption lines. It is possible that a certain degree of interaction between the energy levels of the Pr and Nd ions in each glass seems to be taking place, affecting their relative transition probabilities but not their energies of transition.

The energies of emission states for both Nd and Pr gave rise to a single peak in the range of measurement (800 nm to 950 nm). These measured emissions of Nd and Pr yielded exponential decay lifetimes in the range 1·094–1·225 µs for these borate glasses. The errors associated with these measurements were of the order of the size of the symbols of the data points in Figure 4. These values were similar for an equivalent set of phosphate glasses.[7] The similarity of decay times from all these samples indicates that the luminescent states are only weakly affected by the host structures.

5. Conclusions

Up to 15 mol% of rare earth (La, Pr and Nd) oxide has been incorporated in a B_2O_3-based glass matrix to produce homogeneously coloured, bubble free glasses. From FTIR data, the presence of BO_4 tetrahedra and a continuous random network with B–O–B linkages are both supported.

UV-visible absorption spectra for these glasses show that the energies of the 4f electronic levels of the Nd and Pr ions (at these concentrations) are only weakly linked to the surrounding host glass network,

borate or phosphate. Intensities of optical absorption transitions in the Nd ions were enhanced by the presence of Pr ions in these glasses, possibly due to interaction between the 4f levels of these two ions in the glasses. Thus, while the probabilities of transitions between the 4f levels of these rare earth ions are affected by their environment and neighbours, the energies of transition between the states remain essentially unchanged.

The lifetimes of luminescent emissions from the rare earth ions are similar and of the order of 1 µs for these borate glasses, as well as for a similar set of phosphate glasses. This indicates that the structural surroundings of the rare earth ions do not greatly affect the luminescent levels of these ions.

Acknowledgements

The authors gratefully acknowledge: Ms S. Kumari and Ms N. Pereira's help in preparing the phosphate glasses; UGC-DAE-CSR for support through CRS-M-128; Prof. S. K. Mendiratta, J. Cascalheira, Department of Physics, University of Aveiro, Portugal for help in the luminescence studies.

References

1. Chakraborty, I. N., Shelby, J. E. & Condrate Sr., R. A. *J. Am. Ceram. Soc.*, 1984, **67**, 782.
2. Chakraborty, I. N., Day, D. E., Lapp, J. C. & Shelby, J. E. *J. Am. Ceram. Soc.*, 1985, **68**, 368.
3. Shikerkar, A. G., Desa, J. A. E., Krishna, P. S. R. & Chitra, R. *J. Non-Cryst. Solids*, 2000, **270**, 234.
4. Moorthy, L. R., Rao, T. S., Jayasimhadri, M., Radhapathy, A. & Murthy, D. V. R. *Spectrochim. Acta A*, 2004, **60**, 2449.
5. Moorthy, L. R., Jayasimhadri, M., Radhapathy, A.& Ravikumar, R. V. S. S. N. *Mater. Chem. Phys.*, 2005, **93**, 455.
6. De Sousa, B. & Singh, M. *MSc Project Report*, 2004, Goa University.
7. Kumari, S. & Pereira, N. *MSc Project Report*, 2003, Goa University.
8. Ajroud, M., Haouari, M., Ben Ouada, H., Maaref, H., Brenier, A. & Garapon, C. *J. Phys.: Condens. Matter*, 2000, **12**, 3181.
9. Kamitsos, E. I., Karakassides, M. A. & Chryssikos, G. D. *J. Phys. Chem.*, 1987, **91**, 1073.
10. Lowther, J. E. *J. Phys. C*, 1974, **7**, 4393.

Proc. VI Int. Conf. Borate Glasses, Himeji, Japan, 18–22 August 2008   *Phys. Chem. Glasses: Eur. J. Glass Sci. Technol. B, February 2010,* **51** (1), 59–64

Structural effects of europium on lead borate glasses

Amy Marquardt,[1] *Matt Roberts,*[1] *Steve Feller,*[1] *Steve Singleton*[2] *& Mario Affatigato*[1,*]

[1] *Physics Department, Coe College, 1220 First Av. NE, Cedar Rapids, Iowa 52402, USA*
[2] *Chemistry Department, Coe College, 1220 First Av. NE, Cedar Rapids, Iowa 52402, USA*

Manuscript received 14 February 2008
Accepted 29 July 2008

Studies of the structure of lead metaborate glasses doped with up to 15 mol% Eu_2O_3 are reported. Measurements of the samples using laser ionization time of flight mass spectrometry revealed an isolated role for single Eu^{+3} ions and a preference for sites near B_3O_6 rings. The relative isolation of the europium modifier ions was confirmed by infrared spectroscopy and UV/visible absorbance and fluorescence. The impact on the lead borate network was limited to an increase in the fraction of three-coordinated borons and an increase in lead rich units. Fluorescence measurements suggest a stable (+3) oxidation state for the europium ions, and very little change in the fluorescence emission lifetime. The stable borate sites (near rings) occupied by the Eu and its high field strength are possible explanations of these observations.

I. Introduction

The introduction of europium ions in glasses has been of considerable interest over the past two decades. The value of doping to optical applications has been substantial, and the usefulness of adding rare earth (RE) elements has been well established. The precise structural role of the individual cations, however, has remained an open question. In particular, the environment surrounding the RE ions is known[1] to affect the fluorescence yields and lifetimes, and cross-relaxation between nearby *single* RE ions has also been noted,[2,3] including in borates.[3,4]

In this paper we report on our investigations of the structure of europium doped lead borate glasses using laser ionization time-of-flight mass spectrometry (LITOF-MS), which allows us to look quite accurately[5–8] at ions and charged molecular units desorbed from the glass samples upon laser irradiation. Additionally, infrared and UV measurements were taken. Combined, these methods provide new insights into the structural role of the RE cations, as well as their effect on the larger glass network.

II. Experimental procedures

Sample preparation

Glass compositions were prepared using the following reaction:

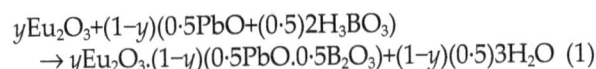

$$yEu_2O_3 + (1-y)(0.5PbO + (0.5)2H_3BO_3)$$
$$\rightarrow yEu_2O_3.(1-y)(0.5PbO.0.5B_2O_3) + (1-y)(0.5)3H_2O \quad (1)$$

where y denotes the molar fraction of europium oxide, which was varied from 0·001 to 0·15. The concentration of lead was held constant at 50 mol%, a very stable base glass. The samples were prepared from high purity (>99·99%) europium oxide, lead

oxide and boric acid to prevent contamination. Glass samples were successfully made at y=0·001, 0·005, 0·01, 0·03, 0·05, 0·08, 0·1 and 0·15. Higher europium contents would not make glass samples.

The components were mixed together in a platinum crucible and heated to a temperature in the range from 1000°C to 1350°C depending on the composition. After an initial heating, weight loss measurements were taken to ensure the stoichiometry of the sample, which had an estimated maximum compositional error of <2%. The glass was reheated to the same temperature, and then the melt was quenched using a twin roller quencher to ensure a high and repeatable cooling rate (approximately 10^5 K/s). For the optical measurements we also used some air cooled samples. The average thickness of the roller quenched samples was 36±1 μm and 3·2±0·05 mm for the air quenched samples.

Glass shards that were free of visible imperfections were selected and ground into a powder using a mortar and pestle. The powdered sample was mounted on the time-of-flight (TOF) insertion tip, and inserted into the vacuum chamber, which was kept at pressures less than 5×10^{-7} Torr.

Laser ionization mass spectroscopy

The instrument used was a Comstock reflectron time-of-flight mass spectrometer. Glass units were desorbed from the sample using a Nd:YAG laser (266 nm, 4 ns pulse duration, 0·5 Hz repetition rate) to ionize a sample area of 1·5 mm². The samples were rotated during acquisition to ensure proper surface sampling. All reported data represent an average of 200 spectra taken with power ranging from 70 to 113 mW and the detector set to similar gains. The positive and negative ion spectra were obtained for the full glass composition range, and a lead metaborate glass

* Corresponding author. Email maffatig@coe.edu

was used to calibrate the LITOF-MS. The resolution of the instrument exceeded $M/\Delta M = 1200$, enough for isotopic separation.

The identification of the ionic species was carried out by comparing the mass multiplets seen in the experimental spectra to simulations. The simulations, which take into account the different appearance of each multiplet due to varying the naturally occurring isotopic ratios, were carried out using IsoPro® modelling software. An ionic species was considered to have been identified only if there was a substantial match to the simulated set of peaks.

Absorption/emission and infrared measurements

A JASCO spectrofluorometer was used to measure the fluorescence of roller quenched and air quenched samples over the compositional range. A strong emission spectrum was observed from 600–700 nm when using an excitation wavelength of 393·4 nm. The area and height of the peaks in the emission spectra were measured over the compositional range.

The Eu^{3+} emission lifetimes were measured by exciting the samples with the 355 nm output of a frequency tripled Nd:YAG laser with a pulse width of 5 ns. The subsequent emission was passed through a 0·1 m monochromator with a bandwidth of 0·5 nm, tuned to 612 nm. The signal was detected with a photomultiplier tube, and sent directly to a digital oscilloscope for analysis. The average of 512 shots was taken and fitted to a single exponential decay to obtain the emission lifetime.

The infrared measurements were obtained using a Nicolet Nexus 670 FTIR instrument. The glass samples were crushed, mixed with KBr powder in a proportion of 1:100, and pressed into pellets. Sixty four scans were then taken in transmittance mode, with a resolution of 4 cm^{-1}. The background was a pellet of pure KBr. The spectra obtained were then deconvoluted using IgorPro® software, with the bands following a Gaussian profile. The calculations of the band areas in the infrared spectra were done using the built-in IgorPro integration routine, and the areas were then divided by the net area of the whole IR spectrum.

III. Results and Discussion

Figure 1 shows a typical negative ion spectrum for a Eu-doped lead borate glass. Figure 1(a) shows the spectrum of the sample $0.05Eu_2O_3.0.95(0.5PbO.0.5B_2O_3)$ over a relevant mass range, while in Figure 1(b) we have enlarged the scale to highlight the $(EuB_3O_6)^{-1}$ mass multiplet (group of peaks) and compare it to its isotopic simulation (shown in Figure 1(c)). For this intermediate dopant concentration, the spectrum in Figure 1 shows multiplets for some pure borate ionic groups

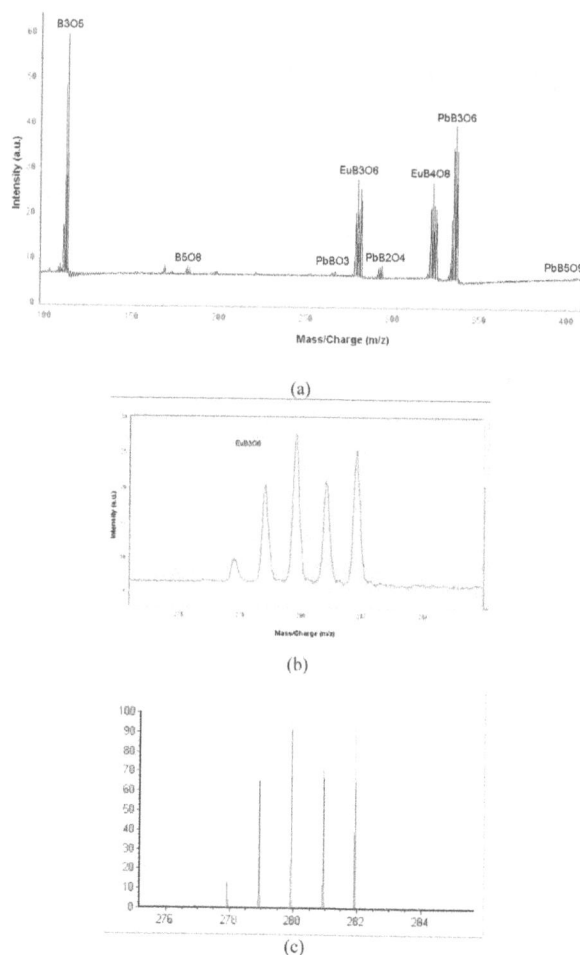

Figure 1. (a) Negative ion spectrum of the sample $0.05Eu_2O_3.0.95(0.5PbO.0.5B_2O_3)$, (b) the region of the spectrum for the $(EuB_3O_6)^{-1}$ mass multiplet, (c) isotopic simulation of the $(EuB_3O_6)^{-1}$ mass multiplet

$((B_3O_5)^{-1}$ and $(B_5O_8)^{-1})$, some lead borate groups $((PbBO_3)^{-1}, (PbB_2O_4)^{-1}, (PbB_3O_6)^{-1},$ and $(PbB_5O_9)^{-1})$, and two europium borate species $((EuB_3O_6)^{-1}$ and $(EuB_4O_8)^{-1})$. Previously,[8] we have identified the $(PbB_3O_6)^{-1}$ unit as a metaborate ring containing an adjoining lead cation, which provides local charge balance. The pure borate species were associated with interconnecting chain fragments, and possibly with some fragmented borate ring structures. Thus, the $(EuB_3O_6)^{-1}$ unit is tentatively identified with a metaborate ring containing a europium ion instead of a lead ion. As will be discussed later, there appears to be a competition between the lead and europium cations for occupation of sites next to the rings. Although the time-of-flight measurements do not yield the oxidation state of the europium ion, the fluorescence measurements show characteristic emission lines[9] for Eu^{3+} in borate glass. The $(EuB_4O_8)^{-1}$ multiplet, on the other hand, is slightly less abundant than the $(EuB_3O_6)^{-1}$ multiplet, and could easily be formed by the addition of a BO_2 unit to the $(EuB_3O_6)^{-1}$ ionic group. We also note that there are no *negative* pure europium oxide (Eu_mO_n) groups.

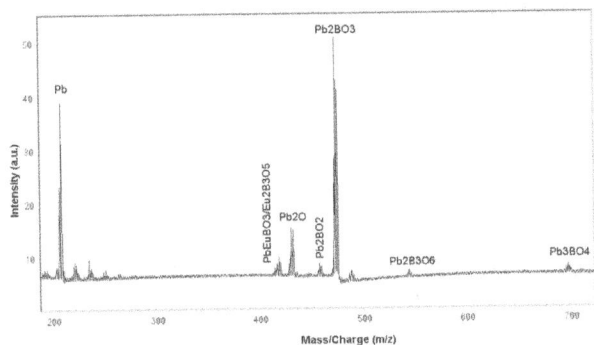

(a)

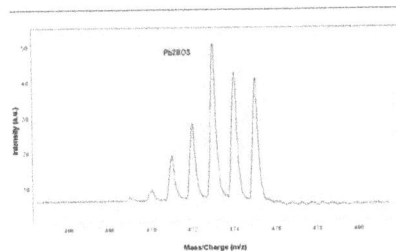

(b)

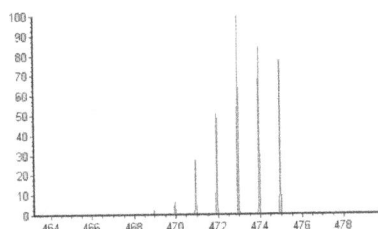

(c)

Figure 2. (a) Positive ion spectrum of the sample $0.05Eu_2O_3.0.95(0.5PbO.0.5B_2O_3)$, (b) the region of the spectrum for the $(Pb_2BO_3)^{+1}$ mass multiplet, (c) isotopic simulation of the $(Pb_2BO_3)^{+1}$ mass multiplet

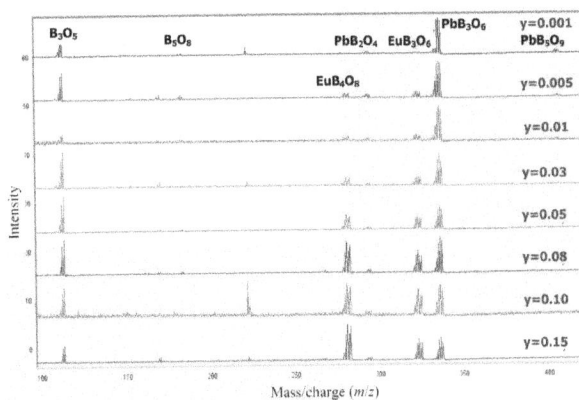

Figure 3. Changes in the negative time of flight mass spectra as a function of the Eu_2O_3 content, y

A typical positive ion spectrum is shown in Figure 2. As before, Figure 2(a) displays the relevant mass spectrum for the $0.05Eu_2O_3.0.95(0.5PbO.0.5B_2O_3)$ sample, while in Figure 2(b) we have enlarged the scale to highlight the $(Pb_2BO_3)^{+1}$ group, and Figure 2(c) shows the IsoPro simulation for this group. Other than the expected lead ions, Pb^{+1}, there are several lead borate species $((Pb_2BO_2)^{+1}, (Pb_2BO_3)^{+1}, (Pb_2B_3O_6)^{+1}$, and $(Pb_3BO_4)^{+1})$, one europium borate group, $(Eu_2B_3O_5)^{+1}$, in small abundance, and, interestingly, one mixed europium lead borate species, $(PbEuBO_3)^{+1}$. The lack of units containing a single lead cation is attributable to the high lead content of the base glass. In our previous work[5,8] we have seen a trend towards groups containing more than one lead cation as the amount of lead oxide in the glass sample is increased; the $0.5PbO.0.5B_2O_3$ base glass already contains enough metal oxide to have few single lead cation units.

The europium borate group, though in small abundance, suggests there are enough europium cations to have two locally associated to the borate fragment $(B_3O_5)^{-1}$; recall that these borate groups were

abundant in the negative ion spectrum. Of greatest interest is the presence of the mixed $(PbEuBO_3)^{+1}$ ionic group in the glass. This appears to arise from the substitution of one of the (very abundant) lead ions in $(Pb_2BO_3)^{+1}$ by a europium cation. If the ionic groups arose from random interactions, we would expect the ratio, z, of the mixed multiplet to the pure lead-borate one to be proportional to the ratio Eu/Pb for the glass composition, which is given by

$$z = \frac{4y}{1-y} \qquad (2)$$

The ratio of areas for the two relevant multiplets in the T-O-F spectrum of the sample with y=0.05 mol% Eu_2O_3 is 0.081, while the value for z predicted by Equation (2) is 0.21. This suggests that the europium cations preferentially occupy the borate sites, displacing the lead ions. This is why the $(PbEuBO_3)^{+1}$ multiplet appears in such prominence in the TOF spectra, even at relatively low doping concentrations.

Figure 3 illustrates the compositional changes that occur in the negative spectra as the Eu_2O_3 content is increased. An obvious change is the rapid decrease of the boroxol/metaborate ring multiplet $(PbB_3O_6)^{-1}$, accompanied by the growth of the two europium borate molecular species, $(EuB_3O_6)^{-1}$ and $(EuB_4O_8)^{-1}$. Similarly, the far less abundant lead borate species, $(PbB_2O_4)^{-1}$ and $(PbB_5O_9)^{-1}$, also decrease to the point of extinction. We again note that the changes are very sensitive to the Eu_2O_3 content over the range y=0.001 to y=0.15. The reasons for the larger impact are likely to involve the higher oxidation state of the europium cation (+3 versus +2 for Pb), which may therefore occupy sites near the ring oxygens more readily than the lead ions. The presence of boron in all europium-containing units may indicate the preferential bonding of the Eu ions to the more negative borate anions. Mixed units (those containing lead and europium cations) only appear after the Eu concentration exceeds 3 mol% Eu_2O_3, while pure lead oxide units $(Pb_{me}O_{ne})$ grow as the concentration of Eu increases. In summary, the europium ions appear to

(a)

(b)

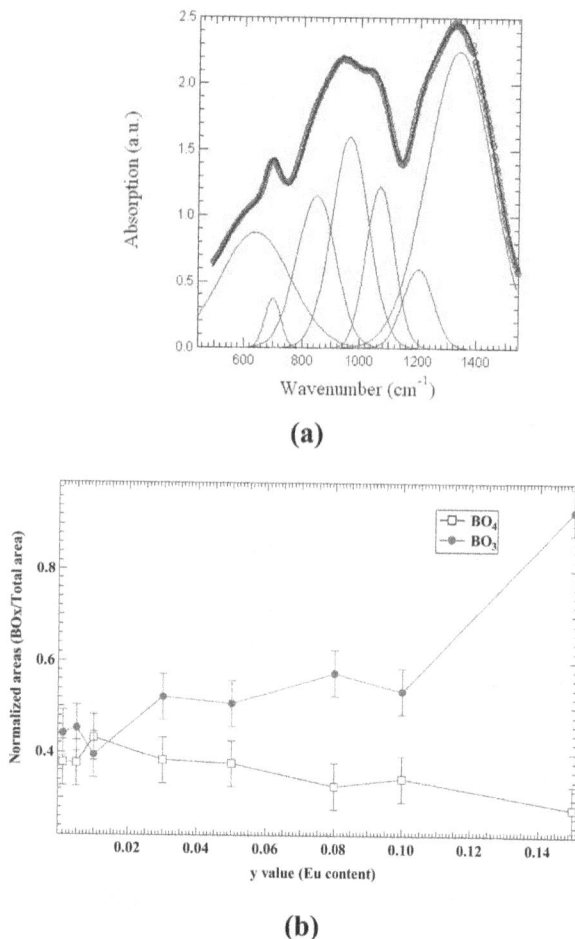

Figure 4. (a) Deconvolution of the infrared absorption spectrum of the sample $0.005Eu_2O_3.0.995(0.5PbO.0.5B_2O_3)$. The data are represented by the small symbols, while the solid line is the fit. (b) Relative quantification of the abundance of tetrahedral BO_4 and trigonal BO_3 units. At y=0.15 the sum exceeds one because of changes in the baseline area of the spectrum

effectively replace the lead ions in their ionic role as borate network modifiers. The lead then creates an oxide-like subnetwork within the glass, an effect that occurs naturally at higher lead concentrations.

The infrared spectra shown in Figure 4 support this hypothesis. Deconvolution of the absorption spectra (Figure 4(a)) shows a variety of borate bands, all of which have already been assigned by Kamitsos & Chryssikos.[10–12] There are three bands near 820 cm⁻¹, 925 cm⁻¹, and 1040 cm⁻¹, which have been assigned to BO_4 stretching vibrations, and two bands near 1180 cm⁻¹ and 1330 cm⁻¹, assigned to stretching modes of trigonal BO_3. The smaller, sharper band at about 710 cm⁻¹ is assigned to bond bending vibrations of B–O–B linkages. No Eu–O bands are observed within the experimental range of the FTIR spectra (400–4000 cm⁻¹). There is some speculation[13] of a BO_4 stretching band at approximately 1310 cm⁻¹, of which we have no evidence in our data. We have therefore assumed that only BO_3 groups contribute to the features in the 1180–1400 cm⁻¹ region.

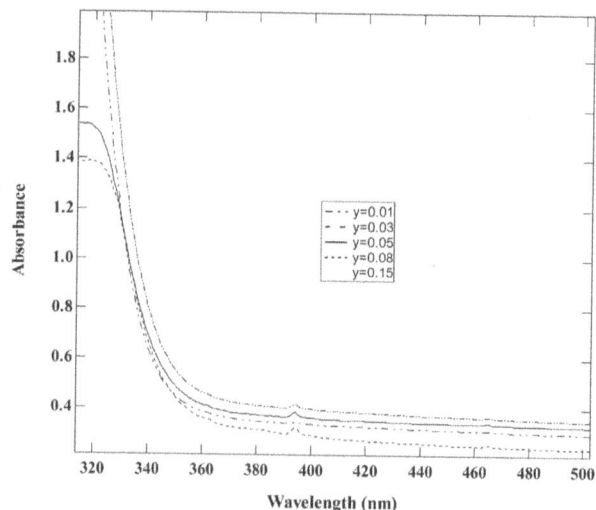

Figure 5. UV/visible absorbance spectra of the samples as a function of Eu_2O_3 concentration

A relative quantification was carried out in which the areas of the BO_3 bands (1180, 1330 cm⁻¹) were added for each Eu_2O_3 concentration and, separately, the areas of the BO_4 bands (820, 925, 1040 cm⁻¹) were also summed. These sums were then normalized by the total spectrum area, and the results are plotted as a function of the Eu_2O_3 content. Figure 4(b) shows that the addition of modifier (PbO plus increasing amounts of Eu_2O_3) decreases the abundance of four-coordinated boron units and increases the fraction of BO_3 units. This behaviour is also in agreement with previous results from RE-doped borates.[13] The continuing decrease in the fraction of BO_4 units also explains why the lead cations are not yet making lead oxide units: the net concentration of modifier is not high enough, and the added oxygen can still be absorbed by the borate network through the $BO_4{\rightarrow}BO_3$ reconversion. We also note that the nature of the BO_3 triangles may well shift toward more of a pyroborate arrangement by the time the composition reaches y=0.15. We did not observe significant changes or trends in the fitted-peak widths in this region, which averaged 258±26 cm⁻¹.

Optical measurements

Figure 5 shows the UV-visible absorption spectra. Clear evidence of the Eu^{3+} ($^7F_J{\rightarrow}^5L_6$) and ($^7F_J{\rightarrow}^5D_2$) absorptions is seen for samples with concentrations greater than y=0.01. These transitions are not readily observed for lower y values due to the thinness of the roller quenched samples.

Figure 6 shows the emission spectra corresponding to the primary Eu^{3+} relaxation process when the samples are irradiated at 394 nm. This excitation wavelength corresponds to the ($^7F_J{\rightarrow}^5L_6$) transition shown in Figure 5. The spectra are consistent between glass samples, and do not point to any significant changes in the electronic environment surrounding

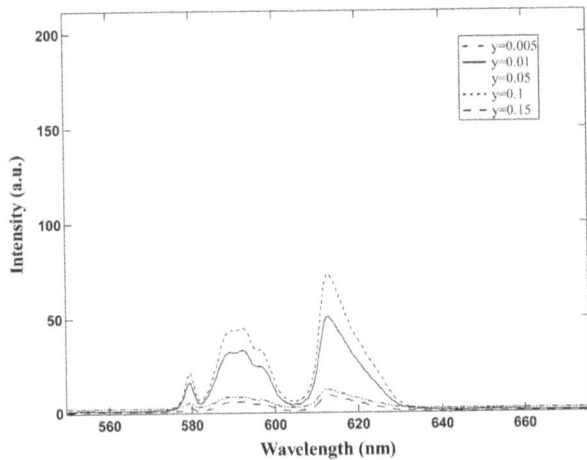

Figure 6. *Emission spectra corresponding to an excitation wavelength of 394 nm. The excitation wavelength corresponds to the ($^7F_J{\rightarrow}^5L_6$) transition*

the europium cations. This is again taken as evidence that europium has an ionic modifier role within the network which does not change with Eu_2O_3 concentration, with europium positioning itself preferen-

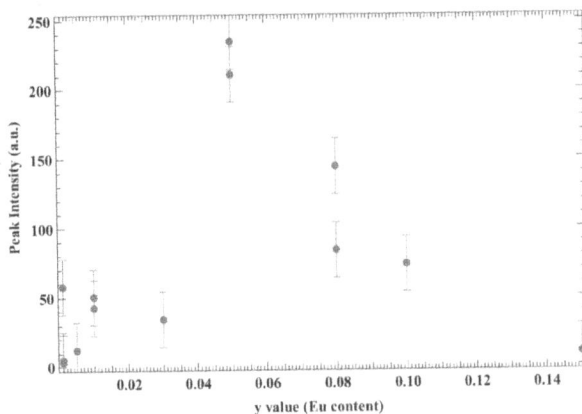

(a)

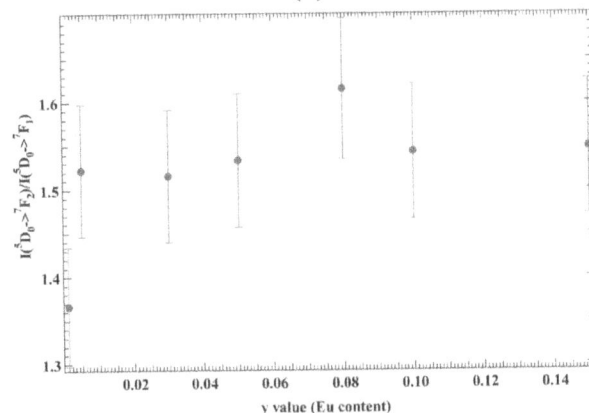

(b)

Figure 7. *(a) Emission intensities of the ($^5D_0{\rightarrow}^7F_2$) peak at 612 nm as a function of y. The intensity of this peak increases to a maximum at 5 mol% Eu_2O_3. The error bars are ±20 a.u. (b) Ratio of the integrated intensities of the $^5D_0{\rightarrow}^7F_2$ and $^5D_0{\rightarrow}^7F_1$ transitions. Errors bars represent an estimated 5% error*

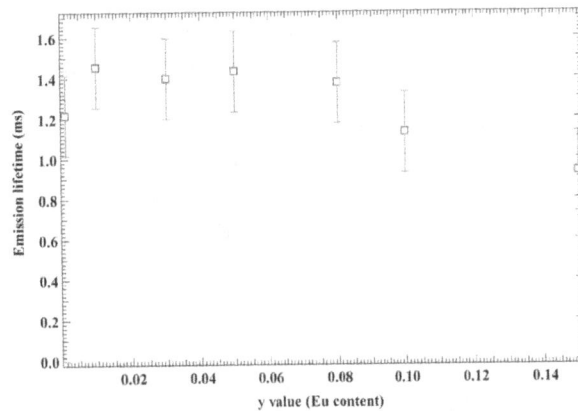

Figure 8. *Emission lifetimes of the 612 nm band as a function of y. The error bars represent 0·2 ms*

tially near borate sites, as suggested by the TOF data. The ions are relatively isolated from the rest of the network, and show no evidence of clustering. The high Dietzel field strength[14] of the Eu^{3+} cation would then result in an insensitivity to changes in the surrounding electronic environment, with the exception of longer range interactions and quenching by other Eu ions as their concentration increases. This agrees with other work[15] that found little change in the Eu emission until strongly electronegative PbF_2 units were added, and a relative insensitivity[16] in the Eu coordination number (N=6) to changes in a sodium borate network.

The variation in intensities of the ($^5D_0{\rightarrow}^7F_2$) peak at 612 nm with *y* are shown in Figure 7(a). The intensity of this peak increases to a maximum at 5% Eu_2O_3 content. Above this concentration, self quenching by the europium ions leads to a decrease in the fluorescence intensity. A traditional way[4,17,18] to look at the effect of the glass network on the Eu cation is to calculate the ratio of the fluorescence intensities for the $^5D_0{\rightarrow}^7F_2$ and $^5D_0{\rightarrow}^7F_1$ bands. The results of our calculation are shown in Figure 7(b), and they indicate very little change in the symmetry of the europium ion sites as a function of the Eu concentration. A lack of rigidity of the lead borate network would allow for the incorporation of the europium cations without any significant localized distortion of the sites (and thus without any large symmetry changes).

Finally, Figure 8 shows the lifetime of the 612 nm emission as a function of *y*. The results were best fitted using a single exponential function. The 1/*e* lifetimes are fairly consistent at about 1 ms, with perhaps a slight decrease over the compositional range. The emission lifetime is not strongly influenced by the europium concentration over the range studied.

IV. Conclusions

Measurements using laser ionization time-of-flight mass spectrometry, UV-visible emission and absorption, and infrared spectroscopy on europium doped lead borate glasses have provided insights into the structural changes occurring in the glass network. These include:

1. Up to the doping limit of the lead borate samples (15 mol% Eu_2O_3), the europium displaces the lead cations, providing local charge balance and attaching preferentially to borate units, especially metaborate rings. The displaced lead cations appear to form lead borate units with multiple Pb ions. In base lead borate glasses with higher lead concentrations (>50 mol% PbO) we previously noted[8] an increase in lead oxide units, but this was not a significant effect in the glasses reported herein. This is likely due to the total amount of modifier present.

2. The europium ions retain the role of ionic modifier in the borate network. The borate network is affected and a back-conversion of tetrahedral BO_4 to trigonal BO_3 units occurs as the Eu_2O_3 concentration is raised. The europium takes the form of single Eu^{3+} ions spread throughout the network.

3. Fluorescence measurements suggest no changes in the oxidation state of the europium ions, and very little change in the fluorescence emission lifetime. The relatively stable borate sites occupied by the Eu and its high field strength are likely explanations.

V. References

1. Zhao, D., Zhang, J., Wu, C., Qin, G., De, G., Zhang, J. & Lu, S. *Chem. Phys. Lett*, 2005, **403**, 129.
2. Jubera, V., Garcia, A., Chaminade, J. P., Guillen, F., Sablayrolles, J. & Fouassier, C. *J. Lumin.*, 2007, **124**, 10.
3. Aleksandrovsky, A. S., Krylov, A. S., Malakhovskii, A. V., Potseluyko, A. M., Zaitsev, A. I. & Zamkov, A. V. *J. Phys. Chem. Solids*, 2005, **66**, 75.
4. Wada, N. & Kojima, K. *J. Lumin.*, 2007, **126**, 53.
5. Affatigato, M., Feller, S., Schue, A., Blair, S., Stentz, D., Smith, G. B., Liss, D., Kelley, M. J., Goater, C. & Leelesagar, R. *J. Phys.: Condens. Matter*, 2003, **15**, S2323.
6. Blair, S., Stentz, D., Feller, H., Leelasagar, R., Goater, C., Feller, S. & Affatigato, M. *J. Non-Cryst. Solids*, 2001, **293–295**, 490.
7. Blair, S., Stentz, D., Goater, C., Feller, S. & Affatigato, M. *J. Non-Cryst. Solids*, 2001, **293–295**, 416.
8. Stentz, D., Blair, S., Goater, C., Feller, S. & Affatigato, M. *Appl. Phys. Lett.*, 2000, **76**, 1.
9. Hirao, K. & Soga, N. *J. Am. Ceram. Soc.*, 1985, **68**, 515.
10. Kamitsos, E. I., Karakassides, M. A. & Chryssikos, G. D. *J. Phys. Chem.*, 1987, **91**, 1073.
11. Kamitsos, E. I. *Phys. Chem. Glasses*, 2003, **44**, 79.
12. Kamitsos, E. I., Patsis, A. P., Karakassides, M. A. & Chryssikos, G. D. *J. Non-Cryst. Solids*, 1990, **126**, 52.
13. Pisarski, W. A., Pisarska, J. & Ryba-Romanowski, W. *J. Mol. Struct.*, 2005, **744–747**, 515.
14. Varshneya, A. K. *Fundamentals of Inorganic Glasses*, Academic Press, San Diego, 1994.
15. Pisarski, W. A., Pisarska, J., Maczka, M. & Ryba-Romanowski, W. *J. Mol. Struct.*, 2006, **792–793**, 207.
16. Shimizugawa, Y., Qiu, J. R. & Hirao, K. *J. Non-Cryst. Solids*, 1997, **222**, 310.
17. Bettinelli, M., Speghini, A., Ferrari, M. & Montagna, M. *J. Non-Cryst. Solids*, 1996, **201**, 211.
18. Oomen, E. W. J. L. & van Dongen, A. M. A. *J. Non-Cryst. Solids*, 1989, **111**, 205.

Proc. VI Int. Conf. Borate Glasses, Himeji, Japan, 18–22 August 2008   *Phys. Chem. Glasses: Eur. J. Glass Sci. Technol. B, August 2009, 50 (4), 237–242*

Structure–property relationships in the mixed glass former system Na_2O–B_2O_3–P_2O_5

Randilynn Christensen, Jennifer Byer, Tobias Kaufmann & Steve W. Martin*

Department of Materials Science and Engineering, Iowa State University, Ames, IA 50011, USA

Manuscript received 15 October 2008
Revised version received 24 June 2009
Accepted 26 June 2009

The structures and properties of glasses in the mixed glass former (MGF) system Na_2O–B_2O_3–P_2O_5 have been investigated. Specifically, the atomic level structures of these glasses have been investigated using IR and Raman vibrational spectroscopies and the volume (density) and glass transition temperature have been studied as representative and important physical and thermal properties. These glasses are from one of the mixed glass former glass (MGFG) systems that exhibit significantly higher alkali ion (Na^+) conductivity, and as such become attractive candidates as solid electrolytes in next generation solid state batteries. In this study, these glasses were found to also exhibit a significant mixed glass former effect (MGFE) in the density and glass transition temperature, where a maximum above the simple mixing rule behaviour was observed near the mid-composition range. The IR and Raman spectra, while not completely evaluated in this first study of these glasses, do show the appearance of vibrational bands that also maximise in intensity near the mid-composition range in these glasses. Both low (35 mol%) and mid (50 mol%) alkali oxide series were investigated, and the MGFE appears to be strongest for the low alkali series, suggesting that the MGFE likely arises from new MGF structures, whose changes appear to be greatest for the higher fraction of glass formers in the system.

1. Introduction and background

Much is known of the composition dependence (typically as a function of alkali oxide modifier content) of the atomic level structure and the physical and chemical properties of binary glass forming systems, classically studied as the progressive addition of a so-called modifying oxide salt to the base and otherwise mostly covalently bonded glass former.[1] The alkali silicates, borates, phosphates, and germanates have been the most studied for the longest time. More recently, new classes of binary glasses are being studied as well, including alkali and to a lesser extent alkaline earth vanadates, molybdates, tellurates, and tungstates.

A class of so-called mixed glass former glasses (MGFGs) that are comprised of more than one base glass former have also been studied, but to a lesser extent.[2] The structural complexity of the glassy state in general makes the detailed study of such multicomponent systems challenging at best. However, there are a few simple ternary MGFGs that are comprised of one modifying oxide and at most two glass formers where more progress can be made. The widely used alkali borosilicate series of glasses is one such example. The favourable and significant changes in structure and properties that B_2O_3 additions make upon the base sodium silicate glass make these MGFGs useful in a variety of applications.[3] The widespread use of these "Pyrex" glasses in residential, commercial and technical applications is due to the well known reduction in the fraction of nonbridging oxygens and the increase

in the fraction of tetrahedral borons that are created upon adding B_2O_3 to the base soda silicate glass. However, a less well known benefit of mixing glass formers to form a ternary glass forming system has more recently been discovered in that in almost all alkali ion containing MGFG systems studied to date, the measured alkali ion conductivity passes through a maximum that can be as much as a few orders of magnitude higher than for either of the end member binary glass forming systems.[4] This markedly increased ionic conductivity is concomitant with the same significantly improved thermal, chemical, and mechanical properties that have been observed for the alkali borate silicate glasses.

With the increasing need for improved, higher energy density and higher recharge cycle life, electrochemical energy storage systems to meet the growing demands for portable electronic devices and to offset declining petroleum based fuels, new materials are needed to enable next generation lithium batteries that are both safer and cheaper. These MGFGs are among such new materials being studied because they offer the possibility of enabling safe, environmentally benign, and affordable lithium batteries to be developed that hold the promise of reaching (or at least getting significantly closer to) the order of magnitude higher energy density that is theoretically obtainable in lithium batteries when lithium metal anodes are used. Since these MGFGs not only have orders of magnitude higher ion conductivities compared to their binary glass system counterparts, and dramatically improved thermal, chemical and mechanical properties, these glasses are significant

* Corresponding author. Email swmartin@iastate.edu

candidates for improved solid electrolytes in next generation lithium batteries.

For these reasons, we have begun the study of the MGFE (mixed glass former effect) in both sulphide and oxide glasses, and here we report the volume and thermal properties, as well as some preliminary IR and Raman spectra, for the MGFG system Na_2O–B_2O_3–P_2O_5. This system was chosen because, while Li^+ ions are smaller and hence more mobile in glass than Na^+ ions, ^{23}Na has accessible radio nuclides, for both fundamental studies of the diffusion coefficient and studies of their magnetic resonance, that will enable detailed studies of the chemical structure around the mobile Na^+ ion in these glasses. In a similar manner, ^{11}B and ^{31}P were selected as the glass forming cations because these nuclei are also very accessible with NMR spectroscopy, and will enable companion detailed studies of the network structure of these glasses.

In this paper, we report the composition dependence of the density, the molar volume, and the glass transition temperature, T_g, as representative physical properties of the glasses, and the IR and Raman spectra as a first level investigation of the atomic structure. We have so far investigated two series of constant alkali oxide modifier, 35 and 50 mol%. In future studies, we will expand our examination of these glasses to include the thermal expansion and the mechanical modulii to more fully characterise the physical properties. We will further expand our study of the structure by using NMR spectroscopy and x-ray and neutron diffraction techniques, which will be further augmented by simulation (molecular dynamics) and modelling (reverse Monte Carlo) techniques. Finally, we will carefully determine the ionic conductivities to help develop improved understanding of the MGFE in these ternary glasses.

2. Experimental methods

Two series of glasses were made, $yNa_2O.(1-y)[xB_2O_3.(1-x)P_2O_5]$, where $y=0.35$ and 0.50, and $0 \leq x \leq 1$, in steps of x of 0.1. Appropriate amounts of the starting materials, Na_2CO_3, H_3BO_3, and $(NH_4)_2HPO_4$, were combined in an alumina crucible and degassed over a Bunsen burner or in a box furnace for ~1.5 h. After degassing, the samples were ground into a powder and were taken into a N_2 glove box with <0.1 ppm O_2 and <5 ppm H_2O. The samples were melted at 1000–1100°C for ~15 min in alumina crucibles in a closed end muffle tube furnace which was hermetically connected to an outside wall of the glove box. The melts were removed from the furnace and quenched between two brass plates at room temperature, or poured into a brass annealing mould, 40°C below T_g. Powdered glass was prepared by crushing the plate quenched glass with an impact mortar and pestle. Glasses were tested for alumina contamination by

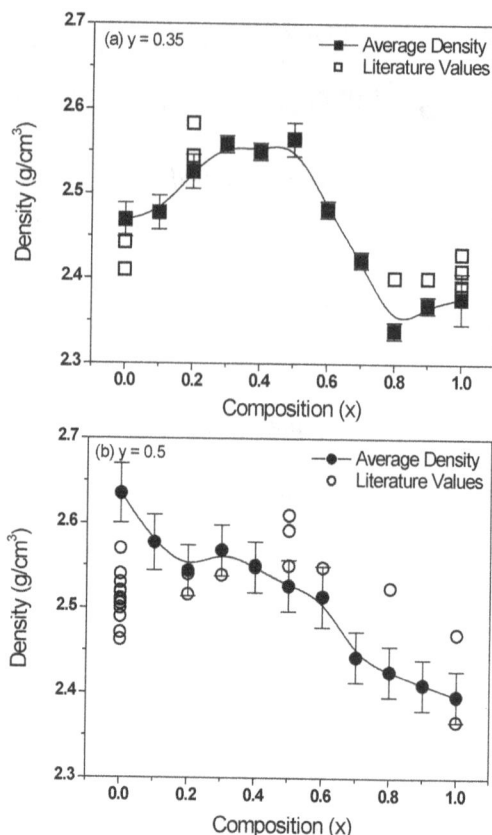

Figure 1. Composition dependence of the density for (a) the y=0·35 series, and (b) the y=0·5 series of yNa₂O.(1–y) [xB₂O₃.(1–x)P₂O₅] glasses. The error bars on the data points were determined by repeated measurements on the same sample, a measure of the measurement precision, and by repeated measurements on samples of known density, a measurement of the measurement accuracy. Literature values from Refs 5–10

EDS (energy dispersive spectroscopy) and contained less than 10 atomic% Al. All glasses were found to be amorphous by XRD.

DSC measurements were run on ~20 mg samples, hermetically sealed in Al sample pans, in a Perkin-Elmer Diamond DSC, that had previously been calibrated on the indium melting point (±1°C). Four scans were run on each sample. First, a survey scan on the as-quenched glass was run at 20°C/min by heating the sample up through T_g until it crystallised. This scan was used to determine the crystallisation temperature, T_c. A second sample was then heated at 20°C/min up to a temperature above T_g but below T_c. The sample was then cooled at 20°C/min and then reheated two times to give two separate measurements of T_g for the glass under a −20/+20°C/min thermal history.

Density measurements were made inside the glove box on bulk plate quenched and annealed samples using the Archimedes' method with dried kerosene as the immersion fluid. The balance was calibrated to ±0·001 g and the density of kerosene was determined using a density flask to ±0·01 g/ml. Repeat measurements on three samples were used and it is believed that the

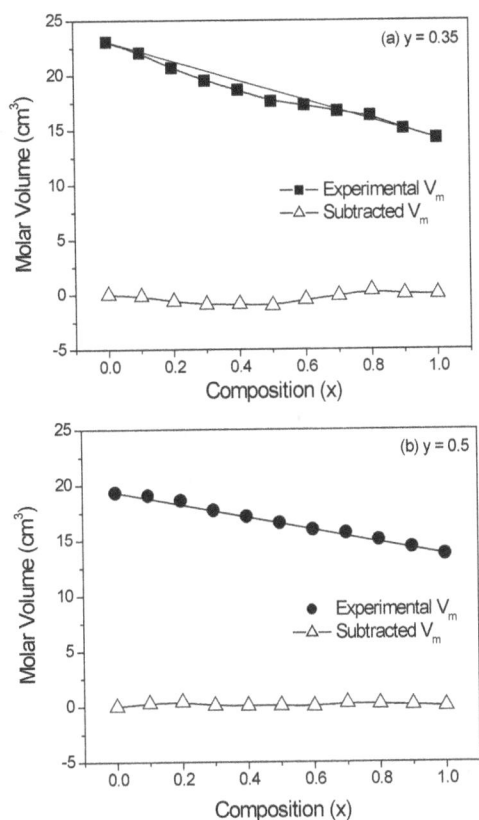

Figure 2. Composition dependence of the calculated molar volumes for (a) the y=0·35 series, and (b) the y=0·5 series of yNa$_2$O.(1–y)[xB$_2$O$_3$.(1–x)P$_2$O$_5$] glasses. The subtracted difference molar volume is the difference between the molar volume of the glass and the linear interpolation, assuming ideal mixing behaviour between the molar volumes of the two end member binary glasses. Literature values from Refs 5 to 11

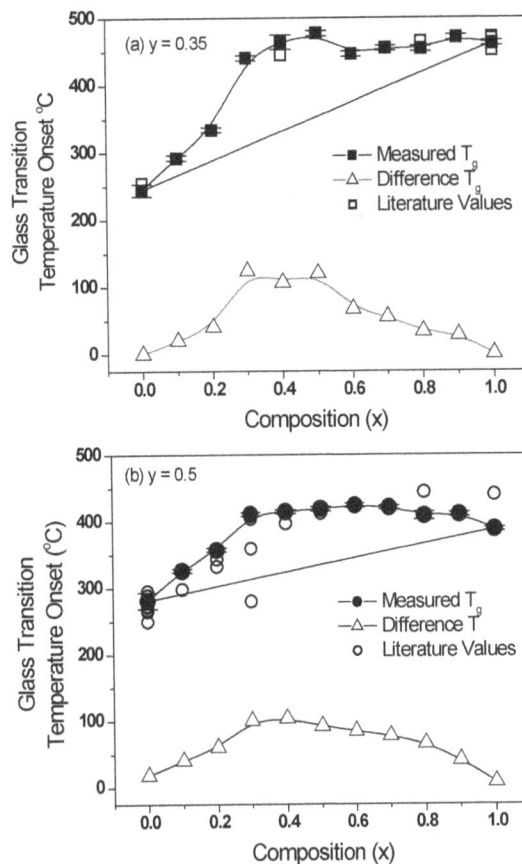

Figure 3. Composition dependence of T$_g$ and difference T$_g$ (determined as the difference between T$_g$ for the MGFG and the interpolation of the values of T$_g$ for the two end member binary glasses) for (a) the y=0·35 series, and (b) the y=0·5 series of glasses. Literature values are taken from Refs 8 and 12

accuracy of the density measurements is ±0·01 g/ml.

IR spectra of the glasses were collected using a Bruker IFS 66/v spectrometer on pressed pellets of 1–3 wt% glass powder and previously dried and milled CsI. Typically, 32 scans at 4 cm^{-1} resolution were used to collect the spectra. The Raman spectra were collected using a Renishaw inVia Raman microscope with a 488 nm Ar$^+$ laser at typically 5–10 mW power for three average collections of 10 s exposures with a 2400 lines/mm grating and a Peltier cooled CCD camera detector.

3. Results

Figure 1(a) and (b) show the composition dependence of the density and Figure 2(a) and (b) show the calculated molar volumes for the two series of glasses studied in this work. While the density values for the y=0·35 series of glasses show a very pronounced MGFE, the density values for the y=0·5 series of glasses show only a weak MGFE.

Figure 3(a) and (b) show the composition dependence of T$_g$ for the y=0·35 and y=0·5 series of glasses,

respectively. As can be seen, T$_g$ shows a pronounced MGFE in the mid-composition range. To better illustrate this behaviour, the difference between the measured T$_g$ and that expected from simple additive ideal mixing behaviour between the two end member glass formers is shown as well. The increase in T$_g$ over that of the ideal mixing behaviour is slightly larger for the y=0·35 series of glasses compared to that of the y=0·5 series of glasses, suggesting that larger changes in the glass structure occur for the y=0·35 series compared to the y=0·5 series.

Figure 4(a) and (b) give the composition dependence of the IR spectra of the y=0·35 and 0·5 series of glasses, respectively, over the wavenumber range of interest. There were no significant bands above 2000 cm^{-1}, except for a minor –OH peak due to slight, <1 wt% –OH contamination. While there are a number of peaks that can be associated with vibrations of the pure end member glasses, there are a few new bands, marked, that grow and disappear across the mixed glass former range. These new peaks, as yet not fully identified, suggest that there are new mixed –O–B–O–P–O structural groups that form in the MGFG that are unique to these glasses.

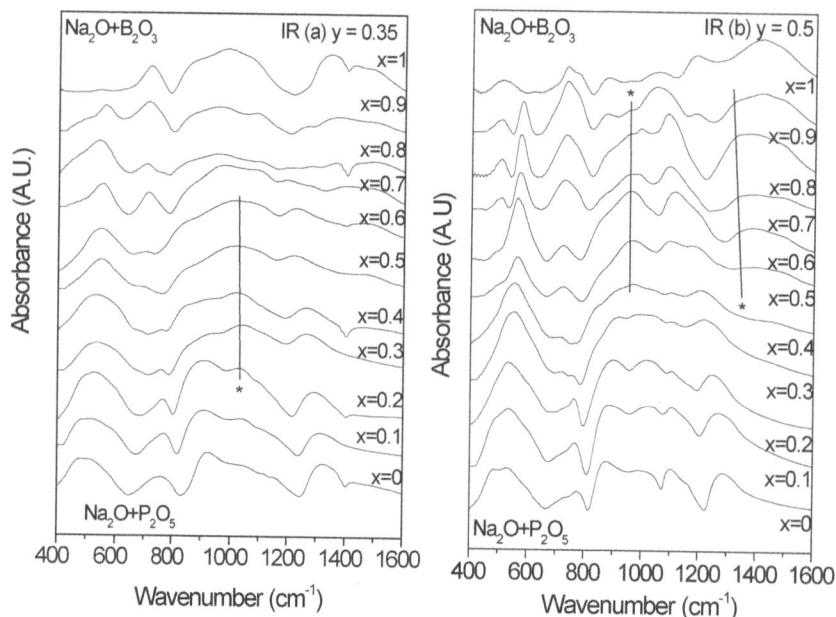

Figure 4. Composition dependence of the IR spectra for (a) the y=0·35 series, and (b) the y=0·5 series of glasses in the mid-IR frequency range

Figure 5(a) and (b) show the composition dependence of the Raman spectra of the y=0·35 and y=0·5 series of glasses. Again, only the frequency range where significant Raman bands were found is shown. As is often the case in comparing Raman and IR spectra, the Raman spectral bands are typically fewer in number and narrower in width. These two characteristics often make the Raman spectra of glass compositions easier to interpret, and since the bands overlap to a smaller degree, they are more sensitive to subtle changes in glass structure. This is the case here, and since it is expected that the changes created by the MGFE are in the second nearest neighbour coor-

dination sphere, new –B–O–P–O– bonds and linkages for example, these changes are expected to give rise to smaller changes in the spectra of these glasses compared to the more dramatic changes caused by changing the level of alkali modifier content. These latter changes bring about significant changes in the short range order bridging and nonbridging structures, and as a result induce larger changes in the vibrational spectroscopy of glass. There are, however, a few systematic changes that occur for the mid-composition glasses, that appear not to be present in either of the binary end member glasses. These are marked by lines in Figure 4 to indicate

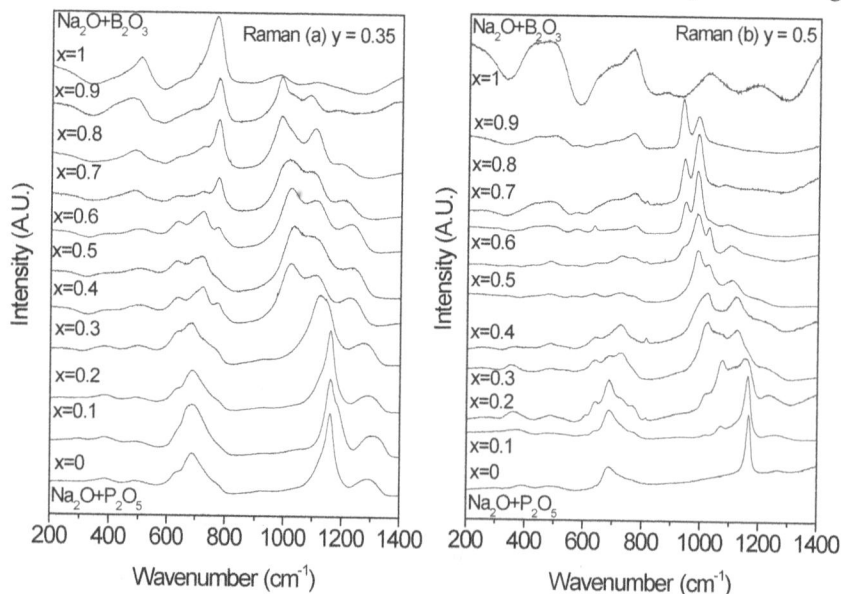

Figure 5. Composition dependence of the Raman spectra for (a) the y=0·35 series, and (b) the y=0·5 series of glasses determined on plates and pieces of glass samples using a Raman microscope with excitation at 488 nm. The notch filter on the spectrometer cuts in at 200 cm⁻¹ shifted from the Rayleigh line. There may be lower frequency modes in this range, but these were not examined at this time

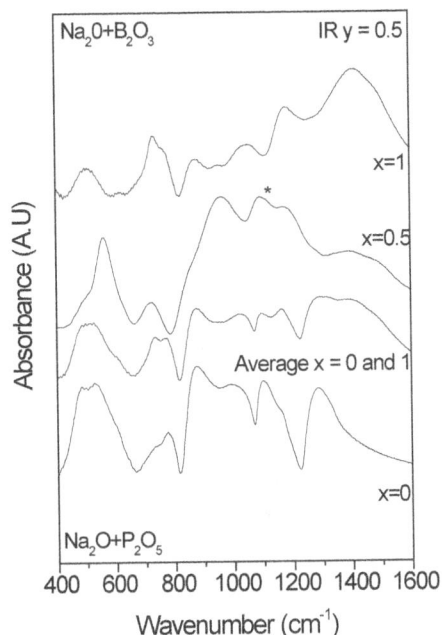

Figure 6. Calculated average IR spectra for the x=0·5 and y=0·5 glass compared to the IR spectra of the two end member glasses and the x=0·5 glass. New bands associated with new –O–B–O–P–O– mixed structures are marked with *

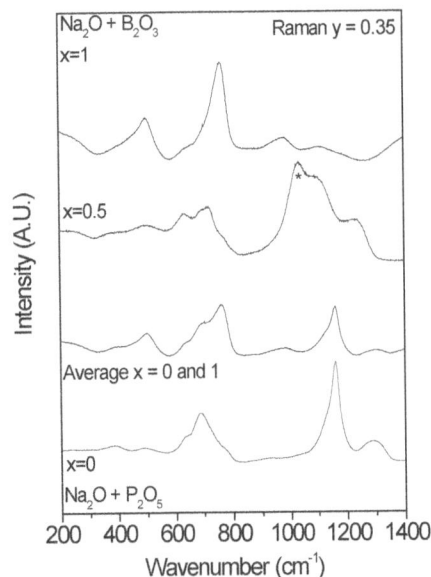

Figure 7. Calculated average Raman spectra for the x=0·5 and y=0·35 glass compared to the Raman spectra of the two end member glasses and the x=0·5 glass. New bands associated with new –O–B–O–P–O– mixed structures are marked with *

the approximate composition behaviour of these IR bands. One vibrational peak, at higher wavenumbers, appears to grow in and disappear for the borate rich glasses and another vibrational peak, at lower wavenumbers, appears to grow and disappear for the phosphate rich glasses. Again, both of these peaks indicate that these glasses have significant structural mixing for the MGFG. Such new bands would not be expected if the glasses were nothing more that simple ideal mixtures of the end member glass compositions.

4. Discussion

The goal of this project was to investigate the volume and thermal properties of the glasses and to investigate their vibrational spectroscopy. This is the first part of a larger study to investigate and understand the highly nonlinear composition dependence of the ionic conductivity in these MGFGs, the so called MGFE. So far we have determined that while the density exhibits a modest MGFE for both compositions the molar volume, on the other hand, appears not to exhibit any MGFE within experimental error. This suggests that since the density can often be dominated by the short range atomic packing of the constituent atoms (and ions) in the glass, the lack of a strong MGFE in the molar volume may indicate that the short range order, the first coordination sphere around each atom (ion) in the glass, is not strongly affected by mixing the glass formers. Indeed, this is to be expected since both end members are oxide materials with the same overall level of alkali oxide modifier and hence mixing of these two similar glasses would

to the first approximation suggest that changes would not occur to the first coordination sphere around each glass forming cation, B and P, Rather, the MFGE might be expected to involve changes in the second coordination sphere. It is in the second coordination sphere where the second glass former cation would be found if there were complete intermixing of the two glass end members; this would involve larger changes in the second coordination sphere compared to the first coordination sphere.

This conclusion that the MGFE involves changes in the second coordination sphere is supported by both the T_g data and the vibrational spectra. Since T_g is a measure of the overall connectively of the glass network, the network being no stronger than the weakest link in the structure, it is the connectivity of the B and P polyhedra, and to a lesser extent the individual nature of the B and P polyhedra, that determine the magnitude of T_g. The fact that T_g reaches a maximum at mid-composition range for each of the MGFG series suggests that there are favourable connections being created by mixing of the glass formers. The hypothesis is that these more energetically stable connections are likely occurring at longer length scales than the first coordination sphere.

The above hypothesis is borne out in the vibrational spectra of these MGFGs. In all four spectral series, the changes in the IR and Raman spectra induced by the mixing of the glass former must be considered as being relatively weak and modest. Indeed, a casual inspection of the spectra would lead one to conclude that the spectra of a MGFG is approximately nothing more than the weighted mixing of the spectra of the end member glasses. This suggests that any changes that do occur must make modest changes to the short range order of these glasses. If larger, more dramatic changes were being made to the short range order of the glasses, then one ought to see more distinct changes in the vibrational spectra which are dominated by short range order bonding effects.

To illustrate this point, Figure 6 compares the IR spectra of the MGFGs for the $y=0.5$ and $x=0.5$ mid-composition glass to that calculated by averaging the spectra of the two end member glasses. In a similar way, Figure 7 compares the Raman spectra of the two end member glasses, $NaBO_2$ and $NaPO_3$, and that of the $y=0.35$ and $x=0.5$ mid-composition glass to that calculated by averaging the spectra of the two end member glasses. As shown in Figures 6 and 7, the actual spectrum has many of the same bands as the average of the two end member spectra with only a few new bands (the strongest of which are marked), representative of mixed –O–B–O–P–O– atomic structures.

From the weak changes in the density and the vibrational spectra compared to the stronger changes in the composition dependence of T_g, it is hypothesised that the MGFE has its greatest effect on the intermediate range order structure of these glasses. This may well be a partial, though incomplete, reason why the MGFE appears to have such a strong influence on the alkali ion conductivity of the MGFG systems. There are important local energetics that control the local (first coordination) sphere energy barriers that the mobile ions experience in their conduction, such as the charge density of the counter anion and the so called "doorway" radius that the mobile cation must diffuse through to enter the next available site. However, it is the long range diffusion of the mobile ions that determines the measured ionic conductivity in these glasses. Long range diffusion of the mobile alkali ions requires that the alkali cations move over extended regions of the glass structure and in doing so, their motions become functions of the longer range structure of the glass, namely, the intermediate range order of the glass. In a similar way, the glass transition (T_g) requires cooperative diffusion of large segments of the glass structure, that comprise the intermediate range order of the glass. Hence, the fact that T_g also shows its maximum effect at the intermediate values of y ($y=0.5$, for example) for both series of glasses suggests that the MGFE is most manifest in changes to the intermediate range order of theses glasses.

In future studies of these glasses, we will focus on this hypothesis that the MGFE is most manifest in the intermediate range order changes to the atomic structure. Specifically, we will use structural techniques that probe the intermediate range order in these glasses, such as multiple quantum NMR techniques, and x-ray and neutron diffraction and simulation and modelling techniques such as reverse Monte Carlo and molecular dynamics. We will also measure physical properties, such as ionic conductivity and tracer and self diffusion,

that are more sensitive to, and dependent upon, the intermediate range order of the glasses.

5. Summary and conclusions

Two sodium borophosphate mixed glass former glass (MGFG) series have been investigated to determine if the mixed glass former effect (MGFE) is manifest in the volume and thermal properties of these glasses. While a strong MGFE was observed in T_g for both MGFG series, the MGFE was much weaker or not present in the molar volume of these glasses. IR and Raman spectra, while not fully interpreted at this time, are consistent with the hypothesis that the mixing of the glass formers in this glass system causes greater changes in the glass structure at longer length scales, leaving the short range order structure, the coordination number and coordinating ligand anion, O, largely unchanged. This hypothesis is further consistent with the observed strong MGFE in the alkali ion conductivity in these glasses that is believed to be dependent upon intermediate range structures in the glass to support long range alkali ion conduction. Future work will explore this hypothesis further by using techniques that are more sensitive to the intermediate range order structure in these glasses.

6. Acknowledgements

This research was supported in part by a grant from the National Science Foundation DMR-0710564.

References

1. Rawson, H. Oxide glasses, *Mater. Sci. Technol.*, 1991, **9**, 279.
2. Pradel, A., Rau, C., Bittencourt, D., Armand, P., Philippot, E. & Ribes, M. Mixed Glass Former Effect in the System 0.3Li₂S-0.7[(1-x)SiS₂-xGeS₂]: A Structural Explanation, *Chem. Mater.*, 1998, **10**, 2162.
3. Kreidl, N. J. Inorganic Glass Forming Systems, in *Glass: Science and Technology, Volume 1: Glass-Forming Systems*, Edited by N. J. Kreidl & D. R. Uhlmann, Academic Press, New York, 1983, Chap. 3.
4. Zielniok, D., Cramer, C. & Eckert, H. Structure/Property Correlations in Ion-Conducting Mixed-Network Former Glasses: Solid-State NMR Studies of the System Na₂O-B₂O₃-P₂O₅, *Chem. Mater.*, 2007, **19**, 3162.
5. Zielniok, D., Cramer, C. & Eckert, H. Structure/Property Correlations in Ion-Conducting Mixed-Network Former Glasses: Solid-State NMR Studies of the System Na₂O-B₂O₃-P₂O₅, *Chem. Mater.*, 2007, **19**, 3162.
6. Feller, S. A., Lower, N. & Affatigato, M. Density as a probe of oxide glass structure, *Phys. Chem. Glasses*, 2001, **42**, 240.
7. Budhwani, K. & Feller, S. A density model for the lithium, sodium and potassium borosilicate glass systems, *Phys. Chem. Glasses*, 1995, **36**, 183.
8. Hudgens, J., Brow, R., Tallant, D. & Martin, S. Raman spectroscopy study of the structure of lithium and sodium ultraphosphate glasses, *J. Non-Cryst. Solids*, 1998, **223**, 21–31.
9. Alekseev, N. E., Gapontsev, V. P., Izyneev, A. A., Zhabotinskii, M. E., Kravchenko, V. B. & Rudnitskii, Yu. P. *Issledovaniya v Oblasti Radiotekhniki i Elektroniki*, 1975, 401.
10. Alekseev, N. E., Gapontsev, V. P., Zhabotinskii, M. E., Izyneev, A. A., Kopylov, Yu. L., Kravchenko, V. B. & Rudnitskii, Yu. P. *Journal?*, 1977, **3**, 99ff.
11. Martin, S. W. & Angell, C. A. Glass formation and transition temperatures in sodium and lithium borate and aluminoborate melts up to 72 mol% alkali, *J. Non-Cryst. Solids*, 1984, **66**, 429.
12. Affatigato, M., Feller, S., Khaw, E. J., Feil, D., Teoh, B. & Mathews, O. The glass transition temperature of lithium and lithium sodium borate glasses over wide ranges of composition, *Phys. Chem. Glasses*, 1990, **31**, 19.

Proc. VI Int. Conf. Borate Glasses, Himeji, Japan, 18–22 August 2008   Glass Tech.: Eur. J. Glass Sci. Technol. A, October 2009, 50 (5), 285–287

Composite materials based on immiscible borate glass and iron spinel nanoparticles

Elena Kashchieva, *Yanko Dimitriev, Vanya Ivanova*

University of Chemical Technology and Metallurgy, 8 Kl. Ohridski Blvd., 1756 Sofia, Bulgaria

Toshka Merodiiska

Institute of Electronics, Bulgarian Academy of Sciences, 72 Tzarigradsko Shausee Blvd., 1784 Sofia, Bulgaria

Manuscript received 18 August 2008
Revised version received 22 December 2008
Accepted 15 January 2009

Composite materials based on a borate glass matrix and iron spinel nanoparticles were prepared by the incorporation technique. The glass composition was selected to be on the phase separation boundary in the system B_2O_3–TeO_2–Fe_2O_3. The batches were melted at 800–1000°C and slowly (100°C/min) or fast (1000°C/min) cooled. Three different types of nanosized particles containing iron (Fe_3O_4, $CoFe_2O_4$ and $CuFe_2O_4$) were obtained by the coprecipitation method in an alkali medium. The composites were produced by a process including: (i) synthesis, powdering, mixing and melting of homogenised 90 wt% $30B_2O_3.60TeO_2.10Fe_2O_3$ glass together with 10 wt% iron spinel nanoparticles; (ii) slow cooling of the composite melts; (iii) heating of the obtained composites at 800°C and slow cooling in an external magnetic field. Using TEM and SEM it was established that the synthesis conditions have an influence on the microstructure of the matrix and composite materials. The magnetic field stimulates the orientation of the heterogeneous formations which also depends on the thermal history of the samples.

Introduction

For the production of new composite materials with combinations of size dependent properties it is necessary to identify a convenient choice of precursors and route of fabrication.[1–3] When the incorporation method is used, the matrix microstructure may be an issue.[4] In this field the application of phase separated glasses is accepted as an attractive non-traditional means of obtaining composites.[5]

In this investigation we aim to produce composite materials by incorporating iron spinel nanoparticles (Fe_3O_4, $CoFe_2O_4$ or $CuFe_2O_4$) in a borate–tellurite based glassy matrix which has a high tendency for immiscibility. The aim is to follow the structural evolution in the composites, depending on the rate of melt cooling, as well as on the presence of an external magnetic field during the cooling. The present study continues our previous works concerning the phase separation and magnetic properties of borate–tellurite glasses containing Fe_2O_3, MnO and CoO,[6,7] and composites based upon them.[8]

The glass used in this study is from the system B_2O_3–TeO_2–Fe_2O_3, in which the immiscibility region covers a much greater range in composition than the glass formation range.[6] The composition of the selected glass, $30B_2O_3.60TeO_2.10Fe_2O_3$, is located on the boundary between the two regions. The presence of boron oxide in the matrix stimulates the creation of microheterogeneous structures, which provoke the segregation of iron ions in separated micro-

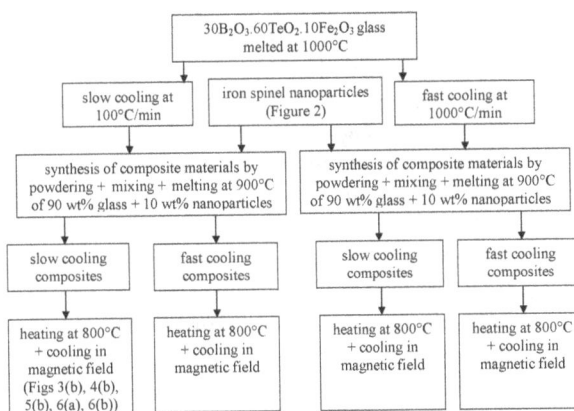

Figure 1. Processing scheme for preparation of composite materials

regions.[7] There is no chemical interaction between B_2O_3 and TeO_2, and thus the phases TeO_2, $4TeO_2.Fe_2O_3$ and $TeO_2.Fe_2O_3$ crystallise in TeO_2-rich samples, while crystals of a-Fe_2O_3 appear in supercooled melts with compositions near the B_2O_3–Fe_2O_3 side.[7] The presence of Fe_2O_3 limits the hydration trend in B_2O_3-rich glasses and stimulates the crystallisation.

Experimental

Composites were produced according to the scheme presented in Figure 1.

The glassy matrix batches were prepared with high purity raw materials: TeO_2, B_2O_3 and Fe_2O_3. Melting was performed at 900°C. Two cooling rates were applied – slow cooling at a rate about 100°C/min, and fast

---

*Corresponding author. Email elena_kaschieva@yahoo.com

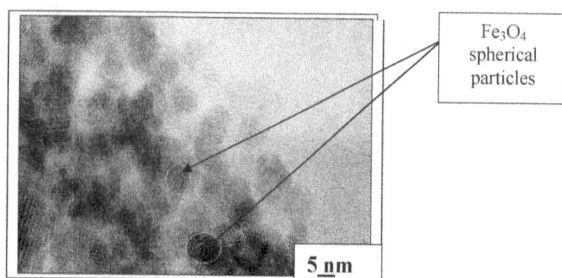

Figure 2. HRTEM micrograph of Fe_3O_4 particles

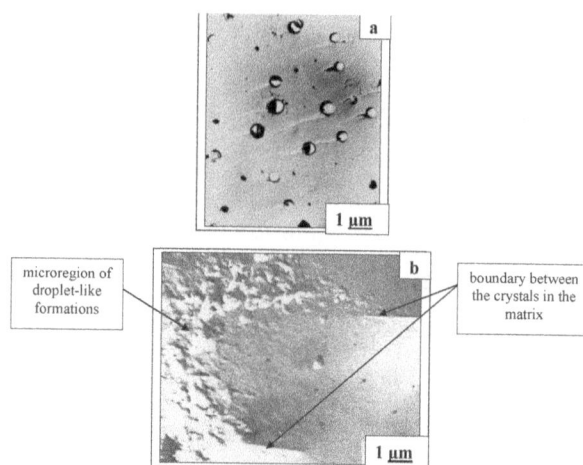

Figure 3. TEM micrographs of a composite formed from 90 wt% glass and 10 wt% Fe_3O_4: (a) slow cooled sample; (b) heated at 800°C and cooled in a magnetic field

cooling at 1000°C/min. The spinel powder particles of Fe_3O_4, $CoFe_2O_4$ and $CuFe_2O_4$ were synthesised at room temperature by the coprecipitation method.[9] Very dilute water chloride solutions of Fe^{3+}, Fe^{2+}, Cu^{2+} and Co^{2+} were mixed in the exact ratio required to give stoichiometric compounds. The high alkalinity of the media, which is necessary for cation coprecipitation, was achieved by adding NaOH solution to the mixed chloride solutions. The black precipitates obtained were washed by distilled water to pH=7 and dried at 50°C in a conventional drier.

The composite was produced from a mixture containing 90 wt% of previously powdered $30B_2O_3.60TeO_2.10Fe_2O_3$ glass and 10 wt% nanoparticles. The mixture was melted at 900°C and slow cooled. Additional heating in silica glass crucibles at 800°C was performed. These melts were free cooled for one hour between the poles of an electromagnet, which produced a magnetic field with induction B=1 T.

The microstructure of the bulk samples was observed by transmission electron microscopy (TEM, Philips EM-400) using the C+Pt replica technique and by scanning electron microscopy (SEM, Philips 525M). All observations were performed on surfaces which were cut in the direction of the applied external magnetic field.

The synthesised spinel powders were investigated by high resolution transmission electron microscopy (HRTEM, JEOL JEM-400 EX) and by x-ray diffraction (XRD) analyses using Co K_α radiation on a TUR-M62 diffractometer with Bragg–Brentano geometry and a computer controlled goniometer, HZG-3.

Results

According to the HRTEM observations of the spinel compounds, they consist of spherical grains with mean diameter D of 10 nm or less: $D_{Fe_3O_4}$=10 nm (Figure 2); $D_{CuFe_2O_4}$=6 nm; $D_{CoFe_2O_4}$=9 nm.[9] Their XRD phase identification, based on JCPDS data,[10] shows that the synthesised materials are mono-phase spinels.

Figure 3(a) illustrates the microstructure of slow cooled composite materials containing 90 wt% $30B_2O_3.60TeO_2.10Fe_2O_3$ glass and 10 wt% Fe_3O_4 nanoparticles. Glassy matrix and well shaped spherical immiscible drops with sizes between 0·1 and 1 μm are observed. After heat treatment of this composite

at 800°C, and its slow cooling in the presence of an external magnetic field, the microstructure is more complex with a pronounced orientation tendency (Figure 3(b)). In composites with $CoFe_2O_4$ or $CuFe_2O_4$ nanoparticles, the external magnetic field leads to an increase in the number of crystals and also in their orientation (Figures 4 and 5). It is established by SEM that the iron is uniformly distributed in the whole composite volume (Figure 6).

Discussion

Recently matrices with different microstructures have been used for fabrication of nanocomposites by the incorporation technique. In many cases they are kinetically stable and amorphous with reduced reaction and crystallisation under heat treatment. Our previous experience[11,12] is related to the preparation of composite materials based on homogeneous lead borate glasses incorporating the $PbMoO_4$ phase with crystal sizes between 10 nm and 50 nm. The synthesis procedure included mixing, pressing, sintering and melting of the starting components. It has been established that the content of uniformly incorporated $PbMoO_4$ crystals has an influence on the

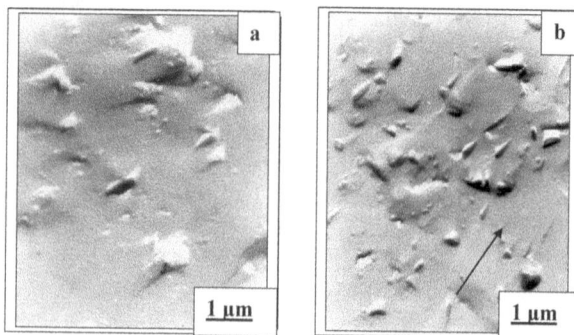

Figure 4. TEM micrographs of a composite containing 10 wt% $CoFe_2O_4$ particles: (a) slow cooled sample; (b) heated at 800°C and cooled in a magnetic field

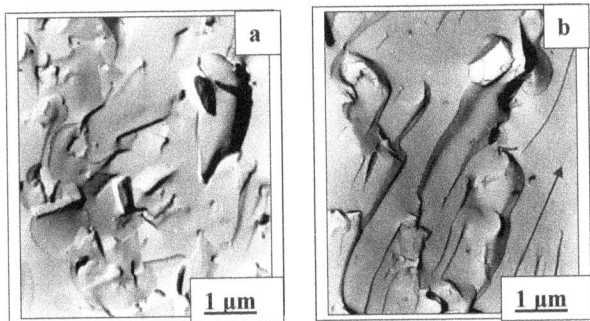

Figure 5. TEM micrographs of a composite containing 10 wt% $CuFe_2O_4$ particles: (a) slow cooled sample; (b) heated at 800°C and cooled in a magnetic field

volume density and specific electrical conductivity of the samples.

A heterogeneous matrix is a possible alternative medium for producing composites by the incorporation route. The presence of different types of inhomogeneities allows the inclusion of micro- and nanoparticles in them, as well as the tracing of the influence of the external factors on the microstructure. For this reason we selected as the matrix for these new composites an immiscible glass containing micron and submicron droplet-like formations. The TEM analysis illustrates that the synthesis conditions have an influence on the structure of this matrix. It has been already established[7] that fast cooling of this glassy matrix leads to formation of a homogeneous structure, while slow cooling allows the evolution of micro- and macrophase separation and crystallisation.

In the composite melts, slow cooling stimulates the development of both droplet-like immiscibility and crystalline formations. According to our preliminary work, the composite containing Fe_3O_4 nanoparticles

shows a microstructure similar to the pure matrix,[7] but some structural inhomogeneities related to the included particles are formed.[8] The EDAX studies[8] showed that the observed immiscibility drops are rich in B_2O_3 and also in both types of iron containing magnetic phases: Fe_2O_3 from the glass, and Fe_3O_4 from the nanoparticles. For this reason, the influence of the magnetic field on the microstructure is clearly pronounced and provokes simultaneously the appearance of oriented matrix crystallisation and segregation in microregions of immiscibility drops in alternation by the microcrystals (Figure 3(b)). This complex structure is due to the superposed influence of the magnetic field on the iron containing phases in the matrix and in the nanoparticles. In composites with $CoFe_2O_4$ or $CuFe_2O_4$ nanoparticles, the influence of the magnetic field is also well expressed. In these two cases the crystallisation ability predominates.

Conclusions

Composite materials have been produced by the incorporation technique based on $30B_2O_3.60TeO_2.10Fe_2O_3$ glass, which has a tendency towards immiscibility, and uniformly distributed nanoparticles of Fe_3O_4, $CoFe_2O_4$ or $CuFe_2O_4$. Additional changes of the structure are realised by the action of an external magnetic field applied during the cooling of the melts. The process of structural evolution is complex and strongly depends on the presence of the magnetic field and the thermal history of the investigated samples: the magnetic field stimulates mainly the orientation of the microstructure, while the heat treatment stimulates increasing crystallisation.

Acknowledgements

The study was performed with financial support of the Bulgarian National Foundation of Science under Grant No. BY-TH-102/2005 and by UCTM, Sofia under Grant No. 10611/2008.

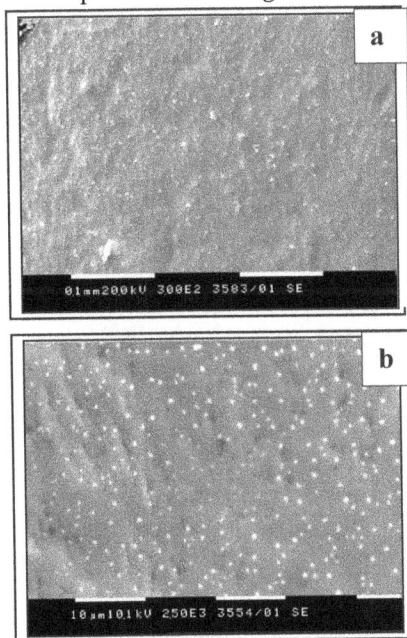

Figure 6. SEM micrographs of composites slow cooled, heated and cooled in magnetic field containing: (a) 10 wt% $CoFe_2O_4$ (×300); (b) 10 wt% $CuFe_2O_4$ (×2500)

References

1. Moriarty, P. Rep. Prog. Phys., 2001, 64, 297.
2. Davis, S. & Klabunde, K., Am .Chem. Soc., 1982, 82, 153.
3. Gleiter, H., Nanostruct. Mater., 1992, 1, 1.
4. Komatsu, T., Shioya, T. & Matusita, K. J. Am. Ceram. Soc., 1993, 76, 2923.
5. Köster, U. Mat. Sci. Forum, 1997, 235, 377.
6. Dimitriev, Y., Kashchieva, E. & Koleva, M. J. Mater. Sci., 1981, 16, 3045.
7. Kashchieva, E., Ivanova, V., S. Stefanova & Dimitriev, Y. Glastech. Ber. Glass Sci. Technol., 2002, 75 (C2), 334.
8. Kashchieva, E., Ivanova, V., Dimitriev, Y., Kordov, A., Nedkov, I. & Merodiiska, T. In: Nanoscience & Nanotechnology, Eds E. Belabanova & I. Dragieva, Heron Press, Sofia, 2002, Vol. 2, p. 94.
9. Nedkov, I., Merodiiska, T., Slavov, L., Vandenberghe, R.E., Kusano, R.E. & Takada, J. J. Magn. Mater., 2006, 300, 358.
10. International Center for Diffraction Data, Alphabetical Indexes, Pennsylvania 19073-3273, 1997, sets 1-86.
11. Kashchieva, E., Ivanova, V., Jivov, B. & Dimitriev, Y. Phys. Chem. Glasses, 2000, 41, 355.
12. Ivanova, V., Kashchieva, E., Jivov, B. & Dimitriev, Y. In: Nanoscience & Nanotechnology, Eds E. Belabanova & I. Dragieva, Heron Press, Sofia, 2001, Vol. 1, p. 30.

Proc. VI Int. Conf. Borate Glasses, Himeji, Japan, 18–22 August 2008   *Phys. Chem. Glasses: Eur. J. Glass Sci. Technol. B, April 2009, **50** (2), 79–84*

An XPS study of the O1s binding energy of sodium borate glasses

Hiroyo Segawa, Tetsuji Yano & Shuichi Shibata

Department of Chemistry and Materials Science, Graduate School of Science and Engineering, Tokyo Institute of Technology, 2-12-1 O-okayama, Meguro-ku, Tokyo 152-8550, Japan

Manuscript received 16 September 2008
Revision received 9 October 2008
Manuscript accepted 10 December 2008

The photoelectron spectra of the oxygen 1s (O1s) orbital of xNa$_2$O.$(100-x)$B$_2$O$_3$ $(0 \leq x \leq 35)$ glasses have been investigated by x-ray photoelectron spectroscopy (XPS). In the O1s photoelectron spectra, only one broad band was observed and the binding energy shifted to lower energy with increasing Na$_2$O content. The spectrum of each sodium borate glass was deconvoluted into three bands, assigned to B^3–$\varnothing$–B^3, B^3–$\varnothing$–B^4 and B^4–$\varnothing$–B^4 (B^3 or B^4 indicates three- or four-coordinated boron). The ratio of these three types of oxygen atoms was calculated from structural group information based on previous NMR data. The O1s binding energies of the three types of oxygen are in the order: B^3–$\varnothing$–$B^3 \approx B^3$–$\varnothing$–$B^4 > B^4$–$\varnothing$–B^4. It was found that B^4–$\varnothing$–B^4 has more negative charge than the other oxygens. The B^4–$\varnothing$–B^4 content is negligibly small for less than 20 mol% Na$_2$O, but for larger Na$_2$O content a small amount of B^4–$\varnothing$–B^4 was observed. The formation of oxygen with high negative charge suggests that the borate anomaly occurs at a composition of about 20 mol% Na$_2$O.

1. Introduction

The coordination number of boron in alkali borate glasses is known to change from three to four with increasing alkali oxide content, resulting in many peculiar phenomena (i.e. the borate anomaly). In Na$_2$O–B$_2$O$_3$ binary glasses, the thermal expansion coefficient[1] shows the anomaly: it increases with increasing Na$_2$O content, shows a maximum at about 20 mol%, and then decreases for higher Na$_2$O content.

From Krogh-Moe's model,[2] NMR[3–5] and Raman spectra,[6] the change in the properties of Na$_2$O–B$_2$O$_3$ glasses as a function of Na$_2$O content is attributable to the formation of various structural groups such as boroxol, tetraborate and diborate. The ratio of these groups varies with glass composition, which causes the borate anomaly.[1] Glassy B$_2$O$_3$ is known to consist of interconnected BO$_3$ triangles ($\varnothing$ denotes an oxygen atom bridging two boron centres) and form planar boroxol rings. The BO$_3$ triangles change into BO$_4^-$ tetrahedra with addition of alkali oxide. For alkali borate glasses, xM$_2$O.$(100-x)$B$_2$O$_3$ (x is in mol% M$_2$O, M is an alkali element), the early studies showed that the fraction of BO$_4^-$ units, N_4, follows the equation

$$N_4 = \frac{x}{100-x} \tag{1}$$

and is independent of the kind of alkali ion for $x<30$. For alkali content above $x=30$, nonbridging oxygen (NBO) atoms are formed on borate triangular units, leading to metaborate triangles B$_2$O, pyroborate dimers B$_2$$\varnothing$O$_4^{4-}$, and orthoborate monomers BO$_3^{3-}$.

Many authors have studied the local structures of various metal ions in alkali borate glasses as follows. Measurements have been made of the absorption coefficient of Co^{2+}[7] and Ni^{2+} ions,[8] the coordination of V^{5+} ion,[8] and the redox equilibrium of Cr^{3+}$\rightleftharpoons$Cr^{6+}[9] and it is reported that these values change markedly over $8<x<28$ mol% alkali oxide content. Duffy *et al*[10] have calculated optical basicities from measurements of the ultraviolet $^1S_0 \rightarrow {}^3P_1$ frequency of p-block probe ions, such as Tl$^+$, Pb^{2+} and Bi^{3+}. Marked increases in the optical basicity occurred over $18<x<28$ mol% alkali oxide content, and were explained as the switching of ions such as Pb^{2+} from low to high basicity sites.[11–13] Kamitsos *et al*[14–16] proposed a two-site model for modifier cations such as alkali and alkaline earth ions which was derived from far-infrared spectroscopy. These abrupt changes in optical absorption showed a dependence on the kind of modifier alkali ion, and the changes occur at lower alkali contents as the radius of the alkali ion increases.[9,11,12] Some researchers used Raman,[17] NMR[18,19] and neutron scattering[20] to investigate how N_4 depends on the kind of alkali ion and it was found that the value of N_4 decreases from Li to Cs. In xLi$_2$O.$(100-x)$B$_2$O$_3$ glasses, N_4 follows Equation (1) in the range $0<x\leq28$, and B$\varnothing_2$O$^-$ units may be formed for $x>28$.[21]

On the other hand, the chemical states of the oxygen atoms can be investigated by x-ray photoelectron spectroscopy (XPS). Especially in alkali silicate glasses, their O1s spectra separate into two bands which are assigned to bridging and NBOs depending on the charges on oxygen atoms.[22–24] In alkali borate glasses, only one band is observed

---

[1] Corresponding author. Email hsegawa@ceram.titech.ac.jp

in the O1s spectra with a wide FWHM (full width at half maximum).[25-27] The broad O1s spectrum of borate glasses originates from the overlap of various structural groups such as boroxol, tetraborate and diborate groups. Since the XPS spectrum gives us information about the binding energy of oxygen in various structural groups, deconvolution of the O1s spectrum into bands assigned to oxygen atoms in different groups is desirable to understand the borate anomaly in particular composition ranges for alkali borate glasses. Using the XPS technique, we can also explain the abrupt change in optical absorption, such as is observed in the optical basicity of borate glasses.

In this paper, we report measurements of the binding energies of O1s photoelectrons for $Na_2O–B_2O_3$ glasses. The broad O1s spectra are deconvoluted into several bands assigned to oxygen in various structural units, and the binding energies of each oxygen atom and the FWHM values are evaluated. On the basis of the results of the deconvolution, we discuss the origin of the borate anomaly for $Na_2O–B_2O_3$ glasses, including the change in optical basicity.

2. Experimental

Batches sized to yield 70 g of $xNa_2O.(100−x)B_2O_3$ $(4.4 \leq x \leq 35)$ glasses were prepared from reagent grade Na_2CO_3 and H_3BO_3. The batches were melted at 1000°C for 2 h in platinum crucibles and then poured. The glasses were crushed and were remelted for another 2 h to improve homogeneity. The glass melts were poured on a carbon plate, and immediately transferred to a preheated annealing furnace at T_g+15°C. After annealing for 1 h, the glasses were cooled at a slow rate.

Pure B_2O_3 glass was prepared from reagent grade H_3BO_3. The reagent was melted in a platinum crucible at 1000°C for 24 h under nitrogen gas flow to eliminate water. The resultant melt was heated up to 1200°C for 10 min, and then poured and annealed.

The glass samples obtained were cut to a size of 2.5×2.5×16 mm to measure XPS spectra. The samples were fractured in ultra high vacuum (<1×10⁻⁹ Torr), and their photoelectron spectra were measured by an XPS spectrometer (ULVAC PHI Inc., Model 5500MT) with monochromatic Al K_α x-rays ($h\nu$=1486.6 eV). High resolution photoelectron spectra were collected by a spherical capacitor at an interval of 0.025 eV and band pass energy of 5.85 eV. A low energy flood gun was used to control surface charging. Binding energies were referenced to C 1s spectra from hydrocarbon (binding energy of 284.6 eV) adsorbed on the analysing surface.

The compositions of these glasses were determined by inductively coupled plasma spectroscopy (SII, Model SRS 1500VR), after the glasses were dissolved with HCl.

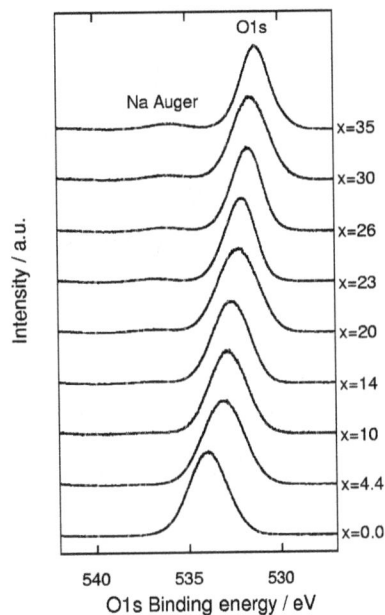

Figure 1. O1s photoelectron spectra of $xNa_2O.(100−x)B_2O_3$ glasses

3. Results

XPS spectra of $xNa_2O.(100−x)B_2O_3$ $(x \leq 35)$ glasses are shown in Figures 1–3. In Figure 1, O1s spectra are shown. Small peaks at about 536 eV were assigned to Na Auger electrons. Each of the O 1s spectra consists of only one broad band, and the spectra shift to lower energy with increasing Na_2O content. These broad spectra can be regarded as containing several bands due to oxygen atoms in different structural groups such as boroxol, tetraborate and diborate. B1s and Na 1s spectra are shown in Figure 2 and 3, respectively.

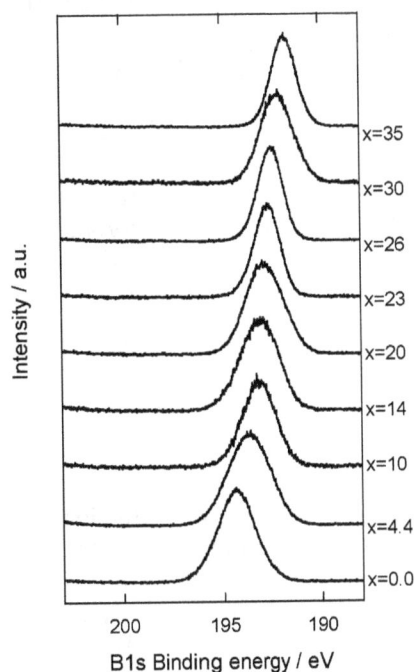

Figure 2. B1s photoelectron spectra of $xNa_2O.(100−x)B_2O_3$ glasses

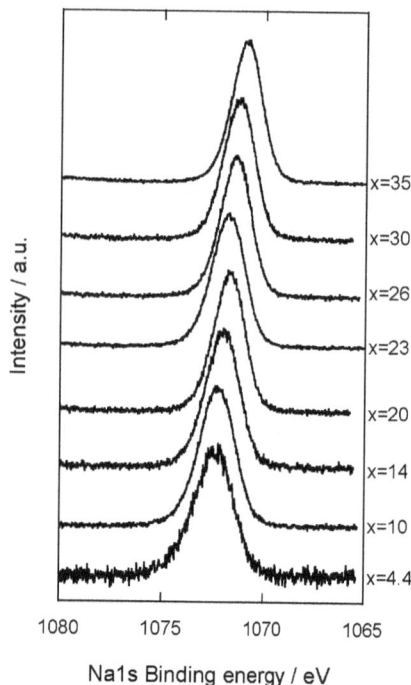

Figure 3. Na1s photoelectron spectra of xNa_2O.
$(100-x)B_2O_3$ glasses

These spectra also consist of one band, which shifts
to lower energy with increasing Na_2O content. The
energy shift of each spectrum is the same as that of
the O1s spectrum. These results mean that the charge
density on each atom increased with increasing Na_2O
content according to the 'charge potential model'.[28]

Using early ^{10}B NMR data measured by Jellison
et al[4] as a basis, the O1s spectra were deconvoluted.
According to the early NMR data, it may be assumed
that NBO atoms are not present in $xNa_2O.(100-x)B_2O_3$
($x \leq 35$) glasses. Thus, the number of moles of all bridging oxygen atoms, $N_{B-\emptyset-B}$, in these glasses is given by

$$N_{B-\emptyset-B} = N_{B^3-\emptyset-B^3} + N_{B^3-\emptyset-B^4} + N_{B^4-\emptyset-B^4} = \frac{300-2x}{100} \quad (2)$$

B^n indicates n-coordinated boron; B^3 and B^4 are
three- and four-fold coordinated borons, respectively.
$N_{B^3-\emptyset-B^4}$ and $N_{B^4-\emptyset-B^4}$ can be calculated from the following equation using Jellison's NMR data;[4]

$$N_{B^3-\emptyset-B^4} = \left(3D + 4T^4\right)\left(\frac{100-x}{100}\right) \quad (3)$$

$$N_{B^4-\emptyset-B^4} = \frac{D}{2}\left(\frac{100-x}{100}\right) \quad (4)$$

where T^4 and D are the molar fractions of tetrahedral coordinated boron in tetraborate and diborate
groups, respectively. The molar ratios of $B^3-\emptyset-B^3$,
$B^3-\emptyset-B^4$ and $B^4-\emptyset-B^4$ ($N_{B^3-\emptyset-B^3}$, $N_{B^3-\emptyset-B^4}$ and $N_{B^4-\emptyset-B^4}$)
are obtained from Equations (2–4). In Figure 4, the
molar ratios of oxygen atoms are plotted as a function of Na_2O content. With increasing Na_2O content,
$N_{B^3-\emptyset-B^3}$ decreases, while $N_{B^3-\emptyset-B^4}$ increases. In glasses

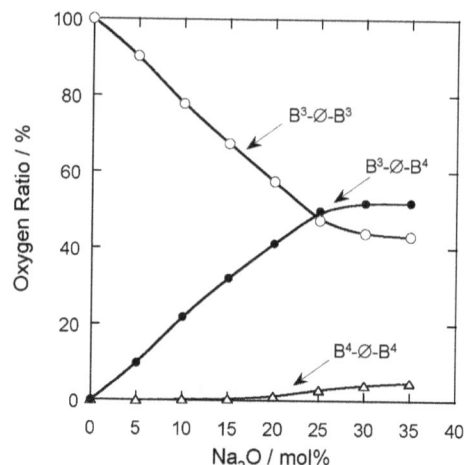

Figure 4. Ratio of oxygen atoms in the different structural
groups as a function of Na_2O concentration; ○ $B^3-\emptyset-B^3$;
● $B^3-\emptyset-B^4$; △ $B^4-\emptyset-B^4$

containing $x \leq 15$, $N_{B^4-\emptyset-B^4}$ is negligible (less than 1%),
but for $x > 15$, a small value was found for this fraction.

The broad O 1s spectra in Figure 1 were deconvoluted into two or three bands, which were assigned
to each type of oxygen atom; $B^3-\emptyset-B^3$, $B^3-\emptyset-B^4$ and
$B^4-\emptyset-B^4$. The spectra were deconvoluted so that the
areas of each band were proportional to the $N_{B^3-\emptyset-B^3}$,
$N_{B^3-\emptyset-B^4}$ and $N_{B^4-\emptyset-B^4}$. The deconvolution was carried
out assuming that the individual bands were Gaus-
sian and that the FWHM of the $B^3-\emptyset-B^3$ band was
2.3 ± 0.2eV (equal to the FWHM of pure B_2O_3 glass, in
which only $B^3-\emptyset-B^3$ is present). Typical examples of
the deconvolution are shown in the following figures.
In the case of $10Na_2O.90B_2O_3$ glass, the O 1s spectrum
was deconvoluted into two bands due to $B^3-\emptyset-B^3$ and
$B^3-\emptyset-B^4$ as shown in Figure 5. The contribution of
$B^4-\emptyset-B^4$ was neglected. Two kinds of deconvoluted
results are shown, because a certain ambiguity was
involved in the deconvolution treatments. In Figure
5(a), the band assigned to $B^3-\emptyset-B^3$ is at higher energy
than that assigned to $B^3-\emptyset-B^4$. In Figure 5(b), $B^3-\emptyset-B^3$
is at lower energy than $B^3-\emptyset-B^4$. The differences
between measured and deconvoluted spectra are
shown under the spectrum as residuals. The residuals
in Figures 5(a) and (b) are essentially the same. This
means that the binding energies assigned to $B^3-\emptyset-B^3$
and $B^3-\emptyset-B^4$ cannot be directly distinguished.

In Figure 6, the deconvolution for $35Na_2O.65B_2O_3$
glass is shown. When $x > 20$, the O1s spectra were
deconvoluted using three bands assigned to $B^3-\emptyset-B^3$,
$B^3-\emptyset-B^4$ and $B^4-\emptyset-B^4$. The binding energy of $B^4-\emptyset-B^4$
was lower than for the other types of oxygen. The fact
that the band assigned to $B^4-\emptyset-B^4$ has lower energy was
used to resolve any ambiguity of the deconvolution.

The binding energy and FWHM values for various
glass compositions, derived from the deconvolutions,
are summarised in Table 1. The average FWHM
values of each band assigned to $B^3-\emptyset-B^3$, $B^3-\emptyset-B^4$
and $B^4-\emptyset-B^4$ are 2.3 ± 0.1, 2.0 ± 0.5 and 1.5 ± 0.5eV, re-

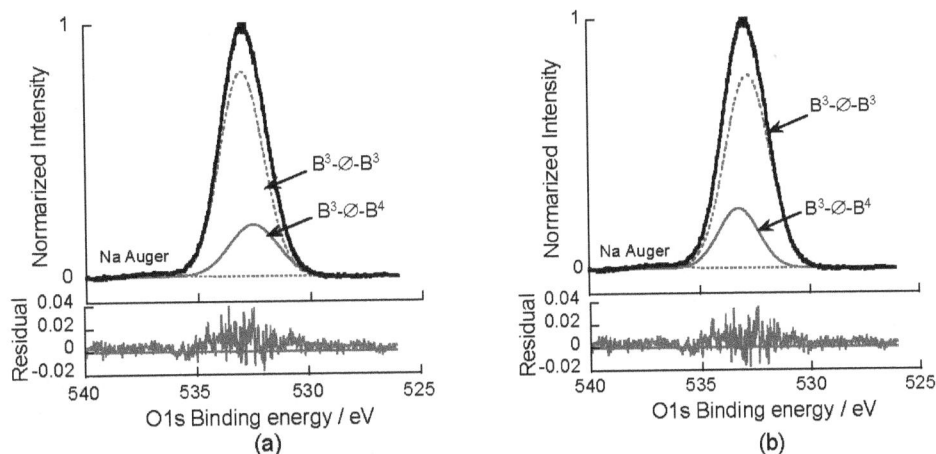

Figure 5. Deconvolution of the O1s spectrum of $10Na_2O.90B_2O_3$ glass; (a)$B^3–\varnothing–B^3>B^3–\varnothing–B^4$, (b)$B^3–\varnothing–B^3<B^3–\varnothing–B^4$

spectively. The FWHM values of the bands assigned to oxygen atoms of different structural units are in the following order: $B^3–\varnothing–B^3>B^3–\varnothing–B^4>>B^4–\varnothing–B^4$.

The binding energies are plotted in Figure 7 as a function of Na_2O content. It was found that the binding energy of $B^3–\varnothing–B^3$ is similar to that of $B^3–\varnothing–B^4$ for a particular Na_2O content, whereas the energy of $B^4–\varnothing–B^4$ is always lower than those of $B^3–\varnothing–B^3$ and $B^3–\varnothing–B^4$. That is to say, the O1s binding energy values due to each oxygen atom are in the order: $B^3–\varnothing–B^3\approx B^3–\varnothing–B^4>B^4–\varnothing–B^4$. The binding energy of each oxygen shifted linearly to lower energy with increasing Na_2O content.

4. Discussion

4.1. O 1s binding energy of different oxygen atoms
One of the most fruitful results in this study concerns oxygen in different structural units. We can estimate the different chemical states of oxygen involved in various structural units in borate glasses, and compare the binding energies with previous theoretical calculations and NMR data.

The binding energies in $Na_2O–B_2O_3$ glasses are in the order: $B^3–\varnothing–B^3\approx B^3–\varnothing–B^4>B^4–\varnothing–B^4$, as shown in Figure 7. Considering 'the charge potential model',[28] in which the binding energy of the inner shell shifts linearly to lower energy with increase of negative

charge, the lowest O1s binding energy for $B^4–\varnothing–B^4$ suggests that negative charge is localised more on $B^4–\varnothing–B^4$ than on $B^3–\varnothing–B^3$ or $B^3–\varnothing–B^4$.

Three-fold coordinated boron in borate glasses has sp^2 hybrid orbital and forms a planar triangular $B\varnothing_3$ unit. The boron has an unoccupied orbital. The interconnected three oxygen atoms have an unshared electron pair. In this $B\varnothing_3$ unit, the unshared electron pair of the oxygen atoms could give an electron to the boron atom, and form $p\pi–p\pi$ bonding between the oxygen and the three-fold coordinated boron. On the other hand, four-coordinated boron forms sp^3 hybrid orbital. In the sp^3 hybrid orbital, the electron, which was used for π bonding in the $B\varnothing_3$ unit, is spatially narrowed and localised,[29] and the $B\varnothing_4^-$ unit has negative charge to compensate the positive charge of the Na^+ ion. The negative charge is distributed on four oxygen atoms in the $B\varnothing_4$ unit and the electron on the $B^4–\varnothing$ bond is increased. Thus, the charge of $B^4–\varnothing–B^4$ is more than that of $B^3–\varnothing–B^3$ or $B^3–\varnothing–B^4$.

Differences in the orbital between boron and oxygen will affect the FWHM values of the bands. The FWHM of pure B_2O_3 glass, 2·3 eV, is larger than that of SiO_2 glass, 1·5 eV.[23] The FWHM values

Table 1. O1s binding energy (BE) and full width at half maximum (FWHM) of the three peaks in the deconvolution of the O1s spectra

Na_2O (mol%)	$B^3–\varnothing–B^3$ BE (eV)	FWHM (eV)	$B^3–\varnothing–B^4$ BE (eV)	FWHM (eV)	$B^4–\varnothing–B^4$ BE (eV)	FWHM (eV)
0·0	533·6	2·3				
4·4	533·2	2·3	532·8	2·0		
10	532·9	2·3	533·0	2·2		
14	532·7	2·3	532·7	2·1		
20	532·4	2·3	532·3	2·1	530·9	1·2
23	532·3	2·4	532·1	1·7	531·4	1·2
26	532·1	2·3	532·1	1·8	530·9	1·7
30	531·7	2·3	531·7	2·1	530·2	1·7
35	531·4	2·3	531·3	1·7	529·8	1·4

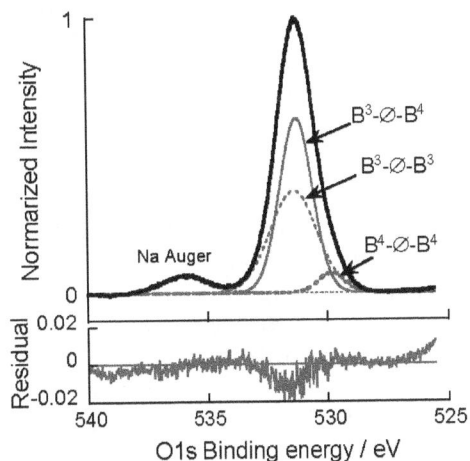

Figure 6. Deconvolution of the O1s spectrum of $35Na_2O$. $65B_2O_3$ glass

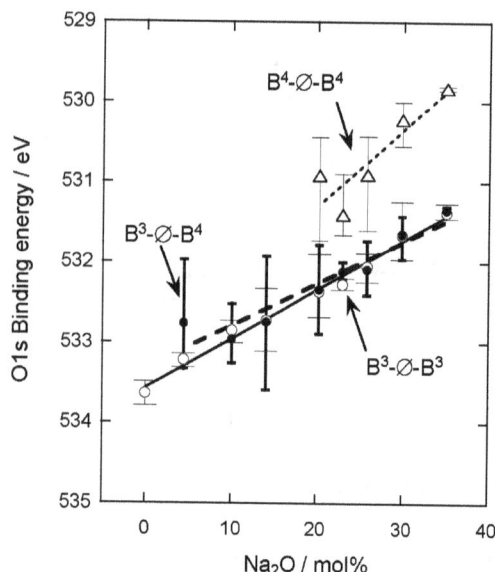

Figure 7. O1s binding energy values of each peak assigned to oxygen atoms in the different structural groups as a function of Na_2O concentration (note the reversed scale of the vertical axis). Values for the different groups are denoted by the following; O; B^3–Ø–B^3, ●; B^3–Ø–B^4, Δ; B^4–Ø–B^4. The lines correspond to the regression line; solid line; B^3–Ø–B^3, thick dashed line; B^3–Ø–B^4, short dashed line; B^4–Ø–B^4

reflect the distribution of the electron density, and the large FWHM of B^3–Ø–B^3 is closely related to the characteristic π-bonding between boron and oxygen atoms, and the occurrence of medium range order in the form of boroxol rings.

Youngman *et al* estimated the fraction of BØ$_3$ in six-membered boroxol rings and in 'loose BO$_3$' units of potassium borate glasses by using ^{11}B dynamic angle spinning NMR and MAS NMR.[30] For example, in pure B_2O_3 glass, more than 65% of boron atoms are in the boroxol rings. A B^3–Ø bond has a π electron between the unoccupied boron orbital and the unshared electron pair of the oxygen. Localisation of an electron on B^3–Ø–B^3 would depend on whether the oxygen is involved in boroxol rings or not. If B^3–Ø–B^3 is in a ring structure, the neighbouring $p\pi$–$p\pi$ orbital become delocalised around the planar ring. On the other hand, for B^3–Ø–B^3 which is not in a ring, the $p\pi$–$p\pi$ bonding is formed between the oxygen and the two neighbouring boron atoms, so that the electron is localised more than in boroxol rings. Thus, the coexistence of three-fold coordinated boron atoms in rings and in loose BØ$_3$ units is expected to result in a large FWHM of the O1s band of B^3–Ø–B^3.

The B^4–Ø–B^4 bonding no longer has a large π-character and the dihedral angle cannot have a large influence on the charge density of oxygen atoms. Therefore, the O1s FWHM for B^3–Ø–B^3 is similar to that of the O1s spectrum of silica glass,[23] in which SiO$_4$ tetrahedral units are bonded each other.

Zhong & Bray[18] reinvestigated NMR data and indicated that the fraction of four-fold coordinated boron atoms arises from a competition between the formation of four-fold coordinated boron and of negative NBO in BO$_3$ units. In our deconvolution of the O1s spectra, we have assumed that there are no NBOs. Strictly, we should consider the existence of NBO in the glass. NBO has more electrons and the O1s binding energy would be lower than for other oxygen. However, considering the equilibrium

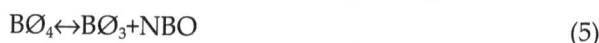

$$BØ_4 \leftrightarrow BØ_3 + NBO \tag{5}$$

the band assigned to B^4–Ø–B^4 may involve a contribution from NBO. If the existence of NBO is not be neglected then the band with lowest binding energy should be assigned to both B^4–Ø–B^4 and NBO.

4.2. Influence of different oxygen bonding on cation sites

Several anomalous phenomena for metal ions incorporated into alkali borate glasses have been reported as an aspect of the borate anomaly. For instance, the ratio of Cr^{3+}/Cr^{6+} for the redox equilibrium in alkali borate glasses increases with increasing alkali content and shows a steep shift towards the high valence state at an alkali content of about 20 mol%. The absorption spectrum of Pb^{2+} in glasses due to the $^1S_0 \rightarrow ^3P_1$ transition, which has been utilised as a probe for scaling the basicity of glass,[10] also shows an abrupt shift towards longer wavelength at around 20 mol% alkali content. The former phenomenon depends on the melting condition at high temperature, while the latter was observed in solid glasses at room temperature. When the type of alkali ions was the same, the drastic changes occurred at around the same alkali content. This indicates that there may be an analogous mechanism that causes the drastic change of the properties in both glass and melt, since the optical properties reflect the oxygen ability to donate an electron (activity of oxide ions) in glass or melt.

The optical basicity measured from the spectrum of Pb^{2+} ions in borate glasses[12] is plotted against Na_2O content together with our O1s binding energy data in Figure 8. The optical basicity increases with increasing Na_2O content and changes abruptly at 20 mol%. The abrupt change of the optical basicity corresponds to the appearance of B^4–Ø–B^4 units with lower O1s binding energy. As mentioned before, the difference in O 1s binding energy shows that oxygen in B^4–Ø–B^4 unit has higher charge density than those in B^3–Ø–B^3 and B^3–Ø–B^4 environments. Duffy *et al* explained that the abrupt shift of the optical basicity around 20 mol% alkali content was caused by ions switching from low to high basicity sites. Based on their model, the appearance of oxygen atoms with high electron density such as B^4–Ø–B^4 or NBO is favourable for the formation of high basicity sites

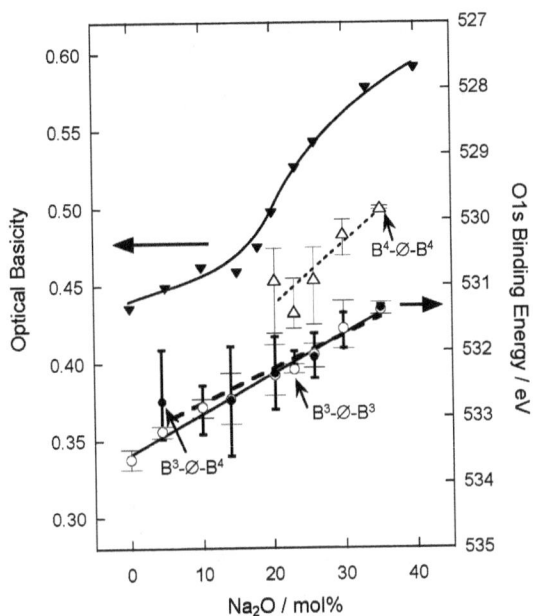

Figure 8. Optical basisity values[28] and O1s binding energy values of each peak assigned to oxygen atoms in the different structural groups as a function of Na_2O concentration

in alkali borate glasses. From Figure 4, we note that the fraction of B^4–Ø–B^4 becomes significant for Na_2O content in the range $x \geq 20$ mol%, although it still remains at relatively modest levels below 10%. The amount of B^3–Ø–B^3 seems to be too low to interact with diluted Pb^{2+} ions of 10^{-2} mol% in glasses unless the ions coordinate preferentially with oxygens in B^4–Ø–B^4 environments ('site preference').

High temperature Raman spectroscopy has been used to investigate the structure of sodium borate glasses and their melts.[31,32] According to these results, when $x<20$ mol%, tetrahedrally coordinated boron $(BØ_4)$ is stable against change of temperature, while in the $x \geq 20$ mol% Na_2O region the fraction of $BØ_4$ changed according to Equation (5) and depended on the melting temperature.[32] For instance, the fraction of $BØ_4$ in the glass with 25 mol% Na_2O decreased from 30 to 11% when the temperature was increased from the glass transition temperature to 1000°C. This means that there are enough NBOs at high temperature to interact with small numbers of metal ions in the melt.

In the redox equilibrium of chromium ions,[9] the increase of NBO in the melt in the region $M_2O>20$ mol% M_2O is sufficient to shift the reaction towards the high valence state. Pb^{2+} in the melt[12] would also interact ionically with a couple of NBOs due to charge compensation. Then as the temperature is lowered, NBO changes to $BØ_4$ as described by the reaction in Equation (5). Therefore, there is a high probability of finding Pb^{2+} near B^4–Ø–B^4 or residual NBO after cooling. Consequently, the site preference of Pb^{2+} for B^4–Ø–B^4 or NBO sites inferred from the optical basicity can be explained as arising from the reaction in Equation (5) occurring whilst the melt is cooled to form the glass.

5. Conclusions

The O 1s spectra of $xNa_2O.(100-x)B_2O_3$ $(x \leq 35)$ glasses were measured by using XPS, and the broad spectra were deconvoluted into different structural units of B^3–Ø–B^3, B^3–Ø–B^4 and B^4–Ø–B^4.

1. The O 1s binding energy values are in the order: B^3–Ø–$B^3 \approx B^3$–Ø–$B^4>B^4$–Ø–B^4, and the FWHM values of each band were decreasing in this order. The differences of binding energy and FWHM among each band were probably caused by the shapes of the molecular orbital of each oxygen atom.

2. From the difference of binding energy values of each type of oxygen, it is concluded that the drastic change of the properties of metal ions at 20 mol% Na_2O concentration, such as optical basicity, originate from the formation of B^4–Ø–B^4 which has a large charge density.

References

1. Gooding, E. J. & Turner, W. E. S. *J. Soc. Glass. Technol.*, 1934, **13**, 32.
2. Krogh-Moe, J. *Phys. Chem. Glasses*, 1962, **3** (4), 101–10.
3. Bray, P. J. & O'Keefe, J. G., *Phys. Chem. Glasses*, 1963, **4** (2), 37–46.
4. Jellison, G. E., Jr. & Bray, P. J. *J. Non-Cryst. Solids*, 1978, **29**, 187.
5. Bray, P. J. Borate Glasses: Structure, Properties, and Applications *Materials Science Research*. Plenum, New York. 1978, **Vol. 12**, P 321.
6. Konijnendijk, W. L. & Stevels, J. M. *J. Non-Cryst. Solids*, 1975, **18**, 307.
7. Paul, A. & Douglas, R. W. *Phys. Chem. Glasses*, 1962, **3** (1), 21–6.
8. Bamford, C. R. *Phys. Chem. Glasses*, 1962, **3** (6), 189–202.
9. Paul, A. & Douglas, R. W. *Phys. Chem. Glasses*, 1968, **9** (1), 27–31.
10. Duffy, J. A. & Ingram, M. D. *J. Non-Cryst. Solids*, 1976, **21**, 373.
11. Duffy, J. A., Kamitsos, E. I., Chryssikos, G. D. & Patsis, A. P. *Phys. Chem. Glasses*, 1993, **34** (4), 153–7.
12. Duffy, J. A. & Harris, B. *Ironmaking Steelmaking*, 1995, **22**, 132.
13. Easteal, A. J. & Udy, D. J. *Phys. Chem. Glasses*, 1973, **14** (6), 107–11.
14. Kamitsos, E. I. & Chryssikos, G. D. *Solid State Ionics*, 1998, **105**, 75.
15. Kamitsos, E. I.,Chryssikos, G. D., Patsis, A. P. & Duffy, J. A. *J. Non-Cryst. Solids*, 1996, **196**, 249.
16. Kamitsos, E. I., Karakassides, M. A. & Chryssikos, G. D. *J. Phys. Chem.*, 1987, **91**, 5807.
17. Chryssikos, G. D. & Karassiides, M. A. *Phys. Chem. Glasses*, 1990, **31** (3), 109–16.
18. Zhong, J. & Bray, P. J. *J. Non-Cryst. Solids*, 1989, **111**, 67.
19. Youngman, R. E. & Zwanziger, J. W. *J. Am. Chem. Soc.*, 1995, **117**, 1397.
20. Majerus, O., Cormier, L., Calas, G. & Beneu, B. *Phys. Rev. B*, 2003, **67**, 024210.
21. Kodama, M. & Kojima, S. In *Borate Glasses, Crystals and Melts*. Edited by A. C. Wright, S. A. Feller & A. C. Hannon. Society of Glass Technology, Sheffield, 1997, Pp 181–8.
22. Brückner, R. Chun, H. -U., Goretzki, H. & Sammet, M. *J. Non-Cryst. Solids*, 1980, **42**, 49.
23. Matsumoto, S., Nanba, T. & Miura, Y. *J. Ceram. Soc. Jpn*, 1998, **106**, 415.
24. Sprenger, D., Bach, H., Meisel, W. & Gütlich, P. *J. Non-Cryst. Solids*, 1993, **159**, 187.
25. Kaneko, Y. *J. Ceram. Soc. Jpn (Yogyo-Kyokai-Shi)*, 1978, **88**, 330.
26. Kaneko, Y., Nakamura, H., Yamane, M. & Suginohara, Y. *J. Ceram. Soc. Jpn (Yogyo-Kyokai-Shi)*, 1982, **90**, 557.
27. Matsumoto, S., Miura, Y., Murakami, C. & Nanba, T. In *Borate Glasses, Crystals and Melts*. Edited by A. C. Wright, S. A. Feller & A. C. Hannon. 1997. Society of Glass Technology, Sheffield, 1997, Pp 173–80.
28. Siegbahn, K., Nordling, C., Johansson, G., Hedman, J., Hedén, P. F., Hamrin, K., Gelius, U., Bergmark, T., Werme, L. O., Manne, R. & Baer Y. *ESCA Applied to Free Molecules*. North-Holland, Amsterdam, 1969.
29. Kawazoe, H., Hosono, H. & Kanazawa, T. *J. Non-Cryst. Solids*, 1978, **29**, 159, 173.
30. Youngman, R. E. & Zwanziger, J. W. *J. Phys. Chem.*, 1996, **100**, 16720.
31. Yano, T., Kunimine, N., Shibata, S. & Yamane, M. *J. Non-Cryst. Solid*, 2003, **321**, 137.
32. Yano, T., Kunimine, N., Shibata, S. & Yamane, M. *J. Non-Cryst. Solid*, 2003, **321**, 147.

Proc. VI Int. Conf. Borate Glasses, Himeji, Japan, 18–22 August 2008   *Glass Technol.: Eur. J. Glass Sci. Technol. A*, October 2009, **50** (5), 280–284

Transformation of boron containing hybrid structures into silicon oxycarbide glasses

Yordanka Ivanova*

University of Chemical Technology and Metallurgy, Kl. Ohridski 8, 1756 Sofia, Bulgaria

Yuliya Vueva

Institute of Inorganic Chemistry, Bulgarian Academy of Science, Acad. G. Bonchev Str. bl. 11, 1113, Sofia, Bulgaria

Manuscript received 18 August 2008
Revised version reeived 27 April 2009
Accepted 21 August 2009

Boron containing hybrid organic–inorganic materials were synthesised by a sol-gel process using co-hydrolysis of methyltriethoxysilane (MTES) and tetraethylortosilicate (TEOS) prepared in two molar ratios, MTES/TEOS=1/1 and MTES/TEOS=2/1. Trimethylborate B(OCH₃)₃ (TMB) was used to introduce B₂O₃ into the hybrid matrix. The resulting gels were pyrolysed in a N₂ atmosphere in the temperature interval 600–1100°C. The structural evolution during pyrolysis of the hybrids was followed by FTIR, ²⁹Si MAS NMR, ¹¹B echo MAS NMR and XRD analysis. The morphology of the multicomponent materials was investigated by scanning electron microscopy. The structures of the hybrid gels and resulting Si–O–C–B glasses depend on the concentration of modifying boron ions and the amount of organic precursor used in the synthesis. The FTIR, ²⁹Si charge polarised (CP) MAS NMR and ¹¹B MAS NMR results show the formation of Si–O–B linkages and Si–C links in the glasses for both systems. FTIR and ²⁹Si CP/MAS NMR spectra obtained after pyrolysis at 1100°C show a profile typical of an oxycarbide structure, in which a small amount of boron is incorporated via B–O–Si linkages.

Introduction

The processing of silicon oxycarbide glasses (SiOC) through pyrolysis of hybrid materials has been established as a novel route to prepare new advanced glasses and ceramics.[1,2] SiOC glasses have attracted attention for their peculiar high temperature stability, good mechanical properties[3] and chemical durability.[4] The incorporation of heteroatoms (such as boron) in these materials provides better thermal stability, as has been observed in similar silicon carbonitride (SiNC) glasses.[5]

Recently, some authors reported that the presence of Si–O–B linkages in the resulting hybrid structure prevents the formation of cyclic or cage siloxane entities and leads to relatively high ceramic yields.[6] This work describes the synthesis of boron containing organic–inorganic hybrid materials derived by a sol-gel process through co-hydrolysis of methyltriethoxysilane (MTES) and tetraethylorthosilicate (TEOS), prepared in two molar ratios MTES/TEOS=1/1 and MTES/TEOS=2/1. Increasing the quantity of MTES molecules is expected to increase the content of hydrophobic organic groups bonded to silicon, which will provide *"in situ"* protection from water attack in the gels, and results in preferential formation of B–O–Si bonds in the gels. The effect of the MTES/TEOS molar ratio and boron load on the gel forming and pyrolysis behaviour were investigated by means of XRD analysis, infrared spectroscopy, ¹¹B echo MAS NMR and ²⁹Si CP MAS (charge polarised magic angle spinning) NMR. Scanning electron microscopy was used to examine the morphology of the hybrid gels and glass materials.

Experimental

The gel hybrids were prepared from a pre-hydrolysed mixture of MTES, TEOS and ethanol (EtOH) (volume ratio 1:1:1, acidic H₂O, pH=2 HCl, pre-hydrolysis time of 30 min). To follow the influence of organic silicon alkoxide MTES on gel forming and pyrolysis behaviour, two molar ratios (MTES/TEOS=1/1 and MTES/TEOS=2/1) were used. Trimethylborate B(OCH₃)₃ (TMB) was used to introduce boron in the glasses in weight ratios MTES/B(OCH₃)₃ of 95/5, 90/10, 80/20, corresponding to the samples MTB5, MTB10, MTB20, and M2TB5, M2TB10, M2TB20 for the MTES/TEOS=2/1 system. After 1 h of stirring, the sols were cast and gelled for several days under air and at room temperature. The hybrids were converted into SiBOC glasses by pyrolysis in a tube furnace at 600°C and 1100°C for 2 h in a N₂ atmosphere, using a heating rate of 10°C min⁻¹.

The structure and properties of the hybrid materials and oxycarbides were studied with FTIR (Mattson 7000) by the KBr disk method over the wavenumber range 4000–400 cm⁻¹, XRD (Rigaku) with Cu Kα radiation in the 2θ-range from 10° to 80°, at a scan rate of

*Corresponding author. Email y.ivanov@uctm.edu

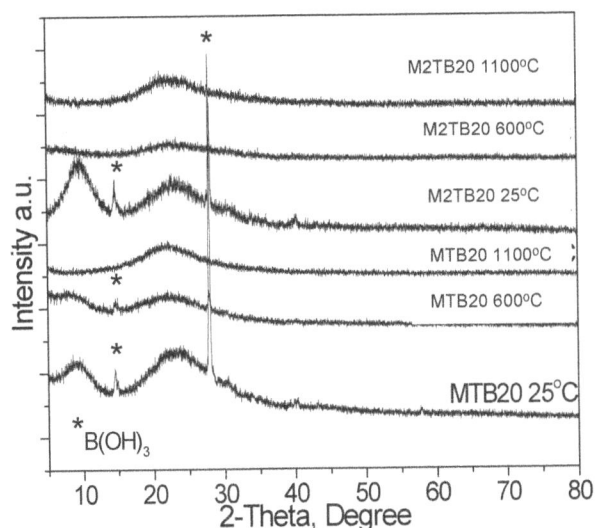

Figure 1. XRD patterns of the MTB20 and M2TB20 samples at different temperatures

$2°\ 2\theta$/min, and with SEM (Hitachi S-4100). ^{29}Si solid state NMR spectra were recorded at 79·49 MHz and ^{11}B solid state NMR spectra were recorded at 128·4 MHz on a (9·4 T) Bruker Avance 400 spectrometer.

Results

Figure 1 shows the XRD patterns at different temperatures (25, 600 and 1100°C) and boron concentration 20 wt% of the gels from both series. Typical XRD patterns of amorphous materials were observed for both gels. Besides the amorphous halo, peaks at $2\theta \approx 15°$ and $2\theta \approx 28°$ were assigned to the boron hydroxide crystalline phase. During the pyrolysis, the characteristic lines of $B(OH)_3$ are reduced but still present to 600°C. After this temperature the samples vitrify and are amorphous. These results are in good

agreement with SEM micrographs (Figures 2 and 3). The MTB5 gel has a grainy amorphous surface (Figure 2(a)), while the surface of M2TB5 gel is smooth (Figure 2(b)). At higher boron contents (20 wt%), crystals of $B(OH)_3$ are visible (Figure 2(c)). During the heat treatment at 600°C, the $B(OH)_3$ crystals are enlarged (Figure 3(b)). At 1100°C the material has a glassy surface (Figure 3(c)).

FTIR and NMR methods were used to investigate the influence of the boron and organic silicon alkoxide on the structural changes in the materials during the heating process. Figures 4(a) and (b) present the IR spectra of the MTB5 and M2TB5 samples, and Figures 5(a) and (b) present the IR spectra of MTB20 and M2TB20. The broad absorption bands around 3450 cm^{-1} visible for all samples are caused by OH related vibrations, and peaks at around 1650 cm^{-1} show the presence of water.[7] The sharp peak at 1280 cm^{-1}, due to $Si–CH_3$,[8] is visible for all samples, and remains in the spectra up to 600°C. Peaks in the region 1500–1300 cm^{-1} due to (ν B–O) are stronger in the spectra of the samples with higher boron concentrations (Figure 5). The absorption at 1195 cm^{-1} is characteristic of (δ B–OH), and is present in the spectra of high boron samples.[9] The region 1500–1300 cm^{-1}, which characterises the B–O bonds, becomes broader as the amount of MTES increases. The band at 1045 cm^{-1} is due to Si–O symmetric stretching in the siloxane network, and the shoulder at 1130 cm^{-1} is due to the asymmetric stretching mode of Si–O bonds.[10] The localisation of the shoulder at 1130 cm^{-1} is different, depending on boron concentration, which suggests that condensation in the siloxane network is different according to the quantity of boron added in the solution and incorporated in the silicon network.

The pattern of the spectra changes when the samples are pyrolysed at 600°C. The band located at 1280

Figure 2. SEM micrographs of (a) MTB5, (b) M2TB5 and (c) MTB20 gel samples

Figure 3. SEM micrographs of (a) MTB20, 25°C, (b) MTB20, 600°C and (c) MTB20, 1100°C samples

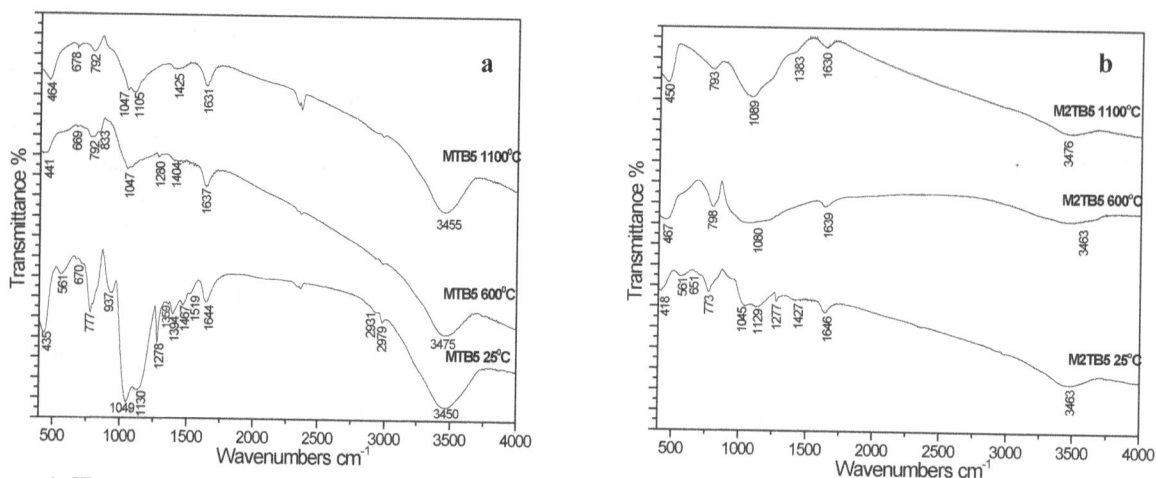

Figure 4. IR spectra at different temperatures of samples: (a) MTB5; (b) M2TB5

cm^{-1}, associated with CH$_3$ symmetric bending of the MTES molecules, decreases with temperature due to depolymerisation and mineralisation reactions. The intensity of this band increases with MTES content. The shoulder located at 1130 cm^{-1}, due to the asymmetric stretching mode in Si–O–Si bonds, increases, broadens and shifts to higher wavenumbers with temperature, due to the increase in the crosslinking of the network. The most important band at 890–880 cm^{-1}, which is due to the formation of Si–O–B linkages, appears in the spectra of both materials.[11] The copolymerisation between boron and silicon proceeds further as the annealing temperature is increased. The peak at 670 cm^{-1}, due to Si–O–B linkages, appears at 600°C,[12] and is still present in the spectra at the highest pyrolysis temperature (1100°C).

At 1100°C the intensity of the peaks around 1400 cm^{-1} decreases due to the loss of boron during the pyrolysis process. Samples with higher boron contents show broadening of the 1400–900 cm^{-1} region, which indicates that boron is being incorporated into the silicon matrix.[13] In the spectra of the heated samples the bands at 1050 cm^{-1} (ν Si–O) and 750 cm^{-1} (SiX$_4$, X=O, C) are still present. The spectra obtained after

pyrolysis at 1100°C show a similar IR profile, typical of an oxycarbide structure.[14]

^{29}Si CP MAS NMR spectra for MTB20 and M2TB20 gels are shown in Figure 6. The solid state ^{29}Si CP MAS NMR spectrum of MTB20 displays resonances centred at –55 ppm (T^2), –92 ppm (Q^2), –100 ppm (Q^3) and –108·7 ppm (Q^4)[15] (where Q^n and T^n respectively indicate trifunctional and tetrafunctional silicon sites with n oxygens which bridge to another silicon). The signals around –55 ppm refer to CH$_3$Si(OC$_2$H$_5$)(OSi)$_2$ and at –92 ppm to [(OC$_2$H$_5$)$_2$Si(OX)$_2$], whilst those at –101 ppm refer to (OC$_2$H$_5$)Si(OX)$_3$ and at –108·7 ppm to Si(OX)$_4$.[16,17] The presence of Q^2 in the MTB20 sample leads to the conclusion that the condensation of the hybrid gel is not completed. In the ^{29}Si CP MAS NMR spectrum of M2TB20 gel, Q^2 structural units are absent, which suggests a better degree of condensation than for MTB20 gel.

The ^{11}B echo MAS NMR spectra of the MTB20 gel (Figure 7) shows boron coordination consisting predominantly of [BO$_3$], with a small amount of [BO$_4$] species. The [BO$_3$] values of the dried MTB20 gel lie at –0·7 ppm, –6·7 ppm, –10·6 ppm and [BO$_4$] at –12·7 ppm (Figure 7(b)).[18] The presence of [BO$_4$] indicates

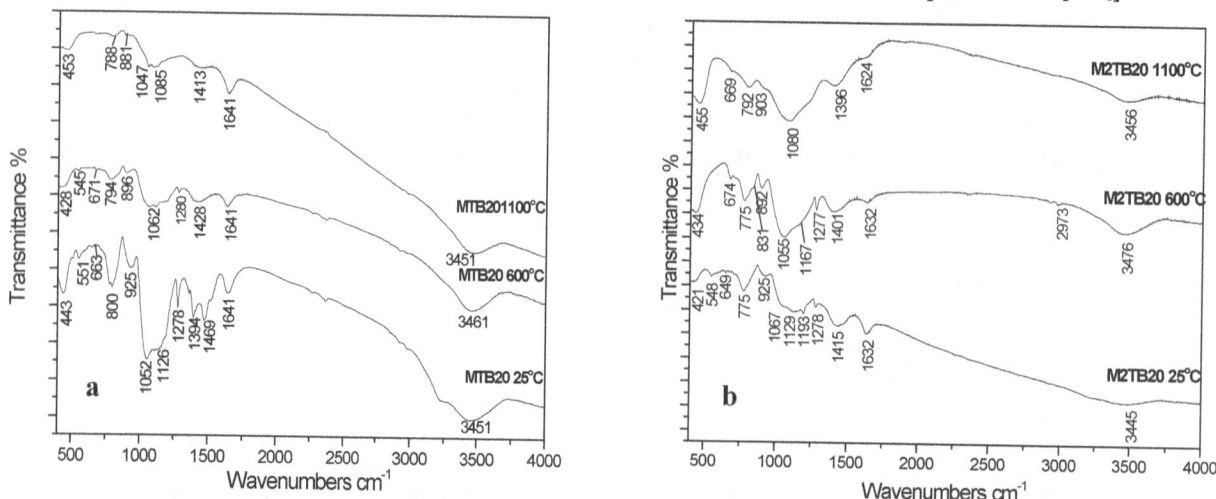

Figure 5. IR spectra at different temperatures of samples: (a) MTB20; (b) M2TB20

Proceedings of the VI International Conference on Borate Glasses, Himeji, Japan, 18–22 August 2008

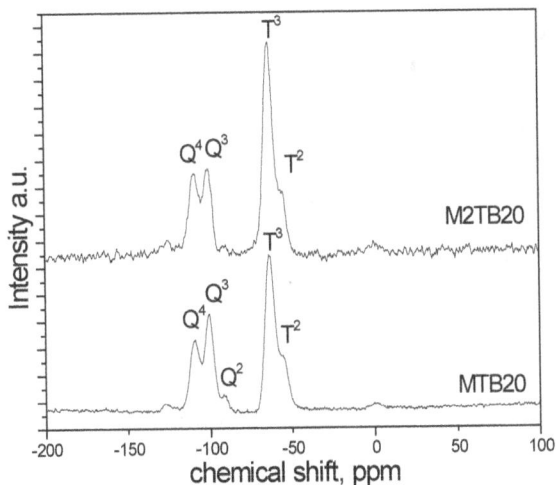

Figure 6. ^{29}Si charge polarised MAS NMR spectra of MTB20 and M2TB20 gel samples

that there are some linkages of the [BO$_4$] tetrahedra to the silica network.

The ^{29}Si CP MAS NMR spectra of the samples pyrolysed at 600°C (Figure 8) show T^3 units and the transformation of Q^3 and Q^2 into Q^4 units, due to the completion of condensation reactions. The spectrum of the glass sample pyrolysed at 1100°C (Figure 9) shows the structural changes involved in the gel to glass conversion. The main peak is now assigned to [SiO$_4$] at −110 ppm, with a smaller [CSiO$_3$] peak be-

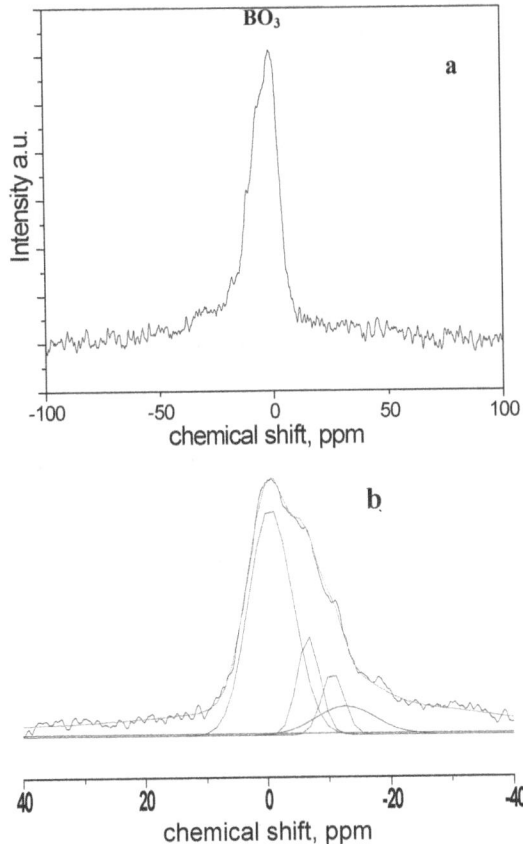

Figure 7. (a) ^{11}B echo MAS NMR spectra of MTB20 gel sample; (b) deconvoluted ^{11}B echo MAS NMR spectra of MTB20 gel sample

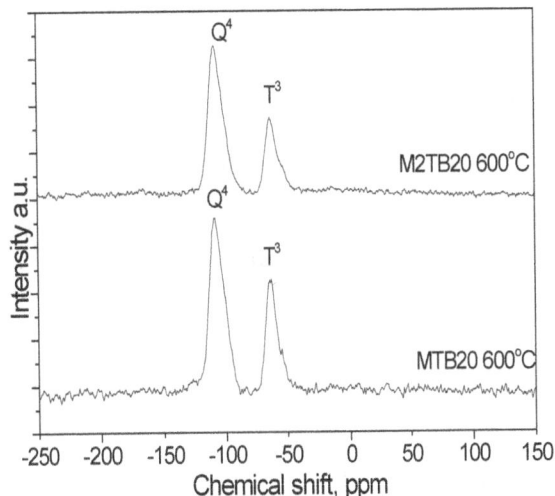

Figure 8. ^{29}Si CP MAS NMR spectra of MTB20 and M2TB20 samples pyrolysed at 600°C

ing apparent at −72 ppm, indicating that Si–C bonds have been formed and retained during pyrolysis, confirmed by FTIR analysis, where a band at 750 cm^{-1} due to Si–C bonds is observed.

Discussion

Boron modified silica hybrid precursors were prepared by the hydrolysis and condensation of MTES, TEOS and TMB. Two different MTES/TEOS molar ratios (MTES/TEOS=1/1 and MTES/TEOS=2/1) were used in order to control the B–O–Si linkages formed in the solution. Increasing the quantity of MTES molecules is expected to increase the content of hydrophobic organic groups bonded to silicon, which will provide *"in situ"* protection from water attack in the gels, and results in preferential forma-tion of B–O–Si linkages in the gels. The FTIR spectra of the MTBx and M2TBx samples had very similar characters. The most important band at 890–880 cm^{-1},

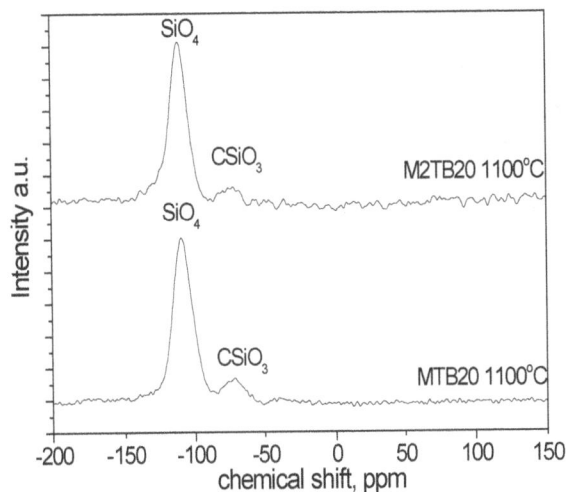

Figure 9. ^{29}Si CP MAS NMR spectra of MTB20 and M2TB20 samples pyrolysed at 1100°C

which is due to the formation of Si–O–B linkages, appears in the spectra of both materials only after heat treatment at 600°C. The band at 650 cm^{-1} due to Si–O–B linkages also becomes prominent at this temperature. The intensity of these bands increases with increasing boron content, but is not strongly influenced by the quantity of MTES. The presence of Si–O–B bands in the structure of silicon oxycarbide glasses at 1100°C suggests that boron has entered into the Si–O network.

^{29}Si CP MAS NMR analysis of MTES/TEOS=2/1 gels showed no Q^2 (SiO$_2$(OH)$_2$) structural units, compared to ^{29}Si CP MAS NMR of MTES/TEOS=1 gels, where Q^2 groups are present. This indicates that MTES-rich gels have better condensation than MTES/TEOS=1/1 gels, which could be due to the increased hydrolysis of organically modified silicon alkoxides in acid conditions, since organic groups decrease the steric effect around Si atoms. Hence an electron containing substitute (–CH$_3$ in CH$_3$Si(OC$_2$H$_5$)$_3$) will stabilise the positive charge around Si, and will increase the extent of hydrolysis (additional influence of inductive effect). The condensation in gels is influenced by steric and inductive effects, as well as by the extent of substitution in organically modified silicon alkoxides, R$'_x$Si(OR)$_{4-x}$. As a result, the condensation increased with increasing substitution.

FTIR analysis was not very informative about the formation of Si–O–B linkages in the gels, due to the broad Si–O–Si band in the region 850–1250 cm^{-1}. ^{29}Si CP MAS NMR analysis is insensitive to Si–O–B linkages, since Si and B^{III} in the second coordination sphere cause the same ^{29}Si chemical shift. The ^{11}B spectrum of gels reveals trigonally coordinated B^{III} species and a small amount of tetrahedrally coordinated B^{IV}, which supposes that boron species are connected to the silicon oxycarbide network by B–O–Si linkages.

High temperature ^{29}Si CP MAS NMR spectra at 1100°C confirmed that carbon atoms are bonded to the glass network in the form of CSiO$_3$ groups.

Conclusions

Boron-containing silicon oxycarbide glasses were synthesised by a sol-gel process from co-hydrolysis of methyltriethoxysilane (MTES) and tetraethylortho-silicate (TEOS), prepared in two molar ratios (MTES/TEOS=1/1 and MTES/TEOS=2/1), and trimethylborate (TMB). The FTIR, ^{29}Si CP/MAS NMR and ^{11}B MAS NMR results show the formation of Si–O–B linkages and Si–C bonds in the gels and glasses for both systems. The amount of Si–O–B linkages in the gel network is affected mainly by the quantity of boron introduced into the solution, rather than the quantity of MTES. Within the Si–O–C–B hybrid network, boron exists predominantly as [BO$_3$] units. FTIR and ^{29}Si MAS NMR spectra obtained after pyrolysis of the gels at 1100°C show a profile typical of an oxycarbide structure, in which a small amount of boron atoms are incorporated via B–O–Si linkages.

Acknowledgment

This work is financially supported by fund "Scientific researches" of the Ministry of Education and Science-Bulgaria, Contract No. TH-1507/05.

References

1. Crouzet, L., Leclereq, D., Mutiu Hubert, P. & Vioux, A. *Chem. Mater.,* 2003, **15**, 1530.
2. Zhang, H. & Pantano, C. G. *J. Am. Ceram. Soc.,* 1990, **4**, 959.
3. Renlund, G. M, Prochazka, S. & Doremus, R. H. *J. Mater. Res.,* 1991, **6**, 2716.
4. Soraru, G. D., Modena, S., Guadagnino, E., Colombo, P., Egan, J. & Pantano, C. *J. Am. Ceram. Soc.,* 2002, **85**, 1529.
5. Riedel, R., Kienzle, A., Dressler, W., Ruwisch, L., Bill, J. & Aldinger, F. *Nature,* 1996, **382**, 796.
6. Gervais, C., Babonneau, D. N. & Soraru, G. D. *J. Am. Ceram. Soc.,* 2001, **84**, 2160.
7. Pouchert, C. J. *The Aldrich Library of Infrared spectra 3rd Ed.,* Aldrich Chemical Co., Wisconsin, 1981.
8. Mendes-Vivar, J. & Mendoza-Serna, R. *J. Sol-Gel Sci. Technol.,* 1997, **8**, 235.
9. Parson, J. L. & Milberg, M. E. *J. Am. Ceram. Soc.,* 1960, **43**, 326.
10. Viart, N. & Rehspinger, J. L. *J. Non-Cryst. Solids,* 1996, **195**, 223.
11. Soraru, G. D., Dallabona, N., Gervaise, C. & Babonneau, F. *Chem. Mater.,* 1999, **11**, 910.
12. Soraru, G. D., Babonneau, F., Gervaise, C. & Dallabona, N. *J. Sol-Gel Sci. Technol.,* 2000, **18**, 11.
13. Peña, A. R., Rubio, J. & Oteo, J. L. *J. Sol-Gel Sci. Technol.,* 2003, **26**, 195.
14. Soraru, G. D., Babonneau, F. F., Maurina, S. & Vicens, J. *J. Non-Cryst. Solids,* 1998, **224**, 173.
15. Brutchey, R. L., Goldberger, J. E., Koffas, T. S. & Tilley, T. D. *Chem. Mater.,* 2003, **15**, 1040.
16. Lotero, E., Vu, D., Nguyen, C., Wagner, J. & Larsen, G. *Chem. Mater.,* 1998, **10**, 3756.
17. Bois, L., Maquet, J. & Babonneau, F. *Chem. Mater.,* 1994, **6**, 796.
18. Prased, S., Clark, T. M., Sefzik, T. H., Kwak, H. T., Gan, Z. & Grandinetti, P. J. *J. Non-Cryst. Solids,* 2006, **352**, 2834.

Proc. VI Int. Conf. Borate Glasses, Himeji, Japan, 18–22 August 2008   *Phys. Chem. Glasses: Eur. J. Glass Sci. Technol. B*, December 2009, **50** (6), 355–357

Boron isotope effect on the high temperature viscosity of sodium borosilicate glasses

Jun Matsuoka,[A,*] *Yusaku Nishida,*[A] *Kunihiro Kimura,*[A] *Satoshi Yoshida*[A,B] *& Toru Sugawara*[A,B]

[A] *Department of Materials Science, The University of Shiga Prefecture, Hassaka 2500, Hikone, Shiga 522-8533, Japan*
[B] *Centre for Glass Science and Technology, The University of Shiga Prefecture, Hassaka 2500, Hikone, Shiga 522-8533, Japan*

Manuscript received 20 August 2008
Revised version received 1 October 2009
Accepted 7 October 2009

Borosilicate glasses are important for many technical applications. However, they contain two kinds of network forming oxides, and so relationships between composition and properties are complicated. Viscous flow is one of the properties for which the mechanism is not yet clarified. Previously we measured the boron isotope effect for high temperature thermal properties of B_2O_3 glass. In the present study, the boron isotope effect on the viscosity of the $33Na_2O.(67{-}y)B_2O_3.ySiO_2$ glass system, with y=5 to 20, was investigated around 10^3 Pa s as a function of the B_2O_3/SiO_2 ratio. The isoviscous temperature of the glass made from ^{11}B is found to be lower than for the glass made from ^{10}B. Mixing of isotopes is found to further decrease the isoviscous temperature. These results suggest that the breaking of B–O covalent bonds dominates the viscous flow of the examined glass compositions.

Introduction

Glass transition phenomena and the mechanism of viscous flow in glass forming liquids are not clarified yet. Many studies have been carried out on the composition dependence of these properties. However, substitution of one atomic element for another causes simultaneous changes in atomic size, electronegativity and mass, and therefore it is not possible to specify the dominant factor which affects the above properties. On the other hand, isotopic substitution of an atom changes atomic mass without changing size or electronegativity, and so it is a powerful tool for studying the atomic mechanisms of the glass transition and viscous flow. However, there have been few studies of isotopic effects on the glass transition [1–6] and viscosity.[7]

B_2O_3 is a common glass forming oxide, and boron has two stable isotopes, ^{10}B and ^{11}B. In our previous study, we have investigated the boron isotope effect on the high temperature thermal properties (heat capacity, fictive temperature and apparent activation energy of the glass transition) of B_2O_3 glass.[4,6,7] Similar effects were also reported for alkali borate glasses by Nagasaki *et al.*[5]

Borosilicate glass systems are important for many technical applications. They are also interesting from the scientific point of view because they contain two kinds of network forming oxides, B_2O_3 and SiO_2. In this study, in order to clarify the role of B–O bond breaking and bond–bond interactions on the viscous flow in the borosilicate glass system, the effect of bo-

ron isotope substitution on $33Na_2O.(67{-}y)B_2O_3.ySiO_2$ glasses with y=5 to 20 and average boron atomic mass of 10·0, 10·5 and 11·0 was investigated in the viscosity range from 10^2 to 10^4 Pa s.

Experimental

The samples were made using isotope enriched H_3BO_3 (Cambridge Isotope Laboratories, Inc., Andover, MA, USA), in which the isotope enrichment was 99% for ^{10}B-enrichment and also 99% for ^{11}B-enrichment, and the chemical purity was more than 98%. Other raw materials were reagent grade Na_2CO_3 and SiO_2 (Sigma-Aldrich Japan, Tokyo). The raw materials were mixed and then melted at 1200°C in a platinum crucible for 2 h to obtain a glass sample. H_3BO_3 with the natural isotope ratio (Sigma-Aldrich Japan, Tokyo) was also used to check the measurement condition.

Viscosity, η, was measured by a parallel plate rotating method [8] from 10^2 to 10^4 Pa s by using a high temperature viscometer PPVM (Opt Corp., Yotsukaido, Chiba, Japan) with a heating rate of about 9 K min^{-1}. This viscosity range is around the working temperature range. The size of the sample was 3·5 mm in thickness and 10 mm in diameter at first, and 2·0 mm and 13 mm at the measurement time. A zirconia ball of 2·0 mm diameter, set at the centre of the parallel plates, was used as a spacer. Measurements were carried out at least four times for all compositions. The obtained data were fitted with the Vogel–Fulcher–Tammann equation. The maximum deviation of the experimental data of log(η/Pa s) from the fitting curve at a given

* Corresponding author. Email matsuoka@mat.usp.ac.jp

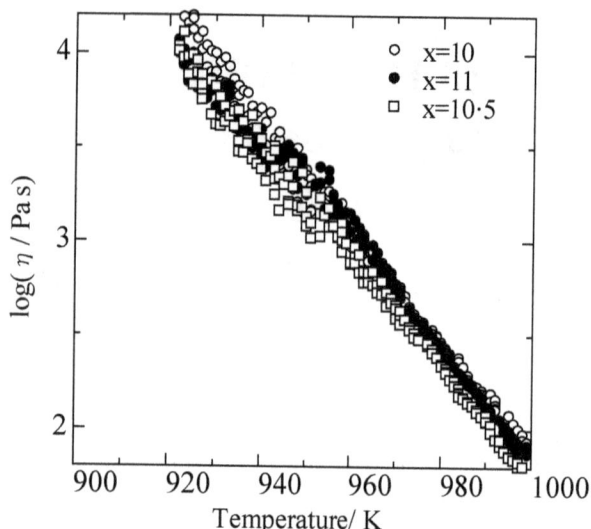

Figure 1. Viscosity of isotope-substituted $33Na_2O.52^xB_2O_3.15SiO_2$ glasses with average boron atomic mass of 10·0, 10·5 and 11·0

temperature is ±0·15 at about $\log(\eta/Pa\ s)=4$, ±0·10 at about $\log(\eta/Pa\ s)=3$, and ±0·05 at about $\log(\eta/Pa\ s)=2$.

Results

Figure 1 shows the viscosity of isotope-substituted $33Na_2O.52B_2O_3.15SiO_2$ glasses with average boron atomic mass of 10·0, 10·5 and 11·0. It can be seen that the viscosity of the glass with average boron mass of 11·0 (^{11}B glass) is somewhat lower than that of 10·0 (^{10}B glass). In addition, the viscosity for average boron

mass of 10·5 ($^{10.5}B$ glass) is much lower than for the two isotopically pure glasses. Similar behaviour was observed for all glasses with different SiO_2 contents.

Figure 2(a) to (d) show the isoviscous temperature at $\eta=10^3$ Pa s. That of ^{11}B glass is about 1 K lower than that of ^{10}B glass independent of the SiO_2 content. On the other hand, the decrease of isoviscous temperature due to isotope mixing strongly depends on the SiO_2 content. The magnitude is about 4 K for $y=5$ and 10 glasses, while that for $y=15$ and 20 glasses is about 1 K. These temperature differences are summarised in Figure 3.

Discussion

The existence of a boron isotope effect means that the breaking of B–O covalent bonds strongly affects the viscous flow process. When we compare the viscosity of the isotopically pure glasses, the value for the heavier isotope glass is lower than that for the lighter isotope glass. Isotopic substitution does not change the energy landscape of the interatomic potential, but it does affect the viscous flow by the following three mechanisms: One is the difference of trial frequency of atomic jump, which is the same as the frequency of atomic vibration, ν, and ν is inversely proportional to the square root of the effective mass of vibration. The second is the difference of zero-point energy, which has the value $h\nu/2$. This affects the effective activation energy of atomic jump in the energy landscape. The third is the difference of the thermal excitation probability of atomic motion at a given temperature

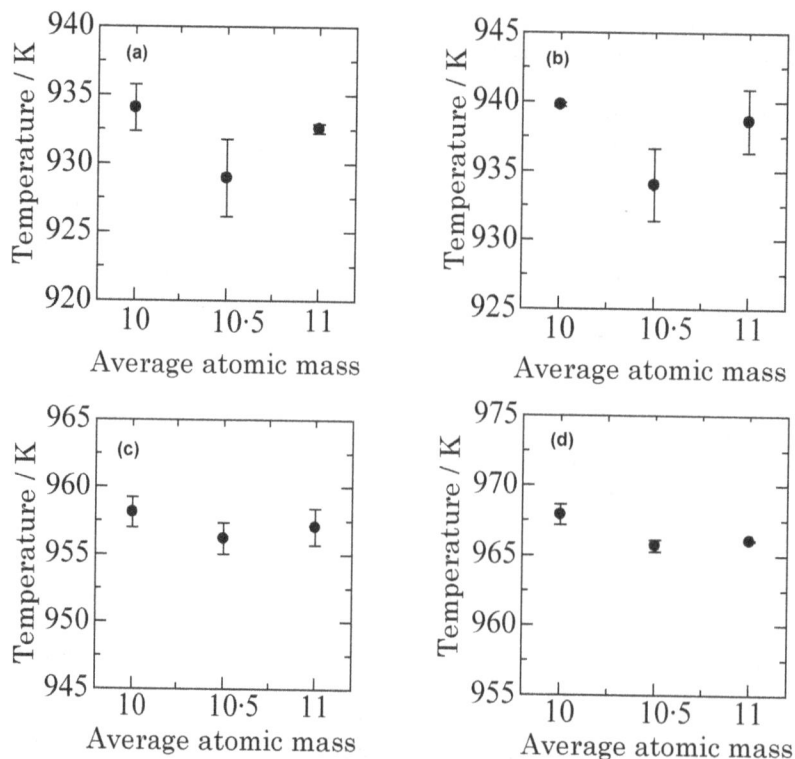

Figure 2. Isoviscous temperature at $\eta=1·0\times10^3$ Pa s, for $33Na_2O.(67-y)^xB_2O_3.ySiO_2$ glasses with x=10·0, 10·5 and 11·0, and y=5 (a), 10 (b), 15 (c) and 20 (d)

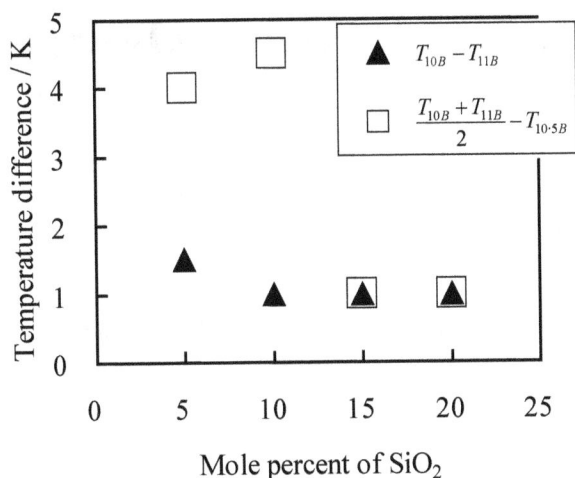

Figure 3. Differences between isoviscous temperatures at $\eta=1\cdot0\times10^3$ Pa s for glasses in the system $33Na_2O.(67-y)$ $^xB_2O_3.ySiO_2$ with y=?5? to 20. Closed triangles show the difference between the isoviscous temperatures of ^{10}B and ^{11}B glasses. Open squares show the difference between the average isoviscous temperature for the two isotopically pure glasses and the value for the isotopically mixed glass

through the difference of Debye temperature. In this effect, lower v caused by heavier isotope gives a lower Debye temperature and hence a larger excitation probability. Among these three effects, the first two give higher viscosity for heavier isotope glasses, whilst the last one (difference of Debye temperature) gives lower viscosity for heavier isotope glasses. Therefore, the experimental result shows that the third effect dominates the isotope substitution effect between the ^{10}B and ^{11}B glasses in this study.

Mixing of ^{10}B and ^{11}B is found to further lower the viscosity. This isotope mixing effect may be due to a difference of Debye temperature, or may be due to a change of the rate of atomic diffusion in which the isotope mixing affects the back diffusion probability or correlation factor of progressive atomic jump. It is not clear yet which effect dominates the mixed isotope effect, but certainly the difference of the dynamic interaction between neighbouring B–O bonds (or neighbouring oscillators of atomic vibration) is the origin of the isotope mixing effect. A similar effect is observed for pure B_2O_3 [7], and clarification of this behaviour will give us the atomic mechanism of viscous flow in oxide glasses.

The isotope mixing effect is large for $y=5$ and 10 glasses. This means that the interaction of two neighbouring B–O bonds is large for these glasses, but it becomes small at $y=15$. The large difference of the magnitude of isotope mixing effect between $y=10$ and 15 agrees with the large difference of isoviscous temperature between the same two compositions.

Conclusions

Substitution of ^{10}B by ^{11}B in the $33Na_2O.(67-y)B_2O_3.$ $ySiO_2$ glass system with $y=5$ to 20 is found to decrease the viscosity, and the magnitude of the decrease is almost constant, independent of SiO_2 content, y. Mixing of the two boron isotopes is found to further lower the viscosity, and the magnitude of this decrease is large for $y=5$ and 10, and small for $y=15$ and 20. The fundamental origin of this behaviour has been discussed.

References

1. Shepley, L. C. & Bestul, A. B. *J. Chem. Phys.*, 1963, **39**, 680.
2. Johari, G. P., Hallbrucker, A. & Mayer, E. *J. Chem. Phys.*, 1990, **92**, 6742.
3. Johari, G. P., Hallbrucker, A. & Mayer, E. *J. Chem. Phys.*, 1991, **95**, 6849.
4. Matsuoka, J., Kamiya, K. & Yamashita, H. *Bol. Soc. Esp. Ceram. Vidrio 31-C*, 1992, **4**, 269.
5. Nagasaki, T., Morishita, R. & Matsui, T. *J. Nucl. Sci. Tech.*, 2002, **39**, 386.
6. Matsuoka, J., Numaguchi, M., Fujino, Y., Matsuo, M., Kurose, S., Yoshida, S. & Soga, N. *J. Non-Cryst. Solids*, 2004, **345&346**, 542.
7. Kurose, S., Yoshida, S. & Matsuoka, J. *Proceedings of the XX International Congress on Glass (CD-ROM)*, 2004, P-07-060
8. Shiaishi, Y. *Met. Technol.*, 1994, **64**, 2. (In Japanese.)

Proc. VI Int. Conf. Borate Glasses, Himeji, Japan, 18–22 August 2008   *Phys. Chem. Glasses: Eur. J. Glass Sci. Technol. B, August 2009, 50 (4), 257–261*

Thermal and optical properties of Bi_2O_3–GeO_2–B_2O_3 glasses

Naoto Yamashita,* Tatsuya Suetsugu, Toshihiko Einishi

Isuzu Glass Co., 6-3-6 Minamitsumori, Nishinari, Osaka, 557-0063, Japan

Kohei Fukumi, Naoyuki Kitamura & Junji Nishii

Photonics Research Institute, National Institute of Advanced Industrial Science and Technology, 1-8-31 Midorigaoka, Ikeda, Osaka, 563-8577 Japan.

Manuscript received 18 August 2008
Revised version received 30 January 2009
Accepted 20 May 2009

The thermal and optical properties of Bi_2O_3–GeO_2–B_2O_3 glasses have been studied. The glass transition and deformation temperatures have a maximum at a Bi_2O_3 content of ~20 mol%. The molar refraction shows additivity with composition, decreases linearly with decreasing Bi_2O_3 content and does not depend on the B_2O_3/GeO_2 ratio. Colourless glasses were obtained in a low Bi_2O_3, high GeO_2 region of composition, as expected from electronegativity. Raman scattering spectroscopy showed that the glass structure depends on the Bi_2O_3/B_2O_3 ratio and on GeO_2 content. By a glass imprinting method, a one-dimensional periodic pattern with a period of 500 nm was successfully fabricated on the glass surface.

1. Introduction

Glasses having sub-wavelength periodic structures on their surfaces show some peculiar optical functions such as antireflection, large angular dispersion and birefringence.[1–3] Recently, a glass imprinting method has attracted attention as a method for fabricating sub-wavelength periodic structures on optical glass surfaces.[3] In this method, a sub-wavelength periodic structure on a mould is transferred to a glass surface by pressing the glass against the mould at high temperature. Glasses with low deformation temperature, high refractive index and good transparency in the region from near-UV to near-IR are required for producing high performance optical elements by the imprinting method. Since Bi_2O_3–B_2O_3 glasses have low deformation temperature and high refractive index,[4] it is expected that Bi_2O_3–B_2O_3 glasses can be used as optical glasses for the imprinting method. It is well known, however, that the cut-off wavelength increases with increasing Bi_2O_3 content. In order to improve the transparency in the near-UV region, it is important to add oxides which include atoms of high electronegativity.[5] GeO_2 has a wide glass forming region in combination with B_2O_3, and has lower deformation temperature and higher refractive index than SiO_2.[6,7] Therefore, it is anticipated that Bi_2O_3–GeO_2–B_2O_3 glass is a candidate optical glass for the imprinting process. In the present study, thermal and optical properties of Bi_2O_3–GeO_2–B_2O_3 glasses have been investigated. Furthermore, the structure of the glasses has been studied by Raman spectroscopy. In addition, a sub-wavelength periodic structure was fabricated on the glass by the glass imprinting process.

2. Experimental procedures

Bi_2O_3–B_2O_3 and Bi_2O_3–GeO_2–B_2O_3 samples shown in Figure 1 and Table 1 were prepared. Bi_2O_3 (Taiyo Koko, 99·76%), GeO_2 (Shinko Chemical, 99·999%) and B_2O_3 (Wako Chemical, 90%) were used as starting materials. The mixture of starting materials was melted in a platinum crucible in an electric heating furnace at 1000°C, and poured onto a carbon plate to form a glass plate with a thickness of ~4 mm and a weight of about 50 g. Then, the glass plate was annealed at around the glass transition temperature and allowed to cool in a furnace. Glass formation was confirmed by eye.

Refractive index, density, optical absorption, glass transition and deformation temperatures and Raman scattering spectra were measured for $x$$Bi_2O_3$.$(1-2x)GeO_2$.$x$$B_2O_3$ glasses (with x in the range from 0·5 to 0·15) along the $(0·5Bi_2O_3.0·5B_2O_3)$–$1·0GeO_2$ compositional line and for $x$$Bi_2O_3$.$(0·8-2x)GeO_2$.$(x+0·2)B_2O_3$ glasses (with x in the range from 0·4 to

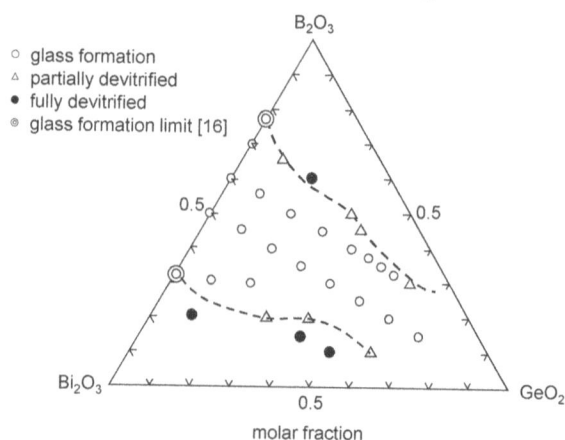

Figure 1. Glass forming region in the Bi_2O_3–GeO_2–B_2O_3 system

* Corresponding author. Email na-yamashita@aist.go.jp

Table 1. Composition of Bi_2O_3–GeO_2–B_2O_3 samples

Nominal composition (molar fraction)			Visual status
Bi_2O_3	GeO_2	B_2O_3	

$xBi_2O_3.(1-2x)GeO_2.xB_2O_3$ glasses

0·50		0·50	glass formation
0·45	0·10	0·45	" "
0·40	0·20	0·40	" "
0·35	0·30	0·35	" "
0·30	0·40	0·30	" "
0·25	0·50	0·25	" "
0·20	0·60	0·20	" "
0·15	0·70	0·15	" "

$xBi_2O_3.(0·8-2x)GeO_2.(x+0·2)B_2O_3$ glasses

0·40		0·60	glass formation
0·35	0·10	0·55	" "
0·30	0·20	0·50	" "
0·25	0·30	0·45	" "
0·20	0·40	0·40	" "
0·175	0·45	0·375	" "
0·15	0·50	0·35	" "
0·125	0·55	0·325	" "
0·10	0·60	0·30	partially devitrified
0·30		0·70	glass formation
0·25	0·10	0·65	partially devitrified
0·20	0·20	0·60	fully devitrified
0·15	0·35	0·50	partially devitrified
0·15	0·40	0·45	partially devitrified
0·60	0·10	0·30	glass formation
0·50	0·20	0·30	glass formation
0·70	0·10	0·20	fully devitrified
0·50	0·30	0·20	partially devitrified
0·40	0·40	0·20	partially devitrified
0·40	0·50	0·10	fully devitrified
0·30	0·60	0·10	partially devitrified

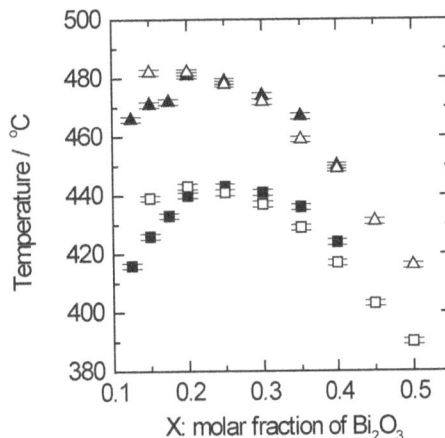

Figure 2. Glass transition and deformation temperatures of $xBi_2O_3.(1-2x)GeO_2.xB_2O_3$ glasses (along the $(0·5Bi_2O_3.0·5B_2O_3)$–$(1·0GeO_2)$ compositional line), and of $xBi_2O_3.(0·8-2x)GeO_2.(x+0·2)B_2O_3$ glasses (along the $(0·4Bi_2O_3.0·6B_2O_3)$-$(0·8GeO_2.0·2B_2O_3)$ compositional line). Triangles and squares show deformation and glass transition temperatures, respectively. Open and closed marks show $xBi_2O_3.(1-2x)GeO_2.xB_2O_3$ and $xBi_2O_3.(0·8-2x)GeO_2.(x+0·2)B_2O_3$ glasses, respectively

0·125) along the $(0·4Bi_2O_3.0·6B_2O_3)$–$(0·8GeO_2.0·2B_2O_3)$ compositional line. In addition, the Raman scattering spectrum was also measured for $0·3Bi_2O_3.0·7B_2O_3$ glass. The refractive index of the glasses was measured at a wavelength of 632·8 nm by a prism coupling method (Metricon, Model 2010 Prism-Coupler). The density was measured by Archimedes' method. The molar refraction was calculated from the refractive index and the density using the Lorentz–Lorenz equation.[5] The optical absorption measurements were carried out for glass samples with a thickness of 3 mm using a conventional spectrophotometer (Hitachi, U4000). Glass transition and deformation temperatures were obtained from the thermal expansion coefficient, which was measured with a thermo-mechanical analyser (SII, TMA/SS6300) at a heating rate of 10°C min^{-1}. Raman scattering spectra of the glasses were measured in backscattering geometry with a Raman spectrometer (Atago Jobin Ybon, T64000) using DPSS (diode pumped solid state) laser light (Coherent, Verdi, λ=532 nm) as an exciting source.

Fabrication of sub-wavelength structure on the glass surface was attempted by the imprinting process. An oblate spheroid glass preform with 11 mm diameter and 4 mm thick was pressed under vacuum at around the glass deformation temperature with a high precision optical glass pressing machine (Toshiba Machine GMP-311V), equipped with a SiC mould on which one-dimensional array of grooves with a period of 500 nm was formed. The fabrica-

tion method of the mould has been described in detail elsewhere.[3] The glass surface after pressing was observed with a scanning electron microscope (Hitachi, S-4700) and an atomic force microscope (Shimadzu, SFT-3500).

3. Results

3.1 Properties of Bi_2O_3–GeO_2–B_2O_3 glasses

Figure 1 shows the glass forming region obtained in the present study. The Bi_2O_3–GeO_2–B_2O_3 ternary system has a wide glass forming region. Figure 2 shows the glass transition temperature and deformation temperature of $xBi_2O_3.(1-2x)GeO_2.xB_2O_3$ and $xBi_2O_3.(0·8-2x)GeO_2.(x+0·2)B_2O_3$ glasses. The glass transition temperature and glass deformation temperature were strongly dependent on Bi_2O_3 content. These temperatures showed maxima at Bi_2O_3 contents of 20 to 30 mol%. The glass transition and deformation temperatures were less affected by the substitution of GeO_2 for B_2O_3.

Figure 3 shows the refractive index, molar refraction and molar volume as a function of Bi_2O_3 content in the glasses. The refractive index and molar volume increase monotonically, but nonlinearly, as the Bi_2O_3 content increases. The substitution of GeO_2 for B_2O_3 causes the refractive index to increase, and the molar volume to decrease. The molar refraction increases linearly with increasing Bi_2O_3 content, and is unchanged by the substitution of GeO_2 for B_2O_3. The molar refraction shows additivity with composition (molar refraction=$32·3f_{Bi_2O_3}+10·0f_{GeO_2}+9·7f_{B_2O_3}$, where $f_{Bi_2O_3}$, f_{GeO_2} and $f_{B_2O_3}$ are the molar fraction of Bi_2O_3, GeO_2 and B_2O_3, respectively).

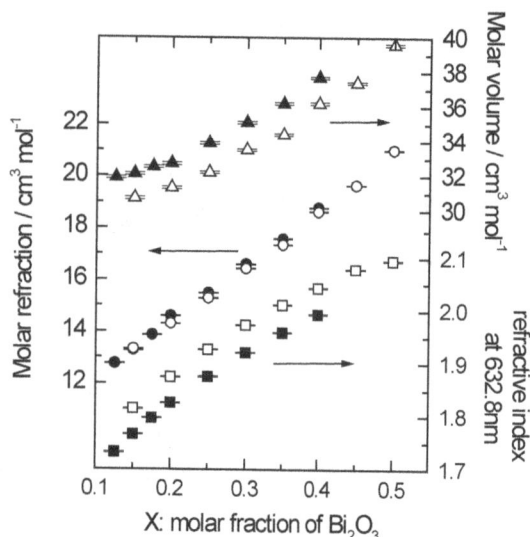

Figure 3. Refractive index (squares), molar refraction (circles) and molar volume (triangles) of $xBi_2O_3.(1–2x)$ $GeO_2.xB_2O_3$ glasses (along the $(0.5Bi_2O_3.0.5B_2O_3)$–$(1.0GeO_2)$ compositional line) and of $xBi_2O_3.(0.8–2x)$ $GeO_2.(x+0.2)B_2O_3$ glasses (along the $(0.4Bi_2O_3.0.6B_2O_3)$–$(0.8GeO_2.0.2B_2O_3)$ compositional line). Open and closed marks show $xBi_2O_3.(1–2x)GeO_2.xB_2O_3$ and xBi_2O_3–$(0.8–2x)GeO_2–(x+0.2)B_2O_3$ glasses, respectively

An inset in Figure 4 shows a typical optical transmittance. No absorption bands were observed in the visible region. The absorption edge is located between 300 and 400 nm. The photon energy where the transmittance of glass is equal to 5% is plotted as an absorption edge energy against Bi_2O_3 content, in Figure 4. The absorption edge energy increases monotonically with decreasing Bi_2O_3 content. The substitution of B_2O_3 and GeO_2 for Bi_2O_3 by 10 mol% increases the absorption edge energy by 0.17 eV and 0.15 eV, respectively. The substitution of GeO_2 for B_2O_3 by 10 mol% decreases the energy by 0.02 eV. Colourless glasses were obtained in the low Bi_2O_3, high GeO_2 region of composition.

3.2 Fabrication of sub-wavelength structure

The Bi_2O_3–GeO_2–B_2O_3 glass so developed was used for the fabrication of sub-wavelength periodic surface structure by an imprinting process. Figure 5 displays

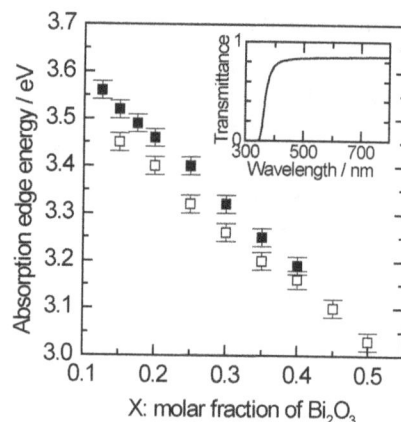

Figure 4. Absorption edge energy of $xBi_2O_3.(1–2x)GeO_2.$ xB_2O_3 glasses (along the $(0.5Bi_2O_3.0.5B_2O_3)$–$(1.0GeO_2)$ compositional line) and of $xBi_2O_3.(0.8–2x)GeO_2.(x+0.2)B_2O_3$ glasses (along the $(0.4Bi_2O_3.0.6B_2O_3)$–$(0.8GeO_2.0.2B_2O_3)$ compositional line). Open and closed marks show $xBi_2O_3.$ $(1–2x)GeO_2.xB_2O_3$ and $xBi_2O_3.(0.8–2x)GeO_2.(x+0.2)B_2O_3$ glasses, respectively. An inset shows the transmittance of $0.15Bi_2O_3.0.50GeO_2.0.35B_2O_3$ glass

the SEM (scanning electron microscope) and AFM (atomic force microscope) images of the glass surface after pressing. It can be seen that a structure with a period of 500 nm and a height of 100 nm was formed on the glass surface. In addition, the glass had a smooth surface after the imprinting process. Colouration was not observable by eye after the imprinting process.

3.3 Structure of Bi_2O_3–GeO_2–B_2O_3 glasses

Figures 6(a) and (b) depict the Raman scattering spectra of the glasses. The Raman scattering intensity was normalised to the intensity of the band at about 150 cm^{-1}. Bands in the regions from 140 to 150 and from 1250 to 1330 cm^{-1} and a shoulder at about 360 cm^{-1} were observed in bismuth borate glasses, as shown in Figure 6(a) and (b). In addition, faint Raman bands were observed in the region from 500 to 1100 cm^{-1}.

In the Raman spectra of $xBi_2O_3.(1–2x)GeO_2.xB_2O_3$ glasses shown in Figure 6(a), bands were observed at about 150 and 420 cm^{-1}, in the regions from 760 to 780 cm^{-1}, and from 1250 to 1330 cm^{-1}. An increase in the GeO_2 content caused the intensity of the 420 cm^{-1} band and 780 cm^{-1} band to increase.

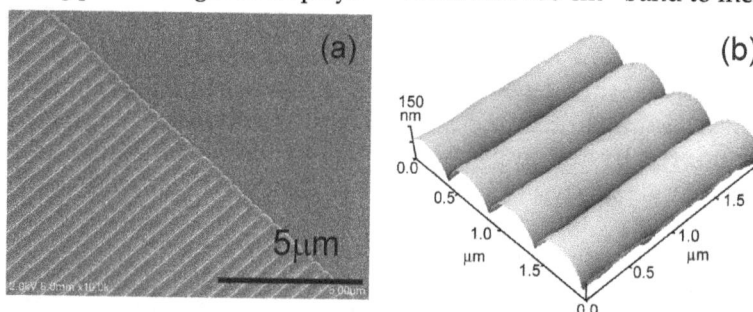

Figure 5. (a) Scanning electron microscope (SEM) and (b) atomic force microscope (AFM) images of the surface of Bi_2O_3–GeO_2–B_2O_3 glass after pressing

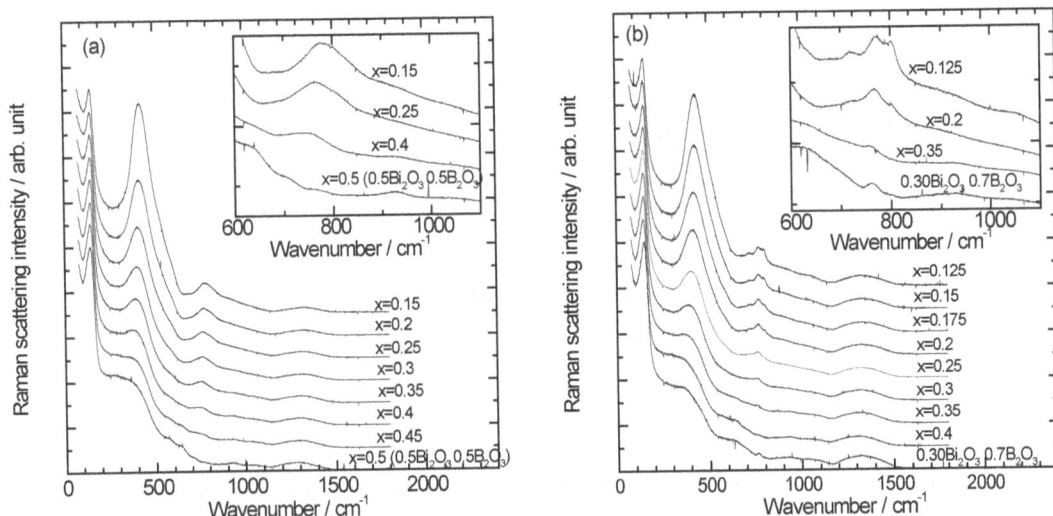

Figure 6. Raman spectra of (a) $xBi_2O_3.(1-2x)GeO_2.xB_2O_3$ glasses (along the $(0.5Bi_2O_3.0.5B_2O_3)$–$(1.0GeO_2)$ compositional line) and (b) $xBi_2O_3.(0.8-2x)GeO_2.(x+0.2)B_2O_3$ glasses (along the $(0.4Bi_2O_3.0.6B_2O_3)$–$(0.8GeO_2.0.2B_2O_3)$ compositional line). The spectrum of $0.3Bi_2O_3.0.7Bi_2O_3$ glass is also shown for comparison

Bands were observed at about 150 and 420 cm^{-1} and in the region from 1250 to 1330 cm^{-1} in the Raman spectra of the $xBi_2O_3.(0.8-2x)GeO_2.(x+0.2)B_2O_3$ glasses (Figure 6(b)), as for the $xBi_2O_3.(1-2x)GeO_2.xB_2O_3$ glasses. A weak but sharp band was observed at about 770 cm^{-1}. In addition, bands were observed at about 720 and 800 cm^{-1} for the glasses containing GeO$_2$ higher than 30 mol%. The 720, 780 and 800 cm^{-1} bands became distinct as the GeO$_2$ content increased.

4. Discussion

4.1. Properties of Bi_2O_3–GeO_2–B_2O_3 glasses

The glass transition and deformation temperatures are strongly dependent on the Bi$_2$O$_3$ content, but are less affected by the substitution of GeO$_2$ for B$_2$O$_3$, as shown in Figure 2. These temperatures show maxima at Bi$_2$O$_3$ contents of 20 to 30 mol%. Since it has previously been reported that the glass transition temperature has a maximum at a Bi$_2$O$_3$ content of about 20 mol% and about 30 mol% in Bi$_2$O$_3$–GeO$_2$ and Bi$_2$O$_3$–B$_2$O$_3$ glasses,[8,9] respectively, it was deduced that such an anomaly occurs for Bi$_2$O$_3$–GeO$_2$–B$_2$O$_3$ glasses too.

The compositional dependence of the refractive index is somewhat different from that of the molar refraction (see Figure 3). The refractive index increases at a decreasing rate as the Bi$_2$O$_3$ content increases, whereas the molar refraction increases linearly. The substitution of GeO$_2$ for B$_2$O$_3$ increases the refractive index, whereas the molar refraction is unchanged by the substitution of GeO$_2$ for B$_2$O$_3$. These differences in compositional dependence between molar refraction and refractive index are due to the effect of molar volume on refractive index. That is, the molar volume increases at an increasing rate as the Bi$_2$O$_3$ content increases, which leads to the increase of refractive index at a decreasing rate. In addition, the molar

volume is decreased by the substitution of GeO$_2$ for B$_2$O$_3$, resulting in an increase of the refractive index.

The absorption edge energy increases monotonically as the Bi$_2$O$_3$ content decreases, as shown in Figure 4. The substitution of B$_2$O$_3$ and GeO$_2$ for Bi$_2$O$_3$ by 10 mol% increases the absorption edge energy by 0.17 eV and 0.15 eV, respectively. Substitution of GeO$_2$ for B$_2$O$_3$ by 10 mol% decreases the energy by 0.02 eV. These tendencies of edge energy shift agree well with those expected from electronegativity.[5]

4.2 Fabrication of sub-wavelength structure

The Bi$_2$O$_3$–GeO$_2$–B$_2$O$_3$ glass so developed was used for the fabrication of sub-wavelength periodic surface structure by an imprinting process. A structure with a period of 500 nm and a height of 100 nm was formed on the glass surface. That is, the periodic structure on a mould was successfully transferred to the glass surface. Since neither roughening nor colouration occurred on the glass surface after the imprinting process, it was confirmed that Bi$_2$O$_3$–GeO$_2$–B$_2$O$_3$ glass could be used for the fabrication of sub-micron structure by the glass imprinting process.

4.3 Structure of Bi_2O_3–GeO_2–B_2O_3 glasses

In the Raman scattering spectra of Bi$_2$O$_3$–B$_2$O$_3$ glasses shown in Figure 6(a) and (b), bands were observed in the regions from 140 to 150 cm^{-1}, from 500 to 1100 cm^{-1}, and from 1250 to 1330 cm^{-1} and a shoulder at about 360 cm^{-1}. Although the band in the region from 140 to 150 cm^{-1} and the shoulder at about 360 cm^{-1} have previously been assigned to Bi^{3+} ion related structures,[10,11] a rigorous assignment of this band and shoulder has not been achieved yet. The band in the region from 1250 to 1330 cm^{-1} has been assigned to BO$_3$ triangles.[12]

In the Raman spectra of $x\mathrm{Bi_2O_3}.(1-2x)\mathrm{GeO_2}.x\mathrm{B_2O_3}$ glasses shown in Figure 6(a), bands were observed at about 150 and 420 cm^{-1}, and in the regions from 760 to 780 cm^{-1} and from 1250 to 1330 cm^{-1}. For germanate glasses, the bands at about 420 and 780 cm^{-1} have previously been assigned to Ge–O–Ge vibration and metagermanate groups, respectively.[13,14] Since it has been reported that the band is located at 420 cm^{-1} in pure GeO$_2$ single glass, and that the bands are located at higher wavenumber in alkali germanate glasses, the band at 420 cm^{-1} in the present glasses was assigned to Ge–O–Ge vibration of GeO$_4$ tetrahedra with no nonbridging oxygen atoms. Therefore, it was deduced that GeO$_4$ tetrahedra with no nonbridging oxygen atoms and with two nonbridging oxygen atoms are present in the glasses. Nevertheless, the presence of the band at 420 cm^{-1} indicates that the fraction of the latter is low compared with that of the former. The formation of metagermanate groups implies that Bi$_2$O$_3$ partially acts as a network modifier. An increase in the GeO$_2$ content causes the intensity of the 420 cm^{-1} and 780 cm^{-1} bands to increase. The ratio in intensities of the 780 cm^{-1} and 420 cm^{-1} bands decreases from 1·4 to 0·2 as the GeO$_2$ content increases from 10 to 70 mol%, due to the decrease of the Bi$_2$O$_3$ content. Bands due to boron related structural units were not observed in the region from 600 to 1000 cm^{-1}.

Bands were observed at about 150, 420 and 770 cm^{-1}, and from 1250 to 1330 cm^{-1} in the Raman spectra of the $x\mathrm{Bi_2O_3}.(0\cdot8-2x)\mathrm{GeO_2}.(x+0\cdot2)\mathrm{B_2O_3}$ glasses (Figure 6(b)). The band at about 770 cm^{-1} was assigned to the triborate group from comparison with the Raman spectra of alkali and alkaline earth borate glasses.[15] The additional bands at about 720 and 800 cm^{-1} observed for glasses containing more than 30 mol% GeO$_2$ were assigned to chain-type metaborate and boroxol groups, respectively.[15] The presence of the 780 cm^{-1} band due to metagermanate groups was unclear, owing to the overlap of the boron related bands. The 720, 780 and 800 cm^{-1} bands, due to boron related structural units, became distinct with increasing GeO$_2$ content. That is, an increase in GeO$_2$ content caused the formation of boroxol, triborate and metaborate groups in $x\mathrm{Bi_2O_3}.(0\cdot8-2x)\mathrm{GeO_2}.(x+0\cdot2)\mathrm{B_2O_3}$ glasses, although the increase in GeO$_2$ content is accompanied by a decrease in B$_2$O$_3$ content. In a previous Raman spectroscopic study on K$_2$O–GeO$_2$–B$_2$O$_3$ glasses, it was shown that the structure depends on the K$_2$O/B$_2$O$_3$ concentration ratio.[14] Since an increase in GeO$_2$ content is accompanied by a decrease in the concentration ratio Bi$_2$O$_3$/B$_2$O$_3$, it is qualitatively reasonable that an increase in the GeO$_2$ content caused the formation of boroxol and triborate groups. However, although the concentration ratio of Bi$_2$O$_3$/B$_2$O$_3$ in 0·15Bi$_2$O$_3$.0·50GeO$_2$.0·35B$_2$O$_3$ glass was the same as that of 0·3Bi$_2$O$_3$.0·7B$_2$O$_3$ glass, boroxol units were observed only in the former glass. This indicates that a small fraction of GeO$_2$ forms nonbridging oxygen atoms and

is associated with Bi^{3+} ions, resulting in a decrease in the amount of Bi$_2$O$_3$ which interacts with B$_2$O$_3$.

5. Conclusions

Two series of bismuth germanoborate glasses were studied, with the compositions $x\mathrm{Bi_2O_3}.(1-2x)\mathrm{GeO_2}.x\mathrm{B_2O_3}$ and $x\mathrm{Bi_2O_3}.(0\cdot8-2x)\mathrm{GeO_2}.(x+0\cdot2)\mathrm{B_2O_3}$. The glass transition and deformation temperatures were in the ranges from 340 to 440°C and from 370 to 480°C, respectively. The glass transition and deformation temperatures had maxima at Bi$_2$O$_3$ contents of 20 to 30 mol%, as observed in bismuth borate and bismuth germanate glasses. The refractive index was in the range from 1·75 to 2·20. Molar refraction decreased linearly with decreasing Bi$_2$O$_3$ content and did not depend on the B$_2$O$_3$/GeO$_2$ ratio. Colourless glasses were obtained in the low Bi$_2$O$_3$, high GeO$_2$ region of composition, as expected from electronegativity. Raman scattering spectroscopy showed that the structure depended on the Bi$_2$O$_3$/B$_2$O$_3$ ratio and the GeO$_2$ content. It was deduced that Ge atoms mainly form GeO$_4$ tetrahedra with no nonbridging oxygen atoms and that the association of Bi^{3+} ions with GeO$_2$ affects the concentration of boron related structural units. By the glass imprinting process, a one dimensional periodic pattern with a period of 500 nm on a SiC mould was successfully transferred to the glass surface. The glass surface was not coloured or roughened after the imprinting process.

Acknowledgement

This work was carried out in Next-generation Nanostructured Photonic Device and Process Technology as part of the Program to Create an Innovative Components Industry, supported by New Energy and Industrial Technology Development Organization.

References

1. Nishii, J., Kintaka, K. & Nakazawa, T. *Appl. Opt.*, 2004, **43**, 1327.
2. Kintaka, K., Nishii, J., Mizutani, A., Kikuta, H. & Nakano, H. *Opt. Lett.*, 2001, **26**, 1642.
3. Mori, T., Hasegawa, K., Hatano, T., Kasa, H., Kintaka, K. & Nishii, J. *Opt. Lett.*, 2008, **33**, 428.
4. Becker, P. *Cryst. Res. Technol.*, 2003, **38**, 74
5. Duffy, J. A. & Ingram, M. D. In: *Optical properties of glass*, Eds D. R. Uhlmann, & N. J. Kreidl, The American Ceramic Society, Westerville, 1991, p. 159.
6. Ma, C., Kiczenski, T., McRae, J., Affatigato M. & Feller, S. A. *Phys. Chem. Glasses*, 2000, **41**, 365.
7. Volf, M. B. *Mathematical approach to glass, Glass science and technology Volume 9*, 1988, Elsevier, Amsterdam, p. 143.
8. Nassau, K., & Chadwick, D. L. *J. Am. Ceram. Soc.*, 1982, **65**, 197.
9. Ehrt, D., *Glass Tech.*, 2000, **41**, 182.
10. Baia, L., Iliescu, T., Simon, S. & Kiefer, W. *J. Mol. Struct.*, 2001, **599**, 9.
11. Bale, S., Srinivasa Rao, N. & Rahman, S. *Solid State Ionics*, 2008, **10**, 326.
12. Yano, T., Kunimine, N., Shibata, S. & Yamane, M. *J. Non-Cryst. Solids*, 2003, **321**, 137.
13. Furukawa, T. & White, W. B. *J. Mater. Sci.*, 1980, **15**, 1648.
14. Chakraborty, I. N. & Condrate, Sr., R. A. In: *Advances in materials characterization*, Ed. D. R. Rossington, R. A. Condrate, & R. L. Snyder, Plenum, New York, 1983, p. 223.
15. Konijnendijk, W. L. *Philips Res. Rep. Suppl.*, 1975, No.1, p. 1.
16. Imaoka, T. & Yamazaki, T. Report of the institute of industrial science, The University of Tokyo, 1972, **22**, 1.

Proc. VI Int. Conf. Borate Glasses, Himeji, Japan, 18–22 August 2008   Glass Tech.: Eur. J. Glass Sci. Technol. A, August 2009, **50** (4), 227–229

Design and operation of a new roller quencher for rapidly cooling melts into glasses

A. J. Havel, S. A. Feller,* M. Affatigato

Coe College, Physics Department, 1220 1ˢᵗ Ave NE, Cedar Rapids, IA 52402, USA

M. Karns & M. Karns

Morgan Meredith Manufacturing, Cedar Rapids, IA 52403, USA

Manuscript received 18 August 2008
Revised version received 21 December 2008
Accepted 6 January 2009

We discuss the design and construction of a new roller quencher that is capable of cooling liquids at rates up to $1.8 \times 10^6\,°C/s$. The new roller quencher is based on refinements to our proven twin roller designs. This new generation of machines will have a number of advantages over the existing ones, including digital control of the speed of the rollers and gap adjustment, the ability to control the atmosphere over the resulting glass samples, ease of cleaning, and size. We also discuss the use of roller quenching at Coe College to form borate, and other glasses.

1. Introduction to previous roller quenchers

Extending the compositional ranges of glass formation has proven fruitful in our research.[1–8] This was made possible by a series of twin roller quenchers of our own construction. Simple by design, these machines rapidly cooled molten samples, readily producing glasses that were difficult to make by other methods.

In our previous design, two rollers, spinning in opposite directions at several thousand RPM, pulled the melt through a 1–700 µm gap. The adjustable gap was achieved by a wedge that could be driven down, forcing the rollers to separate. Though this method worked, it was found to be difficult and cumbersome to use.

2. The design of the improved roller quencher

The new machine has the same fundamental design as the earlier versions, as shown in Figure 1. It retains the twin rollers and adjustable gap, but the internal workings of the machine have been substantially improved. For example, the rollers move apart via two over-eccentric gears (part of this design is shown in Figure 2). This allows for gap adjustment from 0 to 1100 µm with an improved accuracy of 0·5 µm which can be achieved by turning two handles. This is shown in Figure 3. Roller operation is now fully automated by a touch screen control panel that displays the machine's settings, including the rotation rate of the rollers in RPM. Shielding has also been improved to fully enclose the machine and capture the formed glass, see Figure 1. This assists in clean

Figure 1. An image of the sixth generation Coe College Roller Quencher (the COE 102M). This image shows how the user may control the gap by simply turning the two knobs on the edges of the machine. Note the full shielding added to the machine

up, maintenance, and sample purity while simultaneously aiding in user safety.

The machine is also currently being fitted for a fully automated control of the gap width. This will aid in the ease of operation of the machine, as well as giving users the ability to automatically record operating conditions. Another feature currently being designed will allow the user to purge the sample with inert gas, creating dry box conditions. For

* Corresponding author. Email sfeller@coe.edu

Figure 2. The timing placements for the over-eccentric hubs. These gears allow the rollers to separate with precision. Both knobs and rollers drop into the side frames of the machine and are lined up simply by the two indentations on the gear teeth. This ensures the gap is always accurate

Figure 3. Showing the roller quencher without the internal shielding. The rollers and bubble level can be seen

example, this will allow hygroscopic samples to be roller quenched in nitrogen and stored without being subjected to atmospheric water.

One of the greatest improvements in the latest version of the roller quencher was the change in the diameter of the rollers. These were reduced from 15·24 cm to 10·16 cm (as was the case for our fourth generation roller quencher shown in Figure 4). This helps in multiple ways: First, the smaller rollers are 50% lighter, helping to make the solid steel machine less heavy. It now weighs less than 100 kg. Secondly, the smaller rollers allow the machine to be more compact. This decreased the overall size of the machine, from 92×46×61 cm to 30·5×30·5×46 cm. Most importantly, the smaller rollers give a smaller incident angle, which shortens the contact length the melt has with the rollers and yields higher average cooling rates, as shown by Equation (1)

$$\frac{\Delta T}{\Delta t} = \frac{r\omega\Delta T}{y} \qquad (1)$$

where r is the roller radius, ω is the angular velocity, y is the contact distance of the melt and rollers, Δt is the time the melt is in contact with the rollers, and ΔT is the change in temperature of the melt.[9]

The resulting average cooling rates for the new roller quencher are shown in Table 1. The table depicts a linear trend up to the maximum rotation

Table 1. Cooling rates for various speeds of the COE102M roller quencher

RPM	1000	1200	1400	1600	1800	2000	2200	2300 (max)
ω (rads/s)	105	126	147	168	188	209	230	241
Cooling rate ($\times 10^6$°C/s)	0·78	0·93	1·1	1·2	1·4	1·6	1·7	1·8

$\Delta T = 975$°C = 1000°C – 25°C
$r = 5·08$ cm
$y = 0·666$ cm

rate of the machine, 2300 RPM. Compared to the fifth generation roller quencher, whose cooling rate has a maximum of $1·1 \times 10^6$°C/s,[9] the new machine has a calculated cooling rate of $1·8 \times 10^6$°C/s. The smaller roller diameter and accessibility of higher rotation rates gives us a greater cooling range and is likely to allow us to extend glass forming ranges even further than before.

The cooling rates of other quenching methods range from 10^2, 10^3, and 10^4°C/s for air cooling, plate quenching and ice quenching, respectively.[9] A glass may be defined as an amorphous solid with viscosity greater than 10^{13} poise, which does not undergo a change of structure from the liquid state upon cooling.[10] When a melt is cooled quickly enough, an inadequate amount of time is available to create a long range crystalline structure, leaving a solid with only a short to mid range atomic structure (10^{-10}–10^{-9} m length scale). Roller quenching is extremely useful in shortening the time available for structural rearrangement.

3. Previous work with roller quenching

From research completed with the earlier versions of the roller quencher,[1–8] we have shown that roller quenching molten samples can extend the range of glass formation to larger R values for borates, borosilicates, and borovanadates (where R is the molar ratio of modifier oxide to boron oxide) and other glass systems. Kasper et al showed that the compositional range of sodium borate glasses could be vastly increased with the use of roller quenching.[3] Previously, sodium borate glasses have been made in the regions of $R=0$ to $R=0·7$ and from $R=2·0$ to $R=3·5$.[3,7,8] With the use of roller quenching, the glass forming range was

Figure 4. Improved design for the new sixth generation roller quencher (left) compared to the fourth generation roller quencher (right)

extended from $R=0$ to $R=4\cdot9$,[3] apart from the small region $0\cdot8 \leq R \leq 1\cdot2$ near the sodium metaborate crystal composition ($R=1\cdot0$). Similarly, glass formation in lithium borates[8] has been extended over the continuous range $R=0$ to $R=2\cdot8$.

This extension of the glass forming range for lithium borosilicates is described in research performed by Boekenhauer *et al.*[1] The R range was extended, using roller quenching, up to a value of 10 with K values (the relative amount of silica to boron oxide) of up to $K=1$.[1] This corresponds to a molar fraction of lithium oxide

$$X = \frac{R}{1+R+K} = 0\cdot83 \qquad (2)$$

The glass forming range has also been significantly extended for the lithium silicate system. In previous studies it had been found that the glass forming range was limited to $J \leq 0\cdot85$[11] (where J is the molar ratio of alkali oxide to silicon dioxide), but with use of the roller quencher, Peters *et al* were able to extend the compositional range up to $J=1\cdot8$.[4]

The alkaline and alkaline earth vanadate glass forming ranges have also been expanded in our laboratory. Since these are poor glass formers,[6]

roller quenching was our only suitable way of obtaining glass over a large range of R values. We have been able to expand the limit of glass formation to $0\cdot0 \leq R \leq 1\cdot5$.[6] The same glass forming range was found for lead vanadate glasses.[6] Similarly, the bismuth vanadate glass forming range was expanded to $0\cdot0 \leq R \leq 0\cdot4$.[6] Many other systems have enjoyed a similar expansion of the range of glass formation.

4. Conclusion

We have designed and built a roller quencher that surpasses all of our previous versions in regards of user interface, ease of cleaning, size, and defence against contamination. Calculated cooling rates have shown that our sixth generation machine should be able to outperform older versions and allow us to continue expanding compositional ranges of glass formation. Further information on the new roller quencher may be obtained by contacting S. A. Feller at sfeller@coe.edu.

5. Acknowledgments

We would like to thank the Iowa College Foundation (Maytag Innovation Program) for their support and the National Science Foundation for grant DMR 0502051. Corning Inc. is acknowledged for funding the construction of the first of the new generation of roller quenchers. Marc Berman, Tyler Mullenbach, and Maranda Franke are thanked for proofreading the manuscript.

References

1. Boekenhauer, R. *The Density of Lithium Borosilicate Glasses Related to Atomic Arrangement*, Coe College Honors Paper, 1991.
2. Kottke, J. A. *An Examination of the Physical Properties of Rubidium and Cesium Borosilicate Glasses and the Development of a Quantitative Density Model for the Alkali Borosilicate Glass System*, Coe College Honors Paper, 1995.
3. Kasper, J. E., Feller, S. A. & Sumcad, G. L. *J. Am. Ceram. Soc.*, 1984, **67** (4), 71.
4. Peters, A. M., Alamgir, F. M., Messer, S. W., Feller, S. A. & Loh, K. L. *Phys. Chem. Glasses*, 1994, **35** (5), 212–215.
5. Larson, C. J., Doerr, J., Affatigato, M. & Feller, S. *J. Phys.: Condens. Matter*, 2006, **18**, 11323–11331.
6. Basu, A. Lewis, J. O'Brien, C. P. Feller, H. Starns, J. Frueh, J. Feller, S. and Affatigato, M. *Phys. Chem. Glasses*, 2003, **44** (1), 1–4.
7. Lim, H .P. Karki, A. Feller, S. Kasper, J. & Sumcad, G. *J. Non-Cryst. Solids*, 1987, **91**, 324–332.
8. Karki, A., Lim, H. P., Feller, S., Stark, J., Sanchez, C. & Shibata, M. *J. Non-Cryst. Solids*, 1987, **92**, 11–19.
9. Nelson, N. *The Design and Construction of a Rapid Quenching Device*. Coe College Honors Paper, 1998.
10. Kittel, C. *Introduction to Solid State Physics*, John Wiley & Sons, New York, 1986.
11. Bansal, N. P. & Doremus, R. H. *Handbook of Glass Properties*, Academic Press, Orlando, 1986.

Proc. VI Int. Conf. Borate Glasses, Himeji, Japan, 18–22 August 2008   *Phys. Chem. Glasses: Eur. J. Glass Sci. Technol. B, February 2009, 50 (1), 59–62*

Optical properties of ZnO–SnO–B₂O₃ glasses

Naoyuki Kitamura,* Kohei Fukumi & Junji Nishii

National Institute of Advanced Industrial Science and Technology, Ikeda, Osaka, 563-8577, Japan

Manuscript received 18 August 2008
Revised version received 21 October 2008
Accepted 20 November 2008

The optical transmittance, refractive index and reflectivity were studied for ternary (60–x)ZnO.xSnO.40B₂O₃ (x=20–60) glasses. The absorption edge energy varied from 3·05 to 2·82 eV with a change of SnO concentration from x=20 to 60 mol%, and the refractive index n_d increased monotonically from 1·743 to 1·895. According to an analysis using a one-term Sellmeier equation, the resonance energy shifted from 8·2 to 6·7 eV as the SnO concentration increased. The shift of the resonance energy is quite different from that of the absorption edge energy. A broad peak was observed around 4·8 eV in the reflection spectra for all the glasses. The peak position shifted slightly from 4·8 to 4·7 eV as the SnO concentration increased. In addition, a broad shoulder ranging over 3·5–4·3 eV was observed. The intensity of the shoulder as the SnO concentration decreased and it became imperceptible for x=20. Therefore the shift of the absorption edge energy is deduced to be due to the shift of the 4·8 eV band and the change in intensity of the shoulder. The Raman spectra show bands from 620 to 1540 cm⁻¹, which are assigned to vibrations of borate groups. The 180 cm⁻¹ band, which is assigned to Sn–O vibration, decreased in intensity as the SnO concentration decreased and disappeared at x=20. Therefore the shoulder at 3·5–4·3 eV found in the reflection spectra may be due to structure related to this band.

Introduction

Glasses containing tin oxide have high refractive index and low deformation temperature, and so they are candidates for use in diffractive optical elements having sub-wavelength periodic structures on the surface fabricated by a glass nano-imprinting method. Paul et al[1] studied SnO–B₂O₃ glasses by Mössbauer spectra and considered the ionicity of tin ions for different tin concentrations. Nakai et al[2] studied glass formation and thermal properties of SnO–B₂O₃ and LiO–SnO–B₂O₃ systems as low melting ion conducting glasses. Hayashi et al discussed the structure of borate groups and the role of tin ions in SnO–B₂O₃[3] and Li₂O–SnO–B₂O₃ glasses.[4] After the investigation of Li ion conducting glass by Idota et al,[5] studies of SnO–B₂O₃ based glass gave much attention to the electronic properties of the glass.[2–7] However, the optical properties of tin oxide containing borate glasses have not been studied yet. In the present study, we focused on the optical properties such as transmittance, refractive index and reflectivity in ternary ZnO–SnO–B₂O₃ glasses. The effect of chemical composition on the optical properties is discussed in conjunction with the glass structure analysed from Raman spectroscopy.

Experimental procedure

Glasses of the composition (60–x)ZnO.xSnO.40B₂O₃ (x=20, 30, 40, 50, 60 mol%) were prepared by a conventional melt quenching method. Appropriate mixtures of ZnO (Kojundo Chemical Laboratory Co. Ltd., 99·9%), SnO (Wako Chemical, 99%), B₂O₃ (Kanto Chemical, 98%) and 0·5–1 wt% of carbonaceous additives, were melted at temperatures between 900 and 1000°C for 30 min in an alumina crucible under a nitrogen gas flow. The melt was quenched by pouring it into a metal mould in air; subsequently it was annealed for 1 h at the glass transition temperature and then slowly cooled to room temperature. The density of each glass was measured at room temperature by Archimedes' method using distilled water as the buoyancy liquid. The estimated error is ±0·0005 g/cm³. The glass transition temperature (T_g), deformation temperature (A_t) and thermal expansion coefficient (α) were measured using a thermal mechanical analyser (Seiko Instruments Inc., TMA/SS6300) with a heating rate of 10°C/min. The experimental errors on T_g, A_t and α are ±3°C, ±3°C and ±5×10⁻⁷ °C⁻¹, respectively. The sample size for the thermo-mechanical measurement was 3×3×~15 mm. The refractive index of the glass was measured by a prism coupling method (Metricon, Model 2010) at wavelengths of 632·8, 831·8 and 1545 nm and by a V-block method (Shimazu, KPR-2000) at wavelengths

* Corresponding author. Email naoyuki.kitamura@aist.go.jp

Table 1. Density, molar volume, glass transition and deformation temperatures, average thermal expansion coefficient (50–400°C) and refractive index at He-d line of (60–x)ZnO.xSnO.40B$_2$O$_3$ glasses (x=20–60 mol%)

| x | ZnO | SnO | B$_2$O$_3$ | Density | Molar volume | T$_g$ | A$_t$ | α | n$_d$ |
	(mol%)			(g/cm^3)	(cm^3/mol)	(°C)	(°C)	(×10^{-7} °C^{-1})	
20	40	20	40	3·7088	23·548	450	479	50	1·7428
30	30	30	40	3·7450	24·744	427	461	52	1·7809
40	20	40	40	3·7538	26·106	405	455	56	1·8098
50	10	50	40	3·7951	27·227	380	416	63	1·8419
60	0	60	40	3·9005	27·858	359	396	70	1·8948

of spectral lines from He (587·562 nm), Hg (546·07348, 435·83277, 404·6563 nm), Cd (643·847, 479·9912 nm) and H (656·3, 486·1 nm) discharge lamps. The errors on these measurements did not exceed 0·0010 and 0·0001 for the former and latter methods, respectively. The transmittance and reflectivity were measured by a conventional spectrophotometer (Hitachi, U4000) in the wavelength region from 200 to 800 nm and from 200 to 500 nm, respectively. The reflectivity measurement was performed at an incident angle of 10° using an integrating sphere. In order to investigate the relationship between the properties and glass structure, Raman scattering spectra were measured with a Raman spectrometer (Horiba Jobin-Ybon, T64000) equipped with 532 nm DPSS laser (Coherent, Verdi) with output power of 200 mW. Measurement was carried out at room temperature with the backscattering arrangement.

Results

The density and molar volume are listed along with glass transition temperature, deformation temperature and thermal expansion coefficient in Table 1. The molar volume increased monotonically with increasing SnO content. The glass transition and deformation temperatures decreased linearly with increasing SnO concentration. The thermal expansion coefficient increased as the SnO content increased.

The molar extinction coefficient, normalised by SnO concentration, is plotted against photon energy in Figure 1. The absorption edge energy was determined from the intersection of extrapolated lines for the intrinsic absorption edge and reflection loss as shown in Figure 1 and was listed in Table 2. The uncertainty in the estimated edge energy is ±0·02 eV. The absorption edge energy varied from 3·05 to 2·82 eV as the SnO concentration changed from x=20 to 60. Figure 2 shows the wavelength dispersion of the re-

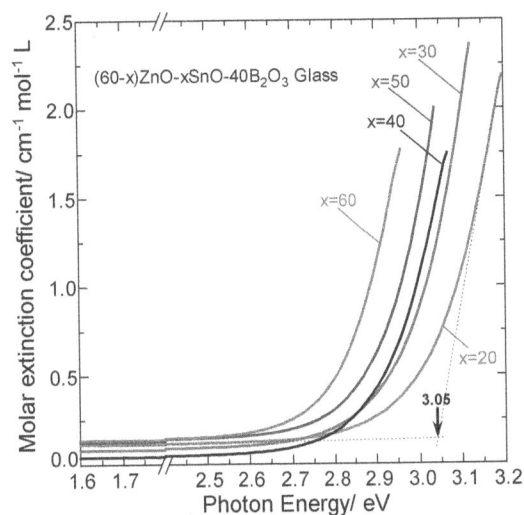

Figure 1. Molar extinction coefficient of the (60–x)ZnO.xSnO.40B$_2$O$_3$ glasses (x=20–60 mol%). The ordinate was normalised by SnO concentration. Dashed lines are the extrapolated lines for estimating absorption edge energy

fractive index for all the glasses. The refractive index, n$_d$, increased linearly from 1·743 to 1·895 as the SnO concentration increased (Table 1). The wavelength dispersion of the index, as shown in Figure 2, was fitted to the following one-term Sellmeier equation[8,9] by the least squares method

$$n^2 - 1 = \frac{A\lambda^2}{\lambda^2 - \lambda_0^2} \tag{1}$$

where n, A and λ_0 are the refractive index at the wavelength λ, the oscillator strength and the resonance wavelength, respectively. The uncertainties on A and λ_0 are ±0·005 and ±0·02 eV, respectively, and optimised parameters are listed in Table 2.

The reflection spectra in the ultraviolet region are shown in Figure 3. A broad peak was observed around 4·8 eV for all the glasses. The peak energy of each sample is listed in Table 2, and the uncertainty

Table 2. Absorption edge energies, reflection peak energies, and parameters analysed from the wavelength dispersion of refractive index of (60–x)ZnO.xSnO.40B$_2$O$_3$ glasses (x=20–60 mol%)

| x | ZnO | SnO | B$_2$O$_3$ | Absorption | Peak | Oscillator | Resonance |
	(mol%)			edge (eV)	energy (eV)	strength	energy (eV)
20	40	20	40	3·05	4·79	1·899	8·20
30	30	30	40	2·98	4·75	2·010	7·72
40	20	40	40	2·92	4·77	2·087	7·33
50	10	50	40	2·89	4·69	2·176	6·98
60	0	60	40	2·82	4·69	2·338	6·73

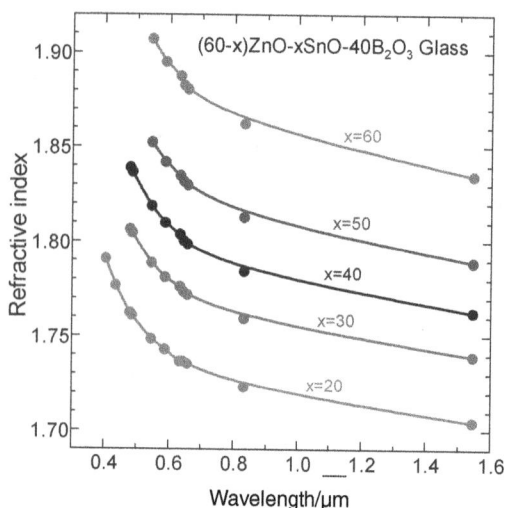

Figure 2. Wavelength dispersion of refractive index for (60–x)ZnO.xSnO.40B₂O₃ glasses (x=20–60 mol%)

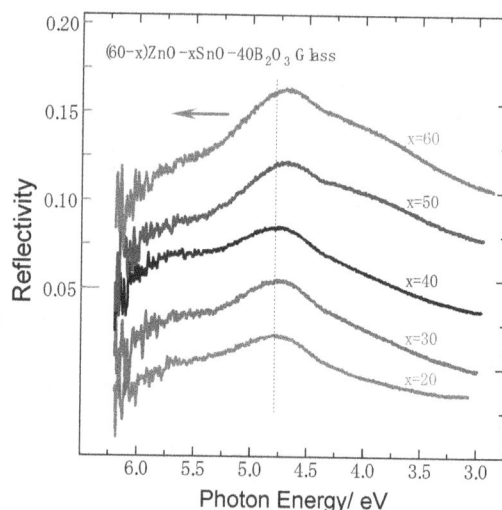

Figure 3. Reflection spectra of (60–x)ZnO.xSnO.40B₂O₃ glasses (x=20–60 mol%). The incident angle for measurement was 10°

in the peak energy is ±0·02 eV. The peak position shifted from 4·8 to 4·7 eV as the SnO concentration increased. In addition, a broad shoulder ranging from 3·5 to 4·3 eV was observed in the reflection spectra. The intensity of the shoulder decreased with a decrease in the SnO concentration and the shoulder was imperceptible at x=20.

The Raman spectra of the glasses are shown in Figure 4. Bands were found in the region of 620–1540 cm⁻¹. In addition, bands were observed in the low frequency region at 180 and 380 cm⁻¹. The bands at 840 and 950 cm⁻¹ are distinct for low SnO (x=20 and 30) glasses and a broad band remains between 820 and 1100 cm⁻¹ for high SnO glasses. The bands at 180 and 380 cm⁻¹ decrease in intensity with as the SnO content decreases, and almost disappear at x=20. Features in the regions of 620–740 and 1160–1540 cm⁻¹ seem to sharpen as the SnO content increases. The intensity of a weak but distinct band at 770 cm⁻¹ increases as

the SnO content increases, while a band 790 cm⁻¹ does not show any appreciable change in band shape with the compositional change.

Discussion

Since the absorption edge shifts toward lower energies as the SnO content increases, as shown in Figure 1, the edge is due to a transition in the electronic state related with tin ions. The edge energy changes by 0·23 eV with the change of SnO concentration from 20 to 60 mol% (Table 2).

The refractive index, n_d, increases linearly from 1·743 to 1·895 as the SnO content increases, as shown in Table 1. The resonance energy, which was calculated from resonance wavelength, shifts monotonically from 8·2 to 6·7 eV, and the oscillator strength also monotonically increases, as the SnO

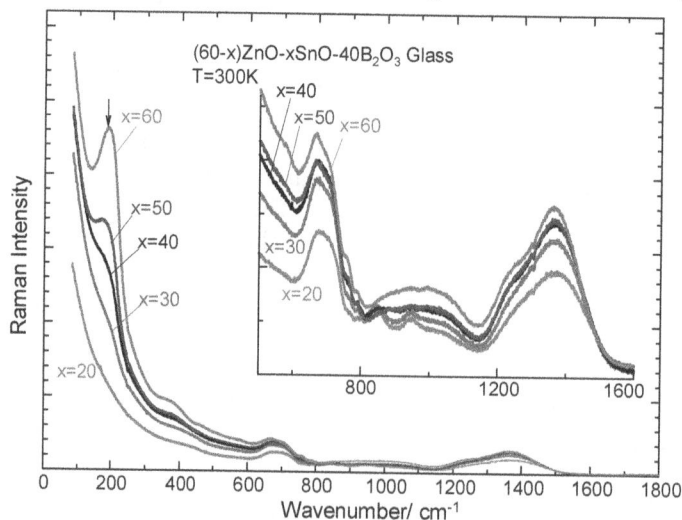

Figure 4. Non-polarised Raman scattering spectra of (60–x)ZnO.xSnO.40B₂O₃ glasses (x=20–60 mol%). The insert shows a magnification of the spectra relating to vibrations of borate groups

content changes from 20 to 60 mol%. The large shift in resonance energy implies a change in the electronic state of the tin ions, but this is not due to a change in shallower levels near the band gap, because the shift of the absorption edge energy was only 0·23 eV, as described above.

In the reflection spectra shown in Figure 3, a broad peak was observed around 4·8 eV for all the glasses, which is related to a transition in the band gap. The peak position shifts slightly from 4·8 to 4·7 eV as the SnO content increases, which is about half of the shift of the absorption edge. In addition, a broad shoulder ranging from 3·5 to 4·3 eV was observed in the reflection spectra. The intensity of the shoulder decreases as the SnO content decreases, and it is imperceptible at $x=20$, as shown in Figure 3. Therefore the slight shift of the 4·8 eV peak and the change in intensity of the broad shoulder, which is at lower energies than the band gap energy, is the origin for the shift of the absorption edge with compositional change.

Raman bands assigned to the vibration of borate groups were found in the region 620–1540 cm^{-1}, as shown in Figure 4. Peaks at 660 and 730 cm^{-1} are assigned to vibrations of ring- and chain-type metaborate groups.[2,10–13] A weak band at 770 cm^{-1} is assigned to the vibration of six-membered borate rings with one or two BO_4 units.[2,10–13] The distinct 790 cm^{-1} band is assigned to the breathing vibration of the boroxol ring.[2,10,12] The 840 and 940 cm^{-1} bands, observed only in glasses with $x=20$ and 30, are assigned to vibrations of pyroborate and orthoborate groups, respectively.[10,12] Meera *et al*[12] have also assigned the 840 and 1240 cm^{-1} bands to the vibration of pyroborate groups. The 1240 and 1370 cm^{-1} bands have been assigned to the vibration of $B-O^-$ in a large borate network.[2,12] However, recently the bands in the region 1100–1600 cm^{-1} were assigned to BO_3 triangles or BO_2O^- units.[14] The 250 cm^{-1} band due to ZnO_4 vibration[15] was not observed in the present glasses. A peak at 180 cm^{-1} is thought to be vibration of Sn–O bonds, because the peak intensity corresponded to the SnO concentration. A peak at 200 cm^{-1} has been observed for crystalline SnO (c-SnO), while peaks at 620 and 770 cm^{-1} have been observed for crystalline SnO_2 (c-SnO_2).[15,16] Geijke *et al*[11] observed a Raman band at 175 cm^{-1} in $SnB_{2.2}O_{4.3}$ glass, and noted the similarity of the band positions in the glass and c-SnO. Therefore, the 180 cm^{-1} peak should be assigned to the vibration of Sn–O in a pyramidal SnO_4 structure which exists only in c-SnO.

The B_2O_3 content does not change in the present glasses. The 840 and 950 cm^{-1} bands, however, develop into a broad feature ranging between 820 and 1100 cm^{-1} for higher SnO content. This may be due to an increase in the interaction between nonbridging oxygens and Sn^{2+} ions, $B-O^- \cdots Sn^{2+}$, due to a decrease in the Sn^{2+} iconicity as the SnO content increases above

20 mol%, as deduced from the chemical isomer shift in the Mössbauer spectra of SnO–B_2O_3 glasses.[1]

On the other hand, as the SnO content increases, there is an increase in the intensity of the 180 cm^{-1} band, which is assigned to pyramidal SnO_4, similar to that in c-SnO. Moreover the change in intensity of the 180 cm^{-1} band is strongly associated with that of the broad shoulder around 3·5–4·3 eV in the reflection spectra. Therefore, the broad shoulder in the reflection spectra is deduced to be due to formation of pyramidal SnO_4.

Conclusions

Monotonic changes in the optical absorption edge energy, refractive index and reflection peak energy for $(60-x)ZnO.xSnO.40B_2O_3$ ($x=20–60$) glasses were observed as the SnO concentration changes. A shift of the absorption edge to lower energy was deduced to be due to the shift of the 4·8 eV reflection peak, which may correspond to the band gap, and also to a change in intensity of the broad shoulder at 3·5–4·3 eV. The change in intensity of the 180 cm^{-1} Raman band follows the change in the shoulder in reflection spectra. This Raman band is assigned to SnO_4 pyramids and hence it is deduced that the shoulder originates from the electronic state of Sn^{2+} in pyramidal SnO_4 structures.

Acknowledgement

This work was carried out as a study of Next-generation Nano-structured Photonic Device and Process Technology as a part of Program to Create an Innovative Components Industry supported by NEDO.

References

1. Paul, A., Donaldson, J. D., Donoghue, M. T. & Thomas, M. J. K. *Phys. Chem. Glasses*, 1977, **18**, 125.
2. Nakai, M., Hayashi, A., Morimoto, H., Tatsumisago, M. & Minami, T. *J. Ceram. Soc. Jpn.*, 2001, **109**, 1010.
3. Hayashi, A., Nakai, M., Tatsumisago, M. & Minami, T. *C. R. Chime*, 2002, **5**, 751.
4. Hayashi, A., Nakai, M., Tatsumisago, M., Minami, T., Himei, Y., Miura, Y. & Katada, M. *J. Non-Cryst. Solids*, 2002, **306**, 227.
5. Idota, Y., Kubota, T., Matsufuji, A., Maekawa, Y & Miyasaka, T. *Science*, 1997, **276**, 1395.
6. Popova, E. & Dimitriev, Y. *J. Mater. Sci.*, 2007, **42**, 3358.
7. Li, H., Lin, H., Chen, W. & Luo, L. *J. Non-Cryst. Solids*, 2006, **352**, 3069.
8. Malitson, I. H. *J. Opt. Soc. Am.*, 1965, **53**, 1205.
9. Kitamura, N., Hayakawa, J. & Yamashita, H. *J. Non-Cryst. Solids*, 1990, **126**, 155.
10. Konijnendijk, W. L. *Philips Res. Rep. Suppl.*, 1975, **30**, 1.
11. Gejke, C., Zanghellini, E., Swenson, J. & Börjesson, L. *J. Power Sources*, 2003, **119–121**, 576.
12. Meera, B. N. & Ramakrishna, J. *J. Non-Cryst. Solids*, 1993, **159**, 1.
13. Konijnendijk, W. L. & Stevels, J. M. *J. Non-Cryst. Solids*, 1975, **18**, 307.
14. Yano, Y., Kunimine, N., Shibata, S. & Yamane, M. *J. Non-Cryst. Soilds*, 2003, **321**, 137.
15. *RASMIN Web*: http://riodb.ibase.aist.go.jp/rasmin/
16. Geurts, J., Rau, S., Richter, W. & Schmitte, F. *J. Thin Solid Films*, 1984, **121**, 217.

Proc. VI Int. Conf. Borate Glasses, Himeji, Japan, 18–22 August 2008   *Phys. Chem. Glasses: Eur. J. Glass Sci. Technol. B*, October 2009, **50** (5), 305–310

Laser induced modification of alkali borate glasses

Ben Franta, Landon Tweeton, Steve Feller & Mario Affatigato*

Physics Department, Coe College, 1220 First Av. NE, Cedar Rapids, IA 52402, USA

Manuscript received 21 August 2008
Revised version received 11 February 2009
Accepted 19 August 2009

We report on the effects of 785 nm light irradiation at 84 and 160 mW on copper-doped alkali borate glasses. Glasses of the chemical formula y(CuO).(1−y)[x(Z₂O).(1−x)(B₂O₃)] were made, with y=0·05, 0·1<x<0·5, and Z representing either Li or Cs. Modification of the surface, including crystallisation, was measured as a function of laser power and sample composition. Laser drawn surface patterning and spheroidisation were also performed. Scanning electron microscopy, atomic force microscopy, micro-Raman spectroscopy, and x-ray fluorescence spectroscopy were used to characterise the changes induced by the laser light.

1. Introduction

The effects of laser irradiation on glasses has been studied for several years. Borates, in particular, have been crystallised[1] and selectively patterned[2] through laser exposure, and numerous glass families have been examined, including $GeSe_2$, zinc tellurite,[3] and PbI_2-doped lead germanate glasses.[4] This paper follows a similar study on vanadate glasses[5] which suggested a thermally driven mechanism for the formation of various crystal forms observed on the surface.

Here, we report on our studies of the laser irradiation of copper-doped alkali borate glasses, which can be modified with 785 nm light at relatively low (84–160 mW) powers. Observed changes due to irradiation include topographical alterations and crystallisation, surface patterning and spheroidisation of particles, and the removal of surface crystals which had formed spontaneously on hygroscopic samples when left in air.

2. Experimental procedure

2.1 Glassmaking procedure

Glasses were prepared according to the following reaction

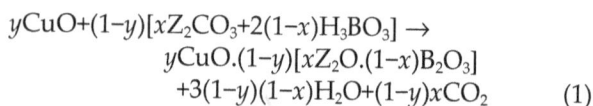

$$yCuO+(1-y)[xZ_2CO_3+2(1-x)H_3BO_3] \rightarrow$$
$$yCuO.(1-y)[xZ_2O.(1-x)B_2O_3]$$
$$+3(1-y)(1-x)H_2O+(1-y)xCO_2 \quad (1)$$

where Z represents either Li or Cs. The starting materials for the glasses were copper (II) oxide (Alfa Products, 99·0%), lithium carbonate (Aldrich, 99%+), caesium carbonate (Sigma-Aldrich, 99%), and boric acid (Sigma-Aldrich, 99·99%). The value of y was held constant at 0·05 (5 mol%) for all samples in order to increase absorption at the laser wavelength (785 nm), and x was varied from 0·1 to 0·5. The powders

were thoroughly mixed in a platinum crucible and heated to 1000°C in an electric muffle furnace. After 10 min the melt was removed from the furnace, allowed to cool, and a weight measurement was taken to ensure composition. The sample was then placed back in the furnace for a further 10 min at 1000°C, and subsequently cooled by splat quenching between two stainless steel plates. The cooling rate was approximately 10000°C/s. Sample thickness was approximately 1·5 mm. Water sensitive samples were stored in a bell jar desiccator.

Crystals of compositions $Li_2O.B_2O_3$, $Cs_2O.B_2O_3$, $Cs_2O.2B_2O_3$, and $Cs_2O.3B_2O_3$ were fabricated through a process of devitrification. Appropriate amounts of chemical powder were mixed and heated as above. Afterwards, the samples were placed in an annealing furnace at 80°C above the glass transition temperature, T_g, and allowed to devitrify for at least 16 h.

2.2. Laser irradiation procedure

The Raman system used for the laser irradiation and analysis was a JASCO NRS-3100 micro-Raman spectrophotometer. A Torsana Starbright 785 nm SLM diode laser with a spot diameter size of 2 μm was used to irradiate the sample at 84 mW and 160 mW for durations ranging from 5 to 40 s. Spectra were then acquired with the laser power attenuated to 16 mW at various sites within the resulting affected area; this power level was not sufficient to cause surface alterations. Changes in the Raman spectrum as a function of distance from the centre of the irradiation site were identified.

2.3. Scanning electron microscopy and x-ray fluorescence spectroscopy

A Tescan Instruments Vega-II Scanning Electron Microscope (SEM) and an Oxford Instruments Inca PentaFET were used for imaging and chemical analy-

* Author to whom correspondence should be addressed. Email maffatig@coe.edu

(a)　　　　　　　　(b)

(c)

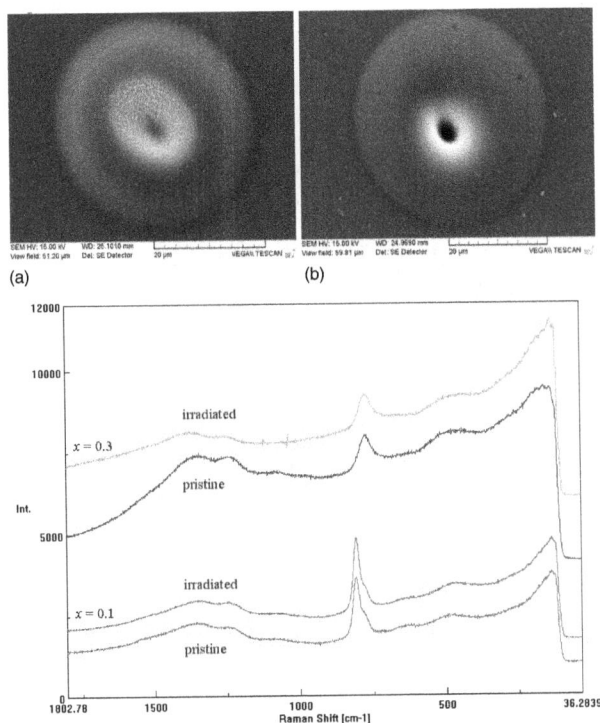

Figure 1. (a) SEM image of lithium borate sample with x=0·1, irradiated at 160 mW for 30 s; (b) Lithium borate sample with x=0·3, irradiated at 160 mW for 30 s. (c) Raman spectra of these samples, pristine and irradiated, indicating that the affected regions are still amorphous

(a)

(b)

Figure 2. (a) SEM image of a lithium borate sample with x=0·5, irradiated at 160 mW for 30 s. Note the crystalline region proximal to the centre. (b) Raman spectra, showing increasing crystallisation towards the centre of the irradiation site

sis. Samples were gold coated with a sputter coater prior to SEM analysis. Images were taken using a side electron detector and with column potentials of 15 and 30 kV. Elemental population spectra were quantitatively optimised using the copper signal of the sample, and relative abundances were calculated using a boron-to-oxygen ratio normalisation. To ensure consistency, multiple readings of areas exhibiting similar surface characteristics were performed and then averaged.

2.4. Atomic force microscopy

A Veeco Multimode Atomic Force Microscope (AFM) was used to analyse the surface morphology of caesium borate glasses in which lines had been drawn with laser irradiation. The AFM tapping mode was used, with the piezoelectric scanner mapping out a 60×60 μm area with a maximum z-range of 5·1 μm. No post-acquisition filtering was carried out.

3. Results

3.1. Crystallisation of sample surface

Figure 1 shows Raman spectra and scanning electron micrographs of lithium borate samples with x values of 0·1 and 0·3, after exposure to 160 mW laser light. Figure 2 shows the same for the lithium metaborate composition (x=0·5), and indicates an increasing level

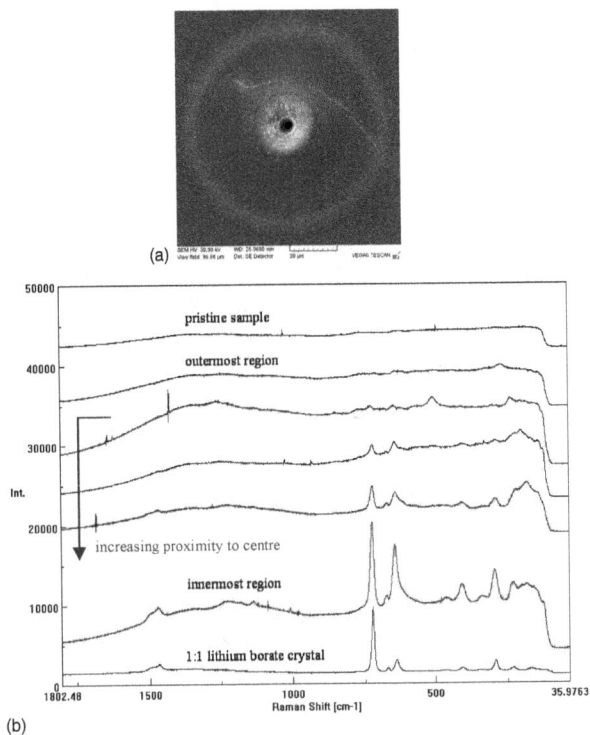

of laser induced crystallisation proximal to the centre of the irradiation site (Raman spatial resolution is approximately 2 μm). X-ray chemical analysis was

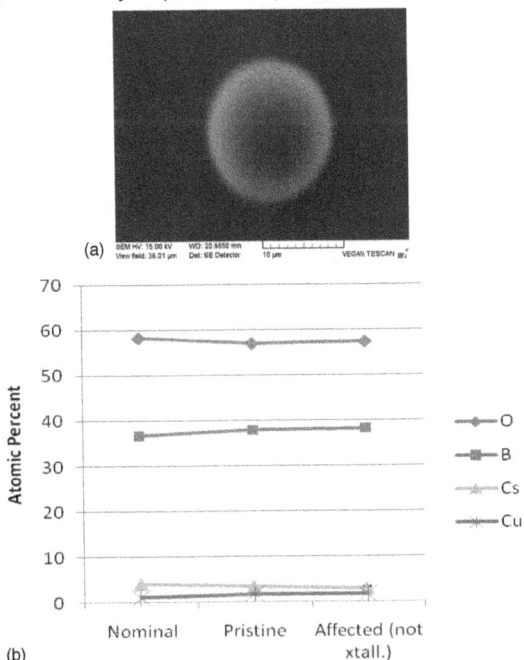

(a)

(b)

Figure 3. (a) Caesium borate sample with x=0·1, irradiated at 160 mW for 30 s; (b) x-ray fluorescence chemical analysis, showing homogeneous composition throughout the region of laser exposure. Interpolating lines serve as a guide to the eye

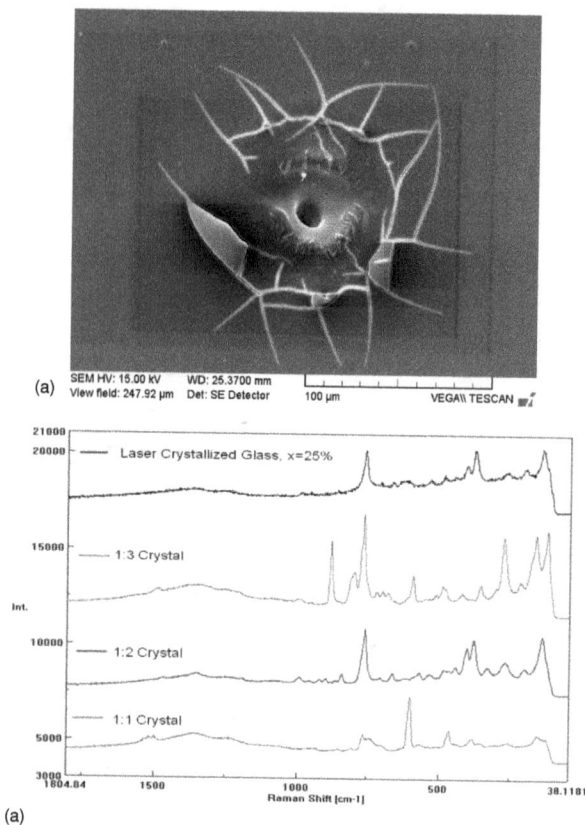

Figure 4. (a) Caesium borate with x=0·25, irradiated at 160 mW for 40 s; (b) Raman spectra, showing a comparison of the crystallised inner ridge formed under laser exposure to the 1:3, 1:2, and 1:1 caesium borate crystals. Note the agreement between the irradiated glass and the 1:2 caesium borate crystal

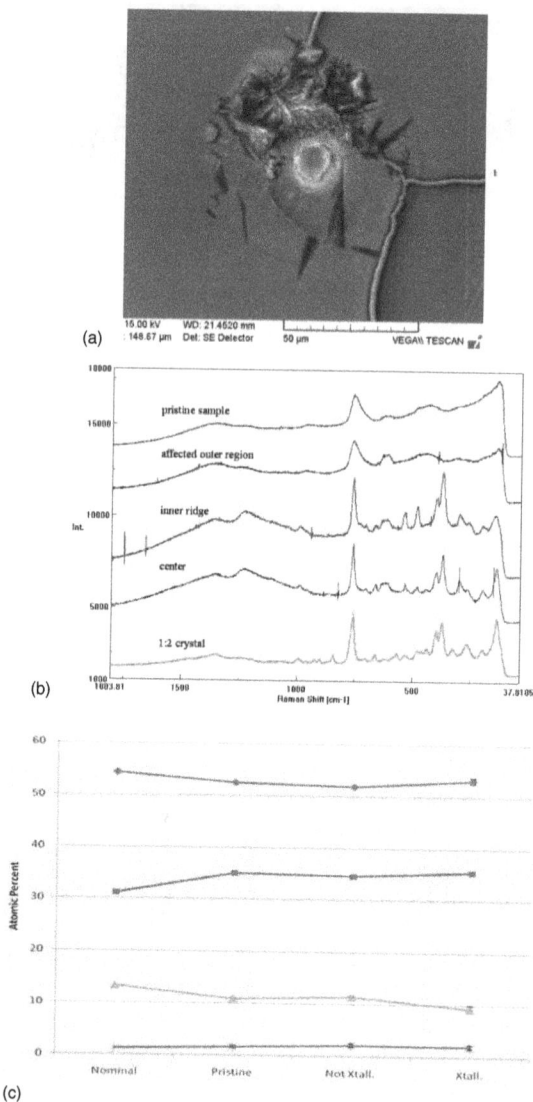

Figure 5. (a) SEM image of caesium borate with x=0·3, irradiated at 160 mW for 30 s; (b) Raman spectra of various regions, showing the presence of $Cs_2O.2B_2O_3$ crystal; (c) chemical analysis of various regions. Interpolating lines serve as a guide to the eye

not performed with these samples as the technique is not well suited to making accurate measurements of lithium.

As shown in Figure 3, low caesium content glasses (x=0·1) displayed limited topographical changes, forming a spot approximately 15 μm in diameter. Analysis with Raman spectroscopy showed this area to remain structurally amorphous. Mid-range compositions (x=0·25) exhibited significant changes in response to the 785 nm irradiation, often including the formation of a crystalline ridge surrounding the centre of the affected area, as shown in Figure 4. Figure 5 displays data pertaining to the x=0·3 composition, which also exhibited a region of crystallisation. The Raman spectrum of this area (Figure 5(b)) compares favourably with that of the prepared $Cs_2O.2B_2O_3$ crystal. Unlike lithium, caesium levels can be accurately measured using x-ray fluorescence spectroscopy, so chemical analyses were carried out on various regions surrounding each irradiation site. Preliminary results were inconclusive, but indicated stable chemical populations throughout irradiated areas which remained amorphous, and suggested slight changes in the caesium content of crystalline areas. We are in the process of refining these measurements.

Caesium borate glasses of composition x=0·4 and 0·5 were water sensitive and crystallised spontaneously when exposed to the atmosphere. Laser exposure had the effect of removing this crystallisation, as shown in Figure 6. Raman data demonstrated that the sample surface was amorphous during irradiation. The Raman spectrum of the crystals themselves could not be acquired, as even modest laser energies caused them to disappear.

3.2. Surface patterning

Caesium borate glasses with x=0·25 were exposed to 785 nm laser irradiation at 84 mW, which caused surface modification but was insufficient to crystallise the sample. During irradiation, the sample position was manipulated to result in a pattern of parallel

(a)

(b)

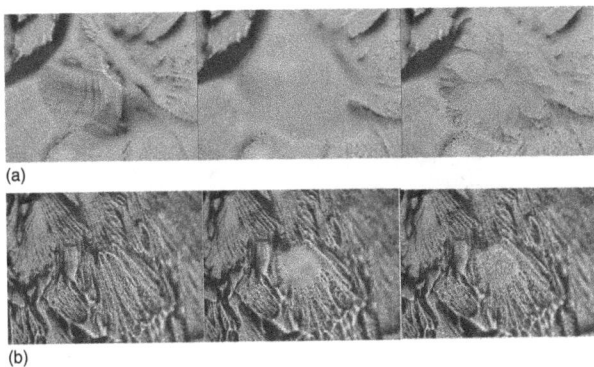

Figure 6. Optical micrographs of caesium borates with x=0·5, showing dispersion of surface crystals under laser irradiation and subsequent reformation (all images are 500 μm across). (a) left: before irradiation; centre: immediately after irradiation at 160 mW; right: 10 s after irradiation. (b) left: before irradiation; centre: immediately after irradiation at 84 mW; right: 5 s after irradiation

lines, as shown in Figure 7. Figure 7(b) shows a cross-sectional analysis of the surface morphology of the sample, indicating a nearly sinusoidal behaviour. The imprinted lines are approximately 375 nm high and have a period of 25 μm.

3.3. Spheroidisation

Caesium borate glass with $x=0·1$ was ground into a fine powder with a mortar and pestle. The powder was placed on a microscope glass slide and individual particles were irradiated at 16 mW, resulting in an immediate spheroidisation of the grains. The spheres, approximately 40 μm in diameter, were adhered to a carbon disk and gold coated before SEM imaging, as shown in Figure 8. The results were nearly spherical; one flat portion remained where the glass particle was in contact with the microscope slide.

3.4. Area of effect

Figure 9 shows a caesium borate glass ($x=0·3$) after the gold coating needed for SEM analysis had begun to flake off, an occurrence commonly observed over a period of days with both lithium and caesium borate glasses. Although laser irradiation produced an initial area of visible effect less than 100 μm in diameter, the gold coating is preserved around the exposure site to a radius of 1 mm. This suggests that areas which are far away from the irradiation site may still be affected, perhaps thermally.

4. Discussion

4.1. Crystallisation of sample surface

Results from Raman spectroscopy and scanning electron microscopy point to significantly different irradiation effects, which are dependent upon the alkali content of the glass. Exposure of lithium bo-

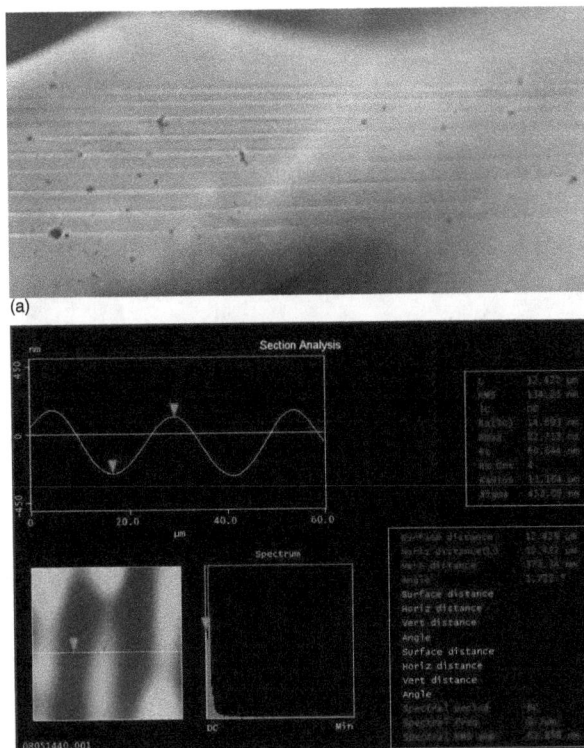

(a)

(b)

Figure 7. (a) Optical micrograph of caesium borate with x=0·25, irradiated with 785 nm laser light at 84 mW focused through a 100× lens. Image acquired under 5× magnification with writing rates of approximately 1–2 mm/s. (b) Computer screenshot of an AFM sectional analysis of the striations, showing a mound-like morphology

rate samples with x values of 0·1 and 0·3 to 160 mW laser light produced topographical changes, but, as evidenced by Raman spectroscopy, did not produce structural changes. This is in agreement with the thermal behaviour: the glass transition temperature increases[6] from 343°C to 496°C over the range in x from 0·1 to 0·3 for undoped lithium borates. The lithium metaborate composition, on the other hand,

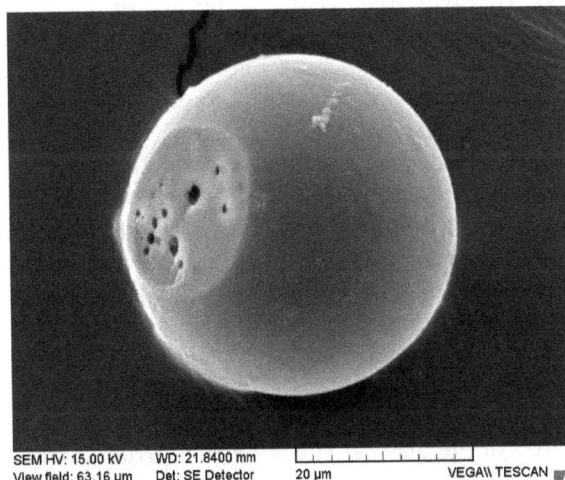

Figure 8. Sphere made from an irradiated powder of x=0·1 caesium borate glass. The flat portion indicates the area which was the bottom of the sphere as it formed

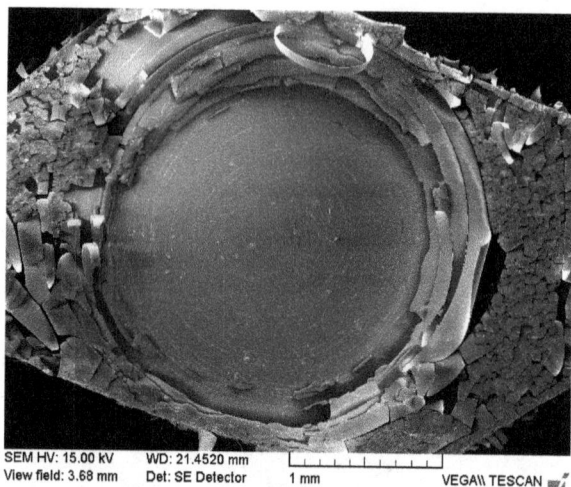

Figure 9. Caesium borate glass with x=0·3, two days after irradiation for 30 s at 160 mW. Original area of effect was less than 100 μm in diameter. As the gold film required for SEM analysis peels from the surface, an area of effect due to laser exposure is observed which is 1 mm in diameter

did show evidence of crystallisation at the irradiation site (Figure 2). As expected, the Raman signal of this laser induced crystal was seen to match that of a laboratory prepared $Li_2O.B_2O_3$ crystal. Exposure duration was not observed to significantly affect the size or overall response of the modified area; however, the affected area of the lithium metaborate composition was significantly larger than that of the other two. A dimple or hole is often observed at the centre of the irradiation site (e.g. Figures 1, 2, and 4), and is likely to be a result of mass transfer due to laser exposure. This hole does not extend completely through the sample.

The response of caesium borate glasses to laser irradiation was, again, dependent upon composition. Low caesium compositions (x=0·1; Figure 3) remained amorphous, while midrange compositions (x=0·25 and x=0·3; Figures 4 and 5) underwent crystallisation. This was somewhat unexpected, as the glass transition temperature actually increases[6] for the midrange composition, reaching 430°C for pure (i.e. y=0) caesium borate glasses. A Raman comparison with three caesium borate crystals (with x=0·25, 0·33, and 0·50) suggested that the laser induced crystalline area was composed of x=0·33 caesium borate crystal (see Figures 4(b) and 5(b)). This indicates that the crystalline areas are caesium enriched during irradiation, and that compositional proximity to the crystal stoichiometry was a key factor in the crystallisation.

For high caesium compositions (x=0·4 and x=0·5), laser exposure resulted in the immediate vitrification of crystals which formed due to the hygroscopic nature of the glasses. The absence of irradiation allowed crystal regrowth to occur over a short period (~10 s; see Figure 6). Because these crystals are most likely formed as a result of hydrolysis, we expect their dispersion by laser irradiation to be caused by

a heating and drying of the surface – essentially a thermal phenomenon.

4.2. Surface patterning

Analysis of the laser drawn lines with atomic force microscopy (Figure 7(b)) indicated that the patterning is formed when the sample material expands under laser exposure, resulting in a continuous, shallow mound.

4.3. Spheroidisation

Figure 8 displays a single vitreous grain (x=0·1 caesium borate) spheroidised by laser irradiation. Numerous microvoids can be seen extending into the interior of the sphere. Again, the spheroidisation process appears to be thermally driven in nature, a result of the glassy grain being heated beyond its glass transition temperature. The sphere remained amorphous after irradiation, indicating the crystallisation temperature, T_x, had not been reached.

4.4. Area of effect

In general, the area of visible surface change due to laser irradiation tends to become greater as alkali content increases. For instance, the x=0·1 caesium borate exhibited an area of effect of approximately 15 μm in diameter, while the x=0·5 sample of the same series showed an affected area which had a diameter of greater than 500 μm. Although visible changes may not be apparent outside these areas, Figure 9 suggests that larger regions of the sample may be affected, nevertheless. For instance, it is possible that, due to these hygroscopic nature of the glasses, a layer of moisture which may normally form on the glass surface evaporated near the irradiation site, allowing the gold coating to bond more strongly to the surface in that area.

5. Conclusions

We have investigated changes in the morphology of copper-doped alkali borate glass upon irradiation with 785 nm laser light. Low alkali content glasses were visibly altered, but not crystallised, at the site of exposure. Higher alkali content glasses showed further surface alterations, such as the formation of crystals. Chemical analysis results indicated that compositional changes were taking place, particularly in regions which appeared to be crystalline. The x=0·4 and x=0·5 compositions of caesium borate exhibited spontaneous surface crystallisation when exposed to air. Laser exposure resulted in vitrification of the site, which then quickly recrystallised in the absence of irradiation.

Other laser induced modifications were produced as well, including the formation of surface striations.

AFM analysis of this surface patterning showed the laser drawn lines to be formed by an expansion of the irradiated material. The surface was shown to be morphologically sinusoidal. The irradiation of small grains was also carried out, resulting in the fabrication of glassy spheres approximately 40 μm in diameter.

Interactions between the glass sample and a gold coating indicated an area of effect from laser exposure which was much greater than the visibly altered site. This supports the suggestion that the laser induced effects are essentially thermal in nature.

Acknowledgments

We would like to acknowledge the financial support of the US National Science Foundation under grants DMR-MRI-0722682, DMR-MRI-0420539, DMR-MRI-0320861, and DMR-0502051. The authors would also like to thank Coe College for its continued support and Dr Norimasa Umesaki and the other organisers of the Sixth International Conference on Borate Glasses, Crystals and Melts.

References

1. Kamitsos, E. I., Karakassides, M. A., Patsis, A. P. & Chryssikos, G. D. *J. Non Cryst. Solids*, 1990, **116**, 115.
2. Honma, T., Benino, Y., Fujiwara, T., Sato, R. & Komatsu, T. *J. Non Cryst. Solids*, 2004, **345&346**, 127.
3. Haro, E., Xu, Z. S., Morhange, J. F., Balkanski, M., Espinosa, G. P. & Phillips, J. C. *Phys. Rev. B*, 1985, **32**, 969.
4. Goutaland, F., Mortier, F., Capoen, B., Turrell, S., Bouazaoui, M., Boukenter, A. & Ouerdane, Y. *Opt. Mater.*, 2006, **28**, 1276.
5. Franta, B., Williams, T., Faris, C., Feller, S. & Affatigato, M. *Phys. Chem. Glasses: Eur. J. Glass Sci. Technol. B*, 2007, **48**, 357.
6. Feller, S. A., Kottke, J., Welter, J., Nijhawan, S., Boekenhauer, R., Zhang, H., Feil, D., Parameswar, C., Budhwani, K., Affatigato, M., Bhatnagar, A., Bhasin, G., Bhowmik, S., Mackenzie, J., Royle, M., Kambeyanda, S., Pandikuthira, P. & Sharma, M. In: *Borate glasses, crystals and melts*, Eds. A. C. Wright, S. A. Feller & A. C. Hannon, Society of Glass technology, Sheffield, 1997, p. 246.

Proc. VI Int. Conf. Borate Glasses, Himeji, Japan, 18–22 August 2008   *Glass Tech.: Eur. J. Glass Sci. Technol. A, August 2009, 50 (4), 217–220*

Mechanochemical synthesis of amorphous solid electrolytes in the system Li$_2$O–B$_2$O$_3$–[EMI]BF$_4$

Akitoshi Hayashi,* Daisuke Furusawa, Keiichi Minami, Kiyoharu Tadanaga & Masahiro Tatsumisago

Department of Applied Chemistry, Graduate School of Engineering, Osaka Prefecture University, Gakuen-cho, Sakai, Osaka 599-8531, Japan

Manuscript received 18 August 2008
Revised version received 28 February 2009
Accepted 8 June 2009

A new type of solid electrolyte was mechanochemically prepared from lithium pyroborate glass as a Li$^+$ ion conductor, and 1-ethyl-3-methyl-imidazolium tetrafluoroborate ([EMI]BF$_4$) as an ionic liquid. It was revealed from FT-IR and ^{11}B MAS-NMR measurements that [EMI]$^+$ cations and BF$_4^-$ anions were not decomposed by high energy ball milling, and were still present in the prepared (100−x)(0·67Li$_2$O.0·33B$_2$O$_3$).x[EMI]BF$_4$ (mol%) glasses. DSC analysis suggests that glass transition temperatures were decreased by the addition of [EMI]BF$_4$ to the lithium borate glass. The ambient temperature conductivity of the glass containing 10 mol% [EMI]BF$_4$ was 10^{-4} S cm^{-1}, which is four orders of magnitude higher than that of 67Li$_2$O.33B$_2$O$_3$ (mol%) glass. The activation energy for conduction was decreased by the addition of [EMI]BF$_4$. The dissolution of an ionic liquid in glass is a novel way to enhance the conductivity of glass electrolytes.

1. Introduction

Glass electrolytes with high Li$^+$ ion conductivity have been studied for use in all-solid-state rechargeable lithium batteries with high safety and reliability. Sulphide glasses in the systems Li$_2$S–P$_2$S$_5$ and Li$_2$S–SiS$_2$ have superior characteristics as solid electrolytes with a high Li$^+$ ion conductivity of 10^{-4}–10^{-3} S cm^{-1} at room temperature.[1–3] In general, oxide glasses are chemically and electrochemically more stable than sulphide glasses, but a big drawback of oxide glass electrolytes is their low conductivity of 10^{-8}–10^{-7} S cm^{-1} at room temperature. Major efforts have thus been focused on improving the conductivity of oxide glasses.[4,5] Increasing lithium ion concentration and adding a lithium salt are conventional techniques to increase conductivity of glass electrolytes.

Lowering the glass transition temperature (T_g) of glasses by incorporating organic oligomer blocks is one strategy for enhancing the ambient temperature conductivity of glass electrolytes.[6,7] The addition of small amounts of 1,4-butanediol as an oligomer into Li$_2$S–P$_2$S$_5$ sulphide glasses has previously been used to increase the conductivity of glass electrolytes by lowering T_g.[7]

Ionic liquids have received much attention as electrolytes in lithium secondary batteries because of their attractive properties, such as high chemical and thermal stability, nonflammability and high ionic conductivity.[8,9] A typical ionic liquid, 1-ethyl-3-methyl-imidazolium tetrafluoroborate ([EMI]BF$_4$), shows a low T_g (ca. −90°C) and high ambient

temperature conductivity of 10^{-2} S cm^{-1}.[9] Ionic liquids are compatible with ion conducting glasses because both materials consist of ionic moieties.[10] Decreasing T_g of oxide glass electrolytes by combining with an ionic liquid is expected to increase their conductivity.

In the present study, new solid electrolytes were prepared from an oxide glass with Li$^+$ ion conductivity and an ionic liquid with low T_g by a mechanochemical technique, using a planetary ball mill apparatus. We used lithium pyroborate glass which has an ambient temperature conductivity of 10^{-8} S cm^{-1}. To increase the conductivity, the borate glass was combined with [EMI]BF$_4$ ionic liquid. The local structure of the obtained materials was analysed by Fourier transform infrared (FT-IR) and solid state ^{11}B MAS-NMR spectroscopies. The thermal and electrical properties of the Li$_2$O–B$_2$O$_3$–[EMI]BF$_4$ glasses were examined.

2. Experimental

Lithium ion conducting borate glasses were prepared via mechanochemistry. Reagent grade Li$_2$O (Furuuchi Chemical Co., 99·9%) and B$_2$O$_3$ (Kojundo Chemical Laboratory Co., 99·99%) powders were used as starting materials. The mechanochemical treatment was carried out for 1 g batches of the mixed materials at the composition 67Li$_2$O.33B$_2$O$_3$ (mol%) in a ZrO$_2$ pot (volume 45 ml) with 160 ZrO$_2$ balls (5 mm in diameter), using a high energy planetary ball mill apparatus (Pulverisette 7, Fritsch GmbH). Fine glass powder was obtained after mechanical milling for 1·5 h at a constant rotation speed of 510 rpm, followed

* Corresponding author. Email hayashi@chem.osakafu-u.ac.jp

by milling for 20 h at 370 rpm. The ionic liquid, [EMI] BF_4, supplied by Tokyo Chemical Industry Co., was used as received. The [EMI]BF_4 was added to the pyroborate glass powder and then mechanochemical treatment was carried out. The milling condition was almost the same as used for the glass preparation, except for the milling periods of 20 h at 370 rpm. All the processes were conducted at room temperature in a dry Ar-filled glove box.

FT-IR spectra of the prepared powders were collected using a FT-IR spectrometer (Spectrum GX, Perkin Elmer Co.). The powders were mixed with anhydrous liquid paraffin, and the obtained slurry was sandwiched between two KRS-5 plates in a dry box. The FT-IR measurements were carried out in a dry N_2 atmosphere. Solid state ^{11}B MAS-NMR spectra were recorded at 96·24 MHz on a NMR spectrometer (Unity Inova 300, Varian Inc.). A $\pi/8$ pulse length of 0·7 μs and a recycle pulse delay of 2·0 s were used. A cylindrical ZrO_2 spinner filled with the powders was rotated at a speed of about 5·5 kHz during the ^{11}B MAS-NMR measurements. Differential scanning calorimetry (DSC) was carried out using a thermal analyser (DSC6200, Seiko Instruments Inc.) for the obtained powders sealed in an Al pan in a dry Ar-filled glove box. A constant heating rate of 10°C min^{-1} was used. Electrical conductivities were measured for pelletised samples obtained by cold pressing (360 MPa) the powders; the diameter and thickness of the pellets were 10 mm and about 1 mm, respectively. Gold electrodes were formed on both the faces of the pellet by vacuum evaporation. AC impedance measurements were carried out in a dry Ar atmosphere using an impedance/gain phase analyser (1260A, Solartron Analytical) in a frequency range of 10 Hz to 8 MHz.

3. Results and discussion

The $67Li_2O.33B_2O_3$ (mol%) glass powder and the [EMI]BF_4 ionic liquid were combined by a mechanochemical method using a high energy ball mill. Fine powders of $(100-x)(0·67Li_2O.0·33B_2O_3).x$[EMI]$BF_4$ (mol%) were obtained with compositions of $x=10$ or less. The prepared powders showed a halo pattern in x-ray diffraction measurements.

Figure 1 shows FT-IR spectra of the prepared $(100-x)(0·67Li_2O.0·33B_2O_3).x$[EMI]$BF_4$ (mol%) glasses. The spectrum of neat [EMI]BF_4 liquid ($x=100$) is also shown in the figure. The bands at 1176 and 1577 cm^{-1} are assigned to the in-plane C–H deformation vibration, and to in-plane C–C and C–N stretching vibrations of the imidazole ring ([EMI]$^+$), respectively.[11] Two bands in the 3100–3200 cm^{-1} region are due to the C–H stretching vibration of the imidazole ring.[12] In the spectra for the glasses with compositions from $x=2$ to $x=10$, the bands attributable to the vibration of the imidazole ring

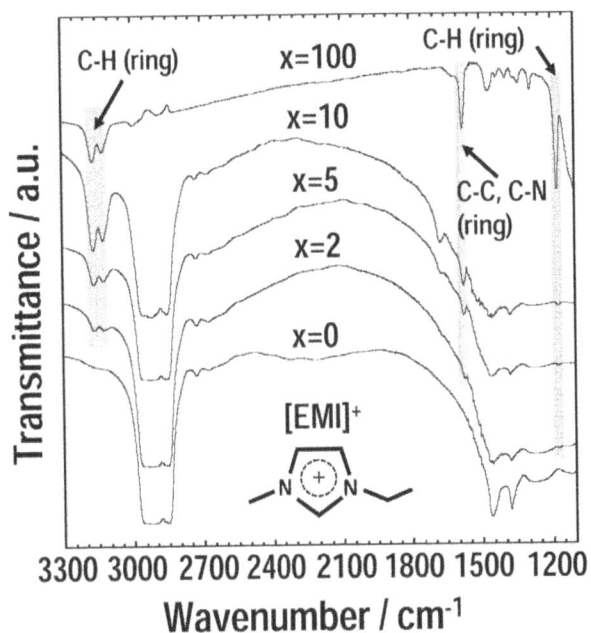

Figure 1. FT-IR spectra of the $(100-x)$ $(0·67Li_2O.0·33B_2O_3).x$[EMI]BF_4 (mol%) glasses, prepared by a mechanochemical technique. The spectrum of neat [EMI]BF_4 liquid ($x=100$) is also shown

are clearly observed; the bands at 1380, 1460 and 2800–3000 cm^{-1} are due to liquid paraffin which was used as a dispersing agent. It is revealed from FT-IR spectra that [EMI]$^+$ cations are not decomposed by high energy ball milling and are present in the $(100-x)(0·67Li_2O.0·33B_2O_3).x$[EMI]$BF_4$ glasses.

Figure 2 shows ^{11}B MAS-NMR spectra of the $(100-x)(0·67Li_2O.0·33B_2O_3).x$[EMI]$BF_4$ ($x=0$ and 5) glasses. Spinning sidebands are observed in the spectra and marked with asterisks. The $67Li_2O.33B_2O_3$

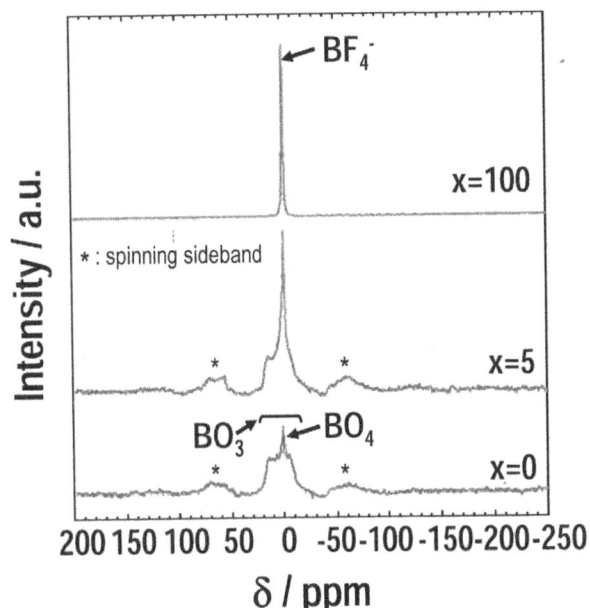

Figure 2. ^{11}B MAS-NMR spectra of the $(100-x)$ $(0·67Li_2O.0·33B_2O_3).x$[EMI]BF_4 ($x=0$ and 5) glasses. The spectrum of [EMI]BF_4 liquid ($x=100$) is also shown

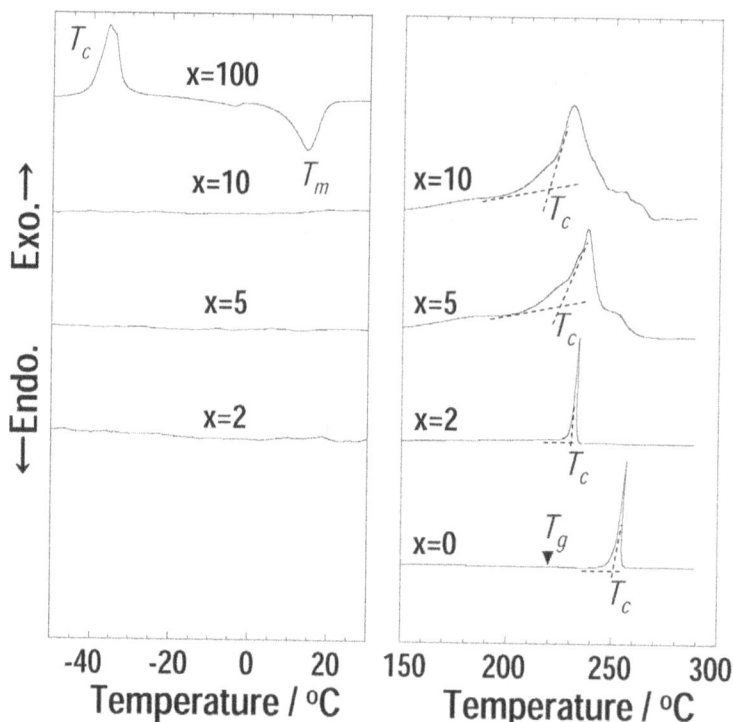

Figure 3. DSC curves of the (100–x)(0·67Li$_2$O.0·33B$_2$O$_3$).x[EMI]BF$_4$ glasses. The DSC curve of [EMI]BF$_4$ liquid (x=100) is also shown

glass (x=0) shows a typical ^{11}B MAS-NMR spectrum,[13] consisting of the peaks attributable to fourfold-coordinated borons in the BO$_4$ group and threefold-coordinated borons in the BO$_3$ group. The BO$_4$ group gives a sharp peak at about 0 ppm. The BO$_3$ group shows a broad peak which overlaps the central BO$_4$ peak because the BO$_3$ group has a larger second order quadrupole broadening than the BO$_4$ group. A strong peak at about 0 ppm is observed for [EMI]BF$_4$ ionic liquid (x=100) in a static NMR spectrum. The peak is attributable to fourfold-coordinated borons in a BF$_4^-$ anion. In the spectrum for the x=5 glass, a sharp peak due to fourfold-coordinated borons and a broad peak due to threefold-coordinated borons are observed; the peaks due to BF$_4$ and BO$_4$ units overlap with each other. The ^{11}B NMR spectra indicate that the glasses prepared from the pyroborate glass and the [EMI]BF$_4$ liquid include BF$_4^-$ anions as well as borate groups of BO$_4$ and BO$_3$.

Figure 3 shows DSC curves of the (100–x) (0·67Li$_2$O.0·33B$_2$O$_3$).x[EMI]BF$_4$ (mol%) glasses. An endothermic change due to a glass transition (at a temperature T_g=220°C) and a sharp exothermic peak due to crystallisation (at a temperature T_c=260°C) are observed in the pristine borate glass (x=0). The addition of [EMI]BF$_4$ to the borate glass decreases the crystallisation temperature. Decreasing glass transition temperatures due to the addition of [EMI]BF$_4$ are thus presumed, although glass transition phenomena are not clearly observed. Moreover, thermal behaviours due to crystallisation and melting (at a temperature T_m) of [EMI]BF$_4$ itself (x=100) are not observed for

the prepared glasses (x=2, 5 and 10). It is thus suggested that [EMI]BF$_4$ was successfully dissolved in pyroborate glass via the mechanochemical synthesis.

Figure 4 shows the temperature dependence of the conductivity for the (100–x)(0·67Li$_2$O.0·33B$_2$O$_3$). x[EMI]BF$_4$ (mol%) glasses. The conductivity data for the ionic liquid [EMI]BF$_4$ [9] is also shown for comparison. The conductivities of the glasses follow the Arrhenius equation. The conductivity of the

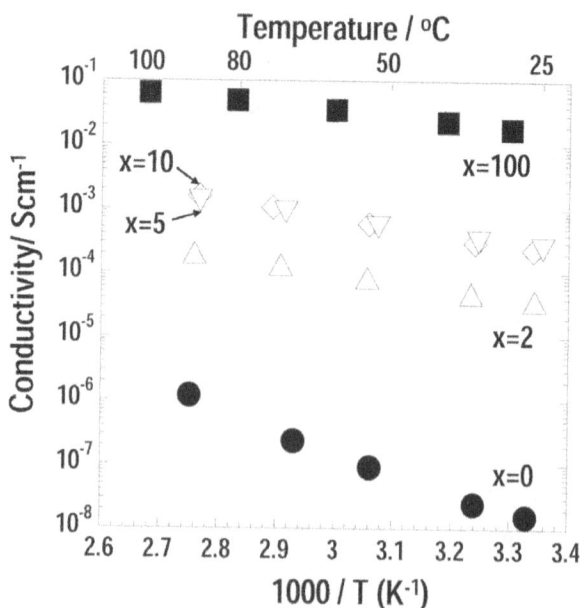

Figure 4. Temperature dependence of the electrical conductivity for (100–x)(0·67Li$_2$O.0·33B$_2$O$_3$).x[EMI]BF$_4$ glasses. The conductivity data for [EMI]BF$_4$ liquid[9] is also shown

67Li$_2$O.33B$_2$O$_3$ glass (x=0) is 1×10^{-8} S cm^{-1} at room temperature. The glasses with the addition of [EMI]BF$_4$ show higher conductivity in the whole temperature range than the pristine borate glass. The conductivity of the glass with 10 mol% [EMI]BF$_4$ is 2×10^{-4} S cm^{-1}, which is four orders of magnitude higher than that of the glass without [EMI]BF$_4$. The activation energy for conduction is decreased from 63 to 29 kJ mol^{-1} by the addition of 10 mol% [EMI]BF$_4$. The ionic liquid [EMI]BF$_4$ has a conductivity of 2×10^{-2} S cm^{-1} at room temperature. However, thermal behaviours due to [EMI]BF$_4$ itself were not detected, as shown in Figure 3. DSC analysis suggested that glass transition temperatures were decreased by the addition of [EMI]BF$_4$ to the 67Li$_2$O.33B$_2$O$_3$ glass. A prospective explanation of increasing conductivity is the increase in Li$^+$ ion mobility in the borate glass when combined with [EMI]BF$_4$. Lithium borate glasses with dissolved [EMI]BF$_4$ liquid were mechanochemically synthesised, and the obtained electrolytes exhibited a high conductivity of 10^{-4} S cm^{-1} at room temperature.

4. Conclusions

A new-type solid electrolyte was mechanochemically prepared from lithium pyroborate glass and [EMI]BF$_4$ ionic liquid. FT-IR and ^{11}B MAS-NMR measurements showed that [EMI]$^+$ cations and BF$_4^-$ anions are not decomposed by high energy ball milling and are still present in the $(100-x)(0\cdot67\text{Li}_2\text{O}.0\cdot33\text{B}_2\text{O}_3).x$[EMI]BF$_4$ glasses. DSC measurements suggested that glass transition temperatures were decreased by the addition of [EMI]BF$_4$ to the lithium borate glass; borate glass with dissolved [EMI]BF$_4$ was successfully synthesised. The ambient temperature conductivity of the glass with 10 mol% [EMI]BF$_4$ was 10^{-4} S cm^{-1}, which is four orders of magnitude higher than that of 67Li$_2$O.33B$_2$O$_3$ glass. The activation energy for conduction was decreased by the addition of [EMI]BF$_4$. The dissolution of ionic liquids in lithium ion conducting glasses is a novel way to enhance the conductivity of glass electrolytes.

References

1. Mercier, R., Malugani, J. P., Fahys, B. & Robert, G. *Solid State Ionics*, 1981, **5**, 663.
2. Pradel, A. & Ribes, M. *Solid State Ionics*, 1986, **18/19**, 351.
3. Minami, T., Hayashi, A. & Tatsumisago, M. *Solid State Ionics*, 2000, **136–137**, 1015.
4. Tuller, H. L., Button, D. P. & Uhlmann, D. R. *J. Non-Cryst. Solids*, 1980, **40**, 93.
5. Tatsumisago, M., Minami, T. & Tanaka, M. *Glasstech. Ber.*, 1983, **56K**, 945.
6. Hayashi, A., Wang, L. & Angell, C. A. *Electrochem. Acta*, 2003, **48**, 2003.
7. Hayashi, A., Harayama, T., Mizuno, F. & Tatsumisago, M. *J. Power Sources*, 2006, **163**, 289.
8. Sakaebe, H. & Matsumoto, H. *Electrochemical Aspects of Ionic Liquids*, Ed. H. Ohno, John Wiley & Sons, Inc., New Jersey, 2005, P 173.
9. Noda, A. & Watanabe, M. *Electrochimica Acta*, 2000, **45**, 1265.
10. Hayashi, A., Yoshizawa, M., Angell, C. A., Mizuno, F., Minami T. & Tatsumisago, M. *Electrochem. Solid-State Lett.*, 2003, **6**, E19.
11. Nanbu, N., Sasaki, Y. & Kitamura, F. *Electrochem. Commun.*, 2003, **5**, 383.
12. Yokozeki, A., Kasprzak, D. J. & Shiflett, M. B. *Phys. Chem. Chem. Phys.*, 2007, **9**, 5018.
13. Bray, P. J. *J. Non-Cryst. Solids*, 1985, **73**, 19.

Proc. VI Int. Conf. Borate Glasses, Himeji, Japan, 18–22 August 2008   *Glass Technol.: Eur. J. Glass Sci. Technol. A, December 2009, 50 (6), 323–328*

A study of the phase and glass forming diagrams of the BaO–Bi$_2$O$_3$–B$_2$O$_3$ system

Martun R. Hovhannisyan*

State Engineering University of Armenia, 105 Teryan st. 0009 Yerevan, Armenia

Rafael M. Hovhannisyan, Bela V. Grigoryan, Hovakim A. Alexanyan

Scientific-Production Enterprise of Material Science, 17 Charents st., 0025 Yerevan, Armenia

Nikolay B. Knyazyan

Institute of General and Inorganic Chemistry of NAS RA, Argutyan st., district 2, 10, 0051 Yerevan, Armenia

Manuscript received 25 August 2008
Revised version received 5 July 2009
Accepted 6 September 2009

The phase diagram of the binary Bi$_2$O$_3$–B$_2$O$_3$ system has been corrected. The eutectic composition has been determined as 48·5Bi$_2$O$_3$.51·5B$_2$O$_3$, between BiBO$_3$ and Bi$_3$B$_5$O$_{12}$, with melting point 665°C. BiBO$_3$ melts congruently at 685±5°C. Six phase diagrams of the pseudo-binary Bi$_4$B$_2$O$_9$–BaBi$_2$B$_2$O$_7$, BiBO$_3$–BaBi$_{10}$B$_6$O$_{25}$, BaBi$_2$B$_4$O$_{10}$–BaBi$_{10}$B$_6$O$_{25}$, BaBi$_2$B$_4$O$_{10}$–BaBi$_2$B$_2$O$_7$, BaB$_2$O$_4$–BaBi$_2$B$_2$O$_7$, BiBO$_3$–BaB$_2$O$_4$ systems have been constructed, and eight pseudo-binary eutectic compositions have been revealed. Two new crystalline ternary compounds, BaBi$_2$B$_2$O$_7$ and BaBi$_{10}$B$_6$O$_{25}$, have been revealed at the crystallisation of the same glass compositions. Both compounds melted congruently: BaBi$_2$B$_2$O$_7$ at 725±5°C, and BaBi$_{10}$B$_6$O$_{25}$ at 690±5°C. Single crystals of the new compound BaBi$_{10}$B$_6$O$_{25}$ were grown from the melt by stoichiometric composition cooling. X-ray characteristics of both compounds were determined. BaBi$_2$B$_2$O$_7$ has an orthorhombic cell with lattice parameters a=11·818 Å, b=8·753 Å, c=7·146 Å, cell volume V=739·203 Å^3, Z=4. BaBi$_{10}$B$_6$O$_{25}$ has orthorhombic crystal symmetry and lattice constants a=6·434 Å, b=11·763 Å, c=29·998 Å, V=2270·34 Å^3, Z=8. The influence of various methods of melt casting on glass forming ability in the ternary BaO–Bi$_2$O$_3$–B$_2$O$_3$ system is investigated. The expanded glass formation area changes from stable glass forming barium borates and binary bismuth borates up to barium bismuthates. All stoichiometric compositions are in the area of stable glasses.

Increased restriction and interdiction on the use of toxic materials (including PbO) in electronics products, due to ecological and regulatory requirements, have led to intensified development of new lead free sealing glasses and glass ceramics with a wide temperature interval of sealing. The solution of this problem requires that the phase diagrams of the new lead free systems are constructed and that low melting glasses are revealed. Intensive studies of lead free low melting systems began in the early 1990s. The binary Bi$_2$O$_3$–B$_2$O$_3$ system is most attractive for this purpose, because of both the presence several low melting eutectics,[1] and the propensity for glass formation.[2] Furthermore, bismuth borate single crystals and glass ceramics have nonlinear optical (NLO) properties and other attractive properties.[3–6] Both these factors are reasons for further study of binary and ternary bismuth borate systems, and the glasses which they form.[7–15]

Zargarova & Kasumova have studied the ZnO–Bi$_2$O$_3$–B$_2$O$_3$ ternary system and constructed its melting diagram.[10] They discovered two ternary compounds,

ZnBi$_4$B$_2$O$_{10}$ and ZnBiBO$_4$. Barbier *et al* reinvestigated this system and revealed a third compound, Bi$_2$ZnB$_2$O$_7$.[12] The uniqueness of the BaO–B$_2$O$_3$–Bi$_2$O$_3$ system is shown by the available sets of compounds and eutectics in the binary BaO–B$_2$O$_3$[16] and Bi$_2$O$_3$–B$_2$O$_3$[1,10] systems, and the four ternary compounds Ba$_3$BiB$_3$O$_9$, BaBi$_2$B$_4$O$_{10}$, BaBiB$_{11}$O$_{19}$ and BaBiBO$_4$ recently found by Egorisheva *et al*[11] as a result of an investigation of phase equilibrium in the BaO–B$_2$O$_3$–Bi$_2$O$_3$ system. Ba$_3$BiB$_3$O$_9$ undergoes a phase transition at 850°C and exists up to 885°C, where it decomposes in the solid state. BaBiB$_{11}$O$_{19}$ and BaBi$_2$B$_4$O$_{10}$ melt congruently at 807 and 730°C, respectively. BaBiBO$_4$ melts incongruently at 780°C.[11] BaBiBO$_4$, or BaBi(BO$_3$)O, a novel borate compound, has also been made by solid state synthesis at temperatures below 650°C from several compositions in the ternary BaO–B$_2$O$_3$–Bi$_2$O$_3$ system by Barbier *et al.*[12] A powder sample of BaBiBO$_4$ had a second harmonic signal with an NLO efficiency equal to five times that of KH$_2$PO$_4$ (KDP). Elwell *et al* investigated the BaO–B$_2$O$_3$–Bi$_2$O$_3$ system by hot stage microscopy and a new ternary eutectic composition, 23·4BaO.62·4Bi$_2$O$_3$.14·2B$_2$O$_3$ (wt%), with a low liquidus temperature of 600°C, was revealed for ferrite

---

*Corresponding author. Email martun_h@yahoo.com

spinel growth.[9] However, the available data are insufficient and the construction of the phase diagram of the BaO–B$_2$O$_3$–Bi$_2$O$_3$ system is essential, first of all to reveal new eutectic and stoichiometric compositions, and to characterise their glass forming ability.

Experimental

About one hundred samples of various binary and ternary compositions in the BaO–B$_2$O$_3$–Bi$_2$O$_3$ system were synthesised and tested. Compositions were prepared from chemically pure grade BaCO$_3$, Bi$_2$O$_3$ and H$_3$BO$_3$ at 2·5–5·0 mol% intervals. Most of the samples were obtained as glasses by various cooling methods, depending on the propensity for crystallisation: casting on metallic plates (bulk samples), and super cooling techniques constructed by our group (tape samples with thickness 30–400 μm). Glass formation was determined visually or by x-ray analysis. The glass melting was performed at 800–1200°C for 15–20 min with a 20–50 g batch in a 20–50 ml uncovered quartz glass or corundum crucible, using an air atmosphere and a "Superterm 17/08" electric furnace. The chemical composition of some glasses was determined by traditional chemical analysis, and the results indicate a good compatibility between the calculated and analytical amounts of B$_2$O$_3$, BaO and Bi$_2$O$_3$. SiO$_2$ contamination from quartz glass crucibles did not exceed 2 wt%, and alumina contamination did not exceed 0·5–1 wt%, according to the chemical analysis data.

Some compositions were studied by solid state synthesis. 15–20 g of material was carefully mixed in an agate mortar, pressed as tablets, located on platinum plates and underwent a first thermal treatment at 450–600°C for 24 h in electrical muffles (supplied by Naber). After regrinding, the powders were tested by DTA (differential thermal analysis) and x-ray methods. Tablets of the pressed samples were subjected to repeated thermal treatment.

DTA of the glasses was conducted using a Q-1500 type DTA in platinum crucibles, with powder glass samples of weight 0·5–0·6 g, a heating rate of 10 K/min, and a sensitivity of 250 μV. The accuracy of temperature measurement is ±5 K.

Powder x-ray patterns were obtained on a DRON-3 type diffractometer with Cu K$_\alpha$ radiation and a Ni filter. Samples for glass crystallisation were prepared with glass powder pressed in the form of rods or tablets. The crystallisation was performed in electrical muffles (supplied by Naber) by a single stage heat treatment. This was done for 24–48 h at around the temperature at which the maximum exothermal effects were observed by DTA on the glasses.

Crystalline phases of binary and ternary compounds formed by both glass crystallisation and solid state sintering were identified by using JCPDS-ICDD PDF-2 release 2008 database.[17]

Results and discussion

The glass forming diagram of the BaO–B$_2$O$_3$–Bi$_2$O$_3$ system (Figure 1) was constructed using different rates of cooling of the melt. The revealed large glass formation area includes all eutectics in the binary BaO–B$_2$O$_3$, Bi$_2$O$_3$–B$_2$O$_3$ and BaO–Bi$_2$O$_3$ systems, and covers the majority of the concentration triangle, reaching up to 90 mol% Bi$_2$O$_3$. All ternary compounds

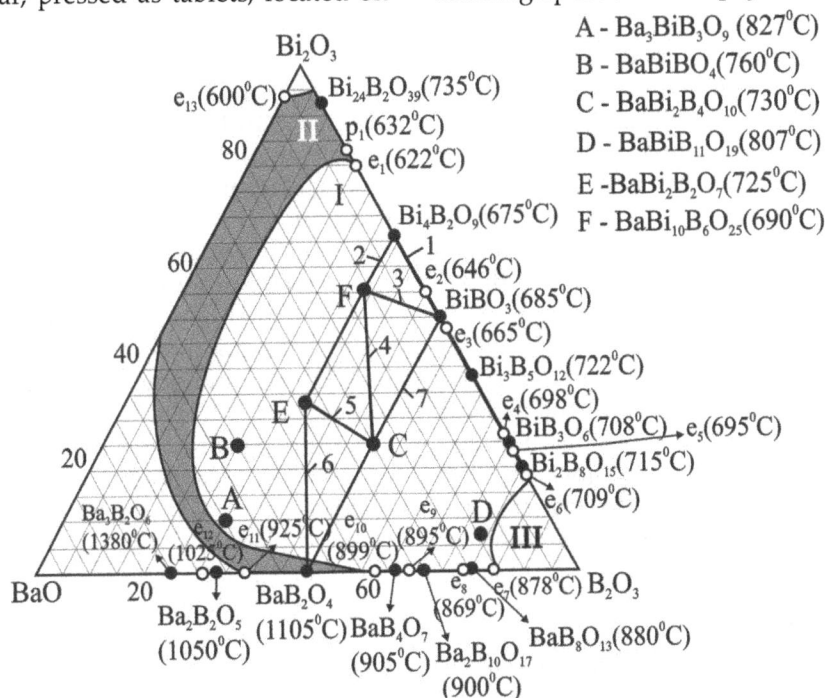

Figure 1. Glass forming diagram of the BaO–Bi$_2$O$_3$–B$_2$O$_3$ system. I indicates the area of stable glasses, II indicates the area of glasses formed by the supercooling method, and III indicates the stable phase separation region

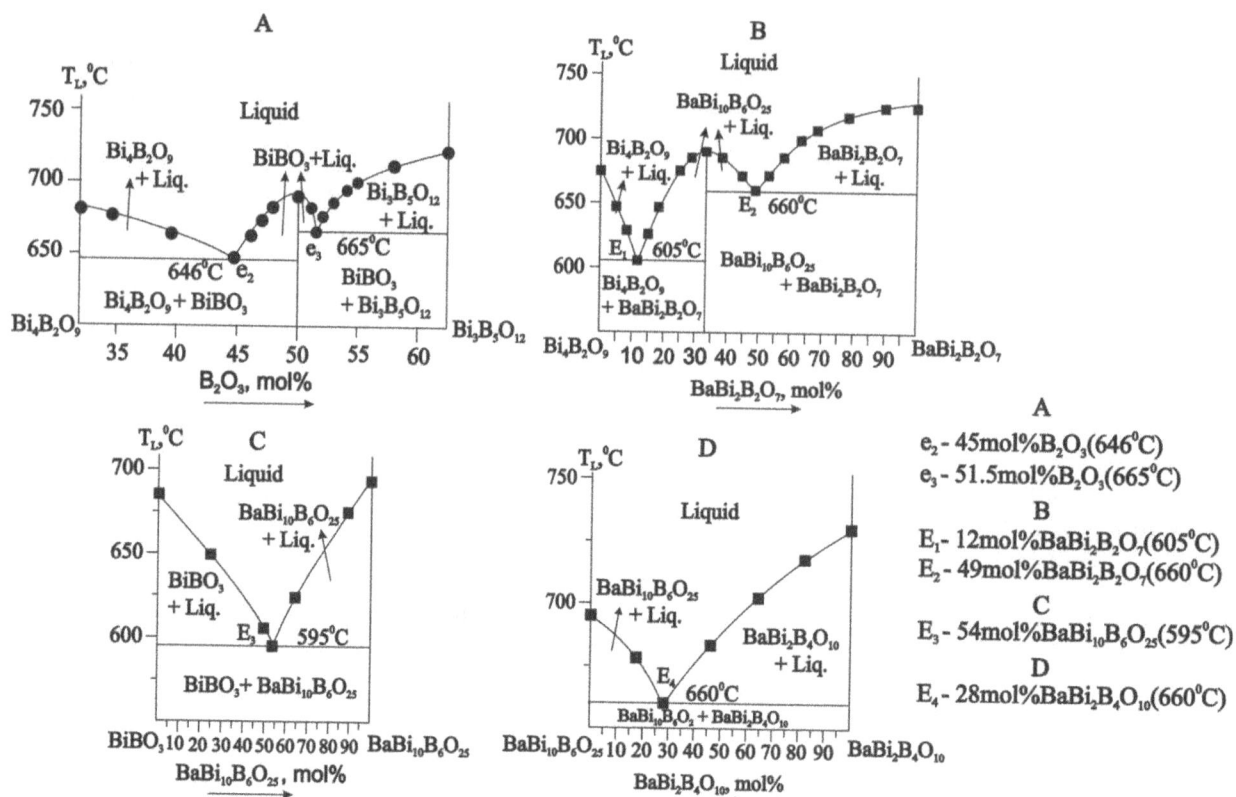

Figure 2. A: The corrected phase diagram of the Bi_2O_3–B_2O_3 system in the interval 30–65 mol% B_2O_3. B–D: Phase diagrams of the pseudo-binary systems: B – $Bi_4B_2O_9$–$BaBi_2B_2O_7$; C – $BiBO_3$–$BaBi_{10}B_6O_{25}$; D – $BaBi_{10}B_6O_{25}$–$BaBi_2B_4O_{10}$

have good glass forming ability, and are in the area of stable glasses (I in Figure 1). A stable phase separation region was also observed for B_2O_3 content more than 84 mol% (III in Figure 1).

The phase diagram of the Bi_2O_3–B_2O_3 system was first determined by Levin & Daniel in 1962 and five crystalline compounds, $Bi_{24}B_{12}O_{39}$, $Bi_4B_2O_9$, $Bi_3B_5O_{12}$, BiB_3O_6 and $Bi_2B_8O_{15}$, were identified.[1] Later, in 1974, Pottier revealed a sixth compound, $BiBO_3$ (bismuth orthoborate),[18] which was missing in the original phase diagram.[1] There are no doubts about the existence of $BiBO_3$ now: transparent colourless single crystals of $BiBO_3$ have recently been grown from the melt and characterised by Becker & Froehlich.[19] Monophase samples of both crystalline $BiBO_3$ modifications were obtained by crystallisation below 550°C of bismuth borate glasses with 50–57 mol% B_2O_3.[19] However, these authors did not correct the phase diagram, and did not determine the melting point of $BiBO_3$ or the eutectic composition between $BiBO_3$ and $Bi_3B_5O_{12}$. The compound $BiBO_3$ and this eutectic point are clearly given on the version of the B_2O_3–Bi_2O_3 phase diagram constructed by Zargarova & Kasumova,[10] without indication of their melting points and the eutectic composition.

First of all, compositions in the Bi_2O_3–B_2O_3 system containing 45–65 mol% B_2O_3 were tested to determine the melting point of $BiBO_3$ and to determine the eutectic composition between $BiBO_3$ and $Bi_3B_5O_{12}$. The compositions used to correct the B_2O_3–Bi_2O_3 phase

diagram were prepared by solid state synthesis at 520°C, with steps of 0·5–1·0 mol% B_2O_3 over the interval 45–55 mol% B_2O_3. As a result, the eutectic composition, 48·5Bi_2O_3.51·5B_2O_3 (mol%), between $BiBO_3$ and $Bi_3B_5O_{12}$, was determined, and its melting point was measured by DTA as 665±5°C (Figure 2A). It was also found that $BiBO_3$ melts congruently at 685±5°C.

Six pseudo-binary systems, $Bi_4B_2O_9$–$BaBi_2B_2O_7$, $BiBO_3$–$BaBi_{10}B_6O_{25}$, $BaBi_2B_4O_{10}$–$BaBi_{10}B_6O_{25}$, Ba-$Bi_2B_4O_{10}$–$BaBi_2B_2O_7$, BaB_2O_4–$BaBi_2B_2O_7$ and $BiBO_3$–BaB_2O_4, were studied by DTA and x-ray analysis, and their phase diagrams were constructed. These phase diagrams are discussed below. Two new crystalline ternary compounds, $BaBi_2B_2O_7$ and $BaBi_{10}B_6O_{25}$, were revealed by crystallisation at the same glass composition. Both compounds, $BaBi_2B_2O_7$ and Ba-$Bi_{10}B_6O_{25}$, melt congruently at 725±5°C and 690±5°C, respectively.

Single crystals of $BaBi_{10}B_6O_{25}$ were obtained by cooling of a melt with the stoichiometric composition. Glass powder of composition 11·11BaO.55·55Bi_2O_3.33·33B_2O_3 (mol%) was heated in a quartz glass ampoule up to 750°C at a rate 10 K/min. After 2 h at high temperature, the melt was cooled at a rate 0·5 K/h. Single crystals with sizes up to 1·66×0·38×0·19 mm³ were grown.

The x-ray characteristics of $BaBi_2B_2O_7$ and Ba-$Bi_{10}B_6O_{25}$ were determined and are given in Tables 1 and 2. The x-ray powder diffraction patterns of

Table 1. X-ray characteristics of the new ternary compound $BaBi_2B_2O_7$, synthesised at the same glass composition crystallisation (640°C, 20 h)

No.	d_{exp}	I/I_0	h k l	No.	d_{exp}	I/I_0	h k l	No.	d_{exp}	I/I_0	h k l
1	6·23	9	1 0 1	20	2·15	25	1 4 0	39	1·34	7	1 2 5
2	5·02	9	1 1 1	21	2·12	5	2 3 2	40	1·33	3	5 4 3
3	4·80	5	2 1 0	22	2·06	24	1 2 3	41	1·28	8	0 3 5
4	4·29	5	0 2 0	23	2·01	6	5 2 3	42	1·24	6	5 6 0
5	4·11	14	1 2 0	24	1·97	25	2 4 1	43	1·21	5	2 6 3
6	3·88	6	3 0 0	25	1·84	15	1 4 2	44	1·21	9	8 4 3
7	3·67	4	0 2 1	26	1·82	7	1 3 3	45	1·20	13	2 7 1
8	3·59	26	3 1 0	27	1·78	6	0 0 4	47	1·19	14	1 4 5
9	3·56	50	1 2 1	28	1·72	52	1 1 4	48	1·173	12	2 4 5
10	3·52	23	2 2 0	29	1·67	23	2 5 0	49	1·17	4	10 1 0
11	3·19	100	1 1 2	30	1·63	34	2 5 1	50	1·14	6	1 2 6
12	3·12	8	2 2 1	31	1·63	5	5 3 2	51	1·11	6	6 6 2
13	3·05	9	2 0 2	32	1·57	4	0 5 2	52	1·10	6	3 2 6
14	2·91	43	0 3 0	33	1·55	10	2 4 3	53	1·09	7	1 8 0
15	2·696	90	1 2 2	34	1·522	8	0 3 4	54	1·042	6	1 4 6
16	2·51	12	2 2 2	35	1·48	23	6 3 2	55	1·021	9	0 6 5
17	2·38	21	0 0 3	36	1·45	6	0 6 0	56	1·018	9	9 4 3
18	2·31	5	4 2 1	37	1·42	10	8 1 1	57	1·01	5	2 7 4
19	2·25	22	1 1 3	38	1·37	11	8 2 1				

$BaBi_2B_2O_7$ and $BaBi_{10}B_6O_{25}$ could be indexed on an orthorhombic cell with lattice parameters as follows: for $BaBi_2B_2O_7$ a=11·818 Å, b=8·753 Å, c=7·146 Å, cell volume V=739·203 Å^3, Z=4; for $BaBi_{10}B_6O_{25}$ a=6·434 Å, b=11·763 Å, c=29·998 Å, V=2270·34 Å^3, Z=8.

$Bi_4B_2O_9$–$BaBi_2B_2O_7$ pseudo-binary system

The introduction of 12 mol% $BaBi_2B_2O_7$ in the pseudo-binary system $Bi_4B_2O_9$–$BaBi_2B_2O_7$ reduced the melting point of initial $Bi_4B_2O_9$, and resulted in the formation of a simple eutectic, E_1, with melting point 605°C (Figure 2B). A maximum of the liquidus with melting point of 690°C is seen at 33·33 mol% $BaBi_2B_2O_7$, which indicates the formation of the new congruently melting ternary compound $BaBi_{10}B_6O_{25}$ (11·11BaO.55·55Bi$_2$O$_3$.33·33B$_2$O$_3$). Further increase of the $BaBi_2B_2O_7$ content (49 mol%) leads to a second eutectic, E_2, with melting point 660°C. Increasing of the liquidus temperature is observed in the post

eutectic region of composition, with a maximum at 725°C. It corresponds to the formation of the new congruently melting ternary compound $BaBi_2B_2O_7$.

$BiBO_3$–$BaBi_{10}B_6O_{25}$ pseudo-binary system

$BiBO_3$–$BaBi_{10}B_6O_{25}$ is a very important system. Initial $BiBO_3$ has a melting point of 685°C. The second maximum in the liquidus curve (Figure 2C) of 690°C is connected with the formation of the new ternary compound $BaBi_{10}B_6O_{25}$. There is a simple eutectic, E_3, between these two compounds at 54 mol% $BaBi_{10}B_6O_{25}$, with a melting point of 595°C.

$BaBi_{10}B_6O_{25}$–$BaBi_2B_4O_{10}$ pseudo-binary system

The $BaBi_{10}B_6O_{25}$–$BaBi_2B_4O_{10}$ system confirms the presence of the new congruently melting ternary compound $BaBi_{10}B_6O_{25}$, with a melting point of 690°C (Figure 2D). $BaBi_2B_4O_{10}$ melts congruently at 730°C.

Table 2. X-ray characteristics of the new ternary $BaBi_{10}B_6O_{25}$ single crystals

No.	d_{exp}	I/I_0	h k l	No.	d_{exp}	I/I_0	h k l	No.	d_{exp}	I/I_0	h k l
1	9·21	3·0	0 1 2	23	2·91	31	0 1 10	45	2·01	5·2	3 2 1
2	6·26	3·0	1 0 1	24	2·8	6	2 2 1	46	1·98	23·7	3 2 3
3	6·02	3·0	0 0 5	25	2·7	75·9	2 0 6	47	1·92	3·0	1 2 14
4	5·01	7·3	0 0 6	26	2·64	3·4	0 4 5	48	1·88	3·0	0 3 14
5	4·89	10·8	1 0 4	27	2·57	3·0	2 0 7	49	1·86	3·0	1 6 2
6	4·63	6·5	0 2 4	28	2·53	4·3	0 4 6	50	1·84	15·1	2 5 4
7	4·19	4·3	0 2 5	29	2·52	5·2	1 4 4	51	1·83	3·0	1 4 12
8	4·18	4·3	1 2 2	30	2·49	6·9	2 3 0	52	1·82	3·0	1 6 4
9	4·11	10	1 1 5	31	2·47	9·5	1 2 10	53	1·81	3·0	2 5 5
10	3·92	6·0	0 3 0	32	2·45	4·7	2 3 2	54	1·79	3·4	1 6 5
11	3·80	3·4	0 3 2	33	2·38	16	0 3 10	55	1·77	3·0	3 2 8
12	3·65	10·8	0 3 3	34	2·35	4·3	0 5 0	56	1·75	3·4	1 4 13
13	3·56	39·7	1 0 7	35	2·34	5·2	0 5 1	57	1·73	55·2	3 4 1
14	3·51	9·5	1 2 5	36	2·33	3·9	1 0 12	58	1·68	16·4	0 7 0
15	3·41	3·4	1 1 7	37	2·31	3·4	1 2 12	59	1·64	31·0	0 7 4
16	3·38	7·3	1 3 0	38	2·25	17·2	2 2 8	60	1·73	55·2	3 4 1
17	3·33	8·6	1 3 1	39	2·21	7·8	1 5 0	61	1·71	4·3	3 4 3
18	3·27	9·1	1 3 2	40	2·19	3·0	0 5 5	62	1·69	3·0	0 6 9
19	3·18	100	1 3 3	41	2·15	21·2	2 3 7	63	1·65	3·0	2 5 9
20	3·04	9·2	2 1 2	42	2·09	3·4	3 1 2	64	1·6	6·0	4 0 2
21	3·07	14·7	2 0 3	43	2·06	21·6	3 0 4	65	1·55	6·5	4 2 1
22	2·94	14·2	0 4 0	44	2·03	3·0	3 1 4	66	1·52	9·5	1 7 7

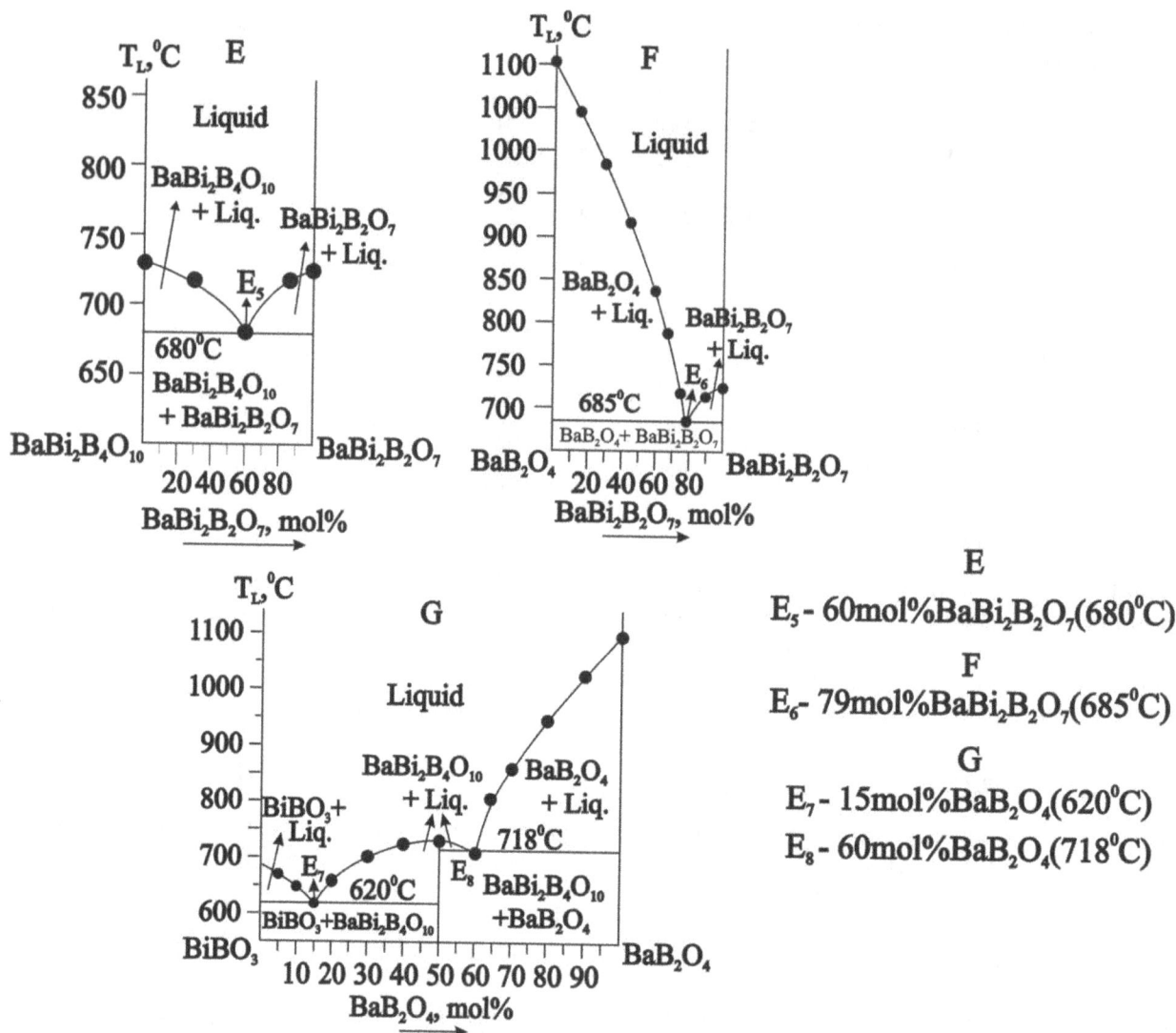

Figure 3. E–G: Phase diagrams of the pseudo-binary systems: E – BaBi$_2$B$_4$O$_{10}$–BaBi$_2$B$_2$O$_7$; F – BaB$_2$O$_4$–BaBi$_2$B$_2$O$_7$; G – BiBO$_3$–BaB$_2$O$_4$

There is a simple eutectic, E$_4$, between these two compounds at 28 mol% BaBi$_2$B$_4$O$_{10}$, with a melting point of 660°C.

BaBi$_2$B$_4$O$_{10}$–BaBi$_2$B$_2$O$_7$ pseudo-binary system

The BaBi$_2$B$_4$O$_{10}$–BaBi$_2$B$_2$O$_7$ system confirms the presence of the new congruently melting ternary compound BaBi$_2$B$_2$O$_7$, with a melting point of 725°C (Figure 3E). There is a simple eutectic, E$_5$, between BaBi$_2$B$_4$O$_{10}$–BaBi$_2$B$_2$O$_7$ at 60 mol% BaBi$_2$B$_2$O$_7$, with a melting point of 680°C.

BaB$_2$O$_4$–BaBi$_2$B$_2$O$_7$ pseudo-binary system

BaB$_2$O$_4$–BaBi$_2$B$_2$O$_7$ is a very important system between congruently melting BaB$_2$O$_4$ and the new ternary compound BaBi$_2$B$_2$O$_7$. This system also confirms the presence of the new congruently melting ternary compound BaBi$_2$B$_2$O$_7$, with a melting point of 725°C (Figure 3F). There is a simple eutectic, E$_6$, between

BaB$_2$O$_4$ and BaBi$_2$B$_2$O$_7$ at 79 mol% BaBi$_2$B$_2$O$_7$, with a melting point of 685°C.

BiBO$_3$–BaB$_2$O$_4$ pseudo-binary system

BaBi$_2$B$_4$O$_{10}$ is a congruently melting compound, with a melting point of 730°C, and it occupies the central area of the BiBO$_3$–BaB$_2$O$_4$ pseudo-binary system (Figure 3G). This system forms two simple binary eutectics, E$_7$ with at 15 mol% BaB$_2$O$_4$, with a melting point of 620°C, and E$_8$ at 60 mol% BaB$_2$O$_4$, with a melting point of 718°C.

Conclusions

The glass forming diagram in the BaO–Bi$_2$O$_3$–B$_2$O$_3$ system has been investigated and constructed. All stoichiometric ternary compositions are in the area of stable glasses.

The phase diagram of the well known binary Bi$_2$O$_3$–B$_2$O$_3$ system has been corrected in the interval between the compounds $Bi_4B_2O_9$ and $Bi_3B_5O_{12}$. The eutectic composition, $48 \cdot 5Bi_2O_3 \cdot 51 \cdot 5B_2O_3$ (mol%), between $BiBO_3$ and $Bi_3B_5O_{12}$, with melting point $665 \pm 5°C$, has been determined. It is shown that the compound $BiBO_3$ is congruently melting with a melting point of $685 \pm 5°C$.

Six pseudo-binary phase diagrams have been constructed for the $Bi_4B_2O_9$–$BaBi_2B_2O_7$, $BiBO_3$–$BaBi_{10}B_6O_{25}$, $BaBi_2B_4O_{10}$–$BaBi_{10}B_6O_{25}$, $BaBi_2B_4O_{10}$–$BaBi_2B_2O_7$, BaB_2O_4–$BaBi_2B_2O_7$ and $BiBO_3$–BaB_2O_4 systems. Eight pseudo-binary eutectic compositions were revealed.

Two new crystalline ternary compounds, $BaBi_2B_2O_7$ and $BaBi_{10}B_6O_{25}$, have been revealed by crystallisation of the same glass compositions. Both compounds melted congruently: $BaBi_2B_2O_7$ at $725 \pm 5°C$, and $BaBi_{10}B_6O_{25}$ at $690 \pm 5°C$. Single crystals of the new compound $BaBi_{10}B_6O_{25}$ were grown by cooling the melt with the stoichiometric composition. X-ray characteristics of both compounds were determined. $BaBi_2B_2O_7$ has an orthorhombic cell with lattice parameters $a=11 \cdot 818$ Å, $b=8 \cdot 753$ Å, $c=7 \cdot 146$ Å, $V=739 \cdot 203$ Å^3, $Z=4$. $BaBi_{10}B_6O_{25}$ has orthorhombic crystal symmetry and lattice constants $a=6 \cdot 434$ Å, $b=11 \cdot 763$ Å, $c=29 \cdot 998$ Å, $V=2270 \cdot 34$ Å^3, $Z=8$.

References

1. Levin, E. M. & McDaniel, C. L. *J. Am. Ceram. Soc.*, 1962, **45**, 355.
2. Mazurin, O. V., Streltsina, M. V & Shvaiko-Shvaikovskaya, T. P. *The properties of glasses and glass-forming melts*. Vol.2, Nauka, Leningrad, 1975.
3. Becker, P. & Bohaty, L. *Phys. Chem. Glasses*, 2003, **44**, 91.
4. Muehlberg, M, Burianek, M., Edongue, H. & Poetsch, C. *J. Cryst. Growth*, 2002, **237**, 740.
5. Becker, P. *Cryst. Res. Technol.*, 2004, **38**, 74.
6. Oprea, I.-I. Optical Properties of Borate Glass – Ceramic. Dissertation zur Erlangung des Grades Doktor der Naturwissenschaften, Osnabruck University, Germany, 2005.
7. Hwang, C.& Fujino, S. XX International Congress on Glass Proceedings, Sept 27–Oct 01, 2004, Kyoto. PDF file No O-07-52.
8. Honma, T., Benino, Y., Fujiwara, T., Komatsu, T. & Sato, R. *Appl. Phys. Lett.*, 2003, **82**, 892.
9. Elwell, D., Morris, A. W. & Neate, B. W. *J. Cryst. Growth*, 1972, **16**, 67.
10. Zargarova, M. I. & Kasumova, M. F. *Zh. Neorg. Mater.* 1990, **26**, 1678 (In Russian).
11. Egorisheva, A., Skorikov, V., Volodin, V., Myslitskii, O & Kargin, Yu. *Russian J. Inorg. Chem.*, 2006, **51**, 1956.
12. Barbier, J., Penin, N., Denoyer, A., & Cranswick, L. M. D. *Solid State Sci.*, 2005, **7**, 1055.
13. Kim, J.-M. & Jung, B.-H. *Mater. Sci. Forum*, 2003, **439**, 18.
14. Kim, J.-M. & Kim, H.-S. *Mater. Sci. Forum*, 2006, **510-511**, 574.
15. Kim, Y.-J., Hwang, S.-J. & Kim, H.-S. *Mater. Sci. Forum*, 2006, **510-511**, 578.
16. Hovhannisyan, R. M. *Phys. Chem. Glasses: Eur. J. Glass Sci. Technol. B*, 2006, **47**, 460.
17. Powder Diffraction File. PDF2 Release 2008. International Center for Diffraction Data, 2008, USA.
18. Pottier M. J. *Bull. Soc. Chim. Belg.*, 1974, **83**, 235.
19. Becker, P. & Froehlich, R. *Z. Naturforsch. B*, 2004, **59**, 256.

Proc. VI Int. Conf. Borate Glasses, Himeji, Japan, 18–22 August 2008   *Phys. Chem. Glasses: Eur. J. Glass Sci. Technol. B, December 2009, 50 (6), 361–366*

Dispersion and fluorescence of Tb³⁺ in CaO–B₂O₃ glass

Noriyuki Wada*

Department of Materials Science and Engineering, Suzuka National College of Technology, Shiroko, Suzuka, Mie, 510-0294, Japan

Misaki Katayama, Kazuo Kojima & Kazuhiko Ozutsumi

Department of Applied Chemistry, College of Life Sciences, Ritsumeikan University, 1-1-1 Noji-higashi, Kusatsu, Shiga, 525-8577, Japan

Manuscript received 18 September 2008
Revised version received 25 November 2008
Accepted 18 September 2009

The effect of Tb³⁺ clustering on the fluorescence properties of $30CaO.70B_2O_3.xTb_2O_3$ glasses (x=0·1–20·0) prepared by the melt quenching method was investigated by measuring the fluorescence spectra, and Tb L_{III}-edge x-ray absorption fine structure and then analysing the local structure, clustering, and florescence properties of Tb³⁺ ions. It was found that the fluorescence bands due to the $^5D_3\rightarrow^7F_J$ transitions (J=5 and 4) of Tb³⁺ ions were caused by the dissolved and dispersed Tb³⁺ ions in the Ca²⁺ site in the CaO-like structure. The Tb–O–Tb and Tb–O–Tb–O–Tb linkages and four-Tb–O–membered rings produce the cross relaxation process, $(^5D_3\rightarrow^5D_4)\rightarrow(^7F_0\leftarrow^7F_6)$ between neighbouring Tb³⁺ ions, therefore it was seen that the fluorescence bands due to the $^5D_3\rightarrow^7F_J$ transitions (J=5 and 4) of Tb³⁺ ions were quenched for x=2, and the fluorescence bands due to the $^5D_4\rightarrow^7F_J$ transitions (J=6, 5, 4, 3, 2, 1, and 0) of Tb³⁺ ions were enhanced in the x range from 0·1 to 12·5. For x>12·5, structures bonding two four-Tb–O-membered rings were formed, and consequently the fluorescence bands were quenched by the cross relaxation process, $(^5D_4\rightarrow^7F_J)\rightarrow3(^7F_J\leftarrow^7F_6)$ (0≤J≤6, 0≤J'≤5).

Introduction

Glasses and ceramics doped with Tb³⁺ ions excited by UV light strongly fluoresce a green colour due to the 4f–4f ($^5D_4\rightarrow^7F_5$) transition, and are utilised as green phosphors for lamps, liquid crystal displays, and plasma displays. The Tb³⁺ green fluorescence has been reported to be enhanced by the cross relaxation process, $(^5D_3\rightarrow^5D_4)\rightarrow(^7F_0\leftarrow^7F_6)$ which has been called resonance energy transfer.[1] In a similar manner to the Tb³⁺ ion green fluorescence, the red fluorescence due to the $^5D_0\rightarrow^7F_2$ transition of Eu³⁺ ions in oxide glasses has been enhanced by the cross relaxation processes, $(^5L_6\rightarrow^5D_{J'})\rightarrow(^7F_{J^*}\leftarrow^7F_{J\#})$ and $(^5D_J\rightarrow^5D_{J'})\rightarrow(^7F_{J^*}\leftarrow^7F_{J\#})$ $(3\geq J>J'\geq0, 6\geq J^*>J^\#\geq1)$, with a 30CaO.70B₂O₃ glass as the optimum host.[2,3]

In general, transition metal and rare earth ions were thought to be homogeneously dissolved into a glass. However, these ions do not always disperse well because of the solubility and site selectivity.[4,5] In addition, the local structure and clustering of rare earth ions are affected by the concentration because the energy transfer between rare earth ions depends on the concentration. However, it has never been reported that the energy transfer was investigated by analysing the clustering of rare earth ions in a glass and that the number of rare earth ions in the second coordination sphere of a rare earth ion was analysed.

In this study, the effect of Tb³⁺ clustering on Tb³⁺ fluorescence properties was investigated by measuring the fluorescence spectra and Tb L_{III} edge x-ray absorption fine structure (XAFS) of 30CaO.70B₂O₃ doped with Tb³⁺ ions and then by analysing the local structure, clustering, and florescence properties of Tb³⁺ ions.

Experimental

To prepare 30CaO.70B₂O₃.xTb₂O₃ glasses (x=0·1–20·0), B₂O₃ (Kojundo, 99·9%), CaCO₃ (Wako, 99·5%), and Tb₄O₇ (Kojundo, 99·9%) powders were used. B₂O₃ and CaCO₃ powders were weighed and mixed, and then the powder mixture was melted at 1200°C for 60 min by using a Pt crucible. The host glass was obtained by quenching the melt on a carbon plate. The host glass was crushed to a powder by using an alumina mortar and a pestle. Appropriate amounts of Tb₄O₇ powder and host glass were combined according to the chemical formula 30CaO.70B₂O₃.xTb₃O₃, so as to produce 10 g of powder which was melted at 1300–1350°C for 60 min. To form a glass, the melt was poured onto a graphite plate. Immediately, the glass was annealed at 650°C for 1 h and then gradually cooled to room temperature.

To prepare samples of thickness 1·5 mm for measuring fluorescence spectra, the glasses were cut and polished. The fluorescence spectra were measured using a fluorescence spectrometer (Hitachi F-2500)

* Corresponding author. Email wadan@mse.suzuka-ct.ac.jp

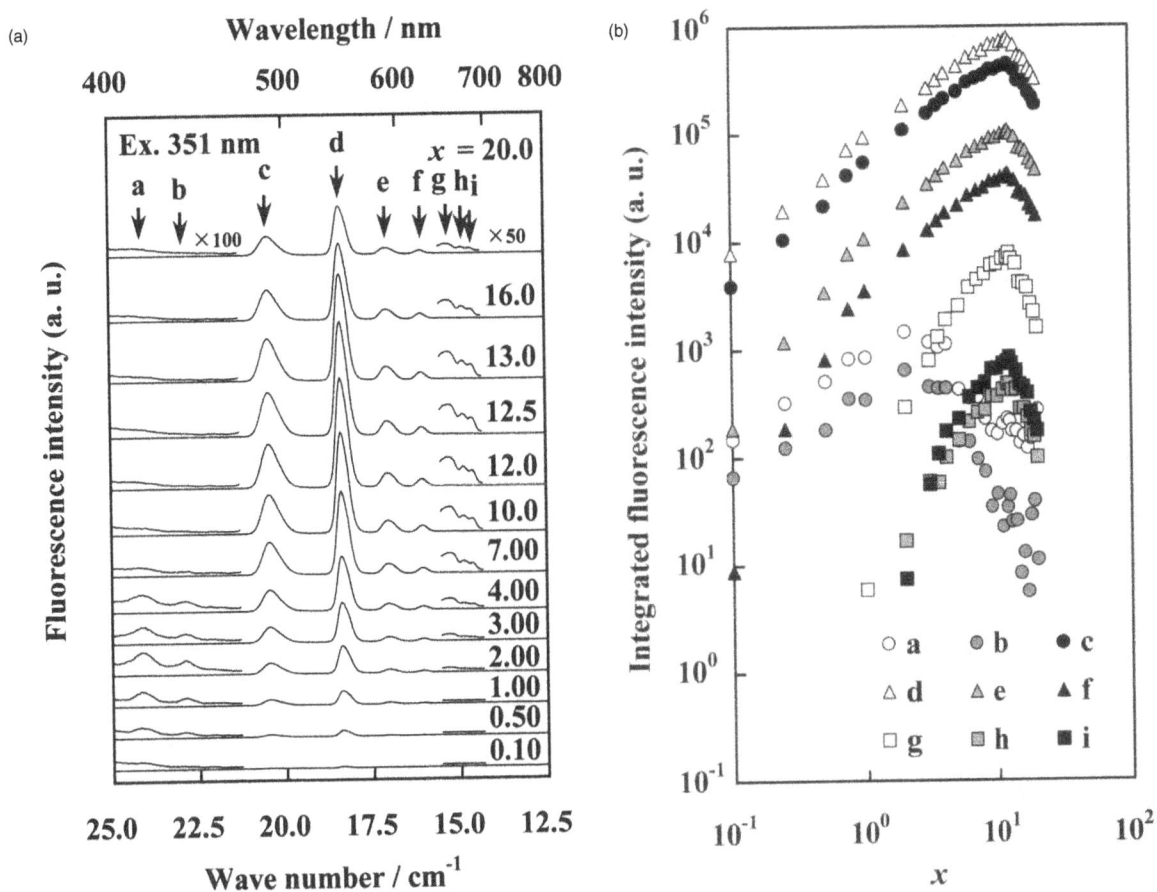

Figure 1. (a) Fluorescence spectra of $30CaO.70B_2O_3.xTb_2O_3$ glasses excited at the wavelength of 351 nm and (b) Tb_2O_3 content dependence of integrated intensities of the fluorescence bands, a–i

with a solid sample holder (measuring size: ø 5 mm) and an excitation wavelength of 351 nm. To prepare samples for measuring XAFS spectra, the glasses were crushed up using an alumina mortar and a pestle and the powder was put on both sides of double sided sticky tape (Sekisui chemical industry). In addition, the reference sample for XAFS spectra, crystalline Tb_2O_3 (Kojundo, 99·9%), was prepared likewise. Tb L_{III} edge XAFS spectra in the range of 7·4–8·4 keV were taken in transmission mode using beam line BL-4 at the SR Centre of Ritsumeikan University. Analysis of XAFS spectra was carried out using REX (Rigaku).

Results

Fluorescence spectra

Typical fluorescence spectra and integrated fluorescence intensities for the $30CaO.70B_2O_3.xTb_2O_3$ glasses are shown in Figure 1(a) and (b), respectively. The fluorescence bands a–i due to the $^5D_3 \rightarrow {}^7F_J$ (J=5 and 4) and $^5D_4 \rightarrow {}^7F_J$ (J=6, 5, 4, 3, 2, 1, and 0) transitions of Tb^{3+} ions appeared at 414, 437, 489, 544, 587, 623, 654, 673, and 684 nm. The fluorescence bands, a and b, with the initial energy level of $^5D_3$, were quenched at a Tb_2O_3 content of x=2, while the fluorescence bands, c–i, with the initial energy level of $^5D_4$, were quenched at a Tb_2O_3 content of x=12·5.

Analysis of XAFS spectra

The XAFS oscillation curve, $\chi(k)$, (where k is the wave number of the photoelectron) was obtained after normalisation by using the cubic spline method and subtraction of the smooth background. The radial structure function, $|F(r)|$, (where r is the distance from the Tb^{3+} ion) was obtained by Fourier transforming the k^3 weighted $\chi(k)$, $k^3\chi(k)$, over the range in k from 35·0 to 90·0 nm^{-1}. The radial structure functions, $|F(r)|$, for the $30CaO.70B_2O_3.xTb_2O_3$ glasses and Tb_2O_3 are shown in Figure 2. For Tb_2O_3, the peaks corresponding to the Tb–O and Tb–Tb interactions appeared at distances of 100–250 and 310–440 pm, respectively. In the glasses, the peak corresponding to the Tb–O interaction appeared at 90–310 pm, similar to Tb_2O_3. In the glasses, the Tb–O interaction peak shifts to longer distance as x changes from zero to 0·75, and then for higher x it shifts to shorter distance. Although the Tb–O interaction peak shifts as x changes, the Tb–O interaction of each glass is similar to that of Tb_2O_3. Hence, the Tb–O bond length, r_{Tb-O}, Debye–Waller factor, σ, and mean free path, λ_{MF} can be obtained by analysing the Tb–O interaction in $|F(r)|$ using the curve fitting method by assuming that the coordination number of the Tb^{3+} ion is six.[3,6,7] The values of σ, λ_{MF}, r_{Tb-O}, and estimated standard deviation, R are shown in Table 1. Judging from the values of R, the values of σ, λ_{MF}, and r_{Tb-O} were

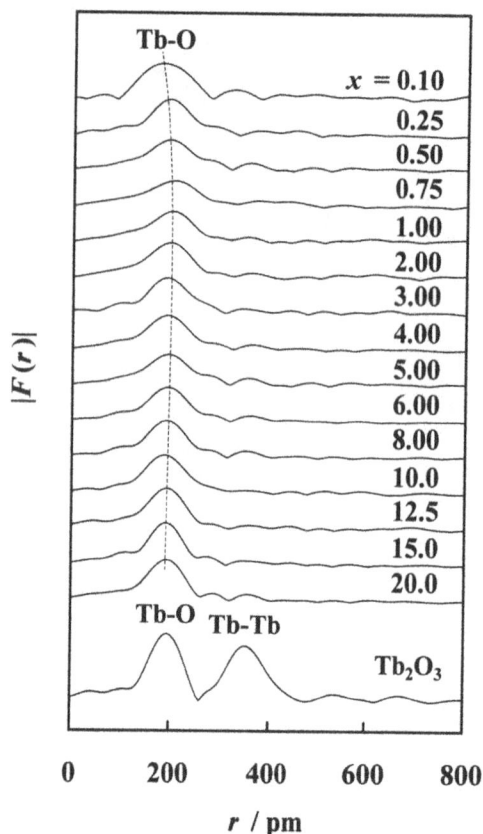

Figure 2. Radial structure functions, $|F(r)|$, for the Tb L_{III}-edge of $30CaO.70B_2O_3.xTb_2O_3$ glasses and Tb_2O_3 standard. The dashed line indicates the trend of the first Tb–O peak with composition

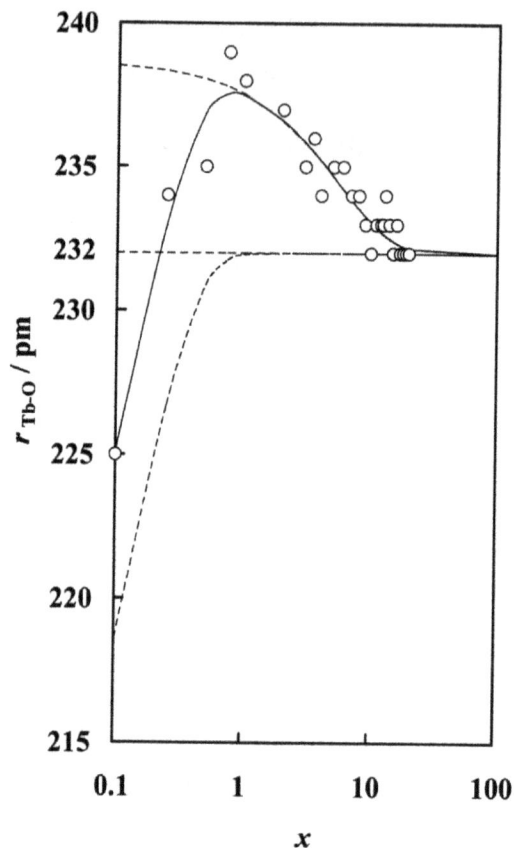

Figure 3. Relationship between Tb–O bond length, $r_{Tb–O}$ and Tb_2O_3 content, x in $30CaO.70B_2O_3.xTb_2O_3$ glasses. Open circles are experimental data points, the solid line is the fitted function of Equation (1), and the dashed lines are the individual terms of this fit

well determined for $x=0.25$ and greater. The values of σ, λ_{MF} and $r_{Tb–O}$ of the glasses are each larger than those of Tb_2O_3 because the uniformity and density of

Table 1. Results of curve fitting analysis of $30CaO.70B_2O_3$. xTb_2O_3 glasses and Tb_2O_3 standard

x	$r_{Tb–O}$ (pm)	σ (pm)	λ_{MF} (pm)	R (%)
0·10	225	13	468	11·3
0·25	234	11	416	3·7
0·50	235	12	396	4·0
0·75	239	13	409	4·7
1·00	238	11	384	3·8
2·00	237	11	397	6·4
3·00	235	12	454	6·1
3·50	236	12	374	4·8
4·00	234	11	430	6·3
5·00	235	12	432	6·0
6·00	235	11	410	5·2
7·00	234	11	419	5·6
8·00	234	11	400	4·9
9·00	233	10	394	4·9
10·0	232	11	421	6·3
11·0	233	10	388	4·0
12·0	233	10	394	5·0
12·5	233	10	388	4·3
13·0	234	11	343	4·9
14·0	233	9	380	3·8
15·0	232	8	348	3·9
16·0	233	10	370	3·5
17·0	232	9	352	3·2
18·0	232	10	413	5·4
19·0	232	9	387	4·0
20·0	232	9	383	4·0
Tb_2O_3	232	7	479	4·9

a glass is generally lower than that of a crystal. The relationship between x and $r_{Tb–O}$ is shown in Figure 3. With increasing x, $r_{Tb–O}$ increases for x less than 0·75, but decreases when x is more than 0·75.

Discussion

Tb^{3+} concentration quenching and clustering

It has been reported that the fluorescence bands, **a** and **b** arise from dispersed Tb^{3+} ions,[8] and that concentration quenching is caused by the cross relaxation process, $(^5D_3 \rightarrow ^5D_4) \rightarrow (^7F_0 \leftarrow ^7F_6)$, due to dipole–dipole and dipole–quadrupole interactions between Tb^{3+} ions in Tb^{3+} clusters.[1,7,8] For clustering of rare earth (Re) ions, MD simulations have revealed the presence of Re–O–Re linkages in various glasses doped with Re_2O_3, because the peaks due to Re–O–Re linkage were observed at 400–600 pm in the Re–Re pair distribution function.[10–14] In addition, the formation of Tb–O–Tb linkage in phosphate glasses has been directly observed by magnetic difference neutron diffraction.[15] The occurrence of Tb–O–Tb linkages increases with increasing x, therefore the fluorescence bands **a** and **b** are easily quenched at higher x. Hence, it is found that the cross relaxation process occurs for

Tb–O–Tb linkages. In a similar manner to the Tb^{3+} ion green fluorescence, the red fluorescence due to the $^5D_0 \rightarrow ^7F_2$ transition of Eu^{3+} ions has been found to be quenched by the cross relaxation process, $(^5D_0 \rightarrow ^7F_J)$ $\rightarrow m(^7F_{J'} \leftarrow ^7F_{J''})$ $(6 \geq J \geq 0, 6 \geq J' > J'' \geq 0)$, between the excited Eu^{3+} ion at the $^5D_0$ level and m ions of Eu^{3+} in the $^7F_{J''}$ levels.[2] Hence, the quenching of the fluorescence bands, **c–i** is caused by the cross relaxation process, $(^5D_4 \rightarrow ^7F_J) \rightarrow 3(^7F_{J'} \leftarrow ^7F_6)$ $(0 \geq J \geq 6, 0 \geq J' \geq 5)$, between the excited Tb^{3+} ion at the $^5D_4$ level and three Tb^{3+} ions in the $^7F_6$ ground level. It is found that the cross relaxation process occurs for a Tb^{3+} ion having three Tb^{3+} ions in the second coordination sphere, i.e. the Tb^{3+}–$(O–Tb)_3$ structure.

Tb–O bond length and Tb^{3+} clustering

Although evidence of Tb^{3+} clustering has been reported, as mentioned above, the Tb–Tb interaction of glasses cannot be observed beyond the Tb–O interaction in Figure 2 because of disorder around Tb^{3+} ions.[16] Therefore, by considering how the Tb–O bond length is changed by the kind and number of ions in the second coordination sphere of a Tb^{3+} ion, the structure of Tb^{3+} clusters was investigated as follows: In a borate glass doped with rare earth ions of a few mole percent, the rare earth ions are considered to be substituted on modifier ion sites which include nonbridging oxygens amongst their neighbours.[5] For $x < 0.75$, the Tb^{3+} ions are considered to be substituted on Ca^{2+} sites in the CaO-like structure (NaCl type) in the glass. A Tb^{3+} ion, which has a smaller ionic radius and a higher valence than a Ca^{2+} ion, has a lower basicity than a Ca^{2+} ion, and accordingly the Tb^{3+} ion attracts oxide ions in its first coordination sphere.[17,18] Therefore, for low Tb^{3+} concentration, r_{Tb-O} decreases, because most Tb^{3+} ions have no other Tb^{3+} ions in the neighbouring modifier ion sites in their second coordination sphere. Because Tb^{3+} ions easily substitute second neighbour sites of Tb^{3+} ions with increasing x, the value of r_{Tb-O} increases, i.e. r_{Tb-O} increases with increasing x because Tb^{3+} ions are substituted into the second coordination sphere of Tb^{3+} ions. On the other hand, for x greater than 0.75, it can be assumed that the Tb^{3+} ions are formed to the simple cubic lattice of the Tb^{3+} octahedron in a CaO-like structure because the Tb^{3+} ions coordinated with six oxide ions are clustering.[1,3] When the Tb^{3+} octahedrons are formed to the simple cubic lattice in a CaO-like structure, considering charge compensation, the maximum number of Tb^{3+} ions in the second coordination sphere of a Tb^{3+} ion is an integer between 3 and 6 because the Ca^{2+} deficiency is formed to 3 or less in the second coordination sphere. The distance r_{Tb-O} decreases with increasing x because the Tb^{3+} ions, the ionic radius of which is smaller than that of the Ca^{2+} ions, are clustering, and are finally approaching a Tb_2O_3-like structure.

Using the fitted results in Table 1, r_{Tb-O} (in units of pm) as a function of x was fitted to give the equation

$$r_{Tb-O}(x) = 232 - 24.95 \exp\left(-\frac{x}{0.1613}\right) + 6.65 \exp\left(-\frac{x}{5.4242}\right) \tag{1}$$

It is seen that the fitted curve as shown by the solid line in Figure 3 corresponds well to the experimental data points. The first term of Equation (1) means that, for high x, r_{Tb-O} of the glass approaches that of Tb_2O_3 crystal (r_{Tb-O}=232 nm), because the Tb^{3+} ions are clustering and finally form a Tb_2O_3-like structure at high x. In addition, the second term indicates that r_{Tb-O} increases with increasing x because the number of dispersed Tb^{3+} ions decreases compared with the number of Tb^{3+} ions coordinated with a Tb^{3+} ion in the second coordination sphere. Moreover, the third term indicates that r_{Tb-O} decreases with increasing x because the Tb^{3+} ions, which have a smaller ionic radius than Ca^{2+} ions, are clustering to form the simple cubic lattice in a CaO-like structure. Therefore, the Tb^{3+} clustering can be investigated by the first and third terms of Equation (1), that is

$$r_{Tb-O}(x) = 232 + 6.65 \exp\left(-\frac{x}{5.4242}\right) \tag{2}$$

The probability that a Tb^{3+} ion substitutes the second coordination sphere of a Tb^{3+} ion, W is determined as follows

$$W = \frac{n_{Tb}}{n_{Tb} + n_{Ca}} = \frac{2x}{2x + 30} \tag{3}$$

where n_{Tb} and n_{Ca} are the relative amounts of Tb^{3+} and Ca^{2+} ions, respectively. As mentioned above, when the simple cubic lattice of Tb^{3+} octahedron is formed in a CaO-like structure, the maximum number of Tb^{3+} ions in the second coordination sphere is an integer between 3 and 6. Therefore, the averaged maximum number of Tb^{3+} ions in the second coordination sphere, N_{Max} is 4.5 (=(3+4+5+6)/4). Thus, the number of Tb^{3+} ions in the second coordination sphere, N_{Tb} as a function of x is

$$N_{Tb}(x) = N_{Max}W = \frac{N_{Max}n_{Tb}}{n_{Tb} + n_{Ca}} = \frac{N_{Max}x}{x + 15} \tag{4}$$

N_{Tb} as a function of r_{Tb-O} is then derived from Equations (2) and (4)

$$N_{Tb}(r_{Tb-O}) = \frac{5.4242 N_{Max} \ln \dfrac{r_{Tb-O} - 232}{6.65}}{5.4242 \ln \dfrac{r_{Tb-O} - 232}{6.65} - 15} \tag{5}$$

When the value of $N_{Tb}(r_{Tb-O})$ is calculated from Equation (5), it is normally a rational number. However, N_{Tb} is the number of Tb^{3+} ions in the second coordination sphere, then the value of N_{Tb} must be one of 1, 2, 3, 4, 5, and 6. Therefore, the value of N_{Tb} cannot be directly calculated from Equation (5). When $N_{Tb}(r_{Tb-O})$

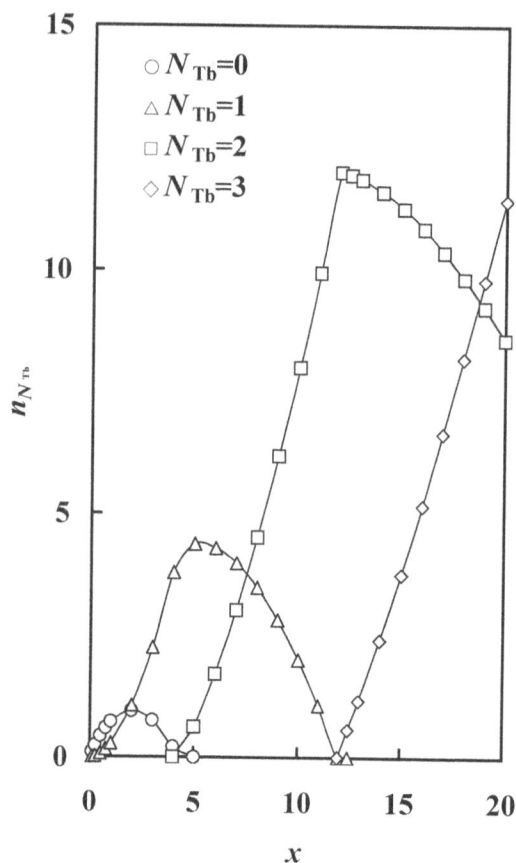

Figure 4. The relative number, $n_{N_{Tb}}$, *of* Tb^{3+} *ions having* N_{Tb} *second neighbour* Tb^{3+} *ions*

is normally a rational number between N_{Tb} and $N_{Tb}+1$, assuming that r_{Tb-O} is determined by the numbers of Tb^{3+} ions having N_{Tb} and $N_{Tb}+1$ second neighbour Tb^{3+} ions, the value of N_{Tb} is defined as functions of r_{Tb-O} and $N_{Tb}(r_{Tb-O})$ by

$$N_{Tb}(r_{Tb-O}) = N_{Tb}R_{N_{Tb}} + (N_{Tb}+1)R_{N_{Tb}+1} \qquad (6)$$

where $R_{N_{Tb}}$ and $R_{N_{Tb}+1}$ ($=1-R_{N_{Tb}}$) are the fractions of Tb^{3+} ions having N_{Tb} and $N_{Tb}+1$ second neighbour Tb^{3+} ions, respectively. The relative number of Tb^{3+} ions having N_{Tb} second neighbour Tb^{3+} ions is then

$$n_{N_{Tb}} = xR_{N_{Tb}} \qquad (7)$$

The relationship between the values of the $n_{N_{Tb}}$ and x is shown in Figure 4. The value of n_0 is largest at $x=2\cdot0$. With increasing x, the value of n_1 increases in the range of x from $0\cdot1$ to $5\cdot0$, and then decreases for x in the range from $5\cdot0$ to $12\cdot0$. The value of n_2 increases with x, for x in the range from $4\cdot0$ to $12\cdot0$, and then decreases for x from $12\cdot0$ to $20\cdot0$. The value of n_3 increases with increasing x.

According to the variation in $n_{N_{Tb}}$ described above, it is found that Tb^{3+} ions cluster as follows: For $x=0\cdot1-2\cdot0$, some of the Tb^{3+} ions cluster with a Tb^{3+} ion in the second coordination sphere, namely, a Tb–O–Tb linkage. The fraction of dispersed Tb^{3+}

ions without a Tb^{3+} ion in the second coordination sphere is a maximum at $x=2\cdot0$. For $x=2\cdot0-5\cdot0$, the value of n_0 decreases and the value of n_1 increases instead, hence Tb–O–Tb linkages are increasingly formed. The number of Tb–O–Tb linkages is a maximum at $x=5\cdot0$. For $x=5\cdot0-7\cdot0$, the value of n_1 decreases and the value of n_2 increases instead, so Tb–O–Tb–O–Tb linkages are formed. For $x=6\cdot0-12\cdot0$, when the condition, $2n_2 \geq n_1$ is satisfied, the Tb^{3+} ions form not only Tb–O–Tb–O–Tb linkages, but also four-Tb–O–membered ($(Tb–O)_4$) rings, because the number of Tb^{3+} ions with n_1 is lacking compared to the number of Tb^{3+} ions with n_2. Therefore, for $x=5\cdot0-7\cdot0$, with increasing x, the number of Tb–O–Tb linkages decreases and the number of Tb–O–Tb–O–Tb linkages increases instead. For $x=6\cdot0-12\cdot0$, with increasing x, the number of Tb–O–Tb–O–Tb linkages decreases, and the number of $(Tb–O)_4$ rings increases instead, and then is a maximum at $x=12\cdot0$. For $x=12\cdot0-20\cdot0$, because the value of n_2 decreases and the value of n_3 increases instead, the $(Tb–O)_4$ ring is bonded to another such ring by –O– bonding, that is, $(Tb–O)_4$–O–$(Tb–O)_4$, $(Tb–O)_4(–O–)_2(Tb–O)_4$, and $(Tb–O)_4(–O–)_3(Tb–O)_4$ clusters are formed.

Relationship between Tb^{3+} clustering and fluorescence properties

It is found that the fluorescence bands, **a** and **b** appear due to the dispersed Tb^{3+} ions on Ca^{2+} sites in a CaO-like structure, because the change of integrated intensities of the fluorescence bands, **a** and **b** in Figure 1 accord with the change of n_0 in Figure 4. Therefore, although the concentration quenching of the fluorescence bands **a** and **b** is caused by the cross relaxation process, $(^5D_3 \rightarrow ^5D_4) \rightarrow (^7F_0 \leftarrow ^7F_6)$, it is found that the quenching is directly caused by the decrease of n_0. This result shows that the cross relaxation has already occurred in the Tb^{3+} concentration before quenching, as reported previously.[1] The cross relaxation occurs by increasing of n_1, n_2, and n_3, instead of decreasing of n_0, therefore the cause of the cross relaxation is the energy transfer between the neighbouring Tb^{3+} ions of the Tb–O–Tb and Tb–O–Tb–O–Tb linkages, the $(Tb–O)_4$ ring, and the structures bonding two $(Tb–O)_4$ rings.

Since the changes of integrated intensities of the fluorescence bands **c–i** in Figure 1(b) accord with the change of n_1+n_2 in Figure 4, it is seen that the fluorescence bands **c–i** are quenched at about $x=12\cdot5$ when n_2 is at a maximum. This concentration quenching occurs by increasing of n_3 instead of decreasing of n_2. Therefore, the enhancement of the fluorescence bands **c–i** is caused by the Tb–O–Tb and Tb–O–Tb–O–Tb linkages and the $(Tb–O)_4$ rings. In addition, it is found that the cause of the quenching is the cross relaxation process, $(^5D_4 \rightarrow ^7F_{J'})$ $\rightarrow 3(^7F_{J'} \leftarrow ^7F_6)$ ($0 \geq J \geq 6$, $0 \geq J' \geq 5$), for Tb ions having three

Tb^{3+} ions in their second coordination sphere, where two $(Tb–O)_4$ rings are bonding.

Conclusions

To investigate the Tb^{3+} clustering, fluorescence, and cross relaxation in $30CaO.70B_2O_3.xTb_2O_3$ glasses (x=0·1–20·0) prepared by the melt quenching method, the Tb L_{III}-edge XAFS and Tb^{3+} fluorescence spectra were measured and analysed. It was found that, for x=0·1–4·0, the fluorescence bands due to the $^5D_3 \rightarrow {}^7F_J$ transitions (J=5 and 4) of Tb^{3+} ions were caused by substituted and dispersed Tb^{3+} ions on the Ca^{2+} sites in the CaO-like structure. For x=1·0–12·5, Tb–O–Tb and Tb–O–Tb–O–Tb linkages, and four-Tb–O-membered rings are formed, and the cross relaxation process, $(^5D_3 \rightarrow {}^5D_4) \rightarrow (^7F_0 \leftarrow {}^7F_6)$, occurs between the neighbouring Tb^{3+} ions; therefore it was seen that by these linkages and ring, the fluorescence bands due to the $^5D_3 \rightarrow {}^7F_J$ transitions (J=5 and 4) of Tb^{3+} ions were quenched and the fluorescence bands due to the $^5D_4 \rightarrow {}^7F_J$ transitions (J=6, 5, 4, 3, 2, 1, and 0) of Tb^{3+} ions were enhanced. Tb^{3+} ions having three Tb^{3+} ions in their second coordination sphere, where two four-Tb–O-membered rings are bonding, caused the cross relaxation process, $(^5D_4 \rightarrow {}^7F_J) \rightarrow 3(^7F_{J'} \leftarrow {}^7F_6)$ ($0 \geq J \geq 6$, $0 \geq J' \geq 5$), and consequently it was found that the fluorescence bands due to the $^5D_4 \rightarrow {}^7F_J$ transitions (J=6, 5, 4, 3, 2, 1, and 0) of Tb^{3+} ions were quenched at x=12·5.

Acknowledgment

A part of this work was conducted at the SR centre of Ritsumeikan University, supported by "Nanotechnology Network Japan" of MEXT, Japan.

References

1. Hayakawa, T., Kamata, N. & Yamada, K. *J. Lumin.*, 1996, **68** 179.
2. Wada, N. & Kojima, K. *J. Lumin.*, 2007, **126**, 53.
3. Wada, N. & Kojima, K. *XXI International Congress on Glass*, 2007, Strasbourg, Abstract M24.
4. Murata, T., Torisaka, M., Takebe, H. & Morinaga, K. *J. Non-Cryst. Solids*, 1997, **220**, 139.
5. Murata, T., Moriyama, Y. & Morinaga, K. *Sci. Tech. Adv. Mater.*, 2000, **1**, 139.
6. Wada, N., Kojima, K. & Ozutsumi, K. *Mem. SR Centre, Ritsumeikan Univ.*, 2005, **7**, 7.
7. Ofuchi, H., Imaizumi, Y., Sugawara, H., Fujioka, H., Oshima, M. & Takeda, Y. *Nucl. Instrum. Meth. B*, 2003, **199**, 231.
8. Silversmith, A. J., Nguyen, N. T. T., Sullivan, B. W., Boye, D. M., Ortiz, C. & Hoffman, K. R. *J. Lumin.*, 2008, **128**, 931.
9. de Graaf, D., Stelwagen, S. J., Hintzen, H. T. & de With, G. *J. Non-Cryst. Solids*, 2003, **325**, 29.
10. Mountjoy, G. *J. Non-Cryst. Solids*, 2007, **353**, 2029.
11. Du, J. & Cormack, A. N. *J. Non-Cryst. Solids*, 2005, **351**, 2263.
12. Corradi, A. B., Cannillo, V., Montorsi, M., Siligardi, C. & Cormack, A. N. *J. Non-Cryst. Solids*, 2005, **351**, 1185.
13. Bernard, C., Chaussedent, S., Monteil, A., Balu, N., Obriot, J., Duverger, C., Ferrari, M., Bouazaoui, M., Kinowski, C. & Turrell, S. *J. Non-Cryst. Solids*, 2001, **284**, 68.
14. Park, B., Li, H. & Corrales, L. R. *J. Non-Cryst. Solids*, 2002, **297**, 220.
15. Cole, J. M., Hannon, A. C., Martin, R. A. & Newport, R. J. *Phys. Rev. B*, 2006, **73** 104210.
16. Peters, P. M. & Houde-Walter, S. N. *J. Non-Cryst. Solids*, 1998, **239**, 162.
17. Morinaga, K., Murata, T. & Wada, N. *Molten Salts*, 1999, **42**, 145. (In Japanese)
18. Morinaga, K., Yoshida, H. & Takebe, H. *J. Am. Ceram. Soc.*, 1994, **77**, 3113.

Proc. VI Int. Conf. Borate Glasses, Himeji, Japan, 18–22 August 2008   *Phys. Chem. Glasses: Eur. J. Glass Sci. Technol. B*, December 2009, **50** (6), 358–360

Thermal conductivity of sodium borate glasses at low temperature

Masashi Tohmori,[A] *Toru Sugawara,*[A,B] *Satoshi Yoshida*[A,B] *& Jun Matsuoka*[A,*]

[A] *Department of Materials Science, The University of Shiga Prefecture, Hassaka 2500, Hikone, Shiga 522-8533, Japan*
[B] *Center for Glass Science and Technology, The University of Shiga Prefecture, Hassaka 2500, Hikone, Shiga 522-8533, Japan*

Manuscript received 18 August 2008
Revised version received 30 September 2009
Accepted 1 October 2009

Thermal conductivity is one of the fundamental thermodynamic properties of materials. However, the temperature dependence below room temperature has been systematically studied only for about twenty glass compositions. We have studied the thermal conductivity of xNa$_2$O.(100–x)B$_2$O$_3$ glasses. Increasing the alkali content is found to increase the thermal conductivity up to x=25, which is mainly due to the increase in heat capacity. The composition dependence of the thermal conductivity becomes small above x=25. This can be attributed to the decrease of phonon mean free path due to the formation of nonbridging oxygens.

Introduction

Thermal conductivity is one of the fundamental thermodynamic properties. The application of glass powders to solders for electronic parts and to glass wools for thermal insulation requires accurate knowledge of it. In addition, this property is also important for the manufacture and use of many glass products.[1]

A number of studies of the thermal conductivity of glasses have been reported. However, there are very few systematic studies of the composition dependence of the thermal conductivity of glass.[2–7] In the pioneering study by Kittel,[3] it was pointed out that the thermal conductivity of amorphous materials is much smaller than that of crystalline materials, due to the shorter phonon mean free path for the disordered structure. However, further study of the composition dependence has not been carried out systematically, and less than 20 glass compositions have been measured below room temperature. Therefore, the temperature or composition dependence of the thermal conductivity of glasses is not yet well understood.

In our previous study,[7] we measured the thermal conductivity of Na$_2$O–SiO$_2$ glasses from 150 K to room temperature. The conductivity was found to decrease with increasing alkali content. This behaviour is attributed to the decrease of the phonon mean free path caused by the formation of nonbridging oxygen (NBO).

Borate glass systems have a composition–structure relationship different from that of silicate systems. Therefore, a different composition dependence of the thermal conductivity is expected and so the thermal conductivity of sodium borate glasses was measured

from 150 K to room temperature in this study. The composition dependence of the thermal conductivity and the phonon mean free path are discussed in relation to the glass structure.

Experimental

The thermal conductivity of glasses in the system xNa$_2$O.(100–x)B$_2$O$_3$ (x=5, 10, 15, 20, 25, 30, 35) has been measured in this study. These glasses were prepared from reagent grade Na$_2$CO$_3$ and B$_2$O$_3$. The obtained glass samples were annealed at a temperature 20 K higher than the glass transition temperature listed in Table 1. The glass transition temperature was measured by using TMA (thermomechanical analysis). The density of the glass samples was measured by Archimedes' method, using kerosene as the immersion liquid.

The thermal diffusivity was measured by a transient heating method.[7] The samples used for the measurements of thermal diffusivity were cut to a size about 10×10×5 mm^3. One side of the sample (10×10 mm^2 surface) was coated with carbon black film to absorb light from a halogen lamp for heating. A thermocouple was attached with silver paste on the opposite side of the sample for measuring the temperature change.

The sample was kept at a temperature from about 150 K to room temperature in a cryostat and under vacuum, and then the thermal diffusivity was estimated by measuring the time–temperature relation due to the irradiation heating process. For a transient heating method adopted in this study, one surface was exposed to a constant and uniform heat flux. The temperature at the opposite side of the sample is given by the following equation, when the dissipation

---

*Corresponding author. Email matsuoka@mat.usp.ac.jp

of heat from the sample can be ignored[8]

$$T-T_0 = \frac{\dot{q}l}{DC_P}\left\{\frac{Dt}{l^2}+\frac{1}{6}-\frac{2}{\pi^2}\sum_{n=1}^{\infty}\frac{(-1)^n}{n^2}\exp\left(-\frac{Dn^2\pi^2t}{l^2}\right)\right\}$$

(1)

where D is the thermal diffusivity (m^2s^{-1}), C_P the heat capacity per volume ($J\,K^{-1}m^{-3}$), $\dot{q}$ the heating power ($J\,m^{-2}s^{-1}$), l the thickness of the sample, T_0 the initial temperature, $T-T_0$ the temperature change, and t the heating time. The exponential term of Equation (1) becomes negligible as the heating time becomes long. In such a case, Equation (1) can be approximated to the simplified equation

$$T-T_0 = \frac{1}{C_P}\left(\frac{\dot{q}}{l}t-\frac{\dot{q}l}{D}\right)$$

(2)

We obtained the thermal diffusivity by fitting this equation to the experimental data. Measurements of the thermal diffusivity were carried out more than 10 times for one condition, and the error was within ±5 %.

The thermal conductivity, κ, was calculated from the thermal diffusivity and the heat capacity per unit volume by using the following equation

$$\kappa = C_P D$$

(3)

The phonon mean free path, λ, was calculated from the thermal diffusivity and mean sound velocity, v, according to

$$\kappa = \frac{1}{3}C_P v\lambda = C_P D$$

(4)

The heat capacity was measured by using a DSC (Seiko, DSC220) from about 150 K to room temperature. Measurements were carried out two or three times, and the error was within ±3%. Mean sound velocity was measured in this study by using the cube resonance method,[9] using a 3×3×3 mm^3 specimen at room temperature. Measurements were carried out nine times, and the error was within ±2% for x=5 and within ±0·3% for other compositions.

Results

The heat capacity and mean sound velocity of xNa$_2$O.(100−x)B$_2$O$_3$ glasses are listed in Table 1, together with the density data. These values increase with increasing Na$_2$O content.

Figure 1 shows the thermal conductivity of xNa$_2$O.(100−x)B$_2$O$_3$ glasses as a function of temperature. By comparing values at the same temperature, it can be seen that the thermal conductivity increases with increasing Na$_2$O content. However, the increase is not linear with composition. The thermal conductivity of the x=5 and 10 glasses is very low, and the compositional change between x=25 and 35 is small. For all compositions, the thermal conductivity increases

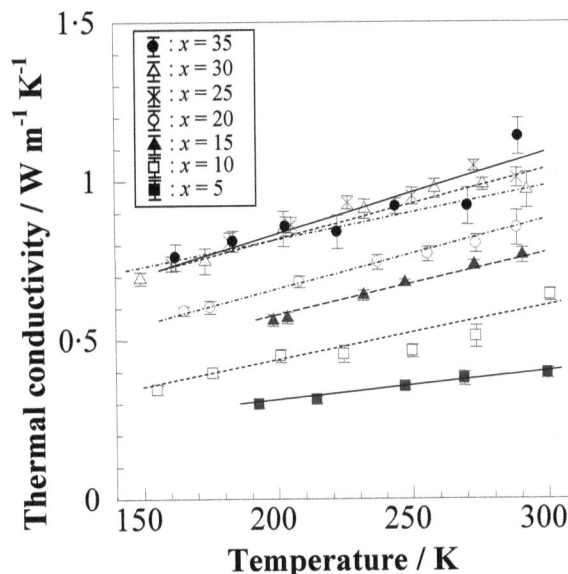

Figure 1. Thermal conductivity of xNa$_2$O.(100−x)B$_2$O$_3$ glasses from 150 K to 300 K. The lines are guides to the eye

with increasing temperature.

Figure 2 shows the phonon mean free path of these glasses. We assumed that the sound velocity is constant with temperature, and the room temperature value was used to calculate the phonon mean free path. The error due to this assumption should be less than 3%, because the temperature dependences of density and elastic constants are small. It is found that the phonon mean free path increases with increasing alkali content from x=5 to x=25. On the other hand, further increase of the alkali content from x=5 to x=35 is found to decrease the phonon mean free path.

Discussion

The thermal conductivity increases with increasing Na$_2$O content, x, but the composition dependence is not linear.

The thermal conductivity of the x=10 glass is only a little higher than that of the x=5 glass. The sound velocity and heat capacity of the former glass are respectively 12% and 6% higher than those of the latter glass. However, the phonon mean free paths

Table 1. Density, mean sound velocity and heat capacity of xNa$_2$O−(100−x)B$_2$O$_3$ glasses

Composition xNa$_2$O.(100−x)B$_2$O$_3$	Density ϱ (10^3 kg/m^3)	Sound velocity v (m/s)	Specific heat C$_{200\,K}$ (10^6 J/m^3K)	C$_{275\,K}$ (10^6 J/m^3K)
5	1·96	2490	1·1	1·37
10	2·04	2799	1·16	1·46
15	2·12	3029	1·28	1·62
20	2·18	3185	1·35	1·68
25	2·27	3312	1·5	1·87
30	2·34	3405	1·57	1·95
35	2·38	3562	1·62	1·97

Maximum error in density is ±0·003×10^3 kg/m^3. The maximum experimental error of sound velocity is within ±2%, and that of specific heat is within ± 3%.

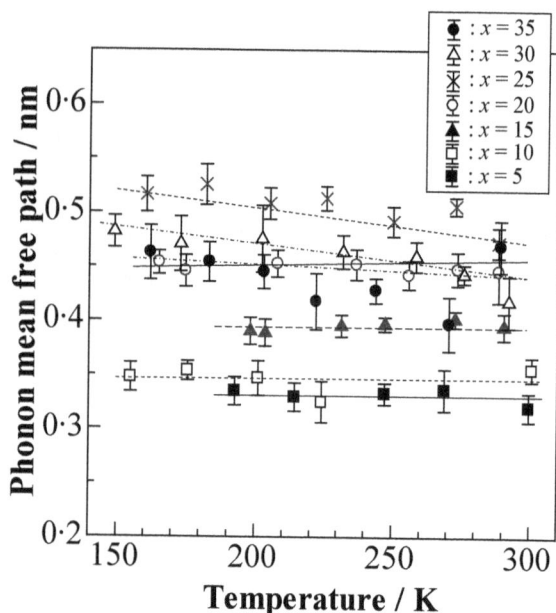

Figure 2. Phonon mean free path of xNa$_2$O.(100–x)B$_2$O$_3$ glasses from 150 K to 300 K. The lines are guides to the eye

of these glasses are almost the same as each other. An increase in the Na$_2$O content in this composition region is known to increase the amount of 4-ford coordinated boron.[10,11] This structural change increases the connectivity of the glass network, and so increases the elastic modulus. However, this increase of connectivity does not increase the phonon mean free path. Pure B$_2$O$_3$ glass consists only of 3-fold coordinated boron,[11] and so the structure should be homogeneous. Addition of Na$_2$O to this glass leads to a fraction of 4-fold coordinated boron, which makes the structure of the glass inhomogeneous, together with the increase of connectivity. The former effect will tend to decrease the phonon mean free path, and the latter effect will tend to increase the phonon mean free path. The similarity of the phonon mean free path of x=5 and 10 glasses is due to a balance between these two competing effects.

From x=10 to 25, the thermal conductivity largely increases with increasing Na$_2$O content, and that of the x=25 glass is two times that of the x=10 glass. In this composition range, heat capacity, sound velocity and phonon mean free path all increase with increasing Na$_2$O content. These changes cause the increase of thermal conductivity with composition. The phonon mean free path of the x=25 glass is about 1·5 times that of the x=10 glass, due to the increase of the network connectivity.

From x=25 to 35, the composition dependence of the thermal conductivity is small. This is due to the decrease of the phonon mean free path, together with the increase in heat capacity and sound velocity. It is

well known that NBO is formed in this composition region.[12] Therefore, the decrease of phonon mean free path with increasing Na$_2$O content is due to the formation of NBO, which reduces the network connectivity. This trend is similar to that for alkali silicate glasses,[7] and is expected to be a common feature in oxide glasses.

The temperature dependence of the phonon mean free path is very small for all compositions in this study. Therefore, the temperature dependence of the thermal conductivity is governed mainly by the temperature dependence of the heat capacity. The constant phonon mean free path is due to the large inhomogeneity of the glass structure at the atomic scale, which governs the scattering tendency of all phonon wavelengths which are significantly excited in this temperature range.

Conclusions

The thermal conductivity of sodium borate glasses, xNa$_2$O.(100–x)B$_2$O$_3$, was measured from about 150 K to room temperature, and was found to increase with increasing alkali content. The dominant factor affecting the behaviour of the thermal conductivity is the composition dependence of the phonon mean free path. The phonon mean free path increases with increasing alkali content from x=5 to x=25. On the other hand, further increases of the alkali content above x=25 are found to decrease the phonon mean free path. The increase of the phonon mean free path up to x=25 is due to the increase of network connectivity arising from the formation of 4-fold coordinated boron. On the other hand, the decrease of the phonon mean free path from x=25 to x=35 is due to the decrease of network connectivity caused by the formation of nonbridging oxygen, as for Na$_2$O–SiO$_2$ glasses.

References

1. Salama, S. N., Salman, S. M. & Gharib, S. J. Non-Cryst. Solids, 1987, **93**, 203.
2. Ratcliffe, E. H. Glass Technol., 1963, **4**, 113.
3. Kittel, C. Phys. Rev., 1949, **75**, 972.
4. Zhu, D.-M. & Kosugi, T. J. Non-Cryst. Solids, 1996, **202**, 88.
5. Freeman, J. J. & Anderson, A. C. Phys. Rev. B, 1986, **34**, 5684.
6. Zeller, R. C. & Pohl, R. O. Phys. Rev. B, 1971, **4**, 2029.
7. Hiroshima, Y., Hamamoto, Y., Yoshida, S. & Matsuoka, J. J. Non-Cryst. Solids, 2008, **354**, 341.
8. Crank, J. The Mathematics of Diffusion, Second edition, Clarendon Press, Oxford, 1975.
9. Goto, T. & Soga, N. J. Ceram. Soc. Jpn., 1983, **91**, 24.
10. Bray, P. J. In: Borate Glasses: Structure, Properties, Applications, Eds. L. D. Pye, V. D. Fréchette & N. J. Kreidl, Plenum Press, New York, 1978, p. 321.
11. Yun, H.Y. & Bray, P. J. Non-Cryst. Solids, 1978, **27**, 363.
12. Shelby, J. E. Introduction to Glass Science and Technology, Second edition, Royal Society of Chemistry, Cambridge, 2005, Chapter 5.

Proc. VI Int. Conf. Borate Glasses, Himeji, Japan, 18–22 August 2008   *Glass Tech.: Eur. J. Glass Sci. Technol. A, August 2009, 50 (4), 214–216*

Compositional dependence of silver ion incorporation into borosilicate glasses through staining for fabrication of graded index optical elements

T. Suetsugu,[1,2] *T. Wakasugi*[1] *& K. Kadono*[1]*

[1] *Division of Chemistry and Materials Technology, Kyoto Institute of Technology, Matsugasaki, Sakyo-ku, Kyoto 606-8585, Japan*
[2] *Isuzu Glass Co., Ltd., Minamitsumori 6-3-6, Nishinari-ku, Osaka 557-0063, Japan*

Manuscript received 20 August 2008
Revised version received 1 December 2008
Accepted 6 January 2009

The staining technique, which is a well known glass colouring method, is of potential use for the fabrication of graded index optical elements based on a glass substrate. We have studied the incorporation of silver into borosilicate glasses, $65SiO_2.20B_2O_3.10R_2O.5Al_2O_3$ (R=Li, Na or K), and also the changes in the optical properties due to silver staining. The dependence of the silver incorporation behaviour on the kind of alkaline ions in the glass was investigated. After staining at 340°C for 12 h, 1.9×10^{18} and 2.7×10^{18} atoms/cm² of silver were incorporated into lithium and sodium borosilicate glasses, respectively. This gave rise to an increase in the refractive index of ~0.03–0.04 at the stained surface of these glasses, although the glasses were colourless and transparent. On the other hand, the amount of silver incorporated into the potassium borosilicate glass was only one third of the amounts for lithium and sodium borosilicate glasses, and the increase in the refractive index was less than 0.02. The difference in the silver incorporation behaviour is discussed from the viewpoint of ionic diffusion.

Introduction

Staining has been long used as a glass colouring technique,[1] in which stains composed of silver or copper inorganic compounds, organic resins, and organic solvents, are applied on the glass surfaces and the glasses are heat treated, resulting in the colouration of the glasses. It has been revealed that some parts of the stained glass windows for cathedrals in medieval times were also coloured by such methods.[2] This technique is still used for colouring glass ware, e.g. drawing the graduation lines or colouring amber for glass containers in laboratories, and a silver stain is currently available commercially.[3]

The staining process includes a reaction in which alkaline ions present in the glass are replaced by the silver or copper ions in the stains.[1,2,4,5] The incorporated ions migrate into the deeper regions of the glass by diffusion. This reaction is regarded as a similar reaction to ion exchange which is usually performed by immersing the glass in molten salt. Therefore, staining can be applied for fabricating graded index optical elements based on glass substrates if the refractive index changes at the glass surface without any absorption change in the visible to near infrared region. We have previously reported that optical waveguides are easily prepared by silver- and copper-staining based on commercially available borosilicate and soda–lime–silica glasses.[6] It is important to control the staining condition in order to obtain moderate index change without any colouration. We have also presented the relationship between staining conditions and optical properties (i.e. absorption and refractive index) after staining, for borosilicate and soda–lime–silica glasses.[7,8] Furthermore, we have investigated how silver incorporation depends on glass composition, and the way in which the optical properties change through silver-staining, for aluminoborosilicate glasses, which are nominally nonbridging oxygen free.[9] It was shown that a much larger amount of silver ions is incorporated into aluminoborosilicate glasses than for borosilicate and soda–lime–silica glasses stained under the same conditions.

Borosilicate is one of the most important glass systems for optical elements because of the good chemical and optical durability, the low thermal expansion, and the wide transparency range. Besides such practical advantages, borosilicate glasses are of interest from the point of view of glass structure. Here, we present studies of the staining of borosilicate glasses, investigating the way in which silver ion incorporation and optical property changes depend upon the kind of alkaline ions.

Experimental procedures

Glass preparation

Borosilicate glasses of the composition $65SiO_2.20B_2O_3.10R_2O.5Al_2O_3$ (mol%), where R=Li, Na or K, were prepared by a conventional melting and cooling method as follows. Reagent grade compounds (Wako Pure Chemical Industries, Ltd.), SiO_2, Al_2O_3, R_2CO_3, and

*Corresponding author. Email kadono@kit.ac.jp

B_2O_3, were used as raw materials. Batch mixtures with desired compositions were melted at 1400°C for 1 h using Pt crucibles. The melts were quenched in a carbon mould and annealed at 580°C. The glasses obtained were cut to a size 20×20 mm² and 3 mm in thickness. Both sides of the substrates were optically polished. Hereafter, the glass substrates with R=Li, Na, or K are referred to as LBS, NBS, or KBS glass, respectively.

Staining and characterisation

Commercially available silver stain (Okuno Chemical Industries Co., Ltd.) was used for the staining. A mixture of the stain with screen oil (50:50 in wt%) was applied on one side of the glass substrate and dried at 200°C for 1 h. The amount of mixture applied to the substrate was 100 mg/cm². The substrate was then heat treated at 300°C, 320°C or 340°C for 12 or 24 h under ambient atmosphere. After the heat treatment, residual stain on the glass substrate was washed off. The amount of incorporated silver was measured by ICP (inductively coupled plasma) elemental analysis. Since the stain contains a mixed salt of silver and sodium, the amount of sodium was also analysed for the LBS and KBS glasses by ICP. The concentration profile of silver as a function of depth from the stained surface was obtained by energy dispersive x-ray (EDX) analysis using a Horiba EMAX super Xerophy S-792XI spectrometer.

The optical absorption spectrum in the range 200 to 900 nm was measured using a Hitachi U-3000 spectrophotometer. The refractive index at 633 nm of the glass surface was measured using a Metricon Model 2010 prism coupler.

Results

The density, glass transition and softening temperatures, and the refractive index at 633 nm for the LBS, NBS, and KBS glasses used in the experiments are summarised in Table 1. The prepared glasses were colourless and transparent in the visible region. No opacity was observed for all the glasses. For the three component systems, SiO_2–B_2O_3–R_2O (R=alkaline metal), the immiscibility regions become larger in the opposite order of the atomic number of the alkaline metals.[10] The SiO_2–B_2O_3–Li_2O system has the largest immiscibility region, which ranges from 20 to 100 mol% of SiO_2, 0 to 65 mol% of B_2O_3, and 0 to 30 mol% of Li_2O at 650°C. Although the composition of

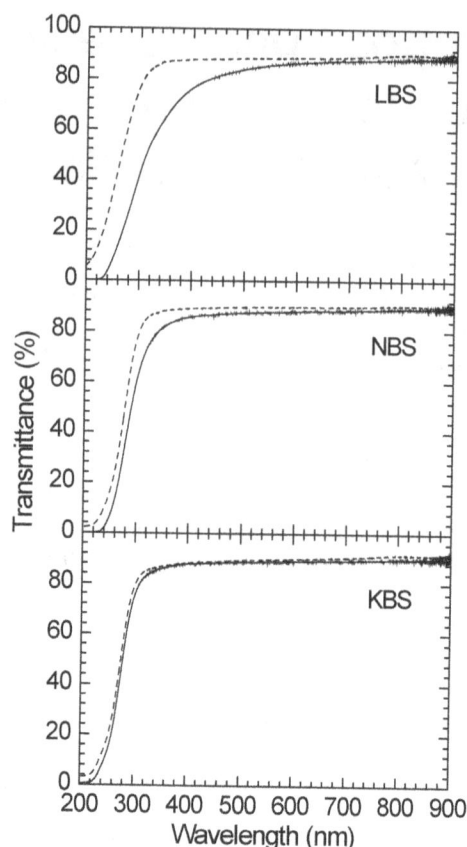

Figure 1. Transmission spectra of the LBS, NBS, and KBS glasses before (dashed line) and after staining at 340°C for 12 h (continuous line)

the LBS glass, except for Al_2O_3, is within this region, phase separation was not observed by the naked eye. This is owing to the suppression effect of Al_2O_3 on phase separation.[11]

Figure 1 shows the transmission spectra of the glasses stained at 340°C for 12 h, compared with those before staining. Since the stained glasses were colourless and had no absorption at longer wavelengths than 400 nm, silver nanoparticles were not formed.[12] The red-shifts observed at the absorption edges were due to absorption assigned mainly to Ag^+.[13]

Table 2 shows the amount of incorporated silver per unit area and the increase in the refractive index at the stained surfaces. The amount of incorporated sodium is also listed for the LBS and KBS glasses. For all the glasses, the amount of incorporated silver

Table 1. Composition, density, glass transition, T_g, and softening temperatures, T_s, and refractive index at 633 nm of the glasses used in the present study

Glass composition (mol%)	Density (g/cm³)	T_g (°C)	T_s (°C)	Refractive index
65SiO₂.20B₂O₃.10Li₂O.5Al₂O₃ (LBS)	2·26	536	627	1·494
65SiO₂.20B₂O₃.10Na₂O.5Al₂O₃ (NBS)	2·32	557	618	1·488
65SiO₂.20B₂O₃.10K₂O.5Al₂O₃ (KBS)	2·29	548	609	1·484

Table 2. Amount of incorporated silver and sodium per unit area (10^{18} atom/cm²) and increase in refractive index at the surface of the glass for various staining conditions

Glass	Staining condition	Temperature (°C)	300	320	320	340
		Time (h)	12	12	24	12
LBS	Amount of incorporated Ag		0·42	0·70	1·42	1·92
	Amount of incorporated Na		1·12	3·02	2·05	2·25
	Increase in refractive index		0·013	0·019	0·027	0·038
NBS	Amount of incorporated Ag		056	0·95	1·39	2·71
	Increase in refractive index		0·012	0·024	0·020	0·035
KBS	Amount of incorporated Ag		010	0·21	0·27	0·65
	Amount of incorporated Na		3·56	4·05	2·28	1·57
	Increase in refractive index		0·002	0·005	0·008	0·016

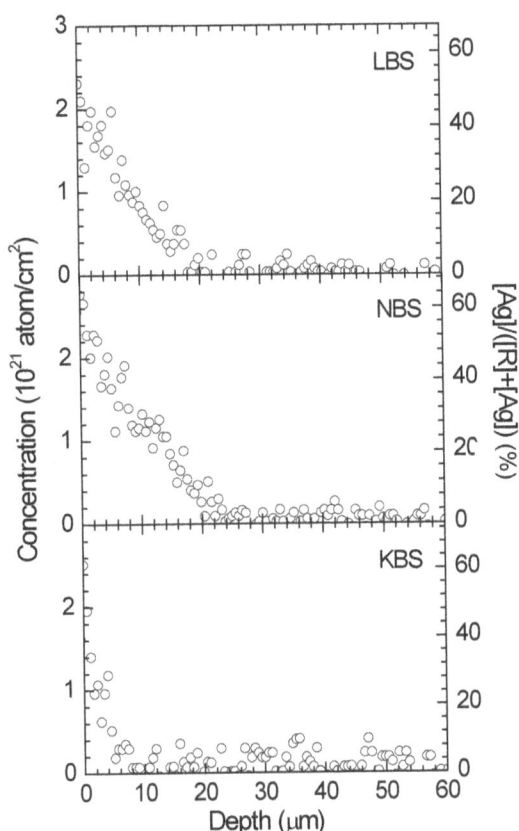

Figure 2. Depth concentration profiles of silver incorporated into the LBS, NBS, and KBS glasses stained at 340°C for 12 h

increases with staining temperature and with staining time at 320°C.

The refractive index increases with the amount of incorporated silver. The increases in refractive index for the LBS and NBS glasses are similar, while that for the KBS glass is less than those of the other glasses.

Figure 2 shows the depth concentration profiles of incorporated silver for the glasses stained at 340°C for 12 h, as obtained from EDX measurements. The diffusion depths for the LBS and NBS glasses are around 20 mm, while that for the KBS glass is only 10 mm.

Discussion

As shown in Table 2, the amounts of incorporated silver for the LBS and NBS glasses are similar, while that for the KBS glass is only one third of those for LBS and NBS. It is well known that the ion exchange process is controlled by the inter diffusion of at least two kinds of ions.[14,15] In the present case, therefore, the diffusion of Ag^+ ions is determined not only by the self diffusion of the Ag^+ but also by the diffusion of the alkaline ions through the inter diffusion coefficients. In particular, the effect becomes stronger when the concentration of Ag^+ increases. Since the diffusion of K^+ ions is slower than those of Li^+ and Na^+, the incorporation of silver into the KBS glass is relatively limited. The amounts of incorporated sodium for the

LBS and KBS glasses are larger than the amount of silver for the same staining condition. This is due to the larger diffusion coefficient of Na^+ than Ag^+.

The changes in the refractive index should be attributed not only to the incorporation of Ag^+, but also to the Na^+ ions contained in the silver stain for the LBS and KBS glasses. Since, however, the refractive index difference between the original glasses is less than 0·01, as shown in Table 1, the increase in the refractive index is mainly attributed to the incorporation of Ag^+ ions which have a large polarisability.

Conclusions

The behaviour and alkali ion dependences of silver incorporation into borosilicate glasses were investigated for silver-staining under various conditions. Silver ions were more easily incorporated into lithium and sodium borosilicate glasses, than for potassium borosilicate glass. The increase in refractive index was ~0·03–0·04 for the lithium and sodium borosilicate glasses after staining at 340°C for 12 h, while it was less than 0·02 for potassium borosilicate glass after the same treatment. The silver ions migrated to a depth of around 20 µm and more than half of the alkaline ions were replaced by silver ions at the glass surfaces for the lithium and sodium borosilicate glasses. On the other hand, the silver diffusion depth was only 10 mm for the potassium borosilicate glass, although the surface concentration of silver was similar to those for the lithium and sodium glasses.

Acknowledgement

This study was performed in the R and D program conducted by Japan Science and Technology Agency (JST). The authors are grateful for financial support from JST.

References

1. Weyl, W. A. *Coloured Glasses*, Society of Glass Technology, Sheffield, 1959, p.409–419 and p.433–435.
2. Jembrih-Simbürger, D., Neelmeijer, C., Schalm, O., Fredrickx, P., Schreiner, M., Vis, K. De., Mäder, M., Schryvers, D., Caen, J. J. *Anal. Atomic Spectrom.*, 2002, **17**, 321.
3. http://www.okuno.co.jp/topindex.html
4. Sakata, S., Ebata, Y. *Osaka Kogyo-gijutsu-shikensho Kihou*, 1955, **6**, 216–223.
5. Rawson, H. *Phys. Chem. Glasses*, 1965, **6**, 81–84.
6. Kadono, K., Suetsugu, T., Ohtani, T., Einishi, T., Tarumi, T. *Jpn. J. Appl. Phys.*, 2006, **45** (2A), 685–688.
7. Zhang, A. Y., Suetsugu, T., Kadono, K. *J. Non-Cryst. Solids*, 2007, **353**, 44–50.
8. Zhang, A. Y., Suetsugu, T., Kadono, K. *J. Ceram. Soc. Jpn.*, 2007, **115** (1), 47–51.
9. Suetsugu, T., Wakasugi, T., Kadono, K. *Mater. Sci. Eng. B*, submitted. Note to author: publication information available yet?
10. Mazurin, O. V., Roskova, G. P., Porai-Koshits, E. A. In: *Phase Separation in Glass*, Eds O. V. Mazurin & E. A. Porai-Koshits, North-Holland Physics Publishing, Amsterdam, 1984, P132.
11. Du, W. F., Kuraoka, K., Akai, T., Yazawa, T. *J. Mater. Sci.*, 2000, **35**, 4865–71.
12. Kreibig, U., v. Fragstein, C. *Z. Physik*, 1969, **224**, 307–23.
13. Borsella, E., Battaglin, G., Garcia, M. A., Gonella, F., Mazzoldi, P., Polloni, R., Quaranta, A. *Appl. Phys. A*, 2000, **71**, 125–32.
14. Doremus, R. H. *J. Phys. Chem.*, 1964, **68** (8), 2212–18.
15. Wakabayashi, H. Ph D thesis, Kyoto University, 1986, P 101–115.

Proc. VI Int. Conf. Borate Glasses, Himeji, Japan, 18–22 August 2008   Phys. Chem. Glasses: Eur. J. Glass Sci. Technol. B, April 2009, **50** (2), 85–88

Conversion kinetics of silicate, borosilicate, and borate bioactive glasses to hydroxyapatite

Steven B. Jung* & Delbert E. Day

Missouri University of Science and Technology, Rolla, Missouri, 65409, USA

Manuscript received 27 August 2008
Revised manuscript received 22 September 2008
Accepted on 6 November 2008

Bioactive 45S5 glass has been studied extensively both in vivo and in vitro, and it is relatively well known that when placed in a phosphate containing solution the glass will react to form the bone-like material hydroxyapatite (HA). In the present work, a kinetic analysis of previously measured weight loss data was done to determine reaction rate constants for four bioactive glasses; one silicate glass, two borosilicate glasses, and one borate glass, via the contracting volume model. The reaction rate increased with increasing B_2O_3 content, with the borate glass reacting nearly five times faster than the silicate 45S5 glass. The three silica containing glasses all deviated from the contracting volume model after approximately 50–70% of the total weight loss; however, when compared to the 3D diffusion model, the normalised data were in good agreement to 100% of the total weight loss. The deviation from the contracting volume model to the slower 3D diffusion model indicates a change in conversion model for the silica containing glasses and can likely be attributed to the formation of a silica rich layer of a certain thickness that began controlling the release of ions from the unreacted glass by diffusion.

1. Introduction

Bioactive glass 45S5 was first reported to bond to bone by Hench in 1971.[1] Ever since, 45S5 has been used extensively in both research and clinical use for the repair of bone and other living tissues. Silicate based bioactive glasses, such as 45S5, and ceramics have served as important biomedical materials due to their ability to bond to surrounding hard and soft tissues and enhance bone formation.[2–9]

Borate glasses have only recently been explored for use in biomedical applications. Richard was the first to investigate replacing SiO_2 with B_2O_3 in the 45S5 glass composition.[10] The borate based 45S5 immersed in a K_2HPO_4 solution at body temperature (37°C) formed a layer of hydroxyapatite (HA), $Ca_{10}(PO_4)_6(OH)_2$, similar to that formed by the silicate based 45S5.[10] The *in vitro* formation of HA from the borate based 45S5 led to further investigation *in vivo*. Particles of the borate based 45S5 glass were placed in a rat tibial defect, and not only promoted bone formation, but did so at a faster rate than the silicate based 45S5 glass.[10]

Possible reaction mechanisms have been described by Huang *et al*[11,12] for the silicate 45S5 glass and the borate analogue of 45S5 (all SiO_2 replaced by B_2O_3). The borate glass fully converted to HA by the glass dissolving, the B_2O_3 and Na_2O going into solution, and the CaO reacting with PO_4^{3-} from the phosphate solution. The silicate glass partially converted to HA, while leaving a sodium depleted core surrounded by a silica rich layer.

According to the results of previous studies,[10–12] the borate based glass most closely follows a contracting volume type of behaviour, where the HA first forms at the outside of the glass particle, and then continually reacts inward toward the centre until completely reacted. The silicate glass forms HA at the outer edge of the particle initially; however it is not clear how the formation of the silica gel layer affects the movement of ions. Therefore, if the silica gel layer becomes the rate controlling mechanism, the weight loss reaction might follow 3D diffusion kinetic behaviour.

The main objective of the following work was to determine if there was a mechanistic difference in the way silicate, borosilicate, and borate bioactive glasses converted to HA. The relative reaction rates of the silicate, borate, and two borosilicate based 45S5 glasses were determined by comparison with the contracting volume model. Finally, the normalised weight loss data was compared to the 3D diffusion model to see if the conversion reaction changed models during the weight loss experiment.

2. Experimental

2.1 Glass preparation and weight loss for the silicate, borate, and borosilicate based 45S5 glasses

The glasses used were based on the bioactive 45S5 glass composition, and the composition of each glass is shown in Table 1. The weight loss data used in this analysis and an in-depth experimental procedure was reported by Huang *et al*.[11,12]

---

* Corresponding author. Email steven.b.jung@gmail.com

Table 1. Nominal composition of glasses used in weight loss experiment[11,12]

Glass	Weight%					Theoretical Weight loss%	Actual Weight loss%	k (CVM)
	Na_2O	CaO	B_2O_3	SiO_2	P_2O_5			
0B	24·5	24·5	0·0	45·0	6·0	56	42	0·0045
1B	24·0	23·9	17·0	29·3	5·8	57	35	0·0065
2B	23·4	23·4	33·1	14·4	5·7	58	42	0·0120
3B	22·9	22·9	48·9	0	5·6	59	57	0·0220

2.2 Kinetic evaluation of the weight loss data

In the previous work by Huang,[11,12] the weight loss for the glass particles was measured for hundreds of hours beyond a measurable change to make sure the experiment has indeed ended, and is shown in Figure 1. The maximum weight losses for the 0B, 1B, 2B, and 3B were 42%, 35%, 42%, and 57% respectively. The weight loss data reported by Huang *et al*[12] were converted to fractional weight change (α) by dividing the measured weight loss values by the maximum weight loss value (Figure 2). The maximum weight loss was determined by the difference of the mass of the starting glass and that of the end product.

The contracting volume model (CVM) is a geometrical model that is used to theoretically describe the reaction rate of a solid sphere of some material and a reactant (gas or liquid). In this experiment, the CVM was used to model the weight loss from the glass particles suspended and the 0·02 M phosphate solution data measured by Huang.[11,12] The values of α versus time for the four glasses were fitted using the CVM shown as Equation (1)[13,14]

$$1-(1-\alpha)^{1/3}=kt \qquad (1)$$

The 3D diffusion model, Equation (2),[14] is a diffusion model that describes the diffusion of material through some rate controlling medium in three dimensions. The 3D diffusion model was compared to the normalised weight loss data just as the CVM to see if a change in the glass conversion model could be detected from the weight loss for any of the four glasses

$$[1-(1-\alpha)^{1/3}]^2=kt \qquad (2)$$

Terms present in Equations (1) and (2) are the normalised weight loss (α), which is calculated by dividing the measured weight loss by the total weight loss for the reaction, rate constant (k), and time (t).

3. Results

3.1 Weight loss of silicate, borate, and borosilicate based 45S5 glasses

The weight loss data measured by Huang *et al*[11,12] for the four glasses tested for the first 1800 h of the reaction is shown in Figure 1. The 0B and 1B glasses reacted at nearly the same rate for the first 200 h while the 2B and 3B glasses reacted significantly faster. At the time the experiment was stopped, the 0B, 1B, and 2B glasses had stopped losing weight, however the

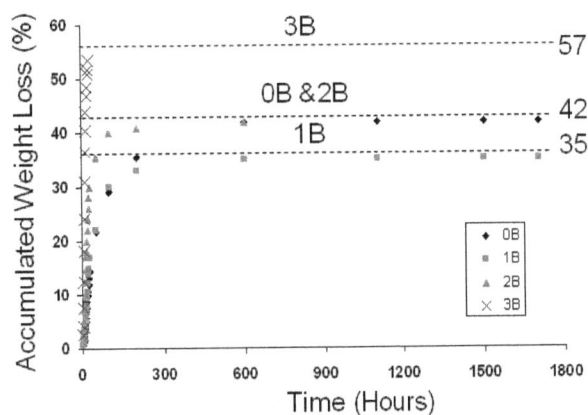

Figure 1. Accumulated weight loss percent[11,12] for the 0B, 1B, 2B, and 3B glasses during the first 1800 h of reaction in 0·02 M K_2HPO_4 solution at 37°C. The dashed horizontal lines denote the maximum weight loss for each glass. The 0B and 2B glasses lost 42%, 1B lost 35%, and 3B lost 57%, respectively

measured weight losses were less than the theoretical weight losses for the three glasses (Table 1). Theoretical weight loss for these glasses was calculated by assuming the end product would be HA, and all the rest of the components (Na_2O, B_2O_3, and SiO_2) would be in solution. Silica rich cores were found at the centre of the 0B, 1B, and 2B glass particles at the end of the experiment, indicating that the silica had not fully dissolved in the phosphate solution.[11] The final (maximum) weight loss of the 3B glass was 57%, which was within 2% of its theoretical limit, and therefore assumed to have gone to completion.

3.2 Isothermal reaction kinetics for the silicate, borate, and borosilicate based 45S5 glasses

The accumulated weight loss percent data shown in Figure 1 were converted to α by dividing the measured weight loss by the maximum weight loss for each glass. The solid lines in Figure 2 show that the normalised weight loss (α) versus time curves for each glass had similar shapes. The k values used to fit Equation (1) to the curves in Figure 2 are listed in Table 1. The k values ranged from as low of 0·0045 for 0B to as high of 0·0220 for the 3B glass, indicating that the 3B glass reacts approximately five times faster than the 0B glass. The CVM model is a good fit for all four glasses for the first 30 h of reaction.

After about 50 h, the normalised weight loss deviated from the CVM model. The reaction rate of the

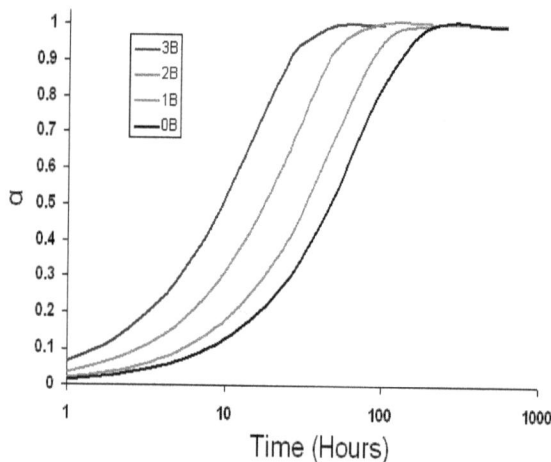

Figure 2. Normalised weight loss (α) versus time for the 0B, 1B, 2B, and 3B glasses[12]

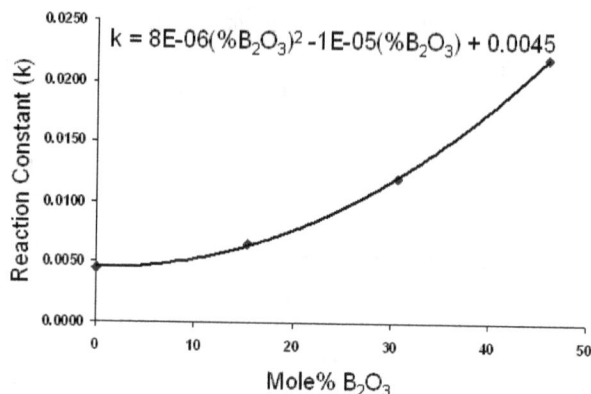

Figure 4. Reaction constant (k) versus mol% B_2O_3 for the 0B, 1B, 2B, and 3B glasses reacted at 37°C in 0·02 M K_2HPO_4 solution

0B, 1B, and 2B particles decreased from that initially described by the CVM. After 50 h, the normalised weight loss data is in good agreement with the 3D diffusion model, indicating a change in reaction model for the three silica containing glasses.

4. Discussion

4.1 Reaction model of the glasses with 0·02 M phosphate solution

The reaction model for the four glasses was determined by plotting α versus time along with fits of Equations (1) and (2) for each glass. A plot with the normalised weight loss data for the four glasses (points) compared to the CVM (solid line) and 3D diffusion equation (dashed line) is shown in Figure 3.

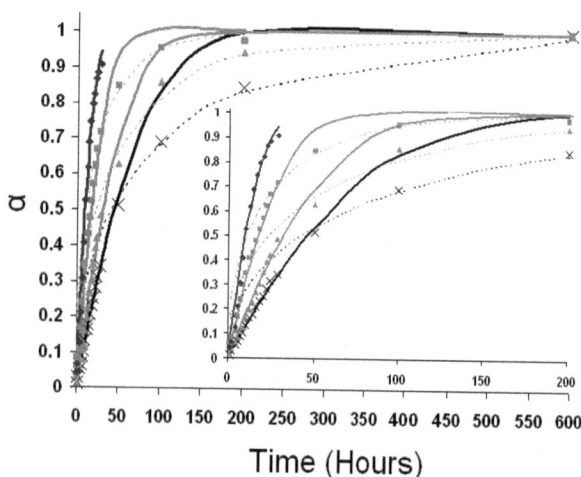

Figure 3. Normalised weight loss (α) versus time for the 0B (crosses), 1B (triangles), 2B (squares), and 3B (diamonds) glasses compared with the Contracting Volume Model (solid lines) and 3D Diffusion Model (dashed lines). The CVM is a good fit of the data for the first 30 h for all four glasses, but after 50 hours, the 3D Diffusion Model is a better fit for the slower reacting 0B, 1B, and 2B glasses (dashed lines)

The first 30 h of data, for all four glasses, were fitted well by the CVM, indicating that the HA formation was due to dissolution of the glass in the phosphate solution.

The normalised weight loss data for the 0B, 1B, and 2B glasses deviate from the CVM after about 50 h. The normalised weight loss after 50 h is well fitted by the 3D diffusion model until the end of the reaction (~600 h). One explanation for this behaviour is that the silica gel layer in the silica containing glasses had reached a thickness such that the movement of ions was controlled by diffusion through the silica gel. The first glass to change from the CVM to the 3D diffusion model, in terms of normalised weight loss (α), was 0B which contained the most silica, and was followed by the 1B and 2B glasses. It is reasonable to suggest that if the change in model was caused by the formation of a silica gel diffusion layer, then the glass containing the most silica (0B) would form a silica gel diffusion barrier first and so on as seen with the 1B and 2B glasses.

4.2 Effect of B_2O_3 content on the reaction rate constant

In previous analysis of the 0B, 1B, 2B, and 3B glasses, the general relationship between higher B_2O_3 content and increased reaction rate was identified,[11] but no mathematical relationship was determined that related the B_2O_3 content of the glasses and the rate of conversion to HA. Figure 4 shows the mol% B_2O_3 plotted versus the rate constant k (Table 1) for each of the glasses. There is a measurable change in k between all four glasses, shown by the solid curved line in Figure 4. This may be a method for accurately determining the effect B_2O_3 is having on the reaction rate constant.

The equation shown in Figure 4 describes the reaction rate constants of the four glasses in the 0·02 M phosphate solution versus the B_2O_3 content. Comparing the phosphate concentration of 0·02 M to the ~0·001 M phosphate concentration in body fluids,[11] the equation would seem irrelevant for practical use as 0·02 M phosphate solution is not a standard solution concentration. The possible importance of the equation shown in Figure 4 comes from finding a relationship between B_2O_3 content and the reaction rate constant k. A similar experiment between the 45S5 glasses and a solution such as simulated body fluid (SBF) could be done to develop a similar equation that is more applicable when trying to model a bioactive glass conversion reaction in a real system, such as in the body.

5. Conclusions

The conversion of the 0B, 1B, 2B, and 3B glasses was described using both the contracting volume model and the 3D diffusion model. The CVM fits the data for the 3B glass over the entire range of α, while the data for the 0B, 1B, and 2B glasses are only fitted well by the CVM for the first 30 h of reaction. The 0B, 1B, and 2B glasses at long times are better fitted by the 3D diffusion model, probably due to the formation of a rate controlling silica gel layer in the glass particles. The relative reaction of the 3B glass was five times faster than the 0B glass, with k values which range from 0·0045 for 0B to 0·0220 for 3B.

The reaction rate constant of 45S5 type bioactive glasses immersed in 0·02 M phosphate solution was shown to depend on the B_2O_3 content, and thus warrants additional work to find a similar relationship for a more common solution such as SBF. Simulated body fluid would more accurately model the *in vivo* environment and could be used to tailor glass compositions for desired *in vivo* reaction rates.

Acknowledgements

The authors would like to thank the National Science Foundation (NSF) (grant #0813608) and the International Materials Institute – New Functionalities in Glass (IMI-NFG) (NSF-DMR-0409588) for their support.

References

1. Hench, L. L., Splinter, R. J., Allen, W. C. & Greenlee, K. *J. Biomed. Mater. Res.* 1991, **2**, 117.
2. Hench, L. L. & Wilson, J. *Handbook of Bioactive Ceramics Vol. 1: Bioactive Glasses and Glass-Ceramics* CRC Press, Boca Raton, FL, 1990, Vol. 1.
3. Hench, L. L., Hench, J. W. & Greenspan, D. C. *J. Aust. Ceram. Soc.*, 2004, **40**, 1–42.
4. Hench, L. L. *J. Mater. Sci.: Mater. Med.*, 2006, **17**, 967–978.
5. Hench, L. L. *Glass Technol.*, 2003, **44**, 1–10.
6. Hench, L. L. & Paschall, H. A. *J. Biomed. Mater. Res.*, 1974, **5** (1), 49–64.
7. Clark, A. E. & Hench, L. L. *J. Biomed. Mater. Res.*, 1976, **10**, 161–174.
8. Boccaccini, A. R., Chen, Q., Lefebvre, L., Gremillard, L. & Chevalier, J. *Faraday Discuss.*, 2007, **136**, 27–44.
9. Kokubo, T., Kushitani, H., Sakka, S., Kitsugi, T. & Yamamuro, T. *J. Biomed. Mater. Res.*, 1990, **24**, 721–734.
10. Richard, M. N. C. MS Thesis, University of Missouri-Rolla, 2000.
11. Huang, W., Day, D. E., Kittraratanapiboon, K. & Rahaman, M. N. *J. Mater. Sci: Mater. Med.*, 2006, **17**, 583–596.
12. Huang, W., Rahaman, M. N., Day, D. E. & Li, Y. *Phys. Chem. Glasses: Eur. J. Glass Sci. Technol. B*, 2006, **47**, 1–12.
13. Bamford, C. H. & Tipper, C. F. R. *Comprehensive Chemical Kinetics, Vol. 22. Reaction is the Solid State*, Elsevier Scientific Publishing Company, New York 1980.
14. Rahaman, M. N. *Ceramic Processing and Sintering – 2nd edition* Marcel Dekker Inc. New York, New York, 2003.

Proc. VI Int. Conf. Borate Glasses, Himeji, Japan, 18–22 August 2008   *Glass Technol.: Eur. J. Glass Sci. Technol. A, December 2009, 50 (6), 319–322*

Fabrication of OCT light source by Bi_2O_3–B_2O_3 glass phosphor doped with Yb^{3+} and Nd^{3+}

Shingo Fuchi,* Ayako Sakano, Ryota Mizutani & Yoshikazu Takeda

Department of Crystalline Materials Science, Graduate School of Engineering, Nagoya University, Furo-cho, Chikusa-ku, Nagoya 464-8603, Japan

Manuscript received 18 August 2008
Revised version received 20 October 2008
Accepted 12 November 2008

We investigated the spectroscopic properties of Bi_2O_3–B_2O_3 based glasses doped with Yb^{3+} and Nd^{3+} in order to find appropriate phosphors for an optical coherence tomography (OCT) light source. We successfully obtained a wideband emission from $3{\cdot}0Yb_2O_3$–$4{\cdot}0Nd_2O_3$–$46{\cdot}0Bi_2O_3$–$46{\cdot}0B_2O_3$–$1{\cdot}0Sb_2O_3$ (nominal molar composition) with a full width at half maximum of 85 nm in the wavelength region around 1000 nm. We achieved a maximum output power of 78 μW (luminescence efficiency of 1·5%) using a green light emitting diode (LED) as a photoexcitation source. The interference signal was measured by using a Michelson interferometer, and a coherence length of 5·4 μm was achieved. This value indicates that the depth resolution of the OCT with this light is twice that of conventional LEDs and super luminescent diodes. Therefore, we conclude that Bi_2O_3–B_2O_3 based glass doped with Yb^{3+} and Nd^{3+} can be a new type of OCT light source.

1. Introduction

Wideband light sources may be used in fibre gyroscopes[1,2] and for optical coherence tomography (OCT), a novel cross sectional imaging technique for biological tissues.[3] Since OCT is based on the Michelson interferometer, a low coherence, wideband light source has several advantages for a high depth resolution. The depth resolution Δz, which is equal to the coherence length in the Michelson interferometer, is calculated using

$$\Delta z = \frac{2\ln 2}{\pi}\frac{\lambda_c^2}{\Delta\lambda} \tag{1}$$

where λ_c is the central wavelength and $\Delta\lambda$ is the full width at half maximum (FWHM) of the Gaussian shaped spectrum. It is obvious from this equation that a shorter central wavelength results in a higher depth resolution. However, the central wavelength is limited to the near infrared region in current applications, leading to penetration depths in biological tissues larger than that at other wavelengths.[3] The water absorption profile has a local minimum at ~1060 nm, and a recent study has shown that imaging biological tissue at a central wavelength around 1000 nm yields minimal OCT axial resolution degradation resulting from water dispersion.[4] Therefore, the central wavelength λ_c should be ~1000 nm. In addition, in order to achieve a higher depth resolution, a wider spectral width is required.

Super luminescent diodes (SLDs) and light emitting diodes (LEDs) in the near infrared region are usually used in the OCT. The spectral width of SLDs and LEDs is approximately 50 nm, corresponding to a coherence length around 10 μm. These light sources are simple and not very expensive. However, this coherence length of around 10 μm is insufficient for some medical applications. Recently, using a super continuum laser, a coherence length less than 10 μm has been achieved. [3] However, this light source is complex, quite expensive, and difficult to use in practical applications in a hospital setting. Zhang *et al* have reported a method for the synthesis of several LEDs to shorten the coherence length.[5] However, synthesis of several LEDs and control of the light intensity of individual LEDs are complicated for practical use.

Therefore, we propose a new type of light source for OCT by combining an LED and a near-infrared emitting phosphor in one package.[6] Since this light source is similar to a white LED, composed of a blue LED and a yellow phosphor, it is simple and useful for practical applications.

To realise a new light source, a phosphor that emits light at around 1000 nm and having a large FWHM Gaussian shaped spectrum is needed. It is well known that the luminescence of Yb^{3+} and Nd^{3+} is located near 1000 nm. Thus, Yb^{3+} and Nd^{3+} were used for the luminescence centre. However, in general, rare earth ions show very sharp luminescences, and therefore a glass was used as a matrix for broadening the luminescence of the rare earth ions. In particular, it has been reported that a high refractive index glass is effective for broadening the luminescence spectrum.[7] In this study, from amongst the various high refractive glasses, we chose a Bi_2O_3–B_2O_3 based glass, because it is of great interest for optical device fabrication due to its low melting temperature, and because of its wide glass formation range in composition.

In this paper, we report the luminescence properties of Bi_2O_3–B_2O_3 based glasses doped with Yb^{3+} and Nd^{3+}. Then, we demonstrate the interference signal.

---

* Corresponding author. Email fuchi@mercury.numse.nagoya-u.ac.jp

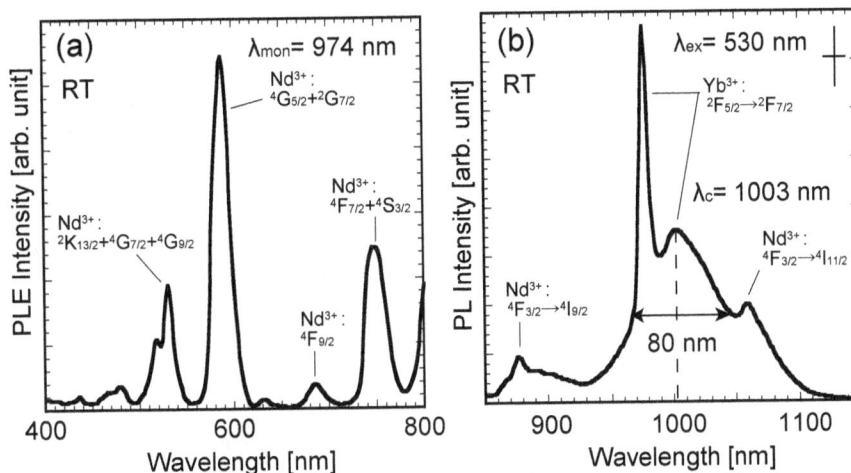

Figure 1. (a) The PLE spectrum monitored at 974 nm, and (b) the PL spectrum, excited by a green LED (central wavelength of 530 nm), of $1{\cdot}0Yb_2O_3$–$1{\cdot}0Nd_2O_3$–$48{\cdot}5Bi_2O_3$–$48{\cdot}5B_2O_3$–$1{\cdot}0Sb_2O_3$ glass phosphor. There are many excitation bands of Nd^{3+} ions in (a). The luminescence of Nd^{3+} ions and Yb^{3+} ions is observed in (b)

2. Experimental

The samples used in this study were synthesised by melt quenching. Powders of Yb_2O_3, Nd_2O_3, Bi_2O_3, H_3BO_3, and Sb_2O_3 were mixed in nominal molar compositions of xYb_2O_3–yNd_2O_3–Sb_2O_3–$(99{-}x{-}y)(50{\cdot}0Bi_2O_3$–$50{\cdot}0B_2O_3)$. The mixed powders were melted at 1250°C in an Al_2O_3 crucible in an electric furnace. Sb_2O_3 was used to suppress the reduction of Bi^{3+}. After waiting for 10 min, the molten liquid was poured between two stainless steel mould plates kept at room temperature for the formation of the glass. We did not observe melting of the Al_2O_3 crucible.

Photoluminescence (PL) measurements and photoluminescence excitation (PLE) measurements were carried out at room temperature. The photoexcitation source for the PL was a green LED (central wavelength of 530 nm). An InGaAs photomultiplier mounted on a 0·20 m grating monochromator was adopted for the lock-in detection. Interference signals were measured by a Michelson interferometer. An InGaAs photodiode was used for the detection of the interference signals.

3. Results

Figure 1 shows (a) the PLE and (b) the PL spectra of $1{\cdot}0Yb_2O_3$–$1{\cdot}0Nd_2O_3$–$48{\cdot}5Bi_2O_3$–$48{\cdot}5B_2O_3$–$1{\cdot}0Sb_2O_3$ glass phosphor. There are many excitation bands in Figure 1(a). These excitation bands correspond to the Nd^{3+} transitions $^2K_{13/2}{+}^4G_{7/2}{+}^4G_{9/2}$, $^4G_{5/2}{+}^2G_{7/2}$, $^4F_{9/2}$, $^4F_{7/2}{+}^4S_{3/2}{\leftarrow}^4I_{9/2}$, as indicated in Figure 1(a). The assignments of these transitions were made according to Ref. 8. From Figure 1(a), it is considered that the green region ($^2K_{13/2}{+}^4G_{7/2}{+}^4G_{9/2}{\leftarrow}^4I_{9/2}$) is one of the candidates for the photoexcitation wavelength. We therefore decided to use a green LED as the photoexcitation source in this study.

In Figure 1(b), luminescence peaks at 876 nm, 976 nm, 1003 nm, and 1059 nm are observed. These peaks correspond to the $^4F_{3/2}$ to $^4I_{9/2}$ (~876 nm) and $^4F_{3/2}$ to $^4I_{11/2}$ (~1059 nm) transitions of Nd^{3+} ions, and the $^2F_{5/2}$ to $^2F_{7/2}$ (~976 nm and ~1003 nm) transitions of Yb^{3+} ions, respectively. Although there is no Yb^{3+} excitation band around 530 nm, luminescence of Yb^{3+} is clearly observed in Figure 1(b). This is due to the energy transfer from Nd^{3+} ions to Yb^{3+} ions.[9] Therefore, increase of Nd_2O_3 concentration is effective for increasing the luminescence of both Yb^{3+} ions and Nd^{3+} ions.

The FWHM of the broader luminescence (peak at 1003 nm) is 80 nm. This value corresponds to a depth resolution of 5·5 µm, which is two times better than that of conventional SLDs and LEDs. However, the sharp luminescence peaked at 976 nm causes side lobes of the coherence function. Therefore, we need to control the luminescence spectrum of Bi_2O_3–B_2O_3 based glasses doped with Yb^{3+} and Nd^{3+} in order to eliminate the sharp luminescence at 976 nm.

It is known that the re-absorption of Yb^{3+} influences the luminescence spectrum of Yb^{3+}.[10] In this mechanism, the sharp luminescence is re-absorbed, and other luminescence is not absorbed. Therefore, it is clear that increasing the re-absorption causes a decrease in the sharp luminescence. Thus, we can expect that increasing the Yb_2O_3 concentration is effective for decreasing the sharp luminescence.

Following these considerations, we investigated the dependence of the luminescence properties on Yb_2O_3 and Nd_2O_3 concentrations. Figure 2 shows the dependence of integrated PL intensity on Yb_2O_3 and Nd_2O_3 concentrations. The integrated PL intensity increases as Yb_2O_3 and Nd_2O_3 concentrations increases. However, concentration quenching occurs when Yb_2O_3 increases from 3·0 to 4·0 mol% or Nd_2O_3 increases from 4·0 to 5·0 mol%. As a result, the maximum integrated PL intensity is achieved at 3·0 mol% Yb_2O_3 and 4·0 mol% Nd_2O_3.

Figure 3 shows the dependence of the output power on the excitation power. The output power increases linearly as the excitation power increases, as shown in Figure 3, and the luminescence efficiency

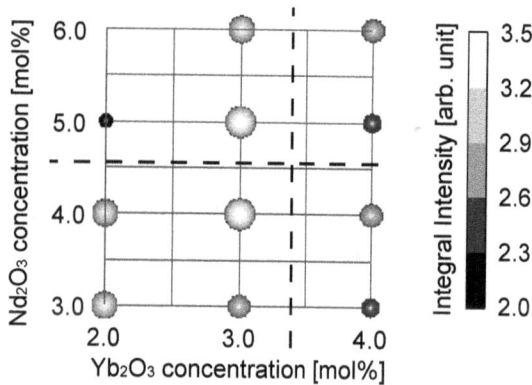

Figure 2. *Dependence of integrated PL intensity on Yb_2O_3 and Nd_2O_3 concentrations. When the Yb_2O_3 concentration increases from 3·0 to 4·0 mol% or the Nd_2O_3 concentration increases from 4·0 to 5·0 mol%, the integrated PL intensity decreases*

of 1·5% is obtained from the slope. We achieved a maximum output power of 78 μW, which is approximately 1/10 of the expected value for practical use.

Figure 4 shows the PL spectrum of $3·0Yb_2O_3$–$4·0Nd_2O_3$–$46·0Bi_2O_3$–$46·0B_2O_3$–$1·0Sb_2O_3$. Luminescence peaks at 976 nm and 1006 nm are observed. In Figure 4, it is obvious that the sharp luminescence peak at 976 nm decreased with increasing Yb_2O_3 concentration (cf. Figure 1(b)). Therefore, the side lobes of $3·0Yb_2O_3$–$4·0Nd_2O_3$–$46·0Bi_2O_3$–$46·0B_2O_3$–$1·0Sb_2O_3$ are expected to be small compared to that of $1·0Yb_2O_3$–$1·0Nd_2O_3$–$48·5Bi_2O_3$–$48·5B_2O_3$–$1·0Sb_2O_3$. The FWHM of the broader luminescence peaked at 1006 nm is 85 nm, which corresponds to a depth resolution of 5·3 μm.

Figure 5 shows the interference signal of $3·0Yb_2O_3$–$4·0Nd_2O_3$–$46·0Bi_2O_3$–$46·0B_2O_3$–$1·0Sb_2O_3$ measured by the Michelson interferometer. In Figure 5, the central packet and side lobes are observed. The FWHM of the central packet is 5·4 μm. The side lobes are due to the sharp luminescence peaked at 976 nm of the PL spectrum shown in Figure 4. This interferometer measurement result suggests that our glass phosphor improves the depth resolution by a factor of about two compared with conventional SLDs and LEDs.

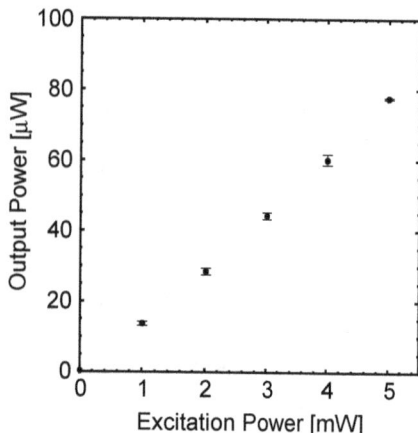

Figure 4. *The PL spectrum of $3·0Yb_2O_3$–$46·0Bi_2O_3$–$46·0B_2O_3$–$1·0Sb_2O_3$. The sharp luminescence peak at 976 nm is decreased due to the increased Yb_2O_3 concentration (cf. Figure 1(b))*

4. Discussion

In the Michelson interferometer, the signal light $u(t)$ and reference light $u(t+\tau)$ interfere. Here, τ is the time difference. Thus, the output of the interferometer $I(\tau)$ is described as

$$I(\tau)=|u(t)+u(t+\tau)|^2$$
$$=|u(t)|^2+|u(t+\tau)|^2+u^*(t)u(t+\tau)+u(t)u^*(t+\tau) \qquad (2)$$

In Equation (2), $|u(t)|^2$ and $|u(t+\tau)|^2$ are the mean (DC) intensities, and the third and fourth terms represent the amplitude of the interference signal. In OCT, the alternating component (AC) is important. Therefore, we focus on the third and fourth terms of Equation (2).

In a glass phosphor, there are many luminescence centres, and individual luminescence centres show a very sharp luminescence. Therefore, (1) we consider the AC of a synthesised luminescence centre (SLC), and then (2) we calculate the AC of an individual luminescence centre. If we use several light sources simultaneously, the light from different sources does not interfere with each other. Then, the AC of

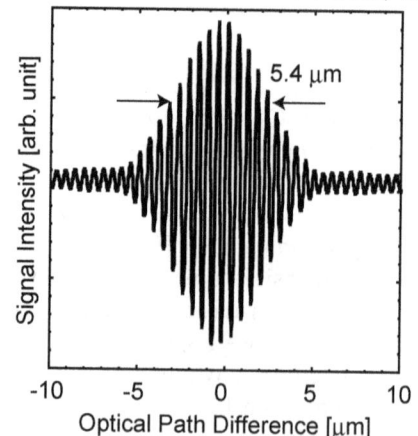

Figure 3. *Dependence of the output power on the excitation power. The output power increases linearly as the excitation power increases*

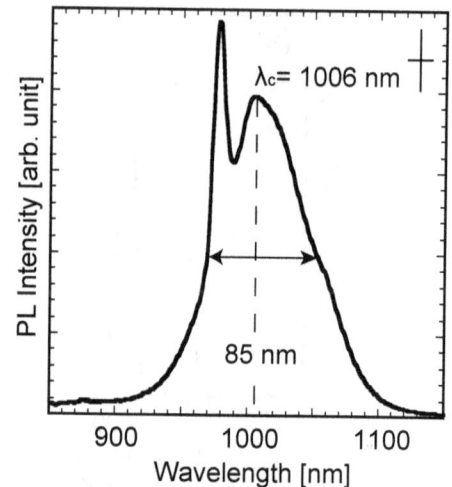

Figure 5. *The measured interference signal of $3·0Yb_2O_3$–$4·0Nd_2O_3$–$46·0Bi_2O_3$–$46·0B_2O_3$–$1·0Sb_2O_3$. The FWHM of the central packet is 5·4 μm*

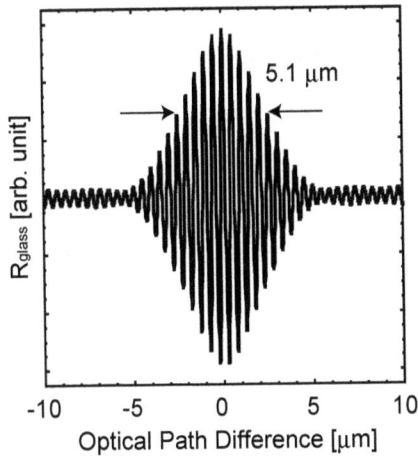

Figure 6. The $R_{glass}(\tau)$ calculated using Equation (6) and the PL spectrum of $3 \cdot 0Yb_2O_3 – 4 \cdot 0Nd_2O_3 – 46 \cdot 0Bi_2O_3 – 46 \cdot 0B_2O_3 – 1 \cdot 0Sb_2O_3$ shown in Figure 4. The FWHM of the central packet is $5 \cdot 1$ μm. The calculation and measurement show a good agreement

the coherence function of synthesised luminescence centres, $R_{SLC}(t)$, has the following form

$$R_{SLC}(\tau) = \sum_n \left\{ u_n^*(t)u_n(t+\tau) + u_n(t)u_n^*(t+\tau) \right\} \quad (3)$$

where n is the total number of luminescence centres.[11] Since the individual luminescence centres generally show very sharp luminescence, we assume that the optical wave of one luminescence centre is represented by

$$u_n(t) = \begin{cases} \exp(i\omega_n t) & 0 \leq t \leq \tau_c \\ 0 & t < 0, \tau_c < t \end{cases} \quad (4)$$

where ω is the optical frequency, and τ_c is the coherence time. Therefore, autocorrelation function of one luminescence centre, $R_n(\tau)$, is described as

$$R_n(\tau) = \int_0^{\tau_c} \left\{ \begin{array}{l} \exp(-i\omega_n t)\exp(i\omega_n(t+\tau)) \\ +\exp(i\omega_n t)\exp(-i\omega_n(t+\tau)) \end{array} \right\} dt \quad (5)$$

$$= 2(\tau_c - \tau)\exp(-i\omega_n\tau)$$

From Equations (3) and (5), when the PL intensity of the sample is represented by $P(\omega_n)$, the AC of coherence function of the sample, $R_{glass}(\tau)$, is represented by

$$R_{glass}(\tau) \propto (\tau_c - \tau)\sum_n P(\omega_n)\exp(-i\omega_n\tau) \quad (6)$$

Therefore, by using Equation (6) and the PL spectrum in Figure 4, we can calculate $R_{glass}(\tau)$.

Figure 6 shows the calculated $R_{glass}(\tau)$ of $3 \cdot 0Yb_2O_3 – 4 \cdot 0Nd_2O_3 – 46 \cdot 0Bi_2O_3 – 46 \cdot 0B_2O_3 – 1 \cdot 0Sb_2O_3$. Note that in Figure 6 the horizontal axis shows the optical path difference, $c\tau$, where c is the speed of light. In Figure 6, the central packet and side lobes are observed. The FWHM of the central packet is $5 \cdot 1$ μm, which is almost the same as the calculation result from the PL spectrum and Equation (1). The side lobes are due to the sharp luminescence peak at 976 nm in the PL spectrum shown in Figure 4. The FWHM of

the central packet, the relative intensity between the central packet and the side lobes, and the shape of the side lobes show a good agreement between the calculation result and the measurement. The FWHM of the central packet for the measurement is $5 \cdot 4$ μm, whilst the calculation gives $5 \cdot 1$ μm. This difference arises because the calculation model, Equation (4), ignores the homogeneous line width, leading to a narrower FWHM.

4. Conclusion

We have investigated the luminescence properties and interference signal of $Bi_2O_3 – B_2O_3$ based glasses doped with Yb^{3+} and Nd^{3+} in order to find appropriate phosphors for an OCT light source. Luminescence peaks of Nd^{3+} ions at 876 nm and 1059 nm, and of Yb^{3+} ions at 976 nm and 1003 nm, are observed when Nd^{3+} ions were excited using a green LED. This is due to the energy transfer from Nd^{3+} ions to Yb^{3+} ions. To increase luminescence intensity and to decrease the sharp luminescence peaked at 976 nm, we studied the dependence of luminescence properties on Yb_2O_3 and Nd_2O_3 concentrations. In this way we have successfully obtained a wideband emission with a FWHM of 85 nm from a glass with nominal composition $3 \cdot 0Yb_2O_3 – 4 \cdot 0Nd_2O_3 – 46 \cdot 0Bi_2O_3 – 46 \cdot 0B_2O_3 – 1 \cdot 0Sb_2O_3$. We achieved a luminescence efficiency of $1 \cdot 5\%$. We conducted both calculation and measurement of the interference signal of the same phosphor, and obtained a good agreement. A coherence length of $5 \cdot 4$ μm was achieved. This value should improve the depth resolution of OCT by a factor of about two compared with conventional SLDs and LEDs.

Acknowledgements

We would like to thank Prof. Nakamura, Assistant Prof. Unuma, and Mr Hirabayashi at Nagoya University for the Michelson interference measurement, PLE measurement, and fruitful discussions. This work was partly supported by the Nippon Sheet Glass Foundation for Materials Science and Engineering.

References

1. Cutler, C. C., Newton, S. A. & Shaw, H. J. *Opt. Lett.*, 1980, **5**, 488.
2. Blin, S., Kim, H. K., Digonnet, M. J. F. & Kino, G. S. *J. Lightwave Technol*, 2007, **25**, 861.
3. Fercher, A. F., Drexler, W., Hitzenberger, C. K. & Lasser, T. *Rep. Prog. Phys.*, 2003, **66**, 239.
4. Wang, Y., Nelson, J., Chen, Z., Reiser, B., Chuck, R. & Windeler, R. *Opt. Express*, 2003, **11**, 1411.
5. Zhang, Y., Sato, M. & Tanno, N. *Opt. Lett.*, 2001, **26**, 205.
6. Fuchi, S., Sakano, A. & Takeda, Y. *Jpn. J. Appl. Phys.*, (accepted for publication).
7. Tanabe, S., Sugimoto, N., Ito, S. & Hanada, T. *J. Lumin.*, 2000, **87–89**, 670.
8. Chen, Y., Huang, Y., Huang, M., Chen, R & Luo, Z. *J. Am. Ceram. Soc.*, 2005, **88**, 19.
9. Batalioto, F., de Sousa, D. F., Bell, M. J. V., Lebullenger, R., Hernandes, A. C. & Nunes, L. A. O. *J. Non-Cryst. Solids*, 2000, **273**, 233.
10. Chen, Y., Huand, Y. & Luo, Z. *Chem. Phys. Lett.*, 2003, **382**, 481.
11. Sato, M., Wakaki, I., Watanabe, Y. & Tanno, N. *Appl. Opt.*, 2005, **44**, 2471.

Proc. VI Int. Conf. Borate Glasses, Himeji, Japan, 18–22 August 2008   *Phys. Chem. Glasses: Eur. J. Glass Sci. Technol. B, June 2009, **50** (3), 189–194*

Borate glasses and glass-ceramics for near infrared luminescence

Joanna Pisarska[1]* & Wojciech A. Pisarski*[2]

[1] *Silesian University of Technology, Department of Materials Science, Krasińskiego 8, 40-019 Katowice, Poland*
[2] *University of Silesia, Institute of Chemistry, Szkolna 9, 40-007 Katowice, Poland*

Manuscript received 18 August 2008
Revised version received 30 October 2008
Accepted 26 March 2009

Near infrared luminescence of Nd^{3+} and Er^{3+} ions in lead borate glasses and transparent glass-ceramics was examined. Selected radiative parameters for rare earth ions in B_2O_3–PbO based glasses were experimentally obtained and theoretically calculated from the Judd–Ofelt framework. Luminescence spectra at 1·06 mm and 1·5 mm due to main $^4F_{3/2}$–$^4I_{11/2}$ and $^4I_{13/2}$–$^4I_{15/2}$ laser transitions of Nd^{3+} and Er^{3+} in lead borate glasses containing PbX_2 (X=F, Cl, Br) and thermally treated samples were also analysed. In both cases, the spectral linewidth and the luminescence decays from excited levels for Er^{3+} differ significantly from those for Nd^{3+} ions.

1. Introduction

Borate glasses with heavy metal oxide and/or halide components such as PbO and PbX_2 (where X denotes F, Cl or Br), have attracted a great deal of interest due to their important physicochemical properties and unique structural behaviour. The structure and properties of lead borate glasses have been studied using theoretical calculations and different experimental techniques.[1–6] The incorporation of PbO and/or PbX_2 in borate glasses results in the creation of tetrahedral BO_4 units as well as a significant increase in their IR transparencies, refractive indices and optical nonlinearities. Moreover, lead borate glass creates an excellent host to incorporate optically active ions. The optical characterisation of lead borate glasses containing several different transition metal or rare earth ions has recently been performed. Quite efficient visible luminescence for Mn^{2+}, Ni^{2+}, Co^{2+},[7] Eu^{3+} and Tb^{3+},[8] Sm^{3+},[9] Ho^{3+},[10] Pr^{3+} and Tm^{3+}[11] ions in oxide lead borate glasses has been found.

Rare earth doped oxide and oxyfluoride glasses based on B_2O_3–PbO and B_2O_3–PbO–PbF_2 have been examined mainly for two reasons. Firstly, lead borate glasses present interesting spectroscopic properties in relation to practical application as tuneable solid state lasers and optical amplifiers. In comparison to other glass systems, Er-doped B_2O_3–PbO glass presents the largest absorption and stimulated emission $\Delta\lambda_{eff}$ in the 1·5 mm transmission window.[12] The broad linewidth spectra for the main $^4I_{13/2}$–$^4I_{15/2}$ (Er^{3+}) near infrared (NIR) laser transition and high fluorescence band efficiency is fundamental for tuneable solid state lasers. Several radiative parameters of rare earth ions, such as the measured luminescence lifetime,

the quantum efficiency of the excited state or the stimulated emission cross section, are considerably higher for B_2O_3–PbF_2 glass than for B_2O_3–PbO glass. For example, the total replacement of PbO by PbF_2 in borate glass results in a two-fold increase in the $^4I_{13/2}$ luminescence lifetime of Er^{3+} ions from 0·4 to 0·82 ms.[6] Secondly, lead borate glasses can be successfully used in the preparation of oxide and oxyfluoride transparent glass-ceramic (TGC) systems. Crystalline phases, such as $PbMnO_4$ in B_2O_3–PbO[13] and cubic b-PbF_2 in B_2O_3–PbF_2–CdF_2,[14] have been identified.

The research in the literature has focused on borate glasses and glass-ceramics singly doped with Nd^{3+} and Er^{3+}. These rare earths are amongst the most widely studied luminescent ions in optical systems, due to the $^4F_{3/2}$–$^4I_{11/2}$ (λ=1·06 µm) and $^4I_{13/2}$–$^4I_{15/2}$ (λ=1·5 µm) laser transitions of Nd^{3+} and Er^{3+}, respectively.

The first part of this paper is concerned with NIR luminescence of rare earths in borate glasses based on B_2O_3–PbO and B_2O_3–PbO–PbX_2 (X=F, Cl or Br). The influence of PbX_2 content on several spectroscopic parameters for Ln^{3+} (where Ln is Er^{3+} or Nd^{3+}) is reported. The emission of Er^{3+} has previously been studied in oxyhalide tellurite glass based on TeO_2–PbO–PbX_2,[15] but has not yet been studied in the B_2O_3–PbO–PbX_2 system.

The second part of this paper contains results for rare earth doped B_2O_3–PbO–PbF_2 glass after annealing. During controlled heat treatment, transparent glass-ceramics can be obtained, in which lead fluoride (PbF_2) particles are dispersed into the borate glass matrix. The transformation from glass to glass-ceramic causes changes in the spectroscopic properties such as a narrowing of spectral lines and elongation in lifetimes of luminescent states.

*Corresponding author. Email Joanna.Pisarska@polsl.pl

2. Experimental

Multicomponent oxide and oxyhalide lead borate based glasses were prepared. The nominal chemical composition of lead borate glass was as follows (in wt%): $18B_2O_3.yPbX_2.(72-y)PbO.6Al_2O_3.3WO_3.1Ln_2O_3$, where X=F, Cl or Br; y=0, 9, 36; Ln=Nd or Er). The small amount of Al_2O_3 increases glass stability, whereas the addition of a low WO_3 content is promising for luminescence, as observed by optical measurements in thulium doped TeO_2–WO_3 systems.[16,17]

Anhydrous oxides and lead halide PbX_2 (99·99% purity, Aldrich) were used as starting materials. To prepare samples, the appropriate amounts of all components were mixed homogeneously together and heated in an atmosphere of dry argon. The mixed reagents were melted at 850°C for 2 h in Pt crucibles, then poured into preheated copper moulds and annealed below the glass transition temperature. After this procedure, the samples were slowly cooled to room temperature. Characteristic temperatures, such as the glass transition temperature and the crystallisation onset, were determined by use of a Perkin Elmer differential scanning calorimeter. The DSC curves were measured with a heating rate of 10°C/min. The x-ray diffraction analysis was carried out using an INEL diffractometer with Cu K_α radiation. Emission spectra were measured using a Continuum Model Surelite I optical parametric oscillator, pumped by a third harmonic of a Nd:YAG laser. Luminescence was dispersed by a 1 metre double grating monochromator and detected with a photomultiplier with S-20 spectral response. Emission spectra were collected using a Stanford SRS 250 boxcar integrator with accuracy of ±0·2 nm. Luminescence decay curves were recorded and stored by a Tektronix TDS 3052 oscilloscope with an accuracy of ±2 ms. All measurements were carried out at room temperature.

3. Results

3.1 Oxide glasses based on B_2O_3–PbO

Figure 1 presents NIR luminescence spectra recorded for Nd^{3+} (a) and Er^{3+} (b) ions in a $B_2O_3.PbO.Al_2O_3.WO_3$ glass. The spectra were measured under excitation by 514 nm (Nd-doped samples) and 488 nm (Er-doped samples) laser lines. Luminescence bands correspond to $^4F_{3/2}$–$^4I_{J/2}$ (J=9, 11, 13) and $^4I_{13/2}$–$^4I_{15/2}$ transitions of Nd^{3+} and Er^{3+} ions, respectively.

The radiative transition probability A_J for excited levels of rare earth ions from an initial state J to a final ground state J' is

$$A_J = \frac{64\pi^4 e^2}{3h(2J+1)\lambda^3} \frac{n(n^2+2)^2}{9} \sum_{t=2,4,6} \Omega_t (<4f^N J\|U^t\|4f^N J'>)^2$$

(1)

where h is Planck's constant and λ is the mean wavelength of the transition. In the analysis, a constant value of 1·92 was used for n, the refractive index of

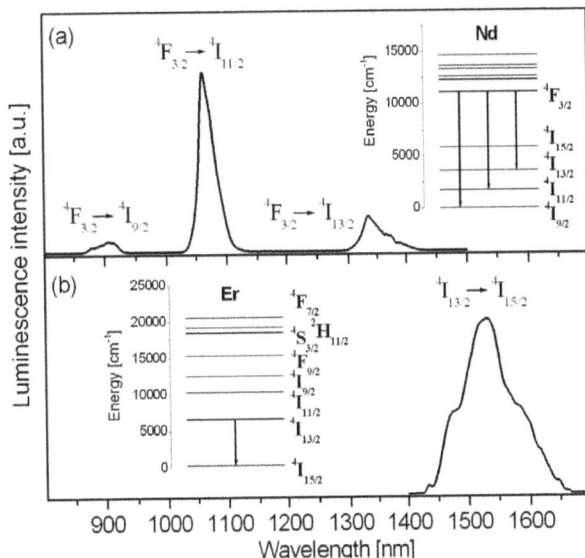

Figure 1. NIR luminescence spectra for Nd^{3+} (a) and Er^{3+} (b) ions in oxide lead borate glasses, under excitation by 514 nm (a) and 488 nm (b) laser lines. The glass composition is $18B_2O_3.72PbO.6Al_2O_3.3WO_3.1Ln_2O_3$, where Ln=Nd or Er. All transitions on the energy level schemes are also indicated

the medium, $\|U_t\|^2$ represents the square of the matrix elements of the unit tensor operator U^t connecting the initial and final states. The squared reduced matrix elements $\|U_t\|^2$ were taken from Ref. 18. The Ω_t (t=2, 4, 6) phenomenological intensity parameters were obtained from the absorption measurements[19,20] and Judd–Ofelt calculations.[21,22] The three Judd–Ofelt intensity parameters for rare earth ions in lead borate glasses are found to be; Ω_2=3·41±0·16, Ω_4=2·88±0·25 and Ω_6=4·03±0·10 (for Nd^{3+}) and Ω_2=3·31±0·26, Ω_4=1·63±0·28, Ω_6=1·29±0·10 (for Er^{3+}) in 10^{-20} cm^2 units.

The total radiative emission probability A_T involving all the intermediate terms is given by the sum of the A_J terms calculated over all terminal states. Thus, the radiative lifetime, τ_{rad}, of an excited level is given by the inverse of the total radiative emission probability

$$\tau_{rad} = \frac{1}{\sum_i A_{Ji}} = \frac{1}{A_T}$$

(2)

The quantum efficiency of excited state η is defined by

$$\eta = \frac{\tau_m}{\tau_{rad}} \times 100\%$$

(3)

where τ_m and τ_{rad} denote measured lifetime and calculated radiative lifetime, respectively. The peak stimulated emission cross-section, σ_{em}, was obtained from the calculated radiative transition probability, A_J, using the following equation

$$\sigma_{em} = \frac{\lambda_p^4}{8\pi c n^2 \Delta\lambda} A_J$$

(4)

where λ_p is the peak emission wavelength, c is the

Table 1. *Spectroscopic parameters for Nd^{3+} and Er^{3+} ions in oxide lead borate glasses*

Spectroscopic parameters	Ln^{3+}	
	Nd^{3+}	Er^{3+}
Transition	$^4F_{3/2}-^4I_{11/2}$	$^4I_{13/2}-^4I_{15/2}$
Peak emission wavelength, λ_P (nm)	1061	1531
Spectral linewidth, $\Delta\lambda$ (nm)	28	100·5
Radiative transition probability,[*] A_J (s⁻¹)	2197	225
Radiative lifetime,[*] τ_{rad} (μs)	233	4400
Measured lifetime, τ_m (μs)	86	400
Quantum efficiency, η (%)	37	10
Peak stimulated emission cross-section, σ_{em} (10^{-20} cm²)	3·58	0·47

[*] obtained from the Judd–Ofelt calculations

velocity of light, n is the refractive index, and $\Delta\lambda$ is the effective linewidth, defined by the full width at half maximum (FWHM).

The radiative parameters for Nd^{3+} and Er^{3+} ions in lead borate glasses were calculated using Equations (1)–(4). The results for the $^4F_{3/2}-^4I_{11/2}$ (λ_p=1·06 mm) and $^4I_{13/2}-^4I_{15/2}$ (λ_p=1·5 mm) main NIR laser transitions of Nd^{3+} and Er^{3+} are given in Table 1.

3.2 Oxyhalide glasses based on B_2O_3–PbO–PbX_2 (X=F, Cl, Br)

Lead halide PbX_2 (X=F, Cl or Br) was introduced to the base B_2O_3–PbO glass matrix. The actual content of X ions has not been estimated. However, the final glass compositions are somewhat different from the nominal starting ones, because of the loss of gas during the melting process, due to the following reaction: $PbX_2 + H_2O \rightarrow PbO + 2HX\uparrow$. Analytical studies of other oxyhalide glasses has shown that the fluorine[23] or chlorine[24] losses can be quite large.

Figure 2 shows NIR luminescence spectra for Nd^{3+} (a) and Er^{3+} (b) ions in oxyhalide lead borate glasses. The luminescence spectra for Er-doped samples were normalised in order to compare the spectral linewidth. Selected parameters such as measured lifetime, τ_m, and spectral linewidth, $\Delta\lambda$, were estimated for Ln^{3+} ions in oxyhalide glasses and then compared to the values obtained for samples without PbX_2. The results are given in Table 2.

3.3 Oxyfluoride transparent glass-ceramics

The oxyfluoride lead borate glasses were also analysed after annealing. Selected Ln-doped samples were thermally treated above the glass transition temperature, T_g (Table 3). The NIR luminescence

Table 2. *Spectroscopic parameters for Nd^{3+} and Er^{3+} ions in oxyhalide lead borate glasses*

9% PbX_2	Nd^{3+}			Er^{3+}		
	$^4F_{3/2}-^4I_{11/2}$			$^4I_{13/2}-^4I_{15/2}$		
	λ_P (nm)	$\Delta\lambda$ (nm)	τ_m (μs)	λ_P (nm)	$\Delta\lambda$ (nm)	τ_m (μs)
[*]	1061	28·0	86	1531·0	100·5	400
X=F	1061	28·5	85	1531·5	52·5	600
X=Cl	1064	30·0	87	1532·5	60·5	435
X=Br	1063	29·0	83	1533·0	80·0	545

[*] without PbX_2

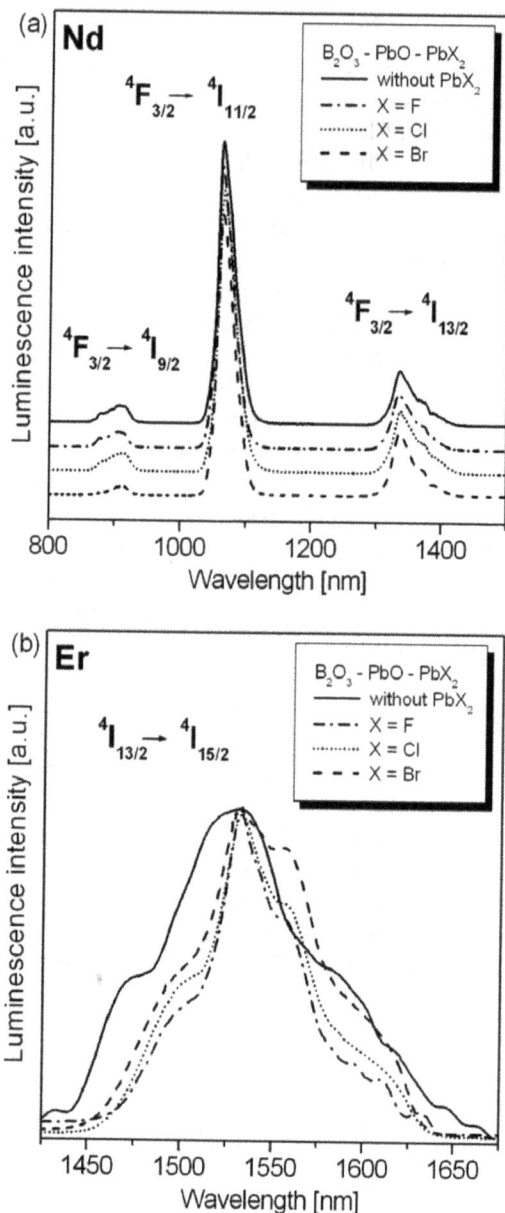

Figure 2. *NIR luminescence spectra for Nd^{3+} (a) and Er^{3+} (b) in oxyhalide lead borate glasses, under excitation by 514 nm (a) and 488 nm (b) laser lines. The glass composition is $18B_2O_3.9PbX_2.63PbO.6Al_2O_3.3WO_3.1Ln_2O_3$, where X=F, Cl or Br; Ln=Nd or Er*

spectra for Nd^{3+} and Er^{3+} ions in B_2O_3–PbO–PbF_2 based glasses before and after annealing (400°C/5h) were measured (Figure 3). Insets to the figure show FT-IR spectra and x-ray diffraction patterns measured for Ln-doped samples before and after annealing. Crystalline peaks due to the orthorhombic PbF_2 phase (PDF-2 card no. P411086) were identified using x-ray diffraction analysis.[25]

4. Discussion

Trivalent neodymium and erbium are the most popular rare earth ions emitting light in the NIR ranges. Many crystalline and amorphous optical materials

Table 3. Thermal parameters and heat treatment conditions for Ln-doped lead borate glasses

Thermal parameters	
Glass transition temperature T_g	353°C
Crystallisation onset T_x	515°C
Thermal stability factor $\Delta T = T_x - T_g$	162°C
Heat treatment conditions	
Annealing temperature T	400°C
Annealing time t	5 h
Phase identification	
PDF-2 card no. P411086	
Crystal	PbF_2
Phase	Orthorhombic
Space group	Pnma
Lattice parameters	$a=6.44$ Å, $b=3.90$ Å, $c=7.65$ Å

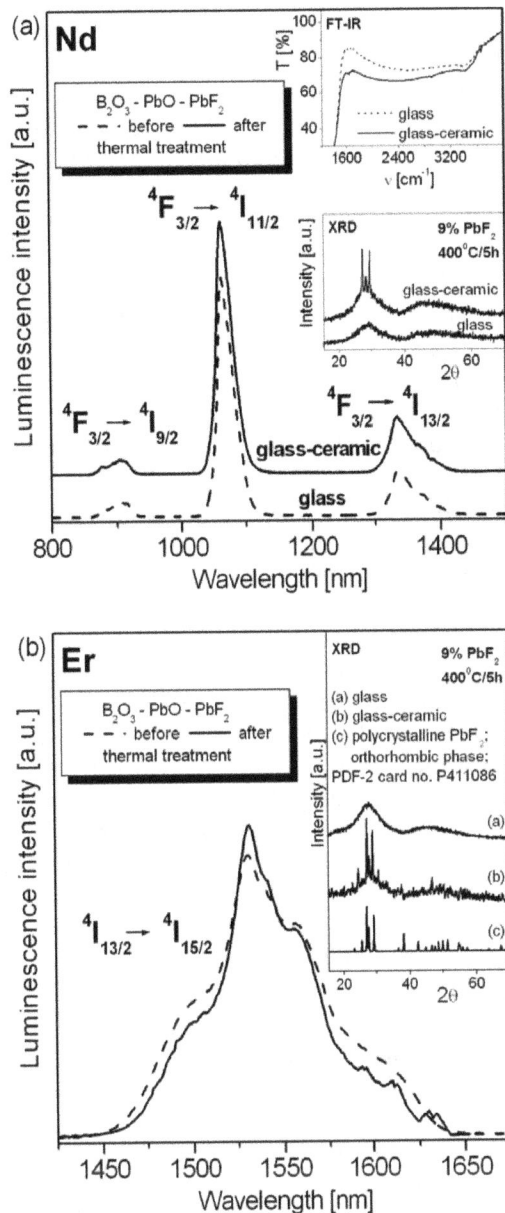

doped with Nd^{3+} and Er^{3+} ions have been studied for NIR luminescence. Several Ln-doped systems (Ln=Nd or Er) were applied as a solid state lasers and optical amplifiers. In particular, glasses or glass fibres containing Er^{3+} can operate as optical amplifiers[26,27] in the standard telecommunication window. It is well established that infrared radiative transition occurs from the $^4I_{13/2}$ excited level of Er^{3+} at about 1·5 mm, which is the low loss optical window of the waveguide. Active Er-doped optical fibre can be quite easily obtained from the precursor borate glass. However, Er-doped borate glasses are rather inefficient fluorescent materials. They are excited into the $^4F_{7/2}$ level. Due to the relatively low energy gaps between the excited levels of Er^{3+} ions and the high phonon energy of the host (~1320 cm^{-1}), borate glass usually eliminates luminescence from the higher lying levels. The high frequencies of the B–O vibrations can quench the Er^{3+} emission at 1·5 mm and degrade the laser performance. For that reason, luminescence has not been observed for various borate based glass systems. It is also noted that the excitation energy transfers nonradiatively very fast from the $^4F_{7/2}$ level to the $^4I_{13/2}$ level, and sometimes very weak NIR luminescence corresponding to the $^4I_{13/2}$–$^4I_{15/2}$ transition of Er^{3+} can be detected. The luminescence of Er^{3+} ions in the lead borate system is quite strong, because the introduction of heavy metal oxide, such as PbO, to the borate glass matrix considerably increases the radiative transition rate, A_J, for the $^4I_{13/2}$ level. For example, the radiative transition rate for the $^4I_{13/2}$ level of Er^{+3} increases from 89 s^{-1} and 117 s^{-1} for B_2O_3–BaO–K_2O and B_2O_3–BaO–Na_2O glasses, to 225 s^{-1} (see Table 1) for our PbO–B_2O_3–Al_2O_3–WO_3 system. As a consequence, the first excited $^4I_{13/2}$ level of Er^{3+} is populated quite well and efficiently. Thus, a surprisingly strong emission from the $^4I_{13/2}$ level of Er^{3+} ions in lead borate based glass was detected and the stimulated emission cross-section is relatively high in comparison to the values obtained for other Er-doped glass systems. For example, the maximum emission cross-section for Er-doped lead borate based glass ($\sigma_{em}=0.47\times10^{-20}$ cm^2 at 1537·5 nm) is nearly the same as the value ($\sigma_{em}=0.52\times10^{-20}$ cm^2 at 1508·3 nm) obtained for Er-doped InF$_3$ based fluoride glass.[20] In particular, quite intense and relatively long-lived NIR emission at 1·5 mm for the studied

Figure 3. NIR luminescence spectra for Nd^{3+} (a) and Er^{3+} (b) in oxyfluoride lead borate glasses before and after annealing, under excitation by 514 nm (a) and 488 nm (b) laser lines. The composition is $18B_2O_3.9PbF_2.63PbO.6Al_2O_3.3WO_3.1Ln_2O_3$, where Ln = Nd or Er. Inset shows FT-IR spectra and XRD patterns for glasses and glass-ceramics

Er-doped lead borate glass, together with large peak stimulated emission cross-section σ_{em} and unusual high spectral linewidth $\Delta\lambda$ nearly close to 100·5 nm (Table 1), is required for tunable solid state lasers and broadband optical amplifiers.

Similar effects were also observed for the studied Nd-doped samples, where nonradiative processes play an important role in relaxation from the excited levels of Ln^{3+}. After excitation by the 514 nm laser line, the energy is transferred very fast to the $^4F_{3/2}$ level by multiphonon relaxation. The energy gap, ΔE, between the $^4F_{3/2}$ and $^4I_{15/2}$ levels is about 5300 cm^{-1}. Only four phonons with energies of 1320 cm^{-1}

are required to bridge the energy gap between the $^4F_{3/2}$ level and the next lowest lying level, $^4I_{15/2}$, so the multiphonon nonradiative process starts to dominate the excited state relaxation. However, the addition of heavy metal oxide (PbO) to the borate matrix increases the radiative transition rate for the $^4F_{3/2}$ level of Nd^{3+} ions from 3386·8 s^{-1} for B_2O_3–Al_2O_3–Na_2O glass[28] to 4290 s^{-1} (Table 1) for our multicomponent B_2O_3–PbO–Al_2O_3–WO_3 system. Consequently, the measured luminescence lifetime, τ_m=86 µs, is found to be much shorter than the radiative lifetime, τ_{rad}=233 µs, calculated from the Judd–Ofelt theory. Thus, the quantum efficiency for the $^4F_{3/2}$ level of Nd^{3+} ions in B_2O_3–PbO based glass is close to η=37%, but this value is relatively high in comparison to the lower values which are obtained for other borate systems without PbO. For example, the quantum efficiency for Nd-doped alkali borate glass is close to η=20%.[28] The introduction of heavy metal oxide (PbO) to the borate matrix increases not only the radiative transition rate, but also the quantum efficiency of the excited states. For this reason, quite intense NIR luminescence at 904, 1061 and 1335 nm was detected. There are three luminescence bands due to transitions from the $^4F_{3/2}$ level to the lower lying $^4I_{9/2}$, $^4I_{11/2}$ and $^4I_{13/2}$ levels of Nd^{3+} ions, respectively. The strong fluorescence band at 1061 nm, due to the $^4F_{3/2}$–$^4I_{11/2}$ transition, has been considered as a potential laser emission of Nd^{3+}. One of the most important radiative parameters for the design of NIR solid state lasers is the peak stimulated emission cross section. An efficient laser transition is characterised by a large value of σ_{em}. Nd-doped lead borate glass which has a relatively large peak stimulated emission cross section is thus useful for laser operation at 1061 nm.[19]

The introduction of lead halide, PbX_2, to the borate glass matrix changes the coordination sphere around Ln^{3+}, and hence partially changes the borate glass network. The anion electronegativities (Br=2·8, Cl=3·0, F=4·0) and ionic type bond character increase in the Br→Cl→F direction. This results in a reduction of the spectral linewidth and an increase of the luminescence lifetime for Ln^{3+} ions (see Table 2). Also, PbX_2 has an important influence on the enhancement of upconversion luminescence of Ln^{3+}.[29] This effect is more significant for Er^{3+} than Nd^{3+} ions. The spectral linewidth for the $^4I_{13/2}$–$^4I_{15/2}$ transition of Er^{3+} in oxyhalide B_2O_3–PbO–PbX_2 based glasses is reduced in the Br→Cl→F direction (Figure 2). All $\Delta\lambda$ values are considerably smaller than that obtained for the sample without PbX_2 (Table 2). In all cases, the $^4I_{13/2}$ lifetime of Er^{3+} ions increases, when PbO is partially replaced by PbX_2, and the highest value of τ_m was obtained for the sample with PbF_2. The F^- ions may have a special effect on the luminescence lifetime.[15] In contrast to the results obtained for Er^{3+}, spectroscopic parameters for Nd^{3+} like the linewidth for the main $^4F_{3/2}$–$^4I_{11/2}$ laser transition at 1·06 mm and the $^4F_{3/2}$ luminescence lifetime,

are nearly independent of PbX_2 content.

Transparent glass-ceramics are of interest as new host materials for Ln^{3+} due to their unique behaviour with fluoride crystal domains in the oxide glassy matrices.[30,31] The narrowing of spectral lines and elongation of lifetimes of fluorescent states of Ln^{3+} ions are the optical consequence of transformation from glass to glass-ceramic. This behaviour can be explained by changes in the environment around Ln^{3+} ions. Depending on glass host and heat treatment conditions, some of the Ln^{3+} ions are incorporated into the crystalline phase during the annealing process. In this way, transparent glass-ceramics containing LaF_3 crystals can be successfully obtained and rare earth ions are well incorporated into the crystalline LaF_3 phase. The LaF_3 crystal is considered to be a more suitable host for rare earth ions than PbF_2 crystal, because the solid solubility of other rare earth ions can be superior, owing to the similar ionic radius and the same valence as the La^{3+} ion.[32] The situation is more complicated when trivalent rare earth ions are substituted for divalent Pb^{2+} ions in transparent glass-ceramics containing PbF_2 crystals. The previously published results indicate that it is experimentally possible to prepare transparent glass-ceramics containing Ln^{3+}:MF_2 (M=Ca,[33] Sr,[34] Ba[35] and Pb[36]) crystals. In the latter system, the Er^{3+} ions were incorporated into the crystallites, forming a $Pb_{1-x}Er_xF_{2+x}$ solid solution. The incorporation of Ln^{3+} ions into MF_2 crystals involves the replacement of divalent M^{2+} cations by trivalent Ln^{3+} ions. In order to compensate the excess positive charge, several processes can be proposed, such as the creation of cation vacancies or the presence of an interstitial F^- fluoride anion.[37] The second mechanism is more likely. In this case, the replacement of Pb^{2+} by Ln^{3+} takes place inside the cubic β-PbF_2 unit cell. Despite the presence of interstitial F^- ions, arising from the charge balance associated with the Pb^{2+}/Er^{3+} substitution, the reduction of the ionic radius from Pb^{2+} to Er^{3+} dominates and explains the decrease of the unit cell parameters.[36]

In order to obtain transparent glass-ceramics, the precursor glass samples are isothermally treated (constant temperature) several times,[38] or heat treated (constant time) at different temperatures.[39,40] In both cases, the glass-ceramics usually lose their transparency with increasing annealing time and/or temperature. The decrease in transparency may be due to light scattering related to larger crystals. In our case, the glass-ceramic samples have quite high transparency, due to the smaller size of the precipitated crystals compared to the visible–near infrared wavelength. Very little difference of transmittance spectra between glass and glass-ceramic can be observed (Figure 3(a)), similar to what has been observed for a TGC system containing LaF_3 crystals.[32]

Selected Ln-doped oxyfluoride samples have been examined before and after heat treatment using x-ray diffraction analysis and luminescence spectroscopy. Independent of PbF_2 content, heat treatment conditions and the kind of Ln^{3+} (Ln=Nd or Er), the orthorhombic PbF_2 phase was identified in the glass-ceramics (Table 3). The $^4F_{3/2}$ luminescence lifetime for Nd^{3+} ions in the studied samples after annealing was the same as in the precursor glass ($\tau_m{\sim}86$ μs). This suggests that Nd^{3+} ions are not incorporated into the crystalline phase, but instead are located in the glass matrix. However, several intense and narrowed diffraction lines were formed during heat treatment, which indicates that transparent glass-ceramics were successfully obtained. A quite different situation was observed for the heat treated Er-doped samples. The luminescence band at 1·5 μm due to the main $^4I_{13/2}$–$^4I_{15/2}$ laser transition of Er^{3+} is more intense and narrow for the glass-ceramic than for the precursor glass (Figure 3(b)). Also, the $^4I_{13/2}$ luminescence lifetime slightly increases from 600 μs (glass) to 670 μs (glass-ceramic). This suggests that some of the Er^{3+} ions are incorporated into the PbF_2 crystalline phase. The previously published results indicate that Er-doped lead borate glasses and transparent glass-ceramics are promising for NIR luminescence and upconversion applications.[41]

5. Summary

Selected Ln-doped lead borate glasses and glass-ceramics have been investigated for near infrared (NIR) luminescence. The results obtained for Nd^{3+} and Er^{3+} in multicomponent oxide and oxyhalide glasses and transparent glass-ceramics lead to the following conclusions:

1. Near infrared luminescence, due to the $^4F_{3/2}$–$^4I_{J/2}$ (J=9, 11, 13) and $^4I_{13/2}$–$^4I_{15/2}$ transitions of Nd^{3+} and Er^{3+} ions, respectively, was detected. Some spectroscopic parameters for both rare earth ions were evaluated. The relatively large emission cross sections are promising for laser applications.

2. The introduction of PbX_2 (X=F, Cl or Br) to the borate glass results in a reduction of the spectral linewidth and an increase of the luminescence lifetime of Er^{3+} ions. This effect is more significant for Er^{3+} than Nd^{3+} ions. The linewidth for the main $^4F_{3/2}$–$^4I_{11/2}$ laser transition at 1·06 μm and the $^4F_{3/2}$ luminescence lifetime of Nd^{3+} are nearly independent of PbX_2 content.

3. Transparent glass-ceramics were successfully prepared. Several diffraction lines were formed after heat treatment. Phase identification reveals that the crystalline peaks can be related to the orthorhombic PbF_2 phase. Samples containing Nd^{3+} show little variation in NIR luminescance properties, but in contrast samples containing Er^{3+} show a relatively more intense and narrow luminescence band at 1·5 μm (due to main $^4I_{13/2}$–$^4I_{15/2}$ laser transition) and a longer decay from the $^4I_{13/2}$ state for the glass-ceramic than for the precursor glass. This indicates that Nd^{3+} ions are more difficult than Er^{3+} to incorporate in the crystalline PbF_2 phase.

Acknowledgment

The Ministry of Science and Higher Education supported this work under research project N N507 3617 33.

References

1. Ushida, H., Iwadate, Y., Hattori, T., Nishiyama, S., Fukushima, K., Ikeda, Y., Yamaguchi, M., Misawa, M., Fukunaga, T., Nakazawa, T. & Jitsukawa, S. *J Alloys Compounds*, 2004, **377**, 167.
2. Sokolov, I. A., Murin, I. V., Mel'nikova, N. A. & Pronkin, A. A. *Glass. Phys. Chem.*, 2003, **29**, 291.
3. Pisarski, W. A, Goryczka, T., Wodecka-Duś, B., Płońska, M. & Pisarska, J. *Mater. Sci. Eng. B*, 2005, **122**, 94.
4. Ciceo-Lucacel, R. & Ardelean, I. *J. Non-Cryst. Solids*, 2007, **353**, 2020.
5. Rada, S., Culea, M., Neumann, M. & Culea, E. *Chem. Phys. Lett.*, 2008, **460**, 196.
6. Pisarski, W. A, Dominiak-Dzik, G., Ryba-Romanowski, W. & Pisarska, J. *J. Alloys Compounds*, 2008, **451**, 220.
7. Lakshminarayana, G. & Buddhudu, S. *Spectrochim. Acta A*, 2006, **63**, 295.
8. Thulasiramudu, A. & Buddhudu, S. *Spectrochim. Acta A*, 2007, **66**, 323.
9. Surendra Babu, S., Jayasankar, C. K., Babu, P., Tröster, T., Sievers, W. & Wortmann, G. *Phys. Chem. Glasses: Eur. J. Glass Sci. Technol. B*, 2006, **47** (4), 548.
10. Pisarska, J., Dominiak-Dzik, G., Ryba-Romanowski, W., Goryczka, T. & Pisarski W. A. *Phys. Chem. Glasses: Eur. J. Glass Sci. Technol. B*, 2006, **47** (4), 553.
11. Pisarski, W. A., Pisarska, J., Dominiak-Dzik, G. & Ryba-Romanowski, W. *J. Phys.: Condens. Matter*, 2004, **16**, 6171.
12. Chen, Q., Ferraris, M., Milanese, D., Menke, Y., Monchiero, E. & Perrone, G. *J. Non-Cryst. Solids*, 2003, **324**, 12.
13. Kashchieva, E. P., Ivanova, V. D, Jivov, B. T. & Dimitriev, Y. B. *Phys. Chem. Glasses*, 2000, **41**, 355.
14. Silva, M. A. P., Messaddeq, Y., Briois, V., Poulain, M. & Ribeiro, S. J. L. *J. Braz. Chem. Soc.*, 2002, **13**, 200.
15. Ding, Y., Jiang, S., Hwang, B.-C., Luo, T., Peyghambarian, N., Himei, Y., Ito, T. & Miura, Y. *Opt. Mater.*, 2000, **15**, 123.
16. Cenk, S., Demirata, B., Öveçoglu, M. L. & Özen, G. *Spectrochim. Acta A*, 2001, **57**, 2367.
17. Özen, G., Aydinli, A., Cenk, S. & Sennaroğlu, A. *J. Lumin.*, 2003, **101**, 293.
18. Carnall, W. T., Fields, P. R. & Rajnak, K. *J. Chem. Phys.*, 1968, **49**, 4412.
19. Pisarska, J., Pisarski, W. A., Dominiak-Dzik, G. & Ryba-Romanowski, W. *J. Molec. Struct.*, 2006, **792–793**, 201.
20. Pisarski, W. A. *J. Molec. Struct.*, 2005, **744–747**, 473.
21. Judd, B. R. *Phys. Rev.*, 1962, **127**, 750.
22. Ofelt, G. S. *J. Chem. Phys.*, 1962, **37**, 511.
23. Fortes, L. M., Santos, L. F., Gonçalves, M. C. & Almeida, R. M. *J. Non-Cryst. Solids*, 2003, **324**, 150.
24. Bueno, L. A., Messaddeq, Y., Filho, F. A. & Ribeiro, S. J. L. *J. Non-Cryst. Solids*, 2005, **351**, 3804.
25. Pisarska, J., Goryczka, T. & Pisarski, W. A. *Solid State Phenom.*, 2007, **130**, 263.
26. Yamauchi, H. & Ohishi, Y. *Opt. Mater.*, 2005, **27**, 679.
27. Tanabe, S. *J. Alloys Compounds*, 2006, **408–412**, 675.
28. Mehta, V., Aka, G., Dawar, A. L. & Mansingh, A. *Opt. Mater.*, 1999, **12**, 53.
29. Yang, J., Dai, N., Dai, S., Wen, L., Hu, L. & Jiang, Z. *Chem. Phys. Lett.*, 2003, **376**, 671.
30. Tanabe, S., Hayashi, H., Hanada, T. & Onodera, N. *Opt. Mater.*, 2002, **19**, 343.
31. Kishi, Y. & Tanabe, S. *J. Alloys Compounds*, 2006, **408–412**, 842.
32. Wang, J., Qiao, X., Fan, X. & Wang, M. *Physica B*, 2004, **353**, 242.
33. Hu, Z., Wang, Y., Ma, E., Chen, D. & Bao, F. *Mater. Chem. Phys.*, 2007, **101**, 234.
34. Qiao, X., Fan, X., Wang, J. & Wang, M. *J. Appl. Phys.*, 2006, **99**, 074302.
35. Qiao, X., Fan, X. & Wang, M. *Scripta Mater.*, 2006, **55**, 211.
36. Dantelle, G., Mortier, M., Patriarche, G. & Vivien, D. *J. Solid State Chem.*, 2006, **179**, 1995.
37. Dantelle, G., Mortier, M., Goldner, P. & Vivien, D. *J. Phys.: Condens. Matter*, 2006, **18**, 7905.
38. Takebe, H., Murakami, T., Kuwabara, M. & Hewak, D. W. *J. Non-Cryst. Solids*, 2006, **352**, 2425.
39. Fan, X., Wang, J., Qiao, X. & Wang, M. *J. Phys. Chem. B*, 2006, **110**, 5950.
40. Qiao, X., Fan, X., Wang, M., Yang, H. & Zhang, X. *J. Appl. Phys.*, 2008, **104**, 043508.
41. Pisarski, W. A., Goryczka, T., Pisarska, J. & Ryba-Romanowski, W. *J. Phys. Chem. B*, 2007, **111**, 2427.

Proc. VI Int. Conf. Borate Glasses, Himeji, Japan, 18–22 August 2008   *Glass Tech.: Eur. J. Glass Sci. Technol. A, October 2009, 50 (5), 273–276*

Electrochemical performance of SnO–B₂O₃–V₂O₅ glasses as negative electrodes for lithium secondary batteries

Akitoshi Hayashi,[1] Hideyuki Morimoto,* Miyuki Nakai, Kiyoharu Tadanaga & Masahiro Tatsumisago

Department of Applied Chemistry, Graduate School of Engineering, Osaka Prefecture University, Sakai, Osaka 599-8531, Japan

*Department of Chemistry and Chemical Biology, Graduate School of Engineering, Gunma University, Kiryu, Gunma 376-8515, Japan

Manuscript received 18 August 2008
Revised version received 27 February 2009
Accepted 8 June 2009

Glasses in the system SnO–B₂O₃–V₂O₅ were prepared using a planetary ball mill apparatus. Amorphous 45SnO.45B₂O₃.10V₂O₅ (mol%) powder was obtained by ball milling for 50 h at a rotation speed of 370 rpm. The glass transition was observed at about 440°C by differential thermal analysis. Electrochemical cells using a conventional liquid electrolyte (1 M LiPF₆ in EC-DEC) were assembled. The cell with the 45SnO.45B₂O₃.10V₂O₅ glass as a working electrode exhibited an initial capacity of 610 mAh g⁻¹, which was smaller than the capacity of the cell using a 50SnO.50B₂O₃ electrode (710 mAh g⁻¹). On the other hand, the glass with V₂O₅ retained a larger capacity of 330 mAh g⁻¹ than the glass without V₂O₅ (180 mAh g⁻¹) after 20 cycles. The addition of V₂O₅ played an important role in enhancing the cycle performance of cells using SnO-based glass electrodes.

1. Introduction

Due to increasing demands for energy storage devices applicable to eco-cars, such as electric vehicles, the development of lithium ion batteries has been strongly encouraged. Negative electrode materials, as an alternative to the graphite electrodes used in commercially available lithium ion batteries, have been extensively studied. SnO-based glasses have a large specific capacity, almost twice as large as graphite electrodes.[1–3] These glasses were prepared not only by the conventional melt quenching technique, but also by mechanical milling.[3–6] Glasses in the system SnO–B₂O₃ have previously been prepared by mechanical milling, and applied as the negative electrode for rechargeable lithium batteries.[6] Mechanochemical synthesis has a favourable feature that fine powders are directly obtained at room temperature without an additional pulverising process of bulk glasses.

The 50SnO.50B₂O₃ (mol%) glass exhibited the largest initial capacity of 700 mAh g⁻¹ in the SnO–B₂O₃ system.[7] The initial capacity of SnO–B₂O₃ glasses is larger than that of SnO–BPO₄ and SnO–P₂O₅ glasses.[8] However, rapid capacity fading of cells using SnO–B₂O₃ glasses is a problem to be solved.

A strategy to improve cyclability of tin borate electrodes involves modification of the glass structure. Glass former components have an effect on electrochemical properties, such as cell potential, capacity and cyclability; SnO–P₂O₅ glasses exhibit better cycle performance than SnO–B₂O₃ glasses.[8] On the other hand, V₂O₅-based glass electrodes were reported to show good cycleability in lithium ion cells.[9]

In the present study, 50SnO.50B₂O₃ glass with the addition of V₂O₅ was mechanochemically prepared in order to improve the cycle performance of lithium cells using SnO–B₂O₃ glasses.

2. Experimental

Reagent grade SnO (Furuuchi Chemical Co., 99·9%), B₂O₃ (Kojundo Chemical Laboratory Co., 99·9%) and V₂O₅ (Wako Pure Chemical Industries) powders were used as starting materials. The mechanochemical treatment was carried out for 1 g batches of the mixed materials at the composition $(100-x)(0.5SnO.0.5B_2O_3)$. xV_2O_5 (mol%) in a stainless steel pot (25 ml in volume) with seven stainless steel balls (10 mm in diameter), using a high energy planetary ball mill apparatus (Pulverisette 7, Fritsch GmbH). The rotation speed was fixed at 370 rpm, and all the processes were conducted at room temperature in a dry N₂ filled glove box.

X-ray diffraction (XRD) measurements (Cu Kα) were performed using a diffractometer (XRD-6000, Shimadzu Co.). Differential thermal analysis (DTA) was carried out using a thermal analyser (Thermoplus 8110, Rigaku Co.) on obtained powders sealed in an Al pan in a dry N₂ filled glove box. The heating rate was 10°C min⁻¹.

Electrode materials were prepared by mixing the

[1] Corresponding author. Email hayashi@chem.osakafu-u.ac.jp

362   *Proceedings of the VI International Conference on Borate Glasses, Himeji, Japan, 18–22 August 2008*

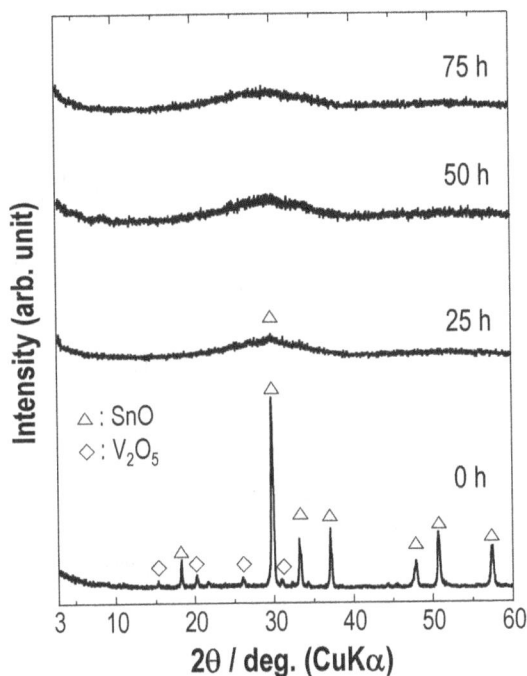

Figure 1. XRD patterns of the powder samples with a nominal composition of $45SnO.45B_2O_3.10V_2O_5$ (mol%) prepared by mechanical milling for several hours. The numbers in the figure denote the milling time

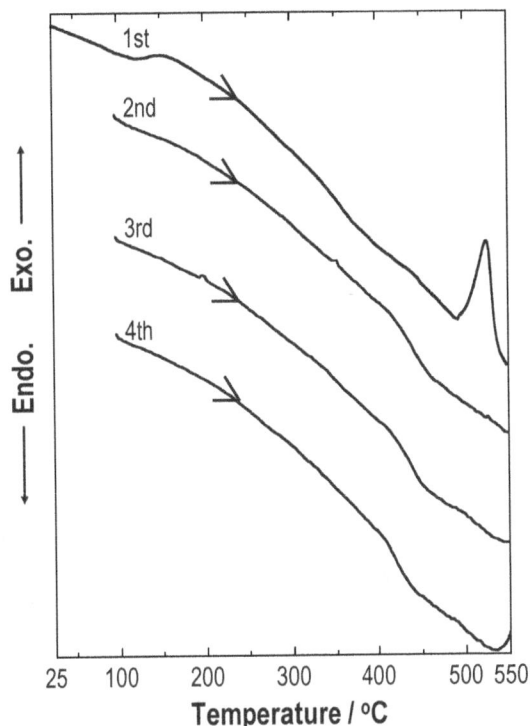

Figure 2. DTA curves for the $45SnO.45B_2O_3.10V_2O_5$ (mol%) amorphous powder prepared by mechanical milling

glass powder (70 wt%), acetylene black (20 wt%), and polyvinylidene fluoride (PVDF, 10 wt%) with an agate mortar in a dry Ar filled glove box. Slurries of the mixture were painted on a Ni mesh as a current collector and then dried. Electrochemical measurements were performed in a simple three-electrode cell, with the obtained electrode as the working electrode, and lithium sheets as counter and reference electrodes. A mixture of ethylene carbonate (EC) and diethyl carbonate (DEC) with 1:1 volumetric ratio containing $1 \ mol \ dm^{-3}$ (1 M) $LiPF_6$ was used as an electrolyte. The charge–discharge performance of the cells was evaluated at a constant current density of $1 \ mA \ cm^{-2}$.

3. Results and discussion

Figure 1 shows XRD patterns of the powder samples with a nominal composition of $45SnO.45B_2O_3.10V_2O_5$ (mol%) prepared by mechanical milling for several hours. XRD peaks due to the SnO and V_2O_5 crystals are observed in the powder mixture before mechanical milling (0 h); the other starting material, B_2O_3, was amorphous. As the milling period increases, the intensity of the SnO and V_2O_5 peaks decreases and the halo pattern becomes dominant. An amorphous sample was obtained after milling for 50 h, and further milling treatment retains the amorphous state (no formation of new crystal phases).

Figure 2 shows DTA curves for the $45SnO.45B_2O_3.10V_2O_5$ (mol%) amorphous powder prepared by mechanical milling. DTA measurements were carried out for four cycles in the temperature range from 100

to 550°C. An exothermic peak due to crystallisation is observed at 480°C and an endothermic change due to a glass transition at about 440°C appears for the first heating process. It was revealed from XRD measurements that VO_2 crystal was precipitated after the first heating up to 550°C. Although a partial crystallisation occurs at the first heating, a glass transition is clearly observed at about 420°C after the second heating process. The amorphous powder prepared by mechanical milling is thus in a glassy state.

Electrochemical measurements were performed in a simple three electrode cell using a liquid electrolyte. Figure 3 shows the first and 20th charge–discharge curves for cells using $(100-x)(0.5SnO.0.5B_2O_3).xV_2O_5$ ($x=0$ and 10) glasses as a working electrode under the cutoff voltage 0–2·0 V. The abscissa shows the charge–discharge capacities per unit weight of the glasses. A charge corresponds to an insertion of lithium ions to the working electrode, while a discharge corresponds to an extraction of lithium ions from the working electrode. Two plateaus at 1·5 V and 0·5 V are observed in the charge curve, and one plateau at 0·5 V is observed in the discharge curves for both glasses ($x=0$ and 10).

The cell with the $45SnO.45B_2O_3.10V_2O_5$ ($x=10$) glass exhibits charge and discharge capacities of 1120 and $610 \ mAh \ g^{-1}$, respectively, at the first cycle. The cell with the $50SnO.50B_2O_3$ ($x=0$) glass shows charge and discharge capacities of 1230 and $710 \ mAh \ g^{-1}$, respectively. The initial capacity of the glass with 10 mol% V_2O_5 is smaller than that of the glass without V_2O_5. On the other hand, the former glass exhibits

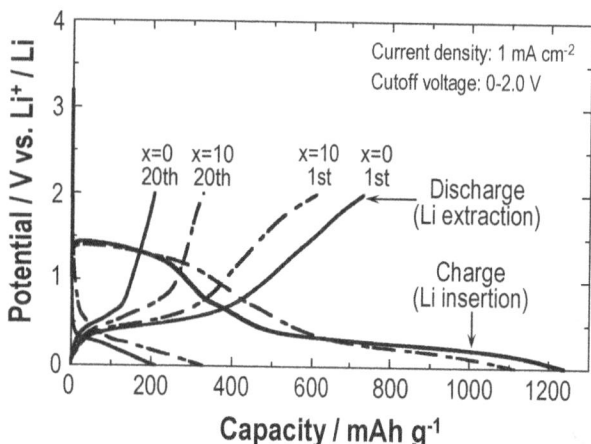

Figure 3. The first and 20th charge–discharge curves for cells using (100−x)(0·5SnO.0·5B$_2$O$_3$).xV$_2$O$_5$ (x=0 and 10) glasses as the working electrode. A mixture of 1 M LiPF$_6$/ EC+DEC was used as an electrolyte. The measurements were carried out at a constant current density of 1 mA cm^{-2} under the cutoff voltage 0–2·0 V

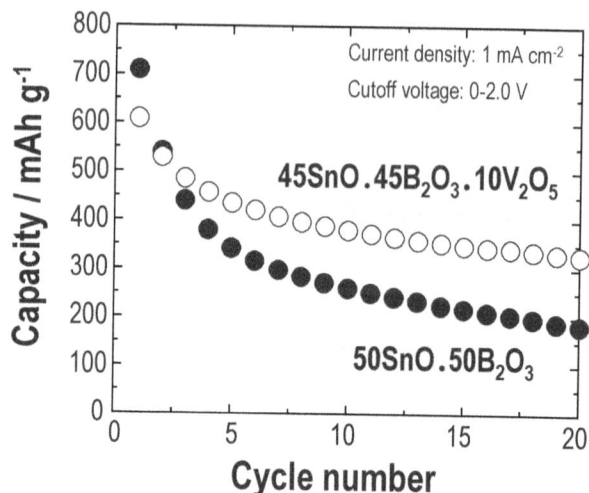

Figure 4. Cycle performance for the cells with the (100−x) (0·5SnO.0·5B$_2$O$_3$).xV$_2$O$_5$ (x=0 and 10) glasses under the cutoff voltage 0–2·0 V

a larger discharge capacity of 330 mAh g^{-1} than the latter glass (180 mAh g^{-1}) at the 20th cycle.

Figure 4 shows the cycle performance for cells with (100−x)(0·5SnO.0·5B$_2$O$_3$).xV$_2$O$_5$ (x=0 and 10) glasses under the cutoff voltage 0–2·0 V. The errors in capacity are within the point symbols. Although the capacities of both glasses gradually decrease during charge–discharge cycling, the 45SnO.45B$_2$O$_3$.10V$_2$O$_5$ glass shows better cycling performance than the 50SnO.50B$_2$O$_3$ glass.

In the charge curve, the first plateau at 1·5 V is due to the formation of metallic Sn^0 (Sn^{2+}→Sn^0) and Li$_2$O–B$_2$O$_3$ glassy matrix, and the second plateau at 0·5 V is due to the formation of Li–Sn alloy (Sn^0→Li–Sn).[1–3] The former reaction is known to be irreversible, and the formation of metallic Sn is the main reason for the large irreversible capacity at the

first cycle. The latter reaction is basically reversible; the formation of Li–Sn alloy (Sn^0→Li–Sn) occurs in the charge process while the formation of Sn (Li–Sn→Sn^0) occurs in the discharge process.

The addition of V$_2$O$_5$ to SnO–B$_2$O$_3$ glass electrode reduces the initial capacity, but improves the cycling performance. The lower Sn content as an active site for electrochemical reaction is responsible for the decreased initial capacity of the 45SnO.45B$_2$O$_3$.10V$_2$O$_5$ glass electrode. The capacity fading during cycling is mainly caused by a large volume change, due to the agglomerated Sn particles.[2] A possible explanation for the improved cycle performance is that a V$_2$O$_5$ domain plays an important role in restricting the agglomeration of Sn particles during charge–discharge cycles.

Further enhancement of cycle performance was achieved by selecting the cutoff voltage for

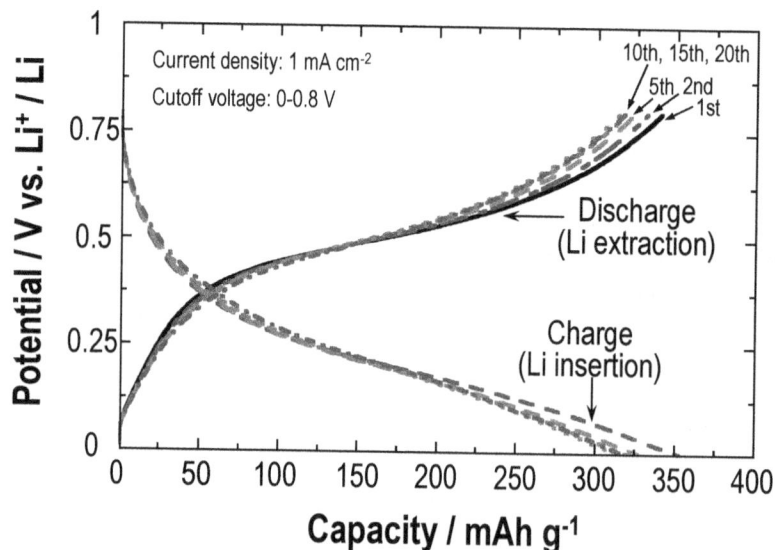

Figure 5. Charge–discharge curves for the cell using the 45SnO.45B$_2$O$_3$.10V$_2$O$_5$ glass under the cutoff voltage 0–0·8 V for 20 cycles

charge–discharge measurements. Figure 5 shows the charge–discharge curves for the cell using the 45SnO. $45B_2O_3.10V_2O_5$ glass under the cutoff voltage 0–0·8 V for 20 cycles. Although the first discharge capacity is smaller than the capacity at the cutoff voltage 0–2·0 V, as shown in Figure 3, the cell maintains rechargeable capacity of about 300 mAh g^{-1} for 20 cycles. It was previously reported that agglomeration of Sn particles is accelerated if the voltage range is larger than 0·8 V for the discharge process.[2] The cycling performance, therefore, is considerably enhanced by reducing the upper cutoff voltage from 2·0 to 0·8 V.

4. Conclusions

Glasses with compositions $(100-x)(0·5SnO.0·5B_2O_3)$. xV_2O_5 (x=0 and 10) were prepared using a planetary ball mill apparatus. Amorphous $45SnO.45B_2O_3.10V_2O_5$ powder was obtained by ball milling for 50 h at a rotation speed of 370 rpm. A glass transition phenomenon was observed at about 440°C in the DTA measurements. Electrochemical cells using a conventional liquid electrolyte (1 M LiPF$_6$ in EC-DEC) were assembled. A cell with $45SnO.45B_2O_3.10V_2O_5$ glass as a working electrode exhibited an initial capacity of

610 mAh g^{-1}, which was smaller than the capacity of a cell using a $50SnO.50B_2O_3$ electrode (710 mAh g^{-1}). On the other hand, the glass with V_2O_5 retained a larger capacity than the glass without V_2O_5 after 20 cycles. The addition of V_2O_5 enhanced the cycle performance of cells using SnO–B_2O_3 glass electrode. Moreover, a decrease of the upper cutoff voltage was effective in improving the cycle performance of a cell using $45SnO.45B_2O_3.10V_2O_5$ glass.

References

1. Idota, Y., Kubota, T., Matsufuji, A., Maekawa, Y. & Miyasaka, T. *Science*, 1997, **276**, 1395.
2. Courtney, I. A. & Dahn, J. R. *J. Electrochem. Soc.*, 1997, **144**, 2943.
3. Tatsumisago, M. & Hayashi, A. *Glass Technol.; Eur. J. Glass Sci. Technol. A*, 2007, **48**, 6.
4. Nakai, M., Hayashi, A., Morimoto, H., Tatsumisago, M. & Minami, T. *J. Ceram. Soc. Jpn.*, 2001, **109**, 1010.
5. Morimoto, H., Tatsumisago, M. & Minami, T. *J. Electrochem. Soc.*, 1999, **146**, 3970.
6. Hayashi, A., Nakai, M., Morimoto, H., Minami, T. & Tatsumisago, M. *J. Mater. Sci.*, 2004, **39** 1.
7. Hayashi, A., Nakai, M., Tatsumisago, M., Minami, T. & Katada, M. *J. Electrochem. Soc.*, 2003, **150**, A582.
8. Hayashi, A., Konishi, T., Tadanaga, K., Minami, T. & Tatsumisago, M. *J. Non-Cryst. Solids*, 2004, **345&346**, 478.
9. Machida, N., Fuchida, R. & Minami, T. *J. Electrochem. Soc.*, 1989, **136**, 130.

Appendix 1
Conference Participants

Prof Mario AFFATIGATO
Coe College, 1220 First Av. NE Cedar Rapids, Iowa 52402, USA
Email maffatig@coe.edu

Mr Takanobu ARAKI
Department of Materials Science and Chemistry, Graduate School of Engineering, University of Hyogo, Shosha 2167, Himeji 671-2280, JAPAN
Email et07x005@steng.u-hyogo.ac.jp

Dr Taro ASAHI
Niihama National College of Technology, 7-1 Yagumo-cyo, Nihama-shi Ehime 792-8580, JAPAN
Email asahi@mat.niihama-nct.ac.jp

Dr Emma BARNEY
ISIS, R1 G62, Rutherford Appleton Laboratory, Harwell Science and Innovation Campus, Didcot OX11 0QX, UK
Email E.Barney@rl.ac.uk

Mr Nathan BARROW
University of Warwick, Physics Department, Coventry CV4 7AL, UK
Email n.s.barrow@warwick.ac.uk

Mr John BERKOWITZ
Physics Department, Coe College, 1220 First Ave NE, Cedar Rapids, IA 52402, USA
Email jaberkow@coe.edu

Dr Lucica BOROICA
National Glass Institute, Th. Pallady 47, Bucharest sector 3 032258 ROMANIA
Email boroica_lucica@yahoo.com

Ms Randilynn CHRISTENSEN
Iowa State University, 2220 Hoover Hall, Ames, IA 50011, USA
Email rbchris@iastate.edu

Dr Laurent CORMIER
CNRS - Institut de Mineralogie, Institut de Mineralogie et Physique des Milieux Condenses, 92140 FRANCE
Email cormier@impmc.jussieu.fr

Dr Yusuke DAIKO
University of Hyogo, 2167 shosha, Himeji, 6712201, JAPAN
Email daiko@eng.u-hyogo.ac.jp

Prof Giuseppe DALBA
University of Trento, Department of Physics, Via Sommarive 14, Trento 38100, ITALY
Email dalba@science.unitn.it

Prof Erwin DESA
Goa University, Department of Physics, Taleigao Plateau, Goa 403 206, INDIA
Email edesa@unigoa.ac.in; erwindesa@gmail.com

Ms Adriana DIACONU
National Institute of Glass, 47th Th. Pallady Av., S3, Bucharest 032258, ROMANIA
Email mail_to_adriana@yahoo.com

Prof Yanko DIMITRIEV
University of Chemical Technology and Metallurgy, blvd."Kl. Ohridski" 8, Sofia 1756, BULGARIA
Email yanko@mail.uctm.edu

Prof Doris EHRT
Otto-Schott-Institut of the University Jena, Fraunhoferstr. 6, Jena D-07743, GERMANY
Email doris.ehrt@uni-jena.de

Ms Margit FABIAN
Research Institute for Solid State Physics and Optics, Budapest, Konkoly-Thege 29-33 1121, HUNGARY
Email fabian@szfki.hu

Prof. Steven FELLER
Physics Department, Coe College, 1220 First Ave NE, Cedar Rapids, IA 52328, USA
Email sfeller@coe.edu

Dr Guillaume FERLAT
IMPMC, Université Pierre et Marie Curie, Paris VI IMPMC, Bât. 7 Campus Boucicaut, 140 rue de Lourmel, 75015, FRANCE
Email ferlat@impmc.jussieu.fr

Ms Maranda FRANKE
Physics Department, Coe College, 1220 First Ave NE, Cedar Rapids, IA 52402, USA
Email mafranke@coe.edu

Mr Benjamin FRANTA
Physics Department, Coe College, 1220 First Ave NE, Cedar Rapids, IA 52402, USA
Email bafranta@coe.edu

Dr Shingo FUCHI
Department of Crystalline Materials Science, Graduate School of Engineering, Nagoya University, Furo-cho, Chikusa-ku, Nagoya 464-8603, JAPAN
Email fuchi@mercury.numse.nagoya-u.ac.jp

Ms Etsuko FUJINAKA
Department of Materials Science and Chemistry, Graduate School of Engineering, University of Hyogo, Yamanoue 684-33, Hiraoka-tyo, Kakogawa-shi, Hyogo 675-0112, JAPAN
Email et07y044@steng.u-hyogo.ac.jp

Mr Shohei FUJIWARA
Graduate School of Engineering, Kobe University, Rokko-dai, Nada, Kobe, Hyogo 657-8501, JAPAN
Email fujiwara@cx2.scitec.kobe-u.ac.jp

Mr Yasuteru FUKAWA
Graduate School of Pure and Applied Sciences, University of Tsukuba, Kojima lab, Tennoudai1-1-1, Tsukuba, Ibaraki 305-8573 JAPAN
Email s-fukawa@ims.tsukuba.ac.jp

Dr Ken-ichi FUNAKOSHI
JASRI/SPring-8 1-1-1, Kouto, Sayo-cho, Sayo-gun, Hyogo 679-5198, JAPAN
Email funakosi@spring8.or.jp

Mr Daisuke FURUSAWA
Osaka Prefecture University, 1-1 Gakuencho, Nakaku, Sakai, Osaka 599-8531, JAPAN
Email furusawa@chem.osakafu-u.ac.jp

Dr Rolly GRISENTI
Dipartimento di Fisica, Università di Trento, via Sommarive, 14 I-38100, ITALY
Email rolly.grisenti@unitn.it

Prof Alex HANNON
ISIS Facility, Rutherford Appleton Laboratory, Chilton, Didcot, Oxon OX11 0QX, UK
Email a.c.hannon@rl.ac.uk

Mr Takumi HARADA
Chiba University, Ka-zaSakura, 202 room, 3-2-23 Kurosunadai Inageku, Chibasi, Chiba 263-0041, JAPAN
Email ikkocukan@yahoo.co.jp

Ms Amanda HAVEL
Physics Department, Coe College, 1220 First Ave NE, Cedar Rapids, IA 52328, USA
Email ajhavel@coe.edu

Dr Akitoshi HAYASHI
Department of Applied Chemistry, Osaka Prefecture University, 1-1 Gakuen-cho, Naka-ku, Sakai, Osaka 591-8023, JAPAN
Email hayashi@chem.osakafu-u.ac.jp

Mr Michael HAYNES
Iowa State University, 2220 Hoover Hall, Ames, IA 50011, USA
Email mjh3sccr@iastate.edu

Dr Yusuke HIMEI
Nippon Electric Glass Co. Ltd., 2-7-1 Seiran, Otsu, Shiga 580-8639, JAPAN
Email yhimei@neg.co.jp

Prof. Kazuyuki HIRAO
Kyoto University, A3-120, Katsura, Nishikyo-ku, Kyoto 615-8510, JAPAN
Email hirao@bisco1.kuic.kyoto-u.ac.jp

Dr Tsuyoshi HONMA
Department of Materials Science and Technology, Nagaoka University of Technology, Kamitomioka 1603-1, Nagaoka 940-2188, JAPAN
Email honma@mst.nagaokaut.ac.jp

Mr Makoto HOSOKAWA
Graduate School of Engineering, Kobe University Rokko-dai, Nada, Kobe, Hyogo 657-8501, JAPAN
Email fujiwara@cx2.scitec.kobe-u.ac.jp

Mr Martun HOVHANNISYAN
State Engineering University of Armenia, 105 Teryan str, Yerevan 0009, ARMENIA
Email martun_h@yahoo.com

Dr Rafael HOVHANNISYAN
Scientific-Production Enterprise of Materials Science, 17 Charents Street, Yerevan 0025, ARMENIA
Email hovhannisyan@netsys.am

Dr Yuka IKEMOTO
JASRI/SPring-8, 1-1-1 Kouto, Sayo, Hyogo 679-5165, JAPAN
Email ikemoto@spring8.or.jp

Dr Seiji INABA
Kyushu University, 744 Motooka, Nishi-ku, Fukuoka 819-0395, JAPAN
Email sinaba@chem-eng.kyushu-u.ac.jp

Prof. Yordanka IVANOVA
University of Chemical Technology and Metallurgy, 8 Kl. Ohridski Blvd., Sofia 1756, BULGARIA
Email y.ivanov@uctm.edu

Prof. Yasuhiko IWADATE
Graduate School of Engineering, Chiba University, Yayoi-cho 1-33, Inage-ku, Chiba 263-8522, JAPAN
Email iwadate@faculty.chiba-u.jp

Mr Steven JUNG
Missouri University of Science and Technology, 902 West 13th Street Rolla, MO 65401, USA
Email steven.b.jung@gmail.com

Prof. Kohei KADONO
Kyoto Institute of Technology, Matsugasaki, Sakyo-ku, Kyoto 606-8585, JAPAN
Email kadono@kit.ac.jp

Dr Akihiko KAJINAMI
Center for Environmental Management, Kobe University, Rokko-dai, Nada, Kobe, Hyogo 657-8501, JAPAN
Email kajinami@kobe-u.ac.jp

Dr Efstratios KAMITSOS
Theoretical and Physical Chemistry Institute, National Hellenic Research Foundation, 48 Vassileos Constantinou Avenue, 116 35 Athens, GREECE
Email eikam@eie.gr

Ms Ol'ga KARAS'
Far East Geological Institute, Prospect 100-letya Vladivostoku, 159 Vladivostok 690022, RUSSIAN FEDERATION
Email okaras@yandex.ru

Prof. Elena KASHCHIEVA
University of Chemical Technology and Metallurgy, 8 Kl. Ohridski Blvd., Sofia 1756, BULGARIA
Email elena_kashchieva@yahoo.com

Mr Yoshinari KATO
Nippon Electric Glass Co. Ltd., 7-1, Seiran 2-chome, Otsu, Shiga 520-8639, JAPAN
Email ykato@neg.co.jp

Mr Mitsuru KAWASHIMA
Graduate School of Pure and Applied Sciences, University of Tsukuba, Kojima lab, Tennoudai1-1-1, Tsukuba, Ibaraki 305-8573, JAPAN
Email m_kawa326@yahoo.co.jp

Dr Atul KHANNA
Guru Nanak Dev University, Department of Applied Physics, Amritsar 143005, INDIA
Email ak.ap@gndu.ac.in;akphysics@yahoo.com

Prof. John KIEFFER
University of Michigan, 2300 Hayward Street, 48109, USA
Email kieffer@umich.edu

Dr Naoyuki KITAMURA
National Institute of Advanced Industrial Science and Technology, 1-8-31, Midorigaoka, Ikeda, Osaka 563-8577, JAPAN
Email naoyuki.kitamura@aist.go.jp

Mr Thomas KLOSS
SCHOTT AG, Otto-Schott-Str. 13, 07745 Jena, GERMANY
Email thomas.kloss@schott.com

Prof. Masao KODAMA
Sojo University (retired) Home: Ikeda 2-32-25, Kumamoto 860-0082 JAPAN
Email masaokodama@arion.ocn.ne.jp

Mr Shunsuke KOIDE
Department of Chemistry and Materials Science, Graduate School of Science and Engineering, Tokyo Institute of Technology, S713 2-12-1 Ookayama, Meguro-ku, Tokyo 152-8550, JAPAN
Email koide.s.aa@m.titech.ac.jp

Prof. Kazuo KOJIMA
Ritsumeikan University, 1-1-1 Noji-higashi, Kusatsu, Shiga 525-8577, JAPAN
Email kojimaka@se.ritsumei.ac.jp

Prof. Seiji KOJIMA
Institute of Materials Science, University of Tsukuba,1 Tsukuba, Ibaraki 305-8573 JAPAN
Email kojima@bk.tsukuba.ac.jp

Prof. Ladislav KOUDELKA
University of Pardubice, Faculty of Chemical Technology, Department of General and Inorganic Chemistry, Studentska 95, Pardubice 53210, CZECH REPUBLIC
Email ladislav.koudelka@upce.cz

Prof. Scott KROEKER
Department of Chemistry, University of Manitoba, Winnipeg, Manitoba R3T 2N2, CANADA
Email Scott_Kroeker@UManitoba.ca

Prof. Gerald LUCOVSKY
North Carolina State University, Department of Physics, Campus

Box 3202, Raleigh, NC 27695, USA
Email lucovsky@ncsu.edu

Mr Seiichi MAMIYA
Graduate School of Pure and Applied Sciences, University of Tsukuba, Institute of Materials Sciences, 1-1-1 Tennodai, Tsukuba, Ibaraki 305-8573, JAPAN
Email mamiya@ims.tsukuba.ac.jp

Dr Tatiana MARKOVA
Physics Department at Saint-Petersburg State Technological University of Plant Polymers, 4, Ul. Ivana Chernykh, St-Petersburg 198095, RUSSIAN FEDERATION
Email T.S.Markova@mail.ru

Ms. Amy MARQUARDT
Physics Department, Coe College, 1220 First Ave NE, Cedar Rapids, IA 52402, USA
Email aemarqua@coe.edu

Dr Steve MARTIN
Iowa State University, 2220 Hoover Hall, Ames, IA 50011, USA
Email swmartin@iastate.edu

Dr Hirokazu MASAI
Department of Applied Physics, Tohoku University, 6-6-05, Aramaki, Aoba, Sendai 980-8579, JAPAN
Email masai@laser.apph.tohoku.ac.jp

Ms Midori MATSUBARA
Niigata University, 8050 Ikarashi 2-no cho, Niigata 950-2181, JAPAN
Email ohtori@chem.sc.niigata-u.ac.jp

Mr Yu MATSUDA
Graduate School of Pure and Applied Sciences, University of Tsukuba, Kojima Lab., Institute of Materials Sciences, Tsukuba, Ibaraki 305-8573, JAPAN
Email hiiro_s721@yahoo.co.jp

Prof. Jun MATSUOKA
The University of Shiga Prefecture, Hassaka 2500, Hikone, Shiga 522-8533, JAPAN
Email matsuoka@mat.usp.ac.jp

Dr Doris MOENCKE
Theoretical and Physical Chemistry Institute, National Hellenic Research Foundation, 48 Vassileos Constantinou Avenue, Athens 11635, GREECE
Email moencke@eie.gr

Mr Tyler MULLENBACH
Physics Department, Coe College, 1220 First Ave NE, Cedar Rapids, IA 52402, USA
Email tkmullen@coe.edu

Prof. Tokuro NANBA
Okayama University, 3-1-1, Tsushima-Naka, Okayama 700-8530, JAPAN
Email tokuro_n@cc.okayama-u.ac.jp

Mr Shuji NISHIMOTO
Mitsubishi Materials Corporation, 1002-14 Mukoyama, Naka-shi, Ibaraki 311-0102, JAPAN
Email nishis@mmc.co.jp

Dr Hideo OHNO
JASRI/SPring-8, 1-1-1 Kouto, Sayo-cho, Sayo-gun, Hyogo 679-5198, JAPAN
Email ohno@spring8.or.jp

Prof. Norikazu OHTORI
Niigata University, 8050 Ikarashi 2-no cho, Niigata 950-2181, JAPAN
Email ohtori@chem.sc.niigata-u.ac.jp

Dr Hiroshi OJI
JASRI/SPring-8, 1-1-1 Koto, Sayo, Hyogo 679-5198, JAPAN
Email oji-h@spring8.or.jp

Dr Toshihiro OKAJIMA
Kyushu Synchrotron Light Research Center, 8-7 Yayoigaoka, Tosu, Saga 841-0005, JAPAN
Email okajima@saga-ls.jp

Dr Armenak OSIPOV
Institute of Mineralogy, UB RAS Russia, Chelyabinsk region, Miass, 456317, RUSSIAN FEDERATION
Email armik@mineralogy.ru

Ms Gillian PARKIN
Immobilisation Science Laboratory, The University of Sheffield, Department of Engineering Materials, Sir Robert Hadfield Building, Mappin Street, Sheffield S1 3JD, UK
Email mtp05gcp@sheffield.ac.uk

Dr Joanna PISARSKA
Silesian University of Technology, Department of Materials Science, Krasinskiego 8, Katowice 40-019, POLAND
Email Joanna.Pisarska@polsl.pl

Dr Francesco ROCCA
IFN-CNR, Institute for Photonics and Nanotechnologies, Via alla Cascata, 56/c POVO (Trento) 38100, ITALY
Email rocca@science.unitn.it

Dr Koichi SAKAGUCHI
Nippon Sheet Glass Co. Ltd., 2-13-12 Konoike, Itami, Hyogo 664-8520, JAPAN
Email KoichiSakaguchi@mail.nsg.co.jp

Dr Bogdan Alexandru SAVA
National Institute of Glass, 47th Th. Pallady Av., S3, Bucharest 032258, ROMANIA
Email savabogdanalexandru@yahoo.com

Dr David SCHUBERT
US Borax Inc., Rio Tinto Minerals, 8051 E. Maplewood Avenue, Greenwood Village, Colorado 80111, USA
Email dave.schubert@riotinto.com

Dr Hiroyo SEGAWA
Tokyo Institute of Technology, 2-12-1 Ookayama, Meguro-ku, Tokyo 152-8550, JAPAN
Email hsegawa@ceram.titech.ac.jp

Prof. Sabyasachi SEN
Department of Chemical Engineering & Materials Science, 3094 Bainer Hall, University of California at Davis, CA 95616, USA
Email sbsen@ucdavis.edu

Mr Kazuyoshi SHINDO
Nippon Electric Glass Co. Ltd., 7-1, Seiran 2-chome, Otsu, Shiga 520-8639, JAPAN
Email kshindo@neg.co.jp

Mr Tatsuya SUETSUGU
Kyoto Institute of Technology, Hashigamicho-1, Matsugasaki, Sakyoku, Kyoto 6068585, JAPAN
Email d7811501@edu.kit.ac.jp

Mr Masanori SUZUKI
Division of Materials and Manufacturing Science, Graduate School of Engineering, Osaka University 2-1 Yamadaoka, Suita, Osaka 565-0871, JAPAN
Email masanori.suzuki@mat.eng.osaka-u.ac.jp

Dr Kiyoharu TADANAGA
Graduate School of Engineering, Osaka Prefecture University 1-1 Gakuen-cho, Naka-ku, Sakai, Osaka 599-8531, JAPAN
Email tadanaga@chem.osakafu-u.ac.jp

Dr Akira TAKADA
Asahi Glass Co. Ltd., 1150 Hazawa-cho, Kanagawa-ku, Yokohama-shi 221-8755, JAPAN
Email akira-takada@agc.co.jp

Mr Masataka TAKAGI
Nippon Electric Glass Co. Ltd., Seiran 2-7-1, Otsu, Shiga 520-8639, JAPAN
Email mtakagi@neg.co.jp

Mr Koji TAKANO
Okuno Chemical Industries Co. Ltd., R&D Laboratory, Inorganic Materials (Osaka Prefecture Univ., Dept. of Applied Chemistry Graduate School of Engineering) 2-1-25, Hanaten-Nishi Joto-ku, Osaka 536-0011, JAPAN
Email k-takano01@okuno.co.jp

Dr Masaki TAKATA
SPring-8/RIKEN/JASRI, 1-1-1 Kouto, Sayo-cho, Sayo-gun, Hyogo 679-5148, JAPAN
Email takatama@spring8.or.jp

Prof. Hiromichi TAKEBE
Ehime University, 3 Bunkyo-cho, Matsuyama, Ehime 790-8577, JAPAN
Email takebe@eng.ehime-u.ac.jp

Mr Yusuke TANAKA
Okayama University, 3-1-1, Tsushima-Naka, Okayama 700-8530, JAPAN
Email tokuro_n@cc.okayama-u.ac.jp

Prof. Toshihiro TANAKA
Division of Materials and Manufacturing Science, Graduate School of Engineering, Osaka University 2-1 Yamadaoka, Suita, Osaka 565-0871, JAPAN
Email tanaka@mat.eng.osaka-u.ac.jp

Mr Yosuke TANIGUCHI
Japan Synchrotron Radiation Research Institute, 1-1-1 Kouto, Sayo-cho Sayo-gun, Hyogo 679-5198, JAPAN
Email yosuke@spring8.or.jp

Prof. Masahiro TATSUMISAGO
Osaka Prefecture University, 1-1 Gakuencho, Nakaku, Sakai, Osaka 599-8531, JAPAN
Email tatsu@chem.osakafu-u.ac.jp

Mr Masashi TOHMORI
Department of Materials Science, The University of Shiga Prefecture, Hassaka 2500, Hikone, Shiga 522-8533, JAPAN
Email t22mtoumori@ec.usp.ac.jp

Mr Satoru TOMENO
Research Center, Asahi Glass Co. Ltd., 1150 Hazawacho, Kanagawa-ku, Yokohama-shi, Kanagawa 221-8755, JAPAN
Email satoru-tomeno@agc.co.jp

Dr Koichiro TSUZUKU
Taiyo Yuden Co. Ltd., 5607-2, Nakamuroda-machi, Takasaki-shi, Gunma 370-3347, JAPAN
Email ktuzuku@jty.yuden.co.jp

Dr Norimasa UMESAKI
Japan Synchrotron Radiation Research Institute, 1-1-1 Kouto, Sayo, Hyogo 679-5198, JAPAN
Email umesaki@spring8.or.jp

Mr Wilson VAZ
Department of Physics, Goa University, Opp. HDFC Bank, Sadar, Ponda Goa 403 401, INDIA
Email wilsonvaz@yahoo.com

Dr Natalia VEDISHCHEVA
Institute of Silicate Chemistry of the Russian Academy of Sciences, Nab. Makarova, 2, St. Petersburg 199034, RUSSIAN FEDERATION
Email natalia@mail.rcom.ru

Dr Noriyuki WADA
Department of Materials Science and Engineering, Suzuka National College of Technology Shiroko, Suzuka, Mie 510-0294, JAPAN
Email wadan@mse.suzuka-ct.ac.jp

Prof. Adrian WRIGHT
University of Reading, J.J. Thomson Physical Laboratory, Whiteknights, Reading RG6 6AF, UK
Email a.c.wright@reading.ac.uk

Mr Shigeru YAMAMOTO
Nippon Electric Glass Co. Ltd., 2-7-1 Seiran Otsu, Shiga 520-8639, JAPAN
Email shyamamoto@neg.co.jp

Mr Naoto YAMASHITA
Isuzu Glass Co. Ltd, 6-3-6 Minamitumiri, Nshinari, Osaka 557-0063, JAPAN
Email nyamashita@isuzuglass.co.jp

Mr Yoshiki YAMAZAKI
Department of Applied Physics, Tohoku University 6-6-05, Aramakiazaaoba, Aoba, Sendai, Miyagi 980-8579, JAPAN
Email yayo@laser.apph.tohoku.ac.jp

Prof. Tetsuji YANO
Department of Chemistry and Materials Science Tokyo Institute of Technology 2-12-1-S7-4 Ookayama, Meguro-ku, Tokyo 152-8550, JAPAN
Email tetsuji@ceram.titech.ac.jp

Mr Hirotoshi YASUNAGA
Tohoku University, Graduate School, Engineering Graduate Course, Department of Applied Physics, Fujiwara Laboratory 6-6-05, Aramaki Aza Aoba, Aoba-Ku, Sendai-Shi, Miyagi 980-8579, JAPAN
Email yasunaga@laser.apph.tohoku.ac.jp

Prof. Tetsuo YAZAWA
Graduate School of Engineering, University of Hyogo, 2167 Shosha, Himeji, Hyogo Pref. 671-2201, JAPAN
Email yazawa@eng.u-hyogo.ac.jp

Dr Yoshinori YONESAKI
University of Yamanashi, Miyamae 7, Kofu, Yamanashi 400-8511, JAPAN
Email yonesaki@yamanashi.ac.jp

Dr Satoshi YOSHIDA
Center for Glass Science and Technology, The University of Shiga Prefecture, 2500, Hassaka, Hikone, Shiga 522-8533, JAPAN
Email yoshida@mat.usp.ac.jp

Dr Randall YOUNGMAN
Corning Incorporated, Science & Technology SP-AR-02-4, Corning, NY 14831, USA
Email youngmanre@corning.com

Author Index

Subject Index

First Announcement

Seventh International Conference on Borate Glasses, Crystals and Melts

Dates: 21–25 August 2011
Location: Dalhousie University, Halifax, Nova Scotia, Canada

Background

The BORATE2011 conference follows the previous meetings held at Alfred, New York, USA (1977); Abingdon, UK (1996); Sofia, Bulgaria (1999); Cedar Rapids, USA (2002); Trento, Italy (2005); and Himeji, Japan (2008).The conference will be dedicated to Professor Tsutomu Minami to honor his achievements in Glass Science, and in particular Borate Glasses.

Scope of the Conference
1. Short and intermediate range order;
2. Structure and physical properties;
3. Computer simulation and modeling;
4. Phase separation and inhomogeneities;
5. New spectroscopic techniques;
6. EPR, XAFS, XPS, IR, NMR, and diffraction;
7. Novel borate glasses and crystals;
8. Borate crystallography;
9. Thermal properties and thermodynamics;
10. Industrial applications of borate and borosilicate glasses;
11. Biomedical applications;
12. Optical properties and materials;
13. Superionic systems and ionic conductivity

Professor Josef W. Zwanziger
Department of Chemistry
Dalhousie University
Halifax
NS B3H 4J3
Canada.
Email jzwanzig@dal.ca

www.ingramcontent.com/pod-product-compliance
Lightning Source LLC
Chambersburg PA
CBHW082304210326

41598CB00028B/4438

* 9 7 8 0 9 0 0 6 8 2 6 3 6 *